REVISED SEVENTH EDITION

DEVELOPMENTAL ANATOMY

A TEXTBOOK AND LABORATORY MANUAL OF EMBRYOLOGY

BY

LESLIE BRAINERD AREY, Ph.D., Sc.D., LL.D.

Robert Laughlin Rea Professor of Anatomy, Emeritus,
Northwestern University

With 638 Illustrations — Some in Color

W. B. SAUNDERS COMPANY — Philadelphia and London

W. B. Saunders Company: West Washington Square
                          Philadelphia, PA 19105

                          1 St. Anne's Road
                          Eastbourne, East Sussex BN21 3UN, England

                          1 Goldthorne Avenue
                          Toronto, Ontario M8Z 5T9, Canada

Developmental Anatomy—*Revised Seventh Edition—1974*   ISBN   0-7216-1377-2

Last digit is the print number;   9   8

To

WEBSTER CHESTER

Professor of Biology, Emeritus, Colby College

*An inspiring teacher, scholarly scientist and true friend of youth
who laid my biological foundations, tendered encouragement
and help in mastering early difficulties and
pointed the way to greater
opportunities*

THIS VOLUME

IS INSCRIBED WITH A DEEP SENSE OF ADMIRATION

AND GRATITUDE

PREFACE TO THE SEVENTH EDITION– *REVISED*

Despite a slow-down in the output of conventional research on mammalian and human development, the past decade has yielded a considerable amount of significant new information. The present volume attempts to incorporate these advances through the alteration of various existing accounts and the insertion of new material. These changes, together with numerous improvements in presentation, affect one fourth of the pages in the main text. Forty-one illustrations have been improved.

Since this book, as a whole, largely retains its former character, it was decided to designate it as the Seventh Edition, Revised. It has been the author's policy not to assign a new number to any edition unless drastic changes had made it obviously a different product.

As in previous editions, a consistent policy has been adopted in the use of large and small type. In large type will be found the basic information for which all students might be held accountable. In small type are presented such additional information and applied data that are deemed important to more mature readers, and especially to medical students and graduates.

Unless the context implies the contrary, the reader may assume that the unfolding of the developmental story in this book is an account of his own formative course. All time references are in accordance with the crown-rump lengths of embryos and fetuses, as tabulated on page 104. Acknowledgment is extended to the W. B. Saunders Company, as publisher, for its coöperation in matters of policy and technology.

Chicago, Illinois   L. B. Arey

CONTENTS

PART III   A LABORATORY MANUAL OF EMBRYOLOGY

PART I. GENERAL DEVELOPMENT

Chapter I. Fundamental Concepts

THE NATURE AND SCOPE OF EMBRYOLOGY

Embryology is the science that treats of the origin and development of the individual organism which, in this period, may be broadly designated an *embryo*. But what is the meaning of 'development' when used in this sense? It is a gradual bringing to completion, both in structure and in function. Its chief characteristic is cumulative change in a progressive direction, in which each component act and result loses significance unless viewed against what precedes and what follows. Although the vital processes employed in the development of an individual may not differ basically from most of those exhibited in the activities of the final organism, their results tend to be permanent rather than transitory. That is, they establish patterns of form and of function rather than merely maintaining what has already been perfected.

An embryo is a functioning mechanism, adequately adapted to its needs and environment at all stages of its development. Nevertheless, most of the advances achieved at any period are in anticipation of functions that do not appear until some later time. Developmental changes, as a whole, thus contrast sharply with the recurring, non-progressive, physiological changes that are concerned solely with the maintenance of stabilized life.

The Developmental Period. The development of many animals is divided by the incident of birth or hatching into prenatal and postnatal periods. For a long time attention was focused on the events taking place before birth, when the most striking advances occur in these animals as a whole. Only gradually was it realized that important changes, beyond mere growth, continue to occur even into the adult state. This broader concept of embryology brings into its range all of the developmental events resulting from sexual reproduction. The total advance is sometimes designated as *ontogeny*.

1

Many animals, including such vertebrates as fishes and amphibians, are capable of an independent existence at relatively immature stages; these free-living forms, with much or most of their development still before them, are called *larvæ*. It is quite otherwise with reptiles, birds and mammals. The human newborn, for example, is fairly complete anatomically, yet is utterly dependent on its elders for food and care. Throughout infancy, childhood and adolescence come the completion of some organs and a gradual remolding of body shape. Only at about the age of twenty-five are the last of these progressive changes finished and the body stabilized in the adult condition.

It is instructive to list the divisions of the life span in man and thus to re-emphasize how many of these entries belong to the developmental period:

Prenatal life
- *Ovum.* Fertilization to end of first week.
- *Embryo.* Second to eighth week, inclusive.
- *Fetus.* Third to tenth lunar month, inclusive.

Birth

Postnatal life
- *Newborn.* Neonatal period; birth to end of second week.
- *Infancy.* Third week until assumption of erect posture at end of first year.
- *Childhood.*
  - *Early.* Milk-tooth period; second to sixth year, inclusive.
  - *Middle.* Permanent-tooth period; 7 to 9 or 10 years, inclusive.
  - *Later.* Prepuberal period; from 9 or 10 years to 12–15 years in females and to 13–16 years in males.

Puberty

Adolescence. The six years following puberty.

Adult.
- *Prime and transition.* Between 20 and 60 years.
- *Old age and senescence.* From 60 years on.

Death

The Fields of Embryology. The general topic of development subdivides conveniently into morphological and functional categories. The morphological division deals with form, structure and relations, and is purely descriptive and comparative in treatment. It traces the formative history of animals from a germ cell of each parent to the resulting, adult offspring. Its objective is to paint the progressive panorama of change that cells, tissues, organs, and the body as a whole undergo in attaining their final states. These unified descriptions of advancing form, structure and relations can be designated by the term *developmental anatomy;* both normal and abnormal (pathological) subdivisions are recognized in this field. The other division of embryology is functional and attempts to explain, on the basis of experiment and analysis, the ways in which development works. How seemingly mysterious happenings can be resolved into familiar physical and chemical phenomena; how parts interact in determining and co-ordinating the evolving embryo; how fetal physiology makes its beginnings and then operates—all these, and more, constitute *developmental physiology*. Much of the effort in this field has centered about an attempt to discover the forces, factors and mechanisms that govern development. This experimental attack on dynamic causation has come to be known as *experimental embryology*. Also important is the renewed assault on the chemistry of developmental processes—*chemical embryology*.

All multicellular animals have certain similarities in their ways of development. It is, however, only in the very earliest stages that all the different kinds of embryos have much in common structurally. Also, among closely related groups the correspondence in the form and method of development is greatest and lasts longest. Thus all vertebrate (*i.e.*, backboned) animals are built about a common anatomical plan and have much the same fundamental style of development. Naturally some variant developmental methods are utilized and some class peculiarities exist, while in the end 'higher' vertebrates achieve greater complexities than do 'lower' ones. Although *comparative embryology*, which deals with these matters, is indispensable for gaining a broad understanding of development, its former importance in supplying missing pages of the human story has diminished greatly. In the experimental field alone is there a high degree of dependence on lower forms. Since mammalian embryos are not readily amenable to experiment, the embryology of mammals and man has not advanced much beyond the descriptive stage.

The Value of Embryology. A general conception of how man, like other animals, develops from a single cell should share in the cultural background of every educated mind. From the philosophical side, embryology is a key that helps unlock such secrets as heredity, the determination of sex and organic evolution. To the medical student, embryology is of primary importance because it supplies a comprehensive and rational explanation of the intricate arrangements of human anatomy. The body does not just happen to be arranged as it is; each end-result is preceded by a definite course of developmental events. Because of this, malformations of various kinds can be explained on the basis of departures from the usual pattern. Embryology is also able to interpret vestigial structures, to explain growth, differentiation and repair, and to throw light on some pathological conditions. For all these reasons it is essential to sound training in anatomy, pathology and surgery. Furthermore, obstetrics is largely applied embryology, while pediatrics and other specialties find it an indispensable tool.

THE HISTORICAL BACKGROUND

Several centuries before our era, Aristotle (384–322 B.C.) wrote the first

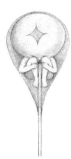

Fig. 1. Human sperm cell, containing a miniature organism, according to Hartsoeker (1694).

treatise on embryology. It was a mighty compendium of observation and argument, so far in advance of his age that for nearly two thousand years almost nothing of significance was added. Aristotle was the first to formulate the alternative that an embryo must be either preformed and only merely enlarging during its development, or it must be actually differentiating from a formless beginning. He decided in favor of the latter interpretation and thereby set off a controversy that extended through the centuries. Although Aristotle discovered many astonishing facts in comparative embryology and followed the general progress of the developing chick, he naturally fell into error on things about which he had to speculate. Thus he credited the popular belief that slime and decaying matter are capable of producing living animals, and he described the human embryo as organizing out of the mother's activated menstrual blood. Such origins were disproved by Redi (1668), although the death blow to the persistent belief in the spontaneous generation of microscopic animalcules and bacteria was dealt only in 1864 by Pasteur. That every living organism comes from a pre-existing, living organism (*omne vivum ex vivo*) and that every cell arises from the subdivision of a pre-existing cell (*omnis cellula e cellula*) are fundamental concepts, so commonplace today that their long struggle for recognition is often overlooked.

Until about the year 1800 it was generally believed either that a fully formed animal exists in miniature in the egg, needing only the stimulus of the sperm to initiate growth and unfolding, or that similarly preformed organisms, male and female, constitute the sperms and these merely enlarge when they get inside the eggs (Fig. 1). To be consistent this doctrine of *preformation* had to admit that all future generations were likewise encased, one inside the sex cells of the other, like so many Chinese boxes. Simple mathematical considerations made such a concept difficult to defend. In modern times a modification of the preformational point of view has been reintroduced into biology, but in a far more subtle form than the original doctrine taught. Beginning in 1910, Morgan and others amassed convincing evidence that the chromosomes of the fertilized egg have, localized in their genes, definite determinative powers over development.

The original preformation theory was virtually destroyed by Wolff (1759–69) who, like Harvey in the preceding century, saw the parts of the early chick embryo take shape as new formations. But Wolff was able to go further and show that the future embryonic region on an egg first consists of 'globules' (*i.e.*, cells), lacking in any arrangement that can be related directly to the form or structure of the future embryo. Only gradually did these 'globules' organize into rudiments which, in turn, took on the characteristics of the various organs of the embryo. This method of progressive development from the simpler to the more complex, through the utilization of building units known as cells, is *epigenesis*. Many years, nevertheless, elapsed before Wolff's views gained proper recognition. The final chapter in the obituary of the original doctrine of preformation was written by Driesch (1900) who proved that in many forms the daughter cells of a fertilized egg (*i.e.*, half- or quarter-eggs), when separated, will develop into complete embryos. The present view on these matters is that development is strictly preformational as regards the genes and their hereditary influences, but rigorously epigenetic in actual constructional activities.

With the overthrow of preformation, scientists sought afresh what it could be that the egg transmits to the next generation. Darwin and others thought that each part of the body might contribute something to the sex cells of an individual, and that these representative tokens could make the operation of heredity physically possible. Weismann (1893) argued convincingly that the facts are quite otherwise, a child in no way inheriting its characters from the bodies of the parents but, rather, from their sex cells alone. These germ cells, in turn, acquired their characters directly from pre-existing germ cells of the same kind. He pictured the '*germ plasm*' as a self-perpetuating, cellular legacy which has existed as an unbroken stream through the ages. At each new generation a temporary body (or *soma*) is built up around it, to serve as a carrier of the germ cells and to hold them in trust for the forthcoming offspring (Fig. 2). The reason, therefore, why offspring resembles parent is because each develops from portions of the same immortal stuff. Modern investigation has shown that the self-perpetuating elements are really the genes, that these occur identically in all cells of the body and that even a body-cell nucleus can substitute for an egg nucleus. Weismann's belief in a fundamental difference between sex and somatic cells has lost much of its original force, but his concept of germinal continuity threw a great light on the nature of the hereditary process.

Harvey (1651) and Malpighi (1672) contributed fundamental descriptions of the stages of the developing chick, as seen with simple lenses. How these observations were refined by Wolff (1759–69) has been told in a previous paragraph. In 1817 Pander demonstrated the three primary germ

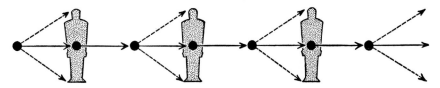

Fig. 2. Diagram showing the concept of the continuity of germ plasm.

layers from which the chick embryo and its constituent parts develop. Von Baer (1829–37) soon afterward broadened this concept to a generalization for all animals; he also determined the origins of the principal organs and made the science of embryology comparative. Exactly 150 years after Leeuwenhoek (1677) first reported the discovery of the sperms of man and other mammals, von Baer (1827) identified the mammalian egg. For these several far-reaching contributions, which influenced all subsequent growth of embryology, he has justly been honored as the 'father of modern embryology.' Cleavage, or subdivision of the egg into the building units of the embryo, was first definitely described by Prévost and Dumas in 1824, but its true meaning had to wait for the recognition of the cell as the structural and functional unit of the organism. This biological landmark was set by Schwann in 1839, and about twenty years later the egg and sperm were recognized as true cells. Hertwig, in 1875, was the first to observe and appreciate the main events involved in the fertilization of an egg by a sperm, while Van Beneden (1883) soon proved that the male and female sex cells contribute the same number of chromosomes to the fertilized egg. That these chromosomes perpetuate themselves by mitosis, maintain a persistent individuality and possess different hereditary qualities was supported by Boveri (1888; 1909). The identification of Mendel's hereditary characters with genes, situated at definite locations in particular chromosomes, was the achievement of Morgan (1912) and his associates.

Like biology in general, embryology has passed through three stages. The first was pure description and fact-gathering. At present it continues chiefly in the program of obtaining a well-rounded account of human development; His and Keibel in Europe and Minot and Mall in America were the original leaders in this endeavor. The second stage was comparison, in which the observations on various animal types were classified and compared, and common trends and principles were sought. The dominance of von Baer in comparative embryology has never been challenged. Description and comparison received a great impetus in the last half of the nineteenth century from the then new theory of evolution; it was hoped that the full evolutionary history of an animal would be revealed in its embryonic development (p. 8). The third stage, experimental and analytical, was pioneered by Roux and Spemann in Europe and by Morgan and Harrison in this country. It is the most vigorous and significant branch of contemporary embryology. Of late there has been a revival of interest in chemical embryology; utilizing new methods of biochemistry and biophysics, the basic mechanisms underlying development are sought.

THE RÔLES OF HEREDITY AND ENVIRONMENT

There is an intimate interrelation between the developing embryo and the immediate environment which its body, organs and tissues encounter. Both the directing force of heredity and the molding influences of environment play important rôles in development. These two factors are, however, unlike in quality. Actually, both are components of a common interacting system, and to argue which is the more important is meaningless. To weigh the value of one against the other is to lose sight of the integrated process of development as a whole: each is essential and important. All of the developmental effects are produced co-operatively by interactions between genic and environmental factors.

Heredity operates through *internal factors,* present in the fertilized egg itself. Chief among these are the *genes,* or hereditary determiners, which are located in the chromosomes of the nucleus; they are contributed equally to the fertilized egg by each parent. The genes direct the development of a host of enzymic reaction-systems that are the responsible agents for the production of the various physical and functional characteristics of an individual. Another possible hereditary mechanism involves the cytoplasm, and some lower animals furnish evidence for a limited operation of such *cytoplasmic inheritance.* Formerly it was urged that the cytoplasm of every egg directs the establishment of those general characteristics that mark an individual as a vertebrate, mammal and human, whereas the genes control only the lesser, and more specific, characteristics. This concept rapidly lost favor as the larger role of the gene became better understood. Many hereditary characteristics, like the blood groups of mammals, can in no way be influenced directly by the environment. Also, through the study of identical twins, reared apart under dissimilar conditions, it is certain that heredity has a much greater influence than environment on the determination of most structural, physiological and mental characters.

Environment supplies the *external factors* that make development possible and allow heredity to find expression. Such environmental factors include adequate warmth, moisture, oxygen, food and various non-nutritive, chemical substances. Moreover, certain general traits (such as weight and, possibly, intangible mental and emotional qualities) can be modified experimentally. Without doubt, environmental factors can condition the appearance of genetic characters; they can modify the developmental expression of inherited characters; and they can so alter the genic constitution of chromosomes as to make possible the creation of new heritable characters (*mutations*). Biochemical genetics suggests that the greater the number of biochemical steps between a gene and its trait, the greater is the opportunity for environmental influences to come into play.

Although an altered environment may induce physical changes in the body proper (*i.e.,* the *soma* as contrasted with the sex cells, or *germ plasm*), there is no proof that such acquired characters can impress themselves on the germ cells and thus become transmitted to subsequent generations. In this regard, the ineffectualness of somatic mutilations, such as circumcision, even though continued through many generations, is too obvious for extended comment. A correlated folklore belief, especially prevalent among animal breeders, is that offspring sometimes inherit characters impressed on the dam by a previous sire. This hypothesis (*telegony*), like other examples of the supposed inheritance of acquired characters, lacks any adequate proof.

ANCESTRAL REPETITIONS

The *theory of recapitulation* long taught that an individual in the course of its development passes through successive stages that approximate the series of adult ancestors from which it is descended. This repetition of ancestral stages was said to be crowded back in development and abbreviated, but nonetheless to present phylogeny in review. The theory would insist, for example, that the embryonic organs and parts of a mammal pass through adult fish-like, amphibian and reptilian phases before the mammalian states are attained. It also asserts that the various adult types of ancestors have been able to leave their imprint on the style of development used by their descendants. In short, during its life history 'every animal climbs up its family tree.' This doctrine goes beyond the facts.

Embryos of different groups do resemble one another in the early stages of their development, but this resemblance tends to diminish progressively as they advance toward their final forms. Moreover, a fish, reptile and mammal do not start alike and pass through the same stages; they are individualistic from their beginnings. The similarities that exist are good proofs of a common origin, while the repetition of like ancestral features in the development of different animals is owing to the presence of the same hereditary factors in the several kinds of fertilized eggs and the development of these eggs under conditions that permit those features to appear. An embryo of a reptile, bird or mammal does not possess gill arches like an adult fish, but only like those of a fish at a corresponding stage of development. All that can be maintained is that the development of any individual may more or less recapitulate the style of development that its ancestor used. Stages may be omitted, sequences altered, larval specializations interpolated and new structures developed.

Some of the structures appearing during human development are apparently useless survivals (*e.g.,* tail), but caution is indicated in judging individual cases since it is doubtful whether any part is retained for long in the evolutionary time scale unless it is either useful or wholly insignificant. For example, the provisional embryonic kidney (mesonephros) of a higher vertebrate functions for a time, and its duct supplies a necessary stimulus to the development of the permanent kidney (metanephros). Certain ancestral organs abandon their original embryonic function, yet are retained and utilized for new purposes (*e.g.,* mesonephric tubules and ducts of mammals become the permanent sex canals of the male). Other parts make their appearance, only to change at once into quite different structures (*e.g.,* gill pouches into thymus and parathyroids). Since these latter are necessary organs it is understandable why in this instance the embryonic pouches appear even though they are never respiratory in function.

Some embryonic organs neither disappear nor take on permanent function, but rather persist throughout life as *vestiges;* more than 100 have been listed for man. Among such are the coccyx, appendix, body hair, wisdom teeth and ear muscles. Many of these vestiges are doubtless on their way toward elimination from the developmental course. Somewhat different are *atavistic characters,* or ancestral reversions. These are features that normally have dropped out of development but may, on occasion, reappear. They result from the perpetuation and inheritance of particular genes that are able to reassert themselves whenever the proper environmental conditions are re-established.

The various ancestral structures that recur in human development represent features that first appeared in lower embryos of the vertebrate stock and have persisted as survivals. Such common characters argue eloquently for common ancestry. However incomplete their developmental review may be, the fact remains that the stages encountered constitute the only record that supplies any significant information as to how the human species may have reached its present state.

THE METHODS OF STUDY

The orderly progress of development is ascertained partly by observing entire embryos and dissections in proper age-sequence, and even more profitably by the microscopical study of stained sections. When sections of an embryo are arranged in the same progressive order as cut, they furnish a related series that permits the changing appearances from level to level to be traced and reconstructed either as mental images or physical models. In doing this, a third dimension, *depth,* must be visualized in addition to the customary dimensions of *length* and *breadth* which are used when inspecting ordinary sections. Still a fourth dimension, *time,* enters into consideration, since it is necessary to recreate the course of development mentally through a period of time. In doing this effectively one should also attempt to correlate the degree of development of any part at any moment with respect both to associated organs and to the condition of the embryo as a whole. Thus, in a sense, embryology becomes a four-dimensional study.

Descriptive and *comparative embryology* cannot throw light on how the events of development are related and why they occur where and when they do. Such questions enter into the field of causation. Their basis is attacked by devising experiments in which living embryos are made to supply data for the information desired. This is the realm of *experimental embryology.*

TERMINOLOGY

In describing development it is necessary constantly to employ words denoting the position of one part with reference to another, or to the body as a whole. The logical usage tabulated here is common to embryology and comparative anatomy. The terms *superior* and *inferior, anterior* and *posterior,* as used in adult human anatomy, are unfortunate choices based on man's erect posture and distinctive locomotion.

A few examples will illustrate the proper application of these terms: The backbone lies *dorsally;* the breast bone is *ventral* to it. The neck attaches to the *cranial* end of the trunk, while the latter extends *caudad* from the neck. The nose occupies the *median* plane; it is *medial* to the cheek which for its part is placed more *laterally.* The wrist is *distal* to the elbow, while the elbow is *proximal* to the wrist. A nerve is traced *distad* toward its ending. (Sagittal, frontal and transverse planes lie with respect to each other as do three adjoining surfaces of a cube.)

TERMS DESCRIBING ANATOMICAL RELATION

ADJECTIVE	ADVERB DENOTING FIXED RELATION	ADVERB DENOTING PROGRESS TOWARD	GENERAL REGION OF BODY REFERRED TO
dorsal	dorsally	dorsad	Back surface.
ventral	ventrally	ventrad	Front surface.
{ cranial	{ cranially	{ craniad	{ Head end. ('Rostral' used mostly in descriptions of the head to indicate snoutward.)
} rostral	} rostrally	} rostrad	{
caudal	caudally	caudad	Tail end.
median	. . . .	. . . .	In the midplane.
medial	medially	medially	Nearer the midplane.
lateral	laterally	laterad	Farther from the midplane.
proximal	proximally	proximad	A more central part.
distal	distally	distad	A more peripheral part.
sagittal	sagitally	. . . .	Any plane parallel to midplane and dividing embyro into right and left portions.
frontal	frontally	. . . .	Any plane parallel to long axis and dividing embryo into dorsal and ventral portions.
transverse	transversely	. . . .	Any plane perpendicular to long axis.

RECOMMENDATIONS FOR COLLATERAL READING

History of embryology:

Meyer. 1939. The Rise of Embryology (Stanford Univ. Press).
Needham. 1959. A History of Embryology (Cambridge Univ. Press).
Oppenheimer. 1967. Essays in the History of Embryology (M.I.T. Press).

Fundamentals of early development:

Balinsky. 1970. An Introduction to Embryology (Saunders).
Kellicott. 1913. A Textbook of General Embryology (Holt).
Wilson. 1925. The Cell in Development and Heredity (Macmillan).

Comparative embryology:

Nelsen. 1953. Comparative Embryology of Vertebrates (Blakiston).
Witschi. 1956. Development of Vertebrates (Saunders).

Human development:

Hamilton, Boyd and Mossman. 1972. Human Embryology (Williams & Wilkins).
Keibel and Mall. 1910–12. Human Embryology (Lippincott).
Keith. 1948. Human Embryology and Morphology (Arnold).
Patten. 1970. Human Embryology (Blakiston).

Human growth and postnatal development:

Scammon. 1923. Vol. I, Chapter III in Abt: System of Pediatrics (Saunders).
Scammon. 1953. Section I in Morris: Human Anatomy (Blakiston).

Human malformations:

Morrison. 1963. Foetal and Neonatal Pathology (Butterworth).
Potter. 1961. Pathology of the Fetus (Year Book Publishers).

Ruben. 1967. Handbook of Congenital Malformations (Saunders).
Schwalbe. 1906–37. Die Morphologie der Missbildungen des Menschen (Fischer).
Willis. 1962. The Borderland of Embryology and Pathology (Butterworth).

Experimental embryology:

Needham. 1942. Biochemistry and Morphogenesis (Macmillan).
Waddington. 1956. Principles of Embryology (Macmillan).
Weiss. 1939. Principles of Development (Holt).
Willier & Oppenheimer. 1964. Foundations of Experimental Embryology (Prentice-Hall).

Chemical embryology.

Brachet. 1950. Chemical Embryology (Interscience Publications).
Brachet. 1960. The Biochemistry of Development (Pergamon Press).
Weber. 1967. The Biochemistry of Animal Development (Academic Press).

Physiological embryology:

Raven. 1954. An Outline of Developmental Physiology (McGraw-Hill).
Windle. 1971. Physiology of the Fetus (Thomas).

Human genetics:

McKusick. 1971. Mendelian Inheritance in Man (J. Hopkins Press).
Stern. 1973. Principles of Human Genetics (Freeman).

Human genetic disorders:

Goodman. 1970. Genetic Disorders of Man (Little, Brown).
Kemp. 1951. Genetics and Disease (Munksgaard).
Sorsby. 1953. Clinical Genetics (Butterworth).

Chapter II. General Features of Development

A multicellular animal begins its development as a fertilized (*i.e.*, activated) egg. Further progress depends upon: (1) *cell proliferation;* (2) *growth;* (3) *differentiation,* which includes *morphogenesis* (the molding of the body and its organs into form) and *histogenesis* (the specialization of cells into tissues); and (4) *integration,* to produce a unified, working organism.

CELL PROLIFERATION

All cells arise from pre-existing cells by cell division. Although a direct fission of the nucleus and cytoplasm into two portions is described for certain old or specialized cells of the body, this style (*amitosis*) plays little or no rôle in development. Several distinctive features characterize the ordinary method of cell division, named *mitosis.* In part, these include the following steps:[1] (1) the synthesis of deoxyribonucleic acid (DNA) at some time during interphase (the so-called resting stage), thus duplicating each loosely arranged chromosome and its genes (Fig. 3 *A*); (2) the reappearance from an uncoiled state of a characteristic number of thread-like *chromosomes* in pairs (*B*); (3) the marked shortening and thickening of such chromosomes (*C*); (4) the arrangement of the double set of chromosomes as an 'equatorial plate' midway of a 'mitotic spindle' (*D, E*); (5) the separation of the two complete single sets and their passage to the opposite poles of the cell (*F*); and (6) the return of the nucleus to the so-called resting state, and the division of the cytoplasm into two masses (*G, H*) which then grow to the characteristic cell size.

It seems like a long span from the egg to the trillions of cells that comprise the completed body of man, yet this prodigious final number could be attained readily by continuously repeated cell division. So effective is the doubling process that some 45 generations (2^{45}) of mitoses would be sufficient.

The complicated events of mitosis serve the purpose of dividing accurately the chromatic substance of the nucleus in such a way that the chromosomes of each daughter

12

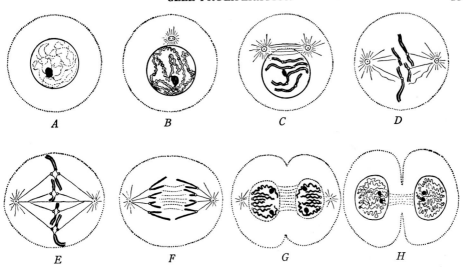

Fig. 3. A diagram of the stages of mitosis. *A*, interphase; *B–D*, prophase; *E*, metaphase; *F*, anaphase; *G, H*, telophase.

cell may be identical, both in number and composition. This is important since it is believed that self-perpetuating particles, or *genes*, in the chromosomes are hereditary determiners, and that these are arranged in definite linear order in particular chromosomes. Before a chromosome splits, each gene in it duplicates itself and one daughter gene goes into each daughter chromosome. It is estimated that the human chromosomes contain genes totaling 30,000 pairs, or more.

Although a gene is too small to be seen, except possibly by the electron microscope,[2] the existence of genes and their constant positions in specific, recognizable chromosomes have been demonstrated convincingly in the fruit fly, Drosophila, through extensive breeding experiments. Remarkable confirmation of these experimental proofs is afforded by the discovery of giant, compound chromosomes in the salivary glands of this insect (Fig. 4). The number and position of distinctive bands (*chromomeres*) in these chromosomes correspond well with charts that plot the positions of genes as deduced from breeding experience; moreover, alterations in appropriate bands accompany new mutations. It is not known, however, what relation may exist between the position of genes and the presence of particular bands or interspaces. The gene has a diameter of about 0.000006 mm., and its important component is a complex molecule of deoxyribonucleic acid (DNA). It acts as an organic catalyst and is said to produce a single, specific, immediate effect. More precisely, its action is interpreted by some as inducing the formation of a particular enzyme in the cytoplasm.[3] Although the gene may have a primary unitary action of this sort, the derived effects can be widespread. The gene is the smallest living thing that is known to grow and reproduce its kind exactly.

Fig. 4. Giant compound chromosomes from the salivary glands of insects (after Painter). *A*, Sex chromosome of the fruit fly ($\times$ 1000); 21 known characters are identified with bands (gene loci) in the spaces indicated by vertical lines. *B*, Segment of chromosome of the black fly, showing the organization of chromatin threads and the alignments that give the appearance of bands ($\times$ 2000).

GROWTH

Growth[4] may be defined as a developmental increase in mass. It is a fundamental property of life and an important factor in development. Without growth no organism could exceed greatly the size of the egg from which it came. Since all living organisms consist basically of cells and these have definite size limitations, increase in bulk during development naturally is conditioned by cell proliferation. This multiplication, by itself, does not result in growth; but it does produce more units to participate in growing. Exceptional is the period of cleavage, which is the initial step in development taken by the fertilized egg. During this period the originally over-large egg subdivides into cells of ordinary size, but significant growth does not enter at all. By the time of birth, a baby is several billion times heavier than the egg from which it came.

The Methods of Growth. Growth is accomplished in several ways. Most important is *protoplasmic synthesis,* by which new living matter (protoplasm) is created from available foodstuffs. In the last analysis, animals depend on plants for their proteins, which are the building materials out of which new protoplasm is constructed. Digestive enzymes split the proteins of food into amino acids, and these products are used by the cells in the processes of synthesis. Growth requires anabolism to exceed catabolism.

A second method of growth involves *water uptake.* The amount of water in a living organism is large. In the early weeks of its development the human embryo is nearly 98 per cent fluid; the adult is 70 per cent water. The colloids within cells and between them have the capacity of imbibing water and swelling. The ability to hold water and release it is governed in part by ionic concentrations.

A third method of growth is by *intercellular deposition.* This results from the manufacture and deposit of nonliving substances. Such material is probably not a direct protoplasmic transformation, but rather a secretion or even a substance brought to existence outside the cell. It is usually located between cells and consists of jelly, fibers or the ground substance of cartilage and bone. A fourth method is *intracellular storage* (e.g., fat).

The Measurement of Growth.   The amount of growth is expressed both in *absolute* and *relative terms,* but comparisons are more easily made when the latter are employed. Thus the absolute gain in weight of a 10-pound baby and a 100-pound youth might be 1 pound each, whereas the relative gains (expressed as percentages of the initial weights) would be 10 per cent, against 1 per cent. It is the same with *growth rates.* The absolute rate, in terms of any chosen unit of time, is the amount of increase during any period divided by the length of that period. But comparisons are more instructive if relative growth rates are employed. The relative rate is the relative increment *(i.e.,* relative increase) per unit of time. It is computed by dividing each absolute rate by the initial value (in weight, volume or length); the result expresses the relative rate in terms of the unit of measurement used. Such computations show that a newborn rabbit and pig, though widely different in weight, grow at the same relative rate. By contrast, a newborn sheep grows eighteen times faster than the human newborn which originally equaled it in weight.

The growth of one part of the body often appears to be quite out of step with the growth of another part, or of the organism as a whole. Yet, in general, such relations of size or weight at any period fit into a simple type of mathematical formula which takes into account the amount of divergence that occurs between the growth progress of each part. For example, the facial region of a baboon outgrows the cranial region so enormously that the newborn and adult skull seem to be unrelated. Yet the dynamics of skull growth is a harmonious process throughout, and a formula may be devised that fits any stage in its development.

Differential Growth. The development of an organism is characterized by a progressive alteration of form and proportions, both externally and internally. It is obvious that uniform growth throughout a body-mass cannot produce these changes. Actually, diversity of form is acquired through *differential rates of growth* operating in various regions and in definite directions. These absolute rates may vary among individuals of any species according to circumstances, but the ratios existing between the growth rates of different parts of the body in that species are relatively constant. It is these fixed relations that produce similar final form in the countless individuals of any animal-kind. And this is accomplished in spite of the fact that the constituent parts of an organism make their appearance and begin to grow at different times. The variance in starting times and growth rates among species is responsible for what may be called their *growth patterns*.

The changes in the proportions of the body and its parts, due to unequal growth, are produced by: (1) local differences in the growth intensity; (2) growth gradients; (3) reduction of the early dominance of cranial over caudal levels; (4) functional demands; and (5) influence of the growth rate

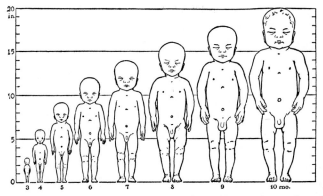

Fig. 5. Diagram illustrating the changes in size of the human fetus when drawn to scale (Scammon and Calkins).

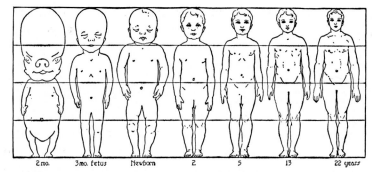

| 2 mo. | 3 mo. fetus | Newborn | 2 | 5 | 13 | 22 years |

Fig. 6. Diagram illustrating the changing proportions of the human body during pre-natal and postnatal growth (Scammon). All stages are drawn to the same total height.

of a neighboring part. The visible way in which differential growth accomplishes the progressive modeling of external and internal form can be illustrated sufficiently through two specific examples. Figure 167 shows stages in the emergence of the limbs from initial, bud-like swellings. Figure 437 illustrates the early form changes undergone by the brain while advancing toward its final shape.

Growth Data. Many pertinent data have been collected concerning the growth rates of the human body and its organs during prenatal and postnatal development. Analyses of these data have brought to light definite growth tendencies and patterns. Some of the more general conclusions, largely due to studies by Scammon, will be summarized in the paragraphs that follow.[5]

1. CHANGES IN SIZE AND FORM. The growth and external changes in a fetus subsequent to the second month are illustrated in Fig. 5. Its growth in length averages about 2 mm. a day. If an adult maintained the chubby newborn shape, his weight would be twice the amount it actually is. Figure 6 shows the proportions of the body at various developmental periods, all drawn to the same height. Note: (a) the greater decrease in the size of the head; (b) the constancy of the trunk length; (c) the early completion of the arms and the tardier growth of the legs; (d) the upward shift of the umbilicus and symphysis pubis; and (e) the downward trend of the midpoint of total length. Figure 7 illustrates the relative proportions of the major organ systems, at birth and maturity, when the paired individuals are drawn to the same height.

Certain of these facts are plainer when tabulated as follows:

CHANGES IN RELATIVE SIZE OF THE PARTS OF THE BODY

(In per cent of the total body volume.)

AGE	HEAD AND NECK	TRUNK	ARMS	LEGS
Second fetal month................	43	52	3	2
Sixth fetal month..................	36	44	7.5	12.5
Birth.............................	32	44	8	16
Two years........................	22	51	9	18
Six years.........................	15	50	9	26
Maturity..........................	10	52	9	29

POSTNATAL INCREASE IN THE SIZE OF THE BODY AND ITS PARTS

(In relation to their sizes at birth as unity; the range indicates the minimal and maximal increase of organs within each group.)

VOLUNTARY MUSCULATURE	GENITAL ORGANS	TOTAL BODY, SKELETON AND LUNGS	LYMPHOID ORGANS	MAJOR VISCERA	ENDOCRINE ORGANS	NERVOUS SYSTEM
38	28–38	18–23	3–21	12–15	2–13	2–5

2. INCREASE IN SURFACE AREA. The relation of surface area to body mass or volume has a profound influence on the rate of both metabolism and heat loss. This relation shifts greatly during the postnatal period. At birth the surface area averages 2200 sq. cm. This is doubled in the first year, trebled by the middle of childhood, and increases rapidly before puberty. At maturity the total postnatal gain is seven-fold. Since, however, the weight of the body has increased some twenty-fold in the same time, it is obvious that there has been a relative loss. Thus, in the newborn there are over 800 sq. cm. of skin per kilogram of body weight, while in the adult there are less than 300 sq. cm. per kilogram.

3. INCREASE IN WEIGHT. During prenatal life weight increases six billion times, whereas from birth to maturity the increment is only twenty-fold. In absolute mass, however, 95 per cent of the final weight is acquired after birth. The ratio of increase during each fetal month to the weight at the beginning of that month is shown in the table on p. 104. It is an astonishing fact that if the body continued to grow even at the greatly reduced rate during the last fetal month, the weight of the adult would be two trillion times that of the earth.

4. INCREASE IN LENGTH. Embryos between four and nine weeks old grow 1 mm. each day; for the rest of intra-uterine life the daily gain in sitting height is about 1.5 mm. Growth in length and in weight have certain features in common, although the relative increase in length is the smaller of the two. This is because weight measures mass which

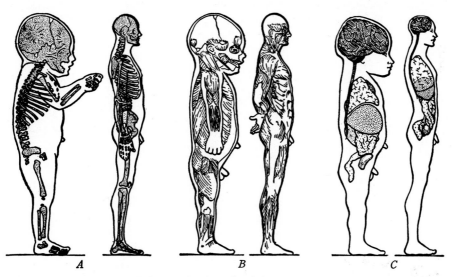

Fig. 7. Lateral views of the newborn and adult, drawn to the same total height to facilitate the comparison of proportions in the body and its systems (Wilner in Morris). *A,* The skeleton; *B,* the musculature, subcutaneous tissue and skin; *C,* the central nervous system and major visceral organs.

extends in three dimensions. The ratio of the increase in length each week or month to the length at the beginning of that period is shown in the table on p. 104. During the first year after birth, length increases 50 per cent. The total postnatal increment is 3.3 times the length at birth. Throughout most of childhood the linear increase is very slow (6 to 7 cm. a year), but at the prepuberal period there is an acceleration; as with weight, this is begun and ended earlier in girls than in boys. Growth in length is complete at about 18 years in females and soon after 20 in males. The body is heaviest in proportion to its length during late fetal life and early infancy (Fig. 7). From the middle of the first year until after puberty there is a decline in this ratio. Thereafter there is an increase in relative mass which may continue throughout life. Except at the puberal period, girls are relatively lighter than boys.

5. GROWTH OF THE ORGAN SYSTEMS. The *skeleton* grows rather slowly until the last two fetal months, whereupon it shows an acceleration. At birth it constitutes from 15 to 20 per cent of the body weight. Postnatal growth of the skeleton apparently parallels that of the body as a whole. The *musculature* likewise grows slowly at first, but represents about 25 per cent of the weight of the newborn and 40 to 45 per cent of the adult. The *blood vessels* show the same general trend. The *central nervous system,* on the other hand, is relatively huge in the young embryo. It decreases from about 25 per cent in the second month to about 15 per cent at birth and 2.5 per cent in the adult. The *peripheral nervous system* likewise undergoes a considerable reduction in relative weight during the postnatal years. The *skin* (including the subcutaneous fat) increases in relative weight up to birth (26 per cent) and shows little change thereafter. As a whole, the *viscera* decrease slowly and steadily in relative weight after the first two embryonic months. In the second prenatal month these organs comprise about 15 per cent of the total body weight; there is a reduction to about 9 per cent at birth and 5 to 7 per cent in the adult.

6. GROWTH OF THE ORGANS. Although the general course of relative growth in the individual organs follows that of the visceral group as a whole, each has its characteristic curve. Every fetal organ tends to increase more or less rapidly to a maximum relative size, and then to decrease throughout its subsequent history even to maturity.

During fetal life the curves of absolute growth are much alike. The various organs have an initial period of slow increase, followed after the fifth month by a terminal phase of rapid growth. This uniformity, however, disappears at birth when most of the organs can be arranged in four main groups; their postnatal growths are shown graphically in Fig. 8.

Factors Controlling Growth. Certain factors make growth possible and control it. Among these, the following require comment:

1. THE CONSTITUTIONAL FACTOR. Every animal species has its characteristic rates and limits of growth. Under identical conditions of development the speed of growth is approximately the same in all individuals of a species, and there is little difference in the final size attained. This is because of inherited qualities that predispose toward a definite basic rate of cell division and growth. It should be emphasized, however, that the original rate undergoes characteristic alterations in different regions of the embryo as the cell strains specialize.

2. TEMPERATURE. Within limits, the growth rates vary with the temperature. Each species has its critical maximum and minimum at which development ceases. Somewhere between these extremes lies the most favorable temperature.

3. NUTRITIONAL FACTORS. New protoplasm has to be created through-

out life, and the amino acids are the building materials out of which this synthesis is accomplished. The human organism can make some of its own amino acids, but others must be obtained in the food proteins. Certain of them favor growth, but not tissue differentiation. Nine amino acids are indispensable for human growth, and the absence of any one of these results in growth failure. The requirements for growth through new tissue-building are more exacting than those that suffice for the maintenance and repair of protoplasm already on hand. Several amino acids are required in the diet of the growing young beyond those that the adult needs in its diet to repair tissue losses due to functioning.

Food must not only be suitable but also adequate in amount if growth is to occur. There is a minimal degree of utilization below which growth fails. Above this level growth accelerates, but it cannot exceed an optimum rate, characteristic of the organism, even if an excess of food is available.

4. GROWTH-PROMOTING FACTORS. Certain substances, which in themselves are not foodstuffs, further the processes that result in the production of new protoplasm.

The Embryonic Factor. Tissues cultivated outside the body thrive better if juices extracted from an embryo are added to the nutrient medium. These extracts increase mitoses and shorten the time taken by each mitosis. There is a strong suggestion that definite cytoplasmic co-enzymes are involved. Since cell proliferation is a prerequisite to the growth of an organism, the presumed similar influence of this factor within an embryo is significant even though indirect.

Hormones. Some of the secretions elaborated by the ductless glands are

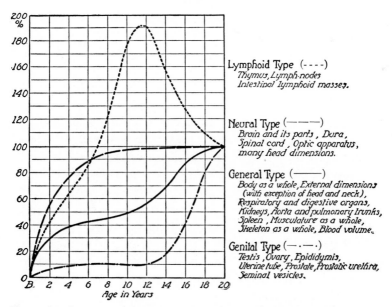

Fig. 8. Chart showing the course of postnatal growth in the various organ types (Scammon). Growth is calculated in relation to adult weights as 100 per cent.

regulators of growth. The thyroid hormone raises the rate of cell metabolism, presumably by acting as a catalyst to increase oxidative processes. It is essential for maintaining a normal level of metabolism. In deficient or excess amounts growth may be affected, but the results vary with the kind of animal and tissue. Thus a young mammal, deprived of its thyroid, remains small and undeveloped in some ways, whereas a tadpole grows slowly. When an excess is fed to a tadpole, only certain parts of the body respond by unusual growth.

One of the hormones produced by the anterior lobe of the hypophysis stimulates the growth of various tissues. Removal of the hypophysis in young animals results in retarded growth, while injections of the growth-promoting hormone into normal animals lead to generalized gigantism. Another hypophyseal hormone stimulates specifically the gonads, while the ovarian follicles, thus made to grow, in turn bring the genital tract to its adult size and also control the cyclic growth of its lining.

Vitamins. These are accessory food substances which, on the whole, animals cannot make and have to obtain in their plant foods. Their actions are after the manner of chemical catalyzers, and the amounts required are insignificant in comparison to the effects induced. In the absence of vitamin A the young animal fails to gain weight, although its skeleton does grow. Vitamin B_2 (riboflavin) exerts a specific influence on growth, and without it growth cannot take place. Vitamins C and D are also essential to normal growth.

5. GROWTH-ARRESTING FACTORS. Birds and mammals cease growing when they have attained a certain characteristic age and size. Even cold-blooded animals, which grow throughout their entire life spans, do so only at greatly reduced rates. Embryonic cells, grown in tissue culture and supplied with adequate food, have an infinite capacity for continued proliferation and growth. Why, then, does growth of the same cells in an organism become slowed or limited?

The total agencies determining growth rates and size limitation are many and unlike, so that a few comments must suffice. There is a fundamental antagonism between cell differentiation and cell proliferation, since the factors that promote differentiation make proliferation increasingly difficult (p. 26). Many cells reach a level of specialization at which they rarely divide, and some never do so; all this is bound up with the general phenomenon of aging. As cell differentation proceeds during development, increasing numbers of them pass beyond the stage where mitosis is easy or even possible. This automatically decreases the rate of relative growth. Another check on growth is cell destruction; the growth of certain organs, such as epidermis, blood and some glands, is offset by cell losses. Still different is the cessation of growth in the long bones of birds and mammals; this is apparently due to hormonal interference.

Abnormal Growth. The mass of an individual is largely set by the size of his skeleton, to which the soft parts tend to conform. Gigantism (*macrosomia*) and dwarfism (*microsomia*) designate conditions that lie outside the normal size range. An individual who exceeds 79 inches in height is rated as a giant, while adults less than 54 inches qualify as dwarfs. The known extremes measured 114 and 19 inches, respectively.

Many giants are well proportioned and are otherwise normal (Fig. 9 *A*). The condition usually starts before birth and the oversized newborn continues to grow at an accelerated rate. On the other hand, there may be an hereditary predisposition toward gigantism and yet it is not until some time during childhood that this tendency is first aroused into action by an infectious disease or other agency. The chief departure from normality in all such giants is the unusual length of the spine and especially of the limbs. The basic cause of this excessive elongation is an overproduction of the growth-promoting hormone secreted by the acidophil cells of the hypophysis. An associated, contributing factor is a delay in the closure of the growing regions near the ends of bones; this extends the period ordinarily allotted to growing. A different kind of overgrowth (*acromegaly*) occurs in adults after the sites of growth in the long bones have closed (Fig. 9 *C*). Under these conditions the new growth results mainly in a thickening rather than an elongation. The bones of the face, hands and feet overgrow in such a heavy manner; the features coarsen and the gonads tend to atrophy. These changes are instigated by an acidophil tumor in the hypophysis. An hereditary influence is usually a factor in the production of giants, and endocrine disturbances tend to show in the family lines. Rarely growth proceeds unequally in the two halves of the body (*D*) and gigantism may even be confined to specific, local regions (*E, F*). It has been suggested that unilateral gigantism may depend on irregular chromosome distribution in the cleaving egg.[6]

Some dwarfs are well proportioned and are normal in all respects except size (Fig. 9 *B*). Such is the *midget* who is small even at birth. Growth proceeds slowly and stops in the various bones at the usual times. The individual merely becomes a miniature adult. This condition is of genetic origin and is frequently transmitted to the offspring. Another type of dwarfism is due to a secondary deficiency of the growth-promoting hormone. The individual at birth is usually normal in size and appearance; growth proceeds normally for a time and then ceases. Such an adult has the uncompleted skeleton of a child of about seven years and often is sexually infantile. Although representing a genetic condition, these individuals may have normal brothers and sisters. A *cretin* is a dwarf whose condition is due to the congenital lack of thyroid tissue (*G*). It has attained an incomplete degree of development but cannot continue its differentiation without the help of the thyroid hormone. A person in such an arrested child-stage has a characteristic facial expression, is mentally defective and shows a dry, thick skin. Sexual development is arrested and most of these dwarfs are sterile. *Achondroplasia* designates a congenital dwarfism,

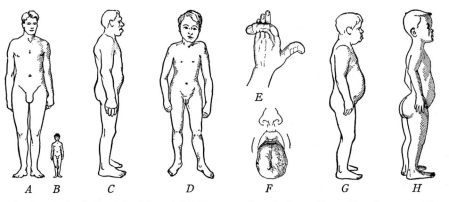

Fig. 9. Growth irregularities. *A*, Well-proportioned giant; *B*, smallest known midget, drawn to scale alongside the largest known giant; *C*, acromegalic gigantism; *D*, unilateral gigantism; *E, F*, localized gigantism; *G*, cretinous dwarf; *H*, achondroplastic dwarf.

characterized by growth cessation at the base of the skull and in the long bones (*H*). Such individuals have a broad, flat face with undershot jaw and possess short limbs. Head-size, trunk-length, intelligence and sexual development are not inferior. The condition is apparently due to a genic mutation; it is definitely hereditary.

DIFFERENTIATION

The fertilized egg straightway subdivides into numerous cells, more suitable in size to serve as the building units of the future embryo. At this point mass movements redistribute the formative cells into three superimposed plates, the *primary germ layers*. From their positions they are termed the *ectoderm* (outer layer; literally, outer skin), *mesoderm* (middle layer) and *entoderm* (inner layer) (Fig. 10 *A*). While the ectoderm and entoderm remain chiefly as sheets exposed on one surface (*i.e., epithelia*), the mesoderm also gives rise to mesenchyme (*B*). This is a diffuse spongework of cells that is a primitive filling-tissue. Such are the materials out of which the embryo differentiates.

Mesenchyme is predominantly derived from the mesoderm, both from the somites and from the sheets of somatic and splanchnic mesoderm (Fig. 71), but some of it comes from the ectoderm and this contribution is often called *mesectoderm*. Such an origin from ectoderm is partly direct and partly indirect; in the latter instance the source is a tissue known as neural crest which also gives rise to the nervous ganglia (Fig. 422). Another limited source is possibly the entoderm.

Position and environment are factors that help determine whether cells take the form of a sheet (*i.e.*, epithelium) or a spongework such as mesenchyme. In tissue culture one type can be transformed into the other by varying the conditions of the experiment. Even in normal organ-development it is known that ectodermal epithelium can, on occasion, give rise to a spongework resembling mesenchyme (*e.g.*, enamel pulp; Fig. 176), and the same is true of entoderm (*e.g.*, thymic reticulum; Fig. 199 *C*). Mesenchyme is an extraordinarily versatile tissue with many potentialities which find expression under the varied conditions offered during the course of development (Fig. 11).

Differentiation has two meanings, both of which imply increasing structural complexity. One refers to a change in the shape and organization of the body and its parts (*morphogenesis*); the other applies to a progressive change in the substance and structure of the cells themselves whereby tissues

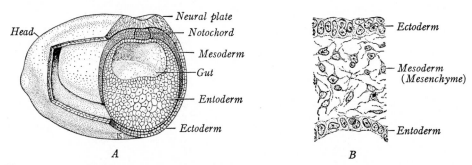

Fig. 10. Germ layers of early embryos. *A,* Stereogram of the head-half of a frog embryo (× 15). *B,* Section from an early human embryo (× 400).

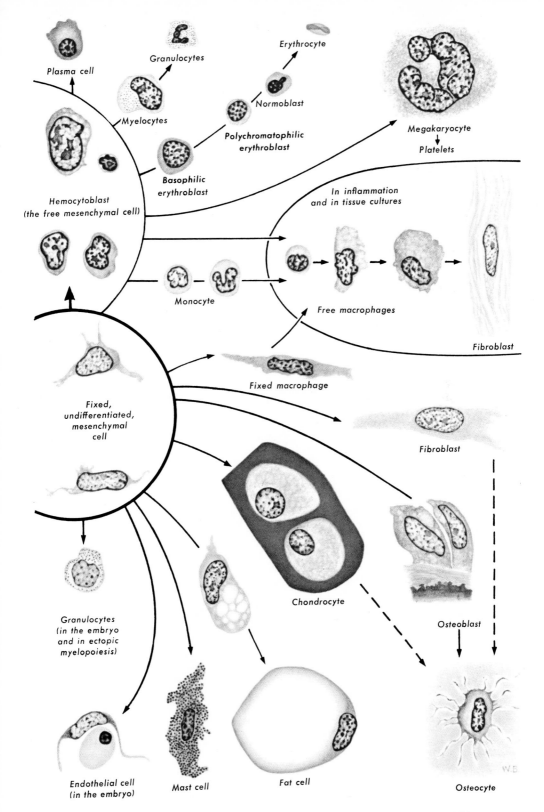

Fig. 11. Derivatives of the primitive mesenchymal cells of man (Bloom). × 720.

23

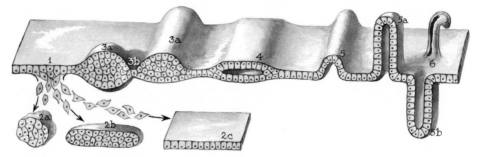

Fig. 12.   Stereogram illustrating the morphogenetic processes. Numbered as in text.

are created (*histogenesis*).[6, 7] Tissue differentiation culminates in the assumption of co-ordinated functional activities. Differentiation, in general, is favored by the thyroid influence and by certain amino acids.

   Morphogenesis.   This aspect of development includes all of the changes that occur during the molding of the body and its organs into form and pattern. The processes employed in morphogenesis are relatively simple acts. Although diverse in nature they occur in orderly and logical sequence, each part merely using whatever method may be appropriate to its needs at the moment. Viewed as a whole, the acquisition of form starts simply, becomes rapidly a scene of seeming confusion as many changes get under way, and then gradually stabilizes as the principal maneuvers are executed. Following the early period of great and varied activity, the later and longer periods of development are characterized by a much more leisurely perfecting of form.

   An underlying requisite to morphogenesis is, of course, cell multiplication and growth. The more important of the specific morphogenetic processes are the following (Fig. 12): (1) *cell migration;* (2) *cell aggregation,* forming (a) masses, (b) cords and (c) sheets; (3) *localized growth,* or its lack locally, resulting in (a) enlargements of various kinds and (b) constrictions; (4) *fusion* and *splitting,* which latter includes the delamination of single sheets into separate layers, the cavitation of cell masses and the forking of cords; (5) *folding,* including circumscribed folds which produce (a) evaginations, or out-pocketings, and (b) invaginations, or in-pocketings; (6) *bending,* which like folding is due to unequal growth; (7) *degeneration,* acting to remove temporary parts or excesses. Differential growth, resulting in enlargements, bendings and folds of all sorts, is the chief process utilized by the embryo in molding its general form and shaping new organs and parts.

   The word *primordium* (or its German equivalent, *anlage*) is a term applied to the cellular beginnings of a future tissue, organ or part before it has taken on its characteristic differentiation and grosser features. Thus the ectodermal thickening in front of the optic cup is the primordium of the lens, as is the arm bud of the arm.

   Histogenesis.   Cell specialization is controlled by genes in the chromosomes (p. 181). The most important constituent of these structures is *deoxyribonucleic acid* (DNA), which is able to transmit coded information to *ribonucleic acid* (RNA) located in the *ribosomes* of the cytoplasm. In this way instructions for specific protein syntheses are imposed on the ribosomes. The transfer of such directions is done by the genes impressing a message on 'messenger' ribonucleic acid (RNA), which acts as an intermediary.

All the cells of a germ layer are at first alike in visible structure and lack of specific shape. Many, nevertheless, already possess qualitatively different cytoplasms, acquired through the cleavage of an egg with regional specializations. Following an invisible chemo-differentiation (p. 176), with the elaboration of unique enzyme patterns, such cells progressively assume distinctive characters which permit their fates to be foretold. At this early period of differentiation in form and structure (histodifferentiation) they are often designated by the suffix -*blast*. Thus a neuroblast will in due time complete its differentiation into a nerve cell, and a myoblast into a muscle cell. The specialization that cells undergo in form and structure are conformable with the particular functions they will perform later and, in fact, anticipate those functions.

Cells of the same specialized type occur in larger or smaller groups and, thus set apart, come to be known as *tissues*. There are four main groups of tissues. Each of the germ layers gives rise to sheet-like *epithelia*. In addition, the ectoderm forms *nervous tissue* while the mesoderm (including mesenchyme) produces the different kinds of *muscle* and the various *connective tissues*. The total process by which cells differentiate into distinctive kinds and assume specific tissue characters is included in the term *histogenesis*. The histogenesis of an individual tissue summates all the departures it has made from the kind of cell it once was. In doing this the specific differences that separate it from other specialized cells which it once resembled are brought into sharp relief.

Illustrative of histogenetic differentiation is the history of the originally single-layered ectoderm. Such embryonic cells have relatively little cytoplasm; it is stained by basic dyes. This reaction reflects the presence of ribonucleic acid (RNA), which is necessary for protein synthesis as the cytoplasm specializes. The ectodermal cells proliferate and gradually change their form and character as they produce the layers of the epidermis (Fig. 401). More specialized epidermal products are the hairs, nails, lens of the eye and enamel of the teeth. Glandular derivatives of the ectoderm vary from the sweat and grease glands of the skin to the more highly organized tissue of the mammary gland, salivary glands and anterior lobe of the hypophysis. Other local specializations produce the sensory epithelium of the organs of smell, hearing and vision, and the smooth muscle elements of the iris. Part of the primitive ectoderm becomes the thickened neural plate, from which both nerve cells and supporting elements arise; a diagram of the lineage of these cells will illustrate a typical course of cell diversification during histogenesis (Fig. 418).

Cell growth and division are an integral part of the complete picture of tissue differentiation as it actually operates. Both daughter cells of a mitosis may continue to divide and grow; both may enter on differentiation; or one may continue as a proliferative cell and the other begin its differentiation. A differentiating cell may, for a time at least, interrupt its specialization and return to cell division. Cell differentiation within an embryo proceeds on different time schedules. Some lines advance steadily and rapidly toward their end stages. Other strains start later, are characteristically slower or indulge in rest intervals. Individual cells of many lines undergo arrest before completing their differentiation and persist indefinitely as *reserve elements*. Their differentiation, and usually division, can be resumed at any subsequent time at the call of an appropriate stimulus. In a tissue such as the epidermis the basal cells continue as proliferative stem-cells throughout life. From them arise cells which move to higher levels and progressively specialize, die and shed. In nervous tissue, on the other hand, all of the neuroblasts differentiate into mature neurons and a subsequent loss can never be replaced.

Differentiating cells may reverse their trend and return to a simpler state. Thus, under a changed environment, cartilage may lose its matrix and its cells come to resemble mesenchyme. This process is called *de-differentiation*. Nevertheless, despite such reversal and apparent simplification, these cells retain their former, specific potentiality. Under suitable conditions they can differentiate again and regain their previous definitive characteristics.

There is a certain antagonism between cell division and cell differentiation. Cells undergoing rapid division are in a state of turbulence which is unfavorable to cytoplasmic specialization. On the other hand, cells that are producing cytoplasmic elaborations of a physical nature tend to lose the plasticity that is requisite to mitosis. The cell types resulting from the processes of differentiation are discrete entities, without transitional forms; for example, an intermediate between a muscle cell and nerve cell is never seen. Neither can one region of a cell specialize in one direction (*e. g.*, toward muscle) and another region in a different direction (*e. g.*, toward nerve). In other words, once a cell becomes committed to any type of differentiation it cannot at the same time engage in another kind; nor can a cell abandon its original line of specialization and change to a different course. Moreover, any particular course of differentiation must be pursued in the distinctive way that characterizes the species to which an embryo belongs.

The path followed by cells during histogenesis shows certain trends which become evident when the conditions at the beginning and at the end of differentiation are contrasted:

Trends from Earlier Stages of Cells Toward Later Stages (Modified after Weiss)

From	Toward
uniformity (of size, structure and capacities)	diversity
irregularity (of shape)	regularity
vagueness (non-distinctive shape and qualities)	definiteness
dispersion (through whole embryo or part)	localization
variability (random arrangements or patterns)	stability
generality (primitive characteristics and qualities)	specialization
plasticity (or adaptability)	rigidity
mobility (ameboid and other shifts of position)	immobility
simplicity (of structure and function)	complexity

INTEGRATION

Morphogenesis and histogenesis are decentralizing processes which resolve the early embryo into a mosaic of organs and parts. During the course of development the organs become independent of former unifying controls, existent from the time of the egg. Although the new organs and organ systems possess structural coherence and unity, they need to be reintegrated into co-operative working mechanisms.[8] This control is supplied in part by the system of *endocrine glands.* Their rôle in activating, synchronizing and co-ordinating, by making use of the body fluids as carriers of their specific chemical substances, is important both among the later developmental phenomena and in ordinary physiological action. The other integrating instrument is the *nervous system* which constitutes the primary mechanism of physiological control and co-ordination.

The supplying of organs with adequate nervous, vascular and hormonal influences is a decisive factor in causing development to pass from a *prefunctional period,* which is preparatory and anticipatory in nature, to a *functional period* of actual (or potential) performance. The time of this transition varies greatly in different organs; moreover, growth and differentiation continue into the functional period.

REFERENCES CITED

1. De Robertis, E. *et al.* 1970. Cell Biology, 5th ed. Chapter 20 (Saunders).
2. Miller, O. L., Jr. & B. R. Beattie. Science, *164,* 955–957.
3. Beadle, G. W. 1945. Chem. Rev., *37,* 15–96.
4. Balinsky, B. I. 1970. An Introduction to Embryology, 3rd. ed. Chapter 18 (Saunders).
5. Scammon, R. 1953. Sect. 1 in Morris: Human Anatomy (Blakiston).
6. Warren, D. C. 1945. J. Heredity, *36,* 227–231.
7. Balinsky, B. I. 1970. An Introduction to Embryology, 3rd. ed. Chapter 17 (Saunders).
8. Weiss, P. 1939. Principles of Development (Holt).

Chapter III.   The Reproductive
Organs and Sex Cells

THE ORGANS OF GENERATION

The *ovaries,* or female sex glands, are ovoid bodies located in the pelvic region (Figs. 13, 32 *A*). Each is supplied with an egg duct, the *uterine tube,* which begins as an open, fringed tube in close relation with the ovary and discharges into the *uterus.* The latter is a muscular organ which serves as a gestation sac; from it the mature fetus is expelled through a combined copulatory and birth canal named the *vagina.*

The *testes,* or male sex glands, are ovoid bodies located outside the general body cavity in a double-chambered scrotal sac (Figs. 14, 285 *A*). Each male duct is given three names at different levels: there is the much coiled *epididymis,* where the male cells are stored, the long *ductus deferens,* and the terminal *ejaculatory duct* through which the male cells are expelled into the urethra and thence out of the body. Three *accessory glands* (seminal vesicles; prostate; bulbo-urethral) supply the bulk of the seminal fluid.

COMMON FEATURES OF SEX-CELL DIFFERENTIATION

The development of a multicellular animal is prefaced by the formation and ripening of the sex cells which will unite and give it origin. These germ cells, or *gametes* are generated within the sex glands of the male and female parents and are termed *spermatozoön* and *ovum,* respectively. The ovum, or egg, is a generalized type of animal cell produced in the ovary. Quite different is the spermatozoön which, differentiating in the testis, is a highly modified and atypical cell. It is the purpose of this chapter to describe how these two kinds of elements develop and mature. The course of events for sex cells in general is termed *gametogenesis.*

Origin of the Sex Cells.   First of all, it is important to inquire how closely Weismann's belief in a separate germ plasm (p. 5) agrees with actual observation. In some lower animals certain cells are set apart early as progenitors of the future sex cells, and it can be shown clearly that every egg or sperm arises from these cells and from no others. For example, when an embryo of the worm, Ascaris, consists of but two cells this specialization into germinal and somatic lines is distinguishable. At the 16-cell stage, one

28

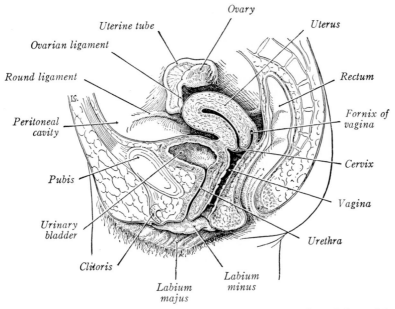

Fig. 13. The female genital system, shown in a sagittal hemisection of the pelvic region (Turner).

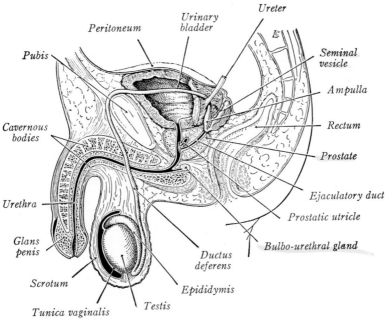

Fig. 14. The male genital system, shown in a sagittal hemisection of the pelvic region (Turner).

cell definitely limits itself to the further formation of nothing but sex cells; it is the first *primordial germ cell*. Similarly, in the early vertebrate embryo there can be recognized large, pale cells (often located at first relatively far from the sex glands) which appear comparable. In man and other mammals they are identified earliest in the yolk-sac entoderm, near the caudal end of the body. From there they migrate forward through the mesenchyme of the mesentery and into the genital ridge which soon becomes the sex gland (Figs. 279 *B*, 286 *A*).[1, 2]

A majority opinion inclines to the belief that these primordial cells become the definitive sex cells.[3, 4] Some, however, deny that the primordial germ cells ever become functional elements. Instead, they derive the definitive sex cells from the proliferation of ordinary cells located in the 'germinal epithelium' which surfaces the sex gland. In the female this proliferation is held to be periodic, month by month, during the potential child-bearing years.[5] From the standpoint of heredity it makes no practical difference which outcome is correct, since all cells of the body contain precisely the same assortment of chromosomes and genes.

The Course of Differentiation. The sex cells of all animals undergo similar histories in achieving maturity. Even the consecutive stages that an egg and sperm pass through in their individual developments are fundamentally comparable in all but the final outcome. The general process of egg-formation is *oögenesis;* that of sperm-formation is *spermatogenesis*. Each shows in succession three equivalent stages (Fig. 15): (1) a period of *cell proliferation,* during which the primitive germ cells divide repeatedly; (2) a period of *growth,* marked by rapid enlargement of the cells so produced; and (3) a period of *maturation,* which involves fundamental nuclear changes and is limited to the final two divisions. At the end of maturation

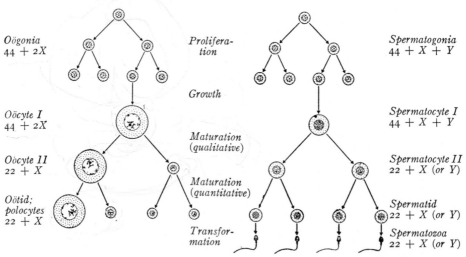

Fig. 15. Diagrams comparing oögenesis and spermatogenesis. The assortment of human chromosomes is indicated at each stage.

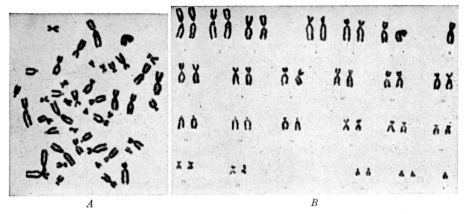

Fig. 16. Human chromosomes from a cultured marrow cell of a male (Ferguson-Smith). ×500. *A,* Ordinary 'squash' preparation. *B,* Chromosomes arranged in pairs according to standard numbering; X and Y are lettered. This is called the karyotype.

the development of an egg is complete and it is ready to function. The male cells, on the other hand, must pass through an additional stage (*transformation*) which converts them from ordinary appearing cells into specialized, motile spermatozoa.

The process of maturation would be of the greatest importance if only for the following reason. Since normal reproduction depends upon the union of male and female sex cells, it is manifest that without some special provision this union would necessarily double the number of chromosomes at each generation. Such progressive increase is, however, prevented by the events of maturation which reduce the number of chromosomes in each sex cell to one-half that characteristic of the species. The details of this process and its full significance will be described in later paragraphs (pp. 36, 46).

Chromosome Numbers. The cells of every animal species contain a definite and characteristic number of chromosomes. This number is identical for all the somatic cells of any animal and for its immature sex cells as well. The smallest possible chromosome assortment is two; it occurs in one form of Ascaris, a round worm. The largest number known is found in a moth, where 224 can be counted. The total number of chromosomes in each human cell has been revised as 46 for both man and woman.[6] It is important to understand that there is a double set of chromosomes in each cell; hence the human assortment contains 23 pairs. With one exception, each member of a chromosome pair is similar to its mate and possesses the same general potentialities. In the female there are 23 different kinds of chromosomes; in the male, however, the members of the sex-chromosome pair are unlike, thereby making 24 different kinds (Fig. 16).

OÖGENESIS

For convenience, the word 'egg' or 'ovum' is often used when referring to any stage in the course of differentiation of the female sex cell during oögenesis. In precise usage, 'ovum' designates the end-stage alone; in order to prevent any possible misunderstanding, the term 'mature ovum' is sometimes employed to designate such a ripe cell.

Origin of the Follicles.   During the fetal period of mammalian development, primordial cells (*oögonia*) proliferate within the cortex of the ovary. Late in fetal life, smaller epithelial cells, of germinal-epithelial origin come to encase the oögonia and so produce *primary follicles* (Fig. 287). Shortly after birth the formation of human oögonia comes to a halt and the cells are then termed *oöcytes*. The total number present at this time and later in each ovary varies considerably. At birth it averages about 370,000; at puberty, 190,000; at 30 years, 26,000.[7] Naturally enough, follicles in various stages of regression (*i.e., atresia*) are abundant at all times. Several years after the end of the childbearing span, follicles are no longer seen.

With occasional futile exceptions, there is no advance beyond the stage of the primary follicle until puberty, which occurs at about the fourteenth year. Thereafter, during the next thirty or more years that constitute a woman's reproductive period, larger follicles in various stages of growth are always to be found. As already stated (p. 30), these growing follicles are probably tardily activated primary follicles that have been dormant since their formation in the fetus, rather than being new structures proliferated by the germinal epithelium of the functional ovary as needed. Actually only a few hundred of the nearly one million oögonia that colonized both fetal ovaries survive the struggle for existence and eventually become liberated as cells ready for fertilization.

Growth of the Follicles.   All sexually mature mammals produce a crop of enlarging follicles during each cycle of ovarian activity (Fig. 17). Most of these follicles, by far, fail to achieve maturity and at some stage of growth succumb to retrograde changes; that is, they suffer *atresia*. At the start a human oögonium measures 0.02 mm. in diameter and its follicular covering consists of a single layer of flattened epithelial cells. Through growth

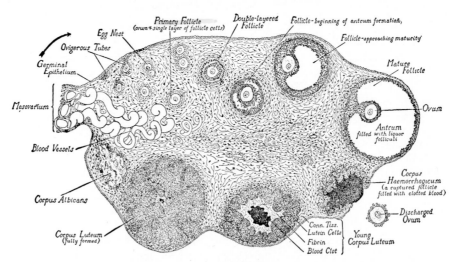

Fig. 17.   Life cycle of an egg and its follicle, shown in a diagram of the mammalian ovary (Patten). Start at the arrow and follow the stages clockwise around the ovary.

the *primary oöcyte,* as it is then called, increases in diameter seven-fold (0.14 mm.). In company with such enlargement on the part of the primary oöcyte, the follicle cells become cuboidal elements; they proliferate and form an encasement, several cells thick. At this time, when the oöcyte is full grown, irregular fluid-filled spaces appear between the follicle cells and then unite into a crescentic cleft (Fig. 17). Progressive enlargement of this cavity converts the original follicle into a definite sac, the *vesicular (Graafian) follicle* whose cavity *(antrum)* is filled with secreted follicular fluid *(liquor folliculi).* This type of follicle is characteristic of mammals alone, but the reason for such specialization is not well understood.

As the growth of the follicle continues, the oöcyte becomes located more and more eccentrically. It is buried in a mound of follicle cells, termed the *cumulus oöphorus* (egg-bearing hillock), situated at any chance position (Fig. 18). The follicle cells as a whole are arranged as a stratified epithelium, named the *stratum granulosum.* Around this epithelial layer the connective-tissue stroma of the ovary has been differentiating a sheath, the *theca folliculi.* The theca is composed of an inner, cellular and vascular *tunica interna* and an outer, fibrous *tunica externa.* In the final phase of marked growth the superficial portion of the follicle approaches closer to the surface of the ovary and raises it into a tense, blister-like elevation.

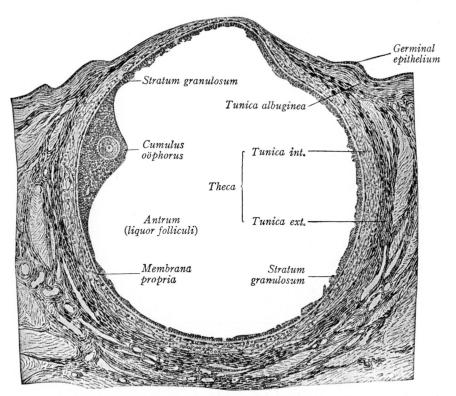

Fig. 18. Vesicular human follicle, with ovum approaching maturity, shown in median section (Bumm). × about 20.

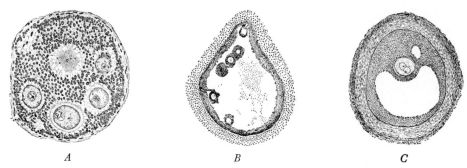

A   B   C

Fig. 19.   Abnormal follicles. *A*, Young, growing follicle of monkey, containing several oöcytes (Prentiss; × 300). *B*, Human vesicular follicle, containing eleven large oöcytes (Arnold; × 15). *C*, Precocious vesicular follicle from a human fetus, containing an oöcyte with two nuclei (von Herff; × 70).

Growth of the follicle is slow at the start, but a rapid increase occurs in the last day or two before rupture. The full-grown, human follicle is millions of times bulkier than was the primary follicle; its final diameter is about 14 mm. The endocrine function of growing follicles will be discussed in Chapter IX.

Abnormal Follicles and Eggs. Probably all mammals develop some follicles that contain more than one egg (Fig. 19 *A*), but this is infrequent in man (*B*). Although such compound follicles conceivably can lead to the production of twins, it is claimed that they usually degenerate.[8] Rarely an egg has a giant nucleus or more than one nucleus (*C*). The development of binucleate eggs in insects does not result in twinning, and presumably this is the case in mammals as well.

Maturation. After a primary oöcyte has finished its growth, the succeeding stages of oögenesis are devoted to the important process of *maturation*. The principal feature of maturation is two specialized nuclear divisions between which a resting nucleus is not restored, as in ordinary mitosis. During the first division each chromosome of a pair passes into a daughter cell; in the second division each remaining chromosome 'splits' in the usual way. Hence each of the four cells finally formed contains the 'reduced' number of chromosomes; that is, a complete single set of chromosomes replaces the double set that characterized the oögonium and primary oöcyte. This process of reduction, by an atypical method of cell division, is named *meiosis*. In the two cell divisions the chromosomes 'split' but once.

Maturation of the egg shows another unusual feature. Although the nuclei of all four cells are equivalent, the cytoplasm is divided very unequally so that the end-products are one large, ripe ovum and three rudimentary ova known as *polar bodies*, or *polocytes* (Fig. 15). The latter are so named because they pinch off at the 'animal pole' of the egg (p. 38). There is an obvious advantage in concentrating on the production of but one large, functional egg: it is destined to enter on a prolonged course of cell division and differentiation, and for this reason should retain all of the cytoplasm and yolk possible. In order to gain this advantage, the prospective ovum develops at the expense of the three polocytes which, having sacrificed their future, soon degenerate. In most animals only two polocytes are actually formed. This is because the first one to be given off fails to subdivide mitotically, although it may fragment as it nears dissolution. This omission is understandable since further multiplication would be a superfluous act.

The extrusion of the polar bodies from the egg of the mouse is illustrated in Fig. 20. At the end of the division that pinches off the first polar body (*A*) the primary oöcyte becomes a *secondary oöcyte.* The separation of the second polar body is shown as stage *B* and again, as photomicrographs, in *C* and *D*. When both polocytes have become free (*E*) the egg nucleus is reconstituted as the so-called *female pronucleus,* considerably smaller than it was before. Such a ripe ovum is technically an *oötid,* although not often called such. At the close of these maturative events the centrosome disappears and the nucleus, with its single set of chromosomes, is ready to unite with the similarly reduced pronucleus brought in by the sperm. Of practical interest is the fact that most animals whose gametes meet inside the body of the female expel technically unripe eggs from the ovary. Only the first polocyte has been cut off when the egg is set free, whereas the second never appears unless fertilization by a spermatozoön follows. This is another instance of economy of effort.

CHROMOSOME BEHAVIOR. The formation of the polar bodies, and especially the distribution of the chromosomes during meiosis, can be explained with the aid of diagrams

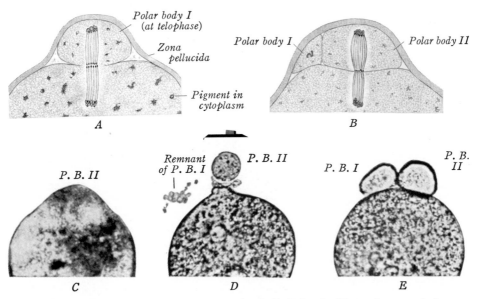

Fig. 20. Maturation of the mouse ovum. *A, B,* Polar bodies and part of the egg proper, sectioned (after Sobotta; × 1250). *C–E,* Formation of the second polar body, photographed from living eggs whose lower hemispheres are omitted. (Lewis and Wright; × 500.)

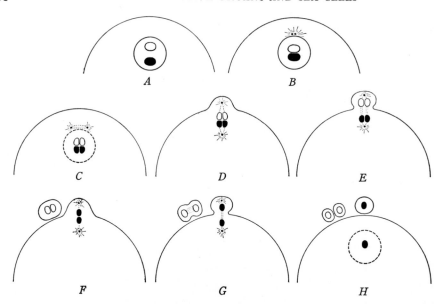

Fig. 21. Diagrams of the maturation of the ovum in an animal with two chromosomes. *A–E,* Budding off of the first polar body. *F, G,* Formation of the second polar body and subdivision of the first. *H,* Mature egg with polar bodies. (Full explanation in the text.)

(Fig. 21). For simplicity only two chromosomes are drawn (*A*). In reality these are a pair, one member (*e.g.,* black) having been inherited from the father of the present individual, the other member (*e.g.,* white) from the mother. Long before the prophase of the first meiotic division, the two chromosomes come to lie side by side (*synapsis; B*). Each chromosome then duplicates itself along its length in such a way that the two halves are identical in their genetic value. Each resulting pair is called a *dyad,* whereas the two dyads comprise a quadruple group known as a *tetrad* (*C*).

The tetrad next undergoes a subdivision which accompanies the budding off of the first polar body (*D, E*). At this division one dyad (the double copy of one original chromosome) passes into the polar body, and the other dyad remains behind in the egg. In the diagram the white (maternal) dyad is shown entering the polar body. In practice, chance alone determines which dyad of each tetrad group leaves, and which one remains. Since this first division separates the original pair of chromosomes of the egg into single (though 'split') chromosomes, it is termed *reductional.* The division is qualitative in nature.

The second meiotic division follows. One half of the remaining dyad (called a *monad*) passes into the second polar body and the other monad stays in the egg (Fig. 21 *F–H*). Meanwhile, in theory at least, the first polar body undergoes a similar division (*G, H*). This type of division is called *equational,* since each monad is the exact duplicate (except for possible cross-over differences) of an original chromosome. Hence the equational division differs in no essential way from an ordinary mitosis. Every chromosome pair of a primary oöcyte exhibits the same behavior during meiosis as did the example whose history has just been traced.

The tetrad, therefore, proves to be a group of four chromosomal elements, peculiar to meiosis. It is formed by the precocious doubling of each chromosome of a joined pair before even the first of the ensuing divisions occurs. It is useful because it anticipates properly two rapidly succeeding divisions, between which the customary resting periods do not occur. The duplicating ('splitting') of individual chromosomes preceding the prophase is also characteristic of an ordinary mitosis, but the mating and separation of whole chromosomes of a pair, combined with this duplication, occur in meiosis alone. The reduction of the original number of chromosomes characteristic of any species to one-half that number is often expressed by saying that the *diploid* (double) number has been reduced to the *haploid* (single) number. Some further information concerning the significance of meiosis in heredity will be presented at the end of this chapter.

HUMAN MATURATION. There are a few observations on the progress of maturation before the human follicle bursts. Figure 22 *A* shows the chromosomes of a primary oöcyte arranged in the meiotic spindle that will lead to the formation of the first polar body. Other ovarian oöcytes, with the first polar body actually cut off (and the metaphase spindle of the second in a state of arrest), have been described.[9] There is reason for believing that all these events take place during the last hours before the egg is set free.

A number of free eggs have been recovered by flushing out the uterine tubes (Fig. 22 *B*). These unfertilized specimens show no advance since leaving the ovary. Hence it seems certain that, as in vertebrates in general, the free egg continues unchanged until it is penetrated by a sperm. This act then furnishes the stimulus for cutting off the final polocyte; only then is the ovum truly mature. Fertilized ova, with both polar bodies formed, have been recovered from the tube,[10] and also produced by fertilization *in vitro*.[11] Since the full number of human chromosomes is 46, the reduced number in the mature ovum is 23 (Fig. 15).

The Mature Ovum. Although always relatively large, the final size of a ripe ovum is correlated with the amount of yolk substance stored in it and not with the size of the animal producing it. The smallest known egg is that of the mouse (0.07 mm.); the largest have a diameter measurable in

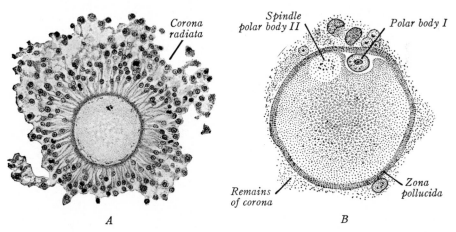

A   B

Fig. 22. Maturation of the human ovum. *A,* Primary oöcyte, with cells of the corona radiata, from a large, unruptured follicle (Stieve; × 155); near the top of the egg are the chromosomes of the first meiotic spindle. *B,* Secondary oöcyte, recovered from the uterine tube (after Allen; ×500); the first polar body has been cut off and the chromosomes left behind are grouped in the final maturation spindle.

inches (birds; sharks). Most animal ova are nearly spherical in form and all possess the usual cell components (Fig. 23). The *nucleus,* also sphe- roidal, is bounded by a *nuclear membrane;* it contains a *chromatin network* (*i.e.,* chromosomes in a diffuse state) and one or more *nucleoli.* The nucleus is essential to the life, growth and reproduction of a cell. Its chromatin bears the determiners of hereditary qualities (genes) by virtue of a content of deoxyribonucleic acid. The nucleolus stores ribonucleic acid, used in protein syntheses. The abundant *cytoplasm* is distinctly granular and con- tains few to many inert *yolk granules;* this nonliving substance has a nu- tritive value. The cytoplasm also contains such 'organelles' as the *mito- chondria* and *Golgi apparatus;* until the egg is finally ready for fertilization there is also a minute *centrosome.* These organelles are living, self-perpetu- ating parts, specialized beyond the general cytoplasm. The mitochondria contain enzymes involved in cell respiration, whereas the Golgi apparatus is related to secretory activity. The centrosome is active during cell division only.

The yolk, or *deutoplasm,* consists of fatty and albuminous substance which has been transported to the cytoplasm and stored as rounded granules. It serves as nutriment for the developing embryo. Since no type of ovum is totally devoid of yolk, this material is useful in classifying eggs. One classification is based on the relative abundance of yolk (*i.e.,* small, medium or large amount). Still more significant in relation to the mechanics of development is the distribution of yolk within the cell: (1) Those ova that contain little yolk tend to have it dispersed rather uniformly, and are accordingly termed *iso- lecithal* (*i.e.,* equal yolk) or *miolecithal* (*i.e.,* little yolk) (Fig. 23 *A*). Examples are found among the invertebrates and in all but the lowest mammals; such embryos have no need for much yolk since they either attain an independent existence quickly or are soon sheltered and nourished within the uterine wall of the mother. (2) As the yolk becomes more abundant it tends to concentrate in one hemisphere, and the ova are then said to be *telolecithal* (*i.e.,* yolk at end) (*B, C*). Various invertebrates and all vertebrates lower than marsupial mammals illustrate this type. If the yolk is moderate in amount, as in amphibians, the egg is often termed *medialecithal* (*i.e.,* medium yolk). The large, yolk- rich eggs of bony fishes, reptiles and birds are frequently termed *megalecithal* (*i.e.,* much yolk). The so-called 'yolk' of the hen's egg is actually a complete cell, taking the form of a highly telolecithal ovum; its huge size and yellow color are due to the enormous amount of stored yolk-substance. Similar in nature is the egg of monotreme mammals, such as the duckbill. (3) Among the arthropods the arrangement of yolk is distinctive. It is massed centrally and is surrounded by a peripheral shell of clear cytoplasm. Such eggs are *centro- lecithal* (*D*).

Eggs possess *polarity* which is made manifest in various ways. The *animal pole* is the site where the polar bodies pinch off. This general region of the egg tends to have the highest activity capacities and thus is more vigorous when development gets under way. At the other end of the polar axis is the *vegetal pole.* Its territory tends to be more sluggish and is concerned with the development of nutritive organs. Cytoplasmic components, such as pig- ment and yolk, are often disposed in a polarized or stratified way. This is well illustrated in telolecithal eggs, whose animal pole is more protoplasmic and whose vegetal pole is more yolk-laden; the nucleus lies nearer the animal pole.

The eggs of most animals become enclosed within protective membranes, or envelopes, which are *primary, secondary* or *tertiary* in character. The delicate *cell* (or *plasma*) *membrane,* innermost in position and elaborated by the egg cytoplasm, is a primary membrane (Fig. 23 *A*). The follicle cells about the ovum usually furnish some kind of secondary membrane; the conspicuous *zona pellucida* of mammalian eggs is assigned to this group (*A*).[12] Extensions of the follicle cells pass through radial canals in the zona and reach the surface of the egg, while microvilli at the periphery of the egg enter part way into the zona.[13] Tertiary membranes may be added by the oviduct as the egg passes through it. The jelly around the frog's eggs (*B*), the albumen about the rabbit's egg, and the albumen and shell of the hen's egg (*C*) are of this sort.

THE HUMAN OVUM.   There is little difference in the size of the eggs formed by the various placental mammals; mouse, man and whale are nearly equal in this respect. Such an egg is small in comparison with many ova; yet when set beside ordinary cells it is truly large, since it is just visible to

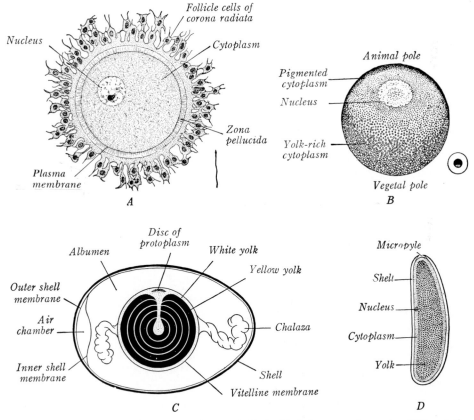

Fig. 23.   Representative types of ova. *A*, Isolecithal human oöcyte (× 200); at the lower right a human sperm cell is drawn to the same scale. *B*, Moderately telolecithal egg of the frog (× 15); at the lower right is a frog's egg surrounded by jelly (× 1). *C*, Highly telolecithal egg of the hen (× 1). *D*, Centrolecithal egg of the fly (× 35).

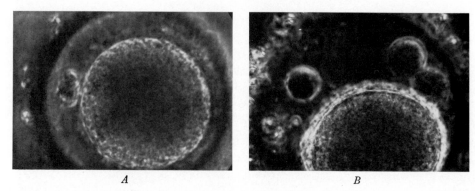

Fig. 24. Living human ova, shown as photomicrographs (Shettles). × 150. *A,* Secondary oöcyte, with first polar body (beginning to subdivide), lying within the space bounded externally by the dark zona pellucida; at the left, spermatozoa lie on the zona. *B,* Ripe egg, said to be artificially fertilized and showing all three polar bodies; unsuccessful spermatozoa are embedded in the locally elevated zona pellucida.

the naked eye as a tiny speck. The diameter of a normal human ovum, freshly discharged, is about 0.10 mm. Calculation shows that the three billion eggs necessary to replace the present population of this world could be contained within the shell of a hen's egg.

The human ovum contains fat globules and conforms closely to the isolecithal mammalian type (Figs. 23 *A,* 24). The plasma membrane is represented merely by the limiting cytoplasmic boundary and is not a distinct envelope in the ordinary sense. Outside the ovum proper lies a thick, tough and highly refractile capsule, the zona pellucida; it increases the total diameter of the egg to about 0.15 mm. Abnormal eggs with giant or double nuclei occur, but they are uncommon (Fig. 19 *C*).

SPERMATOGENESIS

The Course of Differentiation. The sex cells of male vertebrates develop within thread-like testis tubules (Fig. 285 *A*). The latter originate as cellular cords that are derived from the germinal epithelium which covers the sex gland of an embryo. Such a solid cord contains cells of two types (Fig. 25 *A*). The larger are immigrated primordial germ cells which, acting as stem cells, proliferate and become *spermatogonia;* the smaller are so-called indifferent cells. Until the time of sexual maturity these are the only elements to be seen, but then a renewal of proliferate activity and the onset of spermatogenesis advance the testis to its full functional state. Also at puberty the solid epithelial cords first become hollow in man. Two types of cells are then recognizable in the relatively thick wall of a tubule (*B, C*): (1) The male *sex cells,* in various stages of development, arranged in a layered fashion; they are all descendants of the spermatogonia. (2) Tall *sustentacular cells* (of Sertoli), derived from the indifferent cells; they act as columnar supports and apparently serve as nurse cells.

As spermatogenesis gets under way some spermatogonia remain as stem cells, while others enter upon a growth period during which they are called *primary spermatocytes* (Fig. 25 *C*). Up to this stage each cell still contains the full number of chromosomes typical for the species. Next follow the two meiotic divisions which accomplish maturation. Each primary spermatocyte divides into two *secondary spermatocytes,* and each of these, in turn, into two *spermatids.* During these two divisions the cells not only decrease progressively in size, but also the number of chromosomes reduces to half the original number. That is, the double set is reduced to a single set in a way identical with that already described for the egg (p. 36). Finally, all the spermatids become engulfed by the cytoplasm of Sertoli cells, from which they appear to receive nutriment, and gradually transform from typical cells into highly specialized *spermatozoa.* Nothing corresponding to this period of transformation occurs in the development of an egg. When it is complete, the spermatozoa detach and are set free inside the seminiferous tubule. A comparison between oögenesis and spermatogenesis is shown diagrammatically in Fig. 15.

HUMAN SPERMATOGENESIS. The process of sperm formation begins at puberty, extends far past the corresponding time limit in the female, and may persist even into extreme old age. In man, like other animals that do not have a special breeding season, spermatogenesis is continuous. At any

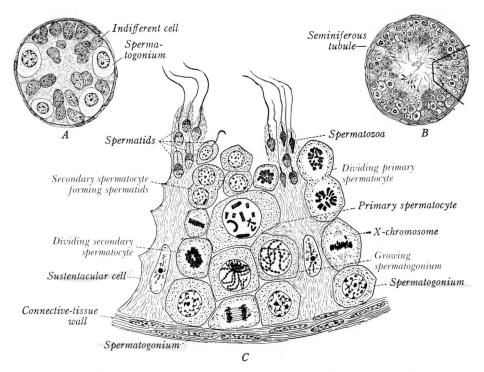

Fig. 25. Human testis tubules, in transverse section. *A,* Newborn ($\times$ 400); *B,* adult ($\times$ 115); *C,* detail of the area outlined in *B* ($\times$ 900).

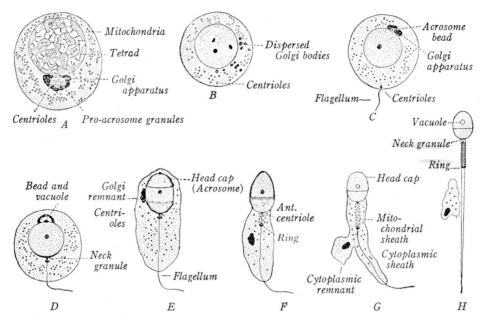

Fig. 26. Stages in human spermatogenesis (adapted after Gatenby and Beams). × 1000–2500. *A,* Spermatocyte; *B,* spermatid; *C–G,* transformation stages; *H,* spermatozoön.

level in a tubule different stages of spermatogenesis occur.[14] A total wave, showing wholly different stages at successive levels, does not progress along the tubule as in some common mammals. The complete course of specialization that produces spermatozoa consumes about two months.[15] The events of human spermatogenesis are typical and agree with the general description already given (Fig. 25 *B, C*).

All the spermatogonia carry the full number of 46 chromosomes. Some of their daughter cells (primary spermatocytes) engage in a period of marked growth. Figure 26 *A* shows such an enlarged cell, with the chromosomes arranged as *tetrads* in 23 mated pairs, preparatory to the first meiotic division; Fig. 36 *A* shows the division in progress. During the division of the primary spermatocytes into secondary spermatocytes, the 23 tetrads separate into two groups, each with a single set of 23 chromosomes already split as *dyads.* Since this division disjoins whole chromosomes of each pair, it is reductional. The secondary spermatocytes then divide equationally into spermatids, each dyad separating into two *monads* (Fig. 36 *B, C*). At the end of meiosis, therefore, each spermatid contains 23 single chromosomes.

The four spermatids derived from each spermatogonium complete their development by undergoing a direct transformation (*spermiogenesis*) into highly specialized spermatozoa.[16, 17] This involves a remodeling of cell shape and a superficial disguising of certain of the typical cell components, as is illustrated in Fig. 26 *B-H.* The nuclear history is simplest, the open-structured nucleus merely condensing greatly and reshaping into the main bulk of the sperm *head.* One of the two centrioles gives rise to a typical *flagellum;* this represents the *axial filament* of the motile *body* and *tail* of the final spermatozoön. Part of the Golgi apparatus becomes the *acrosome,* which constitutes a *head cap* covering the apical half of the head. Mitochondria collect into a spiral *sheath,* located in the body of the spermatozoön. The general cytoplasm is drawn closely around the shaping head and continues as a fairly conspicuous *cytoplasmic sheath* around the body and tail of the spermatozoön.

Details of the participation in the process of transformation by the specialized organ-oids of the cytoplasm are as follows: The Golgi apparatus of the spermatocyte consists of a dark-staining periphery and a paler interior which contains granules within vacuoles (Fig. 26 *A*). In the young spermatid this material, which had become dispersed during the spermatocyte divisions (*B*), reassembles at one point on the surface of the nucleus (*C*) and a distinct bead within a vacuole is again seen in the midst of darker-staining material. The bead applies itself against the nuclear membrane (*D*). It becomes the *acrosome;* the vacuole spreads under the cell membrane and forms the *head cap* that covers the apical half of the head (*E–G*). The peripheral Golgi substance about the vacuole (*D*) detaches, passes down the side of the nucleus (*E*) and finally is cast off along with an unused remnant of cytoplasm (*F–H*).

Meanwhile, the centrioles have migrated from a position between the nucleus and Golgi apparatus (Fig. 26 *A*) to the margin of the cytoplasm (*B*). From the posterior centriole there grows out a threadlike *flagellum,* or future *axial filament* (*C*). Both centrioles and the base of the attached flagellum then move inward to the nucleus (*D*). The *anterior centriole* remains close to the nuclear membrane (*F*). It lies just beneath a neck granule (*D, F*) that causes the nuclear membrane to thicken locally into the so-called *basal plate.* The fate of the *posterior centriole* is uncertain; some believe that it disappears. Meanwhile, a *ring* (or *annulus*) arises near the posterior centriole, seemingly from a dense chromatoid mass (*D, E*). It encircles the axial filament and slips down it for a short distance (*F–H*). Mitochondrial granules, originally scattered throughout the cytoplasm (*A–E*), gather in an end-to-end spiral about the axial filament (*F, G*). This *sheath* extends from the anterior centriole to the ring (*H*).

The Spermatozoön. Rarely (in a few invertebrates only) does the mature male element, or spermatozoön, resemble a typical cell. Most are slender, elongate structures which develop a huge, cilium-like lash; this flagellum whips back and forth to accomplish the active swimming that characterizes the cell. Unlike the egg, which is the largest cell of an organism, the sperm is among the smallest in mass. The extremes of length in animals range from 0.018 mm. in Amphioxus to 2.25 mm. in a toad. Some bizarre types occur, but the commonest shape is that of an elongate tadpole, with an enlarged *head,* short *neck* and thread-like *tail.* The head shows many variations in form (rod; lance; spiral; sickle; spoon; sphere; cone); the tail may bear along its length a fin-like, undulatory membrane.

THE HUMAN SPERMATOZOÖN. At one time the human sperm cells were regarded as parasites, and under this misapprehension the name spermatozoa, or 'semen animals,' was given them. The spermatozoön of man is of average size and shape (Fig. 27). Although its length is nearly one-half the diameter of a human ovum, the relative volume is only as 1:85,000 (Fig.

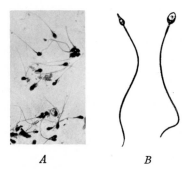

Fig. 27. Human spermatozoa. *A*, Photograph of smear (× 210); *B*, drawings in edge and flat view (× 700).

A　　　　　*B*

23 *A*). All the sperms required to replace the present population of this world could be packed into a spherical vessel having the diameter of 5 mm. The tiny size of spermatozoa makes their structural details difficult to interpret with the ordinary microscope; recent studies with the electron microscope have demonstrated details far more intricate than the main features shown in Fig. 28.[18]

a. The *head* measures nearly 0.005 mm., or one-twelfth the total sperm length of 0.06 mm. It appears oval in flat view, pear-shaped in profile. The interior of the head contains the tightly packed chromosomes of the sperm cell; it is homogeneous in appearance, except for a frequently found vacuole. The anterior half of the head is invested with the *head cap,* a portion of the cell membrane overlying the *acrosome.* Covering the head closely, and also surrounding the remainder of the spermatozoön, is a *cytoplasmic sheath.*

b. The short *neck* begins with a problematical *neck granule,* in contact with the head, and extends to the *anterior centriole.*

c. The *body,* often considered to be a 'connecting piece' belonging to the tail, is slightly longer than the head. Its lengthwise extent is limited by the anterior centriole, located close to the head, and by the *ring,* or *annulus.* The central core (*axial filament*) is surrounded by a double spiral composed of two *mitochondrial strands.*

d. The *tail,* or flagellum, continues the *axial filament* which begins next the anterior centriole and courses the entire length of the tail. It consists of a central pair of fibrils encircled by nine other pairs. The tail shows different portions: (1) The tapering *chief piece* constitutes three-fourths of the total length of the sperm. Here a *fibrillar sheath* encloses the axial filament, and enveloping all is a continuation of the cytoplasmic sheath. (2) The *end piece,* or terminal filament, is a thinner thread since it is merely the termination of the axial filament, clothed with a local portion of the cell membrane.

Abnormal Spermatozoa. Atypical sperm cells occur in all mammals, including man. These may include giant and dwarf forms, badly modeled specimens and elements with more than one head or tail (Fig. 29). Human males of proved fertility have a fair number of their sperms showing such abnormal types. But more than 20 per cent with abnormal form or less than 70 per cent of actively motile sperms is indicative of infertility.

Comparison of the Egg and Sperm. The dissimilar male and female sex cells of animals are admirably adapted to their respective rôles. They illustrate nicely the modifications that accompany a physiological division of labor. Each has the same amount and species-kind of chromatin, although in the sperm it is more compactly stored. Both cells are thus capable of participating equally in heredity, but in certain other respects each is specialized both structurally and functionally.

The ripe egg contains an abundance of cytoplasm, and often a still greater supply of stored food (yolk). As a result, it is large and passive, yet closely approximates the typical cell in all features except that the pre-

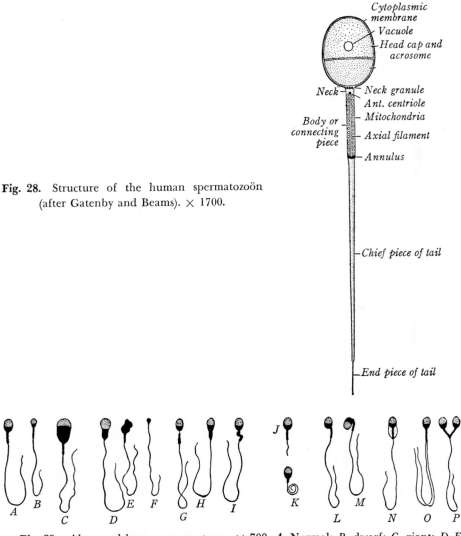

Fig. 28. Structure of the human spermatozoön (after Gatenby and Beams). × 1700.

Fig. 29. Abnormal human spermatozoa. × 700. *A,* Normal; *B,* dwarf; *C,* giant; *D–F,* abnormal head; *G–I,* abnormal neck and body; *J, K,* abnormal tail; *L, M,* abnormal alignment of head and body; *N,* immature (much cytoplasm); *O,* double tail; *P,* double head, neck and body.

viously active centrosome has disappeared. It is only in some invertebrates, however, that the cell by itself is normally capable of cell division and development.

On the other hand, the sperm is small and at casual inspection bears slight resemblance to an ordinary cell. Its cytoplasm is reduced to a bare minimum; it contains a centrosome (in the form of a centriole), but no yolk. Structurally all is subordinated to a motile existence. Functions such as constructive metabolism and cell division are sacrificed. Correlated with the small size of sperms goes an extraordinary increase in numbers, for the greater the total liberated, the more surely will the egg be found. Hence, apart from its rôle in heredity and sex determination, the chief function of the sperm is to seek out the egg and activate it to divide.

COMPARISON OF THE EGG AND SPERM OF ANIMALS IN GENERAL

FEATURES COMPARED	OVUM	SPERMATOZOÖN
Size	Large	Small
Shape	Spheroidal	Elongate
Quantity	Less than sperms (Sometimes but few)	Large numbers (Often millions)
Motility	Lacking	Vigorous (flagellate)
Protection	Egg envelopes	None
Cytoplasm	Bulky	Minimal (mostly in body and tail)
Yolk	Little to much	Lacking
Centrosome	Disappears	Retained (centriole; ax. filament)
Mitochondria	Diffuse	In body only (spiral filaments)
Golgi apparatus	Diffuse	In acrosome only
Nucleus	Open structured	Condensed, as sperm head
Nucleolus	Typical	Indistinguishable
Sex determining rôle	Rarely two kinds (Moths; birds)	Usually two kinds ('Male' and 'female' sperm)

THE SIGNIFICANCE OF MEIOSIS

At meiosis there is a side-by-side association of like chromosomes (one member of each pair having come from the father, the other from the mother of the preceding generation) (Fig. 30 *A*). Each member of a chromosome pair (except the sex-chromosome pair of the male) carries the same general set of hereditary characters as does its mate. The individual genes of any gene-pair within the two chromosomes, however, may be like or unlike in their power of inducing the alternative expressions of a particular character (*e.g.*, for eye color they might be brown-brown, blue-blue or brown-blue). The reducing division of meiosis separates whole chromosomes of each pair, but chance alone governs the actual distribution of the paternal or maternal member of any pair to any particular daughter cell. Reduction obviously halves the chromosome number characteristic for the species, and as a result each daughter cell receives a single (but complete) assortment of chromosomes instead of the double set. This process of reduction necessarily segregates the genes of each pair into separate cells, so that the eggs and sperms of any species are produced with different genic compositions. The second, or equational division of meiosis, not only accomplishes cellular multiplication but also provides an opportunity for the distribution of reshuffled genes as the result of an interchange 'crossing over') of chromosome parts (**Fig. 30 *B***).

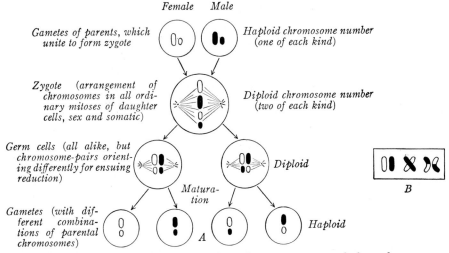

Female   Male

Gametes of parents, which unite to form zygote — Haploid chromosome number (one of each kind)

Zygote (arrangement of chromosomes in all ordinary mitoses of daughter cells, sex and somatic) — Diploid chromosome number (two of each kind)

Germ cells (all alike, but chromosome-pairs orienting differently for ensuing reduction) — Diploid

Matura-tion

Gametes (with different combinations of parental chromosomes) — Haploid

A

B

Fig. 30. *A*, Diagram of the behavior of four chromosomes traced through one generation. *B*, Diagram of stages in a 'cross-over.'

In man nearly seventeen million different final combinations of chromosomes are possible through reduction alone. Vast as this number is, it represents only part of the full picture since the possible recombinations at an ensuing fertilization are measured by the square of seventeen million. A further increase in new hereditary combinations is made possible by the phenomenon of 'crossing over,' already mentioned. This occurs at the stage of chromosomal conjugation in meiosis when the two members of a chromosome pair intertwine and then recombine after an interchange of corresponding parts (Fig. 30 *B*). Reduction and crossing over furnish a basis for variations in the hereditary pattern that find an expression when two cells with such equivalent histories (sperm and egg) meet and pool their single sets of chromosomes in the zygote.

REFERENCES CITED

1. Witschi, E. 1948. Carnegie Contr. Embr., *32*, 67–80.
2. Narbaitz, R. 1962. Carnegie Contr. Embr., *37*, 115–119.
3. Zuckerman, S. 1960. Mem. Soc. Endocrin., No. 7, 63–70.
4. Pinkerton, J. H. N. *et al.* 1961. Obst. & Gyn., *18*, 152–181.
5. Evans, H. M. & O. Swezy. 1931. Mem. Univ. California, *9*, 119–224.
6. Ford, C. E. 1960. Am. J. Human Genetics, *12*, 104–117.
7. Block, E. 1952–53. Acta Anat., *14*, 108–123; *17*, 201–206.
8. Harrison, R. J. 1949. Nature, *164*, 409–410.
9. Hamilton, W. J. 1944. J. Anat., *78*, 1–4.
10. Dickmann, Z. 1965. Anat. Rec., *152*, 293–302.
11. Shettles, L. B. 1958. Am. J. Obstet. & Gyn., *76*, 398–406.
12. Chiquoine, A. D. 1960. Am. J. Anat., *106*, 141–160.
13. Anderson, E. & H. W. Beams. 1960. J. Ultrastr., *3*, 432–446.
14. Clermont, Y. 1963. Am. J. Anat., *112*, 35–45.
15. Heller, C. G. & Y. Clermont. 1963. Science, *140*, 184–185.
16. Gatenby, J. B. & H. W. Beams. 1935. Quart. J. Micr. Sci., *78*, Pt. 2, 1–29.
17. Burgos, M. H., & D. W. Fawcett. 1955. J. Biophys. & Biochem. Cytol., *1*, 287–300.
18. Schultz-Larson, J. 1958. Acta path. et Microbiol. Scand., Suppl. No. 128, 121 pp.

Chapter IV.   The Discharge and Union of Sex Cells

When the eggs and spermatozoa of animals are ripe, they are released from their respective sex glands. In one way or another the two kinds of sex cells are brought into close proximity, whereupon some successful ones meet and unite. The setting free of egg and sperm cells is known as *ovulation* and *semination,* respectively; their union is *fertilization.*

OVULATION

The discharge of the ovum from its follicle comprises *ovulation.* In general, those animals whose offspring attain an independent state with reasonable surety (as the result of development within the mother's body and of parental care after birth) produce far fewer eggs than do those that leave fertilization to chance and development to hazard. The codfish lays 5,000,-000 eggs in a breeding period; an oyster, 50,000,000. By contrast, certain birds and mammals mature only a single egg at a time. Yet the end-result is the same, since all animal types maintain their numbers equally well.

A few animals breed continuously, but most commonly there is a seasonal or annual spawning period. The several mammalian groups show all gradations between ovulation every few days and an annual breeding period. As a whole, lower mammals ovulate spontaneously at the time of sexual excitement, known as the period of heat or *estrus;* only then will the female receive the male. A few, however, such as the rabbit, have 'provoked ovulation', induced only by the act of copulation. Except in the special case of identical twinning, a separate egg is ripened and expelled for each individual developed. It follows that multiple ovulation characterizes many mammals and is an occasional occurrence in man.

In some primates, including man, ovulation is periodic, at intervals of about four weeks, and spontaneous. Such mammals have no restricted period of heat, and hence the urge to mate is not limited to the time of ovulation. The human female begins to ovulate in early adolescence, following puberty (about fourteenth year), and continues until the menopause (about forty-seventh year). Although some large vesicular follicles can be found rather constantly in the ovary between later fetal life and puberty, such precocious follicles eventually degenerate with their contained eggs.

48

As a rule only one follicle (and its egg) matures each month, the ovaries alternating with irregular and unpredictable sequence.[1] Thus, from the hundreds of thousands of potential ova provided, only about 200 ripen in each ovary during the thirty-odd years of sexual activity. There are, nevertheless, many thousands of follicles that reach various degrees of advancement during this period and then, as *atretic follicles,* decline and disappear.

Follicle Rupture. The ripening follicle expands rapidly in the last hours of its existence and, in the human, reaches a diameter of some 15 mm. This expansion causes the surface of the ovary to bulge locally. Here the ovarian wall is stretched thin and at the apex of the area there is a clear, avascular spot named the *stigma* (Fig. 18). Rupture is the final act in a progressive growth process, but the direct antecedent is seemingly not a critically increased pressure by follicular fluid. Rather there is a hormonal factor and local weakening of the follicular wall.[2] The bursting point is at the thinned stigma; as the follicular fluid wells through this gap, the egg becomes torn away from its previously loosened bed in the cumulus oöphorus and rides along in the fluid wave (Fig. 32 *A*). The adhering follicular cells, still investing the egg, constitute the so-called *corona radiata* (Fig. 23 *A*).

The act of ovulation has been observed directly in several mammals, including man (Fig. 31).[3, 4] During the final few minutes of the still intact follicle the thin stigma dilates into a pimple-like cone. Rupture at the tip follows quickly, but it is not explosive. The stretched apex merely opens, the fluid-content of the follicle flows slowly out, often accompanied by a slight hemorrhage, and the follicle gradually collapses as the egg is torn free and swept out. The general course of events resembles somewhat the development and rupture of a boil. The total mechanism of ovulation is not sufficiently understood. Along with the cellular, secretory and vascular changes goes a nicely balanced action of hormones (p. 162).

Egg Transport. Ovulation expels the egg into the peritoneal cavity, but actually the 'cavity' is a restricted space, bounded by the moist surfaces of those organs in the immediate vicinity of the ovary. Some observers have

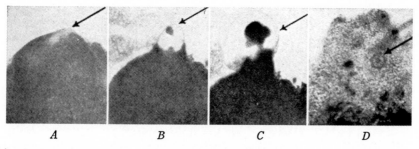

| A | B | C | D |

Fig. 31. Ovulation in the rabbit, recorded in frames from a motion picture (after Hill). *A,* Follicle, with beginning cone (arrow). *B,* Large cone, containing some blood. *C,* Follicular rupture; extrusion of gelatinous material and blood. *D,* Detail of exudate, attached to follicle (below) and containing ovum (at arrow).

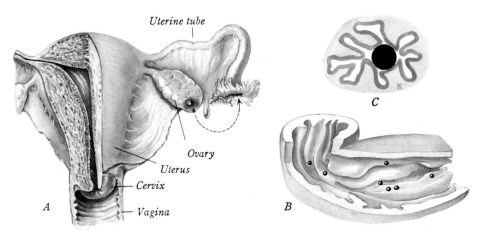

Fig. 32. The parts responsible for egg-transport. *A,* Human ovary, with ruptured follicle, and its associated genital tract (× ⅔); the uterine tube has been displaced from its normal contact with the ovary. *B,* Reconstructed segment of the uterine tube of a rat, opened to show the exact positions of seven eggs on the folded lining (after Huber; × 30). *C,* Lining of the human uterine tube, sectioned transversely at its narrow, lowest level to show an egg drawn to scale (× 42).

watched the fringed end of the uterine tube embrace the ovary and sweep over its surface at the general time of ovulation.[4, 5] It is probable, therefore, that the liberated ovum passes almost directly into the tube without ever gaining the body cavity in any real sense. Both the beating cilia of the tubal lining and the augmented waves in the muscular wall of the tube at this period have been held responsible for directing the egg into the oviduct and then transporting it downward (Fig. 32). Perhaps ciliary activity is more effective in picking up an egg, but there are some reasons for thinking that hormonal control of peristaltic movements is an important factor in moving it down the tube (p. 119).[24]

Passage downward is much slower as an egg reaches the slender, uterine end of the tube, but the reason for this retardation is not surely understood. It is certain that the short-lived egg at times gets across from one ovary to the tube of the other side. This transfer is indicated by the fact that a pregnant tube can occur on one side while the sole corpus luteum (a later stage of a discharged follicle) is located in the opposite ovary. Also there are records of a pregnancy occurring subsequent to the removal of one tube and the opposite ovary. Movements of the pelvic viscera might be thought to accomplish such transport in an accidental manner; on the other hand, several observers have seen a human uterine tube in intimate contact with the surface of the opposite ovary, so the transfer could be quite direct.[6]

The Time of Ovulation. Both human ovulation and menstruation begin with puberty, recur at about 28-day intervals, and discontinue during pregnancy and at the menopause. It is natural that some relation in timing should have been inferred. For many years ovulation and menstruation were supposed to take place synchronously, like 'heat' and ovulation in lower mammals. But when actual data were collected it became apparent that this assumption is untrue. For instance, in 100 patients with normal menstrual cycles, ovulation was found by operative check[7] to occur between days 8 and 19 (counting from the first day of bleeding), and most frequently on days 13 to 15. Additional data obtained by interpreting cervical mucus and an ovulation test set the modal time of ovulation at the thirteenth day.[8] One can conclude that most ovulations occur at the midpoint of an ordinary 28-day cycle, just as they do in the monkey.

It is known that half of all ovulations of the monkey occur on days 12 and 13, yet the full range is from day 8 to 23.[9] Likewise in women, a considerable variation can be expected about the midcycle, average time. How much departure occurs is not surely known—either in any individual or in womankind as a whole. Also some women habitually have cycles shorter or longer than the usual 28-day type. In these persons ovulation takes place sooner or later, respectively, than in the more typical cycle. For this reason many urge that the postovulatory (progesterone controlled) period in the cycle is more constant in length than the preovulatory (estrogen controlled) period, and hence it is more correct to set ovulation at 14 days before the onset of the next menstrual bleeding.

Egg Viability. How long the human egg retains its ability to receive a sperm and then start developing cannot be stated with certainty. In lower mammals this period of competence is brief—a matter of hours rather than days. Similarly, the monkey becomes pregnant only when mated just before its time of ovulation. For the human ovum it is now generally believed that the fertilizable period is less than one day.

If a mammalian egg does not become fertilized, it begins to degenerate while still in the oviduct. In the guinea pig a functional decline enters as early as eight hours after ovulation, and the egg becomes unfertilizable after 20 hours; a physical deterioration is visible within 24 hours. Most of the unfertilized eggs that have been recovered from the uterine tubes of both the monkey and woman showed signs of degeneration.

The Corpus Luteum. Following ovulation the collapsed and folded vesicular follicle transforms into a new ovarian structure, named the *corpus luteum (i.e.,* yellow body) (Fig. 17). It is peculiar to those vertebrates that bring forth living young, and is especially characteristic of mammals. The presence of a corpus luteum is a positive indication of previous ovulation; an expert can estimate the age of a corpus luteum with fair accuracy from its state of development. The so-called *lutein tissue,* which characterizes the corpus luteum, comes chiefly from the enlargement of the stratum granulosum cells of the old follicle (*granulosa lutein cells*) (Fig. 33 *C*). The internal layer of the adjacent theca provides an additional source at the periphery, from which specialized elements (*theca lutein cells*) surround and indent the main granulosa mass (*C*). Within a few days the corpus luteum organizes into a prominent, highly vascularized mass whose structure is typically that of an endocrine gland. There is usually some bleeding

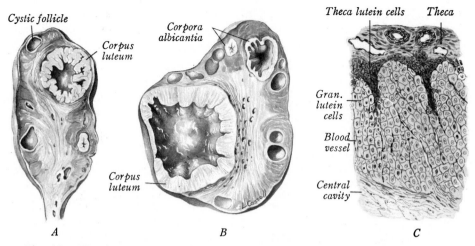

Fig. 33. The human corpus luteum. *A*, Hemisected ovary, containing a mature corpus luteum of ovulation (× 1.2). *B*, Hemisected ovary, containing a mature corpus luteum of pregnancy (×1.5). *C*, Structural detail of a sample region from the wall of a corpus luteum (× 75).

into the collapsed follicle at the time of ovulation, and blood may continue to collect within the central cavity in the remaining days of the menstrual cycle.[10]

The subsequent history of the corpus luteum varies only in the final size and length of life attained. When pregnancy does not supervene, this endocrine body is called a *corpus luteum of ovulation;* in this instance it reaches full size (1–2 cm.) and maturity in eight days (Fig. 33 *A*).[10] Within a few more days secretory exhaustion is followed by a rapid decline. Among the signs of involution is the increase of a fatty pigment which gives the human corpus luteum its characteristic yellow color. Replacement of the regressing corpus luteum by fibrous tissue produces a white scar, the *corpus albicans* (*B*); several months elapse before all traces of it have disappeared. When conception occurs, the transformed follicle, now known as the *corpus luteum of pregnancy,* continues to enlarge until, at eight weeks, it may reach a final diameter of 2 to 3 cm. (*B*). Deterioration is rapid in the second and third months, and again after the completion of pregnancy;[11] conversion into a typical corpus albicans follows quickly. The functional rôle of the corpus luteum will be discussed in Chapter IX.

SEMINATION

In most aquatic animals the eggs and sperm are discharged externally at about the same time and place. Their meeting depends largely upon chance, enhanced by the production of immense numbers of sex cells. Some animals increase the certainty of such cell union by *pseudocopulation;* thus the male frog clasps the female and pours his spermatic fluid over the eggs as they are extruded. On the other hand, many invertebrates and most vertebrates including all reptiles, birds and mammals, have their sex cells unite within the genital tract of the female. This is brought about by the sexual embrace termed *copulation,* or *coitus,* during which the penis of the male serves as a specialized intromittent organ. The biological purpose of coitus is to introduce spermatic fluid into the female, and this deposit constitutes *semination.*

Sperm Storage. At the conclusion of spermiogenesis the spermatozoa of mammals detach from the Sertoli cells. Clusters are moved along the efferent ductules into the epididymis where they separate but remain motionless. The sperms accumulate in the epididymis (which is traversed in about 12 days); any storage in the seminal vesicles is incidental.[12] A physiological ripening, both as regards potential motility and fertilizability, takes place in the epididymis as the sperms are forced onward by newer arrivals. Spermatozoa gradually attain their full functional state, retain it for a limited period, and, if not discharged, then slowly decline in vigor until death supervenes.

Sperm Discharge. At the male climax, during coitus, *ejaculation* occurs. Involuntary muscular contractions forcibly eject the older spermatozoa, along with the secretions of several accessory glands which discharge in rapid sequence.[13] The aggregate mass is the *seminal fluid,* or *semen.* It is a mixture composed chiefly of the secretions of the seminal vesicles, prostate and bulbo-urethral glands, in which are suspended the spermatozoa. The volume of the ejaculate is about 3 ml., and in it swim some 250,000,000 sperms. In his lifetime an average man discharges a billion sperms for every egg that leaves a woman's ovary. Experiments on mammals show that 1 per cent of the normal ejaculate is sufficient to ensure fertilization.

Sperm Transport. The outstanding functional feature of spermatozoa is their lashing, flagellate swimming which resembles that of a tadpole. This property is confined to the tail, and its center of control is apparently located in the body of the sperm (anterior centriole?).[14] Movement is first exhibited after ejaculation when the hitherto quiescent sperm cells are aroused to maximum motility by the combined ingredients of the seminal fluid at body temperature. Forward progress of the human spermatozoön is at the rate of 1.5 to 3 mm. a minute which, in relation to their respective lengths, compares well with the swimming ability of man. On the whole, spermatozoa swim in an aimless fashion; but under certain, perhaps artificial, conditions they orient passively against a feeble current (rheotaxis) and then continue to swim a spiral course upstream.

These innate activities, however, play but little part in the transport of sperms through the female genital tract, since it is largely the muscle in the wall of the several passages that propels the sperms to the vicinity of the ovary.[15] Rabbit spermatozoa complete their upward journey four hours after ejaculation, whereas in the cow and sheep the time spent is about five minutes. In a series of experiments utilizing excised human uteri and tubes, human spermatozoa required 70 minutes to reach their destination (a distance of 7 inches from the uterine cervix).[16] In all mammals the total number of sperms reaching the site of fertilization, in the upper region of a tube, is relatively small; for the rabbit 500 (one out of every 200,000 ejaculated) gain this goal.[17]

The passage of spermatozoa from vagina to uterus is apparently the result of muscular movements of the cervix, and the time occupied is less than a minute in some mammals. The journey through the uterus is similarly accomplished, in some mammals at least, by muscular propulsion; sperms of the dog appear at the ends of the uterine horns within one minute after ejaculation occurs. In the human, sperm ascent through the uterus depends on muscular contraction of the uterine wall;[18] moreover, animal experiments prove that dead sperms pass up the tube as fast as do live ones.

Sperm Viability.   Two important questions arise in connection with the activities of spermatozoa in the female genital tract. One concerns the length of time during which the discharged sperms live and move; the other, and far more important query, is how long such elements actually retain their ability to unite with an egg and activate it. Adequate observations on many animals prove that the ability to fertilize is lost first, so that the mere fact that spermatozoa swim does not necessarily imply that they can still fertilize. In the rabbit, ram and bull they cease to be competent after some 30 hours.

The data for man are based partly on knowledge but mostly on inference. Human spermatozoa have been kept alive *in vitro* for two weeks, and they are said to live even longer in the male sexual ducts following castration. Nevertheless, a comparable length of life and function within the female tract cannot be inferred. Sperms are never present in large numbers in the upper regions of a uterine tube and they disappear within two or three days after coitus.[19] There is no good reason for believing that the duration of fertilizing capacity extends beyond a day or two. That the sperm may lie in wait for the egg, or the reverse, for any considerable period of time is contrary to experience, since impregnation is relatively infrequent in comparison to the number of sperm deposits received by the human female.

When spermatozoa are discharged into the female mammal, they immediately encounter hazards to their longevity. The acidity of the vagina is deleterious or fatal; ingestion by leucocytes is prevalent. Also, these specialized elements undergo a rapid decline in vigor because of their limited amount of potential energy. Previously they were spared from rapid catabolism by their inactivity while stored in the male ducts.

The fertilizing span of sperms within the genital tract of female mammals varies between six hours (mouse) and six days (mare), although the usual time limit is not more than one day. Exceptional are some bats, inasmuch as coitus occurs in the autumn whereas ovulation and fertilization are delayed until the end of hibernation in the spring.[20] Among other vertebrates the hen is known to retain functional sperms in its oviducts for as long as four weeks, while a period of four years has been claimed for the terrapin.

The sparing action of lowered temperature is notable. The use of cooled or frozen spermatic fluid in the delayed, artificial insemination of cattle is now standard practice.

Similarly human semen, frozen in glycerol at −70°C., yields high motility in surviving spermatozoa after many months of storage. Semen, thus preserved for weeks, has even been used for the production of normal babies.[21]

FERTILIZATION

The formation, maturation and meeting of a male and female sex cell are all preliminary to their actual union into a combined cell, or *zygote,* which definitely marks the beginning of a new individual. This penetration of ovum by spermatozoön, and the coming together and pooling of their respective nuclei, constitutes the process of *fertilization* (Fig. 34). In practically all animals fertilization also supplies the stimulus that starts the ovum dividing and thus sets off development in the ordinary sense. The eggs of certain invertebrates (rotifers; crustaceans; insects), however, develop regularly without being fertilized. This method is styled *parthenogenesis* (virgin origin), and in such cases there is often but one polar cell and no reduction in the number of chromosomes. In some instances the sperm of one species is able to fertilize successfully the egg of another species. The hybrid progeny is usually infertile, like the mule, but often possesses greater size and vigor than either parent.

The Events of Fertilization. Both the male and female sex cells have to be in a proper state of maturity if union is to occur. The time when the egg becomes receptive to a sperm varies slightly in different animal types. It may be before maturation begins, after it is completed or at any inter-

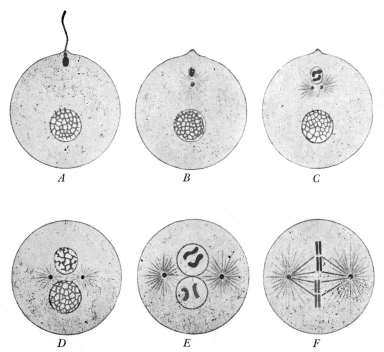

A   *B*   *C*

D   *E*   *F*

Fig. 34.   Semidiagrammatic stages of the events in fertilization (Howell, after Boveri).
The chromatin of the ovum is colored blue, that of the spermatozoön red.

mediate stage. In vertebrates as a whole it is necessary for the first polar body to be extruded; not until then does penetration by the sperm begin. In almost all mammals the second polar spindle is also present but in a state of arrest (Fig. 22 B). Only during the preliminary events of fertilization does this second meiotic division go through to completion. To insure success in its rôle, a spermatozoön must still possess high motility and, like the egg, must be in the functionally potent phase intermediate between under- and over-ripeness. Both egg and sperm must reside in the female tract some five hours before they are fully prepared (*capacitated*) for fertilization.[15]

PENETRATION. Random movements bring the sperm cells in contact with an egg. There is no real proof of any actual chemical attraction, but the secretions of some eggs may serve to trap sperms accidentally entering their sphere of influence. A positive tactile response (thigmotaxis) also tends to keep the sperm head in contact with anything touched. The spermatozoa of mammals propel themselves past the cells of the corona and attach to the surface of the secondary egg membrane (zona pellucida). Only motile sperms are able to make this attachment. The mammalian sperm head penetrates the zona pellucida, after which the lashing movements quickly cease and the one successful male element is passively engulfed by the egg cytoplasm and drawn inward, usually tail and all (Figs. 34 A, 35 A).[22] Some competing, unsuccessful spermatozoa commonly embed in the zona pellucida (Fig. 43 A), or even penetrate into the space between it and the egg proper.

The eggs of mammals and many other animals can be entered at any point by a sperm. On the other hand, sperms avoid the yolk-laden pole of an egg like that of the frog. Even more severe restrictions attend the eggs of fishes, molluscs and insects which are invested with heavy membranes; these egg capsules usually have a definite, funnel-shaped aperture, the *micropyle,* through which the male cell must enter (Fig. 23 D). In many animals, including mammals, only one sperm normally gains entrance into an egg; others, endeavoring to penetrate, are excluded in some incompletely understood way. If accident or reduced vitality admits more than one sperm, the condition is termed *polyspermy;* development then is abnormal and soon ends. On the other hand, some eggs (for the most part, those heavily laden with yolk) regularly exhibit polyspermy. In all such instances, nevertheless, only one sperm actually unites with the egg nucleus; all others perish more or less promptly, without having contributed in any significant way to the main course of development.

Mammalian spermatozoa secrete an enzyme, *hyaluronidase,* that may be localized in the acrosome[23] and thus aid an individual sperm cell in passing through the cells of the corona radiata and penetrating the zona pellucida.[24] The denudation of the zona by the dispersal of corona cells occurs somewhat later through the influence of a tubal factor whose nature is not well understood.[25]

BEHAVIOR OF THE PRONUCLEI. The sequence of events in fertilization proper is illustrated in Figs. 34, 35. Once within the periphery of the egg, the sperm head rotates, end-for-end, and advances toward the center of the egg. During this journey the head swells, becomes open-structured, and converts into a nucleus of typical appearance which is given the special name of *male pronucleus*. At about this time the tail detaches from the rest of the spermatozoön, but it does not disappear from sight until some-what later. Both the mitochondrial granules and the Golgi substance of the sperm disperse into the egg cytoplasm and fragment.[26] During the progress of these events the final maturation division of the egg has been completed and the now smaller, reconstituted egg nucleus (*female pronucleus*) has been made ready for union.

The climax of fertilization is reached when the two pronuclei approach, meet and unite. In some animals they actually fuse and so produce a *cleavage nucleus*. In others, including mammals, each pronucleus loses its nuclear membrane and resolves its chromatin into a complete single set of chromo-somes. Each set then contributes its chromosomes as a unit to the first cleavage division (Fig. 34). Meanwhile a centrosome, presumably the an-terior centriole of the sperm, appears between the chromosome groups and divides into two. Following these preliminaries the first cleavage spindle organizes, with the double set of male and female chromosomes arranged midway as an equatorial plate. The full chromosome number, temporarily halved in each gamete by maturation, is thus restored in the zygote. Fertili-zation is now complete and the egg, freed from its previous restraints,

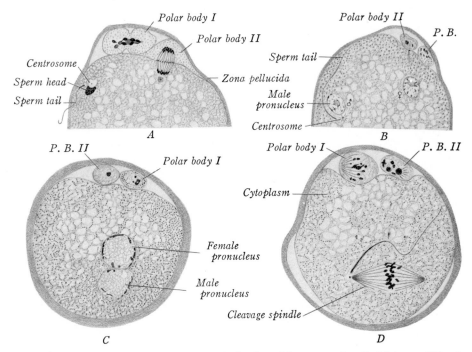

Fig. 35. Fertilization of the ovum of the bat (after van der Stricht). $\times$ 650. *A,* Entrance of spermatozoön, as second polar body cuts off. *B,* Male (♂) and female (♀) pronuclei, about to approach. *C,* Pronuclei, ready to merge; centrosome present at left of pronuclei. *D,* Spindle of first cleavage division.

divides in the ordinary mitotic way. In the rabbit the total events of fertilization occupy 11 hours.

Human Fertilization.   The meeting and union of the human sex cells is believed to take place normally in the upper third of the uterine tube. It is altogether certain that fertilization cannot be delayed until the ovum reaches the uterus, since staleness and degeneration enter rather soon. Presumably it does not take place even in the lowest levels of the uterine tube; at least, degenerating human ova have been recovered from the tube, and the unfertilized eggs of most mammals begin to show visible signs of decline by the time they near the uterus. The final fate of unfertilized eggs is dissolution in the uterus.

Mature eggs have been recovered from the human uterine tube in a stage that shows the discarded tail of a successful spermatozoön and both pronuclei present in the cytoplasm.[27] Attempts to fertilize *in vitro* human oöcytes removed from large ovarian follicles have been successful, followed by cleavage to the morula stage. Even greater success (including living young) has been attained with the rabbit.[28]

The Results of Fertilization.   It is worth while to emphasize again that the male and female cells, merging in fertilization, are each in a sense defective but complementary to the other. Thus the general cytoplasm and yolk are supplied by the egg alone, whereas the sperm probably brings in the functional centrosome (except in parthenogenesis when it must arise within the egg). Both the egg and sperm contribute equally to the requisite nuclear substance; the egg alone may contribute mitochondrial and Golgi bodies. By such pooling of the materials of the two sex cells there results a new, joint product, again characteristic of the species.

The fundamental results of fertilization are: (1) Reassociation of the male and female sets of chromosomes, thus restoring the full diploid number. In this regard, fertilization is the antithesis of maturation which separates the two sets of chromosomes. By bringing together equivalent chromatin contributions from two different parents (and thus restoring the typical number of chromosome pairs), there is furnished a physical basis for bi-parental inheritance and for variation. (2) Determination of the sex to be established in the new individual (p. 59). (3) Activation of the ovum into cell division, or *cleavage*. As the result of the first cleavage mitosis, and all subsequent ones, every cell of the developing body receives a sample of each kind of chromosome that was pooled at fertilization. It should be clearly understood, however, that mitotic activation is not dependent on the presence of two pronuclei or on their union, as natural or artificially induced parthogenesis proves. Such union is, nevertheless, an end and aim of normal fertilization.

Infertility.   Nearly 15 per cent of marriages are barren among couples desiring children; about two-fifths of these failures are chargeable to the husband. Sterility may be congenital: certain anomalies make conception difficult or impossible (*e.g.*, insufficient development of the penis, vagina or uterus); or gonads may be present, but incapable of producing mature, competent sex cells, due to genetic or hormonal causes. Motile spermatozoa may be produced, but in inadequate amounts (less than one-fourth the normal number). Disease may interfere with the proper functioning of the sex organs (*e.g.,* alcoholism; endocrine disturbances), or produce hostile exudates and even inflammatory occlusion of the sexual passages (*e.g.,* gonorrhea; tuberculosis). Psychic factors can exert an influence (impotence or premature ejaculation in the male), but frigidity or failure to experience orgasm by the female is not, of itself, responsible for infertility.

Parthenogenesis. Under normal conditions the eggs of vertebrates do not develop spontaneously to a significant degree. In the ovaries of certain mammals some cell clusters that simulate cleavage may occur, but they probably represent a degenerative fractionation of the egg. At best, stages of this sort do not extend into organized body-building.

Artificial parthenogenesis designates development induced in eggs by chemical or other means. Examples among invertebrates and lower vertebrates can be produced easily as a routine laboratory exercise. The rabbit is most responsive among mammals yet tested; artificial stimulation of oöcytes, then returned to the female genital tract, has resulted in living young.[29]

Superfetation. To fulfill the requirements of superfetation it is necessary that a pregnant female ovulate, conceive and produce a second, younger fetus. In the early months of human pregnancy superfetation is theoretically possible, but an active corpus luteum is known to prevent subsequent fertilization in experimental animals. A few apparent examples have been recorded for lower mammals and at least one suggestive case for man.[30] However, it is difficult to exclude an interpretation of strikingly unequal twins in which one member has experienced retardation in size and differentiation.

Superfecundation. This term designates the impregnation, by successive acts of coitus, of two or more eggs that were liberated at approximately the same time. In those lower mammals that are characterized by multiple births, superfecundation is known to occur; in such instances litter-mates can have different fathers. Its possibility in man cannot be doubted and evidence now exists, from ordinary twins belonging to different blood groups, that cannot be accounted for in any other way.[31]

SEX DETERMINATION

Sexuality is established at the time of fertilization. The decisive factor in the determination of future sex is the particular kind of spermatozoön that fertilizes the egg. The chromosomes related to sex determination in all mammals are a matched XX pair in females and a dissimilar XY pair in males. After halving of the chromosome number has occurred during meiosis, all mature eggs are alike in their chromosomal constitution, and the human assortment consists of 22 ordinary chromosomes and one sex chromosome (22 + X). Spermatozoa, on the other hand, are of two kinds (Fig. 15): in man one group contains 22 + X while the second category, equal in frequency, contains 22 + Y. In the latter instance the Y is a diminutive chromosome that serves as a mate to the X of the male before reduction separates them and distributes each to a different daughter cell (Fig. 36 *A, B*).

Fertilization of any egg (22 + X) by one kind of sperm cell (22 + X) results in a female (44 + 2X). Fertilization by the other kind of sperm cell (22 + Y) produces a male (44 + X + Y). It is the presence of the Y-chromosome in the cells of an embryo that promotes the formation of a testis, regardless of the presence or absence of the X-chromosome. Hence infertile

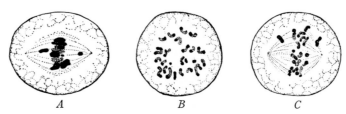

A   *B*   *C*

Fig. 36. Stages of human spermatogenesis showing the behavior of the sex chromosomes during meiosis (after Severinghaus). $\times$ 2000. *A,* Primary spermatocyte in metaphase of first meiotic division; the division is reductional and the X- and Y-chromosomes have separated precociously. *B,* Polar view of the equatorial plate of a secondary spermatocyte, containing the small Y-chromosome (at left and below center), about to divide. *C,* Secondary spermatocyte at the metaphase of the second meiotic division; this division is equational and the two X-chromosomes (monads) have separated and preceded the autosomes.

individuals with such combinations as XXY, XXXY, or XXXXY are still gonadic males. Correspondingly, it is the absence of the Y-chromosome that allows femaleness to assert itself both in the normal female (XX) and in sterile individuals with only a single X.[32, 33]

The two types of sperm cells are equal in number and no significant advantages of size or weight exist between the two kinds.[34] Hence it follows that pure chance should govern which representative will be successful in impregnating any particular egg. For this reason attempts to sway sex determination in either direction have been unsuccessful and offer little promise at present. There is a world-wide preponderance (United States and Europe, 106:100) of live-born males over females, and an even greater disparity among the still-born. The reason for this inequality is not apparent.

In some animals the X-chromosome of the male is without a mate (X O). In birds, certain reptiles and some fishes the sex-determining system is the exact reverse of that already described, inasmuch as the sperms are all alike in chromosomal constitution while the eggs are of two sorts. Actually the details of the sex-determining mechanism are both varied and complex. In some lower organisms the environment is the decisive factor that directs sex into one channel or the other. On the other hand, sex-regulating genes are highly potent in insects. In vertebrates they are, at least, chiefly effective in early stages of development, whereupon their influence is reinforced by gonadal hormones (p. 328).

'Resting' somatic cells of various female mammals, including man, display a small mass of stainable chromatin attached to the nuclear membrane (Fig. 37). This *sex*

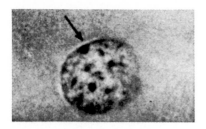

Fig. 37. Sex chromatin (at arrow) in an epithelial cell of a woman (Benirschke). $\times$ 2,000.

chromatin, or *Barr body,* is an inactive X-chromosome. Although normal male cells (X-Y) lack this stainable mass, the sex chromatin does appear in the cells of abnormal males with an XXY constitution. Even though the test tells nothing positive about the presence or absence of the Y-chromosome, it is, nevertheless, a useful criterion in attempting certain diagnoses. It can be employed in embryos before sex is otherwise recognizable (as early as three weeks[35]), for predicting sex before birth, and for establishing the sex of individuals whose external features (including genitalia) are puzzling.[36]

REFERENCES CITED

1. Morse, A. H. & G. van Wagenen. 1936. Am. J. Obst. & Gyn., *32*, 823–828.
2. Blandau, R. J. 1960. Thirty-fifth Pediat. Res. Conf., 39–41.
3. Blandau, R. J. 1955. Fertil. & Steril., *6*, 391–404.
4. Doyle, J. B. 1954. Fertil. & Steril., *5*, 105–129.
5. Westmann, A. 1937. J. Obst. & Gyn. Brit. Emp., *44*, 821–838.
6. Caffier, P. 1936. Zentr. f. Gyn., *60*, 2466–2470.
7. Brewer, J. I. & H. O. Jones. 1947. Am. J. Obst. & Gyn., *53*, 637–644.
8. Corner, G. W. 1952. Brit. Med. J., *2*, 403–409.
9. Hartman, C. G. 1944. Western J. Surg., Obst. & Gyn., *47*, 149–184.
10. Corner, G. W., Jr. 1956. Am. J. Anat., *98*, 377–401.
11. Nelson, W. W. & R. R. Greene. 1950. Am. J. Obst. & Gyn., *76*, 66–90.
12. Stieve, H. 1930. Bd. 7, Teil 2 in Möllendorff: Handbuch (Springer).
13. Mann, T. & C. Lutwak-Mann. 1951. Physiol. Rev., *31*, 27–55.
14. Cody, B. A. 1925. J. Urol., *13*, 175–191.
15. Bishop, D. W. 1961. Sect. D13 in Young: Sex and Internal Secretions (Williams and Wilkins).
16. Brown, R. L. 1944. Am. J. Obst. & Gyn., *47*, 407–411.
17. Braden, A. W. H. 1953. Austr. J. Biol. Sci., *6*, 693–705.
18. Bickers, W. 1960. Fertil. & Steril., *11*, 286–290.
19. Belonoschkin, B. 1939. Arch. f. Gyn., *169*, 151–183.
20. Wimsatt, W. A. 1944. Anat. Rec., *88*, 193–204.
21. Bunge, R. G. *et al.* 1954. Fertil. & Steril., *15*, 520–529.
22. Austin, C. R. & A. Walton. 1960. Chapter 10 in Marshall: Physiology of Reproduction, vol. 1, pt. 2 (Longmans).
23. Colwin, A. L. *et al.* 1957. J. Biophys. & Biochem. Cytol., *3*, 489–502.
24. Blandau, R. J. 1961. Sect. D14 in Young: Sex and Internal Secretions (Williams and Wilkins).
25. Chang, M. C. 1950. Ann. N. Y. Acad. Sci., *52*, 1192–1195.
26. Gresson, R. A. R. 1941. Quart. J. Micr. Sci., *83*, 35–59.
27. Zamboni, L. 1966. J. Cell Biol., *30*, 579–600.
28. Chang, M. C. 1959. Nature, *184*, 466–467.
29. Pincus, G. 1939. J. Exp. Zoöl., *82*, 85–129.
30. Murless, B. D. 1937. Brit. Med. J., 1309–1311.
31. Geyer, E. 1940. Arch. f. Rassenbiol., *34*, 226–236.
32. Carr, D. H. 1961. Anat. Rec., *139*, 214.
33. Russell, W. L. 1961. Science, *133*, 1795–1803.
34. Bishop, M. W. H. & A. Walton. 1960. Chapter 7 in Marshall: Physiology of Reproduction, vol. 1, pt. 2 (Longmans).
35. Park, W. W. 1957. J. Anat., *91*, 369–373.
36. Barr, M. 1959. Science, *130*, 679–685.

Chapter V.   Cleavage and Gastrulation

The Periods of Early Development. There is a high degree of unity in the early development of all multicellular animals. The essential identity of (1) *gametogenesis* (including maturation) and (2) *fertilization* through the entire range of animal groups has already been emphasized. The next phases of early development, which are fundamentally similar in all animals, are named (3) *cleavage* and (4) *gastrulation*. All such sequential 'stages,' of course, blend without sharp limits, but the recognition of characteristic periods of development is highly convenient for descriptive purposes.

Directly after the union of the male and female sex cells, the fertilized egg (the organism-to-be) enters on a series of mitotic cell divisions which provide the building units for the future organism and give the first external sign of development in the ordinary sense of that term. This initial process in the production of a new, many-celled individual is called *cleavage*. By it the egg is subdivided into many smaller cells which typically arrange themselves into a hollow sphere, the *blastula*. An important advance in organization is then accomplished by *gastrulation*, through which the cells of the blastula become redistributed as the *primary germ layers*. These are three in number and from their positions are named *ectoderm, mesoderm* and *entoderm*. They contain the material out of which the embryo and all its parts will differentiate.

The Vertebrate Groups. Although the development of man is the main theme of this book, it is both necessary and desirable to refer from time to time to conditions in lower animals, and particularly in other vertebrates. *Chordates* are animals characterized by the possession of an axial, rod-shaped support known as the *chorda dorsalis* or *notochord*. Most important of the lower chordates to embryology is Amphioxus whose development has furnished considerable fundamental information. The highest group of chordates is the *vertebrates* whose provisional notochord is replaced by a skull and vertebral column. Vertebrates fall into five classes. The three highest groups possess an enveloping embryonic membrane named the amnion, and this feature is the basis of a convenient classification:

A. ANAMNIOTA (amnion absent).
  1. *Fishes:* lamprey; sturgeon; shark; bony fishes; lung fish.
  2. *Amphibians:* salamander; frog; toad; etc.
B. AMNIOTA (amnion present).
  3. *Reptiles:* lizard; crocodile; snake; turtle.

62

4. *Birds:* characterized by wings and feathers.
5. *Mammals:* characterized by hair and mammary glands.
   a. Monotremes: duck-bill; echidna. Primitive mammals possessing a cloaca, like lower vertebrates. They lay large eggs with shells.
   b. Marsupials: opossum; kangaroo; etc. The young are born immature and are sheltered and nourished in a pouch of the abdominal skin.
   c. Placentalia: All other mammals; their unborn young are nourished in the uterus by means of a placenta. The highest order is the Primates with specialized 'nails' (lemur; monkey; ape; man). Since the lemuroid group is considerably different from other primates, the monkey, ape and man are conveniently placed in the suborder of Anthropoids.

CLEAVAGE

Cleavage progressively splits the fertilized egg into smaller cells, termed *blastomeres* (Fig. 38). Cleavage divisions are always mitotic and each daughter cell receives the full double assortment of chromosomes, one set of which came from each parent. The increase in blastomeres tends typically to follow the doubling sequence, 2, 4, 8, 16, etc., although in practice the regularity of this series of divisions is disturbed sooner or later, and thereafter becomes irregular. In most animals the mitoses follow in relatively quick succession. Consequently, at each division the blastomeres are reduced progressively in size until finally the size-relation between the originally overlarge cell bodies and their nuclei is normal. In a strict sense, therefore, cleavage is a fractionating process, which provides building units of convenient size, rather than a process of truly constructive development. The total mass of living substance, available for development, has not increased when cleavage comes to an end. Although there has been no growth in the cleavage cells as a whole, their nuclei do enlarge somewhat and synthesis maintains the proper amount of nucleic acid in the chromosomes.

The cluster-stage of cohering, sticky blastomeres is sometimes called a *morula* from its general resemblance to a mulberry. By this time the blastomeres tend to be arranged about a central, free space. Their continued subdivision produces the *blastula,* whose central, fluid-filled cavity is the *blastocœle,* or cleavage cavity. In its simple, typical form the blastula is a hollow sphere of cells.

Studies indicate that there is a definite organization of the egg that largely governs the pattern and rate of cleavage. The significance of the blastula stage is that it contains specific regions, destined in normal development to give rise to the major components of the future embryo. These districts are located in advantageous positions for participating in the events of gastrulation that follow immediately.

It is the active protoplasm of the egg that accomplishes division. The inert, stored yolk-substance is not involved beyond acting as an impediment which retards the process of mitosis, and even prevents it from extending into overdense regions. In this way, however, the relative amount of yolk and its even or uneven distribution throughout the egg have a profound, deterring influence on cleavage and the mechanics of moving the germ layers into their final positions. Yet, in spite of the different amounts of hindering yolk, the processes at work and the results accomplished are fundamentally comparable in all vertebrate types. The simplest explanation of this basic uniformity is the directing influence of a common inheritance which labors as best it can with eggs variously endowed with yolk.

On the basis of the abundance and distribution of yolk, cleavage is classified as follows:

A. *Total.*   Entire ovum divides; *holoblastic ova.*

     1. *Equal.*   In isolecithal ova; blastomeres are of approximately equal size; *e.g.,* Amphioxus, marsupials and placental mammals.

     2. *Unequal.*   In moderately telolecithal ova; yolk accumulated at the vegetal pole retards mitosis, and fewer but larger blastomeres form there; *e.g.,* lower fishes and amphibians.

B. *Partial.*   Protoplasmic region alone cleaves; *meroblastic ova.*

     1. *Discoidal.*   In highly telolecithal ova; mitosis is restricted to the animal pole; *e.g.,* higher fishes, reptiles, birds and monotreme mammals.

     2. *Superficial.*   In centrolecithal ova; mitosis is restricted to the peripheral cytoplasmic investment; limited to arthropods.

Observations on cleavage bring to light certain general principles which can be formulated as rules. Nevertheless, these should not be regarded as invariable laws because they are occasionally disturbed by other, incidental influences.

1. A mitotic spindle occupies the 'center of protoplasmic density' of the egg or blastomere in which it lies. (Hence, in an isolecithal ovum the spindle is located centrally; in a telolecithal ovum it is nearer the animal pole.)

Corollary: Blastomeres divide into two equal parts unless the yolk is unevenly stored in them.

2. The axis of a spindle occupies the longest axis of the protoplasmic mass in which it lies. (Evident in ovoid blastomeres.)

Corollary: The ensuing plane of division cuts across this long axis, and the daughter cells revert to a more spheroidal shape.

3. Each new division plane tends to intersect the preceding plane at right angles. (Acts to maintain the spheroidal shape of blastomeres.)

4. The speed of cleavage is inversely proportional to the amount of yolk encountered. (In telolecithal ova, blastomeres at the animal pole divide faster than those nearer the vegetal pole.)

An understanding of cleavage and gastrulation is best gained through a comparative approach. In this way the increasing influence of yolk can be appreciated. Also the information obtained from lower vertebrates can be used to explain certain conditions in mammals that are puzzling except on the basis of a yolk-rich ancestry.

Amphioxus.   This fish-shaped animal is a representative of the low chordates. Its early stages of development are similar to those of some invertebrates, including the widely studied echinoderms. The almost microscopic egg (0.1 mm) contains a small amount of yolk which is somewhat concentrated at one end, the vegetal pole. Yet for all practical purposes it can be considered isolecithal. Cleavage is total and nearly equal. About one hour after fertilization the egg divides into two blastomeres, the plane of this first cleavage passing through the egg axis from pole to pole (Fig. 38 *A*). Soon the daughter cells again cleave in a vertical (meridional) plane, but at right angles to the first plane, thus forming four cells (*B*). In the third series of divisions the plane of separation is horizontal (*C*). As the yolk is somewhat more abundant toward the vegetal pole, the four mitotic spindles lie nearer the animal pole; consequently, in the resulting eight-celled stage the upper tier of four cells is slightly smaller than the lower four. A return to cleavage in the meridional plane produces a 16-celled *morula* (*D*). At this time the blastomeres surround a rather definite space at the interior which is the early cleavage cavity, or *blastocœle*. The continuation of practically synchronous cell divisions in alternate planes produces a 32-, 64- and 128-cell stage; during this period the size of the cell is progressively diminished while the central cavity enlarges (*E*). After the seventh set of divisions (128 cells) the regularity of timing is lost. When the embryonic mass is about four hours old it consists of 128 or (theoretically) 256 cells and is a diagrammatic blastula (*F*). This is a hollow sphere; it is made of a single layer of cells which are arranged about a large blastocœle, filled with a watery jelly.

Amphibians. This group represents animals whose eggs have a fair amount of yolk, somewhat concentrated toward the vegetal pole. For this reason these eggs are classified as moderately telolecithal; cleavage is total, but it is necessarily unequal. The egg is 1 to 10 mm. in diameter and contains sufficient yolk to crowd the nucleus and much of the cytoplasm nearer

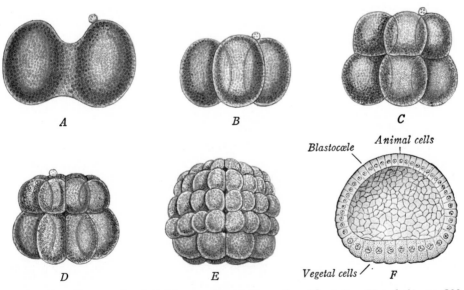

Fig. 38. Cleavage in Amphioxus, viewed from the side (after Hatschek). × 200. *A*, Two blastomeres, separating; one polar body retained. *B*, Four blastomeres. *C*, Eight blastomeres. *D*, Morula, with sixteen blastomeres. *E*, Young blastula. *F*, Older blastula, hemisected, to show cavity of the blastocœle.

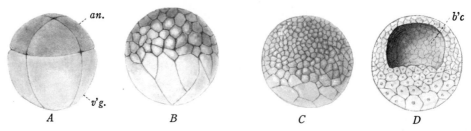

Fig. 39. Cleavage in the frog, viewed from the side. × 12. *A*, Eight blastomeres, with animal (*an.*) and vegetal (*v'g.*) cells; *B*, about 128 blastomeres; *C*, early blastula; *D*, hemisection of *C*, showing blastocœle (*b'c.*).

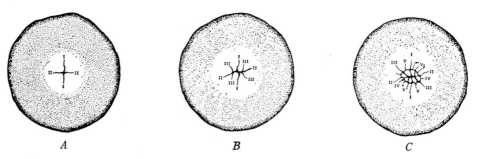

Fig. 40. Cleavage of the pigeon's ovum, viewed from above (Patten, after Blount). ×4. The order of appearance of cleavage furrows on the pale blastoderm is indicated by Roman numerals. *A*, Second cleavage; *B*, third cleavage; *C*, fifth cleavage.

the animal pole (Fig. 23 *B*). As in Amphioxus, the first two divisions subdivide the egg as one would quarter an apple. The spindles for the third cleavage are again nearer the animal pole, but this division takes place in a horizontal plane (Fig. 39 *A*). Hence the upper four cells (*micromeres*), thus cut off, are distinctly smaller than the lower four (*macromeres*). In the further cleavages that follow, the larger yolk-laden cells divide more slowly than the smaller, more purely protoplasmic ones of the animal hemisphere (*B*). After about 32 cells have been formed, tangential divisions (whose separation planes parallel the surface) begin to occur along with the other types already described. Cleavage ends, after about one day, with the completion of a quite typical, hollow *blastula* (*C, D*). The central *blastocœle* is enclosed by blastomeres which are small at the animal pole and larger and fewer at the vegetal pole. The amphibian blastula differs from that of Amphioxus in two regards: (1) the wall is more than one cell thick; (2) the blastocœle is relatively small, and is located above center because of the very thick wall of the vegetal hemisphere.

Reptiles and Birds. The egg is large and contains a great amount of stored yolk. It represents the highly telolecithal type of egg in which the active cytoplasm localizes as a cap at the animal pole, where the nucleus also is situated (Fig. 23 *C*). The huge yolk mass, far more extensive than the vegetal hemisphere of an amphibian egg, is a non-living inclusion and does not participate in cleavage or the formation of the embryo proper. As a result, cleavage is partial and discoidal. Higher fishes and egg-laying mammals also follow the same pattern.

The first two planes of separation are vertical furrows which cross at right angles through the animal pole of the egg but do not extend all the way to the margin of the cytoplasmic cap (Fig. 40 *A*). Succeeding furrows pass first in radial (*B*) and then in circumferential planes (*C*), and the original disc of cytoplasm is transformed into a mosaic of separate nucleated areas, all continuous for a time with the yolk beneath. Following this stage, cleavage divisions also take place in a horizontal plane, thereby producing a certain amount of layering (Fig. 41). The end result, after about one day, is a discoidal plate of cells perched on the surface of the yolk and separated from it by a cleft. At the periphery, the cellular disc progressively gains new cells from a proliferating, syncytial margin which blends into the yolk.

Cleavage thus produces a modified blastula (named a *discoblastula*) in which the cellular cap is termed the *blastoderm*. The space between blastomeres and yolk mass is often called a *blastocœle,* but its strict homology with that in Amphioxus and amphibians is doubtful (p. 76). The massive yolk, which serves as a floor to the blastula cavity, is not contained within cells; hence this floor is fundamentally unlike the vegetal hemisphere of the blastula of those chordate types (Fig. 45 *A-C*).

Mammals. The embryos of all marsupial and placental mammals develop with dependence on the mother. Hence little or none of the inherited yolk has been retained; the eggs are essentially isolecithal and practically microscopic in size (Fig. 23 *A*). Cleavage takes place within the zona pellucida, so the blastomeres have to accommodate themselves to this spheroidal cavity. During this period, while still in the uterine tube, the zona becomes denuded of the cells of its corona radiata through the dispersing influence of an unknown tubal factor.[1] Mammalian cleavage is total and nearly equal, much as in Amphioxus, but the blastula is considerably different both in arrangement and in its subsequent developmental course (Fig. 45 *A, D*). Although subdivision of the mammalian egg begins in the uterine tube, the later stages of cleavage are completed in the uterus. The process has been studied thoroughly in various common mammals; cleavage in the rabbit has

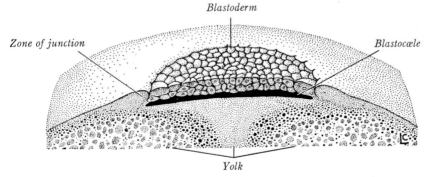

Blastoderm

Zone of junction

Blastocœle

Yolk

Fig. 41. Stereogram of an early blastula of the pigeon, hemisected.

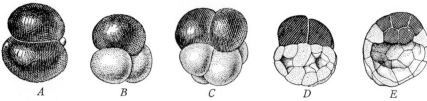

$$A \qquad B \qquad C \qquad D \qquad E$$

Fig. 42. Cleavage stages and blastocyst of the pig (Heuser and Streeter). × 240. The enveloping zona pellucida has been omitted. *A*, Two blastomeres; *B*, three blastomeres; *C*, six blastomeres; *D*, hemisected morula (20 cells), with early cavities appearing; *E*, hemisected blastocyst (30 cells), whose dark cells give rise to the embryo proper and whose light cells become trophoblast.

also been carried out in culture and recorded by motion pictures. The known extremes of time consumed in completing cleavage are 70 hours for the rabbit and 190 hours for the cow.

As in Amphioxus and amphibians, the first two planes are vertical and the third horizontal. The resulting two-, four-, eight- and sixteen-celled stages of the monkey are attained approximately at 36, 48, 72 and 96 hours after ovulation. Nevertheless, some cells tend to divide faster than others, so that the exact doubling sequence often fails (Fig. 42 *A-C*). In some mammals, such as the pig, this difference in the rate of mitosis is regional and is associated with two cell-types which may be recognizable even at the first cleavage. Darker blastomeres, with slower cleavage, are destined to become the embryo proper, whereas the clear cells, with rapid cleavage, differentiate precociously into auxiliary tissue known as the *trophoblast* (Fig. 42). It is possible that a sorting out of cell substances, with different prospective values, accompanies even the earliest steps in blastomere formation. At about the 16-cell (morula) stage the future trophoblast cells begin to flatten against the zona pellucida and there produce a kind of cellular capsule. At the same time pools of clear fluid accumulate between the more centrally located trophoblast cells (*D*), and these spaces soon coalesce into a common, central reservoir (*E*). The fluid, entering from the exterior through the zona pellucida, is secreted against pressure by the trophoblast cells. By the time some thirty cells have formed, the embryo is a definite hollow sac known as the *blastocyst* (*E*). The cells destined to become the embryo-proper constitute an *inner cell mass;* presently this mass flattens and is then the equivalent of a blastoderm (*cf.* Fig. 43 *F, G*). In Fig. 42 *E* trophoblast is lacking above the inner cell mass, but this is merely because the larger cells of this region are laggard in separating off the trophoblast cells that belong there. The completed trophoblastic sac is purely an embryonic adjunct, soon to become associated intimately with the uterus; it is concerned with protective, nutritive and excretory functions.

In certain other mammals, such as the rabbit and monkey, the blastomeres are more nearly equal in size and the trophoblast is already a complete capsule by the time the inner cell mass is recognizable as such (Fig. 43). The young blastocyst of all mammals is spheroidal in shape. It grows rapidly and distends with accumulated fluid; early in this period of enlargement the zona pellucida thins out and disappears.

HUMAN CLEAVAGE. Stages from 2 to 12 cells and blastocysts with 58 and 107 cells have been recovered from the uterine tube and uterus (Fig. 44).[2] In the monkey, cleavage groups, morula and blastocyst have all been studied in detail.[3] Compared with most mammals (e.g., the rabbit, whose blastocyst is 4.5 mm. long when attaching to the uterus) the human blastocyst enlarges slowly. It comes to lie wholly within the uterine wall and, at the time of attachment, is not much larger than the original egg.

It is clear that the thin-walled blastocyst is a specialized blastula, but its proper interpretation is not apparent at first glance. Actually the mammalian blastocyst is comparable to the blastula of the reptile or bird at the completion of the blastodermic overgrowth, but with the yolk removed (Fig. 45 C, D). Stated differently, the cavity of the blastocyst is not a simple blastocœle, like that of Amphioxus and amphibians (A, B); on the contrary, it corresponds to the bird's blastocœle combined with a yolkless yolk-

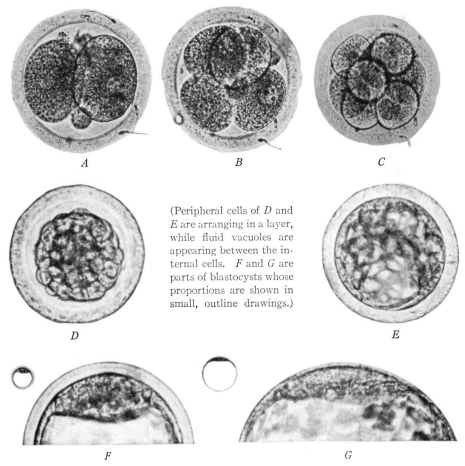

(Peripheral cells of D and E are arranging in a layer, while fluid vacuoles are appearing between the internal cells. F and G are parts of blastocysts whose proportions are shown in small, outline drawings.)

Fig. 43. Cleavage stages and blastocysts of mammals, photographed from life (Lewis, Hartman and Gregory). × 225. A-C, Two-, four and eight-celled stages in the monkey, 30 to 50 hours after ovulation. D-G, Morulæ and blastocysts in the rabbit, 45 to 80 hours after ovulation.

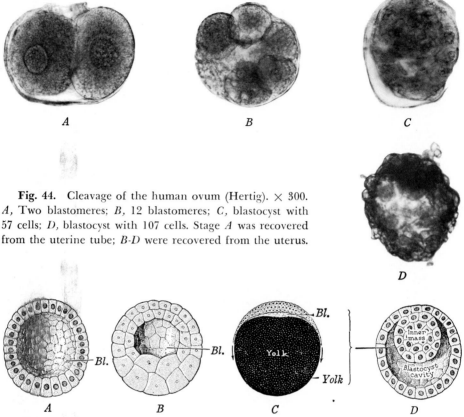

Fig. 44. Cleavage of the human ovum (Hertig). × 300. *A,* Two blastomeres; *B,* 12 blastomeres; *C,* blastocyst with 57 cells; *D,* blastocyst with 107 cells. Stage *A* was recovered from the uterine tube; *B-D* were recovered from the uterus.

Fig. 45. Blastula types among chordates, shown as hemisections. *A,* Amphioxus; *B,* amphibians; *C,* reptiles and birds; *D,* mammals. *A* and *B* are fundamentally comparable; *C* is a discoblastula; *D* is a modification of *C,* with loss of yolk.

Bl. Blastocœle; arrows on *C* indicate the expanding blastoderm.

cavity. The trophoblast represents a precocious development of external cells which, in the bird and reptile, gradually envelop the yolk. The more rapid completion of a trophoblastic capsule in true mammals is a necessary preparation for the early association of the embryo-complex with the tissue of the maternal uterus. Having discarded yolk as a source of nutriment, the mammalian embryo must establish prompt relations (through its trophoblast) with the mother.

In reptiles and birds the embryo-formative region is a superficial blastoderm, while in true mammals its equivalent is the inner cell mass. The higher mammalian ovum, although wholly devoid of yolk, thus develops into a 'blastula' fundamentally resembling the type attained by the yolk-laden eggs of reptiles and birds. That this similarity is real and has an evolutionary significance is attested by the occurrence of typical discoidal cleavage and a discoblastula in the highly telolecithal eggs of present-day monotreme mammals, which lay eggs and incubate them.

GASTRULATION

Gastrulation is the process through which specific regions of the blastula move into certain positions that presage the characteristic body plan. More

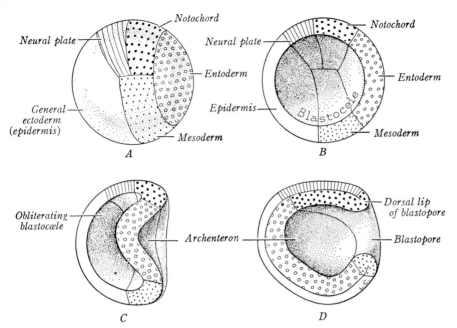

Fig. 46. Stereograms of gastrulation in Amphioxus. × 350. Areas, whose normal fates are known, are marked distinctively. *A,* Blastula; *B,* hemisected blastula; *C, D,* early and later gastrulæ, hemisected, during invagination and involution.

specifically, these rearrangements produce a layered stage whose components are the three *germ layers*. The relation of these layers, one to another, is indicated by their names: *ectoderm* (outer skin); *mesoderm* (middle skin); and *entoderm* (inner skin).

The blastula possesses polarity and bilateral symmetry. It is an important stage practically because it already contains definite cell areas which, in normal development, become the germ layers; these, in turn, give rise to different organs and parts of the embryo. Maps of these prospective regions have been made by staining trial areas on living blastulæ with nontoxic dyes and then discovering what they become. In this way the locations of the prospective ectoderm, mesoderm and entoderm have been mapped, as well as such prospective organs as the neural plate and notochord. These blastula-maps show a fundamentally similar pattern in all of the vertebrate groups.

The progressive steps in gastrulation have been clarified by the staining of these several presumptive regions with dyes of different colors and then following their movements to their later positions. These studies prove the essential similarity of gastrulation in the various chordate groups and have changed certain time-honored interpretations. The chief difference encountered among the several classes of vertebrates is the way in which entoderm becomes segregated; these variations are related to the different physical forms that the blastula assumes.

Amphioxus. Since the animal pole of the blastula corresponds roughly to the front end of the future embryo, Fig. 46 is drawn with the main axis horizontal. Stages *A* and *B* show a late blastula mapped with the cell-territories whose normal fates can be foretold.[4] About the animal and vegetal poles are the future *ectoderm* (epidermis) and *entoderm* (gut), respectively. In between is a girdling zone which is subdivided into prospective *mesoderm, notochord* and *neural plate.* Gastrulation begins about five and one-half hours after fertilization when the blastula contains some 500 cells. An inbuckling (*i.e., invagination*) of the vegetal cells is followed by an inrolling (*i.e., involution*) of cells around the margin of the double-walled cup thus being formed (*C*). The continuation of these movements carries entoderm, mesoderm and notochord to the interior and obliterates the original blastula cavity (*D*). The new, central cavity is the provisional gut or *archenteron,* and its mouth is the *blastopore.* At this period the young embryo is termed a *gastrula* (*i.e.,* little stomach).

Involution took place around the circular margin, or lip of the blastopore. Backward growth of this lip-region next elongates the cup, and unequal growth elevates the blastopore (Fig. 46 *D*). The roof of the archenteron consists of a median strip of notochordal cells, flanked on each side by a strip of mesodermal cells. The sides and floor of the cavity are bounded by entoderm. The *notochord* soon cuts off as a solid, cellular rod (Fig. 47). The mesodermal strips likewise fold off into a series of pouches (future *somites* and other mesoderm) on each side of the notochord. The entoderm then closes in the dorsal defect caused by the loss of the notochord and mesoderm, and thus produces the definitive, tubular *gut.* The cells left on the outside of the gastrula are primitive ectoderm; dorsally they specialize as the *neural plate* in contrast to the general covering of the embryo which will become *epidermis.* These progressive stages in laying down the characteristic ground plan of the vertebrate body can be followed in Fig. 47.

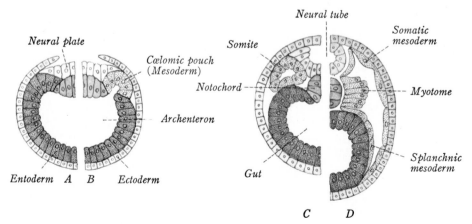

Fig. 47. The acquisition of the body-plan in Amphioxus, as shown in transverse half-sections. × about 200. *A, B,* Early and late gastrulæ; *C, D,* early larvæ.

The mechanics of invagination is incompletely understood; the problem is one that concerns foldings in general.[5] Involution and other shiftings of cell territories in chordates result from mass movements of the cells themselves. Such dynamic, self-contained cell migrations are called *morphogenetic movements* because they give rise to fundamental structures and arrangements. The underlying mechanics is incompletely understood.[22]

An earlier concept of the germ layers can be revised somewhat.[6] Originally it was believed that the blastula is wholly ectodermal, that part of it becomes the entodermal lining of a then two-layered embryo, and that one or the other of these layers next gives rise to mesoderm and so produces a three-layered embryo. As logical as this interpretation once seemed (and to this concept Amphioxus apparently lent support, because for a time the mesoderm is a part of the primitive inner layer of the gastrula, Fig. 47 *A, B*), it is no longer tenable. There is no one-, two- and three-layered stage in the serial sense implied. All three germ-layer territories (as also the future notochord and neural plate) exist potentially in the blastula before gastrulation begins. These regions are then moved to their later positions and superposed as distinct layers through the devices of gastrulation. In accomplishing these shifts the early, presumptive organ-materials are brought into relations that are considered 'normal' or characteristic for that animal group.

Amphibians. Simple invagination of the vegetal hemisphere, as in Amphioxus, is not mechanically possible, so gastrulation is accomplished largely by another method—*involution*. The first indication of gastrulation is a local groove, well below the equator of the blastula (Fig. 48 *A*). This deepening groove is covered by a lip-like fold of the blastula wall. The pocket itself marks the beginning of an *archenteron;* the mouth of the pocket is the *blastopore* and the upper margin of the curved fold is the *dorsal lip* of the blastopore. The early, short groove is extended progressively into a crescent (*B*), then a horseshoe, and finally a circle (*C*). At the sides are the so-called lateral lips of the blastopore and below is the *ventral lip*. Involution takes place at all points along this circular blastoporic lip, but chiefly at its dorsal portion. Cells of the blastula wall move downward along meridians, pass around the lip of the fold in an undertucking manner, and continue to migrate as an internal layer. At the completion of this inrolling, a broad zone girdling the vegetal hemisphere has involuted around the margin of the blastopore and into the interior. To compensate for this loss, the cells of the animal hemisphere spread and overgrow the vegetal hemisphere (*A-D*). In this way the entire surface becomes clothed with primitive ectoderm. This process of ectodermal expansion is often called *epiboly*.

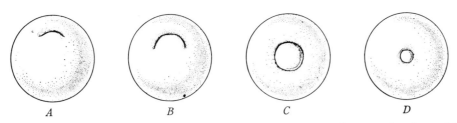

A   *B*   *C*   *D*

Fig. 48. Gastrulation in a tailed amphibian, viewed from the vegetal pole. × 10. Successive stages illustrate the early lip of the blastopore (*A*), its completion (*B, C*) and its overgrowth of the yolk-rich cells (*B-D*).

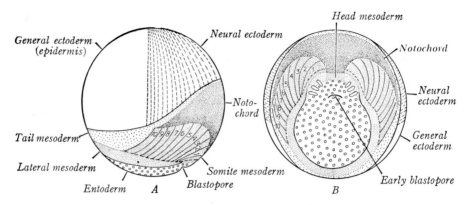

Fig. 49.   Maps of prospective parts of embryos of tailed amphibians at the beginning of gastrulation (after Vogt). *A,* Side view; *B,* view from vegetal pole.

The amphibian blastula has been well mapped for prospective organ-forming regions,[7] some of which are shown in Fig. 49 by differential markings. An important landmark is the sinuous line that demarcates the involuting material (prospective *entoderm, mesoderm* and *notochord*) from the prospective *ectoderm* which does not involute (*A*). The notochordal and mesodermal material (for convenience shortened to 'chorda-mesoderm') forms a broad, girdling band which encircles the yolk field (*B*). It should be noted that the areas that will become axial organs (like the neural plate, notochord and somites) have their greatest extent at this period in a side-to-side direction, in contrast to their ultimate cranio-caudal (roughly, pole-to-pole) orientation.

When *involution* begins at the dorsal lip of the blastopore, the first cells turned in are those of the future entoderm that lie just above the lip (Fig. 49 *B*). Next to follow is head mesoderm and then notochordal material near the midplane. As the blastopore extends, and in succession assumes the shape of a crescent, horseshoe and circle, the more lateral notochordal material and that of the somites and unsegmented mesoderm will be tucked in progressively. During these movements toward the interior, a dye-marked circular area becomes elongate in shape, and this expansion (in a longitudinal direction) continues after involution has finished. Hence *elongation,* which affects practically all parts of the gastrula during gastrulation, is a second basic movement of gastrulation. As the various areas pass around the blastoporic lips, their more lateral parts (*e.g.,* the lateral wings of the notochordal area) move toward more median positions in the interior. This medial *convergence* of more lateral areas, including parts like the neural area that remain on the outside, toward more median locations is the third basic movement of gastrulation. These three types of movement (involution, elongation and convergence) characterize gastrulation in all chordates.

The internal changes and relations during gastrulation can be followed in Fig. 50, which is marked as in Fig. 49. Stages *A-C* show the progressive enlargement of the archenteron and the corresponding obliteration of the blastocœle, the withdrawal of the yolk to the interior and its changing position, and the internal spread of the chorda-mesoderm. Stage *D* is a model, in transverse section, of the caudal half of stage *C*. Gastrulation ends with the following relations established: (1) the general ectoderm (future *epidermis;* now stretched and thinner) and *neural plate* are left on the outside; (2) the *notochord,* extending in a median dorsal location, is flanked by wing-like plates of *mesoderm;* (3) lining the archenteron in front and on the sides is involuted *entoderm;* (4) the floor of the archenteron is the main mass of large-celled entoderm which does not involute but merely elevates into the interior; and (5) the roof of the archenteron is still incomplete at this period, but growth and fusion of its dorsolateral walls will soon fill in this gap (Fig. 10 *A*). Gastrulation in amphibians differs from that in Amphioxus in two important respects: (1) invagination plays no significant rôle; (2) the notochord and mesoderm are not continuous with the entoderm to produce a temporary, composite internal layer.

Reptiles and Birds. The inert yolk mass is proportionately so enormous that it cannot participate, even passively, in gastrulation. For this reason

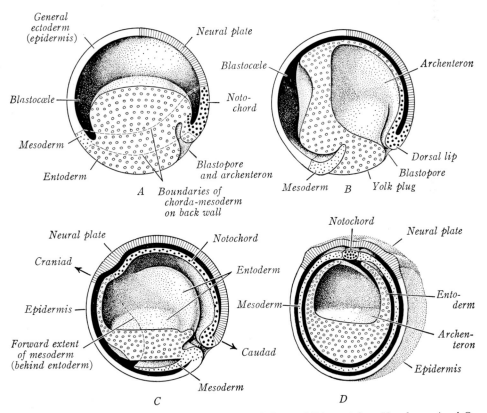

Fig. 50. Stereograms of gastrulation in tailed amphibians (after Hamburger). *A-C,* Early to late stages, showing the movements of areas differentially marked, as in Fig. 49, on the cut surfaces of longitudinal hemisections. *D,* Caudal half of stage *C,* shown by a transverse hemisection.

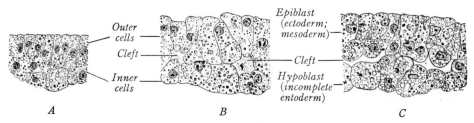

Fig. 51. Entoderm formation in the chick, shown by vertical sections through sample regions of the blastoderm (after Peter). × 340. *A,* Early segregation of the future entoderm. *B, C,* Later stages of actual separation into two layers.

the events of gastrulation are confined to the blastoderm which contains the cells of all three future germ layers. The process, as a whole, takes place in two stages: (1) first the *entoderm* separates from the rest of the disc; (2) then the cells of the *chorda-mesoderm* move into position between the entoderm and the residual outer layer; the latter henceforth is called *ectoderm.* The events in reptiles are similar (except for details) to those in birds, on which the following descriptions are based.

It is now generally agreed that certain cells, richer in yolk, are segregated from others in the blastoderm. Moving to a deeper position they detach, thereby producing the primitive entodermal layer (Fig. 51).[8, 9] Later contributions to the entoderm from cells migrating from the outer layer directly[10, 11, 12] and by way of the primitive streak[13, 14] (see beyond) have also been described.

The cellular plate, overlying the split-off entoderm, is commonly called *epiblast.* It contains not only the future ectoderm (including the neural plate), but also the future mesoderm (including the notochord). Similarly, the early split-off under layer is often called *hypoblast.* Not until it receives further cells from the epiblast and primitive streak is it termed *entoderm.*

The separation of the entoderm from the rest of the blastoderm makes it possible to interpret the resulting stage as a tardy blastula which, though flat, is comparable to those of lower forms.[9] The upper cellular plate would then correspond to the animal hemisphere of a typical blastula, the entoderm to the vegetal hemisphere, while the newly created cleft between these layers would be the blastocœle. Under this interpretation the non-cellular mass of yolk has no counterpart in Amphioxus or amphibians and is a new auxiliary feature. Moreover, the original space over the yolk, produced during cleavage and now located between the entoderm and the yolk, would not be a true blastocœle; neither is it an archenteron in the ordinary sense, since it is not an invagination cavity.

A surface map of the blastoderm, after entodermal delamination has occurred and left an epiblast layer, is shown as Fig. 52 *A.* The relative positions and shapes of the areas occupied by prospective ectoderm (future epidermis and neural plate), notochord and mesoderm, are strikingly similar to those of the amphibian blastula (*cf.* Fig. 49). The only real difference is in the shapes of the areas containing lateral mesoderm, and this difference naturally results from the absence of an entodermal field on the surface of the blastoderm of the bird. The mass movements which take place during gastrulation are also of the same nature as those described for the amphibian: (1) convergence; (2) involution; and (3) elongation.[14, 15] All of the areas swing, or converge, toward the midline (Fig. 52 *B*); the chorda-mesoderm involutes at the midline and then spreads as a middle layer (*C-E*); after these shifts are completed there is considerable elongation of all areas.

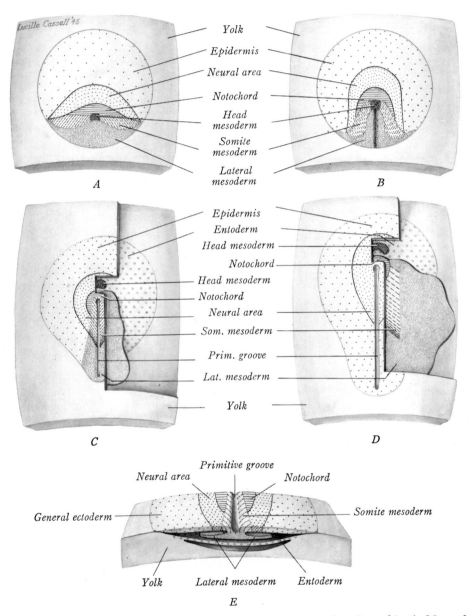

Fig. 52. Gastrulation movements in the chick (largely after Pasteels). *A*, Map of prospective parts, differentially marked on the surface of an early blastoderm. *B*, Formation of the early primitive streak. *C, D,* Passage of chorda-mesoderm to a middle level. *E*, Transection through primitive streak, stage *C*.

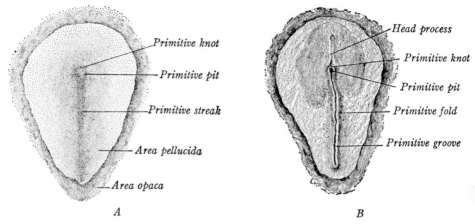

Fig. 53.   Blastoderms of the chick, in surface view. × 16. *A*, Stage of the primitive streak;
B, stage of the head process.

Involution of the chorda-mesoderm takes place through a thickened axial band in the upper layer known as the *primitive streak*. This linear massing of cells is a result of the convergence of mesoderm from each side toward the midline (Fig. 52 *B*). Another result of this convergence is a change of shape of the several surface areas; their lateral halves swing toward the midline and thus become roughly parallel. The future *lateral mesoderm* lies nearest the primitive streak and is the tissue that first gave it origin. This mesoderm continues to migrate and approach the primitive streak from each side; as it arrives at the midline its cells turn downward through the streak and then diverge, right and left, as they spread laterad between the surface layer and the entoderm (*C, E*). A small area of prospective *head mesoderm* (the so-called *prechordal plate*) next turns in, as evidenced by a dimple at the front end of the primitive streak, known as the *primitive pit*. On account of its previous position ahead of the primitive streak, it advances in a forward (anterior) direction (*B-D*). The prospective *notochord* follows, also passing downward through the primitive pit and then forward in the midline. In the region of the primitive streak the paired areas of *somite mesoderm* involute, as did the lateral mesoderm before them (*C, D*). At the completion of gastrulation, the residual outer layer is definitive *ectoderm*. The originally crescentic area representing the *neural plate* converges to become a tear-shaped field located within the general ectoderm, or future *epidermis* (*A-D*).

The primitive streak is a stretched and seam-like blastopore through which the involution of chorda-mesoderm occurs. The fact that it has no open mouth and may not be related to the segregation of entoderm does not alter the homology. The primitive streak acquires a knob at its forward end (Fig. 53 *A*); this is the *primitive knot* (of Hensen) which is said to be originally a separate mass.[16] This knob indicates the immediate source of the tongue of cells that turns downward and forward as the

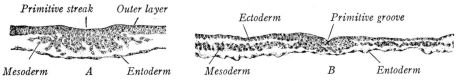

Fig. 54. Involution and spread of mesoderm in the chick, shown in transverse sections through the primitive streak. × 165. *A,* Early streak; *B,* later streak.

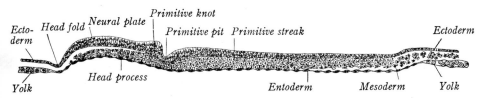

Fig. 55. Head process and primitive streak of the chick embryo, shown in longitudinal section. × 100.

notochord. A shallow *primitive groove* presently courses lengthwise along the middle of the streak and ends at the knot in the *primitive pit* (*B*). The pit and groove result from the active involution of cells. Transverse sections through the streak show the involuting and spreading mesoderm (Fig. 54). A longitudinal section demonstrates the relation of the notochord (also at this period called the *head process*) to the primitive knot and the latter to the primitive streak (Fig. 55). While gastrulation is going on, the original circular blastoderm elongates and acquires a pear-shaped outline (Fig 53 *B*).

Mammals. As in birds, gastrulation occurs in two stages. The first phase takes place when certain cells appear on the under surface of the inner cell mass and arrange themselves into a definite sheet, the *entoderm* (Figs. 56, 57*A*). In monotremes[17] and marsupials[18] these cells are smaller and darker ameboid elements which move out of a common layer to a deeper position. In placental mammals the entodermal cells detach from the inner cell mass. This has been called delamination, but it is possible 'that the process involves a strict segregation in which cells specializing toward entodermal

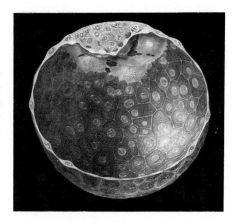

Fig. 56. Blastocyst of the monkey at nine days, hemisected (Streeter). × 200. Entodermal cells have appeared at the under surface of its inner cell mass, and similar cells occur (by spreading) on the nearby back-wall of the blastocyst.

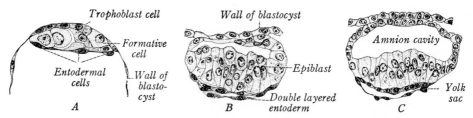

Fig. 57. Entoderm formation in the monkey (after Streeter and Heuser). Only the upper portions of the sectioned blastocysts (through embryonic disc) are shown. *A,* At eight days (× 250); *B,* at eleven days (× 250); *C,* at twelve days (× 200).

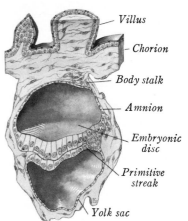

Fig. 58. Reconstruction of the right half of a human embryo of fourteen days. × 105.

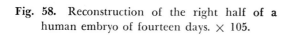

fates are sorted out from others that are prospective ectodermal, mesodermal and trophoblastic elements.[19] In most mammals the entoderm spreads rapidly and lines the blastocyst as a relatively large sac (Fig. 92 *A*). In primates there is a temporary structure which might be interpreted as a similar, large yolk sac (Fig. 64), yet this homology is both affirmed and denied by those who have studied it most.[19] In any event, only the entodermal cells under the inner cell mass persist, and these fashion themselves into a smaller, definitive yolk sac (Figs. 57 *B, C;* 58).

Those cells of the inner cell mass that are neither entoderm nor trophoblast become a plate of *epiblast* containing the progenitors of future ectodermal and mesodermal cells (Fig. 57 *B, C*). Directly beneath is the layer of entoderm that serves as a roof to the yolk sac. These two layers make up the earliest *embryonic disc,* or blastoderm (Fig. 58). The second phase of gastrulation is concerned with the segregation of the mesoderm and notochord as definite parts, located at a middle level. Technical difficulties have not permitted the mammalian blastoderm to be marked with dyes and mapped. Yet there is reason to suspect that the areas of presumptive epidermis, neural plate, mesoderm and notochord, and the movements and involution of mesoderm, are similar to those which have been determined for the chick (*cf.* Fig. 52).

A typical primitive streak appears caudally on the upper surface of the embryonic disc (Fig. 59 *A*). The spread of *mesoderm* through the primitive streak is illustrated in section by Fig. 59 *B* and in surface view by Fig. 60 *A-D*. The appearance of the *primitive knot,* and the growth of the *notochord* (or head process) from it, are indicated in Fig. 60 *D, E.* At the conclusion of these movements of chorda-mesoderm, whereby they come to occupy a middle level, the residual upper layer of the embryonic disc is *general ectoderm* (prospective *epidermis*) and the material of the future *neural plate.*

In most mammals the mesoderm grows rapidly and extends beyond the region of the embryonic disc. Continuing to expand around the wall of the blastocyst, it fills in the space between the trophoblast (usually rated as ectoderm) and the temporarily large sac of entoderm, until its peripheral margins meet and fuse (Fig. 61 *A*). This peripheral tissue is *extra-embryonic mesoderm,* which will clothe such auxiliary structures as chorion, amnion, body stalk and yolk sac; it takes no part in forming the embryo proper. In higher primates, extra-embryonic mesoderm appears precociously before the mesoderm of the embryo becomes recognizable as such. It arises as cells that separate away from the trophoblast of the original blastocyst wall (Fig. 64), and thus has a separate origin from the mesoderm of the embryo itself.[20]

HUMAN GASTRULATION. The youngest gastrulation stage known is a blastocyst in which a layer of entodermal cells has just become segregated from the inner cell mass (Fig. 63).[21] At this stage a cleft is separating a layer of cells, which make up the auxiliary membrane named the amnion, from the rest of the inner cell mass. This leaves a cellular plate (*epiblast*), beneath the cleft, which contains the formative cells that will give rise to the definitive ectoderm and mesoderm of the embryo (Fig. 58). The stages that follow (primitive streak; head process) are well known (Figs. 59, 68 *A*). During this period the cells of the mesoderm and notochord move to their characteristic positions and the three primary germ layers are thereby established.

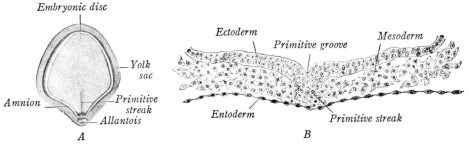

Fig. 59. Human embryo of sixteen days (after Streeter). *A*, Dorsal view of the embryonic disc (× 25). *B,* Transverse section through the primitive streak (× 185).

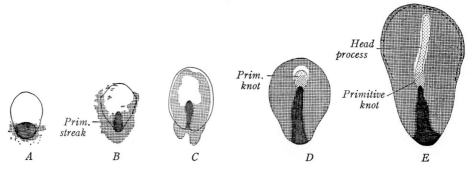

Fig. 60. Embryonic discs of the pig, mapped to show the spread of mesoderm (cross hatched) and the growth of the primitive streak and head process (Streeter). × 25.

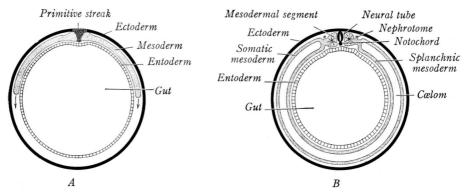

Fig. 61. Spread and differentiation of the mesoderm, shown in diagrammatic transverse sections of a mammalian embryo at different periods (after Prentiss).

EARLY DEVELOPMENT COMPARED IN DIFFERENT VERTEBRATE TYPES

ANIMAL TYPE	TYPE OF EGG	CLEAVAGE	BLASTULA	BLASTULA CAVITY	METHOD OF GASTRULATION	COMMENT ON GASTRULATION
Amphioxus	Isolecithal (little yolk).	Total; nearly equal.	Sphere; wall a single layer.	Spherical; relatively large.	Invagination; involution via blastopore.	Early inner layer is a composite.
Amphibians	Telolecithal (moderate yolk).	Total; unequal.	Sphere; wall layered; thickness varies.	Spherical; relatively small; eccentric.	Chiefly involution via blastopore.	Involuted layers are separate from start.
Birds	Telolecithal (massive yolk).	Partial; discoidal.	Cell-disc; (blastoderm) lies on massive yolk.	Shallow; delayed until entoderm forms.	Delamination; involution via primitive streak.	Occurs in two stages; modified amphibian style.
Mammals (except lowest)	Isolecithal (little yolk).	Total; nearly equal.	Blastocyst, with inner cell mass.	Merged with yolk-free blastocyst cavity.	Delamination; involution via primitive streak.	Occurs in two stages; basically similar to birds.

DERIVATIVES OF THE GERM LAYERS

It was formerly believed that the early germ layers of chordates have predetermined capacities and fixed fates, so that they are, from the first, rigidly restricted to the kinds of differentiation that they ordinarily show. This concept has not withstood experimental tests. A germ layer, as such, or any part of it is not foreordained to carry out specific programs of organ building. Rather, the constitutional plan of an organism assigns the various emerging organs and parts to definite spatial positions, and any cells that chance to lie in a particular region at the proper time will be recruited to participate in the specific job of organ construction characteristic of that site. In normal development, therefore, the germ layers serve merely as advantageously located assembly grounds, out of which the constituent parts of an embryo emerge at precise positions and in orderly sequences. Each early germ layer possesses a potential plasticity that would enable it to do many things other than the activities that actually find expression. Sooner or later, nevertheless, this plasticity is lost when a region receives 'instructions' that determine its fate inexorably; but this acquirement of instructions is wholly a secondary matter. Yet, despite this revision of original concepts, the germ layers remain as eminently useful subdivisions for practical, descriptive purposes.

Since the ectoderm covers the body, it is primarily protective, but it also gives origin to the nervous system and sense organs. The entoderm, on the other hand, lines the primitive digestive canal and has nutritive relations; later it also becomes respiratory. The mesoderm, occupying an intermediate position, naturally is related to skeletal support, muscular movement, circulation, excretion and reproduction. The subjoined table lists the principal derivatives of the three germ layers as they arise in normal development:

THE GERM-LAYER ORIGIN OF HUMAN TISSUES

ECTODERM	MESODERM (*including mesenchyme*)	ENTODERM
1. Epidermis, including: Cutaneous glands. Hair; nails; lens. 2. Epithelium of: Sense organs. Nasal cavity; sinuses. Mouth, including: Oral glands; enamel. Anal canal. 3. Nervous tissue, including: Hypophysis. Chromaffin tissue.	1. Muscle (all types). 2. Connective tissue; cartilage; bone; notochord. 3. Blood; bone marrow. 4. Lymphoid tissue. Epithelium of: 5. Blood vessels; lymphatics. 6. Body cavities. 7. Kidney; ureter. 8. Gonads; genital ducts. 9. Suprarenal cortex. 10. Joint cavities, etc.	Epithelium of: 1. Pharynx, including: Root of tongue. Auditory tube, etc. Tonsils; thyroid. Parathyroids; thymus. 2. Larynx; trachea; lungs. 3. Digestive tube, including: Associated glands. 4. Bladder. 5. Vagina (all?); vestibule. 6. Urethra, including: Associated glands.

REFERENCES CITED

1. Swyer, 1947, Nature, *159,* 873–874.
2. Hertig, A. T. *et al.* 1954. Carnegie Contr. Embr., *38,* 199–220.
3. Lewis, W. H. *et al.* 1941. Carnegie Contr. Embr., *29,* 7–55.
4. Conklin, E. G. 1932. J. Morph., *54,* 69–151.
5. Costello, D. P. 1955. In Willier: Analysis of Development, 213–229, (Saunders).
6. Oppenheimer, J. M. 1940. Quart. Rev. Biol., *15,* 1–27.
7. Vogt, W. 1929. Arch. Entw.-mech. d. Organ., *120,* 384–706.
8. Peter, K. 1941. Ergeb. Anat. u. Entwickl., *33,* 285–369.
9. Pasteels, J. 1945. Anat. Rec., *93,* 5–21.
10. Pasteels, J. 1940. Biol. Rev., *15,* 59–106.
11. Rosenquist, G. C. 1966. Carnegie Contr. Embr., *38,* 71–110.
12. Vakaet, L. 1962. J. Embr. & Exp. Morph. *10,* 38–57.
13. Hunt, T. E. 1937. Anat. Rec., *68,* 449–460.
14. Pasteels, J. 1937. Arch. de Biol., *48,* 381–488.
15. Spratt, N. T. 1946; '47. J. Exp. Zoöl., *103,* 259–304; *104,* 69–100.
16. Spratt, N. T. 1942. J. Exp. Zoöl., *89,* 69–101.
17. Flynn, T. T. & J. P. Hill. 1942. Proc. Zoöl. Soc. London (Ser. A), *111,* 233–253.
18. Hill, J. P. 1910. Quart. Jour. Micr. Sci., *56,* 1–134.
19. Heuser, C. H. & G. L. Streeter. 1941. Carnegie Contr. Embr., *29,* 15–55.
20. Hertig, A. T. 1935. Carnegie Contr. Embr., *25,* 37–82.
21. Hertig, A. T. & J. Rock, 1945. Carnegie Contr. Embr., *31,* 65–84.
22. Trinkhaus, J. P. 1965. pp. 55–104 in DeHaan and Ursprung: Organogenesis (Holt).

Chapter VI. Embryonic and Fetal Stages

The reader, having traced advancing development to the establishment of germ layers arranged in an embryonic disc, is now prepared to survey the full course of progress that a human embryo makes in acquiring its characteristic shape and general body-plan. In this treatment some attention will have to be paid to the origins of several accessory organs (amnion; chorion; yolk sac; allantois; body stalk), although the later histories of these extra-embryonic parts will be deferred to the two chapters that follow.

An initial phase of prenatal life (*period of the ovum*) encompasses the zygote, its subdivision into blastomeres and their arrangement into a hollow blastocyst. It extends through the first week of human development. A second phase (*period of the embryo*) includes the second through the eighth weeks of prenatal life. This period is characterized by rapid growth, the establishment of a placental relationship with the mother, the differentiation of the primordia of the chief organs and the acquisition of the main features of external body form. A final phase of prenatal life includes the third to tenth lunar months (*period of the fetus*). In it differentiation continues and organs become competent to assume their specialized functions. For convenience and clarity in the following descriptions the period of the embryo will be divided further into three subperiods (two-layered, three-layered and somite stages).

PERIOD OF CLEAVAGE (FIRST WEEK)

A previous chapter has described the way in which cleavage breaks down the mammalian egg into cells of suitable size to serve as building units. With the production of a typical *blastocyst* this process of size reduction nears its end. At this stage the cells of the future embryo-proper are segregated as an *inner cell mass,* whereas the capsule-like wall of the blastocyst represents auxiliary tissue, the *trophoblast,* which will establish nutritive and other relations with the uterus.

The first three days of human development are spent in descending the uterine tube, during which time cleavage produces about eight blastomeres. The next two or three days are passed in the uterine cavity as a free morula and blastocyst. By the end of the sixth day the blastocyst attaches to the epithelial lining of the uterus and presently begins to sink into the soft tissue beneath. A few cleavage stages, an abnormal morula and two un-attached blastocysts are the known representatives of the first week of development (Fig. 44).[1] Cleavage stages and the attached blastocyst, as they occur in the monkey, are shown in Figs. 43 *A-C* and 96.

PERIOD OF THE TWO-LAYERED EMBRYO (SECOND WEEK)

The rapid growth of the human embryo, and also of the sac in which it develops, during the second week, can be visualized by referring to Fig. 62 *C-G*. These simplified sections are all drawn to the same scale. It will be remembered that gastrulation segregates the embryo-formative cells of the inner cell mass into three germ layers which are advantageously situated to begin the building of the body and its organs. If cleavage can be compared to the quarrying of building stones, then gastrulation is the cartage of these units to convenient working points on the site of a future edifice. In mammals gastrulation occurs in two stages and it is the first of these, ento-derm formation, that ends the first week and ushers in the second week of human development.

The youngest human embryo of this period is not more than seven and one-half days old.[2] Its inclusion within the lining of the uterus had not been completed, even though a marked collapse of the trophoblastic sac, due to a temporary loss of fluid, had already occurred (Figs. 63, 97 *C*). Several important advances in this specimen mark a stage of development beyond that of the simple, free blastocyst. The *trophoblast* (*i.e.*, the blasto-cyst wall) is much thickened wherever it has come in contact with the con-

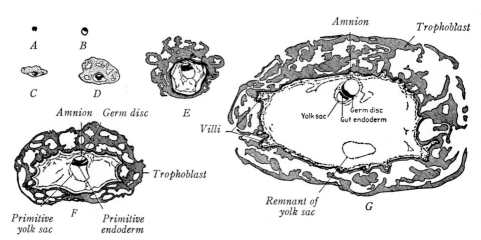

Fig. 62. Human development in the first two weeks, as shown in vertical sections drawn to the same scale (adapted from Hertig). × 25. *A,* Beginning cleavage (two days); *B,* free blastocyst (five days); *C,* early implanting blastocyst (seven days); *D,* nearly im-planted blastocyst (nine days); *E,* fully implanted blastocyst (eleven days); *F,* at twelve to thirteen days; *G,* at fourteen days.

nective tissue of the uterus. Here most of the cells have lost their boundaries and become a syncytium; contrasted are the discrete, pale cells with large nuclei that underlie them. In the region of the inner cell mass a cleft is in the process of separating the *amnion* (which like the trophoblast is auxiliary tissue) from the embryo-formative cells. The latter constitute the *embryonic disc*. It consists of a thicker plate of potential ectodermal and mesodermal cells, not yet recognizable as such, and a definite layer of segregated *entoderm* which faces the collapsed cavity of the trophoblastic sac.

A slightly later specimen, not more than nine days old, lies almost wholly within the uterine lining (Fig. 97 *D*).[2] The chief change is in the *syncytial trophoblast* which has become thick and spongy through the appearance of irregular -spaces; some of these connect with maternal capillaries. The innermost part of the trophoblastic capsule, next the central cavity of the blastocyst, is not syncytial but consists of a layer of individual cells. This constitutes the *cellular trophoblast*.

At 11 days two new features are visible (Fig. 64)[3]: (1) Primitive *mesodermal cells* are differentiating and separating from the inner surface of the cellular layer of trophoblast. These mesodermal elements are all extra-embryonic, the main plate (epiblast) of the embryonic disc not having begun to segregate its cells into layers of definitive ectoderm and mesoderm. (2) The *primary yolk sac* is prominent. Its roof (gut entoderm) is composed of cuboidal cells, but the sac proper, which arose from the gut entoderm by spreading, is very thin.[4]

By the thirteenth day the definitive *yolk sac* is appearing. This is apparently brought about by portions pinching off from the primary sac, thus making a smaller vesicle (Fig. 62 *F, G*).[4] The primitive extra-embryonic mesoderm is consolidating into a definite layer beneath the trophoblast. Some of this mesoderm is beginning to extend as stubby cores into the trophoblastic cords; this marks the beginning of true *chorionic villi* (Fig. 67 *A*). The entire capsule of mesoderm and trophoblast can now be called the *chorion,* and its cavity the *extra-embryonic cœlom.*

The appearance of an embryo at the end of the second week is shown in

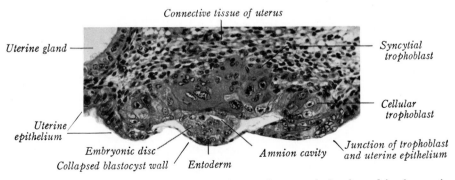

Fig. 63. Section through a human embryo of seven days, partly implanted in the uterine wall (Hertig and Rock). × 200.

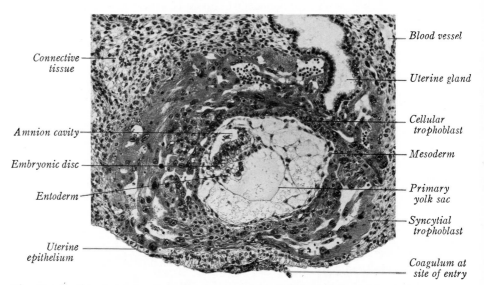

Connective tissue

Amnion cavity

Embryonic disc

Entoderm

Uterine epithelium

Blood vessel

Uterine gland

Cellular trophoblast

Mesoderm

Primary yolk sac

Syncytial trophoblast

Coagulum at site of entry

Fig. 64. Section through a human embryo of eleven days, implanted in the uterine wall (Hertig and Rock). × 110.

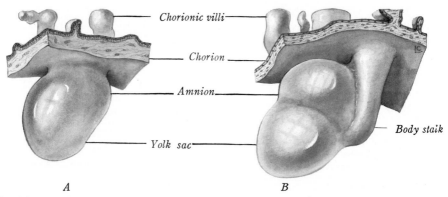

Chorionic villi

Chorion

Amnion

Yolk sac

Body stalk

A   *B*

Fig. 65. Reconstructions of the exterior of early human embryos. *A,* At fourteen days (× 80); *B,* at sixteen days (× 95).

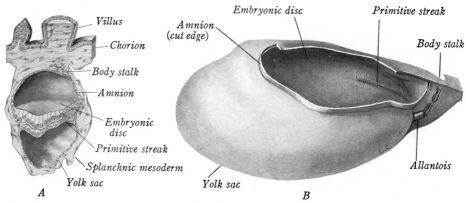

Villus

Chorion

Body stalk

Amnion

Embryonic disc

Primitive streak

Splanchnic mesoderm

Yolk sac

Amnion (cut edge)

Embryonic disc

Primitive streak

Body stalk

Allantois

Yolk sac

A   *B*

Fig. 66. *A,* Right half of a human embryo (Brewer) of fourteen days. × 85. *B,* Human embryo (Mateer-Turner) of sixteen days, viewed from the left and above. × 50.

Figs. 65 *A*, 66 *A*. The circular embryonic disc still lacks separate ectodermal and mesodermal layers, although there are slight indications that a
primitive streak is organizing. At its peripheral margin, the disc is continuous with the amnion above and the yolk sac below. The extra-embryonic
mesoderm can be designated regionally by special names. The layer that
now clothes the yolk sac is *splanchnic mesoderm,* while the layer covering
the amnion and lining the chorion is *somatic mesoderm* (Fig. 67 *A*). The
roof of the dome-like amnion is attached broadly to the chorion by mesoderm. The external covering of the chorion and the epithelial lining of the

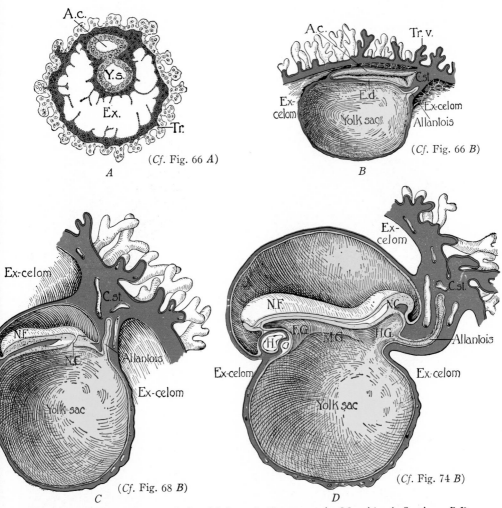

Fig. 67. Human embryos of the third week (Scammon in Morris). *A,* Section; *B-D,*
right halves of models.

A.c., Amnion cavity; *C.st.* body stalk; *E.d.,* embryonic disc; *Ex.,* extra-embryonic cœlom;
F.G., fore-gut; *H.,* heart; *H.G.,* hind-gut; *M.G.,* mid-gut; *N.C.,* neurenteric canal; *N.F.,*
neural folds; *Tr.,* trophoblast; *Tr.V.,* chorionic villi; *Y.s.,* yolk sac.

amnion are usually classified as *extra-embryonic ectoderm,* by analogy with lower forms.

PERIOD OF THE THREE-LAYERED EMBRYO (THIRD WEEK)

At 16 days the location of the margin of the embryonic disc can be identified externally by a zone of constriction located between the amnion and yolk sac (Fig. 65 *B*). The roof of the amnion is now free from the chorion and only a bridge of mesoderm, the *body stalk,* connects the caudal end of the embryo with the chorion. The chorionic villi branch, and blood vessels are appearing in the mesoderm of the villi as well as in the mesoderm of the chorion proper, body stalk and yolk sac. A well-formed *primitive streak* is evident caudally on the surface of the pear-shaped embryonic disc (Fig. 66 *B*). A transverse section shows the mesoderm spreading from the streak as a prominent layer between the ectoderm and entoderm (Fig. 59 *B*). This

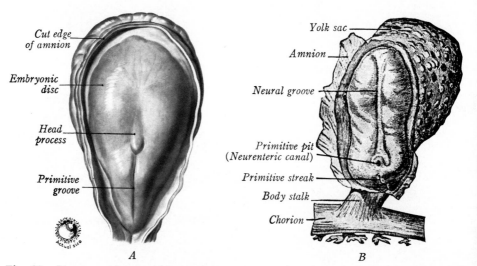

Fig. 68.   Human embryos, with amnion cut away, viewed from above. *A,* At eighteen days (Heuser; × 45); *B,* at nineteen days (v. Spee; × 23).

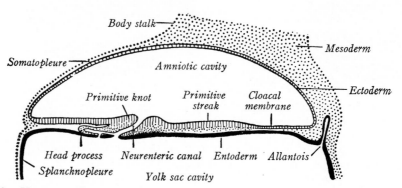

Fig. 69.   Human embryo, at nineteen days, in diagrammatic sagittal section (after Scammon).

segregation of *embryonic mesoderm* signifies that the second phase of gastrulation is in progress. The *allantois* is a slender, entodermal tube which
has extended from the caudal end of the yolk sac into the mesoderm of the
body stalk (Fig. 67 *B*).

Stages at 18 days possess a *head process* which extends forward from the
primitive knot at the front end of the primitive streak (Fig. 68 *A*). The originally solid head process of the previous day has become tunneled by a
notochordal canal and the floor of the canal is disappearing (Fig. 69). As
a result, there is a temporary communication at the site of the primitive pit
between the cavities of the yolk sac and amnion; this passage is known as the
neurenteric canal. At the caudal end of the primitive streak, the ectoderm
and entoderm fuse as the *cloacal membrane*. The primitive streak was the
first landmark that revealed the polarity of the embryonic disc. The head
process extends this defining of the median plane of the future embryo and,
with the primitive streak, divides the embryonic disc into precise right and
left halves. The developmental fates of local regions of a chick blastoderm
at this period, tested by color marking, are indicated in Fig. 70.

An embryo of 19 days ends the presomite period (Figs. 67 *C*, 68 *B*). The
embryonic disc is slipper-shaped in outline and there is slight constriction of
the somewhat convex disc from the yolk sac. Growth has elongated the portion of the embryo ahead of the primitive knot. A median strip of ectoderm
in this region is thickened as the *neural plate;* a definite *neural groove*
courses along its length. The floor of the head process has disappeared, leaving the roof as the *notochordal plate;* this plate soon rounds up into an axial
rod, the *notochord*. The *fore-gut* is beginning to form and there are slight
indications of the future *heart*.

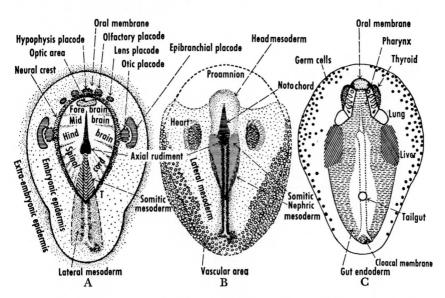

Fig. 70.   Fates of local regions in the top, middle and lowest layer at the head-process
stage, as tested by experiments on the chick (Witschi).

A GRADED SERIES OF PRESOMITE EMBRYOS AND THEIR DIMENSIONS

AUTHOR AND DESIGNATION OF EMBRYO	CHORIONIC SAC		YOLK SAC	EMBRYONIC DISC*	ESTIMATED AGE
	Aver. ext. diam. (mm.)	Aver. int. diam. (mm.)	Max. diam. in (mm.)	Length × width	(In days)

I. Primitive Streak Absent

A. Amniotic cavity; entoderm; solid trophoblast

Hertig and Rock (C.C. 8020), 1945	.22	.15	Unformed	.09 × .08	7

B. Primary yolk sac; extra-embryonic mesoderm; spongy trophoblast

Hertig and Rock (C.C. 7699), 1941	.75	.36	Unformed	.14 × .09	i1

C. Definitive yolk sac; chorionic villi differentiating

Linzenmeier, 1914	1.04	.63	.10	.21 × .11	13

II. Primitive Streak Present

A. Beginning primitive streak

Brewer (U.C. 1496), 1938	2.8	1.5	.2	.21 × .18	14

B. Primitive groove; embryonic mesoderm; allantois; villi branching

Streeter (Mateer), 1919	6.8	4.7	1.5	.92 × .78	16

C. Solid head process; cloacal membrane

Thompson-Brash, 1923	7.2	5.7	.9	.9 × .9	17

D. Notochordal canal tunneling head process

Ingalls (W. R. 1), 1918	7.9	6.7	2.5	2.0 × .8	18

E. Neural folds; neurenteric canal; fore-gut indicated

Spee (Glæ), 1889; 1896	9.3	7.8	2.1	1.5 × .6	19

* The horizontal measurements of the amnion are either the same, or nearly the same, as those of the embryonic disc.

PERIOD OF THE EMBRYO WITH SOMITES (FOURTH WEEK)

Vertebrate Characteristics. Since the embryo is now ready to enter into body building, it is worth while to have in mind some of the chief features that characterize vertebrates in general:

1. A tubular *central nervous system,* wholly dorsal in position.
2. An *internal skeleton,* composed of living tissue.
3. A *mouth,* closed by a lower jaw.
4. A *pharynx,* which differentiates gills or lungs.
5. A ventral *heart,* connecting with a closed system of *blood vessels.*
6. A *cœlom,* or body cavity, which is unsegmented but is divided into compartments for the heart and abdominal organs (and, in higher vertebrates, for the lungs as well).
7. The *limbs,* arranged in two pairs, with an internal skeleton.

The Primitive Body Plan. During the fourth week of human development all of the parts just listed (and many others) make their beginnings. Certain items in this foreshadowing of the future organization of the body require comment and illustration (Figs. 71, 75):

THE NEURAL TUBE. The neural plate folds into a tube which detaches from the general ectoderm and becomes the nervous system. This includes the *brain, spinal cord* and *nerves.*

THE NOTOCHORD. This cord of mesodermal cells runs axially between the neural tube and gut. It serves as a primitive 'backbone' and is later surrounded and replaced by the *vertebral column.*

THE GUT. The roof of the entodermal yolk sac folds into a tubular gut which becomes the *digestive tract* and *respiratory system.* The pharynx of fishes and aquatic amphibians opens to the outside by gill slits; incomplete homologues appear in the embryos of reptiles, birds and mammals.

THE SOMITES. These primitive segments lie alongside the spinal cord in pairs and are a prominent feature of vertebrate embryos. They arise when transverse clefts subdivide the thickened mesoderm next the midplane into block-like masses. Each somite gives rise to a *muscle mass,* supplied by a spinal nerve, while each of these somite-pairs also collaborates in producing a *vertebra.* At the level of any pair of somites lie primitive kidney tubules, and also blood vessels arising from the aorta. This whole group of associated, mesodermal structures is repeated serially throughout much of the embryo's length.

This segmental arrangement brings to mind the serial divisions, or *metameres,* of an earthworm's body. In the worm each metamere similarly contains a ganglion of the nerve cord, a muscle segment, and pairs of nerves, blood vessels and excretory tubules. Such serial repetition of homologous parts is called *metamerism.* Hence the vertebrate embryo is also fundamentally metameric, even though much of its segmentation is lost as development advances. Just as a worm grows by adding new metameres at its tail-end, so the somites and associated structures of the vertebrate embryo appear first in the head region and are added progressively tailward. But there are these differences between the metamerism of a worm and of the vertebrate embryo: in the worm it is complete, and both external and internal; in the vertebrate it is incomplete ventrally, and purely internal.

THE NEPHROTOMES. A short plate of cells extends ventrolaterad from each somite. From these serially arranged plates will develop the kidneys and their ducts.

THE LATERAL MESODERM. The remainder of the mesoderm of the embryo, ventrolateral to the nephrotomes, is not segmented. It splits into two layers, the *somatic* and *splanchnic mesoderm.* From the beginning, the ectoderm and somatic mesoderm are closely associated. They constitute a natural unit, named the *somatopleure,* which produces the lateral and ventral body wall of the embryo and continues beyond the embryo as the amnion and chorion. In a similar way, the entoderm and splanchnic mesoderm combine

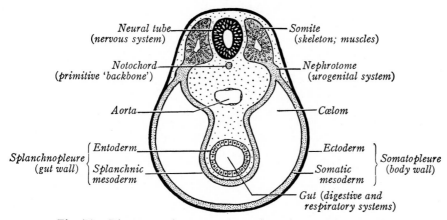

Fig. 71. Diagrammatic transverse section of a vertebrate embryo.

as the *splanchnopleure*. It forms the gut and lungs. The splanchnopleure of the young embryo is continuous with that of the yolk sac.

THE CŒLOM. The space between the split layers of lateral mesoderm is the *cœlom,* or body cavity. In mammals the cœlomic cavity of the chorionic sac (and the mesoderm of the chorion, amnion and yolk sac which faces this cavity) exists before there are similar cavities within the embryo itself. These external representatives of cœlom and mesoderm are designated as *extra-embryonic cœlom* and *extra-embryonic mesoderm.* Until the body wall closes off, there is direct continuity between the cœlom inside and outside the embryo. The original cœlom within the embryo becomes subdivided into separate compartments for the heart, lungs and abdominal viscera. The surface layer of the mesoderm, which everywhere bounds the cœlom, is a specially named epithelium, termed *mesothelium.*

VESSELS. Tiny spaces, appearing within the mesoderm (mostly of the spongy type known as mesenchyme), link into vascular networks which spread rapidly in the chorion, yolk sac and embryo proper. They become the *heart, blood vessels* and *lymphatics.* Their thin, lining layer is an epithelium, named *endothelium.*

Developmental Advances. In the somite-period the neural plate and various other primordia are elaborated further, so that the embryo acquires its general body plan. While the embryo is fashioning a neural tube from the neural plate and is also elaborating other axial structures, it is sometimes called a *neurula;* this designates a stage next beyond the gastrula. During the fourth week there is an average increase in total length from about 2 to 5 mm., but size is too variable among the smaller specimens to constitute a reliable index of development. Better correlated with the degree of development is the number of mesodermal somites. These make their appearance progressively; they begin to appear at the end of the third week and attain nearly their full number (about 42) during the fourth week. Such momentous changes characterize this period that the embryo advances from a simple disc to a relatively complex organism.

Some of the head of an embryo arises from the material of the embryonic disc in the region in front of the early primitive knot. But shortly after the primitive knot is formed, the forward streaming of cells, which produced the head process, ends. The primitive streak then begins to shorten and the knot moves caudad, leaving (or actively paying out?) in its wake most of the notochord and the floor of the neural tube (Fig. 72).[5, 6, 7] Simultaneously with this retreat the somite-pairs appear in steady succession on each side of the notochord, organizing from the appropriate mesoderm brought in by

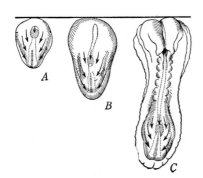

Fig. 72. Diagrams of the caudal growth of the body, partly at the expense of the retreating primitive knot (Streeter). The primitive knot is stippled.

the movements of gastrulation. By the time the majority of somites have formed, the surviving primitive knot overtakes the greatly shortened streak and the two combine as a compact mass of tissue. This mass, located at the caudal end of the embryo, is known as the *end bud* or *tail bud* (Figs. 73 *D,* 76 *A*). In this swelling separate germ layers cannot be recognized as such, but it is apparently not an 'indifferent' material of the sort some have claimed.[8, 9] Actually, there is no real difference between this region and those at more cephalic levels, except that from the first the germ-layer materials are crowded and condensed progressively in a caudal direction in a way that hides their identity. In other words, the gastrulation movements (and the resulting segregation of ectoderm, mesoderm and entoderm) are fundamentally the same with respect to all levels of the future embryo.[6, 7, 10]

The most important maneuver in the establishment of general body form is the transformation of the flat embryonic disc into a roughly cylindrical embryo attached to the yolk sac by a narrower stalk. Three factors co-operate to produce this change: (1) There is more rapid expansion of both the embryonic area and the yolk sac in contrast to a slower rate of growth at the region of transition between the two. The enlarging embryonic area, bound at its more sluggish, inelastic periphery, at first buckles upward and then overlaps the slower growing margin (Fig. 68); the latter becomes a zone of 'constriction' between embryo and yolk sac. Since the growth is particularly rapid at the future head- and tail-ends, the embryo soon becomes elongate (Fig. 74). The entire process can be described as one of internal growth resulting in folding; the embryo enlarges somewhat as does a soap bubble blown from a pipe. (2) In conjunction with the overgrowth just described, there is important underfolding, most evident at the front and hind ends of the embryo.[11] As the neural axis elevates and projects forward beyond the margin of the embryonic disc, the future pharyngeal membrane and the

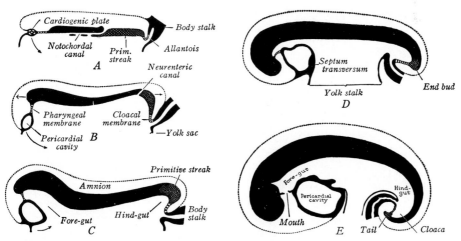

Fig. 73. Sagittal sections of human embryos to illustrate the reversal occurring at the cranial and caudal ends. × 20. *A,* At presomite stage; *B,* at one somite; *C,* at six somites; *D,* at twelve somites; *E,* at twenty-two somites. Arrows show growth directions.

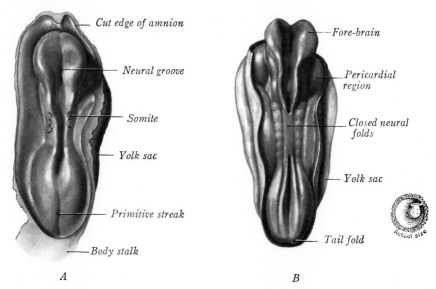

Cut edge of amnion

Neural groove

Somite

Yolk sac

Primitive streak

Body stalk

Fore-brain

Pericardial region

Closed neural folds

Yolk sac

Tail fold

Actual size

A   B

Fig. 74. Human embryos of twenty-one days, in dorsal view (Streeter). *A,* Ingalls embryo of 1.4 mm., with three somites ($\times$ 42). *B,* Payne embryo of 2.2 mm., with seven somites ($\times$ 27).

cardiac area swing beneath, as on a hinge (Fig. 73 *A-C*). In doing this the cardiac area, originally ahead of the pharyngeal membrane, necessarily becomes the more caudal of the two in position, while the amnion and the yolk sac (originally at the rim of the disc) then attach caudal to the pericardium (*D, E*). Caudal growth of the end bud brings about a similar reversal at the caudal end of the embryo (*A-E*). As a result, the cloacal membrane and body stalk turn under onto the ventral side. (3) Finally, a certain amount of true constriction, through growth, purses all these parts at the site of the future umbilicus (*C-E*).

Throughout the entire period during which the body and its parts are being laid down, development and differentiation appear first in the head region and then advance tailward. For this reason, many structures that extend longitudinally for an appreciable distance show a gradation in their development, with progressively advanced stages located at higher levels. The size advantage initially gained by the head-end as a whole is relinquished only slowly. A further tendency toward progressively graded development is expressed from the mid-dorsal line in lateral directions. Such relations are the visible expressions of gradients in growth and differentiation.

Embryos of the Fourth Week. Turning now to actual embryos representative of the early days of the fourth week, it will be seen that rapid, differential growth causes them to take a cylindrical shape (Figs. 74, 75). *Neural folds,* rising high and uniting in a progressive manner, roll up a *neural tube* in which a larger *brain region* becomes plainly indicated. Many preserved specimens of this period show a markedly concave back, ap-

parently produced artificially by shrinkage of the yolk sac (Fig. 76 *A*). Internally the *fore-* and *hind-gut* elongate into blind tubes (Fig. 67 *D*). The
heart becomes conspicuous, a system of paired *blood vessels* is established,
and a functional circulation begins. *Somites* increase rapidly in number
and total a dozen or more.

Later in the week additional characteristics appear (Figs. 76 *B*, 77 *A*).

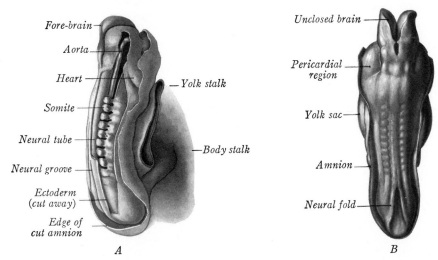

Fig. 75. Human embryos of twenty-two days. *A,* Veit-Esch embryo of 2.3 mm., with
nine somites, partially dissected and viewed from the right side (× 25). *B,* Corner embryo
of 1.7 mm., with ten somites, in dorsal view (Streeter; × 34).

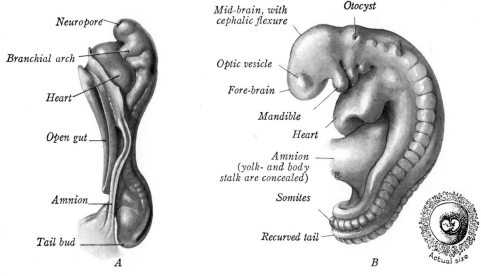

Fig. 76. Human embryos of twenty-four and twenty-six days, viewed from the left
side. *A,* Atwell embryo of 2.6 mm., with nineteen somites (Streeter; × 23). *B,* Embryo of
3.6 mm., with twenty-five somites (× 16).

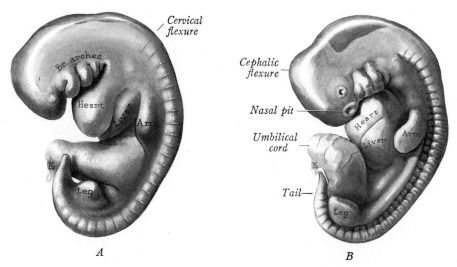

Fig. 77. Human embryos of four and five weeks, viewed from the left side. *A*, At 5 mm. (× 12); *B*, at 8 mm. (× 7.5).

A definite *trunk*, convexly curved, ends in a conspicuous *tail;* the total number of somites (about 42) is acquired. The *head* is flexed ventrally. It bears a sharp bend (*cephalic flexure*) at the level of the mid-brain, while a more gentle curvature (*cervical flexure*) occurs in the region of the future neck. The *heart* is prominent and bulging. *Sense organs* and *limb buds* are indicated and the *branchial arches* become prominent; the first pair of arches bifurcates into primitive *jaws*. The *yolk stalk*, or connection with the yolk sac, is now relatively small and slender. It is the rapidly elongating neural tube, in contrast to the slower growing ventral surface of the embryo, that is responsible for producing the characteristic curves and flexures in the embryo as a whole.

PERIOD OF EMBRYO COMPLETION (FIFTH THROUGH EIGHTH WEEK)

These embryos, ranging between 5 and 23 mm., show marked changes. Their external form, although quite unfinished, comes to resemble more the 'human' condition, and after the second month the developing young is commonly called a *fetus*. This external metamorphosis may be followed by studying Figs. 77–80 *F*. It is due principally to the following factors: (1) Changes in the *flexures* of the body; the dorsal convexity is lost, the head becomes more erect and the body straighter. (2) The *face* develops. (3) The external structures of the *eye, ear* and *nose* appear. (4) The *limbs* organize as such, with digits demarcated. (5) The prominent *tail* of the fifth week becomes inconspicuous both through actual regression and concealment by the growing buttocks. (6) The *umbilical cord* becomes a definite entity, its embryonic end occupying a relatively diminishing area on the belly wall.

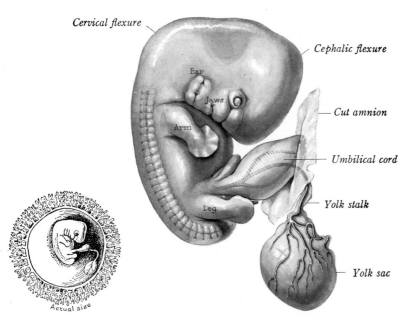

Fig. 78. Human embryo of six weeks (12 mm.), viewed from the right side. × 5.

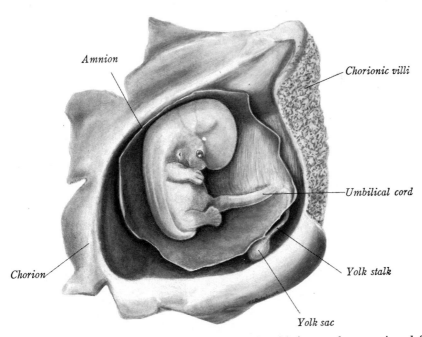

Fig. 79. Human embryo of seven weeks (18 mm.), with its membranes, viewed from the right side. × 2. The chorion has been opened and reflected, and half of the amnion removed.

(7) The *heart,* which was the chief ventral prominence in earlier embryos, now shares this distinction with the rapidly growing *liver;* these two organs determine the shape of the ventral body until the eighth week when the gut dominates the belly cavity and the contour of the abdomen is more evenly rotund. (8) The *neck* becomes recognizable, due chiefly to the settling of the heart caudad and the effacement of the branchial arches. (9) The *external genitalia* appear in their 'sexless' condition. (10) The neuro-muscular mechanism attains sufficient perfection so that spontaneous movements are possible.

Almost all of the internal organs are well laid down at two months; henceforth, until the end of gestation, the chief changes undergone are those of growth and further specialization of the tissues.

PERIOD OF THE FETUS (THIRD THROUGH TENTH MONTH)

During the *third month* (lunar) the fetus definitely resembles a human being, but the head is still disproportionately large (Fig. 80 *G, H*); the previous umbilical herniation is reduced by the return of the intestine into the abdomen; the ears rise to eye level; the eyes are directed forward, rather than laterally, and the eyelids fuse; nails begin forming and ossification centers appear in most of the bones; sex can be distinguished externally. At *four months* the face has a truly human appearance and individual differences become recognizable (*I*); the face is broad and the eyes are still widely separated; the umbilical cord attaches higher on the abdominal wall, above an expanding infra-umbilical region that scarcely existed at two months.

At *five months* downy hairs (lanugo) cover the body and some head hairs show; fetal movements ('quickening') are felt by the mother. At *six months* the eyebrows and lashes become well defined; the body is lean but better proportioned; the skin is wrinkled. At *seven months* the fetus looks like a dried-up old person, with red, wrinkled skin covered with greasy vernix caseosa; the eyelids reopen. At *eight months* subcutaneous fat is depositing; the testes begin to invade the scrotum; infants of this age, born prematurely, may survive with proper care.

At *nine months* the dull redness of the skin fades, wrinkles smooth out, the body and limbs become more rounded. At *full term* (38 weeks) the body is plump; the skin has lost its lanugo hair coat but is still smeared with vernix caseosa; the nails project beyond the fingers and to the tips of the toes; the attachment of the umbilical cord is at the center of the abdomen; the upper limbs are still slightly longer than the lower limbs; the testes are usually in the scrotum; the labia majora are in contact; the average weight is seven pounds and the height is 20 inches.

The embryos shown in Figs. 63-79 are drawn at progressively lower magnifications. In order to obtain a better idea of the actual and relative sizes of embryos at different periods, the series assembled as Fig. 80 should be studied. These embryos are all drawn at natural size.

The tabulation of correlated development, inserted at the end of this chapter, is designed to present an epitome of human development for purposes of study and reference. In the vertical columns the sequential development of each system is listed. Of even greater importance is a study of the horizontal entries which record the correlated changes throughout the embryo at definite periods. It is this picture of parallel development that is otherwise extremely difficult to comprehend.

Anomalies. Grossly abnormal embryos are not infrequent among those obtained by spontaneous abortion or operative intervention. The external body form may show all gradations from mildly faulty modeling to an amorphous mass that is scarcely recognizable as a fetus (Fig. 81 *A, B*). Various pathological alterations in the embryo commonly accompany those morbid disturbances that induce its stunting or death. Degenerative changes are common also in the fetal membranes, although the chorionic sac sometimes continues to grow for a time quite normally; this may occur after the embryo has died and even after it has disappeared. Many of these imperfect specimens of different kinds result from

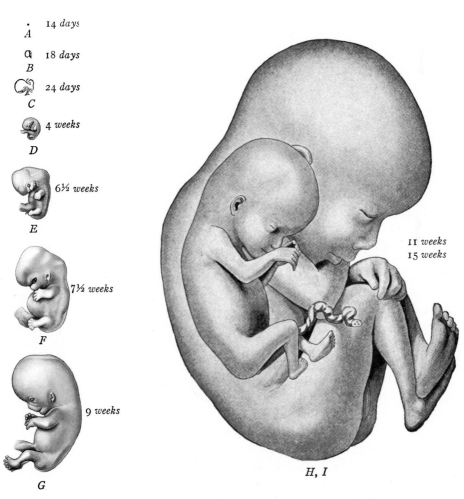

Fig. 80. A graded series of human embryos, at natural size.

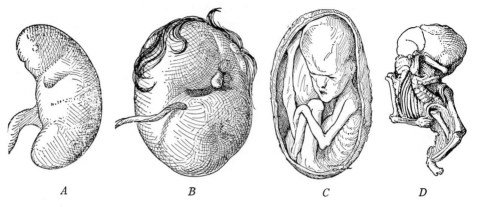

A   B   C   D

Fig. 81. Malformed human embryos. *A,* Stunted embryo, with poorly developed external form ($\times$ 4). *B,* Amorphous fetus, twin to a normal, full-term baby. Externally there was hair and a rudimentary mouth and cyclopic eye; internally there were three vertebræ and the base of a skull, besides much vascular connective tissue and fat and a little muscular and nervous tissue ($\times$ ⅓). *C,* Mummified fetus, within a calcified gestation sac ($\times$ ⅓). *D,* Lithopedion ($\times$ ½). *E,* Fetus papyraceus ($\times$ ⅕).

E

zygotes of such initial poor quality that normal, continued development was impossible. If a dead fetus is retained it usually macerates and is resorbed, but it may mummify (*C*) or even calcify into a *lithopedion* (*i.e.,* 'stone child') and persist indefinitely (*D*). Compression produces a *fetus papyraceus,* or 'paper-doll fetus' (*E*).

Malformations of the more specific parts of the body, externally and internally, will be described in conjunction with the detailed accounts of the development of those regions (Part II of this book).

DETERMINATION OF THE AGE OF EMBRYOS

The determination of the exact age of a recovered human embryo is beset with difficulties. For the practitioner it is fortunate that significant errors concern chiefly the rarer specimens of the early weeks. Development starts with fertilization and if this date could be determined reliably, the age-problem would largely disappear. But the available data, usually supplied by the unaided memory of the patient, too often are either incorrect or, at least, open to alternative interpretation as to the probable time of conception. This lack of a reliable starting date for most pregnancies makes further computation but approximate. On the other hand, the terminal age-date of a healthy embryo whose normal growth has been interrupted by operation is definite, and the true age of such a specimen (whether determinable or not) is the interval since fertilization. But some embryos experience a progressively slowing growth rate prior to their operative removal or spontaneous abortion. Moreover, aborted embryos commonly are not only dead, but have been retained in this condition for a time before extrusion occurs. In neither instance is the recovery date of much value with respect to normal age spans.

It is impossible to establish the precise time of fertilization. Even in fortunate cases when the occasion of an isolated, fruitful coitus is surely known, it can only be assumed that fertilization probably occurred within the day following such coitus. Accordingly, in the common absence of a reliable coital history it becomes important to determine when the effective ovulation most probably took place. If this can be set, the presumptive fertilization date will then be indirectly established within a day—for the reason that the egg loses its fertilizability so rapidly. In a previous discussion of the time relation between ovulation and menstruation (p. 50) it was stated that ovulation occurs most commonly at the middle of the menstrual month. Nevertheless, it is well recognized that this is at best an average, and oftentimes ova are liberated either earlier or later.

Thus it is approximately correct to compute the age of an embryo *from the fourteenth day after the onset of the last menstruation.* There are, however, three practical difficulties which may make such a reckoning unreliable in any specific case. First, deviation from the average time of ovulation; some clinicians even contend that there is no day of the cycle on which instances of conception have not been proved.[12] Second, ovulation earlier or later than the fourteenth day because of short or long cycles. Third, bleeding that resembles menstruation (the 'placental sign') not infrequently appears early in pregnancy, thus obscuring the true time of the last normal menstruation.[13]

For ordinary purposes it is both convenient and reasonably accurate to compare a given specimen with a standard age- and size-table (p. 104). These norms have been established through careful studies on fetuses that were accompanied by fairly adequate clinical histories. It is, however, necessary to understand that such tables merely state averages, whereas the normal size-range varies appreciably above and below the means listed. Furthermore, size alone is rather unreliable as a basis of comparison in the first month;

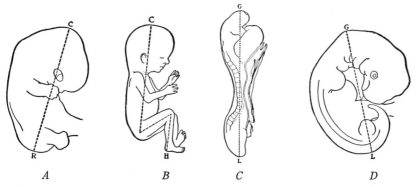

Fig. 82. Measurement of embryos. *A,* Crown-rump length; *B,* crown-heel length; *C,* greatest length; *D,* greatest length is neck-rump length.

more important is the state of structural development which advances in definite, orderly sequence (pp. 88–98).

Embryos are measured in two principal ways (Fig. 82). Commonest is the *crown-rump length* (designated *CR*), or sitting height; this is the distance from vertex to breech (*A*). The other measurement is the *crown-heel length* (*CH*), or standing height (*B*). When young embryos are nearly straight, the *greatest length* (*GL*) is most practical (*C*); in embryos of about four weeks, when the head is greatly flexed, this becomes a neck-rump measurement (*D*). The table on this page lists some statistical averages for human embryos of definite ages. The lengths of embryos between three and eight weeks should be memorized at the outset, because frequent reference will be made to these ages and sizes.

Handy rules for calculating the crown-heel length in inches of an embryo or fetus are as follows:

For the first five months, add the numbers of the previous months.

For the last five months, multiply the number of the month by two.

How well these rules work is shown in the appended table which gives the calculation for each month and the true value for that month in parentheses:

$$
\begin{array}{llll}
1 \text{ mo.} & = 0 & = 0 \text{ in.} & (0.2 \text{ in.}) \\
2 \text{ “} & = 1 & = 1 \text{ “} & (1.2 \text{ “}) \\
3 \text{ “} & = 1 + 2 & = 3 \text{ “} & (2.9 \text{ “}) \\
4 \text{ “} & = 1 + 2 + 3 & = 6 \text{ “} & (6.2 \text{ “}) \\
5 \text{ “} & = 1 + 2 + 3 + 4 = 10 \text{ “} & & (9.4 \text{ “}) \\
6 \text{ “} & = 6 \times 2 & = 12 \text{ “} & (11.7 \text{ “}) \\
7 \text{ “} & = 7 \times 2 & = 14 \text{ “} & (14.0 \text{ “}) \\
8 \text{ “} & = 8 \times 2 & = 16 \text{ “} & (16.1 \text{ “}) \\
9 \text{ “} & = 9 \times 2 & = 18 \text{ “} & (18.0 \text{ “}) \\
10 \text{ “} & = 10 \times 2 & = 20 \text{ “} & (19.9 \text{ “}) \\
\end{array}
$$

RELATIONS OF AGE, SIZE AND WEIGHT IN THE HUMAN EMBRYO

AGE OF EMBRYO	CROWN-RUMP LENGTH (MM.)	CROWN-HEEL LENGTH (MM.) †	EXTERNAL DIAMETER OF CHORIONIC SAC (MM.)	WEIGHT IN GRAMS	AMOUNT OF INCREASE EACH MONTH WHEN VALUE AT START OF MONTH EQUALS UNITY	
					CR LENGTH	WEIGHT
One week..........	0.1*	.....	0.2			
Two weeks.........	0.2*	.....	3			
Three weeks........	2.0	.....	10			
Four weeks........	5.0	.....	20	.02	49.0	40000.00
Five weeks.........	8.0	.....	25			
Six weeks..........	12.0	.....	30			
Seven weeks........	17.0	19.0	40			
Two lunar months..	25.0	30.0	50	1	3.6	49.00
Three lunar months.	56.0	73.0	....	14	1.4	13.00
Four lunar months..	112.0	157.0	....	105	1.0	6.50
Five lunar months..	160.0	239.0	....	310	0.43	1.95
Six lunar months....	203.0	296.0	....	640	0.26	1.07
Seven lunar months.	242.0	355.0	....	1080	0.14	0.69
Eight lunar months.	277.0	409.0	....	1670	0.14	0.55
Nine lunar months..	313.0	458.0	....	2400	0.13	0.43
Full term (38 weeks).	350.0	500.0	....	3300	0.12	0.38

* Total length of embryonic disc.   † Recent study:[15] 7 wk. = 23 mm.; 8 wk. = 32 mm.

The Delivery Date. Of practical interest is the determination of the date of delivery of a pregnant woman. The average time for delivery is ten lunar months, or 280 days, from the beginning of the last menstrual period. This period of 280 days (the so-called *menstruation age*) should not be confused with the duration of pregnancy (*i.e.*, the true age of the fetus or *fertilization age*) which is about two weeks less. Two-thirds of all deliveries vary not more than 11 days above and below the mean of 280. The expected delivery date can be set by counting back three calendar months from the first day of the last period, and then adding a year and one week. This date is, of course, only a forecast based on averages. Since bleeding, which is mistaken for menstruation, sometimes occurs after pregnancy begins and since there is some normal variation in the length of pregnancy, the computation may prove unreliable in any particular case.

For comparison and reference, the gestation periods and average number of young of a few representative mammals are appended. Some of these in a sense are premature at birth; as an example, the newborn rat is blind, hairless and helpless. At the other extreme, the guinea pig is well developed, able to walk and even to eat solid food. The weight of the newborn in comparison to the mother ranges from 0.1 per cent in the polar bear to 33 per cent in the bat.

COMPARATIVE DATA CONCERNING GESTATION IN MAMMALS

ANIMAL	GESTATION PERIOD	NO. IN LITTER	ANIMAL	GESTATION PERIOD	NO. IN LITTER
Opossum...............	13 days	8	Macacus monkey......	24 weeks	1
Mouse; rat.............	20; 22 da.	6; 8	Man; manlike apes.....	38 weeks	1
Rabbit.................	32 days	6	Cow.................	40 weeks	1
Cat; dog; guinea pig.....	9 weeks	4– 6	Mare.................	48 weeks	1
Sow..................	17 weeks	6–12	Rhinoceros...........	18 months	1
Sheep; goat............	21 weeks	1– 2	Elephant.............	21 months	1

VIABILITY AND LONGEVITY

The survival-ability of the protoplasm with which a fertilized egg (and hence the future individual) is endowed varies enormously.[14] Some specimens succumb early in development and others later; in all, about one pregnancy in three is unsuccessful, largely because the embryos are not vigorous enough to reach birth as living individuals. Moreover, this selective elimination does not cease at birth but continues throughout the life span. A person beyond middle age has realized the expected viability of a zygote of average quality; on the other hand, an individual attaining old age comes from one with great initial energy, balance and resistance. But not only do zygotes as a whole differ in endurance, vulnerability and capacity for growth, but the several organs and parts also are similarly variable. Some relatively unimportant organs, such as teeth and hair, suffer a natural early decline, yet this does not matter. When, however, a functionally important organ fails, for whose loss the rest of the body cannot compensate, then life is imperiled. In this instance an otherwise competent human machine meets an untimely death merely because of a single, defective, critical organ, such as the heart or kidney. By contrast, frail individuals frequently totter into old age because they are organically well balanced and have no vulnerable weakness.

This concept of the importance of the quality of the zygotes from which mankind traces origin is fundamental. To a certain degree its implications are fatalistic, yet there are other interacting factors in the total equation besides the innate quality of inherited protoplasm. Chance and a hostile, local environment may cut short a life intended for long performance. On the other hand, an intelligently ordered existence can do much to conserve constitutional endowments to their full expectancy.

REFERENCES CITED

1. Hertig, A. T. & J. Rock. 1948; '49, Am. J. Obst. & Gyn., *55*, 6–17, 440–452; *58*, 968–993.
2. Hertig, A. T. & J. Rock. 1945. Carnegie Contr. Embr., *31*, 65–84.
3. Hertig, A. T. & J. Rock. 1941. Carnegie Contr. Embr., *29*, 127–156.
4. Heuser, C. H., *et al.* 1945. Carnegie Contr. Embr., *31*, 85–99.
5. Streeter, G. L. 1927. Carnegie Contr. Embr., *19*, 73–92.
6. Pasteels, J. 1937. Arch. de Biol., *48*, 381–488.
7. Spratt, N. T. 1947. J. Exp. Zoöl., *104*, 69–100.
8. Holmdahl, D. E. 1951. Z'ts. f. mikr.-anat. Forsch., *57*, 359–392.
9. Peter, K. 1951. Z'ts. f. mikr.-anat. Forsch., *57*, 393–401.
10. Gaertner, R. A. 1949. J. Exp. Zoöl., *111*, 157–174.
11. Grünwald, P. 1941. J. Morph., *69*, 83–125.
12. Weinstock, F. 1934. Zentr. f. Gyn., *58*, 2947–2952.
13. Brewer, J. L. 1937. Am. J. Anat., *61*, 429–481.
14. Streeter, G. L. 1931. Sci. Monthly, *32*, 495–506.
15. Iffey *et al.* 1967. Acta Anat., *66*, 178–186.

E TABLE OF CORRELATED HUMAN DEVELOPMENT

Age in Weeks	Size (C R) in Mm.	Body Form	Mouth	Pharynx and Derivatives	Cœlom and Mesenteries	Urogenital System	Vascular System	Skeletal System	Muscular System	Integumentary System	Nervous System	Sense Organs	Age in Weeks
2.5	1.5	Embryonic disc flat. Primitive streak prominent. Neural groove indicated.	....	....	Extra-embryonic cœlom present. Embryonic cœlom about to appear.	Allantois present.	Blood islands appear on chorion and yolk sac. Cardiogenic plate reversing.	Head process (or notochordal plate) present.	....	Ectoderm a single layer.	Neural groove indicated.	....	2.5
3.5	2.5	Neural groove deepens and closes (except ends). Somites 1–16± present. Cylindrical body constricting from yolk sac. Branchial arches 1 and 2 indicated.	Mandibular arch prominent. Stomodeum a definite pit. Oral membrane ruptures.	Pharynx broad and Pharyngeal pouches ing. Thyroid indicated.	Embryonic cœlom a U-shaped canal, with a large pericardial cavity. Septum transversum indicated. Mesenteries forming. Mesocardium atrophying.	All pronephric tubules formed. Pronephric duct growing caudad as a blind tube. Cloaca and cloacal membrane present.	Primitive blood cells and vessels present. Embryonic blood vessels a paired symmetrical system. Heart tubes fuse, bend S-shape and beat begins.	Mesodermal segments appearing (1–16±). Older somites begin to show sclerotomes. Notochord a cellular rod.	Mesodermal segments appearing (1–16±). Older somites show myotome plates.	....	Neural groove prominent; rapidly closing. Neural crest a continuous band.	Optic vesicle and auditory placode present. Acoustic ganglia appearing.	3.5
4	5.0	Branchial arches completed. Flexed heart prominent. Yolk stalk slender. All somites present (40). Limb buds indicated. Eye and otocyst present. Body flexed; C-shape.	Maxillary and mandibular processes prominent. Tongue primordia present. Rathke's pouch indicated.	Five pharyngeal p present Pouches 1–4 have plates. Primary tympanic indicated. Thyroid a stalked s	Cœlom still a continuous system of cavities. Dorsal mesentery a complete median curtain. Omental bursa indicated.	Pronephros degenerated. Pronephric (mesonephric) duct reaches cloaca. Mesonephric tubules differentiating rapidly. Metanephric bud pushes into secretory primordium.	Hemopoiesis on yolk sac. Paired aortæ fuse. Aortic arches and cardinal veins completed. Dilated heart shows sinus, atrium, ventricle, and bulbus.	All somites present (40). Sclerotomes massed as primitive vertebræ about notochord.	All somites present (**40**).	....	Neural tube closed. Three primary vesicles of brain represented. Nerves and ganglia forming. Ependymal mantle and marginal layers present.	Optic cup and lens pit forming. Auditory pit becomes closed, detached otocyst. Olfactory placodes arise and differentiate nerve cells.	4
5	8.0	Nasal pits present. Tail prominent. Heart, liver and mesonephros protuberant. Umbilical cord organizes.	Jaws outlined. Rathke's pouch a stalked sac.	Phar. pouches gai and vent. divertic Thyroid bilobed. Thyro-glossal duc phies.	Pleuro-pericardial and pleuro-peritoneal membranes forming. entral mesogastrium draws away from septum.	Mesonephros reaches its caudal limit. Ureteric and pelvic primordia distinct. Genital ridge bulges.	Primitive vessels extend into head and limbs. Vitelline and umbilical veins transforming. Myocardium condensing. Cardiac septa appearing. Spleen indicated.	Condensations of mesenchyme presage many future bones.	Premuscle masses in head, trunk and limbs.	Epidermis gaining a second layer (periderm).	Five brain vesicles. Cerebral hemispheres bulging. Nerves and ganglia better represented. [Suprarenal cortex accumulating.]	Chorioid fissure prominent. Lens vesicle free. Vitreous anlage appearing. Octocyst elongates and buds endolymph duct. Olfactory pits deepen.	5
6	12.0	Upper jaw components prominent but separate. Lower jaw-halves fused. Head becomes dominant in size. Cervical flexure marked. External ear appearing. Limbs recognizable as such.	Lingual primordia fusing. Foramen cæcum established. Labio-dental laminæ appearing. Parotid and submaxillary buds indicated.	Thymic sacs, ulti chial sacs and sol thyroids are con and ready to de Thyroid becomes converts into pl	leuro-pericardial communications close. Iesentery expands as intestine forms loop.	Cloaca subdividing. Pelvic anlage sprouts pole tubules. Sexless gonad and genital tubercle prominent. Müllerian duct appearing.	Hemopoiesis in liver. Aortic arches transforming. L. umbil. vein and d. venosus become important. Bulbus absorbed into right ventricle. Heart acquires its general definitive form.	First appearance of chondrification centers. Desmocranium.	Myotomes, fused into a continuous column, spread ventrad. Muscle segmentation largely lost.	Milk line present.	Three primary flexures of brain represented. Diencephalon large. Nerve plexuses present. Epiphysis recognizable. Sympathetic ganglia forming segmental masses. Meninges indicated.	Optic cup shows nervous and pigment layers. Lens vesicle thickens. Eyes set at 160°. Naso-lacrimal duct. Modeling of ext., mid. and int. ear under way. Vomero-nasal organ.	6
7	17.0	Branchial arches lost. Cervical sinus obliterates. Face and neck forming. Digits indicated. Back straightens. Heart and liver determine shape of body ventrally. Tail regressing.	Lingual primordia merge into single tongue. Separate labial and dental laminæ distinguishable. Jaws formed and begin to ossify. Palate folds present and separated by tongue.	Thymi elongating ing lumina. Parathyroids bec beculate and with thyroid. Ultimobranchial fuse with thyro Thyroid becomi centic.	ricardium extended by plitting from body wall. esentery expanding apidly as intestine coils. gaments of liver prominent.	Mesonephros at height of its differentiation. Metanephric collecting tubules begin branching. Earliest metanephric secretory tubules differentiating. Bladder-urethra separates from rectum. Urethral membrane rupturing.	Cardinal veins transforming. Inf. vena cava outlined. Atrium, ventricle and bulbus partitioned. Cardiac valves present. Stem of pulm. vein absorbed into l. atrium. Spleen anlage prominent.	Chondrification more general. Chondrocranium.	Muscles differentiating rapidly throughout body and assuming final shapes and relations.	Mammary thickening lens-shaped.	Cerebral hemispheres becoming large. Corpus striatum and thalamus prominent. Infundibulum and Rathke's pouch in contact. Chorioid plexuses appearing. Suprarenal medulla begins invading cortex.	Chorioid fissure closes, enclosing central artery. Nerve fibers invade optic stalk. Lens loses cavity by elongating lens fibers. Eyelids forming. Fibrous and vascular coats of eye indicated. Olfactory sacs open into mouth cavity.	7
8	23.0	Nose flat; eyes far apart. Digits well formed. Growth of gut makes body evenly rotund. Head elevating. Fetal state attained.	Tongue muscles well differentiated. Earliest taste buds indicated. Rathke's pouch detaches from mouth. Sublingual gland appearing.	Auditory tube panic cavity di able. Sites of tonsil and indicated. Thymic halves become solid. Thyroid follicle	uro-peritoneal communications close. icardium a voluminous c. phragm completed, including musculature. phragm finishes its lescent.'	Testis and ovary distinguishable as such. Müllerian ducts, nearing urogenital sinus, are ready to unite as utero-vaginal primordium. Genital ligaments indicated.	Main blood vessels assume final plan. Primitive lymph sacs present. Sinus venosus absorbed into right atrium. Atrio-ventricular bundle represented.	First indications of ossification.	Definitive muscles of trunk, limbs and head well represented and fetus capable of some movement.	Mammary primordium a globular thickening.	Cerebral cortex begins to acquire typical cells. Olfactory lobes visible. Dura and pia-arachnoid distinct. Chromaffin bodies appearing.	Eyes converging rapidly. Ext., mid. and int. ear assuming final form. Taste buds indicated. External nares plugged.	8
10	40.0	Head erect. Limbs nicely modeled. Nail folds indicated. Umbilical hernia reduced.	Fungiform and vallate papillæ differentiating. Lips separate from jaws. Enamel organs and dental papillæ forming. Palate folds fusing.	Thymic epitheli forming into and thymic co Ultimobranchia disappear as s	cessus (saccus) vagiles forming. estine and its mesenry withdrawn from rd.	Kidney able to secrete. Bladder expands as sac. Genital duct of opposite sex degenerating. Bulbo-urethral and vestibular glands appearing. Vaginal sacs forming.	Thoracic duct and peripheral lymphatics developed. Early lymph glands appearing. Enucleated red cells predominate in blood.	Ossification centers more common. Chondrocranium at its height.	Perineal muscles developing tardily.	Epidermis adds intermediate cells. Periderm cells prominent. Nail field indicated. Earliest hair follicles begin developing on face.	Spinal cord attains definitive internal structure.	Iris and ciliary body organizing. Eyelids fused. Lacrimal glands budding. Spiral organ begins differentiating.	10
12	56.0	Head still dominant. Nose gains bridge. Sex readily determined by external inspection.	Filiform and foliate papillæ elevating. Tooth primordia form prominent cups. Cheeks represented. Palate fusion complete.	Tonsillar crypt invaginate. Thymus formi and becomin ingly lympho Thyroid attai structure.	ntum an expansive ron partly fused with sal body wall. enteries free but exit typical relations. mic extension into bilical cord obliterd.	Uterine horns absorbed. External genitalia attain distinctive features. Meson. and rete tubules complete male duct. Prostate and seminal vesicle appearing. Hollow viscera gaining muscular walls.	Blood formation beginning in bone marrow. Blood vessels acquire accessory coats.	Notochord degenerating rapidly. Ossification spreading. Some bones well outlined.	Smooth muscle layers indicated in hollow viscera.	Epidermis three-layered. Corium and subcutaneous now distinct.	Brain attains its general structural features. Cord shows cervical and lumbar enlargements. Cauda equina and filum terminale appearing. Neuroglial types begin to differentiate.	Characteristic organization of eye attained. Retina becoming layered. Nasal septum and palate fusions completed.	12
16	112.0	Face looks 'human.' Hair of head appearing. Muscles become spontaneously active. Body outgrowing head.	Hard and soft palates differentiating. Hypophysis acquiring definitive structure.	Lymphocytes in tonsils. Pharyngeal to development.	ter omentum fusing transverse meson and colon. duodenum and asling and descending ocolon attaching to wall.	Kidney attains typical shape and plan. Testis in position for later descent into scrotum. Uterus and vagina recognizable as such. Mesonephros involuted.	Blood formation active in spleen. Heart musculature much condensed.	Most bones distinctly indicated throughout body. Joint cavities appear.	Cardiac muscle appearing in earlier weeks, now much condensed. Muscular movements *in utero* can be detected.	Epidermis begins adding other layers. Body hair starts developing. Sweat glands appear. First sebaceous glands differentiating.	Hemispheres conceal much of brain. Cerebral lobes delimited. Corpora quadrigemina appear. Cerebellum assumes some prominence.	Eye, ear and nose grossly approach typical appearance. General sense organs differentiating.	16
20–40 (5–10 mo.)	160.0–350.0	Lanugo hair appears (5). Vernix caseosa collects (5). Body lean but better proportioned (6). Fetus lean, wrinkled and red; eyelids reopen (7). Testes invading scrotum (8). Fat collecting, wrinkles smoothing, body rounding (8–10).	Enamel and dentine depositing (5). Lingual tonsil forming (5). Permanent tooth primordia indicated (6–8). Milk teeth unerupted at birth.	Tonsil structu (5).	nterial attachments pleted (5). al sacs passing into um (7–9).	Female urogenital sinus becoming a shallow vestibule (5). Vagina regains lumen (5). Uterine glands appear (7). Scrotum solid until sacs and testes descend (7–9). Kidney tubules cease forming at birth.	Blood formation increasing in bone marrow and decreasing in liver (5–10). Spleen acquires typical structure (7). Some fetal blood passages discontinue (10).	Carpal, tarsal and sternal bones ossify late; some after birth. Most epiphyseal centers appear after birth; many during adolescence.	Perineal muscles finish development (6).	Vernix caseosa seen (5). Epidermis cornifies (5). Nail plate begins (5). Hairs emerge (6). Mammary primordia budding (5); buds branch and hollow (8). Nail reaches finger tip (9). Lanugo hair prominent (7); sheds (10).	Commissures completed (5). Myelinization of cord begins (5). Cerebral cortex layered typically (6). Cerebral fissures and convolutions appearing rapidly (7). Myelinization of brain begins (10).	Nose and ear ossify (5). Vascular tunic of lens at height (7). Retinal layers completed and light perceptive (7). Taste sense present (8). Eyelids reopen (7–8). Mastoid cells unformed (10). Ear deaf at birth.	20–40 (5–10 mo.)

Chapter VII. The Fetal Membranes

In tracing the development of any vertebrate embryo it becomes apparent that only part of the egg (or of its cleavage-cell mass) forms the actual embryo, whereas other parts lie outside the embryonic territory and for this reason are called *extra-embryonic*. These latter regions are concerned with the elaboration of so-called embryonic or *fetal membranes,* although some of them are not 'membranous' in a strict sense. All are auxiliary organs which have arisen partly for the protection of the embryo, and more especially to provide for its nutrition, respiration and excretion until the time arrives when independent existence can be attempted. As historically recent adaptations they have been evolved with relation to the following structures that were already in existence: the gut (yolk sac); body wall (amnion and chorion); and primitive bladder (allantois). All of these membranes eventually will separate from the fetus by natural methods of amputation or withering, and are then discarded.

The fundamental set of fetal membranes includes the *yolk sac, amnion, chorion* and *allantois.* The *placenta* is a distinctive 'membrane' that has been elaborated by the higher mammals. To this compound organ the chorion (and, to a greater or less extent, the allantois) furnishes a part, while the uterine lining supplies the rest. The *umbilical cord* is a vascular cable that serves as a communication between the fetus and its placenta.

HUMAN MEMBRANES

The beginnings of the yolk sac, amnion, chorion, body stalk and allantois during the second and third weeks of embryonic life have been described in the preceding chapter in conjunction with the embryos of those periods. The later histories of these auxiliary organs (except the chorion; Chapter VIII) and the complete developmental course of another appendage, the umbilical cord, will now be traced.

The Yolk Sac. The entodermal roof of the *yolk sac,* composed originally of taller cells, provides the primary material from which the tubular gut is fashioned by folding (Fig. 67). At first only a slightly narrower region connects the unclosed gut with the yolk sac proper (Fig. 83 *A*). With the further growth of the embryo's body there is progressive constriction of the embryo from the yolk sac. This actual constriction is intensified relatively when both embryo and yolk sac continue to enlarge, whereas their region of union lags.

107

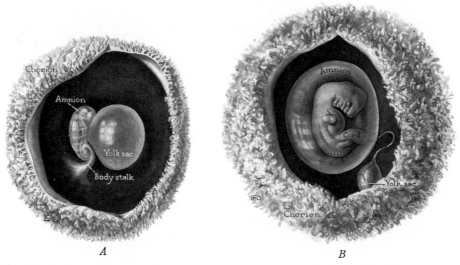

A   *B*

Fig. 83. Early membranes of human embryos, displayed by opening the chorionic sac.
A, At 2.6 mm. (× 5); *B,* at 11 mm. (× 2).

The slenderer connection does, however, elongate greatly to become the thread-like *yolk stalk* which soon is incorporated into the umbilical cord (*B*). The yolk stalk detaches from the gut by the end of the fifth week and presently degenerates.

The yolk sac is a pear-shaped bag which attains an average size of 5 mm. by the middle of the second month. It subsequently shrinks somewhat and converts into a rather solid structure containing flaky debris. The sac usually persists throughout pregnancy and can sometimes be found in the after-birth, beneath the amnion and near the attachment of the umbilical cord to the placenta (Fig. 117 *A*). Although the human yolk sac is not functional in the sense of storing an appreciable amount of yolk, it may well play a significant rôle in the transfer of nutritive fluid to the embryo from the young trophoblast by way of the extra-embryonic cœlom. In the early embryo the epithelial lining becomes specialized, and blood cells and blood vessels differentiate within the mesodermal covering of the sac (Fig. 78). These vessels are of interest as a survival of a necessary nutritional pathway in many vertebrates (p. 117).

Anomalies. In 2 per cent of all adults there is a persistence of the proximal end of the yolk stalk to produce an intestinal pouch, Meckel's *diverticulum of the ileum.* On the average, it arises 30 inches above the beginning of the colon. Although usually a blind sac and less than 10 centimeters long (Fig. 84 *A*), the diverticulum may continue as a solid cord or band to the region of the umbilicus (*B*). Still more rarely it opens at the navel as a completely pervious duct through which intestinal contents escape; this condition constitutes a fecal *umbilical fistula* (*C*). The saccular diverticulum is important surgically since it can telescope into the intestinal lumen and obstruct it. In other instances a loop of the small intestine may become caught and strangulate. This is when the diverticulum extends to the umbilicus, when its free end fuses to an adjacent peritoneal surface, or when there is a supporting band of mesentery.

The Amnion. The margin of the early *amnion* is attached to the periphery of the embryonic disc, the latter serving as a floor to the amniotic cavity (Fig. 68). As the embryonic disc grows and takes the form of a tubular embryo, the amniotic margin also follows the underfolding (Fig. 67). In this way the line of attachment becomes limited to the ventral body wall (Fig. 76 *A*) and then decreases in relative extent as it bounds the constricting umbilical area (*B*). With the development of the umbilical cord, the portion of the amnion near the umbilicus applies itself to the cord as an external covering layer (Figs. 78, 86). The amnion arises by cavitation (p. 117; Fig. 93).[11]

The amnion becomes a thin (but tough), transparent, nonvascular membrane. The lining of the sac, bordering on the amniotic cavity, is a single layer of ectodermal epithelium; the external covering is mesodermal connective tissue. The amniotic cavity enlarges rapidly as the fast-growing amnion expands at the expense of the extra-embryonic cœlom (Fig. 83) and, at the end of the second month, it fills the chorionic sac (Fig. 86 *C*). The amnion then fuses loosely with the chorionic wall, the two fibrous layers combining. This necessarily results in the obliteration of the extra-embryonic body cavity (Fig. 115). Clear, watery *amniotic fluid* fills the sac. There is a circulation of the fluid, which in late pregnancy replaces its water-component at the rate of 500 ml. each hour.[1] In its origin and disposal, the amnion, umbilical cord and fetus are all involved, but the exact pathways are not known completely.[2]

During the early months of pregnancy the embryo is actually suspended in amniotic fluid by its umbilical cord. Thus immersed, the flabby embryo maintains its shape successfully and is able to mold further its body form. Also throughout gestation the amniotic fluid performs several mechanical functions: it serves as a protective water cushion which absorbs jolts, equalizes pressures, prevents adherence of the amnion, and permits change of fetal posture. Amniotic fluid is swallowed by the fetus at least as early as the fifth month.[3] At childbirth the amnio-chorionic sac acts as a hydrostatic wedge to help dilate the neck of the uterus. During the early stages of

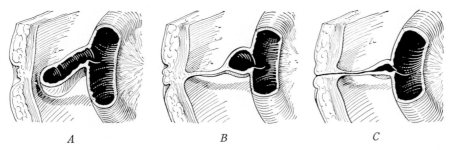

A   B   C

Fig. 84. Meckel's diverticulum of the ileum. *A*, Ordinary, blind sac. *B*, Diverticulum continued to umbilicus as a cord. *C*, Diverticulum, with fistulous opening at umbilicus.

Fig. 85. Vesicular, or hydatidiform mole (DeLee). × ½.

childbirth the membranes usually rupture and nearly a liter of amniotic fluid escapes as the 'waters.' If the tough amnion fails to burst, the head is delivered enveloped in it and this cap is then known popularly as the 'caul.'

Anomalies. When the amount of amniotic fluid exceeds two liters the condition is designated *hydramnios.* A volume less than one-third liter constitutes *oliogohydramnios,* and a marked deficiency may allow the amnion to adhere to the embryo and cause injury. It should be emphasized, however, that the fibrous *amniotic bands,* so-called, which at times interconnect the amnion and fetus appear to result from the breakdown of fetal tissues, rather than being amniotic derivatives that cause specific injuries by attaching to the fetus.[4]

The Chorion. Previous descriptions have traced the differentiation of the primitive capsule of trophoblastic tissue into a shaggy sac which encloses the embryo and all other fetal membranes (Fig. 83). The further differentiation of the *chorion* is related to its increasing association with the lining of the uterus. Such partnership reaches its highest specialization in the region that becomes the *placenta.* This organ acts as an intermediary in the functions of nutrition, respiration and excretion. These topics are so important that they will be deferred to the separate chapter that follows.

Anomalies. Occasionally the embryo blights and the chorionic sac persists, the branching villi transforming into clusters of fluid-filled vesicles (Fig. 83). Such bladders range up to a grape in size. This transformation constitutes a *vesicular* or *hydatidiform mole* which is ordinarily a benign tumor that may attain huge size. Much more rarely the chorionic trophoblast becomes a malignant, invasive tumor known as a *chorio-epithelioma.*

The Allantois. The entodermal component of the *allantois* is a tube that outpouches from the early yolk sac into the body stalk and thereby gains a covering of mesoderm (Fig. 67 *B, C*). Presently, when the hind-gut folds under, it comes off from the floor of that region (*D*). The human allantois is relatively tiny and never becomes a dilated, important sac, such as occurs in many mammals. The entodermal tube extends far toward the chorion and, when the umbilical cord organizes, the allantois becomes a conspicuous component of it. In the second month, however, growth of the tube ceases, and interruption and obliteration follow (Fig. 86).

Physiologically the allantoic tube is a superseded, dispensable rudiment. Nevertheless, blood vessels, coursing in the mesodermal covering of the allantois, continue onto the chorion and vascularize it (Fig. 310). The only significant feature of the organ, as a whole, lies in these allantoic (future umbilical) vessels. Through the development of a placenta, they will put the embryo into close physiological relation with the maternal circulation.

Anomalies. The allantois apparently lies wholly within the umbilical cord and is not responsible for anomalies of the urachus sometimes charged against it.

The Umbilical Cord. The unclosed area on the ventral surface of an embryo is at first expansive (Fig. 76 *A*). As the embryo enlarges, it becomes relatively smaller and even undergoes some constriction (*B*). This unclosed region, at the junction of embryonic and extra-embryonic territories, is the primitive *umbilicus*. At its margin the amnion and the somatopleuric belly wall become continuous, while through the unclosed umbilical ring extend both the yolk stalk and the body stalk with its included allantois (Fig. 86 *B*).

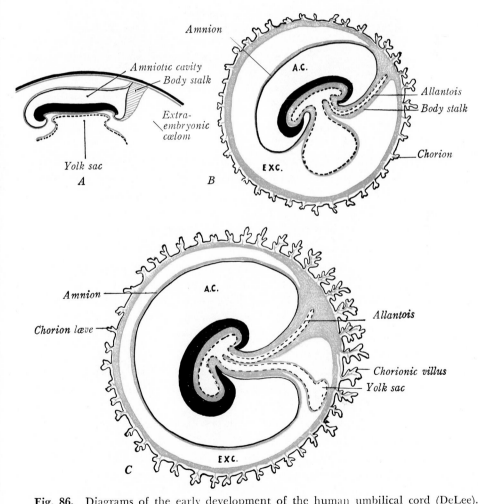

Fig. 86. Diagrams of the early development of the human umbilical cord (DeLee). *a.c.*, Amniotic cavity; *exc.*, extra-embryonic cœlom; solid black, ectoderm; red, mesoderm; dotted black, entoderm.

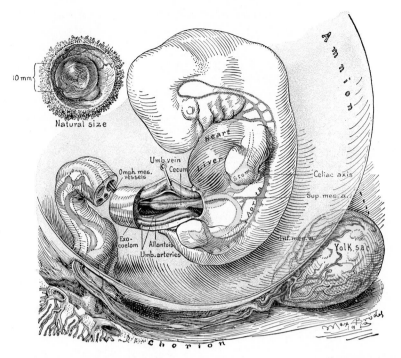

Fig. 87. Relations of the umbilical cord in a human embryo of six weeks (Cullen). × 6. Near the embryo, the cord has been opened and cut across.

Early in the fifth week a cylindrical structure, the *umbilical cord*, comes into existence through the expanding amnion applying itself about the body stalk and yolk stalk as it crowds them together (Figs. 83, 84).[4, 5] Hence the umbilical cord is combined from three auxiliary organs; it is not an outgrowth or extension of the body wall. In addition to the components already mentioned, a part of the extra-embryonic cœlom' is enclosed within the cord for a time. The portion of this cœlom nearest the body of the embryo enlarges greatly and during the seventh to tenth weeks contains coils of the intestine which herniate into it (Fig. 87). After the intestine is withdrawn, the cavity of the cord obliterates by the encroachment of the mesodermal tissue of the cord. Such obliteration marks the final disappearance of the last remnant of the extra-embryonic cœlom. Until the end of gestation the umbilical cord continues to connect the fetus with the part of the chorion that constitutes the fetal side of the placenta. It serves the fetus as a physiological 'life line.' The appearance of the cord during the fetal period, when its vessels are full, is quite unlike that of the collapsed cord, after birth (Fig. 88).[6]

The umbilical cord is covered with the mostly single-layered epithelium of the enveloping amnion. Originally it contains, embedded in mucous tissue, the following structures (Fig. 88): (1) the yolk stalk; (2) a pair of vitelline blood vessels; (3) the allantois; and (4) the allantoic, or umbilical blood vessels (two arteries and, by fusion, a single vein). The *mucous tissue*

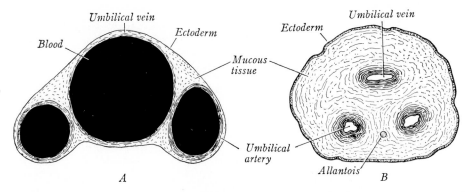

Fig. 88. Transverse sections of the human umbilical cord at full term ($\times$ 3). *A*, Vessels normally distended; *B*, Vessels collapsed.

or jelly of Wharton, peculiar to the umbilical cord, differentiates from the mesenchyme included in the cord at the time of its formation; it is rich in mucoid jelly, poor in fibers and contains neither intrinsic blood vessels nor nerves. In the early months of pregnancy remnants of the yolk stalk, vitelline vessels and allantois are to be seen; the latter strand may continue even to birth.

The mature cord is about one-half inch in diameter and attains an average length equal to that of the full-term fetus (nearly two feet). Its insertion on the placenta is nearly central (Fig. 117 *A*). A spiral twist soon appears, which may finally number as many as forty turns. Several explanations have been proposed to account for this spiraling.[7, 8] The blood vessels frequently curl in loops (by stronger local growth or perhaps, in part, by a local unwinding); these cause external bulgings known as *false knots*.

Anomalies. The extremes of length for the human cord range from almost nothing to seven feet. Abnormal shortness leads to practical difficulties at the time of delivery, and extreme shortness can cause distortion of the fetus. The production of atrophy and amputation through the cord winding about the neck or extremities of a fetus is often alleged, but without convincing proof (Fig. 89 *A*). The umbilical cord may attach to the margin of a placenta (*B*), or even on the adjoining membranes (*C*). Sometimes during the third month the fetus slips through a looped cord in such a way as to produce

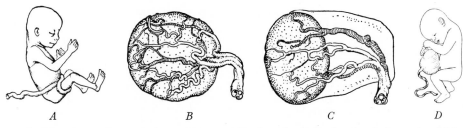

Fig. 89. Anomalous relations of the umbilical cord. *A*, Cord twisted about the thigh; once credited with causing amputation. *B*, Marginal attachment on placenta. *C*, Velamentous attachment on membranes adjoining the placenta. *D*, Omphalocele.

a simple *true knot* (Fig. 117 *A*). Failure of the intestine to retract from its temporary location in the cord results in *omphalocele* (Fig. 89 *D*). Loops of the bowel then occupy a thin sac at the base of the cord; composed solely of the transparent amniotic covering of the cord and peritoneum, it is continuous with the abdominal skin.

COMPARATIVE MEMBRANES

The eggs of fishes, amphibians, reptiles, birds and monotreme mammals are laid, whereupon they undergo development at a suitable temperature in water, earth or air. It is quite otherwise with marsupials, and especially with placental mammals, since the embryos of these animals develop within the uterus of the mother. Such wide differences in environmental conditions, faced by vertebrate embryos, are correlated with considerable diversity in both the number and nature of their fetal membranes.

The embryos of fishes and amphibians develop rapidly to free-swimming, larval stages. Because of this precocity, and their immersion in water, they need no auxiliary aids other than a supply of yolk sufficient to last until independent foraging can be carried on. In amphibians and some fishes the yolk is contained in large, entodermal cells that make up the thick floor of the gut. In sharks and bony fishes the gut-wall and body-wall come to enclose a large, noncellular, bulging yolk mass. Early respiration is performed by the exterior of the embryo or by the vascularized covering of the yolk sac. The lack of physical protection to individual embryos of these groups is offset by the production of lavish numbers, so that adequate replacements survive enemies and accidents.

Reptiles, birds and mammals are in a relatively advanced state of development at hatching or birth. Several auxiliary organs are produced that are of use during the prenatal period alone. Yet the function of no one of these organs is fixed unalterably; only the amnion is relatively stable. Especially in higher mammals has the abandonment of yolk for a physiological dependence on the mother led to the greatest elaboration of these structures. The embryos of all amniote vertebrates are produced in small numbers; they gain protection by such means as a heavy shell, parental incubation and development within the body of the mother.

Reptiles and Birds

The history of the fetal membranes in these amniotes is correlated with the presence of an enormous mass of yolk and an embryonic life spent within a shell. Although the original blastoderm is a small disc, it spreads by peripheral growth and eventually covers the entire surface of the egg (Fig. 90). But only the most central region is directly concerned with the formation of the embryo proper. All the remainder of the blastoderm is *extra-embryonic,* and it is this portion that furnishes most of the fetal membranes. The extra-embryonic blastoderm consists of *somatopleure* (ectoderm and somatic mesoderm) and *splanchnopleure* (entoderm and splanchnic mesoderm), separated by a space which is *extra-embryonic cœlom* (*cf.* Fig. 61 *B*). These components are continuous with their counterparts in the embryonic territory of the blastoderm.

The Yolk Sac. As the embryo enlarges, its circular connection with the extra-embryonic blastoderm grows at a slower rate. This produces a 'constriction' of the splanchnopleure where it joins the rapidly elongating gut. The region of constriction soon lengthens into a tubular *yolk stalk,* whereas the remainder of the extra-embryonic splanchnopleure encloses the massive yolk as the *yolk sac* (Fig. 91). Vitelline blood vessels, arising in the splanchnic mesoderm, ramify on the surface of the yolk sac, and through them absorbed yolk substance and some oxygen are conveyed to the chick during the incubation period (Fig. 90). Shortly before hatching occurs, the shriveled yolk sac slips through the navel into the belly cavity and the body wall closes behind it.

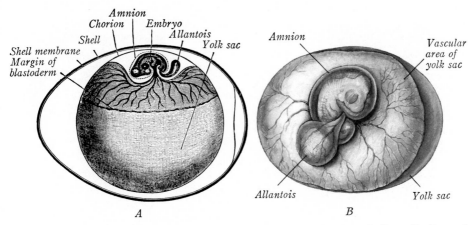

Fig. 90.   Fetal membranes of the chick. *A,* At four days (after Marshall; × 1); *B,* at six days (× 1).

The Amnion and Chorion.   These membranes are concentric sacs which arise by folding of the extra-embryonic somatopleure. The double-layered somatopleure is first thrown up into two crescentic folds. The earliest fold to appear is located just in front of the embryo; later, a second fold arises just behind the embryo (Fig. 91 *A*). These two folds advance like two hoods drawn over the head and the caudal region, respectively (*B*). When they meet, the completed circular fold closes in from all sides over the embryo, as would a bag pursed by draw-strings. The concluding step is the fusion of the several layers located at the margin of the fold (*C*). The result is the production of two separate, compound membranes (*D*).

The inner membrane is the *amnion.* It is lined with ectoderm and covered externally with somatic mesoderm. As the body of the embryo takes form, the amnion is continuous with the belly wall at the rim of the umbilicus. The amnion is a thin, transparent sac which soon fills with a fluid transudate (Fig. 90). Within this *amniotic fluid* the embryo is suspended, immersed in an equivalent of the aquatic habitat used by the embryos of fishes and amphibians. It is protected against drying; well isolated, it is able to develop unimpeded and can change position. The amnion lacks blood vessels, but muscle fibers differentiate in its mesodermal layer. These produce rhythmic contractions which agitate the embryo gently and perhaps help prevent adhesions.

The outer sac of somatopleure is the *chorion* (sometimes called the *serosa*), whose component layers are in reverse order to those of the amnion (Fig. 91 *D*). That is, ectoderm is the covering layer, whereas mesoderm furnishes the lining. The chorion lies next to the shell, encloses both the embryo and all its other fetal membranes, and is separated from them by the extra-embryonic cœlom. The functions of the chorion can best be explained in connection with the allantois, with whose later history it is so closely associated.

The Allantois.   This accessory organ was primarily evolved by reptiles and birds as a temporary sac for urinary storage. It arises as an outpouching of the ventral floor of the gut, near its hind end (Fig. 91 *A*). Since the gut wall is splanchnopleure, this diverticulum necessarily consists of the same layers (entoderm and splanchnic mesoderm) as it pushes outward into the extra-embryonic cœlom. There it forms a dilatation which develops rapidly into a large *allantoic sac,* connected to the hind-gut by the narrower *allantoic stalk* (*B, C*). The expanding sac flattens and spreads throughout the extra-embryonic cœlom (*D*) until, like the chorion, it finally underlies the entire shell. Fusion of the outer wall of the allantoic sac with the overlying chorion produces a functionally common membrane in contact with the porous shell. The allantoic blood vessels ramifying in the

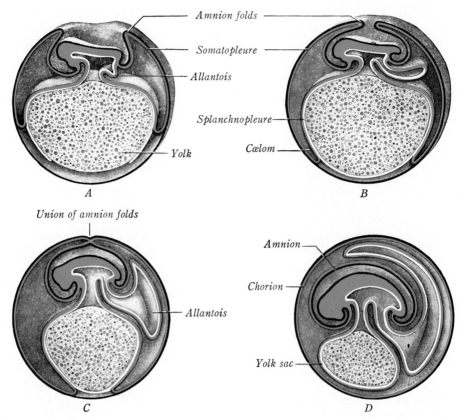

Amnion folds

Somatopleure

Allantois

Splanchnopleure

Cælom

Yolk

A   B

Union of amnion folds

Amnion

Chorion

Allantois

Yolk sac

C   D

Fig. 91. Stages in the development of the fetal membranes of the chick. The diagrammatic stereograms are sagittal hemisections. Ectoderm, black; mesoderm, red; entoderm, white.

combined mesodermal layer of these two membranes are situated favorably to serve as intermediaries in vital gaseous interchanges. Accordingly, the allantois becomes the functional 'lung' of the embryo through which oxygen is delivered to the blood and carbon dioxide is extracted from it. The allantoic cavity continues to act in its primitive capacity as a reservoir for the excreta of the kidneys; protein break-down products are stored there as insoluble uric acid. In addition, part of its wall assists also in the absorption of egg albumen. Shortly before the time of hatching, the allantois dries up and detaches.

Mammals

Since the mammalian embryo depends on the mother for food and oxygen and must provide for the elimination of its own wastes, the fetal membranes begin to develop even while the uterine relation is being established. There is considerable variety in the size, relations and functional rôle of each of these membranes.[9, 10] Some even show differences in the manner of origin. Only the monotremes, whose conditions of development are similar, follow the exact pattern established by reptiles and birds.

The Yolk Sac. Marsupial and placental mammals lack an actual yolk mass, yet a typical, stalked *yolk sac* appears and produces a complete vitelline circulation in quite young embryos. The early history and relations of this organ vary. In the majority of mammals the entoderm spreads just beneath the trophoblastic capsule and for a time lines it as a relatively large sac (Fig. 61); when the extra-embryonic mesoderm and cœlom appear between the two, the entoderm becomes clothed with the splanchnic mesodermal layer (B). After a time the growth of this yolk sac slows, and it then reduces in relative size as the allantois comes into prominence. In sharp contrast, the definitive yolk sac of primates is small from the first and remains a comparatively diminutive vesicle within the large chorionic sac (Figs. 62 G, 79).

The splanchnic mesoderm, surfacing the yolk sac, is the layer that bears the vitelline blood vessels. Many embryos with a highly developed yolk sac establish an intimate association with the uterus by means of a continuous nutritive path which is brought into existence through the union of the yolk sac and chorion. In this way there is formed a *yolk-sac placenta* which, however, is usually transitory.

The Amnion and Chorion. Many mammals produce an *amnion* in a somewhat leisurely manner by folding, but the details of the process vary. In some (rabbit; carnivores) the early trophoblast overlying the embryonic disc disappears. The exposed disc is then a plaque of special formative cells inset into a spheroidal sac of trophoblast (Fig. 92 *A*). Soon the mesoderm appears and its somatic layer combines with the ectodermal trophoblast to produce the extra-embryonic somatopleure. The amnion presently arises by the simple folding of this somatopleure, as in reptiles and birds (B). Also as in these animals, the amnion is important chiefly as a container of the buoyant *amniotic fluid*. The *chorion* is merely all of the original trophoblastic capsule and its lining of extra-embryonic mesoderm, other than those portions used in making the amnion. The region above the embryo is closed in by the outer layer of the amnion fold (B). The chorion often enlarges rapidly; in the pig it reaches the astonishing length of one meter before the embryo advances much beyond the stage of the primitive streak. The chorion has a diverse history, but in all mammals above marsupials it becomes functionally important by differentiating *chorionic villi* and participating in the production of a placenta.

Certain other mammals (guinea pig; hedgehog; anthropoids) acquire an *amnion* quite simply and at a very early stage. The primitive amnion cavity arises as a cleft that separates the inner cell mass into two parts: one is the prospective embryo; the other is non-embryonic, auxiliary tissue (Fig. 93). Thus the floor of the enclosed space is the main plate of the embryonic disc; the sides and roof comprise the thinner 'ectoderm' of the membranous amnion. When soon a layer of somatic mesoderm covers this ectodermal dome of the early amnion (Fig. 67), the structural outcome is identical with the type of amnion derived by folding. The *chorion* of this group of animals is merely a later stage of the original trophoblastic capsule (non-embryonic or subsidiary 'ectoderm') to which a lining of somatic mesoderm has been added (Fig. 67 *A*).

Still other mammals (pig; deer; rat) combine the two methods of amnion formation already described. The inner cell mass first hollows and then its roof ruptures; after this the definitive amnion develops by folding.

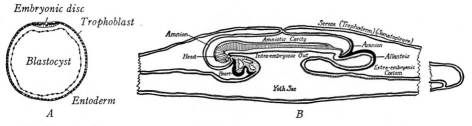

Fig. 92. Fetal membranes of pig embryos, shown in sagittal section. *A*, Blastocyst, at eight days (× 80). *B*, Diagram, at eighteen-somite stage (Patten; × 8); most of the long chorionic (or serosal) sac has been omitted.

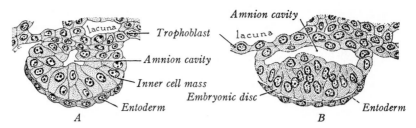

Fig. 93. Amnion formation in the monkey, shown in sections (after Heuser and Streeter). × 330. *A,* At ten days; *B,* at eleven days.

The Allantois. Many mammals, like reptiles and birds, produce a prominent *allantois* by the sacculation of gut-splanchnopleure into the extra-embryonic cœlom (Fig. 92 *B*). For example, in carnivores and ungulates it becomes very large, lines the chorionic sac and fuses with it (Fig. 119 *B*); a goat embryo of two inches has an allantois two feet long. The further history of such conspicuous sacs is a part of the story of the placenta, and will be discussed under that heading (p. 146). Here it need only be mentioned that the urinary wastes actually collect in the allantoic sac, as in reptiles and birds, and are not excreted through the placenta.

By contrast, edentates, rodents and primates tend to have a vestigial allantois. In anthropoids a tiny, entodermal tube pushes into the *body stalk* even before the hind-gut develops (Fig. 67). The body stalk is a bridge of mesoderm which from a very early period connects embryo to chorion in seeming anticipation of the arrival of the tubular, entodermal component of the allantois complex. Since the stalk represents mesoderm across which the cœlom has failed to pass, the fundamental relations are similar to those in other mammals that evaginate a free allantois into the extra-embryonic cœlom. Blood vessels accompany the allantois and extend to the chorion which becomes vascularized through their branches. The entodermal diverticulum itself is functionless and soon regresses, but the blood vessels persist. They become the important umbilical arteries and veins which connect the fetus with the chorionic component of the placenta (Fig. 310).

The Umbilical Cord. Except for the presistence of the allantoic stalk and allantois throughout the fetal life of many mammals, the formation and history of the umbilical cord do not differ significantly from the account already given for man (p. 111).

REFERENCES CITED

1. Plentl, A. A. 1959. Ann. N.Y. Acad. Sci., *75,* 746–761.
2. Bourne, G. L. 1951. The Amnion and Chorion, Chapter 12 (Year Book Company).
3. Reifferscheid, W. & R. Schmiemann. 1939. Zentr. Gyn., *63,* 146–153.
4. Streeter, G. L. 1936. Carnegie Contr. Embr., *22,* 1–44.
5. Politzer, G. & H. Sternberg. 1930. Z'ts. Anat. u. Entw., *92,* 279–379.
6. Chacko, A. W. & S. R. Reynolds 1954. Carnegie Contr. Embr., *35,* 135–150.
7. Wirtinger, W. 1941. Wien. klin. Woch's., *54,* 139–144.
8. Edmonds, H. W. 1954. Am. J. Obst. & Gyn., *67,* 102–120.
9. Grosser, O. 1927. Frühentwicklung, etc. (Bergmann).
10. Mossman, H. W. 1937. Carnegie Contr. Embr., *20,* 51–80.
11. Luckett, W. P. 1973. Anat. Rec., *175,* 375.

Chapter VIII.   Placentation

The subject of *placentation* includes all the events that are related to the following: (1) implantation and establishment of the embryo within the uterus of the mother; (2) the differentiation of the uterine lining into a specialized decidual membrane; (3) the development of a placenta; and (4) the fetal-maternal association throughout pregnancy.

TRANSPORT OF THE OVUM AND BLASTOCYST

The fertilized and cleaving egg is propelled down the central cavity of the uterine tube, guided by the longitudinal folds of its lining membrane (Fig. 94). Muscular contractions, controlled by balanced ovarian hormones, are important agents in this transport, as can be shown by direct observations on rodents.[1, 2] Such muscular force is directed primarily against the fluid, secreted by the tube, that suspends the zygote. In almost all mammals the tubal journey requires three to four days, regardless of the different lengths and calibers of tubes (e.g., mouse; cow). Much of the journey is completed quickly, whereas the cleaving egg lingers at the lower end of the tube. This delay serves to retain the cellular cluster until the uterus is prepared properly to receive it.

The developing human egg is believed to enter the uterine cavity early on the fourth day after ovulation. The period spent in the uterus, as a free morula and blastocyst, varies considerably among mammals; in man it is two to three days. During this interval there is further transport to the site where attachment and embedding will occur. It is suspected that muscular activity may again be the factor responsible for this transfer.[3] While free in the uterine cavity the blastocyst imbibes fluid from the uterine secretion, in which it is immersed, and expands to a diameter of almost 0.2 mm. This secretion, expelled from the uterine glands, also supplies oxygen and some nutriment to the blastocyst.

PREPARATION OF THE UTERUS FOR THE EMBRYO

There is no information as to what determines the site 'selected' by the blastocyst for attachment. If it is not a matter of chance there is, at least, no visible sign of special preparation or predetermination in the area finally used. Instead, the entire lining of the uterus, except in the region of its

119

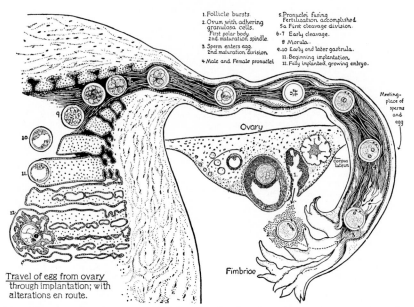

Travel of egg from ovary
through implantation; with
alterations en route.

Fig. 94. Diagram showing stages of the developing human ovum, at various levels in its journey from ovary to uterus (Dickinson).

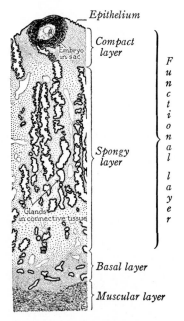

Fig. 95. Vertical section of the human endometrium, with an implanted blastocyst of eleven days (after Hertig and Rock). × 15.

neck, is in a state of special preparation for the reception of the embryo. This favorable condition is a late phase (premenstrual) in the monthly cycle of change controlled by the ovarian hormones. The details of these correlated phenomena will be discussed in the next chapter. For the present it is sufficient to state that the lining membrane has thickened markedly (5 mm., more or less) and undergone specialization. It is well vascularized, and the long, dilated glands contain glycogen and other secreted material. The superficial level, known as the *compact layer,* is the site where the embryo embeds and its placenta develops.

The uterine lining is a mucous membrane named the *endometrium* (Fig. 95). Its exposed surface is covered with a single-layered epithelium which dips inward at intervals to produce the tubular *uterine glands.* Beneath the surface epithelium and between the glands is a soft, cellular connective tissue. The appearance of the endometrium changes somewhat at different depths. Nearest the uterine cavity is the *compact layer,* through which pass the slender necks of the glands. Next deeper is the thick *spongy layer,* characterized by the dilated portions of the glands. Deepest of all is the thin *basal layer* which contains the blind ends of the glands; it does not participate to any extent in the glandular and other changes characteristic of the menstrual cycle and pregnancy. For this reason, the compact and spongy layers are often considered as making a unit; they are then spoken of as the *functional layer* of the endometrium.

IMPLANTATION OF THE EMBRYO

Implantation includes the attachment of the blastocyst to the epithelial lining of the uterus, the penetration of the blastocyst through the epithelium, and its invasion and embedding in the compact layer of the endometrium.

Attachment and Penetration. The follicular cells that adhere to the freshly discharged ovum, as the corona radiata, are lost during the journey down the tube. This dispersion is aided by an unknown factor, emanating from the uterine tube.[4] Shortly before implantation begins, the stretched and thinned zona pellucida disappears as well. This release permits the blastocyst to come into direct contact with the uterine epithelium and makes further growth possible. The human blastocyst probably begins to attach late in the sixth day after ovulation.

Stages in the attachment and beginning penetration of the human blastocyst are lacking, but substitute information is available from similar stages in the monkey that have been studied thoroughly.[5] The sticky, somewhat swollen blastocyst first adheres to the uterine epithelium (Figs. 96, 97 *A*).

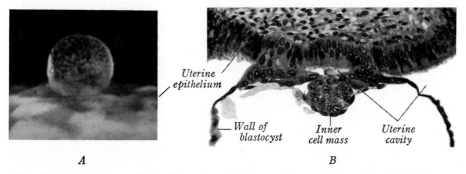

Uterine
epithelium

Wall of
blastocyst

Inner
cell mass

Uterine
cavity

A   *B*

Fig. 96. Attachment of the blastocyst of the monkey to the uterine epithelium at nine days (Heuser and Streeter). *A,* Total view, from side ($\times$ 50). *B,* Median section of the same specimen, showing fusion at two points ($\times$ 200); most of the blastocyst wall is omitted.

In this region of contact, usually between the mouths of glands, the trophoblastic wall of the blastocyst thickens as the result of proliferation; its more
superficial cells lose their boundaries and thus become a syncytium. At the
same time the cells of the uterine epithelium in the area of attachment begin
to break down, apparently as the result of some influence (digestive enzyme?)
exerted by the trophoblast. The injured cells are taken up by the trophoblast
and digested. This erosion creates a gap in the epithelium through which
the invading trophoblast advances and comes into relation with the connective tissue beneath (Fig. 97 *B*).

At this point the series of human specimens begins. The youngest was
slightly more than seven days old; implantation had been in progress for
some 24 hours.[6] This blastocyst had flattened somewhat because of a temporary loss of fluid. It had advanced through the epithelial gap, was pushing
into the soft, fluid-swollen tissue beyond, but was not yet covered in by
uterine epithelium (Figs. 97 *C*, 98 *A*). Wherever the trophoblast had made
contact with maternal tissues it was mostly syncytial and remarkably thickened (Fig. 63). The sac is not rejected though half paternal in origin.

The next known stage, nearly two days older, lay virtually buried within
the compact layer of the endometrium (Fig. 97 *D*.)[6] Rapid growth of the
trophoblast had produced a thick, spongy shell. Whereas the original wall
of the blastocyst was composed of distinct cells, the trophoblastic shell now
contains but a relatively thin layer of such cells (*cytotrophoblast*), located
next the cavity of the blastocyst. Outside of this is a very thick peripheral

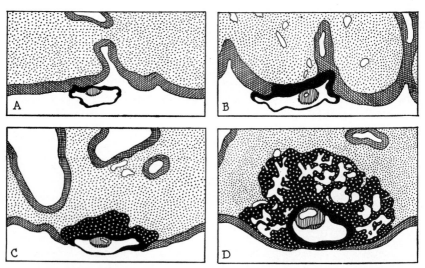

Fig. 97. Stages of implantation, shown by sections (× 90). *A, B,* Attachment of
blastocyst and epithelial erosion in the monkey at nine and ten days (after Wislocki and
Streeter). *C, D,* Advance of human blastocyst through epithelial gap and into uterine
connective tissue at seven and nine days (after Hertig and Rock).

Cellular trophoblast, solid black; syncytial trophoblast, mottled black; epithelium,
crosshatched; connective tissue, stippled. The embryo-proper is at the top of the blastocyst
cavity: ectoderm, thin vertical stripes; entoderm, thick vertical stripes.

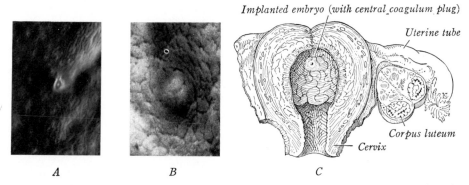

Implanted embryo (with central coagulum plug)

Uterine tube

Corpus luteum

Cervix

A   B   C

Fig. 98. Implantation sites of human embryos, in surface view. *A,* Incomplete implantation, at seven days (Hertig and Rock; × 8); *cf.* Fig. 63. *B,* Complete implantation, at eleven days (Hertig and Rock; × 8); *cf.* Fig. 64. *C,* Hemisected uterus, with implanted embryo, at seventeen days (after Ramsey; × ⅗).

layer *(syntrophoblast),* in which nuclei lie embedded in a common cytoplasmic mass (Fig. 100). The cellular trophoblast is the parent tissue that produces the syncytium by cell division, loss of cell membranes and change in cytoplasmic character.

Implantation Completed. The appearance of a completely embedded embryo, five to six days after attachment began, is shown in Figs. 64, 95, 98 *B*).[7] The original point of entry into the endometrium is sealed with a fibrinous and cellular plug known as the *closing coagulum;* the processes of wound healing, already begun, will cause it to disappear in less than a week. It is plain that a blastocyst is definitely oriented during its penetration and afterward. The side bearing the inner cell mass, or future embryo proper, is the surface that attaches, leads the way in penetration and lies deepest when implantation is completed. All these relations seem to follow naturally from the primary circumstance that the surface of the blastocyst, just overlying the inner cell mass, is more sticky than other regions of the wall and hence is the area that adheres to the uterine lining.

Figure 98 *C* illustrates the internal appearance of a hemisected uterus, containing an implanted embryo of 17 days (11 days after attachment), at three-fifths natural size. Although the site of implantation varies somewhat from specimen to specimen, it is usually at a high level in the uterus and somewhat more often on the back wall than on the front wall.

Anomalies. If the fertilized egg fails to reach the uterus, but implants and develops elsewhere, the condition is known as an extra-uterine or *ectopic pregnancy* (Fig. 99 *A*). The commonest ectopic site is the uterine tube *(tubal pregnancy; B).* Direct attachment to the peritoneum *(primary abdominal pregnancy)* and the development of an unexpelled egg within its ruptured follicle *(ovarian pregnancy; C)* are known also. Continued ectopic growth to maturity is rarely achieved because of the unsuitability of the locations chosen and the inadequacy of the placental arrangements developed at these sites. Pregnant tubes frequently rupture; in some instances the chorionic sac is expelled between folds of the broad ligament, or into the peritoneal cavity, where it may become a *secondary abdominal pregnancy.*

Occasionally the embryo locates in the right or left marginal lining of the uterus. Still less frequently it implants near the neck of the uterus, so that the expanding placenta more or less covers the cervical canal (Fig. 99 *D*). This latter condition is called *placenta prævia* (*i.e.*, placenta leading the way); local detachment and bleeding occur when the region near the canal stretches in the later months of pregnancy.

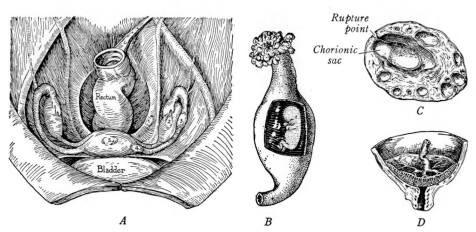

Fig. 99. Atypical implantation sites in woman. *A,* Diagram, in ventral view, of the locations in which pregnancies may occur: 1, abdominal; 2, ovarian; 3, tubal; 4, interstitial; 5, uterine. *B,* Tubal pregnancy, with a window cut in the tubal wall ($\times$ ½). *C,* Ovarian pregnancy, displayed in a halved ovary (after Mall and Cullen; $\times$ ¾). *D,* Placenta prævia, at the cervix of a uterus whose lower portion has been hemisected ($\times$ ⅕).

ESTABLISHMENT OF THE EMBRYO

Even as the blastocyst is becoming implanted, its trophoblastic wall (*i.e.,* the future chorion) starts on a course of specialization which will put the invader in intimate relation with the uterine blood. This phase of implantation (*establishment*) is not a one-sided adjustment, since the maternal tissues, as well, adapt themselves to the new relations and demands. Both sets of co-ordinated changes lead to the production of a specialized organ, the *placenta,* which is the medium for physiological interchanges between the mother and fetus throughout the remainder of pregnancy. The implanted blastocyst is located superficially in the compact layer of the endometrium (Fig. 95). Although it expands greatly as development proceeds, it does not encroach into the spongy layer (Fig. 103).

Trophoblastic Lacunæ. A prominent feature of the syncytial trophoblast is the appearance within it of vacuoles which merge and produce irregular cavities, or *lacunæ* (Figs. 97 *D,* 100).[7] This process produces a communicating labyrinth which adds to itself progressively by incorporating new lacunæ as fast as they appear in the growing, expanding syncytium. This system of channels, which gives the trophoblast its spongy texture, is the beginning of the future *intervillous space* of the placenta. In the week following implantation this labyrinth becomes well developed.

Vascular Relations. Connections are quickly established between the trophoblastic lacunæ and the uterine blood vessels (Fig. 100). Although some minute branches of the spiral arterioles are tapped, the chief communications during this early period are with enlarged capillaries (which connect arterioles and venules) and with the venules themselves.[7] Within a week after implantation there are prominent sinus-like venules beneath the embedded sac (Fig. 103). At first the blood within the trophoblastic lacunæ is small in amount and relatively stagnant. Later the number of arterioles tapped directly increases, and the flow of blood improves. An outstanding characteristic of all trophoblast is that, like endothelium, it does not provoke blood into clotting. The uterine connective tissue in the vicinity of the young chorionic sac is edematous and contains extravasated blood.

Endometrial Erosion. The trophoblast is an invasive tissue which extends progressively into the superficial maternal tissues. (Fig. 100). Paralleling this invasion goes a certain amount of destruction of the endometrial tissues; such erosion is a characteristic of the *border zone* where trophoblast and endometrium meet. Some dissolution of the uterine tissue is apparently related to inadequacies of the blood supply, similar to the necrosis that precedes menstruation. Other destruction is due to a cytolytic enzyme seemingly produced by the cellular trophoblast.[8] The erosive processes are declining in intensity by the end of the third week. At all times erosion is a mild process, under control, and at most the invasion extends but a few millimeters into the endometrium. The relationship of embryo to uterus is in perfect balance, much like that of a parasite to its natural host.

Both the cellular and syncytial trophoblast have the capacity of ingesting maternal tissue, although most of this tissue has undergone a certain degree of necrosis before phagocytosis occurs. Blood cells of all kinds and reticular fibers are the elements that can be identified most commonly within young trophoblastic cells.[9] In addition, there is granular and amorphous material in the process of digestion. Such substance (*histotroph,* or *embryotroph*) is presumably utilized for nourishment in the early period of establishment,

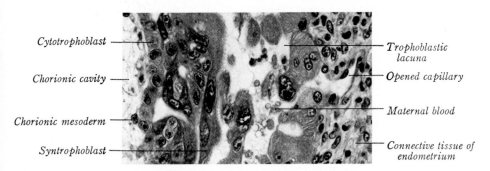

Cytotrophoblast — Trophoblastic lacuna

Chorionic cavity — Opened capillary

— Maternal blood

Chorionic mesoderm —

Syntrophoblast — Connective tissue of endometrium

Fig. 100. Detail of the relations exhibited at the margin of an implanting chorionic sac of eleven days (Hertig and Rock). $\times$ 320. The entire specimen, in section, is shown as Fig. 64.

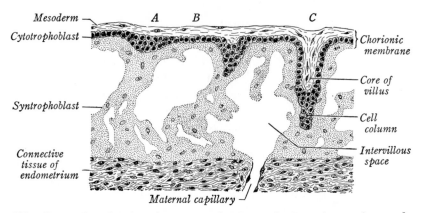

Fig. 101. Composite, showing three stages in the projection of secondary and tertiary villi from an implanted human chorion of two weeks. ✕ about 100.

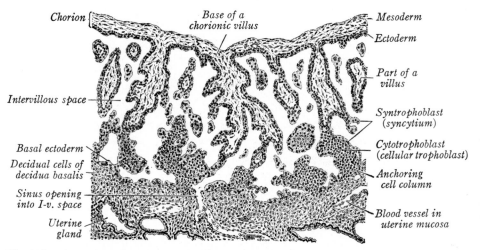

Fig. 102. Part of the human placental site, at seventeen days, in vertical section. ✕ 65.

as is also blood plasma and tissue fluid. It seems likely that the syncytium is a site of synthesis of the proteins required by the embryo at this time.[8]

The Development of Villi. The cellular trophoblast of the differentiating chorionic sac tends to be single-layered, except where proliferation has produced local masses. These masses extend a short distance into the plates and strands of syntrophoblast and indicate a first stage *(primary villi)* in the development of *chorionic villi* (Fig. 101 *A, B*). Just as the cytotrophoblast is said to differentiate the general mesodermal lining (future connective tissue) of the chorion, so also it differentiates a core of vascular connective tissue within villi.[10, 11] At first this tissue is only a stubby center at the base of a villus (B). Nevertheless, as fast as the cellular trophoblast proliferates, perforates the shell of syncytial trophoblast and sends *cell columns* outward, the axial connective tissue extends progressively into them (*C*). Continued growth and branching bring a tuft-like chorionic villus into being (Fig. 102). All such early tufts lack blood vessels and are known as *secondary villi*. During the fourth week the young villi become well formed and vascularized; they are then known as *tertiary villi* (Fig. 109). Such villi increase greatly the fetal-maternal blood relations.

Establishment Completed. Early in the fourth week the essential arrangements have been accomplished which make possible the physiological interchanges between mother and fetus that will characterize the remainder of pregnancy (Figs. 103, 109). Blood vessels (allantoic arteries and veins) pass from the embryo through the body stalk to the connective tissue of the chorion, and then extend into the chorionic villi. This vascularized connective tissue of the chorion and its villi is everywhere covered with trophoblast, which consists of an inner cellular layer and an outer, or superficial, syncytial layer (Fig. 109). Trophoblast also forms a carpet over the eroded surface of the endometrium. Here many of the cell columns pierce the shell of syncytium; coming into contact with the endometrium, they serve to anchor the related villi (Fig. 102). All this trophoblast bounds the *intervillous space* as a complete, common lining. In the labyrinthine channels of the space maternal blood circulates and bathes the villi. The passage of nutritive substances from the circulating blood of the mother to that of the fetus within the vessels of the villi is often spoken of as *hemotrophic nutrition*. It is contrasted with the early *histotrophic nutrition* in which the damaged maternal tissues, containing extravasated blood, and the stagnant blood in the trophoblastic lacunæ are taken up by the trophoblast.

The fetal-maternal relations during the fourth week are illustrated in Fig. 103. By this time the setting is complete for the establishment of a definite *placenta,* but its history and detailed structure must be postponed until the endometrium of pregnancy has been described as a whole.

The association of cellular with syncytial trophoblast is an histological characteristic that is diagnostic of pregnancy (Figs. 100, 111 *A*). Whether found in sections of passed

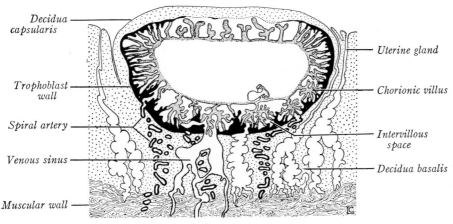

Fig. 103. Vertical section of an implanted human embryo of twenty-two days, showing the relations of the chorionic sac to the endometrium (after Ortmann). × 2.

fragments, of uterine scrapings, of sacs suspected of having harbored blighted embryos, of blood clots, or of an invasive tumor (chorio-epithelioma), the presence of these two types of cells is positive proof that a pregnancy exists or existed.

The fetal trophoblast is a foreign tissue to the uterus, since paternal genes contribute to its constitution. It is remarkable that the uterus does not reject the blastocyst in the same way that immunity reactions cast off homotransplants in general.

THE DECIDUÆ

The mucosal lining *(endometrium)* of the uterus, already specialized in anticipation of pregnancy and utilized as a nesting place by the implanting embryo, rapidly acquires some additional characteristics distinctive of pregnancy and then persists throughout the gestation period. Naturally enough, the greatest disturbance of typical relations within the membrane occurs in the region where the embryo lies. Yet the remainder of the uterine lining becomes involved after a time and then experiences characteristic alterations as well.

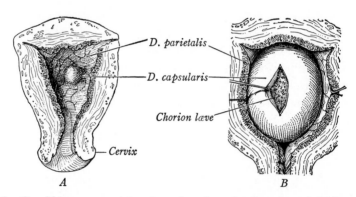

Fig. 104.   Gravid human uteri, hemisected to show the elevation of the human decidua capsularis by the expanding chorionic sac. *A,* At nearly four weeks ($\times$ ½); *B,* at ten weeks ($\times$ ⅖).

The fusions that take place between the entire endometrium and the expanding chorion (which eventually comes in contact with it everywhere) lead to a general splitting off of the uterine lining at birth. The endometrium of the pregnant uterus is, therefore, named the *decidua (i.e.,* that which falls off). Its preparation for pregnancy, the long deferred loss at delivery and the subsequent repair after childbirth extend and exaggerate the events of an ordinary menstrual cycle. The total decidual membrane is actually a direct continuation and further elaboration of the premenstrual (also called 'progravid') type of endometrium.

Even when the early chorionic sac lies embedded within the endometrium, three different regions of this thickened membrane can be recognized (Figs. 104, 105 *A*): (1) The *decidua parietalis,* the general lining of the uterus exclusive of the area occupied by the embryo. (2) The *decidua capsularis,* a region covering the chorionic sac and interposed between the sac and the uterine cavity. (3) The *decidua basalis,* a region underlying the chorionic sac and situated between it and the muscular wall of the uterus. The decidua parietalis, a membrane not directly involved in lodging the embryo, is for some time a typical, undisturbed part of the endometrium of pregnancy. Quite different are the other two decidual membranes. Since the chorionic sac has, in a sense, split the endometrium into two parts, neither the decidua capsularis nor the decidua basalis contains all the levels of a typical endometrium.

Mention has been made previously of the distinctive vascular and glandular specializations that the endometrium builds up in preparation for pregnancy. Another conspicuous specialization is certain greatly enlarged connective-tissue cells of this membrane which appear early in pregnancy and become known as *decidual cells* (Fig. 107). They are characteristic constituents of the deciduæ and are diagnostic of pregnancy.

The decidual cells occur chiefly in the compact layer. Their course of specialization begins during implantation and, although declining in size and numbers in the later months, many remain throughout pregnancy. The decidual cells are large, rounded to angular elements which store glycogen and lipids and may contain more than one nucleus (Fig. 105 *B*). Their size and proliferative increase help account for the thickness of young deciduæ; as a result, the surface of the early decidua parietalis folds characteristically (Fig. 104 *A*). The full significance of the decidual cells is not understood. They also occur at the site of a tubal implantation.

The Decidua Parietalis.   All of the uterine lining of the pregnant uterus, except that in immediate relation with the chorionic sac and the cervical

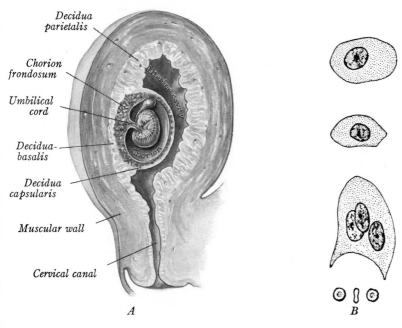

Decidua
parietalis

Chorion
frondosum

Umbilical
cord

Decidua-
basalis

Decidua
capsularis

Muscular wall

Cervical canal

A

B

Fig. 105. *A,* Gravid human uterus of five weeks, hemisected to show the decidual relations to an implanted chorionic sac. × 1. *B,* Decidual cells (× 450); below are red blood corpuscles, drawn to scale.

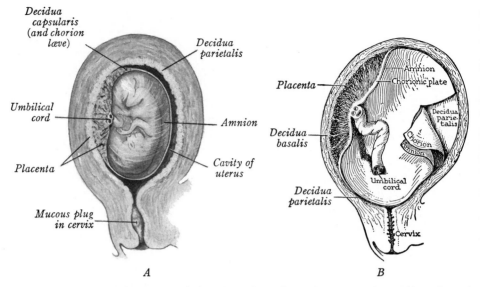

Fig. 106. Gravid human uteri, hemisected to show the progressive obliteration of the uterine cavity. *A*, At three months ($\times$ ½). *B*, At seven months; fetus removed and flaps of the amnion and chorion cut and reflected ($\times$ ⅕).

canal, comprises the *decidua parietalis* (Fig. 105 *A*). The more superficial, *compact layer* of this membrane becomes thick; it contains the slenderer neck regions of the uterine glands, embedded in large quantities of decidual cells (Fig. 107 *A*). Its surface epithelium has usually disappeared by the end of the third month, at which time contact with the expanding decidua capsularis takes place (*B*). The deeper *spongy layer* is characterized for a while by the enlarged and sacculated portions of the uterine glands, continued from their progravid state and even more distended with secretion. Later, as the uterus expands, the glands become stretched to horizontal slits (Fig. 107). After the third or fourth month the membrane thins and shows regressive changes.

The uterine cervix does not elaborate a decidua. Its glands, however, do enlarge and secrete a *mucous plug* which closes off the uterus during pregnancy (Fig. 106 *A*).

The Decidua Capsularis. The superficial portion of the compact endometrium that originally covers the chorionic sac and faces the uterine cavity is the *decidua caspsularis*. Growth of the chorionic sac causes the capsularis to elevate into a progressively expanding dome (Fig. 104). In the earlier stages of pregnancy, blood vessels and some glandular traces occur in the substance of this layer, while its surface epithelium is continuous with that of the decidua parietalis. As the chorionic sac expands, the capsularis grows thin and atrophic (Fig. 107 *A*). At the end of the third month its full surface comes into contact with the decidua parietalis with which it fuses, thereby obliterating the uterine cavity (Fig. 106). During the next two months the capsularis degenerates and disappears as such; this leaves the chorion free to

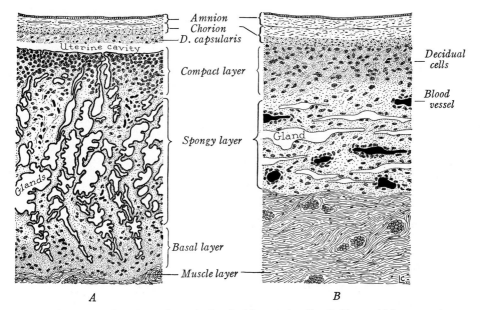

Fig. 107. Vertical sections through the decidua parietalis of the gravid human uterus. *A*, In the third month, before the fusions occur that obliterate the uterine cavity (× 18). *B*, At seven months, obliteration already long completed (× 24).

become adherent to the decidua parietalis for the remainder of pregnancy (Fig. 107 *B*). Long before this, the amnion has fused loosely with the chorion. At term the combined thickness of all these fused membranes has reduced to 2 mm. or less.

Subsequent to the obliteration of the chorionic cavity at two months and of the uterine cavity at three months, the only cavity within the uterus for the remainder of pregnancy is that of the amniotic sac (Fig. 115).

The Decidua Basalis. At the site of implantation the chorionic sac lies within the compact layer of the endometrium; beneath the chorion is the deeper portion of the compacta and the spongiosa (Fig. 103). These two endometrial components specialize in pregnancy as the *decidua basalis*. On the whole, the changes in this portion of the deciduæ parallel those of the parietalis, but the basalis becomes the maternal part of the placenta and thus acquires certain other distinctive components (p. 135).

THE PLACENTA

The first steps leading to the development of a *placenta* have already been described. These include the development of: (1) chorionic villi; (2) a cavernous intervillous space; and (3) a decidua basalis with blood vessels that supply the early intervillous space.

The shape of the human placenta is determined by the form of the patch of villi finally left on the chorionic sac. In the early weeks the villi cover uniformly the entire surface of the chorion and may reach 1000 in number.

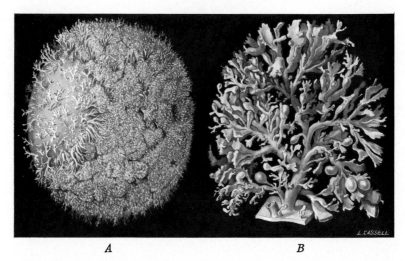

A   *B*

Fig. 108. Human chorionic vesicle at nine weeks. *A,* Entire sac, showing the early distinction between the chorion læve and frondosum (× 1). *B,* Detail of a chorionic villus (× 7).

But with continued growth of the chorionic sac, the villi next the stretched decidua capsularis become compressed and their vascularity is reduced. Atrophy produces a perceptibly bare polar spot at two months (Fig. 108 *A*), while in the fourth month about half of the chorion has become naked. The area of chorion thus lacking in villi is called the *chorion læve* (*i.e.,* smooth chorion). The villi that are associated with the decidua basalis, on the other hand, persist and give the name *chorion frondosum* (*i.e.,* bushy chorion) to this deeper portion of the sac. The area of persistent villi is normally somewhat circular in form, so the human placenta naturally takes the shape of a disc. Since the umbilical cord passes from the embryo to the deep, frondose portion of the chorion, it follows that when this latter region becomes a part of the placenta, the cord then attaches to the fetal side of this organ—and usually near the midpoint (Fig. 105 *A*).

No adequate conception of the placenta is possible without a clear recognition of its composite nature, based on a double origin. The chorion frondosum (both the membrane and its villi) is the *fetal portion,* while the decidua basalis (or, better, what remains of its eroded and altered compact layer) is the *maternal contribution* (Fig. 105 *A*). The *intervillous space,* which to a large degree separates these two components, is an expansion of the cavities arising within the early chorionic trophoblast. The general plan of the placenta is that of two parallel plates (the chorionic membrane and the decidua basalis) between which is a blood sinus (the intervillous space) containing an enormous number of branches belonging to the chorionic villi (Fig. 109). In the paragraphs that follow, these components will be described in detail.

The Fetal Placenta. The chorionic membrane in the placental region comes to be known as the *chorionic plate* (Fig. 109). The inner surface of the plate, bordering the intervillous space, is covered with trophoblast which has a composition and history like that presently to be described for the chorionic villi. During the last half of pregnancy it is replaced largely by fibrinoid material (p. 135). Supporting the trophoblast there is a layer of connective tissue which contains blood vessels radiating from the umbilical cord. These belong to the umbilical (allantoic) system of vessels, and the chorionic plate distributes them to the villi. At the end of the second month the expanding amnion comes everywhere into contact with the connective tissue of the chorion. The ensuing fusion in the placental area causes the amnion to attach to the fetal surface of the placenta (Fig. 106), and this relation persists throughout pregnancy (Fig. 112).

The *chorionic villi* are the most important part of the placenta because they furnish the means by which all kinds of interchanges take place between the mother and fetus. The early villi are compact, bush-like tufts with but few branches, and these are short and plump. Their main stems arise from the chorionic membrane and almost all of the ends (*anchoring villi*) attach and mingle with the eroded surface of the compact decidua basalis. Side branching begins in the second month and produces many *free villi* as well (Figs. 103, 108, 109). During the middle and later months of pregnancy the villi become much more tree-like, with prominent trunks and main branches. The long and slender branches bear innumerable small twigs. Many main-line, terminal branches fuse with the decidua basalis at some distance from their ends, and then recurve into the intervillous space in a J-shaped manner.[12]

All parts of the villous tree have the same structural plan (Figs. 109–111). At the center is a connective-tissue core, which embeds blood vessels. These begin as arterioles and venules, but taper to prominent capillaries

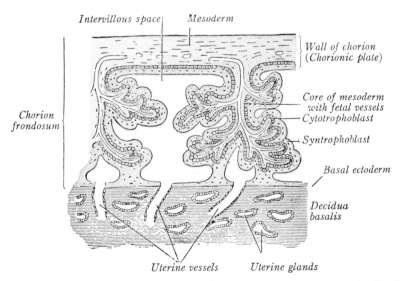

Fig. 109. Diagrammatic section through a human placenta (Bryce in Gray).

Fig. 110.   Tip of a mature chorionic villus, cleared to show the relations between arterial (black) and venous (gray) vessels. × 130.

which continue to the villous tips where they complete a continuous system of closed vessels. The connective-tissue core is covered with a double layer of trophoblast. Inside, next the connective tissue, is the single-layered *cytotrophoblast* with its separate cuboidal cells sharply defined; it is also known as the layer of Langhans. This cellular layer gives rise to syncytium, the *syntrophoblast,* which clothes the villi externally. During the first third of pregnancy the cytotrophoblast of the villi becomes progressively more interrupted and hence scarcer. This is because it is almost completely used up in the forming of syncytium.[13] In the last half of pregnancy only occasional flattened representatives of the cellular layer persist, and the non-mitotic syncytial trophoblast is then the sole covering of the villi that is easily recognizable (Fig. 111 *B*).

The territory of a rapidly enlarging villous tree is in time marked off by *placental septa* into a distinctive lobule known as a *cotyledon.* A septum is a sheet of tissue that projects from the floor of the placenta far toward the chorionic plate (Fig. 114). Originally consisting of a plate of decidual tissue, it becomes compressed and replaced largely by trophoblast.[14] Each cotyledon is a natural vascular unit containing several villous trees; each distributes its branches and twigs throughout that particular lobule. In all there are 14 to 30 cotyledons, incompletely separated from each other by the thin septal partitions.

Although the placental septa become mostly replaced by trophoblastic tissue, in part (and especially basally) they still retain decidual elements.[15, 16]

The core of a villus contains, among other cell types, some special large cells (of Hofbauer) apparently tissue macrophages (Fig. 111 *A*). Embedded in the connective tissue are scattered smooth-muscle fibers.[17] After the second month more and more of the capillaries come to lie close beneath the basement membrane of the trophoblast which here is thinned locally (*B*).[18] The free surface of the syncytium bears microvilli, which increase greatly its absorptive ability (*A*). At intervals, especially among the older villi, the syncytium aggregates into characteristic protuberances with numerous nuclei; these are the so-called *syncytial knots* (*B*), whose special significance is not apparent. They may detach; quantities occur in maternal blood in late pregnancy.

Represented in young stages and increasingly abundant in older placentas are irregular masses of stainable substance known as *fibrinoid material*. It occurs as incomplete layers in the chorion and decidua basalis and as irregular patches on the villi (Fig. 111 *B*). This peculiar material, usually associated with deposits of *fibrin,* has a complex origin to which trophoblast and degenerating decidua contribute.[18] It is an inert by-product of placental aging.

The Intervillous Space.   This cavernous space is primarily of fetal origin (trophoblastic lacunæ), but it grows also at the expense of eroded decidual substance. It rapidly becomes a huge blood sinus, which is bounded above by the chorionic plate and below by the decidua basalis; the space contains all persisting chorionic villi (Fig. 112). The total space is subdivided incompletely by the placental septa into subordinate compartments, each of which belongs to its respective cotyledon (Fig. 114). Just beneath the chorionic plate there is a region, relatively free of villi, which is known as the *subchorial space.* At the periphery of the placenta this space becomes continuous with irregular channels that comprise the so-called *marginal sinus.* For some time the entire intervillous space is lined with trophoblast (Fig. 109); later, part of this lining becomes replaced by fibrinoid material.

Although containing a forest of villi, the intervillous space of a mature placenta has a capacity of about 350 ml., or one-half the volume of the total organ. Its blood supply comes from about 200 arterioles that traverse the decidua basalis and open as nozzles into each cotyledon (Fig. 114).[19,20] Similarly, collecting veins drain each cotyledon, and the marginal sinus as well.

The Maternal Placenta.   The decidua basalis, as the maternal contribution to the placenta, contains representatives of both the compact and the spongy layer of the progravid endometrium (Fig. 112). The glands of the spongiosa become stretched into horizontal clefts by the third month, but there seems to be much variability in their size, shape and even in their persistence during the later months. The portion of the decidua basalis that is most intimately incorporated into the placenta is the *basal plate.* In part this is merely another name for the compact layer of this region. Actually the

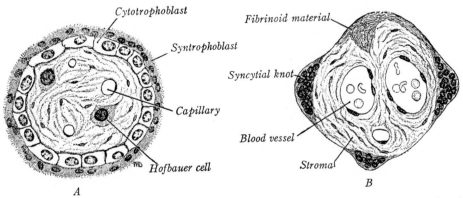

Fig. 111.   Human chorionic villi, in transverse section. × 265. *A,* In the early weeks of pregnancy; *B,* at full term.

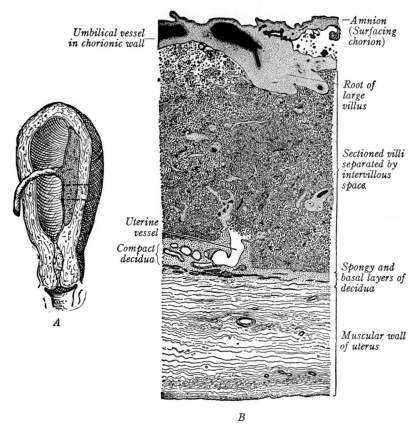

Umbilical vessel
in chorionic wall

—Amnion
(Surfacing
chorion)

Root of
large
villus

Sectioned villi
separated by
intervillous
space.

Uterine
vessel

Compact
decidua

Spongy and
basal layers of
decidua

Muscular wall
of uterus

A

B

Fig. 112. Gravid human uterus at seven months, in section. *A,* Hemisection, with placenta *in situ. B,* Vertical section through the placental site (Minot; × 3.5); the area shown corresponds to the rectangle in *A.*

plate is composed additionally of fibrinoid material and trophoblast. Some of the latter belonged originally to anchoring villi; the remainder represents the residue of the peripheral shell of trophoblast that made junction early with the eroding endometrium (Fig. 109). For a time trophoblast covers the basal plate; it decreases in amount as pregnancy continues and at term has a discontinuous distribution. Spiralling uterine arteries pass through the basal plate obliquely, branching and losing their accessory coats as they proceed (Fig. 114).

Growth and Maturity. Growth of the fetus is roughly paralleled by the enlargement of the uterus and placenta. The placenta continues to increase in area throughout pregnancy; during most of this time it occupies about 30 per cent of the internal surface of the steadily expanding uterus. Increase in thickness results chiefly from the elongation of villi and not from more invasion of maternal tissues. Such thickening is largely completed by the middle of fetal life.[21] The placenta is six times heavier than the fetus at one month, the same weight at four months, and only one-seventh the fetal weight at birth.

IMPORTANT FACTS CONCERNING HUMAN FETAL AND MATERNAL MEMBRANES

ORGAN	ORIGIN	COMPOSITION	LOCATION	FATE	FUNCTION
A. Fetal membranes:					
1. Yolk sac	Fashions from early entodermal layer.	Entoderm and splanchnic mesoderm.	Yolk stalk: within umbilical cord. Yolk sac: between amnion and placenta.	Disconnects from gut early. Stalk disappears early. Sac may persist.	Roof forms the gut. Forms blood cells and vessels. Early absorptive function?
2. Allantois	Diverticulum of hind-gut region of the yolk sac.	Entoderm and splanchnic mesoderm.	First lies within body stalk; later within umbilical cord.	Epithelium disappears early (except traces). Blood vessels persist.	Vessels connect fetal circulation with the placenta.
3. Amnion	Cavitation where inner cell mass joins trophoblast.	'Ectoderm' and somatic mesoderm.	Encloses embryo and umbilical cord. Attaches to embryo at the umbilicus.	Persists until birth. Fuses with chorion. Covers fetal surface of placenta and afterbirth.	Contains fetus, immersed in amniotic fluid. Sole cavity of later pregnant uterus.
4. Chorion	Trophoblastic capsule of blastocyst.	'Ectoderm' (trophoblast) and somatic mesoderm (of trophoblastic origin).	Encloses embryo and all other fetal membranes.	Frondose part becomes fetal placenta. Smooth part fuses with decidua parietalis. Cast off after birth.	Placental area is the fetal organ for nutrition, respiration and excretion.
5. Umbilical cord	Amnion wraps about yolk stalk and body stalk.	Chiefly, allantoic vessels and connective tissue enveloped by the amnion.	Connects belly wall with the fetal side of the placenta.	Cut off after birth and lost with placenta. Stump withers and detaches.	Vascular pathway between fetus and placenta.
B. Maternal membranes:					
1. D. parietalis	Progravid endometrium continued into pregnancy.	Compact and spongy layers (*i.e.*, the 'functional layer' of endometrium).	Lining of uterus, except in placental area and cervical canal.	Stretches but persists. Fuses with d. capsularis and then with chorion. Splits off in spongiosum as part of the afterbirth.	Potential, but unused, placental site. Contributes to growing placental margin?
2. D. capsularis	A superficial part of the endometrium of pregnancy, split and elevated by the implanted chorionic sac.	A more superficial part of the compact layer.	Between chorionic sac and cavity of the uterus.	Pressed into union with d. parietalis. Soon becomes unrecognizable.	Covers and holds implanted chorionic sac in place. No lasting function.
3. D. basalis	Endometrium of pregnancy, located beneath the implanted chorionic sac.	Deeper part of compact layer and all of the spongy layer.	Between chorion and the muscular wall of uterus.	Splits in spongiosum and is lost with rest of placenta.	Supplies maternal blood to the placenta.
C. Placenta	A local association of fetal and maternal tissues.	Chorion frondosum and decidua basalis.	Usually on front or back wall of the uterus.	Cast off as unit after birth. (Continuous with d. parietalis and chorion leave at the placental margin.)	Vital intermediary organ between fetus and mother. Produces hormones and, probably, enzymes.

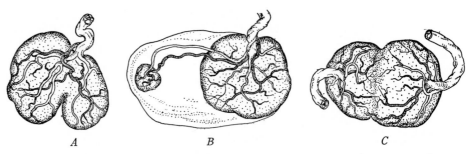

A   *B*   *C*

Fig. 113.   Anomalous placentas. *A,* Bilobed placenta; *B,* main and accessory placenta; *C,* fused placentas of ordinary twins.

The mature placenta makes a prominent, circular patch upon the interior of the greatly enlarged uterus (Fig. 115). Most of its bulk is due to chorionic villi and blood in the intervillous space (Fig. 112). When cut into, the organ is dark-red and spongy. It shows various indications of degenerative changes and declining efficiency, but is still a competent organ functionally. The increasing occurrence of fibrinoid substance in older specimens has been mentioned (p. 133); its presence on villi decreases their absorptive surface and when aggregated constitutes *white infarcts. Red infarcts,* caused by massive coagulation of blood in the intervillous space, also are characteristic.

Anomalies.   The umbilical cord may attach atypically in relation to the placenta (Fig. 89). Departure from a circular shape by the placenta is quite common, ranging from an oval contour to other variant forms (*i.e.,* spindle; pear; heart; crescent; ring) which are more rarely encountered. The placenta may be notched, lobed or even divided completely (Fig. 113 *A*). Occasionally there are one or more *accessory placentas,* of smaller size than the main placenta (*B*). A *circumvallate placenta* bears a circular fold of amnion and chorion on its free surface, owing to an undermining peripheral growth of a too small villous patch. All specimens of atypical shape or number are referable either to irregularities in the shape or growth of the chorion frondosum, or to the persistence and independent development of more than one patch of chorionic villi. *Fused placentas* result when ordinary twins become too closely implanted (*C*); but even the closest union is marked by a plate of compressed tissue at the plane of junction. Quite different is the condition in true twins who share a common placenta that was single from the start (Fig. 151 *B*).

PHYSIOLOGY OF THE PLACENTA

Circulation.   The blood of the fetus and that of the mother circulate independently in totally separate channels. Fetal blood is pumped into the umbilical artery in the umbilical cord, and is distributed by way of the chorionic plate to the villi (Fig. 114). After passing through the capillary mesh of the villi, the blood returns to the fetus by way of the umbilical veins. Each cotyledon of the placenta contains a main villous trunk whose apical branches extend to the basal plate, and may embed there. Maternal blood enters a cotyledon through spiral terminations of the uterine arteries (Fig. 114). Several nozzles open directly into the intervillous space of each cotyledon; yet they do not pass blood simultaneously. Each acts intermittently and irregularly, sending a jet of blood high up toward the chorionic plate. After bathing the chorionic villi, encountered along the course of such a fountain-like spurt, the maternal blood disperses laterally and then mostly downward in the same cotyledon. It is moved passively (forced along by the blood of subsequent jets), and escapes through the most conveniently located venous outlets.[20, 22] About 600 ml. of blood flows through the inter-villous space each minute.

The separation of the fetal and maternal circulations is normally complete. Communication is only through diffusive interchange. The barrier consists of the trophoblastic covering of a villus, the connective-tissue core, and the endothelium of embedded capillaries (Fig. 111). As pregnancy advances, there is marked thinning of the barrier until it may be only 0.002 mm. in thickness.[18] Along with thinning goes an astonishing increase in the transfer rate in both directions. For example, in later pregnancy the fetus receives 4000 times as much water as it incorporates in its tissues.

Functions.   The placenta serves primarily as an organ that permits the interchange of materials carried in the blood streams of the mother and fetus. The functions of the placenta fall into several classes: (1) *Nutrition.* Carbohydrates, proteins and fats, as well as water and inorganic salts, pass in one form or another from the mother's blood to the fetus. (2) *Respiration.* Oxygen, furnished by the mother, diffuses into the fetal blood while carbon dioxide passes in the reverse direction, the placenta thus acting like a lung. (3) *Excretion.* Fluid waste-products of fetal metabolism escape through the placenta, which exerts a kidney-like function. (4) *Barrier.* The placenta is impermeable to particulate matter, such as bacteria and even

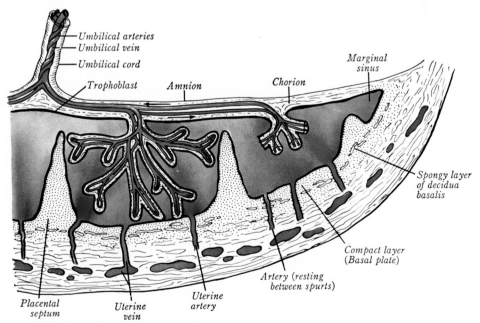

Fig. 114.   Diagram of the fetal and maternal blood flow through the human placenta.

to most large molecules. Yet the barrier is selective in certain cases, permitting some substances of large molecular structure to pass and allowing a retention and storage of some other substances contrary to ordinary osmotic expectations. (5) *Synthesis.* Hormones (estrogen; progesterone; gonadotrophin) are produced in important amounts; certain foodstuffs are synthesized and local enzymes are employed.

The trophoblast is the living membrane that is chiefly important in placental interchanges, both as regards permeability and barrier functions. To a considerable degree it acts like an ordinary semipermeable membrane, and the distribution of substances between mother and fetus is governed by the physical laws of diffusion. Substances that pass through by diffusion include: gases; mineral salts of the body; simple sugars; amino acids; some hormones and vitamins; most drugs; and waste products such as urea. These are examples of materials of relatively small molecular size, in solution. But other dissolved substances fail to be transmitted if their molecules are too large. For example, complex blood proteins do not enter the villi as such, but are first broken down into simpler products, such as amino acids. This requires a rebuilding by the embryo of its more complex proteins from these transferable components. Also fats, for the most part, probably do not pass the barrier as such. Rather, they are split, transferred and then rebuilt by the fetus.

There has been much debate as to whether the trophoblast, besides acting as a physical membrane, also exerts some selective regulation over what passes, like that occurring in glandular secretion. Some evidences of such vital control exist which suggest that simple physical processes are not the only mechanisms involved in the passage of substances from mother to fetus.[23] Also, conditions of transport vary with the age of the placenta. Actually there is a large safety factor in the amount of materials supplied to a fetus. Thus, in late pregnancy 99.9 per cent of the water and sodium reaching the fetus is returned to the maternal circulation.

Since, in general, the placenta is impermeable to particulate matter, even of ultramicroscopic size, it serves as an efficient barrier against the transmission of bacteria. By contrast, the viruses of smallpox and some other diseases pass readily, as do antibodies such as diphtheria antitoxin and anti-Rh agglutinins (produced by an Rh-negative mother). Induction of the Rh response depends on some fetal erythrocytes leaving injured chorionic villi and mingling with maternal blood in the intervillous space.[24, 25] It seems probable that cytoplasmic engulfment (*pinocytosis*) by the exposed border of the syncytium accounts for the uptake of substances of large molecular size, like antibodies and viruses. Nerves are completely lacking in the villi, chorion and cord; there is no possibility of 'maternal impressions' affecting an unborn babe.

As pregnancy advances, the continuous loss in relative weight of the placenta in comparison to the faster growing fetus places mounting demands on the placenta. These are met by producing more and smaller villous twigs (thus increasing the total surface area of the villi exposed to the maternal blood) and by thinning the membrane interposed between the two circulations (p. 134). The total absorbing surface of the chorionic villi at the end of pregnancy is about 140 square feet.[26] This is 50 times the surface area of the skin of a newborn. The thin villous membrane, characteristic of late pregnancy, is correlated with increased permeability; for instance, sodium is transferred 70 times faster than at nine weeks.[27]

CHILDBIRTH

During pregnancy the uterus enlarges into a huge sac whose muscular coat increases in bulk some twenty-four fold and whose capacity becomes over 4000 cc. This enlargement is paralleled by an astonishing increase in the size of individual muscle fibers and by the addition of new fibers from indifferent cells of mesenchymal nature.[28] During the third month the uterus begins to rise out of the pelvis; at six months its upper end has reached the level of the navel, and at the end of pregnancy it is not far below the breast bone. The fetus assumes the characteristic attitude illustrated in Fig. 115. At the time of birth the head is commonest directed downward, but the buttocks may be presented first or the baby may even lie crosswise.

Delivery. Childbirth, or *parturition,* occurs on the average at the time of the tenth missed mensis following conception—that is, 280 days after the last menstrual onset. The factors that induce 'labor' are obscure, but the process consists of a protracted series of involuntary muscular contractions of the uterus, termed 'pains,' combined with reflex as well as voluntary contractions of the abdominal muscles. These bring about a dilatation and effacement of the uterine cervix, the bursting of the bulging fetal membranes ('bag of waters'), and cause the extrusion ('delivery') of the child.

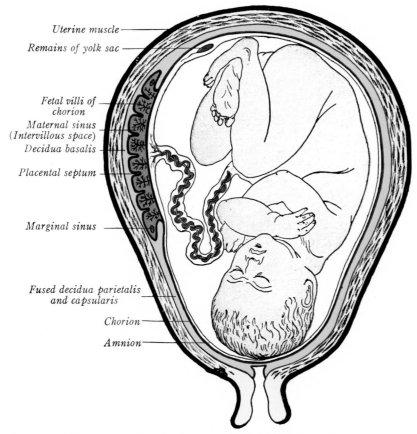

Uterine muscle

Remains of yolk sac

Fetal villi of chorion

Maternal sinus (Intervillous space)

Decidua basalis

Placental septum

Marginal sinus

Fused decidua parietalis and capsularis

Chorion

Amnion

Fig. 115. Diagrammatic longitudinal section of the uterus, illustrating the relation of an advanced fetus to the placenta and other membranes (Ahlfeld).

With the rupture of the membranes the amniotic fluid is expelled but the fetal membranes themselves remain behind, still attached to the deciduæ. Within some minutes after the birth of the baby the pulsation in the exposed cord slows and the umbilical vein collapses, signifying that up to 100 ml. of fetal blood has been restored to the vessels of the newborn child. The cord is then tied and severed. The stump of the cord shrivels and detaches after a week, leaving a depressed scar, which is the *umbilicus* or navel.

The rapid reduction in size of the emptied uterus leads to the detachment of the placenta, whereupon the pull of the placenta progressively helps detach the rest of the deciduæ (Fig. 116 *A*). The plane of separation of these membranes lies usually in the spongy layer where there are only thin-walled partitions between the stretched glands (*B*). At a variable time after the birth of the baby there enters a second series of uterine contractions. Through them the placenta and its associated membranes (the '*afterbirth*') are forced out. Because of the firmly contracted uterus at this time, there is remarkably little bleeding. Restoration and repair of the endometrium proceeds rapidly, and from the deep spongy and basal layers regeneration is nearly completed within a week. The speed of involution of the uterine muscle is astonishing; in a few weeks the weight of the emptied uterus reduces from 1000 grams to 50.

The Afterbirth.   The expelled placenta is a thick, circular disc which averages seven inches in diameter, one inch in greatest thickness, and weighs a little over one pound (Fig. 117). Its entire margin is continuous with the ruptured and cast-off sac that formerly contained the fetus. This latter mem-

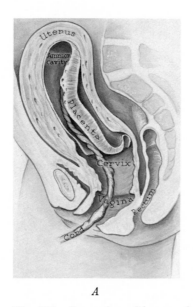

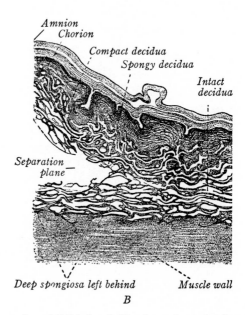

A   B

Fig. 116.   Separation of human deciduæ after childbirth. *A*, Hemisected trunk, showing placental detachment. *B*, Partially separated decidua parietalis and adherent membranes, in vertical section (Broman; × 5).

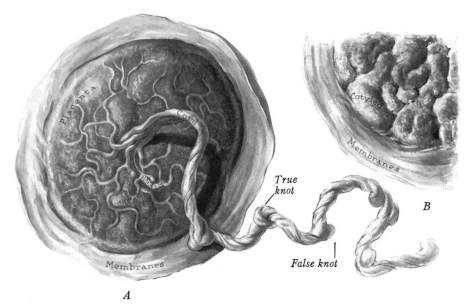

Fig. 117. Mature human placenta and its associated membranes, after expulsion. × ⅓. *A,.*Fetal surface; *B,* quadrant of the maternal surface.

brane is the joint product of the fusions that led to the obliteration of the uterine and chorionic cavities (p. 130). Its constituent parts, from within outward, are: (1) the non-placental amnion; (2) the chorion læve; (3) the decidua capsularis (no longer recognizable as a layer); (4) the decidua parietalis (in the form of shreds). The placenta likewise presents an amniotic or fetal surface, and a uterine or maternal surface. The fetal surface was primarily chorionic but, like the other deciduæ, it is now covered with smooth, glistening, adherent amnion. Usually near but not quite at the center of the fetal surface is attached the umbilical cord, already described at length (p. 112); its main vessels radiate on the fetal surface to the placental margin. The torn, maternal surface of the placenta is irregularly rough, reddish gray in color and bears blood clots. It exhibits 14 to 30 convex areas, of irregular shape, which correspond to the cotyledons. Such lobules are surfaced with tissue of the torn decidua basalis and are separated by shallow grooves. The latter indicate the positions of the bases of placental septa, demarcating cotyledons internally.

Anomalies. Childbirth may be early or late by as much as fifty days, but most of the apparent wide departures from the average duration of pregnancy are due to miscalculations based upon faulty data. The termination of pregnancy at still younger stages, when the fetus is not viable, is designated an *abortion* or *miscarriage*. It probably occurs in one-third of all pregnancies, if all early examples are included. Abortion may follow as a natural consequence upon the death of the fetus through poor placentation or disease. These factors, and others, may induce bleeding and uterine contractions, and thus cause the detachment of a living fetus. Sometimes the aborted deciduæ are discharged as a whole, and in younger specimens they may retain the shape of the uterine cavity (Fig. 118).

A

B

Fig. 118. Aborted fetus within intact membranes. *A,* At two months (× ⅗); the deciduæ have been cast off as a unit. *B,* At four months (× ⅓); the placental area, showing cotyledons, is below while elsewhere the sac is thin.

COMPARATIVE PLACENTATION

The elaboration of a placenta is associated with giving birth to living young (viviparity). When embryos develop within the body of the mother, instead of within laid eggs, they outlast any supply of yolk stored in the egg and have to establish functional relationships with the maternal tissues. Such an intermediate organ is a *placenta*. Some viviparous fishes and reptiles develop a placental relationship of this kind with the maternal tissues. This usually involves the yolk sac and chorion, but in a few reptiles a fairly complex organ results from the fusion of the allantois, chorion and uterine lining.

The egg-laying monotremes naturally forego a placenta. The marsupials, after a brief gestation period, give birth to immature young. Their yolk sac is relatively large and in some forms it unites with the chorion, apparently to complete a transitory *yolk-sac placenta*. In the group, as a whole, the chorion never advances beyond a smooth membrane in close apposition with the vascular uterine lining. Yet in a few marsupials (with longer gestation periods), a kind of chorio-allantoic placental relation does develop. Nevertheless, both monotremes and marsupials are customarily classed as aplacental mammals.

In all higher mammals (*placentalia*) the chorion bears vascular villi, and these engage the uterine mucosa in a more or less intimate functional relation which persists throughout pregnancy. Usually a large allantois fuses with the chorion, thus producing a composite membrane which serves as the fetal component of the so-called *chorio-allantoic placenta* (Fig. 119). When the allantois is insignificant or lacking, the organ is often designated as a *chorionic placenta;* this terminology deliberately ignores the important contribution supplied by the umbilical (originally allantoic) blood vessels. If the chorionic villi lie in apposition with the uterine lining, but do not fuse with it, the association is described as a *semiplacenta*. Since the separation of these two components at birth does not involve any shedding of the uterine lining, this type of placentation is called *nondeciduate*. Much different is the *placenta vera* (true placenta), whose chorionic villi fuse with the eroded uterine mucosa. Since, on the discharge of the placenta at birth, the uterine lining also tears away with some bleeding, this type of placentation is *deciduate*.

Types of Implantation. The relation of the chorionic sac to the uterine wall varies greatly among placental mammals. In general three types of implantation may be distinguished, although transitional conditions occur (Fig. 120): (1) *Superficial.* Growth of the sac brings it into contact with the lining of the main uterine cavity; this type is also known as central implantation (ungulates; carnivores; monkey). (2) *Eccentric*. The sac lies for a time in a fold or pocket which closes off from the main cavity (beaver; rat; squirrel). (3) *Interstitial*. The sac penetrates into the substance of the uterine lining (hedgehog; guinea pig; some bats; ape; man).

Placental Shape. The shape of any placenta and its size relative to the whole chorionic sac are governed by the final distribution-pattern of villi. On this basis four main types have been recognized (Fig. 121): (1) *Diffuse*. Lemurs (among primates) and some ungulates (sow; mare) retain villi over the entire chorion, and their placentas are correspondingly expansive (*A*). In the sow the 'villi' prove on closer inspection to be irregular folded elevations. In the lemur and horse there are short, branched villi separated by smooth interspaces. (2) *Cotyledonary*. True villi occur also in the ruminant (cud-chewing) group of ungulates, such as cattle, sheep and deer. These villi are grouped in well-spaced, prominent rosettes, known as *cotyledons*, which are separated by stretches of smooth chorion (*B*). (3) *Zonary*. The villi of carnivores occupy a girdle-like band about the middle of the chorionic sac. (*C*). (4) *Discoid*. In general, the villi of insectivores, bats, rodents and primates are limited to one or two disc-shaped areas (*D*).

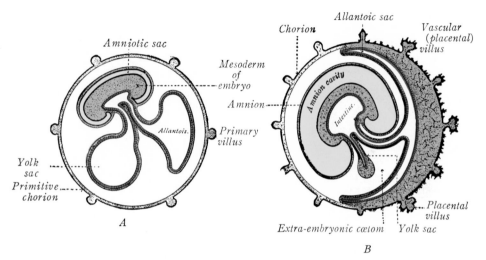

Fig. 119. Diagrams of the fetal membranes of most mammals, in sagittal section (Heisler, after Roule). *A*, Early stage, with relations much as in the chick. *B*, Later stage, with the fetal basis of an allantoic placenta.

Ectoderm, black; mesoderm, red; entoderm, green; amniotic fluid, yellow.

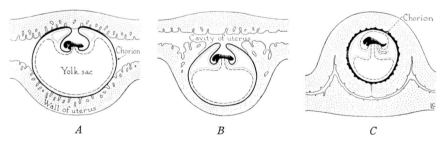

Fig. 120. Variations in the depth of implantation among mammals (after Mossman). *A*, Superficial (dog); *B*, eccentric (ground squirrel); *C*, interstitial (hedgehog).

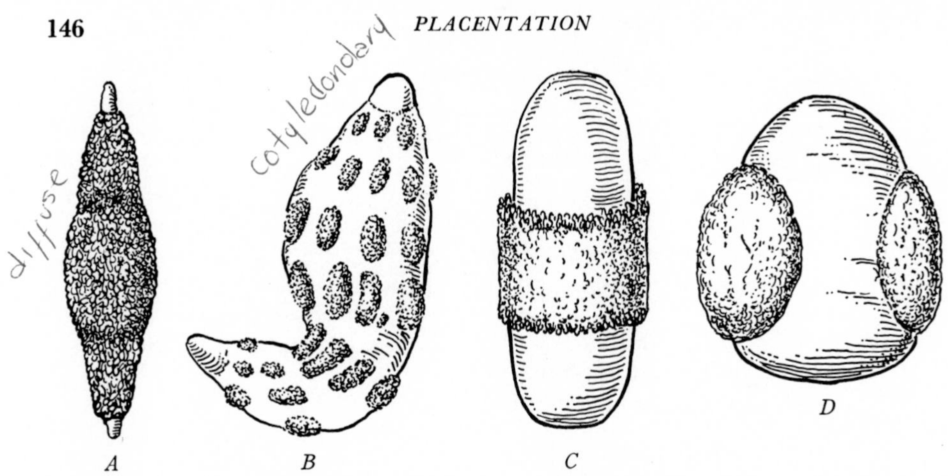

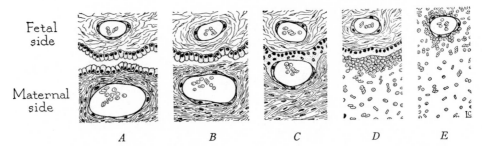

diffuse *cotyledonary*

Fig. 121. The ultimate distribution of chorionic villi as the basis of placental shape in mammals. *A*, Diffuse distribution in the pig; *B*, scattered rosettes (cotyledons) in the calf; *C*, zonary girdle in the puppy; *D*, discoid patches in the monkey.

Fetal side

Maternal side

A   B   C   D   E

Fig. 122. Placental types, arranged in a series to show the progressive elimination of barriers between the maternal and fetal circulations (mostly after Flexner and Gellhorn). *A*, Epithelio-chorial; *B*, syndesmo-chorial; *C*, endothelio-chorial; *D*, hemo-chorial; *E*, hemo-endothelial.

Structural Types. Placentas can also be arranged in a series that is based partly on the degree of contact or gross union existing between the chorion and the uterus, but more particularly upon the minute histological relations established at the zone of junction of these two components. In some forms the placental membrane becomes thinner as pregnancy advances and the structural relations show a progressive shift in type.

1. EPITHELIO-CHORIAL. The least modified placental condition is illustrated by some ungulates (sow; mare) and by lemurs.[29] The allantois and chorion unite and become jointly vascularized by the allantoic vessels. This composite membrane comes into relation with the uterine lining, the simple chorionic villi fitting into corresponding pits in the mucosa of the uterus. The relationship is purely one of apposition, the chorionic ectoderm merely applying itself against the uterine epithelium (Fig. 122 *A*). In the cleft between the two surfaces of epithelial contact is found '*uterine milk,*' composed of secretions and transudates. Nutrient substances and oxygen from the maternal blood must pass out of the uterine vessels and through both layers of connective tissue and epithelium before entering the allantoic vessels. Over the same path gaseous wastes from the embryo travel in the reverse direction. The allantois has, therefore, become important not only as an organ of respiration and excretion (including the storage of urinary wastes), as in reptiles and birds, but also as a participant in nutrition. Since the placenta has taken

over the function belonging to the yolk sac of lower vertebrates, the omission of yolk material from the eggs of higher mammals is understandable.

2. SYNDESMO-CHORIAL. The general type of ungulate placenta, just described, is modified slightly by an advance in the subgroup of ruminants (Fig. 122 *B*). In these mammals the prominent villi of the rosettes occupy deeper pits in the uterine lining. More important still, in some elevated portions of the uterine mucosa, between the villi, there is a local destruction of the uterine epithelium which allows the chorionic ectoderm to come into direct contact with the vascular maternal connective tissue. At the end of gestation, however, the chorionic villi of both types of ungulate placenta are merely withdrawn, and the maternal mucosa is not torn away and does not bleed.

3. ENDOTHELIO-CHORIAL. In carnivores the fetal-maternal union in the region of the villous girdle shows a marked advance in intimacy (Fig. 122 *C*). Erosion of the uterine mucosa practically bares the endothelium of its blood vessels, and the syncytial chorionic epithelium then packs about these maternal vessels. At birth there is destructive separation of the placenta, through which the fetal layer with its enclosed maternal vessels splits off from a deeper, regenerative zone of maternal tissue.

4. HEMO-CHORIAL. A still more intimate placental relation occurs in lower rodents, insectivores, bats and anthropoids; it is characterized by a more thorough erosion of the superficial uterine mucosa. One type (*labyrinthine*) is like the endothelio-chorial placenta of carnivores, except that the endothelium of the uterine vessels is lost and the maternal blood circulates in channels within the fetal syncytium. In the anthropoid type (*villous*) each chorionic villus is, in large measure, a freely branching tuft; these dangle in cavernous spaces and are directly bathed by maternal blood issuing from opened vessels (Fig. 122 *D*). The fusions between chorionic and uterine tissues are such that at birth the placenta tears away as a unit.

5. HEMO-ENDOTHELIAL. In higher rodents (rat; guinea pig; rabbits) is found the nearest approach to actual intermingling of the blood of the two circulations (Fig. 122 *E*). Chorionic villi lose their layers to such a degree that, in most places, the essentially bare endothelial lining of their vessels alone separates the fetal blood from the maternal sinuses.[30]

It is clear that the chorion serves many mammals (*e.g.,* ungulates; carnivores) chiefly by bringing the allantois into close relation with the uterine wall. Sharply contrasted is the condition in anthropoids and rodents where the chorion assumes all the placental functions, while the superseded allantoic sac becomes vestigial or even lacking. Of course, it may be argued that the allantoic vessels are the most important component of the allantois, that these vascularize the chorion of placental mammals in general, and hence that a placenta is fundamentally chorio-allantoic even though the allantois, as a sac, is insignificant.[30]

There is evidence that the rate of transfer of substances from the blood of the mother to that of the fetus increases as the number of layers to be passed decreases. The ascending order of efficiency in permeability is the same as the order in which the structural types of placenta have just been discussed. For example, the human placenta is 250 times more efficient in the transfer of sodium than is the sow's placenta.[27] It is also natural to assume that this structural series, with a progressive thinning of the fetal-maternal barrier, indicates the evolutionary sequence. Yet this may not be the case, since the epithelio-chorial type is widely scattered and tends to occur in mammals highly specialized in other respects. Arguments have even been advanced in favor of erosive placentation as the primitive type.[31] It must be understood, moreover, that permeability is only one kind of efficiency that seems to be correlated with the rapidity of the growth and differentiation of an embryo. Other advantageous qualities may be associated with the thicker types of placental membrane.

SUMMARY CONCERNING PLACENTAL RELATIONS IN MAMMALS

TYPE OF MAMMAL	YOLK SAC; YOLK-SAC PLACENTA	ALLANTOIS; CHORIO-ALLANTOIC PLACENTA	CHORIONIC VILLOUS PATTERN	DEGREE OF FETAL-MATERNAL FUSION *	INTIMACY OF UTERUS AND CHORION	LOSS OF MATERNAL TISSUE AT BIRTH
Monotremes	Huge sac;* None†	Large sac;* None†	Avillous	None (Egg is laid)	Non? (Egg laid in shell)	None (No uterus)
Marsupials	Large sac; Present (In some)	Small sac; Present (In a few)	Avillous	No placenta (Temporary semi-placenta in some)	Epithelio-chorial (Temporary)	None (In some, fetal tissue is retained)
Ungulates in general	Small sac (Large early); Present (Early)	Large sac; Present	Diffuse (Spheroid)	Semiplacenta (Apposition)	Epithelio-chorial	None (Non-deciduate)
Ruminant ungulates	Small sac (Large early); Present (Early)	Large sac; Present	Cotyledonary	Semiplacenta (Slight fusion)	Syndesmo-chorial	Slight (Semideciduate)
Carnivores	Medium sac (Large early); Present (Early)	Large sac; Present	Zonary	True placenta (Fusion)	Endothelio-chorial	Moderate (Deciduate)
Anthropoids	Small sac; None	Vestigial; Not typical‡	Discoid	True placenta (Fusion)	Hemo-chorial	Extensive (Deciduate)
Higher rodents	Large sac (Specialized); None	Small or lacking; Not typical‡	Discoid, cup or spheroid	True placenta (Fusion)	Hemo-endothelial	Moderate (Deciduate)

* The first of each pair of entries in this column concerns the first item in the heading above.
† The second of each pair of entries in this column concerns the second item in the heading above.
‡ Chorionic vessels (primitive allantoic) are the only contributions from the allantois.

REFERENCES CITED

1. Burdick, H. O., *et al.* 1942. Endocrin., *31*, 100–108.
2. Alden, R. H. 1942. Anat. Rec., *84*, 137–169.
3. Hartman, C. G. 1944. Western J. Surg., Obst. & Gyn., *52*, 41–61.
4. Chang, M. C. 1950. Ann. N.Y. Acad. Sci., *52*, 1192–1195.
5. Heuser, C. H. & G. L. Streeter. 1941. Carnegie Contr. Embr., *29*, 15–55.
6. Hertig, A. T. & J. Rock. 1945. Carnegie Contr. Embr., *31*, 65–84.
7. Hertig, A. T. & J. Rock. 1941. Carnegie Contr. Embr., *29*, 127–156.
8. Wislocki, G. B., *et al.* 1948. Obst. & Gyn. Survey, *3*, 604–612.
9. Brewer, J. 1937. Am. J. Anat., *61*, 429–481.
10. Hertig, A. T. 1935. Carnegie Contr. Embr., *25*, 37–82.
11. Wislocki, G. B. & G. L. Streeter. 1938. Carnegie Contr. Embr., *27*, 1–66.
12. Romney, S. L. & D. E. Reid. 1951. Am. J. Obst. & Gyn., *61*, 83–98.
13. Uchida, K. 1957. Acta anat. Nippon., *32*, 287–294.
14. Wislocki, G. B. 1951. Anat. Rec., *109*, 359.
15. Ramsey, E. M. 1960. Chapter 3 in Villee: The Placenta and Fetal Membranes (Williams & Wilkins).
16. Ser, D. M., *et al.* 1958. J. Obst. & Gyn. Brit. Emp., *65*, 774–777.
17. Arey, L. B. 1948. Anat. Rec., *100*, 636.
18. Wislocki, G. B. & H. S. Bennett. 1943. Am. J. Anat., *73*, 335–449.
19. Boyd, J. D. 1956. In: Gestation (2nd Conference; Macy Foundation), 132–194.
20. Ramsey, E. M. 1962. Am. J. Obst. & Gyn., *84*, 1649–1663.
21. Stieve, H. 1940. Anat. Anz., *90*, 225–242.
22. Hamilton, W. J. & J. D. Boyd. 1960. J. Anat., *94*, 297–328.
23. Needham, J. 1931. Chemical Embryology, vol. 3 (Macmillan).
24. Javert, C. T. & C. Reiss. 1952. Surg., Gyn. & Obst., *94*, 257–269.
25. Keenan, H. & W. H. Pearse. 1963. Am J Obst & Gyn., *86*, 1096–1098.
26. Christoffersen, A. K. 1940. Acta Path. & Microbiol. Scand., *17*, 348–374.
27. Flexner, L. B., *et al.* 1948. Am. J. Obst. & Gyn., *55*, 469–480.
28. Stieve, H. 1926. Z'ts. f. mikr.-anat. Forsch., *6*, 351–397.
29. Hitzig, W. H. 1949. Acta Anat., *7*, 33–88.
30. Mossman, H. W. 1937. Carnegie Contr. Embr., *26*, 129–246.
31. Wislocki, G. B. 1929. Carnegie Contr. Embr., *20*, 51–81.

Chapter IX. Reproductive Cycles and Their Hormonal Control

Previous chapters have described the periodic maturing of eggs and their subsequent reception and care by the well-prepared uterus. The purpose of the present chapter is to examine these and other co-ordinated, reproductive phenomena and the ways by which hormones control them.

REPRODUCTIVE CYCLES

Periodic breeding among animals is familiar to everyone. The egg-laying rhythm varies from the daily regularity of the hen to the annual periodicity of various animal kinds. Many animals time their breeding with respect to the seasons in order to take advantage of favorable temperature and food supply. Some marine forms have their reproductive cycles tuned to the tides and even to definite phases of the moon. In a hibernating female mammal, such as the raccoon, the increasing ratio of light to darkness as Spring approaches is the determining factor; this sets into action a chain of events in the retina, brain and hypophysis that ends with ovulation. Other animals, although not co-ordinated with such demonstrable influences, are nonetheless cyclic, and this is true of the females of various domesticated mammals and of primates. The reproductive rhythm of female rats and mice repeats at intervals of four to five days; the guinea pig, every 15 days; the sow, mare and cow, every three weeks; the monkey and human, every four weeks; the chimpanzee, every five weeks; the cat and dog, twice or thrice a year.

The Estrous Cycle. Among mammals, other than primates, the recurring periods of sexual excitement in the adult female are known as 'heat.' The biological term is *estrus,* and the correlated phenomena of the reproductive system from one period of estrus to the next make up an *estrous cycle.* These events can be understood better by recounting what takes place during the typical cycle of a specific mammal, the sow. For 18 days she follows her ordinary routine and shows no interest in the boar. But during the next three days she becomes restless and sexually excited. If a boar is present, he is accepted; the mating normally results in pregnancy, where-

150

upon the cycles cease until after the young are born. If there is no mating, or an infertile one, the cycles continue at the usual intervals of three weeks throughout the year.

Examination of the ovaries of the sow on different days of the estrous cycle shows that a definite series of events takes place. These constitute what is called an *ovarian cycle* (Fig. 123). During the days of sexual inactivity, known as the *diestrus,* the ovarian follicles are all small. About two days before estrus numerous follicles begin to grow, and on the first day of estrus there are large vesicular follicles with maturing eggs. Late in the second day of estrus the follicles rupture, expel the eggs and straightway begin to transform into corpora lutea. These structures attain full development by the seventh day after ovulation. If the eggs are not fertilized, the corpora lutea retain their functional state for seven more days and then begin to degenerate. While involution is still in progress, the development of a new crop of follicles is under way and the cycle repeats. If the eggs are fertilized, the corpora lutea continue on into pregnancy and the cycles cease until after the young are born or are far along with nursing.

Examination of the uterus during an estrous cycle reveals another regular sequence of events. These remarkable happenings comprise the *uterine cycle* (Fig. 123). During the growth and ripening of the ovarian follicles there is growth and change in the endometrium (follicular phase). During the period of the developing and mature corpus luteum the endometrium specializes further in ways that will fit it for pregnancy (luteal phase). If the eggs are not fertilized, the endometrium returns to its original condition as the corpora lutea decline, only to repeat the whole series of changes in the next cycle. If the eggs are fertilized, the endometrium retains its specialized state which is both favorable to pregnancy and necessary for its occurrence.

These several phenomena are all co-ordinated in perfect timing. The eggs are matured just when the sow becomes sexually excited and receptive to the boar; this ensures fertilization and a start on development. The

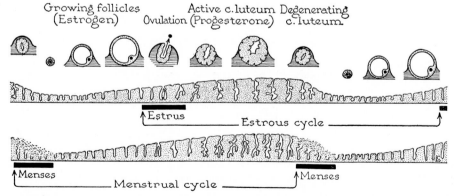

Fig. 123. Diagram showing the correlation of events in the ovarian follicle, in the endometrium of most mammals (with estrous cycle), and in the endometrium of primates (with menstrual cycle); (partly after Corner).

uterus is prepared at the right time to receive and nourish the free blasto-cyst; this ensures continued development. If pregnancy occurs the cyclic happenings are suppressed, since it is necessary that ovulation cease, the corpus luteum persist and the favorable uterine state be retained. If preg-nancy does not occur, the cycle repeats again and again.

The conditions existing in the sow illustrate the fundamental plan of the cycle among mammals. There are minor pecularities in each mammalian type, and there are also major variations. For example, a few animals, such as the cat, ripen follicles in every cycle, but the follicles do not burst unless mating occurs. The rabbit exceeds this by not even completing its follicles in the absence of mating; thus it remains in a state of prolonged estrus, during which it continues to be receptive to the male.

The Menstrual Cycle.   The greatest departure from the basic plan of the mammalian reproductive cycle is found in man, apes and the Old World monkeys. In these forms the periodic growth of the follicle and corpus luteum is quite typical, as are the correlated changes in the endometrium. But an outstanding peculiarity is that the regression of the corpus luteum is accompanied by a destructive breakdown of the endometrium, with hemorrhage, about two weeks after ovulation. This periodic loss of tissue and blood is *menstruation* (Fig. 123). It is the most conspicuous feature of the cycle, and it has no counterpart in lower mammals. Since, however, the estrous cycle of lower mammals has an equally prominent event, estrus, during which ovulation occurs, it was natural to think that the two hap-penings are similar and that ovulation in the higher primates occurs at the time of menstruation. Only in recent decades have these errors been corrected.

The human cycle averages about 28 days in length, but some individuals run to shorter cycles and others to longer ones. There is a popular impres-sion that the cycle is normally regular, but this is far from the truth. If two-thirds of all the cycles of any individual keep within a range of two to three days above or below her average, it is as good a performance as can be ex-pected.[1] No instance of perfect regularity for any considerable period of time has ever been reported. In view of the complex biological mechanism (inter-related hormones) that controls the menstrual cycle no such case is to be expected.

The events of a menstrual cycle can be described as occurring in four stages, although each stage actually blends into the next (Fig. 124). These phases are timed from the start of visible flow, which rates as day one. Former accounts of the cycle have been modified considerably,[2, 3] partly because of direct observations made on pieces of endometrium grafted onto the iris of the eye of a monkey and observed through the transparent cor-nea.[4] The periodic rebuilding of a fresh uterine lining, which is partly lost at each menstrual sloughing, can be viewed as a monthly routine of preparation to make the uterus suitable for the reception of a prospective guest (the blastocyst). In continuance of this metaphor, active menstruation is 'a violent demolition of the premenstrual edifice some days after the expected tenant fails to arrive.'

1. RESURFACING (days 4–6). During the five days of actual menstruation the compact layer, at least, is lost by sloughing.[3] But even before all bleeding has ceased, some local regions have already undergone repair (Fig. 124 *E*). In this process epithelial cells leave the remnants of the glands located in the basal and spongy layers; they glide over the denuded surface and epithelize it anew.

2. FOLLICULAR PHASE (days 7–15). Also known as the *proliferative stage,* this phase completes and extends the postmenstrual repair (Fig. 124 *A*). It coincides with the growth of a new set of follicles in the ovary. The glands proliferate, lengthen rapidly and produce a thin secretion. Connective-tissue cells also multiply and differentiate a new mesh of reticular fibers. The endometrium increases from 1 mm. in thickness to 2 mm. or more.

3. LUTEAL PHASE (days 16–28). Other terms are the *secretory* or *progestational stage.* It parallels the growth and functional life of the corpus luteum. Although the glands no longer proliferate, they elongate further, swell and become tortuous. Throughout much of their length they show sacculations, distended with a thicker mucoid secretion rich in glycogen (Fig. 124 *B*). Peculiar spiral arterioles, continuing their upward growth, break up into capillaries that supply the superficial two-thirds of the endometrium; the basal one-third is supplied by ordinary, straight branches of the uterine arteries. By the end of this period the endometrium has nearly doubled its previous thickness, owing largely to the increase of secretion and to edema fluid.

4. MENSTRUATION (days 1–5). Hours before the onset of active menstruation, the spiral arterioles constrict one by one and cause the endometrium to blanch. Accom-

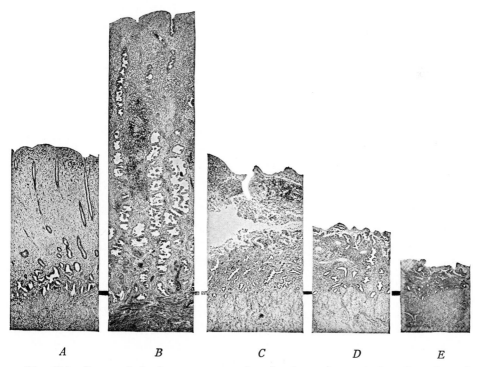

A B C D E

Fig. 124. Stages of the human menstrual cycle, shown by vertical sections through the endometrium (Bartelmez). × 18. *A,* Follicular phase (day 11); *B,* luteal phase (day 23); *C, D,* menstruation (days 1, 2); *E,* resurfacing (day 4).

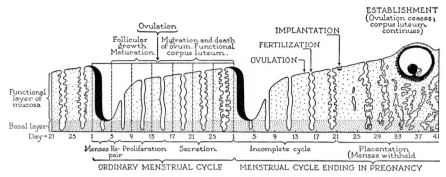

Fig. 125. Graphic presentation of the relations existing between the cyclic events of the ovary and uterus, and their relation to pregnancy (modified after Schröder).

panying this local anemia the mucosa shrinks and death of the blood-deprived tissues follows. At some time each spiral arteriole relaxes for a few minutes, blood escapes from the injured vessel, and sooner or later this noncoagulating blood discharges into the cavity of the uterus. Fragments of the disintegrating endometrium slough away down into the levels supplied by the spiral arterioles (Fig. 124 *C, D*).

Figure 125 brings together many of the timed relations in an ordinary ovarian and uterine cycle, and in a similar cycle that terminates in pregnancy.

Anovulatory Cycles. At times the ovarian follicle fails to rupture and expel its egg. In this event there is, of course, no corpus luteum and no luteal phase of endometrial upbuilding. Yet in spite of this omission, menstruation follows the ending of the follicular phase. It is typical in all respects except that it occurs in an endometrium brought only through the proliferative stage. Such menstruation without ovulation occurs most frequently in girls beginning their menstrual career, in women approaching the end of their reproductive capacity (menopause) and during the nursing of infants. Its incidence in normal adults is about 10 per cent.

Temporary Amenorrhea. When the cyclic phenomena are becoming established (*menarche*) at puberty and adolescence, and when these activities are waning in anticipation of their final failure (*menopause*), there is marked irregularity in menstrual performance; especially are there unusually long intervals (*amenorrhea*) between successive menses. After childbirth there is a delay of several months before ovulation and menstruation are re-established in the non-nursing mother; with nursing there is still further delay. Emotional states, such as anxiety, the influence of change in vocation or climate, and systemic disease may all inhibit the cyclic rhythms for a time.

Estrous and Menstrual Cycles Compared. The estrous and menstrual cycles are alike in having (Fig. 123): (1) a period of growth of the ovarian follicle and ripening of the egg, accompanied by heightened hormone (estrogen) output from the ovary and proliferative growth of the endometrium; (2) ovulation which ends this period; and (3) a period of the corpus luteum, with the production of a luteal hormone (progesterone) and the resulting terminal specialization of the endometrium that makes it suitable for pregnancy. The two cycles are unlike in several respects: (1) in the estrous cycle the animal has a definite period of heat, during which ovulation occurs and the male is accepted; (2) in the menstrual cycle, ovulation is not accompanied by heat or such limited mating, and the end of the luteal phase is signalized by tissue loss and bleeding; and (3) the two cycles, as described, occupy different positions with respect to the sequence of ovarian and uterine changes.

The Vaginal Cycle. The changes that characterize the female sexual cycles are not confined to the ovaries and uterus. The epithelial lining of the uterine tubes and vagina undergoes cyclic alteration as well. Clearest and most noteworthy are the rhythmic changes observed in the vagina of rodents. Migrating leucocytes, present at diestrus (Fig. 126 *A*), disappear just before estrus, and the epithelial cells proliferate and begin to specialize (*B*). During estrus the picture changes still further by the epithelial cells cornifying and shedding in great numbers (*C*). By examining vaginal washings the exact stage of the cycle can be determined at any time. Such tests are of the utmost importance in experimental work because these animals give no other clear signs of their estrous state, as, for example, do the sow by excitement or the bitch by genital swelling.

In primates the vaginal changes are less extensive and distinct.[5, 6] Normal ovarian activity as against inactivity can be detected, and the differences between follicular and luteal stages are distinguishable.

The Mammary Cycle. In the monkey, as in lower mammals, growth changes and regression take place in the mammary glands in co-ordination with the ovarian cycle (Fig. 127).[7] Controlled observations on periodic epithelial growth in the human are lacking, but the breasts do tend to become larger and firmer preceding menstruation. Also, some descriptions

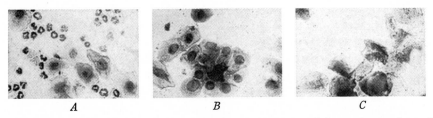

| *A* | *B* | *C* |

Fig. 126. Vaginal cycle of the rat, as shown by smears (Fluhmann). × about 500. *A*, At diestrus; *B*, just before estrus; *C*, during estrus.

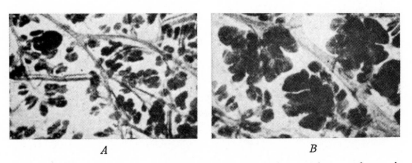

| *A* | *B* |

Fig. 127. Cyclic changes in the mammary gland of the monkey as shown in gross biopsy specimens (after Speert). × 8. *A*, During much of the cycle; *B*, toward the end of the cycle.

exist of correlated glandular changes.[8] During pregnancy there is remarkable growth of the ducts and secretory end-pieces. After the nursing period, regression returns the glandular elements to the so-called resting state.

Male Periodicity.   Testicular activity in the majority of mammals is a seasonal event, as is the associated period of sexual excitement known as *rut.* An annual period of spermatogenesis is followed by an interval of inactivity in which the testes (interstitial tissue as well as tubules) shrink and may be withdrawn from the scrotal sacs to the interior of the body. At this time the accessory sex organs, dependent on the male hormone, also undergo regression. In some such animals (*e.g.,* ferret) artificial lengthening of the daily allotment of light brings the inactive male into a sexually competent state months ahead of the normal schedule.[9] Spermatogenesis of the ground squirrel occurs only in the cooler months, and especially during hibernation; this period may be extended throughout the year by maintaining animals at 40° F.[10] In man and most domesticated and semidomesticated mammals seasonal fluctuations are not apparent.

THE HORMONES CONCERNED WITH REPRODUCTION

The preceding paragraphs have presented the main events of the reproductive cycles in a purely descriptive way, scarcely hinting at the forces that impel and control them in their orderly and well co-ordinated sequences. These agencies are *hormones*—chemical substances produced in endocrine (*i.e.,* ductless) glands. They are distributed through the blood stream and are capable of arousing into action certain target tissues that come under their influence. It is to the hormones concerned with reproduction that attention must now be directed.

The Pituitary Hormones.   The hypophysis, or pituitary body, is a small gland attached to the underside of the brain (Fig. 637). It consists of two main parts, or lobes (Figs. 189 *B*, 190 *A*). The *anterior lobe* is epithelial in structure; specialized cells, distributed through it, secrete the *gonadotrophic hormones.* There is rather good evidence for the existence of three distinct

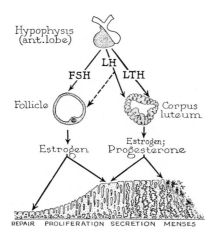

Fig. 128.   Diagram of the hormonal influences of the hypophysis on the ovary, and of the ovary on the uterus.

hormones with gonad-stimulating qualities (Fig. 128).[11] Their changing amounts in relation to days of the menstrual cycle are shown in Fig. 129.

THE FOLLICLE-STIMULATING HORMONE.   This secretion, often abbreviated to FSH, is a product of certain basophilic cells in the lobe. It causes the normal growth of the ovary and brings about its coming into function at puberty. The periodic development of small follicles to the point where they contain full-sized eggs (primary oöcytes) is a self-managed activity. But it is the follicle-stimulating hormone that makes the solid follicles grow into vesicular follicles. It is also a basic factor in producing ovulation and causing the ovary to secrete the *estrogenic hormone.* In the male a hormone, identical with FSH, causes the testis to grow; it also plays a large rôle in governing spermatogenesis. In view of an influence on the gonads of both sexes, *gametokinetic hormone* would be a more appropriate, general designation than FSH.

THE LUTEINIZING HORMONE.   A second pituitary product (LH) is probably secreted by other basophils, specialized in a different direction. Acting in precise balance with FSH this hormone co-operates in bringing about the final stages of follicle development, maturation of the egg and ovulation.[12] Under its influence alone, the emptied vesicular follicle transforms (*i.e.,* luteinizes) into a corpus luteum; but this new glandular mass is not yet capable of secretion. In the male the same hormone induces the development of interstitial cells and the secretion of the male sex-hormone (*testosterone*) by them. Hence *interstitial-cell stimulating hormone* (ICSH) is frequently employed for this pituitary hormone of the male. ICSH and testosterone seem to aid FSH in effecting the complete stimulation of spermatogenesis.[13]

THE LUTEOTROPHIC HORMONE.   A third pituitary product (LTH) (secreted by a particular kind of acidophil) is the agency actually responsible for activating the built-up corpus luteum to release progesterone, and then for maintaining the gland in a functional state.

Proofs of these several effects have been supplied through experiments in which the hypophysis has been removed, transplanted or used in the form of extracts as a substitute for normal pituitary control. This endocrine organ, therefore, is the seat of control without which the ovaries and testes cannot function. Through the gonads it also governs secondarily the cyclic phenomena of the female reproductive tract, the functional state of the male accessory glands (prostate and seminal vesicles) and the secondary sex characters

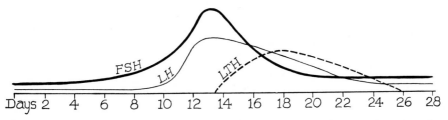

Fig. 129. Graphic presentation of the amounts of gonadotrophic hormones produced during a menstrual cycle.

of both sexes. These pituitary hormones are proteins, and their isolation in pure, or highly concentrated, forms has been accomplished.

The Ovarian Hormones.   These hormones are two in number, and their sources are fairly well established.[14] One, *estrogen,* is almost certainly produced by the internal thecal cells of follicles and corpora lutea. The other, *progesterone,* is seemingly a product of the granulosa cells of the corpus luteum (Fig. 128). The changing amounts of these hormones produced during a menstrual cycle are shown in Fig. 130.

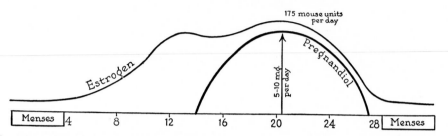

Fig. 130. Graphic presentation of the amounts of ovarian hormones produced during a menstrual cycle, as determined by assays of their end-products in the urine.

ESTROGEN.   In the puberal period of mammals the uterine tubes, uterus, vagina and mammary glands, all of which have remained small and relatively undeveloped since birth, grow rapidly to nearly the adult size and the secondary sex characters of the body become established (Fig. 131 *A, B*). Finally the uterine, vaginal and mammary cycles begin. These phenomena are dependent on the ovaries, as appropriate experiments prove. For example, following the castration of adult mammals the mature state of the reproductive tract and mammary glands is not maintained, and marked atrophy ensues. But the castrate condition can be prevented or corrected by ovarian transplants or by the administration of ovarian extract. Similar treatment of an immature mammal brings about precocious growth and maturity of the reproductive tract and mammary glands.

The hormone responsible for the several effects just mentioned is called *estrogen* because it also produces most of the characteristic features of the estrous or menstrual cycle. Its specific effect on the endometrium is in directing the proliferative stage, also known as the follicular phase (Fig. 128). Estrogen occurs in the ovary in general (including the corpus luteum), and there is evidence that it is made by the internal theca of the follicles.[15, 16] Estrogen is finally lost from the body by being excreted in the urine. Actually 'estrogen' is a collective term because a considerable group of related, chemical substances produces similar effects. They are all organic compounds, belonging to the sterols, which can be isolated in pure crystalline form. *Estradiol* is apparently the actual substance secreted by the ovary. Chemical compounds, with estrogenic qualities, have been synthesized in the laboratory; the most important of these products is *stilbestrol.*

As is characteristic of hormones, tiny amounts of an estrogen produce large effects.[17] For example, *estrone* is an estrogen end-product recoverable from pregnancy urine. The administration of 0.00001 milligram of estrone daily for three days can produce the characteristic estrous changes in a castrated mouse. Six million of these doses would equal a small postage stamp in weight. The daily output of the more potent estradiol from both ovaries of an adult woman has the same effectiveness as 30,000 mouse doses of estrone.

Like other hormones, the group of estrogens is selective in its action. Their influence is directed chiefly at the reproductive tract, mammary glands and secondary sex characters of the female. Injection of estrogen into an immature mammal or an adult that has undergone atrophy of the reproductive tract, following castration, causes the blood vessels of the uterus to dilate, cell proliferation to increase and the glands and muscle to enlarge. Hence, at puberty, estrogen acts on the uterus by bringing it to the full adult condition and then maintaining it there (Fig. 131 *A, B*). It is responsible for the proliferative stage of the uterine cycle (Fig. 128). The vagina also responds by growth and epithelial specialization (Fig. 126). At puberty the nipples and duct system of the mammary gland respond to estrogen by marked growth (Fig. 135, *A, B*).

PROGESTERONE.   The growth and change in the uterus that characterize the full estrous or menstrual cycle are not due wholly to the influence of estrogen. On the contrary, the culminating events (chiefly tissue differentiation) that routinely make the estrogen-primed endometrium suitable for pregnancy are directed by the corpus luteum (Fig. 128). Its characteristic hormone, *progesterone,* implies by its name that it causes the specialization that prepares for gestation (*i.e.,* pregnancy); the last part of the word also indicates that the hormone is chemically a sterol. This progestational development comprises the final, progressive changes in the endometrium which have already been described as the secretory stage or luteal phase (Fig. 131 *C*). Removal of the developing corpus luteum prevents this portion of the uterine cycle from appearing. Under these conditions pregnancy cannot occur. For one thing, the endometrium fails to become sensitized in such a manner that it can react to the presence of an embryo by collaborating in the formation of a placenta. Actually, however, there is no invading embryo in this type of experiment because the blastocysts die while still free in the uterus, awaiting the time for implantation. They die

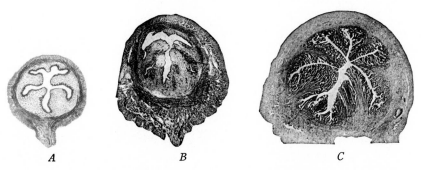

A   *B*   *C*

Fig. 131.   Influence of ovarian hormones on the uterus of the rabbit, shown by transverse sections (*B, C,* Bouin and Ancèl). × 9. *A,* Immature rabbit. *B,* Growth to adult state, caused by estrogen. *C,* Progestational (secretory) changes, caused by progesterone.

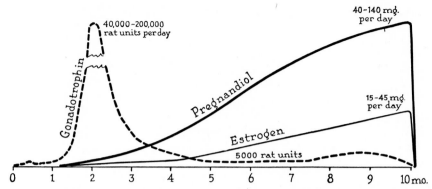

Fig. 132. Graphic presentation of the amounts of placental hormones produced during pregnancy, as determined by assays of their end-products in the urine.

for the lack of chemical substances secreted by the glands of the endometrium under the influence of progesterone. Hence, in the normal course of events, the corpus luteum prepares and sensitizes the endometrium for pregnancy; endometrial secretions nourish and protect the embryo before implantation occurs; and the uterus reacts to the stimulus of implantation by forming a placenta. In some mammals the corpus luteum is also necessary for the maintenance of pregnancy, and its removal leads to resorption or abortion of the fetuses. Such is not the case in primates, once the embryo is well established; this is because the placenta generates substitute hormones, as will be explained in the next topic.

Pure progesterone in crystalline form can be isolated from corpora lutea. Properly used with estrogen, it is able to bring the uterus of an immature mammal or of a mature castrate to its full functional state. When an animal is castrated just after mating, the injection of progesterone can replace the action of the corpora lutea, and embryos can then be carried successfully to birth. Progesterone has not been synthesized from simpler substances, but it is being made by altering other sterols such as cholesterol. The daily output of progesterone is much greater than that of estrogen but the latter is far more potent, weight for weight, in producing its characteristic effects.

The Placental Hormones. During pregnancy the mammalian placenta specializes as an endocrine organ. It takes over certain pituitary and ovarian functions and thus enables the period of gestation to extend beyond the limits of an ordinary uterine cycle. In doing this, the embryo and placenta not only achieve independence from the mother's hormones but also become a self-contained, developing unit. The quantities of the hormones present throughout pregnancy, as indicated by their urinary end-products, are shown in Fig. 132.

CHORIONIC GONADOTROPHIN. A gonadotrophic hormone begins to be formed by the trophoblast by nine days, even while the blastocyst is implanting in the uterine lining. The concentration of this substance in the blood of a pregnant woman rises rapidly and reaches a high level by the seventh week; it has largely subsided by the twelfth week (Fig. 132). This time-range matches the growth, flowering and decline of the cellular tropho-blast, and there is evidence that this tissue is the actual source of the hormone.[18] The secretion is, therefore, named *chorionic gonadotrophin*. It is a specific kind of gonadotroph, different chemically and functionally from the pituitary products, LH and LTH, which it chiefly resembles. This chorionic hormone causes the corpus luteum to outgrow and outlive all ordinary expectations. This extension of luteal function precludes menstrual sloughing and insures the continuance of the characteristic protective actions until the placenta becomes capable of secreting estrogen and progesterone in amounts sufficient to maintain pregnancy. Another result is that the secretion of pituitary gonadotrophins is suppressed, so that follicle growth and ovulation cease promptly.

Chorionic gonadotrophin is excreted in the urine of pregnant women. Its concentration there is sufficiently high so that the injection of a small amount of pregnancy urine into a mature rabbit will induce ovulation (ordinarily dependent on coitus) within less than one day. This response constitutes the Friedman test for pregnancy, which is highly reliable even in the fourth week (Fig. 133). Similarly, the Aschheim-Zondek test makes use of the ability of pregnancy urine to bring quickly to maturity the ovaries of immature mice. Still other tests, quicker and perhaps more reliable, are based on the stimulation of ovulation in the clawed toad or on sperm release by the common frog.

Chorionic Estrogen and Progesterone. The enlarged corpus luteum of human pregnancy is not necessary after the first five weeks, as is proved by experiment. The corpus, or even both ovaries, may then be removed, yet the pregnancy will continue to a normal conclusion. The explanation lies in the fact that the two 'ovarian hormones' are also produced by the placenta and continue their several effects. The transfer of the site of formation of estrogen and progesterone from the ovary to the placenta is probably gradual and is completed at about three months. Indications point to the syncytial trophoblast as the source of these products.[19] Both hormones increase steadily in amount as pregnancy advances, as does also the parent chorionic tissue. They reach a maximum a few days before parturition (Fig. 132); why such large amounts of hormone are present at this time is not known. The placental estrogen is *estriol*, while the elimination product is *estrone*. The excreted form of progesterone, either ovarian or placental, is *pregnanediol*.

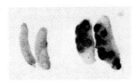

Fig. 133. Friedman test for pregnancy (Fluhmann). At left, ovaries of a control rabbit; at right, multiple ovulation (as evidenced by ruptured, hemorrhagic follicles) induced by pregnancy urine.

It is probable that estrogen is necessary throughout gestation primarily to maintain the capacity of the uterus to grow. Progesterone, however, is the specific agent that is responsible for the large increase in the number of smooth-muscle cells very early in pregnancy;[20] these elements make possible a uterine wall thick enough to withstand subsequent stretching. It is known that estrogen promotes rhythmic contractility in the uterine musculature of some mammals, whereas progesterone effectively inhibits such motility. Opinions concerning the human uterus in this respect do not agree, but it seems probable that progesterone serves to keep the uterus quiet in early pregnancy and thus insures safe implantation to an embryo.

The Testicular Hormone.   Through the centuries it has been known that castration of the prepuberal male suppresses the development of the secondary sex characters and accessory reproductive organs. Similarly castration of the adult leads to the atrophy of these accessory organs and in vertebrates in general, including most mammals, to loss of the sex drive. The aggressive bull and docile ox illustrate the castrate change. The relation of these conditions to a hormone first produced in the testis during puberty is firmly established. The weight of evidence favors the interstitial cells, located in groups between the seminiferous tubules, as the secretory agents.[21] *Testosterone* has been prepared from testis tissue, and the slightly different and much less potent *androsterone* from the urine. Both have been purified in crystalline form and both have been prepared artificially by the degradation of cholesterol. The collective term for these hormones is *androgen,* meaning 'promoting masculinity.'

Many substances with androgenic properties are known, most in the long list being artificial creations of the laboratory chemist. When testis extracts or pure androgens are administered to immature mammals they stimulate the growth of the accessory reproductive organs; in castrated adults they restore these atrophied organs and the mating urge. There is no sharp line of distinction between androgens and estrogens. Their chemical constitution is closely similar and in large doses each can produce certain effects of the other. Both are present in the urine of each sex.

THE HORMONAL CONTROL OF REPRODUCTIVE CYCLES

Not only do the several hormones produce characteristic effects on the reproductive system, but also these hormones are linked in action. Some of them complicate the system of control by acting reflexly on the hypophysis. The reproductive cycles, in all their essentials, can be imitated by injecting hormones into experimental mammals.

The Ovarian Cycle.   The rhythmic action of the ovary, with its sequence of maturing follicles and developing and waning corpora lutea, is believed to be due to an interplay or see-saw action of the gonadotrophic hormones of the hypophysis and the ovarian hormones (Fig. 134).[22, 23] The pituitary hormone, FSH, influences solid follicles to become hollow, and these enlarging, vesicular follicles secrete estrogen. The rising estrogen production stimulates the formation of LH and LTH by the hypophysis. FSH, now in

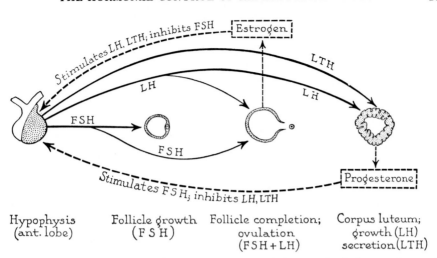

Fig. 134. Diagram of the hormonal inter-relations and control of the ovarian cycle. See full explanation in the text.

full production and aided by an increasing output of LH, brings the large follicles to the end of their growth phase. These same hormones also preside jointly over the terminal events that bring about follicle rupture. LH next transforms the empty follicle into a corpus luteum, but it is the pituitary hormone, LTH, that actually stimulates the secretion of progesterone. The release of these gonadotrophins is activated through neurosecretions arising in the hypothalamic region of the brain; they are transported to the hypophysis by a portal system of blood vessels.

The reverse phase of the see-saw then takes command. When the concentration of estrogen reaches a certain level, it restrains the output of FSH and thereby is responsible for checking its own further production. As the concentration of progesterone rises, it likewise suppresses the output of LH and LTH, and so the corpus luteum suffers regression. A renewed release of FSH to start a fresh cycle may begin when estrogen has declined to a level at which its restraining action on FSH is lost. But some believe that progesterone, building up late in the cycle, actually stimulates the release of FSH anew.

This summary, complex and not wholly agreed upon, fails to account satisfactorily for some peculiar cycles like those of the rabbit or the annual cycle of many wild mammals. A nervous factor acting on the hypophysis also plays a part, and its relative importance varies in different species. Proper amounts and proportions of the pituitary hormones are necessary to produce the typical growth, rupture and luteinization of the vesicular follicles. A proper reactiveness of the follicle is also an important factor.

The Uterine Cycle. The rhythm of the uterus is controlled directly by the ovarian hormones (Fig. 128). During the follicular phase of the ovary (estrogen formation) the proliferative growth occurs in the endometrium.

During the luteal phase of the ovary (progesterone formation) the secretory or progestational changes take place. If pregnancy does not occur, the corpus luteum declines and disappears after a relatively brief existence (for reasons already explained), and the uterine cycle repeats. If pregnancy does occur, the corpus luteum persists longer and the endometrium remains in a specialized state, but ovulation ceases. These pregnancy effects involve the action of chorionic gonadotrophin in preserving the corpus luteum for a time. Thereafter, placental estrogen and progesterone replace their ovarian counterparts.

The peculiar feature of the higher primate type of uterine cycle is the destruction, with bleeding, that brings the progressive changes to an end. Why this occurs in some primates alone and what purpose it performs are unknown, but an explanation of its cause can be given. In an anovulatory cycle the endometrium is brought only through the follicular (proliferative) stage; when the supply of estrogen then reduces to a certain level (through a reflex repression by estrogen on FSH) the endometrium is no longer able to maintain itself and menstruation occurs. In an ordinary ovulatory cycle there is no bleeding during the proliferative phase because the ovaries are supplying estrogen; there is no bleeding in the secretory stage because the corpus luteum is supplying progesterone. With the decline of the corpus luteum, however, the hormonally sustained endometrium is left without adequate support and consequently breaks down.

The Vaginal Cycle.   The proliferation and cornification of the epithelium are dependent on the rise of estrogen in the body, corresponding to the final period of rapid follicular growth. Desquamation and leucocytic infiltration occur as the estrogen declines and its effect wears off, thus returning the lining to its 'inactive' diestrous condition.

The Mammary Cycle.   Before puberty the mammary glands are in a rudimentary state of development, even the ducts being nothing more than short, little-branched sprouts (Fig. 135 *A*). During the puberal period the increasing estrogen in the blood stimulates the nipples to enlarge and the ducts to grow and branch into the tree-like system that characterizes the mature, non-pregnant mammal (*B*). At best there is only quantitative ad-

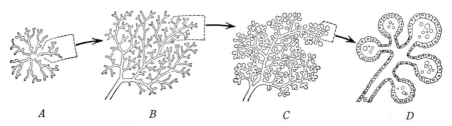

A   *B*   *C*   *D*

Fig. 135.   Development of the mammary gland (after Corner). *A*, Immature state. *B*, Effect of estrogen in bringing the duct system to the adult, virginal state. *C*, Further branching of ducts and budding of secretory end-pieces in pregnancy. *D*, Prolactin effect in causing secretion. The enclosed areas of *A*, *B*, *C* grow into stages *B*, *C*, *D*.

vance beyond this point in response to the ordinary cyclic increase in estrogen.[7, 8]

In pregnancy, however, this degree of differentiation is exceeded strikingly. Then the terminal twigs branch further and bud off secretory endpieces (Fig. 135 *C*). This response in many animals requires the action of progesterone, while in a few species estrogen stimulation alone is adequate. In still other mammals progesterone is not necessary, but apparently is a helpful aid; the human, like the monkey, may prove to be in this group. The actual secretion of milk is brought about and maintained by a hormone, *prolactin,* produced by acidophil cells in the anterior lobe of the hypophysis (*D*). Prolactin is a protein which has been isolated in pure form; it is identical with the luteotrophic hormone (LTH).

The glands are capable of secreting milk in the second half of pregnancy. The reason why the flow is withheld until after childbirth is seemingly because prolactin formation is inhibited by the large amount of estrogen and progesterone that is secreted by the placenta. Suckling sets up a neurohumeral reflex, with the release of the hormone *oxytocin* from the neurohypophysis. This results in the contraction of mammary alveoli and ducts and the consequent ejection of milk. When nursing and the suckling stimulus cease, and milk is no longer withdrawn, the secretory end-pieces undergo a degeneration and the gland returns essentially to the virginal state.

Male Periodicity. Many intergrades exist between animals with a short season of testicular activity and those in which spermatogenesis and mating are uninterrupted. Although light and temperature are factors in certain forms, the gonadotrophic hormones, FSH and ICSH, are the most direct and important regulatory agents. The fluctuations in testicular activity (tubular and interstitial) that occur in seasonal breeders are directly dependent on corresponding variations in the release of the gonadotrophic hormones. Nevertheless, the factors in control of these variations are not known, although it is certain that the gonadal hormones do inhibit the production of FSH and ICSH in mammals in general. Fluctuations in the state of the accessory sex organs and the sex urge stand in direct relation to the amounts of testosterone secreted in the periodic growth and recession of the interstitial tissue.

The Total Reproductive Period. The growth and coming into function of the gonads and accessory reproductive organs, and the development of the secondary sex characters are all phenomena of *puberty.* The several changes depend on the gaining of a requisite degree of sensitivity by the target organs to hormones that have been progressively increasing during the puberal period. But what actuates the gonadotrophic hormones of the hypophysis to increase in amount and become effective at a certain time cannot be said. It is a part of the larger plan of growth and maturing of the body, and the reproductive powers are wisely withheld until the prospective parents are relatively far along in their own developmental courses. The normal loss of the reproductive capacity comes earlier in female mammals than in males. Known as the *climacteric,* it is marked by the cessation of all cyclic, reproductive activities. The most evident indicator is the cessation of menstruation (*menopause*). Involution and atrophy of the ovaries, reproductive tract, external genitalia and mammary glands follow the withdrawal of estrogen support, and the sex desire ultimately wanes. In the male the decline of spermatogenesis and a decrease in the secretion of testicular hormone are individually variable and more gradual than the corresponding phenomena in the female. Yet there is sometimes a male climacteric that is abrupt and disturbing.

REFERENCES CITED

1. Arey, L. B. 1939. Am. J. Obst. & Gyn., *37,* 12–29.
2. Hisaw, F. L. 1961. Sect. C 9 in Young: Sex and Internal Secretions (Williams & Wilkins).
3. Bartelmez, G. W. 1957. Am. J. Obst. & Gyn., *74,* 931–955.
4. Markee, J. E. 1940. Carnegie Contr. Embr., *28,* 219–308.
5. de Allende, I. L. C., *et al.* 1945. Carnegie Contr. Embr., *31,* 1–26.
6. Papanicolaou, G. N., *et al.* 1948. The Epithelia of Woman's Reproductive Organs (Commonwealth Fund).
7. Speert, H. 1948. Carnegie Contr. Embr., *32,* 9–65.
8. Geschichte, C. F. 1948. Diseases of the Breast (Lippincott).
9. Bissonnette, T. H. 1935. J. Exp. Zoöl., *71,* 341–373.
10. Wells, L. G. 1940. Am. J. Anat., *66,* 429–447.
11. Purves, H. D. & R. O. Greep. Sect. B in Young: Sex and Internal Secretions (Williams & Wilkins).
12. Velardo, J. T. 1960. Science, *131,* 357–359.
13. Randolph, P. W., *et al.* 1959. Endocrin., *65,* 433–441.
14. Young, W. C., 1961. Sex and Internal Secretions, Sect. C 7 (Williams & Wilkins).
15. Stafford, W. T., *et al.* 1942. Anat. Rec., *83,* 193–207.
16. Dempsey, E. W. & D. L. Bassett. 1943. Endocrin., *33,* 384–401.
17. Corner, G. W. 1943. The Hormones in Human Reproduction (Princeton Univ. Press).
18. Jones, G. E. S., *et al.* 1943. Bull. Johns Hopkins Hosp., *72,* 26–38.
19. Wislocki, G. B., *et al.* 1948. Obst. & Gyn. Survey, *3,* 604–612.
20. Crandall, W. D. 1938. Anat. Rec., *72,* 195–210.
21. Pollock, N. F. 1942. Anat. Rec., *84,* 23–27.
22. Everett, J. W. 1961. Sect. C 8 in Young: Sex and Internal Secretions (Williams & Wilkins).
23. Lloyd, C. W. 1961. Chapter 7 in Young: Sex and Internal Secretions (Williams & Wilkins).
24. Heller, C. G. & G. B. Myers. 1944. J. Am. Med. Assn., *126,* 472–477.

Chapter X.   Experimental Embryology

Descriptive and comparative embryology offer no explanation as to how and why the steps in development happen when and as they do. Such information comes from experiment. This youngest and most vigorously prosecuted field of embryology is commonly called *experimental embryology;* other terms are developmental mechanics, causal embryology and analytical embryology. Its contributions to an understanding of dynamic causation are already impressive, even though much still remains unexplained.

The present account will deal with the basic, general principles currently recognized as partial explanations of mechanisms underlying development and differentiation. Part II of this book contains more specific information concerning the details of causal development as they apply to the origin and specialization of individual organs and parts. Such material follows the descriptive account of each organ or larger group and is designated by the subheading *Causal Relations.*

THE METHODS OF ATTACK

A biological experiment consists in altering some condition under which a state exists or a process proceeds and then studying the results. The object of experiments in the field of development is to reduce causation to its simplest terms. These results provide, bit by bit, the elements upon which synthesis and comprehensive understanding are based.

One line of experiment tests the dependence of the developing egg or embryo on its environment. The several physical and chemical environmental agents are called *external factors.* Their effects are naturally seen best in those organisms that are directly exposed to their influences; the embryos of viviparous animals, and especially mammals, are less subject to variations of the environment. The external factors of chief importance are the following: (1) *Mechanical,* such as pressure, gravity and centrifugal force. (2) *Physico-chemical,* such as pH and osmotic pressure. (3) *Radiational,* such as heat, ultraviolet rays and x-rays. (4) *Chemical,* through ionic effects. It is not necessary to itemize the detailed effects of these factors which have been tested by adding or subtracting them, one at a time, from the normal environment. Some are requisite to normal development; most, either in excess or deficiency, cause abnormal development of various kinds. These agents also become useful as tools in conducting experiments. For example, the centrifuge disarranges the constituents of the egg and redistributes them into layers according to their specific gravities; heat, ultraviolet rays or x-rays can be used to kill local regions of the cytoplasm; x-rays or radium emanations can destroy the nucleus within a cell; excess sodium chloride can prevent the closure of the spinal cord; calcium-free sea water causes cleavage stages to separate into their component blastomeres.

A second line of attack tests the dependent relations existing between the embryo as a whole and its parts, and the influence of component parts one on another. These interactions and influences are called *internal factors.* Their effects are tested by subtracting parts, by altering relationships through interchange or substitution, and by adding parts or wholes. The terminology and procedure are as follows: (1) *Isolation.* This is carried out by removing a part or region and allowing it to develop in its natural medium (*e.g.,* sea water) or, as an *explant,* in an aseptic, nutrient medium (*i.e.,* tissue culture). When introduced into a reasonably indifferent environment provided by another embryo (*e.g.,* belly cavity; chorio-allantoic membrane) it is called an *interplant.* (2) *Recombination.* Blastomeres can be displaced into strange positions. The grafting of an excised part into the place left by the removal of another part is *transplantation.* When the substitution or exchange is from one individual to another, the embryo supplying the transplanted part is the *donor;* the one receiving it as a graft is the *host.* (3) *Defect.* The egg or embryo, after the excision or destruction of a local region, becomes a defect experiment. An egg nucleus can be destroyed by x-rays or sucked out with a micropipette. Cytoplasmic areas can be destroyed by local pricking, heat or ultraviolet treatment. A fine hair can be employed to cut or constrict an egg or cleavage group. Tiny knives, glass needles and scissors are used for excising the parts of embryos. Some regions can be eliminated by their differential susceptibility to toxic substances. (4) *Addition.* In this category *implantation* is the addition of a part, as a supernumerary structure, to an embryo already complete in every way. *Fusions* of whole eggs or early cleavage stages can be accomplished; the surgical union of older embryos is called *parabiosis.*

Eggs and embryos of various kinds have served as experimental material, but echinoderms and amphibians remain as prime favorites. The embryos of birds can be used to some extent, whereas the development of mammals within a uterus and the unfavorable stickiness of their tissues at operation present formidable difficulties.

AN INTERPRETATION OF EARLY STAGES

Organization of the Egg.   The ripe egg exhibits a degree of organization. Fundamentally important is its *polarity,* with a main axis connecting the two poles. The animal pole possesses higher activity capacities and tends to be near the future apical or anterior end of the embryo; these capacities decrease in gradient fashion to the vegetal pole. Polarity, or axiation, is impressed from without on the ovarian egg, and the animal pole is that end of the egg which was most active in physiological exchanges during oögenesis. Also of great significance is the establishment of *bilateral symmetry.* Innumerable planes, which could divide the egg into physiological (future right and left) halves, pass through the primary axis of the egg that connects the two poles. Not all meridians are exactly equivalent and a certain one comes to possess a slight advantage over the others, owing to influences impressed on it while in the ovary. The existence and localization of such a median plane can be revealed by susceptibility experiments. This differential in favor of bilateral symmetry is not often over-ridden by factors acting at the time of fertilization and afterward, but there is evidence that the point of entry of the sperm or altered gravitational forces may shift the still labile plane of symmetry in the egg of the common frog to a new position.

The cytoplasmic cell body is not homogeneous. There is a greater concentration of pure protoplasm (building material) at the animal pole, whereas reserve materials (such as nutritive yolk) favor the vegetal pole. Such 'animalizing' and 'vegetalizing' factors are responsible for the establishment of dorsoventrality in the embryo. Moreover, the interior of an egg differs from the surface. The internal core of cytoplasm is semifluid; the peripheral shell, more gelatinous. A gel condition favors regional differentiations, and such specialized territories not only occur but even become visible in some eggs. They are characterized by pigmentation, different colloidal consistency or other features. These visible stratifications and distributions, however, are the result and not the cause of polarity. Upon their disturbance by centrifuging the cortical layer of living cytoplasm remains in place and the original axis of polarity still governs further development (Fig. 136). On the other hand, experiments indicate that definite chemical specializations do exist in local regions of the egg cytoplasm. These chemodifferentiations are shown as *morphogenetic substances*. They will be distributed unequally during cleavage and gastrulation, and will then furnish local environments that are differentially favorable for the expression of particular genic potentialities.

The Initiation of Development. The free, unfertilized egg undergoes progressive 'aging.' A demonstrable result is an increasing coarseness and aggregation of cell colloids in contrast to their previous finely dispersed state; with this congealing tendency, or gelation, goes a reduction in vigor and plasticity. *Fertilization* reverses these trends, thus rescuing the egg from impending senescence and death and also bringing about cell rejuvenation. Within limits, an over-ripe egg will still receive a sperm and develop. But as staleness advances, development is progressively poorer; malformations increase and viability of the embryo decreases.[1]

It is obvious, however, that fertilization must do more than rejuvenate;

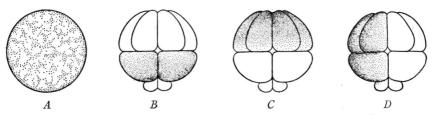

A   B   C   D

Fig. 136. Persistence of polarity in centrifuged eggs (Morgan and Spooner). *A,* Distribution of yolk and pigment granules in the normal egg of the sea urchin. *B–D,* Retention of main axis, with tiny blastomeres at the original vegetal pole, in 16-cell cleavage stages, regardless of the varied dislocation of the heavier granules and lighter clear oil and protoplasm.

it must also activate. During maturation the egg seems to have developed some kind of block that prevents it from advancing beyond a certain stage. Fertilization removes this block, but what the inciting factor may be is not clear. It seems likely that an enzyme is liberated from the sperm that unmasks some important reactive groups in the cortex of the egg.[2] An action is also exerted on the more internal cytoplasm. Among the physico-chemical changes induced are the following: the cytoplasmic colloids become more viscous and stable (especially in the cortex); there is some shrinkage from loss of fluid containing osmotically active substances; permeability increases (except for the zona pellucida); and there is a regulatory effect on the oxidation rate.

One set of chromosomes, from either parent, is adequate for development. That the female pronucleus is not essential to cleavage can be proved easily. Spermatozoa will activate eggs from which the nucleus has been removed or in which the nucleus has been made degenerate through radium treatment. Moreover, even an enucleate, cytoplasmic fragment of some eggs will receive a spermatozoön and develop into a larva. The male pronucleus is equally dispensable. A sperm, with its nucleus fatally damaged by radium, ultraviolet rays or certain chemicals (so that it is inactive, like a foreign body), is still able to enter an egg and stimulate it to develop. Even in the absence of sperms, the eggs of many invertebrates and vertebrates can be made to develop readily through chemical or other stimulation (*artificial parthenogenesis*). Adult frogs have been reared from eggs induced to develop parthenogetically by pricking with a needle; normal rabbits have been born from eggs stimulated by various artificial means.[3, 4] These several facts show that the ability to develop is a fundamental property of a ripe egg and that the actual union of the male and female pronuclei is not the significant factor that sets off development. Neither does the entering sperm supply a specific substance that is necessary for egg activation. Like the various agents that are capable of activating the egg in artificial parthenogenesis, it only releases those reactions within the egg upon which development depends. The egg, therefore, as the essential basis of the future embryo, contains all the substances and factors that are necessary for development and differentiation. It is indispensable, whereas the sperm is dispensable.

Cleavage. By means of *cleavage* the egg is subdivided into smaller building units. The pattern of cleavage depends on the position and orientation of the mitotic spindle at each division, on the rate of mitosis in different regions, and on shifts of the blastomeres after they are cut off. The physical constitution of the cytoplasm and spindle, the amount and distribution of yolk and the influence of surface tension—all these are known factors in determining pattern. Deviations from the general rules of cleavage (p. 64) doubtless have logical explanations, based on asymmetrical, protoplasmic organization or forces, and on local differences in viscosity and other qualities.

Cleavage, no matter how orderly it may be, is not a mechanism primarily designed to distribute particular qualities to the blastomeres which, carrying out irrevocable assignments, then give rise to particular parts of the embryo. The idea that specific qualities are distributed through nuclear divisions during cleavage can be disproved by experiment. For example, a half egg, made to receive but one of 16 cleavage nuclei, develops into a normal individual; hence the several nuclei cannot be qualitatively unlike (Fig. 137). Even more strikingly, when a single nucleus from the gut of a frog tadpole is substituted for the nucleus of an egg, a normal embryo can result.[5] That cleavage is not a specific device for subdividing the egg cytoplasm in such a way that the various blastomeres receive portions with rigidly different, developmental qualities follows from certain facts. For instance, in many animals one of the first two blastomeres is able to develop into a whole embryo, while later blastomeres can have their normal fates swerved to other ends. Again, although the blastomeres sometimes do receive cytoplasmic allotments of different character, it is the organization of the total cytoplasm that governs these distributions, and the cleavage pattern as well. The latter is a somewhat incidental instrument which leads to the organization of the embryo, rather than being the primary cause of its organization.

Gastrulation. Blastomere shifts, during cleavage, that go beyond the adjustments that are made in conformity with local surface tensions are akin to those mass movements that take place in the multicellular blastula during *gastrulation*. The latter rearrangements result from a 'flowing' of cell groups in which the individual cells are mere passengers. Such streaming movements during gastrulation are functions of local regions of the blastula, and can proceed even when these regions are isolated from their normal surroundings. The results of gastrulation are eminently practical, since the cells of the blastula thereby segregate at convenient levels as the germ layers. In some lower animals, gastrulation distributes cells already specialized and with closed fates. But, in vertebrates, all blastomeres lack

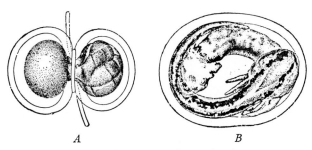

A   *B*

Fig. 137. The equivalence of cleavage nuclei in development (Spemann). × 13. *A,* Constriction of the egg of the newt had confined the nuclei (and cleavage) to the right half; at this point one nucleus was permitted to pass into the left half and the ligature was tightened to produce separation of the halves. *B,* Later identical larvæ, derived from the two halves, prove that cleavage nuclei are equal in quality and potential.

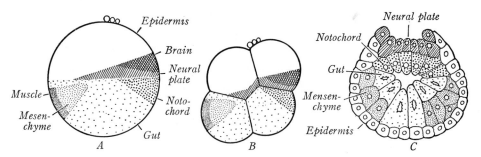

Fig. 138. Organ-forming areas in the egg of the tunicate, Styela (after Pasteels, Vanderbroek and Conklin). *A,* Map of fertilized egg, with the locations of prospective parts marked differentially. *B, C,* Eight-cell cleavage stage in side view and larva in transverse section, showing the fates of the mapped regions.

specific and irrevocable assignments as yet, and merely acquire positions that are advantageous in a future program of specialization and morphogenetic building.

Organ-forming Regions. In the normal course of development, the egg breaks up into blastomeres and these become first the germ layers, and then definite organs and parts of the embryo. Complete *cell lineages* can be followed in some lower forms with a rigid style of cleavage and few and distinctive cells. In vertebrates, it is otherwise; the blastomeres and cell groups of the blastula can be traced surely to their destinations only by marking them with vital dyes that stain the cytoplasm without injuring it. This continuity, though not exact spatial correspondence, between definite territories of early and late stages may even be presaged to a certain extent in some fertilized eggs by localized substances, different in color or texture. In these cases a distinctive area may prove to be the precursor of a specific portion of the later embryo (Fig. 138). It should be emphasized, however, that not every structural differentiation within such egg cytoplasm has functional significance for organ formation. In fact, visible granules, like pigment and yolk, are not primary factors to this end as dislocation of them by centrifuging proves (Fig. 136). Such substances, nevertheless, may be associated with local regions that in normal development do become definite parts of the embryo; in this way they serve as early 'markers' of the location of the materials for these future parts. For these reasons, and since so little is known about actual protoplasmic differentiations, it is better to avoid the term 'organ-forming substances' when designating early differentiations within the cytoplasm.

Any correspondence between specialized regions of the egg and later organs is, of course, not preformation in the gross sense of that term (p. 4). It is merely a *prelocalization* of distinctive regions whose normal (*i.e.,* presumptive) fates can be foretold. For the moment, the existence of such regions in the egg or cleavage group can be considered as having only a topographical and descriptive meaning. Whether the developmental possibilities of these regions just equal their routine performance during development, or whether they may transcend such normal performance will be considered in the discussion of 'potency' which follows.

In addition to local specializations of the cytoplasm, *gradients of activity* represent another factor involved in the formation of organ rudiments. These extend decreasingly from both the animal and the vegetal pole. Although antagonistic, they are interactive and normally are in equilibrium, thus controlling the degree of 'animalization' and 'vegetalization' of the embryo.

THE CONCEPT OF POTENCY

Potency refers to the total range of developmental possibilities that an egg, blastomere or part is capable of realizing under any imposed condition, either natural or experimental. In some animals (*e.g.,* tunicates; molluscs; annelids) cleavage follows a precise pattern (*determinate cleavage*) and each blastomere has its characteristic position and unalterable fate (Fig. 138). A cleavage group or blastula of this sort is a mosaic in which the component blastomeres have received assignments according to an inflexible plan already completed at the time of fertilization. This parceling-out process means that each blastomere at some period becomes the precursor of a definite part of the embryo. An isolated blastomere of the two-cell stage develops into a half-larva, while the destruction of even a part of a blastomere may result in a defective larva (Fig. 139). Later blastomeres have still more restricted possibilities. This is *mosaic development;* the potencies of the blastomeres just equal the fates which they achieve, and blastomeres cannot be swerved away from those fates.

In other animals (*e.g.,* vertebrates) the plan of cleavage is less rigid (*indeterminate cleavage*). Although normal development demonstrates a general relation between blastomere position and fate, still the blastomeres possess more capabilities than they ordinarily show. For example, the de-

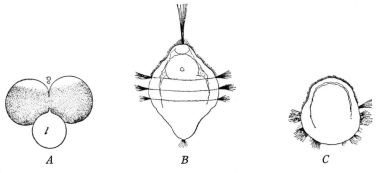

Fig. 139. Mosaic development in the mollusc, Dentalium (Wilson). *A,* Egg, with so-called polar lobe (*l*) protruded, in the first cleavage division. *B,* Normal larva. *C,* Defective larva resulting from the removal of the polar lobe at stage *A;* the apical organ and region below the main zone of cilia are lacking.

struction of a blastomere with a certain presumptive fate or its dislocation to a strange position is followed by readjustments and substitutions that produce a normal embryo. A first cleavage blastomere of a mammal, when isolated, can alter its usual destiny and develop into a perfect (but small) embryo,[6] while two fertilized eggs, made to cohere like a two-cell stage, can produce a single, giant embryo.[7] Even a right or left half of an amphibian blastula will give rise to a whole embryo. This is *regulative development;* the potencies of the blastomeres are greater than their normal performance would lead one to suspect. In comparison to the time-schedules of embryos with mosaic development, a vertebrate embryo does not acquire its general set of organ-specific districts until relatively late. In an amphibian this is at the neurula stage.

All gradations exist between highly determinate, mosaic eggs and indeterminate, regulative eggs. But in every case, development eventually attains the unalterable, mosaic state, so that the differences observed between the eggs of different animals are those that accompany an early or late loss of regulative plasticity. Even the mosaic egg of a tunicate is regulative before fertilization occurs; halves, obtained by a meridional cut and then fertilized, become complete larvæ. The existence of regulation, as shown by the production of a whole embryo from a half egg, a single blastomere, a half blastula, or from the union of two eggs, has an important implication. It furnishes absolute proof that such an embryo is not preformed in the egg but rather that it develops epigenetically (p. 5).

THE PROBLEM OF DETERMINATION

At some time during development every embryonic region loses the more extensive potentialities that it once possessed and becomes limited to a specific line of action and structural differentiation. This fixation of fate by the assumption of an irrevocable assignment for the future is known as *de-*

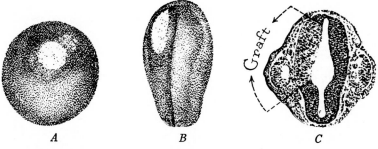

A   *B*   *C*

Fig. 140. Plasticity of parts prior to their determination (Spemann). *A,* Early gastrula of a pigmented amphibian containing, within its prospective brain region, a substitute graft from the prospective belly epidermis of a pale species (× 20). *B,* Resulting embryo, with graft, at neurula stage (× 20). *C,* Transverse section through the head of the later embryo; the transplant has been influenced by its surroundings to differentiate into an inset of pale brain tissue (× 60).

termination. It may already be settled in the fertilized egg, so that cleavage becomes a matter of parcelation, and later development is a mere program of realization (extreme mosaic eggs); or determination may only be getting into swing during gastrulation (regulative eggs, including the eggs of vertebrates). The progress that determination has made at any period is learned by testing *potencies*. For example, when areas of prospective epidermis and neural plate are exchanged at the beginning of amphibian gastrulation, each differentiates into tissue appropriate to its new site (Fig. 140). But a similar interchange, slightly later, at the neurula stage results in the neural tube containing an insert of epidermis, and the skin an island of neural tissue (Fig. 141). Hence, between the early gastrula (when even portions of germ layers are interchangeable) and the early neurula stage, developmental plasticity and the capacity of adaptation have been lost to these parts; that is to say, they have been determined.

Experiment proves that determination appears in different regions at different times; it is established gradually and always proceeds from the general to the particular. *Determination is a 'receiving of an irrevocable instruction,' whereas differentiation of form and substance is the visible carrying out of this assignment.* Only rarely among animals are early cells, even though fully determined, recognizable structurally from their neighbors. Even so, a later visible differentiation has already been presaged by an invisible, chemical differentiation within the cytoplasm of such cells. This first materializes as a unique enzymic pattern.

Following the gastrula stage of an amphibian, the embryo is a mosaic of self-differentiating regions, which are *organ-forming fields*. The developmental history of a limb of an amphibian illustrates the course of determination and its results within such a typical field: The egg and blastula gain an animal and vegetal hemisphere and bilateral symmetry. Cell groups move into position as segregated germ layers, whose cells at first are undetermined and interchangeable. Presently a diffuse, fore-limb field becomes established in the mesoderm. It gains a definite polarity, and determination expressed

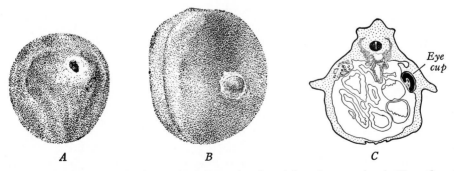

Fig. 141. Loss of plasticity after determination (after Spemann). *A,* Neurula of toad, from which the right eye-region of the brain has been removed. *B,* The portion taken from *A* has been transplanted into the flank of another neurula of equal age. *C,* Transverse section through the resulting tadpole; the transplant has become an eye cup in strange surroundings.

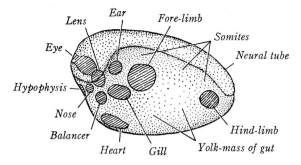

Fig. 142. Amphibian embryo, at the neurula stage, showing the locations of various organ-fields.

as an invisible chemo-differentiation. Centrally within this larger, potential limb district the actual limb bud appears as a localized mass. Although irreversibly determined as a limb (as proved by its self-differentiation after explantation into tissue-culture medium), the bud still possesses regulative ability; if it is halved, two fore limbs develop; if an extra bud is grafted in, the materials merge and make one limb. Moreover, the field retains for a time reserve capacities; if the bud is removed, the field becomes a whole once more and another limb bud can develop in it. At this stage, the limb is determined as an organ, but its cells still lack specific assignments; their determination is an affair of local subfields in which they happen to lie. Overlapping, competing subfields are established for the upper arm, forearm and hand. Mesenchymal condensations then signify the definite localization of these parts. Finally, progressive form and tissue differentiation serve as visible indices of the terminal determinations. Of similar nature is the general behavior of other fields associated with the development of the lens, ear, heart and other organs (Fig. 142).

How is determination brought about? It has already been shown (p. 171) that the cleavage pattern and qualitative differences in the cell nuclei must be excluded from consideration. Determination is the result of progressive change within the cytoplasm and has invisible chemical differentiation (directly or indirectly under the influence of genes) as its basis (p. 181). This internal chemo-differentiation sets the fates of the cells within a local region, whereupon these cells self-differentiate in an undeviating manner. But such a cell group may also emit a chemical substance that affects specifically an adjoining region so that this part becomes determined in a way it otherwise never would; in this instance differentiation is dependent on an outside stimulus. Self- and dependent differentiation are not mutually exclusive; an organ that gets its start through dependence soon acquires the power of self-differentiation and becomes independent. Critical study continues to reveal additional cases of seemingly independent differentiation to be dependent in their earlier periods; yet not all organs (*e.g.,* entodermal derivatives) get their start in a dependent way.

There is regrettable ignorance of the details through which particular differentiations are brought about. In fact, it is poorly known just when cells receive their future assignments. There is, nevertheless, good reason to believe that the ultimate control resides in genes, and that their influence is commonly exerted through the creation of enzymes (p. 181). The stages of differentiation are four: (1) generalized state (lacking any specificity); (2)

chemo-differentiation (internal and invisible); (3) physical differentiation (visible histogenesis); and (4) functional competence.

EMBRYONIC INDUCTION

When one embryonic part transmits a chemical stimulus that influences another part to produce a structure that otherwise would not come into being, then this morphogenetic effect is called an *induction*. The part exerting such an influence is an *inductor* or *organizer,* and the chemical substance emitted is an *evocator*. Induction is an important and widespread mechanism of determination; through it a control of the pattern of protein synthesis is imposed and new kinds of cells, with new behaviors, are created. It occurs especially in organs assembled as a composite from different sources and is useful in bringing about orderly development and the correct timing and fitting together of parts. The proof of an emitted inductive substance has been furnished by cultivating inductor-tissue in a nutritive drop. Characteristic responses follow when bits of appropriate tissue are introduced into this medium, which had acquired significant nucleoproteins.[8]

Local Induction. Inductive effects have been studied most thoroughly in amphibians.[9] The first part to exhibit organizer activity is the chorda-mesoderm tissue (*i.e.,* future notochord plus axial mesoderm) that rolls around the dorsal lip of the blastopore. This tissue, passing to the interior of the gastrula, underlies the dorsal ectoderm like a tongue and is in contact with it (Fig. 50). In this region of contact, the ectoderm first thickens into the neural plate and then folds into the neural tube. That it has been subject to induction can be proved by experiment: if contact between the ectoderm and chorda-mesoderm is prevented, the neural plate fails to develop; if dorsal-lip tissue is implanted under strange ectoderm, it brings about the formation of a neural plate there. This neuralizing effect, however, has further consequences, and it will be instructive to follow one particular sequence of provable inductions to its end. The swelling forebrain, reacting to the influence of the head mesoderm, produces a pair of lateral bulges (the eye vesicles) which become the stalked eye cups. Each vesicle induces the adjacent ectoderm of the head to thicken into a lens plate (Fig. 496 *A*) which then folds into a lens vesicle (*B, C*) and pinches off. The lens, in turn, causes the pigmented epidermis over it to clear and to become the corneal epithelium (Fig. 497). In this sequence of inductions there are, at least, inductors of the first, second and third order.

After the extirpation of an inductor (*e.g.,* eye vesicle), the usual response (*e.g.,* lens formation) fails. This type of experiment, however, does not show that the inductive effect is more than an arousing into action of a tissue already prepared to respond in a particular way. That the eye vesicle can actually 'instruct' adjacent ectoderm is proved by implanting a vesicle beneath the ectoderm of the belly, or by substituting belly ectoderm for the

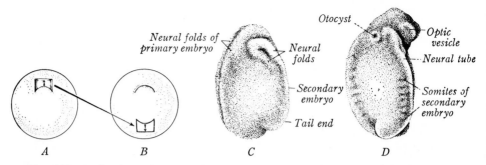

Fig. 143. Induction of a secondary embryo by an extra, primary organizer. *A, B,* Transplantation of the dorsal lip of an early gastrula of the newt to the future belly region of another gastrula. *C, D,* Stages of the resulting primary embryo, with a secondary embryo attached to it (Bautzmann; × 13).

normal lens ectoderm. In both experiments a lens can differentiate from strange ectoderm that normally would never display this activity. That the normal, presumptive lens ectoderm is not at first predisposed exclusively toward lens formation can be proved by supplying it with ear- or nose inductor, whereupon it responds by forming these parts instead of a lens. Once again it may be emphasized that the testimony of these several experiments in favor of the epigenetic mode of development, in contrast to preformation or predetermination, is unanswerable.

In order to produce an induction, the tissue of the inductor must be in virtual contact with the region stimulated. Presumably this is because the chemical substance (the *evocator*) that mediates the response passes by diffusion from cell to cell; it cannot act effectively at a distance. An inductor acquires inductive power and retains it for a limited time only. Strange tissue, grafted into an inductor, becomes imbued with the particular power then residing in that region. So does an induced early organ, which may be made to induce the formation of another organ like itself. Nevertheless, a target region of the embryo is also only capable of responding to a particular inductor for a limited time. During this reactive period, when determination can be established and the subsequent course of differentiation becomes fixed, the responsive tissue is said to possess *competence*. Although an inductor gives a cue or order, the competent tissue carries it out in its own way. For example, when body ectoderm of a frog is grafted onto the future mouth region of a newt, a mouth is induced; but it is a frog's type of mouth, with horny jaws instead of the teeth of a newt.

The Organizer. There is a higher order of induction, with more widespread results, than that already discussed. The prospective chorda-mesoderm tissue is the original focus (*organization center*) with respect to which the entire embryo integrates. This can be proved by grafting the dorsal lip of a gastrula into the belly region of another gastrula (Fig. 143 *A, B*). It sinks beneath the surface and self-differentiates into a mesodermal axis (notochord and segmentally arranged somites, together with kidney tubules

and lateral mesoderm). A neural axis is induced from out of the host ectoderm that is above the notochord and axial mesoderm (*C*), while additional
mesodermal organs may be induced from host tissue and an associated gut
may form from the host entoderm beneath. As development proceeds, an
harmoniously arranged secondary embryo arises which remains attached to
the primary embryo, derived from the host gastrula (*D*). It is this master
organizing-power (self-differentiation combined with complex inductions)
that makes the primordial chorda-mesoderm be recognized as the *primary
organizer* of the embryo. Given the opportunity and materials, it will cause
a whole embryo to be formed.

An organization center has also been discovered in some invertebrates
and in fishes and birds. The primitive streak of a bird is a modified blastopore through which the prospective chorda-mesoderm passes during gastrulation. When transplanted to another blastoderm, its cranial half induces the formation of a neural plate, and can even organize a secondary
embryo, as in amphibians. The entoderm, which precedes the primitive
streak in time of appearance, is the inductor of the primitive streak.[10] Little
is known of the mammal beyond the facts that a chick primitive streak will
induce a neural plate in a rabbit blastoderm and a rabbit primitive streak
acts similarly in a chick blastoderm.[11] These reciprocal activities illustrate
the fact that inductions in general are not species specific.

When various adult tissues (and, to a degree, parts of the blastoporic lip) are tested
as inductors, each tends to favor the formation of one of the following parts: fore- and
mid-brain (including nasal placodes and optic cups); hind-brain and ear vesicles; spino-
caudal structures; and mesodermal parts. Some tissues, killed by heat, are still inductive, but they produce only the fore- and mid-brain parts; their active neuralizing
principle is thermostable ribonucleoprotein.[12] Other tissues, when killed by heat, become inactive though normally inductors of mesodermal parts; their active mesodermalizing principle is a thermolabile protein uncoupled with nucleic acid.[13] Although
these chemical agents are widespread in the tissues of the adult body, existing there
in an inactive or bound state, both are normally present and are set free only by the
primary organizer (chorda-mesoderm plus the prechordal plate of head mesoderm) of
the embryo.[14] This plate of tissue imposes different pattern-forming influences along
its length, as is proved by appropriate transplantation experiments.[15]

An interpretation of these qualitative results on neural-tube differentiation is that
the regional instructions assigned by the living chorda-mesoderm depend on the combinations locally of different concentrations of the neuralizing and mesodermalizing
agents, already mentioned.[16] The neuralizing factor is strongest mid-dorsally; the mesodermalizing factor, caudally. Each declines progressively, in gradient fashion, as the
distance from its center increases. The varying combinations of inductive strengths on
midline ectoderm call forth appropriate regional specializations that become the serial
divisions of the brain and spinal cord. Also from this mid-dorsal strip of strongest influence on ectoderm, progressively weaker levels extend in lateroventral directions.
Here they induce successively the formation of neural crest, sense organs and epidermis.
The chorda-mesoderm itself is most potent along a dorsomedial strip where it becomes
the notochord. Laterally, on each side, a weaker influence is responsible for somite
formation, and still farther laterad and ventrad a further lowering of the gradient
level results in simple plates of lateral mesoderm.

Although the early body plan thus traces origin to the primary organizer, the course of inductions does not end with the conversion of chorda-mesoderm into definitive tissues. These products (notochord, somites and lateral mesoderm) also initiate later inductions, and still other chains of inductive processes continue to arise and act well beyond the stage of the embryo. The organizer is an important factor in producing the plan and differentiation of the body, but it is not the sole factor in achieving this end. The entoderm, for example, apparently differentiates its organs without recourse to inductors. Moreover, when various organ fields are once established, they proceed to organize and regulate themselves.

The evidence is somewhat conflicting as to whether the factors responsible for specificity in induced reactions lie within the responsive tissue or wholly without it in the inductor.[17] A particular response might result from the triggering or activating of innate powers by a nonspecific stimulus,[18] or it might be owing to the introduction into the cytoplasm of a specific morphogenetic substance. An organizer not only initiates cellular differentiation, but also it exerts control of what ensues. It has been proposed that the organizer effect is not simply differentiation as the result of the release of a series of positive inductors; rather, there is a release of a general stimulus upon which is superimposed a progressive series of specific restrictions.[19]

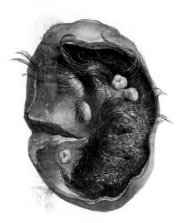

Fig. 144. Teratoma of the human ovary. $\times$ ⅖. Hair, sebum, teeth and a tiny 'tongue,' with papillæ, were visible grossly.

Anomalies. Peculiar growths, known as *teratomas,* occur in various locations, but most frequently in the ovary and sacro-coccygeal region (Fig. 144). They are tumors containing multiple tissues of kinds foreign to the region where found. An immature type with poorly differentiated tissue (*embryoma*) is highly proliferative and usually malignant. The simplest and commonest type of mature tumor is frequently called a *dermoid cyst,* yet these contain more than dermal derivatives. In fact, teratomas usually possess representatives of all three germ layers, but conspicuously absent is any orderly layering or arrangement of the tissues and parts. There is inadequate support for the common belief that teratomas represent one end of a series of gradations that extend through suppressed, included twins to unequal, attached twins.[20] Transitional stages do not exist, while all specimens interpretable as twinning lack neoplastic qualities.

The proper explanation of these chaotic assemblies seems to be that certain plastic, embryonic tissue escapes from the influence of the primary organizer during early development and somehow remains competent to respond to a lower-grade inductive influence that acts at an abnormal time and place. The chief deficiency is the capacity to organize as a whole; histogenesis is satisfactory, whereas morphogenesis is faulty and axiation and metamerism are lacking. Some experimental inductions have been performed on embryos that offer instructive parallels.[21]

OTHER DIRECTING FACTORS

The Rôle of Genes. The cytoplasm provides the immediate material for bringing about development, whereas the genes act as directing and controlling agents. At least one chromosome of each kind must be present in embryonic cells if continued development is to take place. It is probable also that at least one gene of each kind is necessary for normal development, and that every category of development is primarily instigated and controlled by genes. The specializations within the cytoplasm of the egg, whereby it becomes regionally different while still resident in the ovary, are controlled by maternal genes only. It is in this way that the cleavage cells acquire cytoplasms with diverse qualities. An influence of paternal genes may become manifest, shortly after fertilization, in early cleavage and morula stages. Since the same complete assortment of chromosomes and genes is present in every cell of the embryo, it is plain that these elements, by themselves, are incapable of initiating the processes of morphogenesis, differentiation and growth. Hence a gene can produce differential effects only when it comes to lie in a cytoplasm with which it can interact; the opportunity for this reciprocal relationship is already operative when cleavage subdivides the egg into blastomeres with cytoplasms of different qualities. Each newly created cytoplasmic diversity sets up a fresh series of reactions between the genes and the local cytoplasmic substrate, and so differentiation unfolds progressively.

Most genes produce multiple effects in development, and several to many genes may co-operate in producing one final effect. Although the primary genic action is possibly always unitary, this is followed by secondary, multiple results. In the stages of development when details are being worked out, such mechanisms are complex. For example, wing differentiation in the fruit fly has been analyzed into 16 separate processes, under the control of some 40 genes. There is a belief that each gene controls a single biochemical reaction, and it is postulated that this is accomplished by the creation of a single cellular enzyme.[22] These agents act by controlling the processes of cellular metabolism (including the synthesis of proteins and other biological compounds) and by affecting the rates at which reactions go on in the embryo. Catalysts, inhibitors, inductors, antigens and hormones are produced or restrained; are masked or unmasked. Not all agree that genes have a single primary activity, that every gene influences the production of a corresponding enzyme in one step, or that all genes are related to enzyme production. By contrast, others contend that "differentiation is essentially the production of unique enzyme patterns."

Among the fundamental influences exerted by genes are their effects on organizers, on the processes that bring tissues into a competent state and on the fields where differentiation and pattern are being worked out. The reciprocal reactions between genes and conditioned cytoplasm are interlocked and self compensating. There is evidence that some genes act only at certain stages of development or in certain cell types. Yet many basic features of the reciprocal processes remain unexplained. How are proper cytoplasmic substrates prepared? How does the cytoplasm influence the nucleus so as to stimulate or inhibit genic activity? How do genes impress their code on synthesizing cytoplasmic molecules? How do genes control a proper timing (and time order) in the steps of differentiation? How are these steps of determination arranged in space so as to provide a correct developmental pattern? It is desirable for experimental embryology and genetics to join forces, since in the quest for such answers they now have a common goal. There is certainty that the unrolling of development rests fundamentally upon a genic basis and that the solution of the mechanisms involved not only will illuminate the mysteries of development but also will go far toward solving the riddle of life itself.

Hormonal and Nervous Influences. *Hormones* play a rôle in development, and especially in its later phases. They, however, are not primary, creative factors of development. On the contrary, when certain parts of the fetus have arrived at a proper state of differentiation (including hormone sensitivity) they merely react to the chemical stimulation supplied by the hormone substance. Still different is *nervous excitation,* which is not a factor in the differentiation of tissues and organs. For example, a muscle or even a whole limb can develop fully in the complete absence of nerve. Through its inductive power, however, the nervous system does influence the development of the special sense organs and cartilaginous spine.

Gradients. Decreasing intensities of physiological activity extend away from primary centers located in the egg, blastula, embryo and embryonic fields. These gradients may take a course that lies in a craniocaudal, dorsolateral or mediolateral direction. The theory that such quantitative expressions of metabolic activity are the fundamental instruments responsible for determination, pattern and general organization[23] has not withstood criticism well. Axial gradients do indicate relative activities, and such represent a factor to be counted on in development. But these differences (in oxidative values, for instance, as proponents advocate) fall short of furnishing a satisfactory basis for an inclusive explanation of the intricacies of development.

REFERENCES CITED

1. Blandau, R. J. & W. C. Young. 1931. Am. J. Anat., *64,* 303–329.
2. Monroy, A. 1950. Sci. Am., *183,* 46–49.
3. Pincus, G. 1939. J. Exp. Zoöl., *82,* 85–129.
4. Pincus, G. & H. Shapiro. 1940. Proc. Am. Phil. Soc., *83,* 631–647.
5. Gurdon, J. B. 1962. Devel. Biol., *4,* 256–273.
6. Seidel, F. 1960. Archiv f. Entwickl. d. Organ., *152,* 43–130.
7. Nicholas, J. S. & B. V. Hall. 1942. J. Exp. Zoöl., *90,* 441–460.
8. Niu, M. C. 1956. Chapter 7, in Rudnick: Cellular Mechanisms, etc. (Princeton University Press).

9. Holtfreter, J. & V. Hamburger, 1955. Sect. VI, Chapter 1, in Willier: Analysis of Development (Saunders).
10. Waddington, C. H. 1933. Arch. Entw.-mech. d. Organ., *128*, 502–522.
11. Waddington, C. H. 1937. Arch. de Biol., *48*, 273–290.
12. Hayashi, Y. 1956. Embryologia, *3*, 57–67.
13. Yamada, T. 1958. Experientia, *14*, 81–87.
14. Takaya, H. 1955. Proc. Imp. Acad. Japan, *31*, 366–371.
15. Mangold, O. 1933. Naturwiss., *21*, 761–766.
16. Toivonen, S. & L. Saxen. 1955. Ann. Acad. Sci. Fenn., Ser. A, *30*, 1–29.
17. Holtfreter, J. 1951. Growth, *10*, 117–152.
18. Eakin, R. M. 1947. Science, *109*, 195–197.
19. Rose, S. M. 1952. Am. Natur., *86*, 337–354.
20. Willis, R. S. 1935; 1937. J. Path. & Bact., *40*, 1–30; *45*, 49–65.
21. Holtfreter, J. 1934. Arch. Entw.-mech. d. Organ., *132*, 307–383.
22. Horowitz, N. H. 1950. Advances in Genetics, *3*, 33–71.
23. Child, C. M. 1941. Patterns and Problems in Development (Univ. of Chicago Press).

Chapter XI.   Teratology

The subscience of *teratology* is the branch of embryology that deals with all features of abnormal development, including the structure of its end products. *Teratogenesis* is narrower in scope; it is restricted to considerations of the ways and means by which abnormalities arise. It is well recognized that every individual displays some departures from the average in the size, arrangement or composition of his component organs and parts. Especially is this individuality increasingly notable in minor details. There are no rigid boundaries to the extent of 'normal variation,' and any experienced observer is free to set his own standards of normality. When, however, an organ or organism clearly oversteps the reasonable limits in any range of variation, then the condition is known as an *abnormality, anomaly or malformation*. If a fetus differs so markedly from the normal that it is grotesque, and usually nonviable, then the term *monster* or *monstrosity* is sometimes applied to it in a popular, unprecise way. A series of intergrades may connect the normal, slightly abnormal and severely abnormal.

Since the roughing out of the human body and its parts occurs largely in the early weeks of development, almost all physical anomalies date from that time. Yet some functional changes occur at birth (*e.g.*, shifts of circulation) and the development of certain other parts takes place after birth (*e.g.*, permanent teeth; bony epiphyses). These events may also deviate from the normal developmental course. Whereas the most obvious anomalies are visible aberrations of structure, there are others in which enzyme deficiency blocks the course of intermediary metabolism and results in abnormal chemical functioning. Such 'inborn errors of metabolism' involve proteins, carbohydrates, lipids and pigments. The abnormal products may be stored or excreted.

Contrasted against *congenital anomalies*, existing at birth and, indeed, mostly long before birth, are those *acquired defects* that appear secondarily in well-formed parts as the aftermath of mechanical weakness or disease. The latter are traumatic accidents that do not qualify in a discussion of teratology.

Incidence.   One infant in fourteen that survives the neonatal period bears a malformation of some kind and degree, whereas those dying earlier have a much higher incidence.[1, 2] Two-thirds of the malformed still-births show more than one anomaly; this is nearly five times the frequency of

184

multiple defects in malformed live-born. One child in forty is born with a structural defect that demands treatment. Problems that require more exact study include: the chances of the recurrence of the same anomaly in later babies of the same parents (probably not high, if nongenetic in nature); the chances of any anomaly occurring in subsequent babies (somewhat increased); the effect of advancing maternal age (a factor in some types of anomaly, *e.g.,* mongolism); and the increased incidence in later births (expectation increases after the fifth or sixth child, but is not yet dissociated from increased maternal age).

Statistics on incidence vary with races and even in different populations of the same race. They may be taken as approximations, subject to further refinements. Examples of anomalies that are commoner in live-born males than females are: pyloric stenosis (4:1); cleft lip and palate (5:3); and hydrocephalus (5:4). Other types are more frequent in females than males: dislocated hip (7:1); birthmarks (2:1); and brain and cord (6:5).

The incidence-range of specific malformations, expressed as ratios in relation to the general population, is illustrated in the following examples: communicating atria of the heart (1:5); diverticulum of the ileum (1:50); horseshoe kidney (1:600); imperforate anus (1:5000); hemophilia (1:50,000); and absence of appendages (1:500,000). The incidence of the very rare anomalies (such as double nose, fused ears, double penis, and true hermaphroditism) cannot be stated reliably; some run into one in millions and, possibly, billions. Two-thirds of all defects start in early embryonic weeks.

The frequency with respect to organ systems is given as follows: musculo-skeletal (38%); integumentary (20%); central nervous (14%); cardio-vascular (9%); gastro-intestinal (9%); genito-urinary (9%); and respiratory (2%).

Classification. The total group of physical anomalies can be assembled into categories that denote the ways in which imperfect development departs from normality:

1. DEVELOPMENTAL FAILURE (AGENESIS). (*a*) *Organic:* limb; kidney. (*b*) *Cellular:* thyroid cells; ganglion cells of colon.

2. INCOMPLETE DEVELOPMENT. (*a*) *Growth:* dwarfism; infantile uterus. (*b*) *Union:* cleft palate; double uterus; lobed spleen; nonfixation of mesenteries. (*c*) *Subdivision:* of heart chambers; of body cavities. (*d*) *Migration:* undescended testis; unascended kidney. (*e*) *Metabolic processes:* sickle cell anemia; alkapton excretion; pigment lack.

3. DEVELOPMENTAL EXCESS. (*a*) *Overgrowth:* gigantism; cystic kidney. (*b*) *Exaggerated histogenesis:* thick epidermis; general hair coat. (*c*) *Increased numbers:* digits; nipples; twins. (*d*) *Union:* kidneys; fingers; obliterated sigmoid mesocolon. (*e*) *Subdivision:* cleft ureter; double gall bladder.

4. EMBRYONIC SURVIVAL. *Examples:* anal membrane; Meckel's diverticulum; thyro-glossal cyst; double vena cava.

5. MISPLACEMENT (BY ABNORMAL TRANSPORT OR SITE OF ORIGIN). *Examples:* transpositions; aberrant parathyroid; palatine teeth.

6. ATYPICAL DIFFERENTIATION. (*a*) *Tissue substitution:* ligaments from precartilage. (*b*) *Incorrect histogenesis:* achondroplasia; osteogenesis imperfecta. (*c*) *Congenital tumors:* blastomas; teratomas.

7. ATAVISM (P. 8).   *Examples:* azygos lobe of lung (as in quadrupeds); elevator muscle of clavicle (as in climbing primates).

Another type of classification can be based on abnormal functional factors, operating either locally or affecting the organism as a whole. The abnormal nature of each factor may involve its absence, subnormality, excess or perversion. The categories in which these grades of deviation occur are the following: *(1) stimulus; (2) tissue responsiveness; (3) growth capacity; (4) degenerative tendency; (5) enzymic function;* and *(6) hormonal function.*

TERATOGENESIS

Causative Agents.   In ancient times monstrous development was attributed to supernatural agencies, and in later centuries other irrational and superstitious views were still maintained. Even in modern times there remains the world-wide superstition that a pregnant mother may 'mark' her unborn babe for good (by a planned cultural regimen), or for ill (by experiencing fright, dietary upsets or bodily injury). Early in the present century, before direct experiments on teratogenesis became common, such factors as mechanical pressure, disease and faulty implantation were still in high favor as potent causative agents.

In any consideration of causative factors it is necessary to draw a distinction between the genetic constitution (*i.e.,* qualities of the genes) of a given individual and his somatic appearance (which represents the sum total of characters derived from genic action). For the genetic constitution, the term *genotype* is employed; for the 'external' appearance the term *phenotype* is used. The genotype is fixed at the time of fertilization and its phenotype results from the interaction of genes and environment. Since, however, the environment can be varied, the phenotype is also potentially capable of variation. Altered physical or chemical conditions during development may result in a change in form or function similar to that produced ordinarily by a different genetic inheritance. Such an imitative product is known as a *phenocopy.* Since its modified appearance is purely somatic, a phenocopy does not transmit these acquired alterations to the next generation.

No longer can partisans urge the sole efficacy of intrinsic, hereditary mechanisms or of extrinsic, environmental influences in producing anomalous development. Either can play a primary rôle as a causative agent, and both may co-operate in producing a common result. It has been suggested that 20 per cent of all congenital anomalies are genetic in origin, 10 per cent are caused by abnormal distributions of chromosomes and 10 per cent may spring from viral infections early in pregnancy, whereas the remaining 60 per cent are without a known cause at the present time.

A. HEREDITARY FACTORS.   By mutation, genes can produce hereditary malformations.[3] For example, human spontaneous mutation, responsible for hemophilia, occurs once in 30,000 births. Abnormally changed genes may come to lie at corresponding places in both chromosomes of a pair. In this *homozygous* state they may be either *dominant* or *recessive*. Or an abnormal gene may be matched with a normal gene (*heterozygous* state). In the latter event either the normal or abnormal gene may dominate over the other. Mutations that occur in nature are of undetermined origin and are spoken of as 'spontaneous.'

Some abnormal conditions are controlled by a dominant gene, and are transmitted from one generation to the next wherever the abnormal gene occurs. Examples are fusion of fingers and congenital cataract. When both mutant dominants are present in the same individual the effect is often intensified, and may be lethal. Other abnormalities are governed by recessive genes, and only come into being when two recessive genes have joined forces in a fertilized egg. Among these abnormal results are congenital deafness and albinism; some expressions of the homozygous, recessive condition are lethal. Certain recessive genes lie in an X-chromosome, in which instance daughters act as carriers of the abnormality and only their sons are afflicted. Well-known examples are hemophilia and red-green color blindness.

Mutations also have been induced experimentally in the sex cells of animals (including small mammals) and plants through exposure to radiation.[4] It is believed that similar effects in man may have accompanied the bombings in Japan, although clear-cut proofs of an influence on later progeny are meager.

There is still another means of abnormal development chargeable to chromosomes. Certain characteristic derangements in man are now known to be associated with abnormal assortments of chromosomes distributed to cells during meiotic divisions. So-called *mongolism,* which includes retarded mentality among its characteristics, is the result of a mutation in which a certain somatic chromosome occurs in triplicate, producing a set of 47 instead of the customary 46. The *Turner syndrome* designates a genetic female with a sterile ovary and imperfect feminine habitus; her chromosome count is 45, owing to the presence of but a single X-chromosome. *Klinefelter's syndrome* depicts a genetic male with small, subnormal testes and substandard secondary sex characters; his chromosome count is 47, owing to a combination of sex chromosomes represented by XXY. Other atypical assortments of somatic and sex chromosomes are known.

B. Environmental Factors. Various influences can affect the developing embryo (phenotype) adversely, thereby producing phenocopies of developmental errors that ordinarily are products of the hereditary mechanism.[5] Such causative factors are called *teratogenetic agents* (Fig. 145).

1. *Physical Agents.* (a) Mechanical. Unlike lower animals that are directly susceptible to injury, placental mammals are well protected by the uterus and amniotic fluid from mechanical insult. The once-emphasized amniotic folds and bands have lost favor as factors responsible for intra-

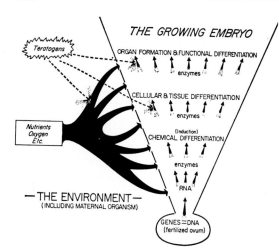

Fig. 145. Diagram illustrating the continuing dependence of an embryo on its environment, and also that the production of a specific malformation by teratogenetic agents coming from the environment is limited to periods of visible, cellular differentiation (Wilson).

uterine amputations and other injuries (p. 110). It is, however, conceivable that too little amniotic fluid in the later months of pregnancy might set up conditions leading to the distortion of so-called compression babies. But, in general, mechanical pressure and kindred forces are no longer considered to be significant factors in teratogenesis.

(b) Irradiation. The mammalian embryo *in utero* is influenced adversely by x-rays applied in appropriate dosage. The primary effect is on mitosis. The secondary, and obvious, effect is the resulting development of malformed parts. As with induced malformations by other agencies, the kind of result obtained depends on the time of administration with respect to the precise state of differentiation existing in various susceptible regions of the embryo. Human embryos are similarly susceptible, and thorough investigation has been made of defects occurring in those born after the atomic bombings in Japan.

2. *Chemical Agents.* (a) Hormones. Deficiencies of the fetal and postnatal endocrine glands produce well-known clinical entities such as cretinism (thyroid), dwarfism (hypophysis) and false hermaphroditism (gonads). The several hormones, administered in excess to the mother, act variously as teratogenetic agents.

(b) Other Chemicals. Most used experimentally is the dye, trypan blue, injected into pregnant mothers. Salts of certain heavy metals and certain drugs (*e.g.*, thalidomide) are among the proved teratogens.

3. *Metabolic Agents.* (a) Vitamins. The absence of any one of several vitamins from the maternal diet produces a deficiency that may result in abnormalities.

(b) Growth Inhibitors. Certain foreign chemicals (nitrogen mustard, urethane, etc.) when used in experiments, act in one way or another to interfere with the growth process, thereby inducing malformations.

4. *Infectious Agents.* In man, the virus of German measles is the sole virus known to be a competent teratogen; to be effective, the disease must

be incurred early in pregnancy. Other viruses have been used successfully in inducing maldevelopment of the chick, pig and guinea pig. A protozoan parasite (Toxoplasma) produces an infectious process in the human mother that induces malformations of the fetus.

5. *Immunity Agents.* Reliable experiments are limited, but the Rh positive human fetus furnishes a natural one. The maternal antigen that is induced by it passes the placenta, reaches the fetus and interferes with the successful termination of its erythropoietic process.

Generalizations. An intensified experimental attack on teratogenesis in recent decades has been fruitful to the extent that certain basic generalizations may be stated with considerable confidence, even though much still remains to be checked and interpreted. Pioneering conclusions, advanced decades ago,[6] have recently been refined and extended as follows:[5]

1. When the dosage of an effective teratogenetic agent reaches a critical threshold-level, there is a response by the embryo that entails irreparable changes leading to maldevelopment, or even death.

2. Each agent seems to act by exerting a specific influence on a definite aspect of cellular metabolism. The pattern of responses aroused by any agent is characteristic, although not necessarily unique to that agent. If different defects are evoked by different agents acting at corresponding times, it may then be inferred that their actions are exerted on wholly different phases of embryonic metabolism.

3. It follows that the same abnormality may be induced by different kinds of agents acting at the time of a particular phase of differentiation. Contrariwise, the same agent can induce different kinds of abnormalities when acting at different phases in embryonic differentiation.

4. Susceptibility to maldevelopment depends upon the exact stage of differentiation taking place within the tissue of a particular region, rather than upon the stage of development of the embryo as a whole.

5. Each organ and organ system has its characteristic susceptible period. This portion of time occurs early in the course of visible, cellular differentiation—*i.e.,* subsequent to invisible chemo-differentiation (Fig. 145). After a brief duration susceptibility declines as organogenesis advances, and becomes negligible in fetal stages.

Mechanisms:

MUTATION. The fifth part of all malformations that arose as mutations, and operates on an hereditary basis, follows the laws of genetics. A wide variety of these congenital conditions are transmitted in regular mendelian fashion. Whatever biological mechanism the genes control, be they unitary enzymic creations or not, this mechanism the abnormal genes disrupt in some way. A minimal physical derangement, with disproportionate functional effects, is illustrated by sickle-cell anemia in which there is a single amino-acid substitution among the 557 amino-acids that make up the peptide chains of the huge, hemoglobin molecule.

HEREDITARY CONTROL. The genetic situation exhibits various complications that tend to obscure ordinary mendelian principles. For example, essentially the same series of effects may be produced in a given species by wholly independent dominant or recessive mutations. Again, both the incidence of affected individuals and the intensity of the effect may be altered by both genetic and environmental factors. Still further, environmental factors may alter the expression of a given mutant. For these reasons there are many instances in which the mode of transmission of a suspected hereditary character is uncertain.

CHROMOSOMAL ABERRATIONS. Unusual assortments of chromosomes result from mishaps during meiotic divisions. Such abnormal behavior includes nondisjunction, translocation, inversion, duplication and deletion.

ENVIRONMENTAL AGENTS. Experimentally induced malformations, brought forth by environmental means, are not entirely free from suspicion of genic involvement. A growing sentiment inclines to the view that these phenocopies assume their forms because of an underlying genic instability.[7] Some even advocate that a majority of congenital defects arises from a combination of circumstances involving the interaction of a number of genetic and environmental factors.[8]

For many years the preferred explanation of the manner of origin of malformations rested on the hypothesis of properly timed developmental arrests.[6] The effect of an environmental agent was said to bring about a temporary retardation or arrest at a critical, susceptible moment in the development of an organ or part; following this inhibition there was a failure of proper recovery and regulation. With the amassing of new detailed analyses, most specialists in this field have abandoned the concept of arrest *per se* as the prime environmental factor in producing abnormal development. Yet the general concept of the differential susceptibility of a part at a particular critical period in a standardized schedule of organ advances remains as an important principle. Other parts may or may not be so scheduled as to be sensitive, at that particular moment, to competent disruptive influences. A part that is sensitive and susceptible at one particular time is immune to the same influence at earlier or later periods of development.

Seemingly there are several ways in which teratogenetic agents can affect susceptible embryonic cells. Enzymes may be blocked or destroyed with consequent crippling or destruction of cells. Some chemicals may substitute in a chain of reactions within cells, but these are incompetent to act so as to meet the metabolic needs of the cells. Other chemicals act by altering certain radicals on the nucleoproteins in cell nuclei; this especially damages cells with high mitotic rates. Whatever the manner of interference may be, the final result is probably either cell impairment or death, or a changed rate of growth. In the first instance a cellular deficiency occurs in the definitive organ or part. As to growth, this activity can be affected by halting it for a time, by slowing it or by accelerating it. Any one of these measures puts local growth out of step with adjoining parts and upsets the co-ordinated schedule of development. Continued, unrestrained growth is a condition leading to the development of congenital tumors.

(The array of anomalous organs and parts will be described and illustrated in subsequent chapters, in conjunction with the serial presentations of normal development.)

TWINNING

Parts of an embryo may show subdivision or *duplication* other than normal. When the replication involves the embryonic axis, and some or all of the embryo proper repeats itself, then the process is commonly called *twinning*. Actually, these words are also used loosely to designate similar repetitions in triplicate, quadruplicate and so on.

Duplications. Supernumerary parts are not an uncommon result of atypical development. Although they represent an abnormal occurrence, these repetitions may approach or achieve structural normality. The basis for duplication resides in the organ- and field primordia, which possess plasticity and a high regulative capacity for a time. A portion of such a primordium contains all the factors necessary to the formation of a whole and, given the opportunity, tends to produce a whole. For example, the heart of an amphibian arises as two plates that normally meet in the mid-plane to form a tube. But if the fusion of these bilateral primordia is prevented, each forms a separate, complete heart (Fig. 146 *A*). On the other hand, a single primordium can subdivide; when a local area that would produce a limb bud is divided, and the halves are prevented from reuniting, two perfect limbs are obtained (*B*). Subdivision of an emerging organ field or the transplantation of a sample area may prove competent even before the field is recognizable as such. Thus a transplant from the early eye field may form an eye in addition to the one produced from the undisturbed, residual tissue.

Experimental Twinning. Organization into an embryo requires the presence of the primary organizer. The formation of separate embryos from isolated, early blastomeres of regulative eggs depends on whether or not these blastomeres contain (or are able to differentiate) organizer substance. In the sea urchin any one of the first four blastomeres includes a sample of the egg from pole to pole and will form a perfect, but small, larva. Subsequent cleavage restricts the organizer material to the vegetal blastomeres, and especially to tiny 'micromeres' at the vegetal pole (Fig. 136); even blas-

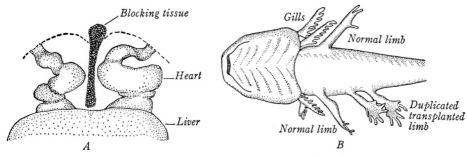

Fig. 146. *A,* Duplication of the heart in a frog whose bilateral, cardiac primordia were prevented from fusing medially (after Ekman). *B,* Duplication of a limb in a salamander; the primordium was split and transplanted to the flank of a host (after Swett).

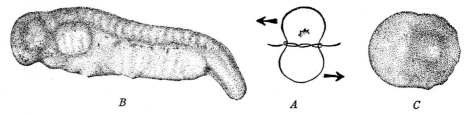

Fig. 147. Dependence of organization on the primary organizer (Spemann). *A,* Gastrula of newt, about to be separated into halves by a constricting hair. *B,* Well-proportioned embryo derived from dorsal half containing the organization center. *C,* 'Twin,' without exterior differentiation, derived from ventral half.

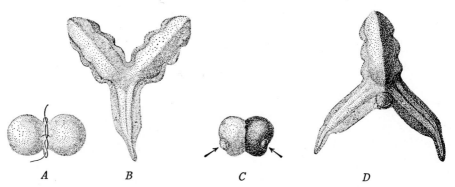

Fig. 148. Joined twins, produced experimentally in the newt. *A, B,* Anterior duplication, resulting from a constriction that causes the primary organizer to split into two tongues. *C, D,* Posterior duplication, resulting from the fusion of parts of two gastrulæ (pale and pigmented) whose normal, main axes are indicated by arrows (after Spemann).

tula-halves that contain both animal and vegetal cells will regulate into complete larvæ. In amphibians, the site of the primary organizer is indicated by a less pigmented territory that appears above the equator of the egg just after fertilization. If the first cleavage furrow transects this *gray crescent,* each of the resulting blastomeres, when isolated, can form an embryo; otherwise the blastomere containing the crescent is the only one that so develops. It follows that only when a blastula or early gastrula is halved in such a manner that each part contains some of the primary organizer (presently the dorsal blastoporic lip) will twin embryos result (Fig. 147).

A double 'monster' forms when the primary organizer of an amphibian is made to separate partially into two chorda-mesodermal tongues (Figs. 148 *A, B*). A similar effect can also be produced by grafting together two half gastrulæ, each containing its blastopore (*C, D*). Depending on the angle the two dorsal lips make with the new main axis, such embryos can be made to have two heads or two tails. A most interesting cruciate (cross-shaped) type results when the dorsal lips face each other directly. The two developing axes then meet head on. Since no further advance forward can be made, each axis splits and the halves move sideways, as follows: − −, ⋖⋗, ⊣⊢.

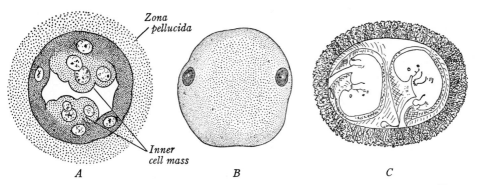

Fig. 149. Identical twin embryos, produced from two inner cell masses. *A,* Hypothetical stage of a sectioned mammalian blastocyst. *B,* Total blastocyst of a sheep, with two embryonic discs (after Assheton; × 15). *C,* Opened chorionic sac, with human embryos of seven weeks (× 1).

The result is a double embryo with fused heads, two faces and partly separate bodies (*cf.* Fig. 153 *A*). But each face is a joint product, the left half being supplied by one individual and the right half by the other; and the midplane of the heads is at right angles to that of the facing bodies. Other types of crossed doubling are produced when a fertilized egg of a frog is inverted during the first cleavage. The heavier yolk then settles, and traces of it interfere with gastrulation.

Twinning can be enhanced in fishes, amphibians and birds by depressing certain environmental factors (temperature, oxygen, etc.) at the time of gastrulation. This tends to abolish the supremacy of the original axis, and leads to the development of other independent or conjoined axes. In mammals an attack on experimental twinning was accomplished by separating the first two blastomeres of the rat and returning them to the uterus. A few of these somewhat traumatized cells developed into early embryos before succumbing.[9] When one blastomere of the two-cell stage of a rabbit was punctured, the unharmed blastomere in two instances developed into living young.[10]

Spontaneous Twinning. The possibility of natural forces separating the early blastomeres of some aquatic animals, and their subsequent development into separate organisms, must be admitted. In vertebrates in general, however, the protective rôle of the egg membranes makes this method of twinning less credible, while the yolk-rich eggs of higher fishes, reptiles, birds and monotreme mammals present an insuperable obstacle to such a procedure.

In true mammals several possible methods of single-egg twinning must be considered:

DIVISION OF CLEAVAGE MASS. It is commonly stated that separation at the two-cell stage is accountable for a fair number of the twins that derive from one egg. A practical obstacle, however, is the thick and tough zona pellucida, which ordinarily does not disappear until the early blastocyst

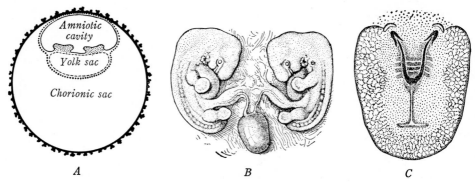

Fig. 150. Identical twin embryos, produced from two embryonic axes on the same embryonic disc. *A,* Hypothetical stage of a sectioned mammalian blastocyst. *B,* Human embryos of six weeks, attached to the same yolk sac by individual yolk stalks ($\times$ 2.3). *C,* Duplicate budding from the primitive axis of a chick embryo ($\times$ 8).

consists of many cells. Of course, under special circumstances it might release the cleaving egg considerably earlier. An alternative theory of total division is that an early blastocyst, at the time when it becomes free and is segregating an inner cell mass from future trophoblast (Fig. 44 *C, D*), divides in such a way that the halves receive portions of both components.

DIVISION OF INNER CELL MASS.   Normally the mass of cells that will form the embryo segregates at one pole of the hollow blastocyst (Fig. 45 *D*). If, during this period, some accident of development establishes two inner cell masses, there is then set up the basis for twin development within the same chorionic sac (Fig. 149).

DUPLICATION OF EMBRYONIC AXIS.   A third method, later in time of origin, occurs when two primitive streaks (and then two embryos) arise on the same embryonic disc as the result of double gastrulation. (Fig. 150 *A, B*).

SUBDIVISION OF EMBRYONIC AXIS.   Observations on lower vertebrates and on the Texas armadillo support the conclusion that a single organization center within an embryonic disc may subdivide by fission or budding (Fig. 150 *C*). Conjoined human twins represent such incomplete subdivision of an embryonic axis (Fig. 153).

DUPLICATION AND SUBDIVISION.   The Texas armadillo, which gives birth to quadruplets regularly, first produces two embryonic axes, and each then buds off another. The Dionne quintuplets are believed to have followed a similar sequence, but with two budding episodes.

Human Twinning.   Man and some other large mammals habitually bear only a single offspring at a time. The frequency of multiple births to total births varies markedly in different countries and races (high in Negroes; low in Japanese). In the United States, a twin pair occurs once in 87 total births. Although only an approximation,[11] triplets have a frequency of $1:(87)^2$ births and quadruplets $1:(87)^3$. Six appears to be the maximum number of spontaneous births that is well authenticated.

A distinction must be drawn between false and true twins. The simultaneous birth of two or more human babies is most commonly owing to the development of a corresponding number of eggs which were discharged from separate follicles at approximately the same time, became fertilized by different sperms and implanted individually in the uterus (Fig. 151 *A*). Such unlike or *fraternal twins*, triplets, etc., are contained within individual chorionic sacs which maintain their separate identity even though, like their placentas (Fig. 114 *C*), they may come into close apposition. The technical designation is *dizygotic twins*. The individual members may be of the same sex or not, as chance happens; they have only the general degree of family resemblance as occurs in brothers and sisters of different ages. Properly speaking they are not twins at all, but merely litter-mates.

Quite different are the true, or *identical twins*, triplets, etc., which are always of the same sex and so strikingly similar in physical, functional and mental traits that only rarely is their diagnosis at all difficult. Doubtful cases are resolved by comparison of such details as blood types, finger prints, eye color, etc. This close duplication is enforced by a derivation from a single fertilized egg (hence *monozygotic twins*), whereby each member acquires the same chromosomal constitution and half of the cytoplasm. Because they develop from a single blastocyst, most human identical twins are contained within a common chorionic sac and have a common placenta (Fig. 151 *B*). The umbilical cords, however, are separate and the amnions are also individual except in rare instances. Accumulated data indicate that one-fourth of all twins possessing individual chorionic sacs trace origin to a single egg. This can result only when the original embryonic mass subdivides before implantation occurs.[12] Hence a single chorionic sac is a useful criterion in assigning twins to the identical category, but it is not an inclusive indicator for all identical twins.

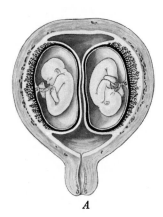 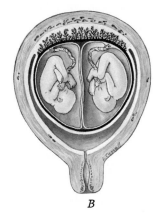

       A   *B*

Fig. 151. Condition of the fetal membranes in the two types of human twins (after Bumm). *A*, Ordinary, double-egg twins with individual chorions and placentas. *B*, True or single-egg twins, with a single chorion and placenta.

Cut surfaces: deciduæ, thick white; amnions, thin white; chorions, black.

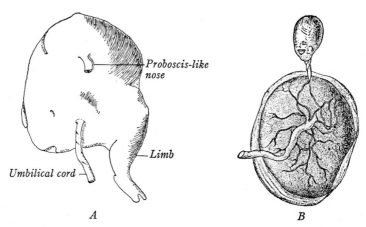

Proboscis-like
nose

Limb

Umbilical cord

A   B

Fig. 152. Acardiac human fetuses, sharing a placenta with a full-term, normal twin.
A, Nearly amorphous specimen, but with some organization. × ¼. *B,* Specimen pictured
in the older literature (after Barkow). × ⅙.

Triplets or higher degrees of multiple births can consist wholly of identical individuals, wholly of fraternal individuals, or any combination of these
two types. The ratio of monozygotic twin births to total births is nearly
constant the world over, whereas the frequency of twinning by multiple
ovulation varies with country and race. Thus, nearly three-fourths of all
American twins are of the two-egg type; by contrast, the Japanese figure is
little more than one-fourth. A tendency toward two-egg twinning runs in
some family lines, but opinions differ concerning the hereditability of one-
egg twinning. [13, 14] As mothers get older, the tendency toward multiple
ovulation increases, but one-egg twinning is uninfluenced by age.[14] If the
first delivery brings twins, a repetition at the next pregnancy entails nearly
five times the risk existent in the general population. In later sequences the
chance decreases to three times.[15]

Some transposition (mirror-imaging) in one or more organs or parts of
the body, and especially in minor details such as handedness and occipital
hair whorl, occurs in nearly one-half of all one-egg twins. Gross transposition of the viscera is rare in completely separate identical twins, but is more
frequent than in ordinary twins or singletons. The incidence of transposition increases markedly in conjoined twins and is frequent in those joined
side by side. The reversal is far commoner in the right than the left member of a conjoined pair. The extensiveness of mirror imaging is presumably
correlated with the degree of progress already achieved in establishing right-
left differentiation when the twinning moment occurred.[16] Conjoined twins
probably represent a relatively late, imperfect twinning episode that takes
place after the symmetry-asymmetry pattern has been rather well established.

Anomalies. It is believed that one member of a pair of identical twins frequently
succumbs before birth and may then macerate and resorb; it may also become compressed
or mummify (Fig. 81 *C-E*). Sometimes an identical twin is much smaller than its mate

and incapable of separate existence after birth. The heart is then either rudimentary or lacking, and there is a corresponding degree of dependence on the normal twin for some or all of the blood supply. This supply may be direct, because of union between the two bodies (Fig. 154), or indirect through the medium of a common placenta (Fig. 152).

Such a twin, with a deficient heart, is a *hemicardius;* if there is no heart at all, it is an *acardius.* The latter type varies structurally from moderate deficiencies to complete lack of ordinary organization (amorphous fetus). A few cases are recorded of a fetus, reasonably well organized and possessing a backbone, that was located outside the abdominal peritoneum but within the trunk of another individual.[17] Such parasitic, included twin specimens (*fœtus in fœtu*) are not to be confused with unorganized teratomas, which are a kind of developmental tumor (p. 180).

Rarely identical twins are conjoined as a 'double monster.' The degree of union may be slight or extensive, and the possession of a single or double set of internal organs varies with the intimacy of the fusion at any level. Union is by the heads, upper trunks or lower trunks; the joining may be by the dorsal, lateral or ventral surfaces (Fig. 153). Sometimes there is a marked disparity in the size of the two components; in such instances the smaller is called a *parasite* (Fig. 154).

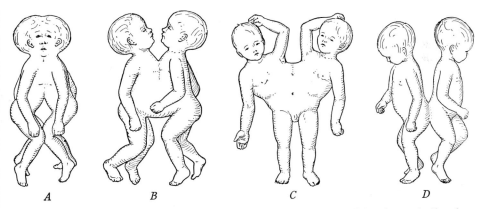

A B C D

Fig. 153. Symmetrical, conjoined twins. *A,* Ventral union of heads; a similar face occurs on the other side of the head. *B,* Ventral union of thoraces. *C,* Lateral union of lower bodies. *D,* Dorsal union in sacral region.

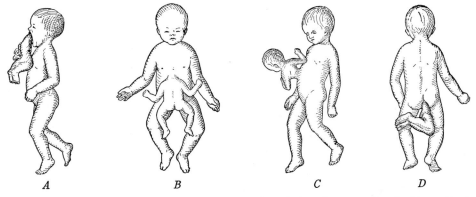

A B C D

Fig. 154. Unequal conjoined twins, one a parasite. *A,* Attachment to head of host; *B, C,* attachment to middle trunk; *D,* attachment to rump.

The general principles of twin production, as established by experiment (p.191), have a special application in birds and mammals. Doubling of the head and upper trunk is due to the chorda-mesoderm splitting into two streams during the forward movements of gastrulation through the primitive streak. Doubling of the lower trunk results somewhat later when the caudally retreating primitive streak and knot produce a forking divergence. Yet it is wholly possible that some conjoining may also result from the merging of closely contiguous embryonic centers.

REFERENCES CITED

1. McIntosh, R., *et al.* 1954. Pediat., *14,* 505–522.
2. Warkeny, J. & H. Kalter, 1961. N. E. J. Med., *265,* 993–1001; 1046–1052.
3. Grüneberg, H. 1952. Bibl. Genet., *15,* 1–650.
4. Russell, W. L. & C. Stern. 1958. Chapters 8, 9 in Claus: Radiation Biology and Medicine (Addison-Wesley).
5. Wilson, J. G. 1959. J. Chr. Dis., *10,* 83–151.
6. Stockard, C. R. 1921. Am. J. Anat., *28,* 115–277.
7. Landauer, W. 1957. J. Exp. Zoöl., *136,* 509–530.
8. Fraser, F. C., *et al.* 1957. Pediat., *19,* 782–787.
9. Nicholas, J. S. & B. V. Hall. 1942. J. Exp. Zoöl., *90,* 441–459.
10. Seidel, F. 1952. Naturwiss., *39,* 355–356.
11. Guttmacher, A. F. 1953. Obst. & Gyn., *2,* 22–35.
12. Corner, G. W. 1955. Am. J. Obst. & Gyn., *70,* 933–951.
13. Dahlberg, G. 1951. Sci. Am., *184,* 48–51.
14. Greulich, W. W. 1934. Am. J. Phys. Anthrop., *19,* 391–431.
15. McArthur, N. 1954. Ann. Genet., *18,* 203–210.
16. Newman, H. H. 1940. Human Biol., *55,* 298–315.
17. Lord, J. M. 1956. J. Path. & Bact., *72,* 627–641.

PART II. SPECIAL DEVELOPMENT

Chapter XII. External Body Form

Tissue combination in definite patterns creates still higher units of organization, the *organs*. Groups of organs associate as *organ systems* within the *organism,* or embryo as a whole. The development of an organ is brought about by the co-operative activities of morphogenesis and histogenesis (pp. 24–27). It is usual to refer to these joint efforts as *organogenesis.*

An organ (*e.g.,* the stomach) has one tissue predominantly important (*i.e.,* its lining epithelium), while the others (*i.e.,* muscle, connective tissue, etc.) are accessory. Whenever an organ is said to originate from a certain germ layer, only its primary tissue is meant; the stomach, therefore, is entodermal. A few organs, like the teeth and suprarenal glands, have equally important parts derived from two germ layers.

A systematic examination of the developmental history of the various organs and parts that make up the human body will comprise Part II (Chapters XII-XXVII) of this book. It includes the various topics that cover regional details (*special development*) in contrast to the more general aspects of development treated in previous chapters. For example, the assumption of the vertebrate type of organization by the embryo and its general shaping into human form were traced in Chapter VI, whereas the present chapter will consider the subject of external form more intimately.

THE HEAD AND NECK

Body-building begins in the future head region, where it gains an early advantage and acquires a favored blood supply. For a long time the head is disproportionately large. This is illustrated in Fig. 77 which shows the highest somites (future base of the head) located midway along the embryo. The gradual adjustment of size relations may be traced in Fig. 6.

The Head as a Whole. The cephalic end of an embryo is composed of two portions almost from the start. One is *neural* in nature and includes the

199

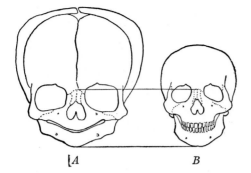

Fig. 155. Skulls of the newborn (*A*) and adult (*B*), drawn to the same face-height to illustrate the relative loss in size of the neural skeleton (Scammon).

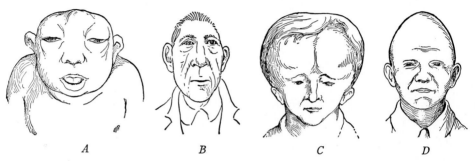

Fig. 156. Malformations of the human cranium. *A*, Cranioschisis, or acrania; *B*, microcephaly; *C*, macrocephaly, or hydrocephaly; *D*, conical cranium.

brain and its associated organs of special sense, as well as their supporting structures. The other is the *visceral* part; it contains the upper ends of the alimentary and respiratory tracts. The neural portion is much the larger in young embryos and this supremacy is never lost completely, although the subsequent differentiation and growth of the nose and jaws succeeds in reducing the early, great disparity in size (Fig. 155).

Anomalies. Only those malformations that involve the external cranial region will be considered at this time. *Cranioschisis,* or open-roofed skull, is usually associated with virtual absence of the brain (Fig. 156 *A*); another designation is *acrania*. *Microcephalus* describes a small cranium housing an undersized and underdeveloped brain (*B*). At the other extreme is an abnormally large head (*macrocephalus*) which accommodates a brain swollen by the excessive accumulation of cerebro-spinal fluid; *hydrocephalus* also designates the same condition (*C*). Various distortions of the normal-sized cranium (asymmetrical; conical; wedge-shaped) depend upon the premature closure of some sutures, while growth continues as usual along other bony margins (Figs. 156 *D*, 384 *C*).

The Branchial Arches and Neck. The construction of jaws and a neck is closely bound up with the history of the *branchial arches*. These are bar-like ridges, separated by grooves, which appear on each ventrolateral surface of the embryonic head during the fourth week (Figs. 76, 77). They correspond to the gill-arches of fishes and some amphibians. In these animals the arches actually bear gills and are separated by clefts (*gill slits*) through which respiratory water flows. Such an arch contains a cartilaginous or bony core, and a main blood vessel (*aortic arch*) which interconnects the dorsal and ventral aortæ; in addition, there are appropriate muscles and nerves. The branchial arches of amniote embryos do not acquire gills; only occasionally are the arches fully separated by transitory clefts.

The human embryo develops five such arches, separated by four ectodermal *branchial grooves*. At the same levels as these external grooves the entoderm of the pharynx pushes aside the mesenchyme and bulges outward to become the *pharyngeal pouches* (Fig. 195). The ectoderm of each groove and the entoderm of its complementary pouch then meet and unite. As a result, a typical arch is separated from the one ahead and behind it by a thin epithelial plate only. The first branchial arch on each side bifurcates into a *maxillary* and a *mandibular process* (Fig. 158 *A*). The last arch lies caudal to the fourth cleft and is poorly defined along its caudal margin.

During the sixth week the second arch overlaps the next three and obscures them (Fig. 78), the more caudal arches then sinking into a triangular depression called the *cervical sinus* (Fig. 604 *A*). In this change the ectodermal grooves become drawn out into *branchial ducts* (Fig. 157). At least that part of the sinus which contains the fourth and fifth arches closes off,[1] whereupon its ectodermal-lined cavity promptly detaches and obliterates. Thus, after a short existence of two weeks, the branchial arches largely disappear as such and the resemblance to the ancestral gilled condition comes to an end. The backward growth of the second arch, and the concealment of the other arches by it, correspond to the covering over of the other gill arches by an *operculum,* as it occurs in bony fishes.

Various muscles, bones and blood vessels differentiate from the mesenchymal cores of the early arches, while their epithelial covering and lining have other, distinctive fates. Moreover, the entodermal pharyngeal pouches, whose later sites can be observed in Fig. 159 *A*, give rise to important derivatives. The completion of these several transformations marks the appearance of a *neck,* which is a characteristic of land vertebrates. This part of the body results from an elongation of the region between the first branchial arch and the body wall enclosing the heart (Fig. 158). The second and

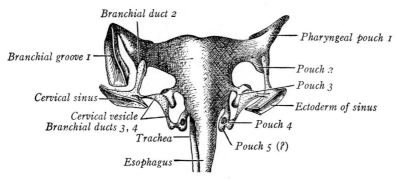

Fig. 157. Human pharynx and cervical sinus, at six weeks, modeled in dorsal view (after Hammar). × 17.

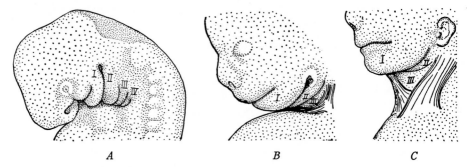

Fig. 158. Human stages, illustrating the relation of the branchial arches (numbered) to the ventral surface of the neck. *A,* At five weeks (× 8); *B,* at seven weeks (× 3.5); *C,* at three months (× 2).

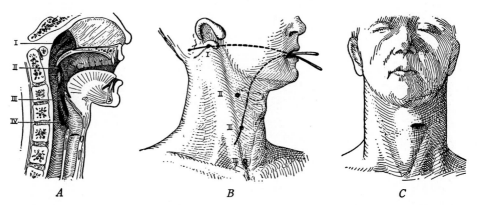

Fig. 159. Anomalies of the primitive branchial apparatus in man. *A,* Left half of an adult head to indicate the final sites of the embryonic pouches, *I-IV* (after Corning). *B,* Common superficial relations of cervical cysts, diverticula and fistulæ. *C,* Cervical fistula, of second pouch origin, communicating with pharynx.

third arches, at least, can be seen to contribute to its ventral and lateral surfaces.[1] The heart itself is left behind, but all the structures that connect head with thorax (*i.e.,* vessels; nerves; muscles; digestive tube; respiratory tube) participate in the elongation. The detailed histories of all derivatives from the branchial-arch system, except the jaws, will be treated in later chapters; at present only the final results will be summarized in a reference table (p. 203).

Causal Relations. Development of the first (mandibular) branchial arch depends on the presence of the ectodermal mouth invagination. All other arches are dependent on the presence of corresponding pharyngeal pouches.

Anomalies. Imperfect obliteration of a branchial 'cleft' leads to the formation of a cervical (branchial) cyst or fistula (Fig. 159 *B*). It is customary to blame the second groove-pouch complex, as also the cervical sinus, for most of these abnormalities.[1, 2] *Cervical cysts* are closed, epithelial sacs which arise in the region of a closing plate. They derive either from an ectodermal groove or its complementary, ectodermal pouch. They are located lateral to the midplane, near the border of the sterno-mastoid muscle.

DERIVATIVES OF THE BRANCHIAL REGION OF THE EMBRYO

ARCH OR POUCH	ECTODERMAL DERIVATIVES		ENTODERMAL DERIVATIVES		MESODERMAL DERIVATIVES		
	ECTODERMAL BRANCHIAL GROOVE	ECTODERMAL COVERING OF ARCH	ENTODERMAL LINING OF ARCH	ENTODERMAL PHARYNGEAL POUCH	SKELETON	MUSCLES (AND THEIR NERVES)	AORTIC ARCHES
I	Ext. auditory meatus. Epithelium of: Meatus. Tympanic membrane (external surface).	Epidermis of auricle (ventral half). Maxillary process: Epidermis of upper lip and cheek. Enamel; parotid gland. Mandibular process: Epidermis of lower lip and jaw. Enamel. Submaxillary gland. Sublingual gland. Epithelium of: Vestibule; palate. Body of tongue.	Epithelium of: Some of sides and floor of mouth.	Cavity and epithelium of: Tympanic cavity. Tympanic membrane (internal surface). Mastoid cells. Auditory tube. (Thyroid arises from floor at about this level.)	Maxillary process: Upper jawbone. Palate. Dentine; cementum. Mandibular process: (Meckel's cartilage.) Lower jawbone. Dentine; cementum. Antimalleolar ligament. Spheno-mandibular ligament. Malleus; incus.	Mastication. M. digastricus (anterior belly). M. tensor palati. M. tensor tympani. (Nerve V innervates this group.)	Degenerates.
II	(Anomalous cysts or fistulæ.)	Epidermis of: Auricle (dorsal half). Upper neck.	Epithelium of: Root of tongue. Pharynx. (In part.)	Palatine tonsil (?): Fossa. Epithelium of: Surface and crypts.	(Reichert's cartilage.) Stapes. Styloid process. Stylo-hyoid ligament. Hyoid (lesser horns).	Expression. Auricular. Epicranial. M. digastricus (post. belly). M. stylo-hyoideus. M. stapedius. (Nerve VII.)	Degenerates.
III	Obliterates in cervical sinus.	Epidermis of: Middle neck.	Epithelium of: Root of tongue. Pharynx; epiglottis. (In part.)	Inf. parathyroid. Thymus. Reticulum. Thymic corpuscles.	Hyoid (body and greater horns).	Pharynx (in part). M. stylo-pharyngeus. Constrictors (some). (Nerve IX.)	Stem of internal carotid.
IV	Obliterates in cervical sinus.	Obliterates in cervical sinus.	Epithelium of: Root of tongue. Pharynx; epiglottis. (In part.)	Sup. parathyroid. Rudimentary thymus.	Thyroid cartilage. Hyo-thyroid ligament (?). Cuneiform cartilage.	Pharynx (in part). Constrictors (others). Larynx (in part). (Nerve X.)	Left: arch of aorta. Right: subclavian (in part).
V	Not formed.	Epidermis of: Lower neck.	(Lungs arise from floor at about this level.)	Ultimobranchial body ('lateral thyroid').	Thyroid cartilage (?). Corniculate, arytenoid and cricoid cartilages.	Larynx (in part). (Nerve X.)	Pulmonary artery. D. arteriosus. (Is it arch 5 or 6?)

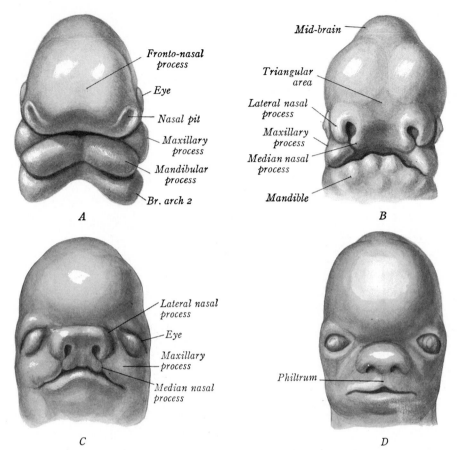

Fig. 160. Development of the human face (adapted after Peter). *A,* At 6 mm. (× 14); *B,* at 10 mm. (× 11); *C,* at 15 mm. (× 10); *D,* at 20 mm. (× 7.5).

Incomplete reduction of the branchial 'clefts' give rise to *cervical fistulæ,* which are of two types: (1) A complete fistula is an open communication between the pharynx and the external surface of the neck (Fig. 159 *C*); its patency is due to the failure of an open second cleft to close. Such a fistula includes the cervical sinus and becomes drawn out into a long tube (Fig. 157). The deep end opens at the tonsillar fossa, while the cutaneous orifice shifts considerably caudad of its original site. (2) Incomplete fistulæ are blind *cervical diverticula,* also called *sinuses,* leading outward from the pharynx or, most commonly, inward from the skin of the neck. They represent a persistent entodermal pouch or ectodermal groove, respectively.

The Face. The site of the *face* is indicated, above, by the region just in front of the bulging fore-brain (future forehead) and, below, by the first pair of branchial arches (future jaws) (Fig. 160 *A*). The emerging eyes and nasal (olfactory) pits are also intimately concerned; in fact, just as the snout constitutes most of the face of low vertebrates, so in mammals a first step in face-construction is the development of the nose. Another important preliminary is the subdivision of each first branchial arch ventrally into a maxillary and mandibular process (Fig. 158 *A*).

An early stage of the face is shown in Fig. 160 *A,* where the expansive *fronto-nasal process* represents much of the front of the head. The *nasal pits* are present and the first branchial arches have bifurcated into *maxillary* and *mandibular processes*. Each nasal pit is soon bounded by a prominent, horseshoe-shaped elevation whose two limbs are named the *median* and the *lateral nasal process* (*B*); at this period the nasal pits communicate by a groove with the mouth cavity, just as in some fishes.

In describing the method of face-construction it is convenient to say that the several 'processes' meet and unite. Actually these components are mere elevations, or ridges, that indicate local centers of mesenchymal proliferation. They are covered by a continuous sheet of folded epithelium. As the mesenchyme spreads from these centers, the furrows between them smooth out and mergers take place.[3, 4] Only to this extent has there been union or fusion.

The development of the human face occurs chiefly between the fifth and eighth weeks. The *lower jaw* is the first to come into being (Fig. 160 *A, B*). This is accomplished by the simple union of the ventral ends of the two mandibular processes. The *upper jaw* is more complex; it is the product of mergers between the two maxillary processes and the two median nasal processes, which latter become compressed toward the median plane (*B-D*).[4] The *forehead* corresponds to the main expanse of the fronto-nasal process. The *nose* has a compound origin (*B-D*). A downward continuation of the fronto-nasal process is the so-called *triangular area;* it elevates slowly into the dorsum (bridge) and apex of the nose. The sides and wings of the nose are furnished by the lateral nasal processes; each of these merges with the maxillary process of the same side. The relations of the several components to the completed face are indicated in Fig. 161 *A*.

When first formed, the nose is broad and flat, with the nostrils set far apart and directed forward (Fig. 160 *C*). In later fetal months the bridge of the nose is elevated and prolonged into the apex, and the nostrils point downward (Fig. 161 *B*). Accompanying this relative narrowing of the nose,

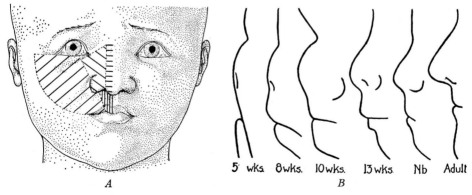

5 wks. 8 wks. 10 wks. 13 wks. Nb Adult

A   *B*

Fig. 161. *A,* Definitive contributions of the facial components, indicated as differentially hatched territories. *B,* Profiles, illustrating the changes in the form and proportions of the face throughout the life span (Scammon).

A   *B*

Fig. 162. Junction of the median nasal and maxillary processes, as marked by angles in the adult upper lip (Lewis). *A,* Mouth open and relaxed; *B,* mouth puckered.

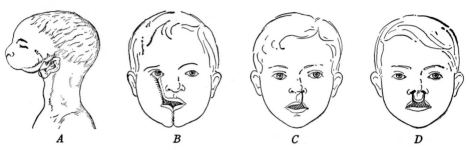

A   *B*   *C*   *D*

Fig. 163. Malformations of the human face. *A,* Agnathia; *B,* oblique facial cleft and median cleft of lower lip; *C,* unilateral cleft (or hare) lip; *D,* bilateral cleft lip.

the head broadens behind the *eyes* and causes them to be directed forward; in this way binocular vision becomes possible in primates. The *eyelids* arise as simple folds of the skin (Fig. 495). The boundary zone between the median nasal processes is evident as the permanent *philtrum,* or median groove of the upper lip. This median region often continues downward into a distinct *labial tubercle* (Fig. 161 *A*). The lateral extent of the median nasal processes is indicated by angular indentations of the upper lip, best seen when the mouth is either open and relaxed (Fig. 162 *A*) or when puckered (*B*). The *lips* begin to split away from the gum regions of the jaws in the seventh week. The original lateral extent of the *mouth opening* is at the point of bifurcation of the maxillary and mandibular processes, near the ear (Fig. 160 *B*). Later this broad slit is reduced markedly in its relative lateral spread; in this way the *cheeks* are established and the lips become pursed. The *chin* is a median projection, grown forward from the fused mandibular processes. Progressive modeling of the face continues throughout childhood and even until the individual becomes full grown. Figure 161 *B* indicates the striking changes that occur in profile-contour.

Anomalies. The fairly frequent occurrence of malformations of the face is correlated with the complex modeling of this region. General failure of the usual transformations results in a featureless face (*aprosopia*). The lower jaw may be retarded (*micrognathia*) or even absent (*agnathia;* Fig. 163 *A*). The primitive mouth slit sometimes fails to reduce normally (*macrostomia;* Fig. 494); on the contrary, the normal degree of closure may be exceeded (*microstomia;* Fig. 517 *C*) and even complete atresia (*astomia*) is known. Fetal or infantile nose shapes are not infrequently retained.

A median defect of the upper lip or jaw, through incomplete growth and union of the median nasal processes, is rare (Fig. 494). Still more unusual is imperfect fusion at the midplane between the mandibular processes which fashion the lower lip and jaw (Fig. 163 *B*). *Oblique facial cleft* describes a slanting furrow that extends from the mouth up the cheek to the eye (*B*). It is usually interpreted as an incomplete union between a maxillary process and the adjoining nasal processes (*cf.* Fig. 161 *A*). Among the commoner human anomalies is *cleft lip,* or *cheiloschisis,* also inappropriately called *harelip.* This defect is usually unilateral, and more often on the left side (Fig. 163 *C*), but it may be double (*D*). The defect is commonly limited to the fleshy lip alone, but it may involve the bony upper jaw as well. The cause lies in a faulty spread of mesenchyme into the normally merging maxillary- and median nasal processes. Virtual absence of mesenchyme at the line of junction can lead to actual separation of these parts.[5, 6] Sometimes the condition of cleft lip is also combined with cleft palate (Fig. 188 *C*).

The Sense Organs. The eye, ear and nose will be considered in detail in Chapter XXVII. The development of the external *nose* has been described in preceding paragraphs dealing with the face. The *eye* makes its appearance in the early weeks, and by the second month lids are present (Fig. 160). For a time the eyes are placed laterally and relatively far apart, but gradually this disproportion is reduced by the differential broadening of the head. The *external ear* is developed around the first branchial groove by the appearance of small tubercles that combine as the auricle (Figs. 78, 80 *F-H*). The groove itself deepens into the external auditory meatus. The ears shift to a position relatively higher and more lateral than their sites of origin (Fig. 516 *A*).

Anomalies. See Chapter XXVII: Nose (p. 524); eye (p. 529); ear (p. 541).

FATES OF THE FACIAL COMPONENTS (EXCEPT EYES)

EMBRYONIC PART	FLESHY DERIVATIVES	BONY DERIVATIVES
Frontal process...............	Forehead.	Frontal.
Triangular area..............	Dorsum and apex of nose.	Nasal.
Median nasal processes.........	Fleshy nasal septum. Median part of upper lip (deep) and gum; incisive papilla.	Ethmoid (perpendicular plate). Vomer. Premaxilla (incisive bone).
Junction of median nasal processes	Philtrum; frenulum. Labial tubercle.	
Lateral nasal process...........	Side of the nose. Wing of the nose.	Maxillary (frontal process). Lacrimal (?).
Junction of lateral nasal and maxillary process........	Naso-lacrimal duct.	
Maxillary process.............	Most of upper lip and gum. Upper cheek region.	Maxillary. Zygomatic.
Mandibular processes..........	Lower lip, gum and chin. Lower cheek regions.	Mandible.

THE TRUNK

The early embryonic region is a layered plate but folding soon rolls it into a hollow cylinder, most of which becomes the *trunk*. This process of

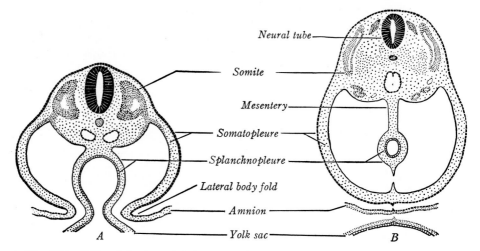

Fig. 164. Rôle of the lateral folds in producing a tubular body, shown semidiagrammatically by transverse sections of human embryos. *A,* At three weeks (× 75); *B,* at four weeks (× 40).

underfolding proceeds beneath the head and caudal end (Fig. 73), and also from both sides (Fig. 164). It is the folded and combined somatopleures that form the sides and ventral surface of the trunk, while the inrolled splanchnopleure becomes separated as the gut and its mesentery. More dorsally, the previously inrolled neural tube and the somites (future axial skeleton and muscles) dominate. The various true fusions of folds and other parts in a developing embryo restore continuities much after the manner of simple wound healing.

In young embryos the trunk is like a cylinder that has been flattened somewhat by lateral compression (Fig. 80 *E*). Its external contour is made irregular by the bulging heart and liver. During the early fetal period these visceral organs become less dominant, and the muscles and skeleton of the trunk also appear. The trunk then acquires an ovoid form, circular in section (*G, H*). From the third fetal month through early infancy there is relatively little change in the trunk proportions. In the middle fetal period the abdominal region between the pubis and umbilicus acquires its characteristic expanse. When erect posture is assumed, the dominance of the thorax and abdomen is reduced and the lumbar region gains in prominence and relative length. The thorax of the newborn is rather conical and thickest below, due to the ribs being more horizontal. In childhood the thorax becomes barrel-shaped—that is, broadest at its middle.

The C-shaped curvature of the fetal body straightens in the newborn (Fig. 165 *A-D*). The permanent curves of the spinal column appear, partly through the pull of the muscles, and are not pronounced until posture becomes erect (*E, F*). The embryonic tail is at its relative maximum at the end of the fifth week when it is one-sixth the length of the embryo. During the succeeding four weeks it disappears from external view, partly through

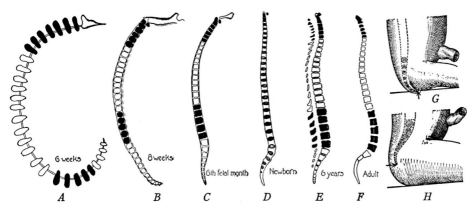

Fig. 165. *A–F,* Spinal curvatures, at various ages, viewed from the right side (Scammon). *G, H,* Recession of the coccyx (arrow) between the tenth fetal week and birth (Schultz).

actual regression; moreover, the coccyx, which represents the remnant of a tail, recedes to a higher position in relation to the buttocks (*G, H*). The *coccygeal fovea,* or postanal pit, of a newborn (Fig. 166 *B*) marks the site where a vestige of the caudal tip of the early spinal cord attaches to the skin (*cf.* Fig. 429).[7]

Anomalies. A grave defect results from faulty closure of the body wall along its midventral line. This is known as *gastroschisis,* or, if the thorax is involved as well, *thoraco-gastroschisis.* The viscera may protrude nakedly through a cleft (Fig. 166 *A*); they may occupy a membranous sac that represents the defective body wall; or the wall may be defective, yet able to resist herniation more or less successfully (Fig. 352 *A*). A kindred malformation is cleft spine (*rachischisis* or *spina bifida*), consequent on the failure of the vertebral column to close normally (Fig. 368 *B, C*). In this instance the neural tube may also herniate or remain open (Fig. 166 *C*).

The embryonic tail has been known to persist and grow (Fig. 166 *B*). Specimens as long as 3 inches have been recorded in the newborn, and one was reported to have be-

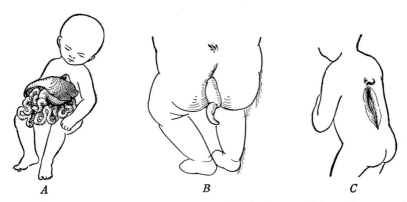

Fig. 166. Malformations of the human trunk. *A,* Gastroschisis, with protrusion of the abdominal viscera. *B,* Tail of an infant (× ¼); this specimen was 'soft' but slightly mobile, and is shown in its contracted state; above is a coccygeal fovea. *C,* Rachischisis, or cleft spine, with gaping skin and exposed spinal cord.

come 9 inches long at 12 years. Most of these tails are soft and fleshy, but a few have contained skeletal elements. Some tumors of the coccygeal region are attributed to the abnormal activity of residual primitive-knot tissue (the end bud).

THE LIMBS

The *limb buds* appear late in the fourth week as lateral swellings but, owing to the early dominance of the head-neck region, the arm buds seem to be located far down the body (Fig. 77). The distal end of a limb bud flattens (Fig. 167 *A, E*) and a constriction divides this paddle-like portion from a more proximal, cylindrical segment (*B, F*). Later, a second constriction separates the rounded part into two further segments (*C, G*); the three divisions of arm, forearm and hand (or thigh, leg and foot) are then respectively marked off. Radial ridges, separated by grooves, first foretell the location of digits (*C, G*). These elongate into definitive fingers or toes, and rapidly project beyond the original plates; the latter, by a slower rate of growth, become restricted to webs between the basal ends of the digits (*D, H*). The thumb early separates widely from the index finger, and the same is true of the great and second toes.

Of the two sets of limb buds the upper pair appears first, begins its differentiation sooner and is earlier in attaining its final relative size. Not until the second year of postnatal life does the leg equal the arm in length; its continued faster elongation throughout childhood is a conspicuous feature of postnatal development (Fig. 6).

The limbs as a whole undergo several changes of position. At the very start they point caudad (Fig. 77 *B*), but soon project outward almost at right angles to the body wall (Fig. 79). Next, they bend directly ventrad at the elbow and knee, so that the elbow and knee then point outward (laterad) and the palm and sole face the trunk; at this stage the thumb (radial) side of the arm and the great-toe (tibial) side of the leg constitute the cranial borders of their respective limbs (Fig. 80 *F*). Finally, both sets of limbs undergo a torsion of 90° about their long axes, but in opposite directions, so that the elbow points caudad and the knee points craniad (Fig. 80 *G—I*).

As a result of the changes just outlined, the straightened limbs of erect man have

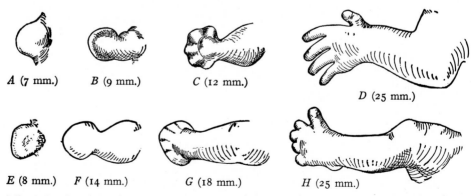

A (7 mm.)   *B* (9 mm.)   *C* (12 mm.)

D (25 mm.)

E (8 mm.)   *F* (14 mm.)   *G* (18 mm.)   *H* (25 mm.)

Fig. 167.   Stages in the development of the human limbs between the fifth and eighth weeks. × 6. Upper row, upper limb; lower row, lower limb.

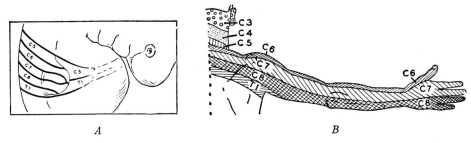

Fig. 168. The segmental (sensory) innervation of the human upper limb (after Keegan and Garrett). *A,* At five weeks; *B,* in adult.

opposite relations with respect to their medial and lateral borders. The elbow faces dorsad, the thumb (radial) side of the arm becomes the outer border (when radius and ulna are parallel), and the palm faces ventrad. Conversely, the knee faces ventrad, the great-toe (tibial) side of the leg is the inner border, and the sole of the foot (in tip-toe position) faces dorsad. By following through these changes it will be seen that the radial and tibial sides of the arm and leg are homologous, as are the ulnar and fibular sides; similarly homologous are the palm and sole, elbow and knee. Since the lateral surface of the upper limb and the medial surface of the lower limb represent the cranial halves of the primitive limb buds, it is only natural that these surfaces are innervated by nerves higher in the spinal series than those nerves that supply the opposite surfaces of the limbs (Figs. 168, 403).

Causal Relations. A limb bud can be induced at will by implanting an ear vesicle or nasal pit in the mesoderm of the lateral plate of an amphibian embryo. Yet the normal source of induction is not obvious and it can only be assumed that the evocating substance is liberated at a proper time by the somatic mesoderm. Experiments on amphibian and chick embryos have illuminated greatly the general facts of morphogenesis that are obtainable by direct observation. When a limb bud first appears, as such, it already is supplied with the materials for the proximal parts of the limb. The capacity to form the distal parts resides in a limited mesenchymal region at the apex of the elongating bud. This tissue progressively lays down the materials for these future parts in a proximodistal sequence and in their definitive spatial pattern (*cf.* Fig. 399). The tip of each limb bud is capped with a prominent epithelial thickening, underlaid by a refractile membrane (Figs. 385, 620). Removal of this tissue suppresses the nicely coordinated activities just mentioned, and the limb becomes limited to a proximal stump. The limb primordium, even when barely discernible as a slight swelling, is capable of self-differentiation when isolated and without nerve supply. It is a mosaic with axes determined and with different regional potentialities; yet each segment is still totipotent as regards its own regulation, and a piece of that region can form a complete segment. If a limb bud is transplanted in any rotated position it will undergo reverse rotation, as a whole, and thus effect regulatory recovery.

Anomalies. The limbs may either fail to develop or, at best, become mere stubs (*amelia;* Fig. 169 *A*). Sometimes the proximal segments of an extremity are normal while the distal portion is deficient and tapers to a stump (*hemimelia; A*). The reverse condition has at least the proximal segment missing, whereupon the hand or foot seems to spring directly from the trunk, like a seal's flipper (*phocomelia; B*). More or less complete union of the legs produces various degrees of the siren or mermaid condition (*sympodia; B*). Fusion results from the developmental failure of a median, wedge-shaped mass of trunk-mesenchyme, which ordinarily keeps them apart. Rarely the hands or feet have missing digits. Extreme in this group is the split or *'lobster-claw'* type, which also usually includes some fusion of fingers; the two main components are separated by an

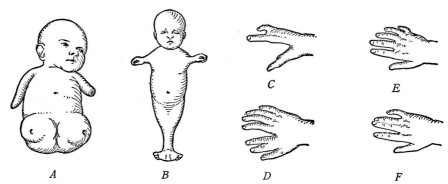

Fig. 169. Malformations of the human limbs. *A,* Hemimelia in arms and amelia in legs; *B,* phocomelia in arms and sympodia in legs; *C,* cleft, or 'lobster-claw' hand; *D,* dichiria, or double hand; *E,* polydactyly; *F,* syndactyly.

abnormally long cleft (*C*). Opposite in nature is a partial duplication (*dichiria; D*), which may approximate a hand or foot in mirror-image. *Polydactyly* (*E*) designates the presence of a supernumerary digit beyond the little finger. The bony fusion or fleshy webbing of digits (*syndactyly; F*) favors the union of the middle and third digits. Abnormal shortness of the digits is *brachydactyly;* it is due either to the omission or marked shortness of phalanges (usually the second). The opposite tendency is *hyperphalangism,* in which supernumerary phalanges are interpolated in the customary digital series. All of these malformations of hands and feet tend strongly to be heritable.

Clubhand or *clubfoot* is said by some to result from primary defects in the differentiating limb buds; others urge that clubfoot is essentially a retention of a transitory condition normal to the fetus and permanent in apes.[8] Congenital elevation of the shoulder results from an arrested descent of the upper limb from its cervical, embryonic position. Congenital dislocation at the hip joint comes from a failure of the outgrowths that normally produce a brim about the socket floor, combined with a failure to make a socket of proper shape.[9] Intra-uterine amputation of an appendage (at any level) sometimes occurs. The cause is intrinsic, due to focal deterioration of the tissues themselves; it is not the result of constriction by a looped umbilical cord or by amniotic bands, as was long believed.[10]

REFERENCES CITED

1. Frazer, J. E. 1926. J. Anat., *61,* 132–143.
2. Frazer, J. E. 1923. Brit. J. Surg., *11,* 131–136.
3. Streeter, G. L. 1948. Carnegie Contr. Embr., *32,* 133–203.
4. Warbrick, J. G. 1960. J. Anat., *94,* 351–362.
5. Stark, R. B. 1954. Plast. & Reconst. Surg., *13,* 20–39.
6. Töndury, G. 1950. Acta Anat., *11,* 300–328.
7. Willis, R. A. 1962. The Borderland of Embryology and Pathology, 2nd ed. (Butterworth).
8. Bohn, M. 1929. J. Bone & Joint Surg., N. S. *11,* 229–259.
9. Rohlederer. 1950. Verh. deuts. orthopäd. Gesel., 1950, 58–67.
10. Streeter, G. L. 1930. Carnegie Contr. Embr., *22,* 1–44.

A. ENTODERMAL DERIVATIVES

Chapter XIII. The Mouth and Pharynx

THE PRIMITIVE DIGESTIVE CANAL

The primary tissue of the entire digestive system is entoderm. This epithelial layer originally lines the whole yolk sac, but a regional difference in the shape of the entodermal cells is apparent from the first (Fig. 66 *A*). Those that underlie the embryonic disc (and serve as a flat roof to the early yolk sac) are taller than the rest; they are the ones that are destined to become gut-entoderm. When, at the twentieth day, the rapidly expanding embryonic disc begins to fold into a cylindrical embryo (p. 95), its gut-entoderm participates as a component layer. Folding first into the head end and then into the hind end of the elongating embryo, this entoderm necessarily takes the form of two internal, blind tubes (Fig. 170). The open end of each tube, where it becomes continuous with the yolk sac, is called an *intestinal portal*, while the tubes themselves are named the *fore-gut* and *hind-gut*. An intermediate region, open ventrally into the yolk sac through the narrower yolk stalk, is sometimes termed the *mid-gut*, but its existence in man is brief since the yolk stalk constricts rapidly during the fourth week and detaches from the gut at the end of the fifth week. Both the fore-gut and the hind-gut elongate and broaden by interstitial growth, so as to keep pace with the growth of the embryo as a whole.

The primitive, tubular gut differentiates into three chief segments: the mouth, pharynx and digestive tube. The latter division includes the esophagus, stomach, small intestine and large intestine; it lies mostly in the body cavity and is suspended or held in place by mesenteries (Fig. 234). The fore-gut specializes into part of the mouth, all of the pharynx and into the digestive tube to a point far along the small intestine. The hind-gut becomes the rest of the small intestine and all of the colon and rectum. Throughout its length the entodermal digestive canal gives rise to numer-

213

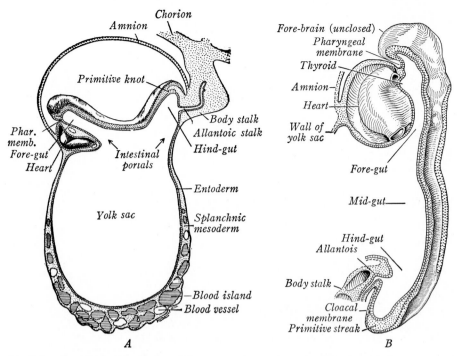

Fig. 170. Entodermal tract of early human embryos, in sagittal section. *A,* At seven somites
(Prentiss, after Mall; × 23). *B,* At ten somites (after Corner; × 30).

ous derivatives, chief of which are the respiratory tract and the thyroid,
parathyroids, thymus, liver and pancreas. The entoderm furnishes merely
the epithelial lining of the digestive and respiratory tracts, and the characteristic epithelial parts of the other organs. The various glands, both large
(such as the liver and pancreas) and small (like the gastric and intestinal
glands), are subordinate growths that push out from the lining epithelium.
All of the accessory coats of the alimentary canal, such as muscle and connective tissue, are secondary investments. They differentiate from the splanchnic
mesoderm nearby (Fig. 71). The epithelium that lines all hollow viscera
of the body is moist and, in many instances, slimy. Together with its underlying connective tissue and glands, it comprises a unit named the *tunica mucosa* or mucous membrane.

At each end of the primitive gut-tube there is a region free of mesenchyme. Here entoderm comes into direct contact with the ectoderm. The
fused plates, thus produced, are the *pharyngeal* (or *oral*) *membrane* and
the *cloacal membrane* (Fig. 172 *A*). The pharyngeal membrane makes a
floor to an external depression known as the *stomodeum*. This pit is
bounded by the fronto-nasal, maxillary and mandibular processes (Fig.
171) and is brought into existence by the overjutting of these parts as
growth progresses. Midway in the fourth week (2.5 mm. embryos) the
pharyngeal membrane ruptures and the stomodeum and fore-gut merge

(Fig. 172 *B*). The stomodeum develops into part of the mouth which is, therefore, ectodermal.

The caudal end of the entodermal tube becomes the temporary *cloaca*, or common vent. It communicates with the allantois (Fig. 172), which preceded it in time of origin, and soon receives the urinary and genital ducts

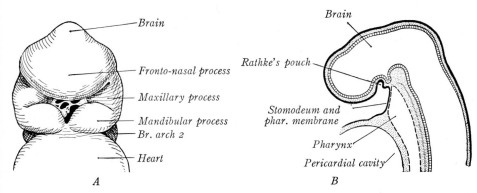

Fig. 171. Human stomodeum and its floor membrane. × 30. *A*, Boundaries of the stomodeum and a partly perforated pharyngeal membrane, shown in front view at 2.5 mm. *B*, Relation of ectoderm (full line) and entoderm (broken line) in this region, illustrated by a sagittal section at 2.5 mm.

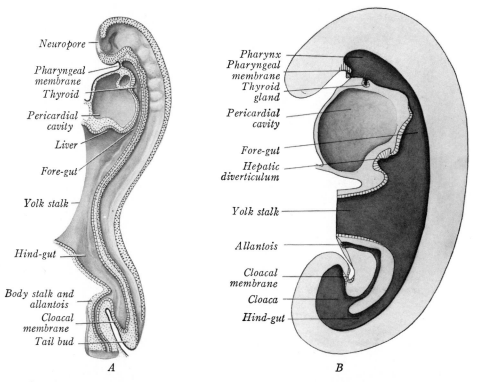

Fig. 172. Entodermal tract, shown in hemisections of human embryos. *A*, At 2.5 mm., with eighteen somites (× 35); *B*, at 2.5 mm., with twenty-three somites (after Thompson; × 38).

(Figs. 195 *A,* 206). Even before these latter connections are complete, the cloaca begins to subdivide into a dorsal *rectum* and a ventral *bladder* and *urogenital sinus* (Fig. 209). By the end of the seventh week the cloacal membrane has separated into an anal and urethral region (Fig. 210), whereupon the two membranes rupture and disappear. Each of the new canals (rectum and urogenital sinus), so formed, acquires thus simply its individual opening to the outside. The end of the digestive tube is then lined for a short distance with ectoderm, and this portion (the *proctodeum*) constitutes some of the future anal canal. It will be noticed that the primitive entodermal tube extends caudad a little beyond the cloacal membrane (Figs. 195 *A,* 206); this *tail-gut* dwindles during the fifth week and soon disappears. How rapidly all these changes occur may be appreciated by comparing embryos of four weeks (Fig. 208) with those two or three weeks older (Figs. 209, 210).

THE MOUTH

After the loss of the pharyngeal membrane there are no exact landmarks to the line of junction between ectoderm and entoderm in the mouth. A considerable caudal displacement has shifted this line in the roof region back into the naso-pharynx, but keeping in front of the palatine tonsils and the entrance to the auditory tubes. Hence the nasal passages, palate, front part of the tongue, and the vestibule are considered to be lined with ectoderm; the enamel of the teeth and probably the salivary glands are likewise ectodermal derivatives. Although these various structures do not belong among the entodermal organs, it is simplest to describe them now along with the systems of which they are functional parts. A further derivative of the ectodermal stomodeum is a dorsal inpocketing, known as Rathke's pouch, which becomes the epithelial lobe of the hypophysis (Fig. 171 *B*). Its point of origin marks the most caudal extent of ectoderm in the completed mouth (Fig. 190 *B*).

Causal Relations. Experiments on the amphibian neurula show that only when pharyngeal entoderm is in contact with the presumptive mouth ectoderm will the pharyngeal membrane rupture and a mouth, with teeth, form. An entodermal sample from a mid-trunk level is unable to induce the formation of a mouth or mouth parts. Yet after a pharyngeal membrane once arises normally, the mouth can develop even when the fore-gut has been removed.

Anomalies. The failure to provide a mouth opening is *astomia.*

Lips and Cheeks. Until the end of the sixth week the primitive jaws are solid masses which do not show any subdivision into lip and gum regions, thus resembling the permanent condition in animals below mammals. The separation of each lip from its respective gum is foreshadowed by the appearance of a thickened band of epithelium (Fig. 174). This *labial lamina* grows from the ectodermal covering of the primitive jaw into the mesenchyme beneath. Following the contour of the jaw, it makes a long, curving band which becomes a deep partitioning plate (Fig. 173 *A*). Progressive disintegration of the more central cells causes each plate to split into two sheets (Fig. 174 *A, C*). In this manner the *lips* become separate from the *gums* by the tenth week, and the epithelial-lined labial groove, so formed, deepens into the *vestibule*. In the midplane of each lip the splitting is less extensive, thereby leaving a soft fold known as the *frenulum*.

The *cheeks* come into existence chiefly through a reduction in the extent of the originally broad mouth opening; this results from progressive fusion of the lips at their lateral angles. The labial and buccal muscles differentiate from mesenchyme of the second branchial arches which migrates between the epidermal covering and mucosal lining of these parts.

The Teeth. Historically the teeth are products of the skin, and both the epidermis and dermis (corium) contribute to their formation. A tooth is a greatly modified connective-tissue papilla that has both undergone a peculiar ossification into *dentine* externally, and become capped by a hard *enamel* elaborated from the epidermis. In addition, the base encrusts with *cementum*, a bony deposit. The homology of a tooth primordium with a dermal papilla and its covering epithelium is somewhat obscured by an early ingrowing that results in the whole primordium coming to lie deep within the gum.

The teeth have a double source of origin in the embryo: the enamel is from ectoderm; the dentine, pulp and cement are mesodermal. There are two generations of teeth in man and most other mammals, but no essential difference exists between the development of the temporary (milk) teeth and the permanent ones. Since the primordia of the *primary dentition* arise first, they will be described first and in greater detail.

The Dental Lamina and Early Tooth Buds. The first indication of oncoming tooth development is an epithelial plate, the *dental lamina,*

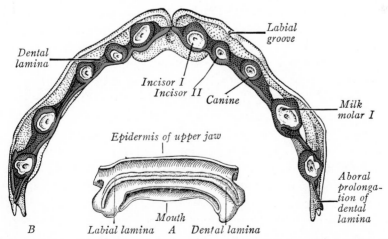

Fig. 173. Isolated epithelium of the human jaws, showing the labial and dental laminæ (after Röse). *A,* At two months (× 8); *B,* at three months, with primordia of milk teeth (× 9).

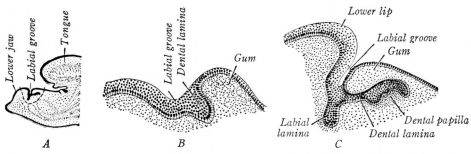

Fig. 174. Development and relations of the labial and dental laminæ, demonstrated by sections through the human lower jaw. *A,* Sagittal section, at nine weeks, to explain the areas included in *B* and *C* (× 10). *B,* Labio-dental lamina, at seven weeks (Röse; × 90). *C,* Detail, at nine weeks, of the area set off in *A* by a broken line (Röse; × 45).

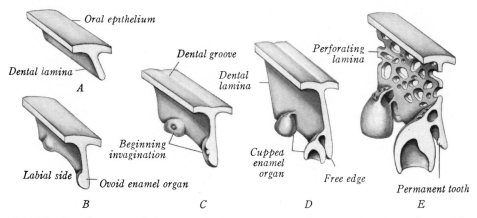

Fig. 175. Development of the enamel organs, at two to four months, shown by models (after Eidmann).

which arises during the seventh week just gumward of the labial lamina, already described (Figs. 173 *A,* 174 *B*). The dental lamina soon becomes a horizontal shelf which projects perpendicularly from the labial lamina and extends well into the substance of the primitive gum (Figs. 174 *C,* 178 *A*). This is brought about as the gum, upgrowing, buries the dental lamina deeper and deeper. Each dental lamina thereby courses alongside the curving labial groove and lies just gumward of it (Fig. 173 *A*).

At intervals along the epithelial lamina there develops simultaneously a series of knob-like thickenings called the *enamel organs,* which will both produce the enamel and serve as the molds for the future teeth (Fig. 175 *A, B*). Early in the third month the deeper side of each enamel organ presses against a dense accumulation of mesenchyme (Fig. 174 *C*). The epithelial surface of contact both buckles inward (*i.e.,* invaginates) and grows around the mesenchymal mound until the whole enamel organ is hollowed like a thick cup (Fig. 175 *C-E*). The concavity, formed in this manner, is occupied by the condensed mesenchymal tissue of the *dental papilla* which is destined to differentiate into dentine and pulp (Figs. 176, 177). An enamel organ and its associated dental papilla are the developmental basis of each tooth. Ten such primordia of the *deciduous,* or *milk teeth* are present in each jaw of a ten-weeks' fetus (Fig. 173 *B*). Later the stalk connecting the enamel organ with the dental lamina, and most of the lamina itself, break down (Fig. 175 *E*). However, the actual free edge of the lamina persists longer and gives rise to the primordia of the permanent enamel organs.

THE ENAMEL ORGAN. This primordium not only deposits enamel but also assumes the directing rôle in tooth development. During the third month it becomes a double-walled sac, composed of an outer, convex wall (*outer enamel layer*) and an inner, concave wall (*inner enamel layer*) (Fig. 176). Between the two is a filling of looser ectodermal cells which transform into a stellate reticulum named the *enamel pulp.* The enamel organ first encases the crown portion of the future tooth, molds its shape and deposits enamel there. Later the enamel organ elongates and similarly models the root portion of the dental papilla, which organizes in response to its influence. This extension is called the *epithelial sheath* of the root (Fig. 179 *A*).

Neither the outer enamel cells nor the enamel pulp contributes directly to tooth development, although the building materials of enamel must pass from the nearby blood vessels through their loosely arranged tissue. By contrast, in the region of the future crown of the tooth the cells of the inner enamel layer become columnar and are designated *ameloblasts* (enamel formers), for they produce *enamel* at the ends that face toward the dental papilla (Fig. 178). This deposit takes the form of parallel *enamel prisms,* one for each ameloblast. The enamel is deposited upon a layer of dentine, which precedes it in origin.

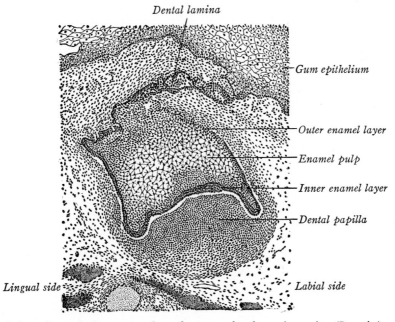

Fig. 176. Primordium of a human tooth, at three months, shown in section (Prentiss). × 70.

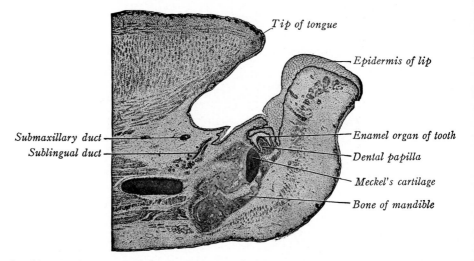

Tip of tongue

Epidermis of lip

Submaxillary duct
Sublingual duct

Enamel organ of tooth

Dental papilla

Meckel's cartilage

Bone of mandible

Fig. 177. Sectioned human lower jaw, at three months, with a tooth primordium *in situ* (Prentiss). × 14.

The enamel substance arises first as a cuticular secretion from the end of an ameloblast; calcification of this 'Tomes process' is secondary. Continued enamel formation produces elongate *enamel prisms,* and these become cemented together (Fig. 179 *C*). As the enamel layer thickens, the ameloblasts are displaced in an outward direction until eventually the internal and external layers of the enamel organ meet. The laying down of an enamel prism by an ameloblast can be compared to the band of toothpaste that is left behind when the tube is squeezed and at the same time drawn away. Enamel is first deposited at the apex of the crown (Fig. 178 *A*); the process then spreads downward in a progressive manner so that the ameloblasts of the neck region are the last to become active (Fig. 180 *B*). Molar teeth have a separate cap of enamel for each cusp; these eventually meet and merge into a compound crown (Fig. 180 *A*). Long before a tooth cuts, its crown is finished and the enamel organ proper undergoes regression. The remains of the enamel organ constitute the transient *dental cuticula* (Nasmyth's membrane).

The *epithelial sheath* of the root is directly continuous with the active part of the enamel organ, but it differs from it both structurally and functionally. The inner layer of epithelial cells remains cuboidal and never produces enamel, while the pulp constituent of the typical enamel organ is lacking (Fig. 179 *A*). Possibly the absence of the latter is significantly correlated with the failure of this region to form enamel.

THE DENTAL PAPILLA.   At the end of the fourth month the superficial cells of the dental papilla arrange themselves in a definite layer that simulates a columnar epithelium (Figs. 178, 179 *A*). These specialized, connective-tissue cells are named *odontoblasts* (tooth formers); actually, 'dentinoblasts' describes them better. Even so, it has not been demonstrated convincingly that odontoblasts are solely responsible for all the dentine substance. The fibrils of the early, jelly-like ground substance that constitutes the *predentine* are continuous with those coursing within the papilla as a whole (Fig. 179 *B*).[1] This soft tissue then calcifies into the definitive *dentine,* or dental bone. Whether the odontoblasts lay down the predentine fibrils is debatable;[2] perhaps they are chiefly concerned with the subsequent deposit of calcium about them. In any region of the tooth, dentine formation precedes slightly the appearance of enamel.

The more central mesenchyme of the dental papilla, internal to the odontoblast layer, differentiates into a soft core-substance (Fig. 179 *A, B*). This is composed of a framework of reticular tissue which binds together blood vessels, lymphatics and nerve fibers. Together with the odontoblast layer it constitutes the *dental pulp,* popularly known as the 'nerve' of the tooth.

As with enamel, the dentine layer is laid down first at the apex of the crown (or cusp of a molar) and then progressively toward the root (Fig. 180 *B*). As the layer thickens, the odontoblast cells retreat before it and so always maintain a more central position. Yet, during the recession, a thread-like process of each odontoblast is spun out and remains behind in the dentine. This *dentinal fiber* (of Tomes) occupies a tiny *dentinal tubule,* essentially like any process of an osteoblast and its canaliculus in ordinary bone (Fig. 178 *B*). The whole odontoblast layer persists throughout the life of a tooth and intermittently lays down dentine. The crowns of the various milk teeth are not completed until 2–11 months after birth, and only then does root development begin. As a preliminary the epithelial sheath elongates, and within this tube the primitive connective tissue is stimulated to condense and organize as it did in the crown. The epithelial sheath of a premolar or molar tooth branches and hence the root comes to have fangs.

THE DENTAL SAC. The mesenchymal tissue surrounding the developing tooth is continuous with that of the early dental papilla. Outside the tooth it differentiates into ordinary connective tissue which constitutes the so-called *dental sac* (Fig. 179 *A*). In the region of the future root the dental sac takes on three important functions: (1) Beginning at the time of eruption its inner cells differentiate into a layer of *cementoblasts.* With the progressive disintegration of the epithelial sheath in a downward direction, these cells deposit upon the dentine an encrustation of specialized bone,

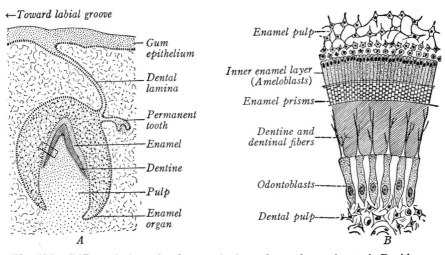

Fig. 178. Differentiation of a human incisor, shown in sections. *A,* Deciduous and permanent primordia, at seven months (× 40). *B,* Detail of the area indicated by a rectangle in *A* (Tourneaux; × about 300).

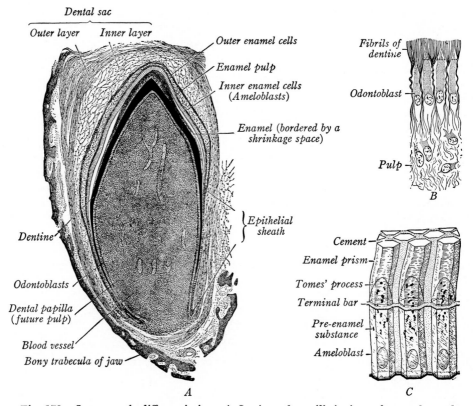

Fig. 179. Later tooth differentiation. *A,* Section of a milk incisor of a newborn dog *in situ* (Stöhr, × 29). *B,* Relation of pulp fibrils to the fibrillar matrix of dentine (after v. Korff; × 350). *C,* Model of enamel prisms and their relation to ameloblasts (× 1000).

known as *cementum.* Deposition proceeds from the neck region downward. (2) During the period of tooth development there has been steady progress in the ossification of the jaw bone. In the region of the teeth the external surfaces of the dental sacs become active in producing bone. Thus each tooth comes to be surrounded with spongy bone, except over its top, and occupies its individual compartment or socket. As a tooth is cut and its root grows to full length, this bone-lined socket, or *alveolus,* reaches a definitive state. (3) The fibrous sac itself consolidates into the thin *periodontal membrane* which holds the tooth in place by embedding some fibers in the cementum and still others in the bony wall of the socket. It is a specialized periosteum.

GROWTH, ERUPTION AND SHEDDING. Progressive growth of the root and other little-understood factors combine in pushing the crown of a milk tooth out of its bony socket, through the overlying portions of the dental sac and gum, and so to the outside. The periods of 'cutting' (*i.e., eruption*) of the various *milk-* or *deciduous teeth* vary with race, climate and nutritive conditions. The permanent teeth in their development eventually press the intervening tissues against the milk teeth (Fig. 181 *A*). The roots of the latter then undergo partial resorption, whereupon their dental pulp is liberated. The combination of tissue loosening and pressure from the permanent teeth leads to the shedding of the milk teeth. Usually the milk teeth are begun, cut, completed and shed at the following times:

	CALCIFICATION OF CROWN BEGINS	TOOTH ERUPTS	CALCIFICATION OF ROOT ENDS	TOOTH SHEDS
Central Incisors....	4　fetal months	6–7½ months	1½ years	7 years
Lateral Incisors....	4½ " "	7– 9 "	1¾ "	8 "
Canines..........	5 " "	16–18 "	3¼ "	12 "
First Molars.......	5 " "	12–14 "	2½ "	10 "
Second Molars.....	6 " "	20–24 "	3 "	11 "

THE PERMANENT TEETH. During childhood and adolescence each jaw elongates considerably at its ends. It can then accommodate three molar teeth on each side that had no counterparts in the provisional set. Otherwise the *permanent dentition* develops essentially like the temporary set. The enamel organs of those permanent teeth that correspond to the milk dentition arise between the fifth and tenth fetal months in another series along the free edge of the disintegrating dental lamina (Figs. 175 *E*, 178 *A*). Located at the same intervals as the deciduous teeth, they come to lie on the lingual side of them and within the same dental sac. In addition, there are the molars (12 in all) not represented in the primary dentition, since the so-called milk molars really correspond to permanent premolars. These permanent molars develop on both sides of each jaw from a backward-growing, free extention of the dental lamina (Fig. 180 *A*).

The primordia of the first permanent molars are present at four months, those of the second molars at birth, while indications of the third perma-

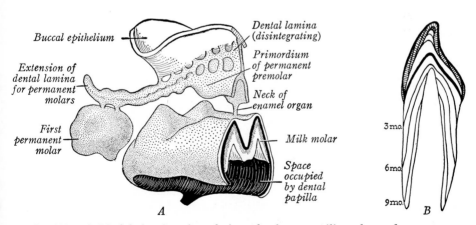

Fig. 180. *A,* Model showing the relation of a human milk molar and a permanent premolar to the dental lamina (adapted after Röse; × 7); at left, a permanent molar is developing from a backward extension of the dental lamina. *B,* The order of deposit of enamel and dentine in a milk incisor is shown at selected intervals after birth; a heavy line delimits the upper portion of the tooth completed at birth.

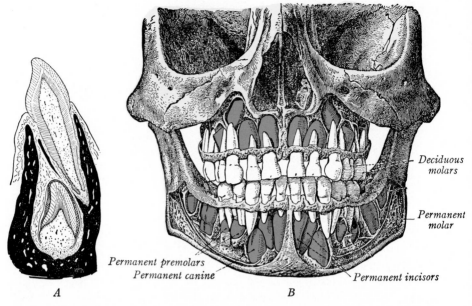

Fig. 181. Relations of the human deciduous and permanent teeth. *A,* Milk canine, with eroding root, and a permanent canine, at four years, sectioned *in situ* ($\times$ 3). *B,* The two dentitions, at five years, in a partly dissected skull (Sobotta).

nent molars, or 'wisdom teeth,' are not found until the fourth to fifth year. Provision for the permanent dentition of 32 teeth is then complete. The permanent teeth develop slowly for a while; later they progress more rapidly and, toward the sixth year, each jaw may contain 26 teeth (Fig. 181 *B*). The enlarging, permanent teeth press against the corresponding milk teeth and aid in their shedding (*A*). The permanent set is then free to remodel the alveoli as needed, to occupy them solely and to erupt.

Deposits of enamel and dentine are rhythmic; from time to time major disturbances, such as birth, weaning, illness and puberty, leave visible records in a series of superimposed *growth lines.*[3] Teeth, like bones, do not follow ordinary organs by continuing to grow after their developmental period ends. For example, a crown has attained full size before its eruption, and its enamel cannot increase further since the ameloblast layer is no longer present. By contrast, dentine is capable of continued development, but only in an inward direction at the expense of the pulp; such intermittent growth may eventually obliterate the root canal. Usually the permanent teeth calcify, erupt and are completed at the following times:

	CALCIFICATION OF CROWN BEGINS	CALCIFICATION OF CROWN ENDS	TOOTH ERUPTS	CALCIFICATION OF ROOT ENDS
Central Incisors....	3– 4 months	4– 5 years	6– 8 years	9–10 years
Lateral Incisors....	3–12 "	4– 5 "	7– 9 "	10–11 "
Canines..........	4– 5 "	6– 7 "	9–12 "	12–15 "
First Premolars....	1½– 2 years	5– 6 "	10–12 "	12–13 "
Second Premolars..	2–2½ "	6– 7 "	10–12 "	12–14 "
First Molars.......	At birth	2½– 3 "	6–7 "	9–10 "
Second Molars.....	2½– 3 "	7– 8 "	11–13 "	14–16 "
Third Molars......	7–10 "	12–16 "	17–21 "	18–25 "

The teeth of vertebrates are homologues of the placoid scales of elasmobranch fishes (sharks and skates) and have a similar development. The teeth of the shark resemble enlarged scales, and many generations of such teeth are produced in the adult fish. Likewise in most reptiles teeth are shed and replaced periodically. The primitive teeth of mammals were of the canine type, and from this conical tooth the incisors and molars have arisen. Just how the cusped tooth differentiated—whether by the fusion of originally separate units or by the development of cusps on a single primitive tooth—is debated.

SUMMARY OF RELATIONS IN TOOTH DEVELOPMENT

```
                ⎧ Main lamina → Degenerates (including necks of enamel organs)
                ⎪                 ⎧ Outer enamel layer ⎫ Pathway for enamel- ⎫ Dental
                ⎪                 ⎪ Enamel pulp......  ⎬   forming materials  ⎬ cuticula
                ⎪ Enamel organs   ⎪                                ↗ Remnant ⎭
Dental lamina ⎨  of deciduous  ⎨ Inner enamel layer → Ameloblasts ⟨ Enamel
                ⎪    teeth         ⎪ Determines shape and size of crown
                ⎪                 ⎪ Organizes papilla and its odontoblast layer
                ⎪                 ⎪              ⎧ Determines shape and size of root
                ⎪                 ⎩ Epithelial sheath ⎨ Organizes root papilla and its odontoblasts
                ⎪ Free edge → Enamel organs of permanent teeth (except molars) ⎫ Fates as
                ⎩ Backward extension → Enamel organs of permanent molars...    ⎭  above

                ⎧               ↗ Dental pulp
                ⎪ Mesenchyme ⟨ Fibrillar basis of dentine (?) ⎫ Dentine
Dental papilla ⎨                                              ⎭
                ⎪                 ↗ Calcifies dentine (?)
                ⎩ Odontoblast layer ⟨ Dentinal fibers (of Tomes)

            ⎧ Outer region → Osteoblasts → Alveolar bone of jaw   ⎫ Alveolar wall
Dental sac ⎨ Middle region → Fibroblasts → Periodontal membrane ⎬
            ⎩ Inner region → Cementoblasts → Cementum
```

Causal Relations. The dependency of an ectodermal mouth, with teeth, on the formative influence of the associated entoderm has been mentioned (p. 216). It is, however, the enamel organ (including the epithelial sheath of the root) that exerts the molding influence under which the entire tooth takes shape and gains size. Through its inductive power the peripheral mesenchyme of the dental papilla is activated to produce the odontoblast layer and to differentiate dentine. A reverse influence is also exerted, since enamel will not form over the crown in the absence of odontoblasts or dentine. An early tooth primordium, transplanted into skin or muscle, will continue to grow and differentiate there. For a time a halved primordium of a molar tooth, grown in culture, is able to regulate and develop as a whole tooth, even producing the normal cusp-pattern.

Anomalies. Developmental abnormalities comprise irregularities in number, size, shape, structure, singleness, position and eruption. *Anodontia* designates a congenital absence of teeth; its mildest expression is the agenesis of a single tooth like the third molar; most extreme is the total suppression of tooth development which is very rare and then is associated with defective skin and its derivatives. Opposite in nature is *hyperdontia* which denotes a production of more than the normal number of teeth, either inserted into the regular series, added (an atavism?) as fourth molars beyond the wisdom teeth, or even occupying abnormal locations outside the dental arch (Fig. 182 *A*). The two regular dentitions are exceeded when representatives of an extra set either precede the milk teeth or follow the permanent set. Persistence of more or less of the milk dentition may be associated with a corresponding failure in the development of the permanent dentition.

Aberrations of form include distortion (Fig. 182 *B*), extra cusps or roots, branched or fused roots (*C*), double crowns (*D*) and fused teeth (*E*). Disturbances (acute diseases; rickets; syphilis) that interfere with enamel deposition produce pits, grooves or even irregular mulberry crowns (*F*); total absence of enamel is known. *Enamel drops,* formed in vesicles of dental-lamina or enamel-organ origin, may occur near a tooth or attached to its root (*G*); *'pearls'* and *cysts,* located in the gums, have a similar origin.

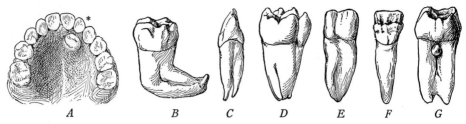

Fig. 182. Human dental anomalies. *A,* Supernumerary milk incisor (*), and an ectopic molar (on palate); *B,* distorted root; *C,* branched root; *D,* double crown; *E,* fused teeth; *F,* faulty enamel deposition; *G,* enamel drop (or pearl).

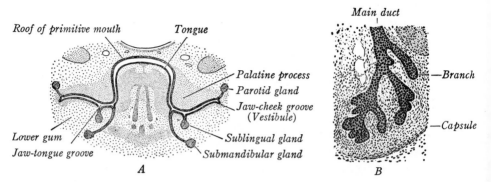

Fig. 183. Origin and early growth of the human salivary glands. *A,* Sites of origin, shown by a diagrammatic frontal section across the jaws at about two months ($\times$ 15). *B,* Detail of the branching submandibular gland, at two months ($\times$ 70).

The Oral Glands. True salivary glands are distinctive characteristics of mammals, the only animals that chew their food. They are usually regarded as ectodermal derivatives, although the site of origin of the submandibular and sublingual glands with respect to the vanished oral membrane is not surely known. All of the salivary glands have a common plan of origin and development.[4] The primordium arises as an epithelial bud and grows by branching into a bush-like system of solid *ducts,* whose end-twigs round out into berry-like, secretory *acini* (Figs. 183 *B,* 184 *A*). Secondary hollowing of the whole system and specialization of the acinal cells complete the epithelial differentiation. A dense mass of mesenchyme, in which the epithelial primordium lies, furnishes an enveloping *capsule* and subdivides the gland into *lobules.* The major salivary glands attain this general plan of organization in the third month. Acinal cells and a canalized duct system are present at six months. As with other glands, acinal differentiation is not complete until some time after birth even though the salivary enzyme is produced by the fetus. The points of origin of the major salivary glands are indicated in Fig. 183 *A,* and their positions and relations at birth in Fig. 184 *B.*

The paired *parotid glands* are the first to appear. In the sixth week (10 mm.) a keel-shaped, epithelial flange has been observed, near each angle of the mouth, growing away from the groove that will divide cheek from gum (Fig. 183 *A*). The flange elongates and, in embryos of seven weeks, separates from the parent epithelium. A tube is then formed by hollowing and this grows backward toward the ear. It soon branches and organizes into the body of the gland, while the stem portion of the tube becomes the *parotid duct* opening into the vestibule (Fig. 184 *B*).

Each *submandibular gland* also arises at the end of the sixth week (12 mm.) as an epithelial ridge, located in the groove between the lower jaw and the tongue and at one side of the midplane (Fig. 183). The caudal end of the ridge soon begins to separate from the epithelium and extend backward and ventrad in the mesenchyme beneath the lower jaw; here it enlarges and branches into the gland proper (Fig. 184). The main stalk, separating in a rostral direction, persists as the *submandibular duct* and opens at the side of the frenulum of the tongue.

Each *sublingual gland* appears during the eighth week as a series of solid buds of epithelium growing downward from the groove between the lower jaw and tongue (Figs. 177, 183 *A*). This group, located just lateral to the submandibular primordium, consists of the sublingual proper, with its major duct (of Bartholin), and of about ten equivalent smaller glands, each with a minor duct. Growth is slower than in the submaxillary gland. The glands lie alongside the tongue and beneath it (Fig. 184 *B*). The major duct opens just lateral to the submandibular duct, or it may join it.

The smaller oral glands (*labial, buccal* and *palatine*) are aggregates that arise at about three months from multiple epithelial buds in their respective locations.

Causal Relations. The system of branching in the submandibular gland is known to be induced by the local mesenchyme of that region.

Anomalies. There may be an absence of any (or, rarely, all) of the salivary glands. Accessory glands occur, as do imperforate ducts that lead to retention cysts. Displacement and mutual fusion are more apt to involve the submandibular and sublingual glands.

The Palate. The mammalian palate is a device for separating the mouth from the nasal respiratory passages, and thus making it possible for the young to suck (and, later, to chew) and breathe at the same time. The two nasal cavities are at first represented by olfactory pits which quickly enlarge into blind sacs, as in adult sharks. The floor of each deepening sac then comes to overlie the roof of the front part of the primitive mouth, and is separated from it by an *oro-nasal membrane* only (Fig. 185 *A*). Both thinning membranes rupture during the seventh week and so create two

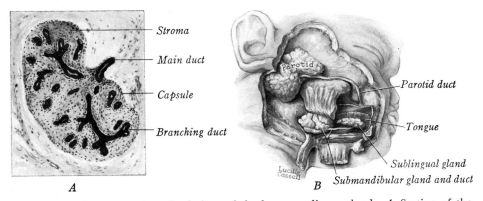

A   *B*

Fig. 184. Later growth and relations of the human salivary glands. *A,* Section of the submandibular gland, at ten weeks (× 40). *B,* Location of the major salivary glands in a newborn (× ½).

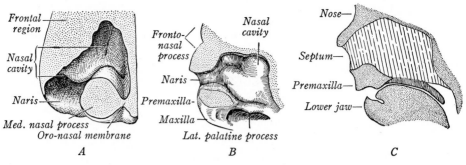

Fig. 185. Separation of the human nasal and oral cavities by the palate. *A,* Medial half of the left nasal sac, at 12 mm. (after Shaeffer; × 45). *B,* Lateral half of the right nasal sac, at seven weeks (× 12). *C,* Relation of the fetal palate and nasal septum, demonstrated by a median section (after Frazer).

internal nasal orifices, the *primitive choanæ* (Fig. 187 *A*). For a short time the two choanæ open directly into the primitive oral cavity whose roof is merely the basal covering of the skull; this simulates the permanent condition in amphibians. Presently, however, the nasal passages are extended backward, becoming separate from the mouth and opening into the pharynx. This is accomplished by means of a partition, the *palate,* that subdivides the primitive mouth cavity horizontally (Fig. 186 *B*). The portion of the primitive mouth, thus added to the olfactory sac, communicates with the pharynx by secondary, definitive *choanæ*. The details of these changes will next be described.

The primordia of the palate are two shelf-like projections that grow from each maxillary-process region of the upper jaw toward the midplane of the mouth cavity (Figs. 185 *B;* 187 *A*). In their growth medially during the seventh and eighth weeks these *lateral palatine processes* encounter the tongue, which rises high at this period, and are forced to bend downward (Fig. 186 *A*). A little later the tongue is withdrawn, because of growth changes, and the lateral palatine processes are then pushed upward to the horizontal plane (*B*). The halves of the palate unite, first with each other and then with the nasal septum. Beginning in the ninth week, this midplane fusion progresses rapidly from the gum-region backward (Fig. 187 *B*). Coincidently bone appears in the front part and forms the *hard palate.* Farther back (where union with the nasal septum does not occur; Fig. 185 *C*) ossification fails. This region constitutes the *soft palate;* the halves of its free apex, the *uvula,* are commonly still notched at birth (Fig. 187 *C*).

Transverse ridges (to aid in the grinding of food) are developed in the mucosal covering of the hard palates of most mammals; their reduced state in man (more so in the adult than in the fetus) is perhaps correlated with the soft nature of his food (Fig. 187 *C*). The folds of the soft palate are invaded from behind by tissue from the third branchial arches; this is responsible for those backward prolongations of the palate, known as the *palatine arches,* which delimit oral cavity from pharynx. From the same source comes the mesenchyme that differentiates into the muscles of the palate. The completed palate shows a median seam, or *raphé,* indicative of the plane of fusion of its two parts.

The median nasal processes, which participate so conspicuously in the formation of the face, also develop so-called *median palatine processes* (Fig. 187 *A*); the latter do not contribute to the palate itself but become the premaxillary portion of the upper jaw (Fig. 185).[5] Fusion between the median palatine processes and the palate is incomplete,

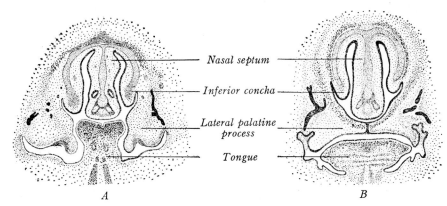

Fig. 186.   Formation of the palate, demonstrated by frontal sections through fetal heads. × 11. *A,* At eight weeks (after Keibel); *B,* at ten weeks (after Kallius).

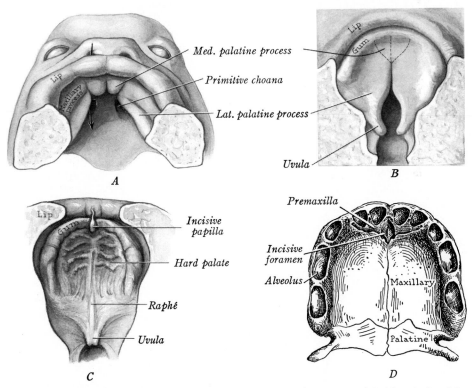

Fig. 187.   Development of the human palate, viewed from the oral side. *A,* At eight weeks (after His; × 9); *B,* at nine weeks (after Peter; × 9); *C,* at birth (× ¾); *D,* bony basis of the hard palate, as shown in the skull of an infant (× 1.5).

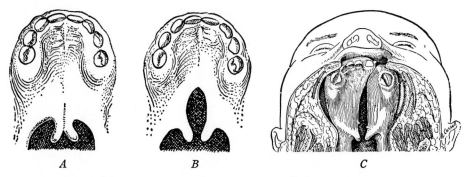

Fig. 188.   Types of human cleft palate. *A,* Cleft uvula. *B,* Cleft involving total soft palate.
C, Cleft involving total soft and hard palates; also combined with severe cleft lip.

so that in the midplane there is a gap, the *incisive foramen,* flanked by the *incisive canals* (of Stenson) (Fig. 187 *D*). These ordinarily become covered with mucous membrane (*incisive papilla; C*) although they sometimes are still open at birth, as is the permanent condition in most mammals (*cf.* Fig. 493 *A*).

Anomalies.   The lateral palatine processes occasionally fail to unite properly, thereby producing a malformation known as *cleft palate,* or *uranoschisis.* The extent of the defect varies considerably. In some persons it involves the soft palate alone and then is median in position (Fig. 188 *A, B*). By contrast, clefts in the hard palate tend to lie at one side of the midline. Both the hard and soft palate may be involved in the same individual (*C*). Cleft palate often accompanies cleft lip (*C*); when double, the premaxilla (formed by the median nasal processes) tends to protrude prominently (Fig. 163 *D*). Epithelial strands may become buried where the edges of the palatine processes fuse, and take the form of 'epithelial pearls' in the newborn or later.

The Hypophysis.   The *hypophysis,* or pituitary body, is an endocrine gland of double origin. One part, obviously glandular in nature, is epithelial; it develops from the ectodermal roof of the stomodeum, which in early stages is adherent to the floor of the fore-brain (Fig. 206).[6] The other component, not so plainly secretory, is a specialized extension from the brain wall. During the subsequent growth of these parts, and the filling-in of mesenchyme between them, both associated regions (stomodeal roof and brain floor) become drawn out into hollow extensions (Fig. 171 *B*).

The stomodeal pocket, known also as *Rathke's pouch,* is located originally just in front of the intact pharyngeal membrane. In embryos about 3 mm. long this pouch is a distinct, shallow sac which quickly enlarges and flattens against the tubular projection from the floor (infundibulum) of the fore-brain (*cf.* Fig. 601). This projection is the future *neural lobe* of the hypophysis. Meanwhile the connection of Rathke's pouch with the oral epithelium has elongated into a slender stalk which lags in development and vanishes by the end of the second month (Fig. 189 *A*). The front wall of the epithelial pouch thickens greatly and becomes the important secretory mass known commonly as the *anterior lobe* (*B*). The back wall of this pouch (the *pars intermedia*) remains thin and lacks functional significance in mammals. The cavity of the closed Rathke's pouch becomes the *residual lumen* of the mature gland; the final condition in man is unique, since after childhood this cleft becomes reduced to cysts or even obliterates completely. The neural lobe differentiates certain specialized cells, but its secretion arises in neurosecretory cells of the brain and drains down the infundibular stalk.[7] Combined with the pars intermedia, the neural lobe makes a complex commonly designated as the *posterior lobe*. During the third and fourth months the hypophysis attains its characteristic shape and arrangement. Somewhat later, cellular differentiation brings about the typical regional organization of this organ (Fig. 190 *A*).

The thick rostral wall of the pouch (technically, the *pars distalis*) differentiates into cords containing highly specialized cells; these cords are interspersed between abundant sinusoids (Fig. 190 *A*). By the tenth week of fetal life all specialized cell types are distinguishable. Later the growth hormone, and still later the gonadotrophic hormones, become detectible when used experimentally. The thin, caudal portion of the pouch-wall, lying between lumen and neural lobe, is named the *pars intermedia;* in man it forms epithelial cysts but is not prominent (Fig. 190 *A*). Another glandular region, the *pars tuberalis,* wraps about the infundibular stalk (Fig. 189 *B*). It develops from the fusion of a pair of early wing-like lobes that bud off laterally from the main pouch. The mostly solid *infundibular stalk* connects the brain (diencephalon) permanently with the swollen end of the stalk which is officially designated the *pars nervosa*. This solid mass is com-

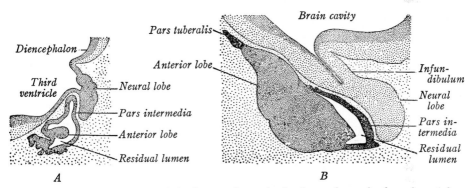

Fig. 189. Early development of the human hypophysis, shown by sagittal sections (after Atwell). × 35. *A*, At eight weeks; *B*, at eleven weeks.

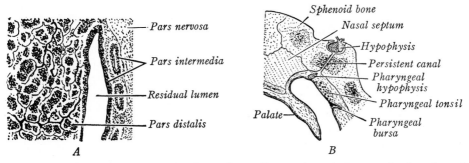

Fig. 190. The later hypophysis, shown in median section. *A*, Area at the junction of the two lobes, at eight months (× 45). *B*, Relations of the main gland and pharyngeal hypophysis in the newborn. × 1.

posed of neuroglial tissue and nerve fibers. Branching cells (*pituicytes*) are modified neuroglial elements.

Growth of the facial region causes the jaws to jut forward, thus deepening greatly the originally shallow stomodeum and making its contribution to the total mouth substantial indeed. After this merger and subsequent growth changes have been accomplished, the original site of Rathke's pouch (which was virtually external) comes to be located well back on the roof of the primitive mouth (Figs. 209, 210). This spot lies at the dorsal and caudal border of the nasal septum (Fig. 190 *B*).

Causal Relations. The cranial end of the notochord is said to induce the development of the neural lobe, while the neural lobe and the brain wall together induce the differentiation of the anterior lobe. The pars intermedia is determined as such by contact with the floor of the fore-brain (diencephalon).

Anomalies. Except when the brain is severely underdeveloped, a poorly differentiated or absent hypophysis is almost unknown. The original course of the stalk of Rathke's pouch is sometimes marked by a canal in the sphenoid bone (Fig. 190 *B*). This *craniopharyngeal canal,* however, is a secondary feature which develops after the stalk has degenerated;[8] it usually obliterates before birth. Residues of the stalk are rare, except for a constant mass located between the nasal septum and the pharyngeal tonsil.[9] It is known as the *pharyngeal hypophysis* (Fig. 190 *B*).

THE PHARYNX

The fundamental importance of the early pharynx would scarcely be suspected from its adult simplicity and unspectacular rôle as a common corridor for the crossing pathways of air and food. Nevertheless, the primitive pharynx is the source of numerous organs and its developmental history is correspondingly complex. Most of these activities occur during the transitional period when the mammalian embryo passes from a stage in which the pharynx is arranged as for branchial respiration to a stage in which the breathing of air is anticipated. Naturally the parts altered most profoundly are the branchial arches and pharyngeal pouches themselves. In addition to the remodeling that is necessary to provide for the mammalian method of chewing and swallowing, various other conversions occur.

The Branchial Arches. In a previous chapter (p. 200) were described the general relations and significance of the branchial arches. These ancestral gill arches are converted into numerous things—including the neck, jaws, face and external ear, already discussed (p. 201 *ff.*), and various arteries, muscles, cartilages and bones to be considered in later chapters. On the floor of the pharynx they contribute especially to the *tongue.* The nearby *larynx,* also of branchial origin, belongs to the respiratory system. Origins and fates in the branchial regions are tabulated on p. 203; inspection of these lists will re-emphasize how important this territory really is.

THE TONGUE. This organ is primarily a pharyngeal derivative. Developmentally it is a mucous-membrane sac which becomes stuffed with voluntary muscle. The tongue arises from the ventral ends of the branchial arches. It consists of two different parts—one oral in origin, the other pharyngeal (Fig. 191). The oral portion, comprising most of the *body* of the tongue, occupies the definitive mouth cavity. It arises from the mandibular arches, in front of the oral membrane, and hence is covered with ectodermal epithelium. This part of the tongue bears papillæ and is concerned with mastication. The pharyngeal portion is the *root* of the tongue. It develops primarily from the union of the second branchial arches, but receives important contributions from the third and, apparently, the fourth arch as well. The epithelial covering is entoderm, beneath which is an infiltration of lymphoid tissue. The root of the tongue is mainly concerned with swallowing. The junction-line between ectoderm and entoderm is apparently in front of the row of vallate papillæ, whereas the body and root are demarcated by a **V**-shaped groove, the *terminal sulcus,* just behind these papillæ (Fig. 192 *D*).

In embryos of four weeks (5 mm.) the body of the tongue is indicated by three primordia (Fig. 192 *A*). These are: (1, 2) paired *lateral swellings* of the first branchial arches, which latter have fused already as the mandible; and (3) the median, somewhat triangular *tuberculum impar (i.e.,* unpaired tubercle) wedged in between the lateral swellings. At the same time the future root is represented by a median elevation, brought into existence by the union of the bases of the second branchial arches; for this reason it is named the *copula (i.e.,* a yoke). Between the tuberculum impar and the copula is the point of origin of the thyroid diverticulum (Fig. 191). This site becomes secondarily depressed by lingual growth-stresses into a prominent pit which occupies the apex of the terminal sulcus and constitutes the permanent landmark known as the *foramen cæcum* (Fig. 192 *D*). The progressive merger of the several lingual components can be followed in the stages shown as Fig. 192. Continued expansion of the tongue in both length and breadth brings into existence a deep, ∩-shaped furrow at the front and along the sides which will make this organ partly free and highly mobile. At the same time (seventh week), the tongue elevates and assumes prominence through the differentiation of voluntary muscle internally (Fig. 177). The body of the tongue acquires *papillæ* of several kinds, whereas the root is the site of numerous pits known as the *lingual tonsil* (Figs. 192 *D*, 193).

The individual components of the tongue are in process of fusion during the sixth week. The lateral swellings of the first arches increase rapidly in size, unite with the

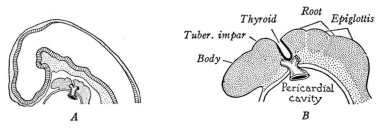

A   *B*

Fig. 191.   The tongue, at about four weeks, in diagrammatic sagittal section. *A,* General relations; *B,* detail of *A.*

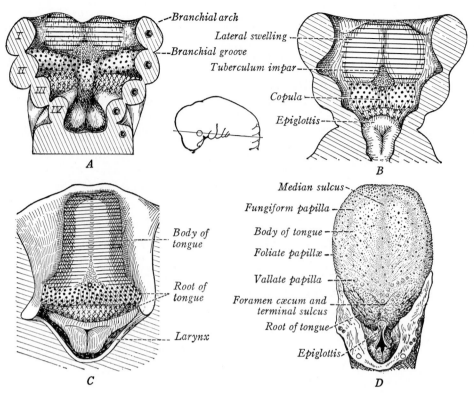

Fig. 192. Development of the human tongue, viewed from above. *A,* At 6 mm. (key figure shows section planes of *A–C*); *B,* at 9 mm.; *C,* at 15 mm.; *D,* at birth (× 1). Distinctive markings indicate the contributions of the branchial arches. The tuberculum impar is designated by circles, and at its base a larger circle indicates the foramen cæcum.

tuberculum impar, and nearly enclose it (Fig. 192 *B, C*). In sharp contrast, the tubercle itself lags in development and becomes progressively less conspicuous. In the end, it furnishes little to the *body* of the final organ.[10, 11] The plane of union of the two lateral swellings is indicated superficially by the *median sulcus* and internally by the fibrous, median *septum*. The copula, together with the adjacent portions of the second branchial arches, enlarges greatly to produce the early *root* of the tongue (*B*). Eventually, however, this territory seems not only to be encroached upon by the third and fourth branchial arches (which primarily form the epiglottis; *C*), but there is a slipping forward of their mucous membrane as well. This conclusion is supported circumstantially by the fact that the sensory portions of the trigeminal and facial cranial nerves (*i.e.,* the nerves of the first and second branchial arches) ultimately supply the epithelium of the body of the tongue, while the glossopharyngeal and vagus nerves (*i.e.,* the nerves of the third and fourth arches) supply the root.

The striated *musculature* of the tongue is innervated by the twelfth, or hypoglossal nerves, and both nerves and muscle belong ancestrally to the occipital region of the head, just caudad of the branchial arches. It is believed that during phylogeny the tongue migrated craniad and invaded the branchial region, retaining its nerve supply the while. Such an invasion would also explain satisfactorily the forward dislocation of the mucosa, already mentioned. In present-day embryos this migration is difficult to trace, although it has been demonstrated in the chick and kitten.[12, 13] Except for slight indications suggestive of migration,[14] the muscles of the human tongue appear to arise *in situ* from the mensenchyme of the arches that make up the floor of the mouth.[15] The temporary elevation of the tongue to a position between the developing halves of the palate (Fig. 186 *A*) is corrected by a withdrawal, due to growth change in the lower jaw; the tongue is then able to broaden greatly (*B*).

The early cuboidal epithelium that covers the tongue is changing to a stratified type at two months. Lingual papillæ are confined chiefly to the oral, or masticatory part of the tongue (Fig. 192 *D*).[16] In fetuses of 9 and 11 weeks, respectively, the *fungiform* and *filiform papillæ* may be distinguished grossly as elevations of the mucosa (Fig. 193 *A*). The *vallate papillæ*, which are entodermal, develop along a V-shaped epithelial ridge just in front of the terminal sulcus. At intervals there appear about nine elevations (*B*). In the tenth week a thickened, epithelial ring delimits each elevation; this ring then grows downward and takes the form of a hollow cylinder (*C*). During the fourth month circular clefts split apart such epithelial downgrowths (*D*), thus separating the sides of the papilla proper from the surrounding wall and forming the trench, or valley, from which this type of papilla derives its name (*E*). At the same time lateral outgrowths arise from the bases of the downgrown ring; they hollow into the *glands of Ebner*. The *foliate papillæ* become distinguishable as parallel folds, along the sides of the tongue, during the third month (*cf.* Fig. 192 *D*). *Taste buds* begin to be indicated at about eight weeks; the details of their development are given on p. 523. One taste bud precedes the appearance of each fungiform papilla and then occupies its top surface (Fig. 193 *A*). Several buds occur also on the dome of an early vallate papilla (*C, D*); these vanish before birth and are replaced by definitive buds on the sides and on the trench wall (*E*).

The *lingual tonsil* is foreshadowed in the fifth month by an infiltration of lymphocytes into the root of the tongue, whereas the pit-like crypts do not differentiate until the time of birth (Fig. 204).

Anomalies. The tongue may be abnormally large (*macroglossia*) because of its oversized muscle mass (Fig. 9 *F*). Smallness (*microglossia*) or absence (*aglossia*) is referable to a failure of growth or development, respectively. Developmental arrest in the body of the tongue

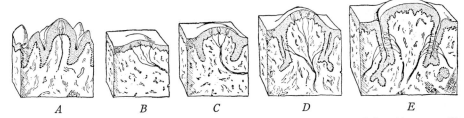

     A   *B*   *C*   *D*   *E*

Fig. 193. Models of developing lingual papillæ. *A*, Filiform and fungiform papillæ, at eleven weeks; *B–E*, vallate papilla, at two to five months. Taste buds are present; stages of Ebner's glands occur in *D, E*.

Fig. 194. Anomalous trifid tongue.

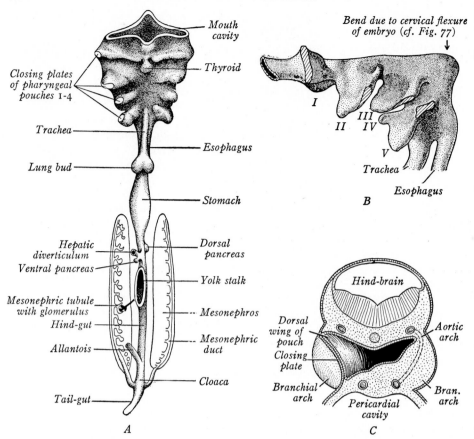

Fig. 195. Early human pharynx, shown as models. *A,* Entodermal tract, at 5 mm., in ventral view (Prentiss; × 25). *B,* Pharynx, at 7.5 mm., viewed from left side (after Kingsbury; × 40). *C,* Pharynx, at 8 mm., sectioned obliquely to show a pouch, at left, and an arch at right (after Frazer).

leads to a bifid organ, as in snakes; even a trifid tongue (unfused lateral components and tuberculum impar) has been known (Fig. 194).

The Pharyngeal Pouches.   The lateral walls of the entodermal pharynx give rise to a series of paired sacculations that extend outward toward corresponding ectodermal grooves. The early relations between these *pharyngeal pouches* and the *branchial grooves* (and branchial arches) are illustrated in Fig. 192 *A*. The pairs of pouches arise in succession in a caudalward direction. Toward the end of the fourth week (4 mm.) five sets have been formed, the last pair being atypical and attached to the fourth (Fig. 195 *A, B*). Meanwhile the pharynx has flattened dorsoventrally and broadened at its cranial end; as a result, it is triangular in outline.

Each typical pouch develops a dorsal and a ventral wing (Fig. 195 *B, C*).[17] Also, in expanding, the pouch pushes aside the intervening mesenchyme and comes into contact with the ectoderm of the corresponding branchial groove. The two layers meet and thus produce a *closing plate* (*C*). Although the plates become perforate in human embryos only occasionally, each combination of pouch and groove, nevertheless, is homologous to a functional branchial cleft of fishes and tailed amphibia; their transitory appearance is an illustration of an unerased, ancestral imprint. The existence of an actual plate is brief, because invading mesenchyme again separates the two component layers. The first and second pharyngeal pouches of each side soon extend off from a broad, lateral expansion of the pharynx (Fig. 157 *B*). The third and fourth pouches, growing laterad, continue to communicate with the pharyngeal cavity through narrow ducts (196 *B*). The questionable fifth pouch is merely a blind diverticulum that shares a common duct with the fourth pouch.

The fates of the entodermal pouches are varied and spectacular (Fig. 196). Although the several derivatives do not continue as parts of the digestive apparatus, their embryonic relations justify their inclusion in the present chapter. The first pouch retains its lumen and differentiates into the *auditory (Eustachian) tube* and the *tympanic cavity* of the middle ear. The second is greatly reduced and becomes the fossa and covering epithelium of the *palatine tonsil*. The third, fourth and fifth lose all trace of a lumen and give rise to a series of specialized organs; these are the *thymus, para-thyroids* and *ultimobranchial bodies*.

Causal Relations. A branchial groove arises only when a corresponding entodermal pouch develops and makes adequate contact with the ectoderm.

THE AUDITORY TUBES AND TYMPANIC CAVITY. The first pair of pouches is the only one that retains its embryonic relations in a recognizable manner. Each main pouch is drawn out during the eighth week into an *auditory tube,* whereas its dorsal angle expands into the *tympanic cavity*

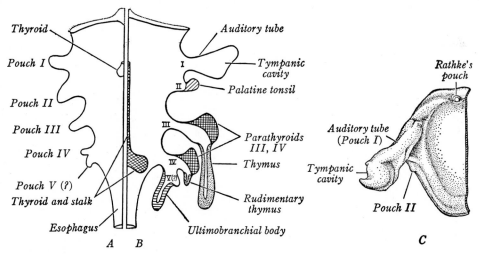

Fig. 196. The human pharynx and its derivatives. *A*, Right half, in ventral outline, at four weeks (× 40). *B*, Left half, in ventral outline, at six weeks (× 25). *C*, Model of left half, in dorsal view, at eight weeks (after Hammer; × 14).

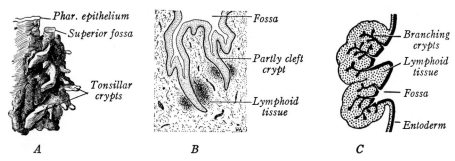

Fig. 197. Development of the human palatine tonsil. *A,* Model of the crypt system, at fourteen weeks (× 18). *B,* Vertical section, at fourteen weeks (× 35). *C,* Diagrammatic vertical section, at five months (× 7).

of the middle ear (Fig. 196 *B, C*). An unconfirmed observation interprets the second pouch, as well, as becoming absorbed into this tract.[18] Simultaneously with the transformation of the first entodermal pouch, the overlying ectodermal groove deepens to produce the *external acoustic meatus.* This canal leads inward to the closing plate which regains a middle layer of mesoderm and persists as the *tympanic membrane* (Fig. 514). Additional details on the development of these derivatives are given on p. 548.

THE PALATINE TONSILS. By the growth and lateral expansion of the pharynx, the second pouch of each side is largely absorbed into the pharyngeal wall (Fig. 196 *C*). It would seem that part of the pouch persists as the *tonsillar fossa,* while its entoderm furnishes the *epithelium* covering the tonsil and lining its crypts. The *crypts* arise progressively in fetuses of three to six months as solid ingrowths from the surface epithelium (Fig. 197 *A*). They branch and hollow secondarily (*B, C*); many of the branches degenerate and reform after birth.[19] Lymphocytes appear near the epithelium in the third month (*B*) and organize as nodules after the sixth month; the arrangement of *lymphoid tissue* makes the tonsil temporarily bilobed (*C*). An enveloping *capsule* and an internal connective-tissue *framework* arise at five months from mesenchyme in the immediate vicinity of the developing organ. The permanent location and relations of the palatine tonsil are shown in Fig. 204.

Some studies have cast doubt on the direct and continuous relation of the second pouch to the palatine tonsil. Rather, it is urged that the tonsil develops at the general site of the second pouch merely because this is a neutral, favorable position in a region of marked growth shiftings.[20] Similarly, the anterior and posterior pillars of the tonsil represent merely the general positions of parts of the second and third branchial arches (Fig. 159 *A*).

Anomalies. The palatine tonsils are highly variable, ranging from stalked, protruding organs to 'buried' types lying in a deep sinus. Cervical cysts, blind diverticula and complete fistulæ occur lateral to the midplane (Fig. 159 *B, C*). They are usually related to the second branchial clefts, and particularly to the cervical sinus, whose drawn-out portion may then open at the tonsillar fossa (p. 204; Fig. 157).

THE THYMUS. Toward the end of the sixth week each third pharyngeal pouch shows a pronounced ventral sacculation (Fig. 198 *A*), and the entire pouch is set free in the week following (*B, C*).[21] At first hollow, these thymic primordia rapidly become solid epithelial bars. The lower ends enlarge and unite superficially during the eighth week, thereby foreshadowing the definitive organ; yet the thymus never loses wholly its paired nature (Fig. 199 *A*). The two lower ends are attached to the pericardium and gradually sink with the latter to a permanent position in the thorax (Fig. 200). During this descent the upper ends become drawn out and finally vanish.

By the tenth week the cells of the thymic epithelium are transforming into stellate elements, and the whole comes to resemble a sparsely fibrous reticular tissue (Fig. 199 *B, C*). This spongework furnishes a delicate support for the entire organ. Some of entodermal cells aggregate into tiny ball-

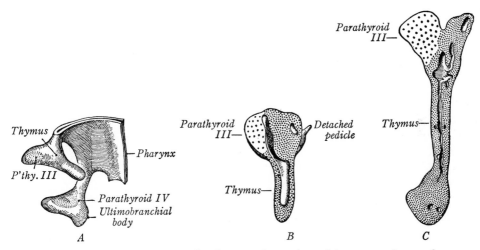

Fig. 198. The third and fourth pharyngeal pouches of human embryos, shown as models (Weller). *A*, At 10 mm., in ventral view (× 65). *B, C*, Detached third pouches, at 14 mm., hemisected lengthwise (× 75).

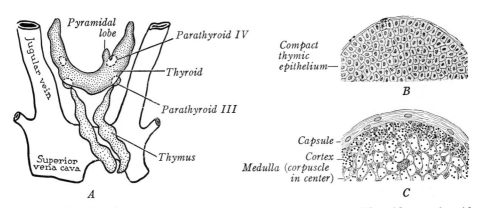

Fig. 199. Glandular derivatives of the human pharynx. *A*, Thyroid, parathyroids and thymi, at two months, in ventral view (after Verdun; × 15). *B*, Thymus, at seven weeks, in section (× 200). *C*, Thymus, at three months, in section (× 150).

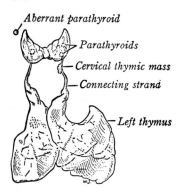

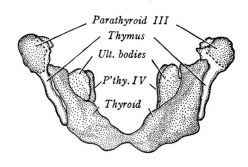

Fig. 200. Thyroid, parathyroids and thymi of a human newborn, in ventral view. × ⅔. An aberrant parathyroid and accessory thymic tissue also are seen.

Fig. 201. Glandular derivatives of the human pharynx, at 14 mm., shown as a model in ventral view (after Politzer and Hann). × 45.

like masses known as *thymic corpuscles* (of Hassal) (*C*).[22, 23] Toward the end of the third month the thymus is becoming increasingly populated with lymphocytes and is differentiating into a denser *cortex* and looser *medulla* (*C*). The organization of a *capsule* and *trabeculæ* from surrounding mesenchyme, and the partial subdivision of the organ into many lobules, complete the process of development. The thymus enlarges steadily until puberty; it then begins to regress, although persisting in reduced form even into old age.

The epithelium of the cervical sinus attaches to the third pouch, but there is no clear evidence that this ectoderm contributes to the definitive organ.[24] An origin of the characteristic small cells (so-called *thymocytes*) of the thymus from entodermal reticulum has been reasserted of late.[25, 26] This conclusion stands in opposition to the view that these elements are migratory lymphocytes that invade the organ from without. In addition to the main thymus, other small thymic masses are frequently encountered in embryos. They arise tardily in the region of the fourth pouches (ventral diverticula of them?), but their exact relations have not been surely traced.[27]

Anomalies. The slender upper ends of the early thymus sometimes persist. They either continue the thymus to the level of the thyroid gland or form separate accessory lobes there, depending on whether they persist wholly or in part (Fig. 200).

THE PARATHYROID GLANDS. The dorsal wing of each third and fourth pouch thickens into a solid mass of cells which is prominent at 10 mm. (Fig. 198 *A*). Each is the primordium of a parathyroid gland. A few days later the two pairs of globular parathyroids (designated III and IV) are set free from the pharynx, although they still remain connected for several more days with the simultaneously detached thymi and ultimobranchial bodies, respectively (Fig. 198 *B, C*). The pair from the third pouches is drawn down by the migrating thymic primordia (Fig. 201) to the level of the caudal border of the thyroid gland and become the *inferior parathyroids* of adult anatomy (Fig. 199 *A*). The pair from the fourth pouches does not shift its position appreciably and remains at the cranial thyroid border as the *superior parathyroids.*[28] Even at their earliest appearance, the glands are differentiating into their distinctive, clear cells; acidophilic cells do not appear until late childhood. All four glands embed superficially in the thyroid capsule and gradually become well vascularized.

Anomalies. Variations in the number, size and location of the parathyroid glands are common (Fig. 200). Both regular and accessory glands may locate at some distance from the thyroid; parathyroids III sometimes even follow the thymus into the thorax.

THE ULTIMOBRANCHIAL BODIES. In the fifth and sixth weeks these sacs are often classified as rudimentary fifth pouches, although there is some doubt as to their true status (Fig. 198 *A*).[29] At the beginning of the seventh week (13 mm.) each ultimobranchial body, joined with the adjacent parathyroid IV, is set free from the pharynx. Meanwhile growth of the thyroid brings its two lobes into contact with the ultimobranchial bodies (Fig. 201). Each of the latter then loses its cavity and is incorporated into the thyroid. Some believe that the final fate of the ultimobranchial bodies is degeneration;[27, 29] others describe an actual conversion ('induction') of ultimobranchial into thyroid tissue, apparently through the dominating influence of a thyroid environment on a plastic, implanted tissue.[28, 30] Still others trace these bodies into the parafollicular (calcitonin) cells of the thyroid.[36, 37]

The Thyroid Gland. The thyroid is the earliest glandular structure to appear.[31, 32] Even an embryo 2 mm. long (six somites) shows an external

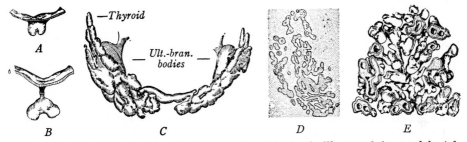

Fig. 202. Development of the human thyroid gland, illustrated by models (after Weller and Norris). *A*, At 4.5 mm., showing the bilobed primordium attached to the floor of the pharynx (× 40). *B*, At 8 mm., showing the thyro-glossal duct (× 40). *C*, At seven weeks (× 40). *D*, At nine weeks, showing (in section) follicular cavities beginning to appear in beaded portions of the epithelial plates (× 40). *E*, At three months, showing the subdivision of epithelium into follicles, some of which are cut across (× 110).

bulge on the ventral floor of its fore-gut that indicates the site of thyroid origin (Fig. 170 *B*). A distinct entodermal outpocketing, the *thyroid diverticulum,* soon protrudes and lies between the first pair of pharyngeal pouches (Fig. 195 *A*). This sac attaches to the pharynx by a narrower neck which is known as the *thyro-glossal duct* (Fig. 202 *B*). It is so named because it is hollow at first and connects the primitive thyroid with the tongue which is organizing from the pharyngeal floor at the same time; here it opens at the aboral end of the tuberculum impar (Fig. 191). The duct presently becomes a solid stalk and then fragments in the sixth week, but its point of origin

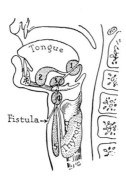

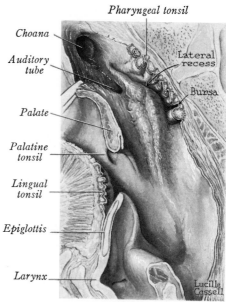

Fig. 203. Diagram showing the course of the thyro-glossal stalk. Along it are indicated the commonest sites (1–5) for cysts and for a fistula (after Chemin).

Fig. 204. Right hemisection through the pharynx of a newborn. × 2.

on the tongue is indicated permanently by an enlarged pit named the *foramen cæcum* (Fig. 192 *D*).

The thyroid sac quickly becomes a solid mass which lies against the primitive aortic stem (Fig. 191). It is bilobed at an early stage (Fig. 202 *A, B*) and, when set free by the atrophy of its stalk, the thyroid begins to be converted into an irregular mass of epithelial plates (*C*). Early in the seventh week the gland becomes crescentic in shape and settles to a transverse position, with a lobe on each side of the trachea. Actually its shift is illusory and is caused by the forward growth of the pharynx, which leaves the aortic trunk and thyroid behind. During the seventh week the enlarging ultimobranchial bodies come in contact with the thyroid primordium and fuse with it (Fig. 202 *C*), thus forcing the thyroid to part company with the aorta and pericardium. As discussed in a previous paragraph, the ultimobranchial bodies rapidly lose their original identity and possibly transform into thyroid tissue. In the eighth week discontinuous cavities begin to appear in swollen or beaded portions of the solid thyroid plates (*D, E*); these represent the beginning of *follicles* which acquire colloid in the third month and soon afterward become functional. By the end of the fourth month this conversion into follicles ends; thereafter new follicles arise only by the budding and subdivision of those already present. A *capsule* and vascular *stroma* differentiate from the local mesenchyme.

Anomalies. Absence of the thyroid results in the cretin type of dwarfism (Fig. 9 *G*). Persistent portions of the thyro-glossal stalk give rise to accessory thyroids, cysts or even, secondarily, to a median fistula opening on the neck (Fig. 203). Cysts lie in the midplane and at, or just below, the hyoid bone. *Accessory thyroids* may also be derived from detached portions of the main primordium. The variable pyramidal lobe of the thyroid, leading upward from the gland, results from the retention and growth of the lower end of the thyro-glossal stalk (Fig. 199 *A*). Failure to descend properly leaves the thyroid located in the base of the tongue.

The Pharynx Proper. This funnel-shaped passage is the residual product after the transformation of its roof, floor and side walls is finished. The original epithelium becomes pseudostratified to stratified. As in the mouth and upper esophagus, the neighboring mesenchyme differentiates a coat of striated (voluntary) muscle. Besides the organs already described, there are a few additional derivatives (Fig. 204):

THE PHARYNGEAL TONSIL. The entrance to the definitive pharynx is in a sense encircled by a lymphoid ring. In addition to the lingual tonsil below and the paired palatine tonsils at each side, there is still another tonsil mass that lies in the dorsal wall. This *pharyngeal tonsil* starts development in the fourth month; its lymphoid accumulation is a response to local vascularity and freedom from growth tensions. The so-called crypts are merely epithelial folds, wrinkled by the stresses of this region,[33] and peculiarly dilated ducts of mucous glands.[34]

The *pharyngeal bursa* is a pit located just below the pharyngeal tonsil. It results from the ingrowth of epithelium along the course of the degenerating tip of the notochord.[35] Until the end of the second month the latter is fused with the epithelium of the pharynx at this point. *Seessel's pouch* is merely the dorsal, blind end of the entodermal fore-gut which, after the loss of the pharyngeal membrane, persists for a short time as a sort of pit (Fig. 590). It has no further significance. The lateral *pharyngeal recess* (of Rosenmüller) is a secondary formation, not related to the second pharyngeal pouch, as formerly claimed. The *piriform recess,* at each side of the entrance to the larynx, marks the site of the earlier third and fourth pouches (Fig. 159 *A*).

DERIVATIVES OF THE PHARYNX AND ITS POUCHES

REGION	LEVEL OF POUCH I	LEVEL OF POUCH II	LEVEL OF POUCH III	LEVEL OF POUCH IV	LEVEL OF POUCH V
Roof:	Caudal end of soft palate?	Pharyngeal tonsil. Pharyngeal bursa.			
Sides: Dorsal wing of pouch.	Tympanic cavity: Lining of drum. Mastoid cells. (Rest of pouch: Auditory tube.)	Palatine tonsil. Fossa. Epithelium of: Surface. Crypts. (Tonsil develops at	Inferior parathyroid gland.	Superior parathyroid gland.	An atypical pouch. Derivative: Ultimobranchial body (lateral thyroid).
Sides: Ventral wing of pouch.	Obliterates as such.	site of pouch, but perhaps not from it.)	Thymus: Reticulum. Corpuscles.	Rudimentary thymus in some specimens.	
Floor: (Arch relations.)	Tongue body (I, II). Thyroid gland. Foramen cæcum.	Tongue root (II, III). Lingual tonsil.	Epiglottis (III, IV).	Larynx (IV, V). Trachea. Lungs.	

REFERENCES CITED

1. Bevelander, G. 1941. Anat. Rec., *81,* 79–97.
2. von Korff, K. 1928. Anat. Anz., *64,* 383–395.
3. Massler, M., *et al.* 1941. Am. J. Dis. Child., *62,* 33–67.
4. Thoma, K. H. 1919. J. Dent. Res., *1,* 95–143.

5. Noback, C. R. & M. L. Moss. 1953. Am. J. Phys. Anthrop., N. S. *11*, 181–187.
6. Gilbert, M. S. 1935. Anat. Rec., *62*, 337–359.
7. Palay, S. L. 1963. Am. J. Anat., *93*, 107–141.
8. Arey, L. B. 1950. Anat. Rec., *106*, 1–16.
9. Boyd, J. D. 1956. J. Endocr., *4*, 66–77.
10. Guthzeit, O. 1921. Ueber die Bedeutung des Tuberculum impar. Inaug.-Diss. Univ. Leipzig, 45 pp.
11. Hammar, J. A. 1901. Anat. Anz., *19*, 570–575
12. Hazelton, R. D. 1970. J. Embr. & Exp. Morph., *24*, 455–466.
13. Bates, M. N. 1948. Am. J. Anat., *83*, 329–355.
14. Frazer, J. E. 1926. J. Anat., *61*, 132–143.
15. Lewis, W. H. 1910. Chapter 12 in Keibel and Mall: Human Embryology (Lippincott).
16. Hellman, T. J. 1921. Upsala läk. Förhandl., *26*, 1–72.
17. Politzer, G. & F. Hann. 1935. Z'ts. Anat. u. Entw., *104*, 670–708.
18. Frazer, J. E. 1914. J. Anat. & Physiol., *48*, 391–408.
19. Minear, W. L. & L. B. Arey. 1937. Arch. Otolaryng., *25*, 487–519.
20. Ramsay, A. J. 1935. Am. J. Anat., *57*, 171–203.
21. Norris, E. H. 1938. Carnegie Contr. Embr., *27*, 191–208.
22. Dearth, O. A. 1928. Am. J. Anat., *41*, 321–352.
23. Kostowiecki, M. 1930. Bull. Internat. Acad. polon. de Sci. de Cracovie Sc. Nat. (Ser. B), 589–628.
24. Garrett, F. D. 1948. Anat. Rec., *100*, 101–113.
25. Auerbach, R. 1961. Dev. Biol., *3*, 336–354.
26. Good, R. A., *et al.* 1962. J. Exp. Med., *116*, 733–796.
27. Van Dyke, J. H. 1941. Anat. Rec., *79*, 179–209.
28. Norris, E. H. 1937. Carnegie Contr. Embr., *26*, 247–294.
29. Kingsbury, B. F. 1939. Am. J. Anat., *65*, 333–359.
30. Politzer, G. 1936. Z'ts. Anat. u. Entw., *105*, 429–432.
31. Sgalitzer, K. E. 1941. J. Anat., *75*, 389–405.
32. Norris, E. H. 1916; '18. Am. J. Anat., *20*, 411–448; *24*, 443–466.
33. Snook, T. 1934. Am. J. Anat., *55*, 323–341.
34. Arey, L. B. 1947. Am. J. Anat., *80*, 203–223.
35. Snook, T. 1934. Anat. Rec., *58*, 303–319.
36. Pearse, A. G. E. & A. F. Carvalheira. 1967. Nature, *214*, 929–930.
37. Stoeckel, M. E. & E. Porte. 1970. Zts. Zellforsch. u. mikr. Anat., *106*, 251–268

Chapter XIV. The Digestive Tube and Associated Glands

THE DIGESTIVE TUBE

The digestive canal proper (esophagus, stomach and intestine) exhibits a rather uniform developmental history, except in such details as size, shape, position and glandular specialization.[1] It consists originally of: (1) an internal tube of entoderm, which is the primary tissue that becomes the *epithelial lining* (including glandular ingrowths; and (2) an investing layer of splanchnic mesoderm that specializes into the thick, supporting wall (Fig. 205, *A, B*). In accordance with the general progress of development in a cranio-caudal direction, higher levels of the digestive tube begin specialization sooner than lower levels and maintain this advantage for some time. As development proceeds, the mucosal lining expands faster than the outer wall and so becomes thrown into folds (*C, D*); these anticipate a future distention by food and also provide additional secretory and absorptive surface. The mesenchymal investment about the lining entoderm differentiates into the *lamina propria, submucosa, muscularis* and *serosa* (or *adventitia*). A thin layer of smooth muscle (*muscularis mucosæ*) arises deep in the lamina propria, demarcating it from submucosa. Of the two chief muscle coats, the circular layer uniformly develops earlier (6–10 weeks) than the longitudinal layer (10–14 weeks). Additional descriptions of tissue specializations

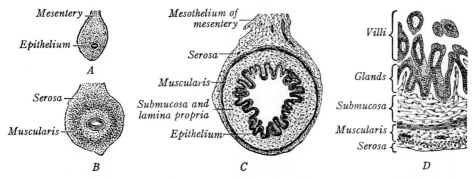

Fig. 205. Organization of the human digestive tube (specifically, the small intestine), shown in transverse sections. *A,* At six weeks (× 33); *B,* at eight weeks (× 33); *C,* at three months (Johnson; × 33); *D,* at four months (Johnson; × 50).

245

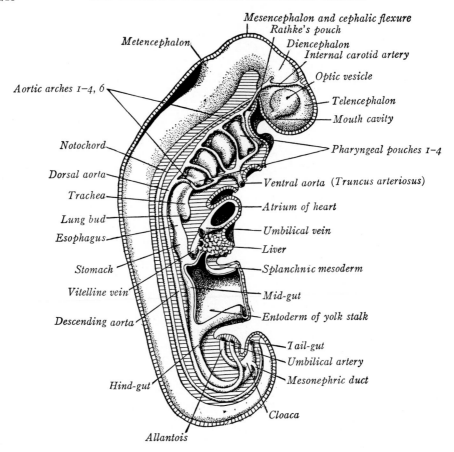

Fig. 206.   Entodermal tract of a 4 mm. human embryo, exposed in a lateral reconstruction
(Prentiss, after His). × 25.

will be confined largely to the distinctive differentiations undergone regionally by the epithelial lining of the primitive tube.

   The Esophagus.   Embryos of about 2.5 mm. lack a definite esophagus
(Fig. 172), as in fishes, and at four weeks it is still a short tube extending
from pharynx to stomach (Fig. 206). Nevertheless, the esophagus soon elongates rapidly, keeping pace with the differentiating neck and the growing
heart and lungs alongside (Figs. 209, 210).

   At five weeks the epithelium has acquired two layers of nuclei, and in the
seventh week vacuoles begin to appear in it so that presently there is an
increase in the size of the lumen (*cf.* Fig. 214 *A-D*). In this way the lining
becomes channeled for a while, but at no time is it totally occluded like the
fetal esophagus of reptiles and birds. The epithelium begins to acquire cilia
at ten weeks and it is not until into the fifth month that a stratified squa-

mous epithelium starts replacing it. At birth the epithelium numbers ten layers, but may still include some ciliated patches. *Superficial glands* are developing in the fifth month, whereas the *deep glands* arise mostly after birth. As a component of the mediastinum, the esophagus never acquires a typical mesentery or serosal tunic (Fig. 230).

Anomalies. There may be *stenosis* (narrowing) or local *atresia* (no cavity) (Fig. 232 *A*). As the latter condition usually involves the trachea as well, it will be easier to explain after that organ has been discussed (p. 271). The partial epithelial occlusion, normally a transient feature, predisposes toward all these abnormalities. A short esophagus occurs when the stomach fails to descend below the diaphragm.

The Stomach. The stomach is discernible in embryos of 4 mm. as a spindle-shaped enlargement of the fore-gut, somewhat flattened on its lateral surfaces (Fig. 206). Originally the stomach lies in the future neck region, but by the end of the seventh week a 'descent' has been completed through a distance of 16 segments to the permanent location in the abdomen (Figs. 208–210). During the period of descent (4–7 weeks) the stomach undergoes certain changes in shape and orientation (Fig. 207): (1) the entire organ increases in length; (2) the dorsal border grows faster than the ventral wall and so produces the convex *greater curvature* in contrast to the passively concaved *lesser curvature;* (3) the *fundus* arises as a local bulge near the cranial end; (4) the stomach rotates 90 degrees about its long axis until the greater curvature (primitive dorsal wall) lies on the left and the lesser curvature (primitive ventral wall) is on the right (Fig. 334 *A*); and (5) the rotating stomach is displaced by the enlarging liver until it extends obliquely from left (above) to center (below).

The epithelium differentiates distinctive *gastric glands* and the mesenchyme produces three incomplete layers in the muscular coat. Rotation causes the original right surface of the stomach to become dorsal and the original left surface to become ventral. This explains why the vagus nerves, coursing originally right and left, assume dorsal and ventral positions in relation to the lower esophagus and stomach.

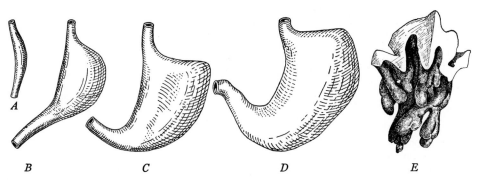

Fig. 207.   Models of the human stomach. *A,* At 5.5 mm. (× 25); *B,* at 9 mm. (× 25); *C,* at 15 mm. (× 25); *D,* at 23 mm. (× 15); *E,* gastric glands, at seven months (after Johnson; × 150).

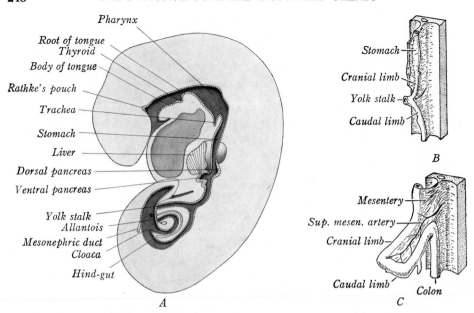

Fig. 208. Early stages of the human entodermal canal. *A,* Right hemisection, at 5 mm. (Prentiss, after Ingalls; × 14). *B, C,* Models of the intestinal loop, at 5 mm. (× 15) and 15 mm. (× 10), showing rotation.

Although usually described as a shift caudad, either by active growth or secondary crowding, 'descents' of the stomach, heart, lungs, and diaphragm are probably better interpreted as relative rather than actual. They result from the forward growth of the head-end of the embryo which leaves these organs behind. Especially are the more dorsal regions of the body concerned in this forward overgrowth, owing primarily to the rapid elongation of the neural tube. The simultaneous, but passive, transport cephalad of the dorsally located somites would explain the apparent descent of other relatively fixed organs, since the somites are customarily used as reference points.

Rotation of the stomach is, at least chiefly, the result of active, unequal growth by the gut wall, although the development of the omental bursa may aid rotation (Fig. 333 *A*).[1, 2, 3] The enlarging liver is responsible for the slanting position of the stomach. It displaces the freely movable cranial end of the stomach to the left, whereas the caudal end is relatively anchored by the short ventral mesentery and bile duct; the vitelline artery also aids by acting as a block.[4]

A pyloric region becomes distinguishable in the third month, but the *pyloric sphincter* is still weakly developed at birth. The mucous membrane shows two early folds that course along the lesser curvature from esophagus to pylorus. These ridges delimit a groove, recognizable permanently as the *gastric canal*. The early epithelium is a simple sheet. Pits, or *foveolæ,* are indicated in the mucosa at seven weeks, and at 14 weeks *gastric glands* begin to bud off from them (Fig. 207 *E*). Both continue to increase manyfold between birth and maturity until they reach a total in the millions. Enzymes are secreted by the fifth month, but the secretion of hydrochloric acid is perhaps delayed until considerably later. Gastric motility occurs at four months.

Anomalies. Transposition to the right side of the abdomen, as in a mirror-image, sometimes occurs in conjunction with the reversal of other asymmetrically placed organs

(Fig. 232 *B*). Some or all of the stomach may be located above the diaphragm (*thoracic stomach*). It is accompanied by a short esophagus and represents an arrested descent. Stenosis results from an overdevelopment of the pyloric sphincter.

The Intestine. In embryos of four weeks (5 mm.) the intestine is a simple tube, beginning at the stomach and ending in the cloaca (Fig. 208 *A*). It occupies the median plane and parallels the curving neural tube; midway along its course the intestinal tube bends slightly ventrad and becomes continuous with the now slender yolk stalk. For convenience the segments of the early intestine above and below the attachment of the yolk stalk are designated as the cranial and caudal *limbs of the intestinal loop* (*B*). At this stage the location of the future *duodenum* is recognizable by its relations to the stomach and the primordial liver and pancreas; the remainder of the cranial loop will become the *jejunum* and much of the *ileum,* while the caudal loop represents the lower ileum and all of the *colon* and *rectum.* The intestine is supported from the dorsal body wall by the *dorsal mesentery;* a *ventral mesentery* exists only in the duodenal region (Fig. 234).

In the fifth week (5–8 mm.) the intestine elongates faster than the trunk, and the intestinal loop becomes a prominent flexure (Fig. 209). By the end of this period the yolk stalk detaches from the apex of the loop. The loss of this marker on the intestinal wall is offset by the acquisition of a more useful one; a bulge in the caudal limb indicates the *cæcum,* and consequently

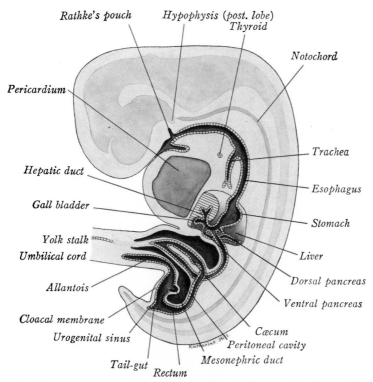

Fig. 209. Human entodermal canal, in hemisection at 9 mm., showing early protrusion of the intestinal loop into the umbilical cord (Prentiss, after Mall). × 9.

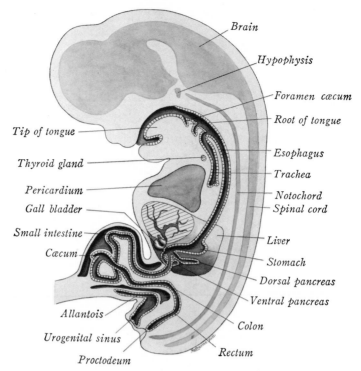

Fig. 210. Human entodermal canal, in hemisection at 17 mm., showing full herniation of the intestine into the umbilical cord (Prentiss, after Mall). × 5.

denotes the boundary between the *small* and *large intestine*. Succeeding gross changes involving the intestine include its torsion, herniation into the umbilical cord and coiling there; next follow a withdrawal into the abdomen, final placement and fixation. Structurally, the differentiation of *villi* and *glands* is a characteristic feature.

ROTATION. While the intestinal loop is developing it undergoes a torsion; in this rotational displacement the superior mesenteric artery, which courses in the mesentery between the two limbs, may be considered as the axis. The twisting is so executed that the cranial limb is carried from the midplane to the right; conversely, the caudal limb shifts to the left (Fig. 208 *B, C*). In other words, there has been anticlockwise rotation as one views the embryo from the ventral side. The torsion is said to result from a change of position of the enlarging left umbilical vein which forces the cranial limb of the bowed gut to the right,[5] but the primary factors stem from an asymmetry determination that is established much earlier (p. 254).

HERNIATION AND COILING. During the sixth week the elongating intestinal loop can no longer be contained wholly within the slower growing abdomen and it begins to escape into the umbilical cord, still retaining the rotated positions of the two limbs (Fig. 87). This protrusion constitutes a temporary (but normal) *umbilical hernia,* the intestine then occupying a cavity continuous with the embryonic cœlom. Thickenings of the mesen-

tery (so-called *retention bands*) occur at the kinked lower end of the duodenum and at the site of the future left (splenic) flexure of the colon. Presumably these prevent the duodenum and future descending colon from
entering the cord, but the loop proper is not so restrained (Fig. 211 *A*). The
loop represents what some authors call the *mid-gut;* it consists of the future
jejunum, ileum, ascending and transverse colon. Continued elongation of
the herniated small intestine leads to extensive coiling; by contrast, the
large intestine and its associated mesentery grow relatively little at this
period (Figs. 210, 211 *A*).

RE-ENTRY AND PLACEMENT. In embryos of ten weeks the abdominal
cavity has increased sufficiently in size both absolutely and relatively (the
latter due particularly to a decline in the growth rate of the liver), so that
the intestine can again be accommodated. The exact cause of withdrawal
from the umbilical sac is not well understood.[1, 4, 6] But whatever the underlying forces may be, it is plain that the return, once begun, is completed
quickly and the cœlom of the cord promptly obliterates. The small intestine is the first to re-enter the abdomen. It does this in a progressive manner,
the proximal portions of the jejunum leading. The returning coils first fill
the available space on the left side of the abdomen, thereby pressing the
non-herniated (future descending) colon also to the left (Fig. 211 *B*), whereas
the later coils (ileum) to return locate in the right half of the abdominal
cavity.[6]

Perhaps because of the prominent cæcal swelling, the large intestine is
the last to leave the umbilical cord and re-enter the abdominal cavity. Its
tendency to straighten then carries this limb slantingly across to the right

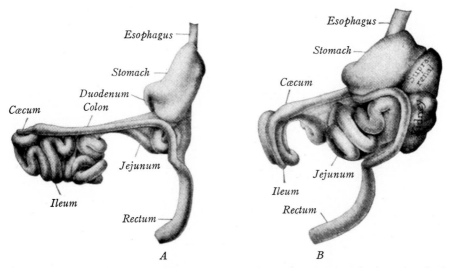

Fig. 211. Reconstructions of the human digestive tube, showing the intestine during
herniation (after Bardeen). *A,* At nine weeks, when coiling is maximum (× 5). *B,* At ten
weeks, after a partial return of the small intestine (× 4).

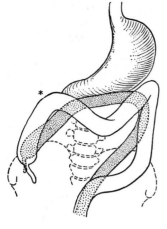

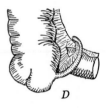

Fig. 212. Position of the colon in human fetuses. In stipple, the relations at ten weeks on return from the umbilical cord; in full line, the relations during the seventh month with the ascending colic limb and right colic flexure (*) evident.

Fig. 213. Development of the human cæcum and vermiform process. *A,* At two months; *B,* at three months; *C,* in the newborn; *D,* at five years.

side, thus completing a rotation that totals 270 degrees (Fig. 242 *A*). In this placement the cæcum lies close to the crest of the ileum and here it becomes fixed in its permanent position.[7] From this point the colon passes obliquely upward to the left of the stomach, where it recurves sharply (*left colic,* or splenic, *flexure*) into the future descending colon which maintains its displaced position on the left side (Fig. 212; in stipple). For several months these conditions persist and there are no definite ascending and transverse limbs of the colon.

COMPLETION.   The *duodenum* acquires an early, characteristic curve in relation to the growing pancreas (Fig. 221). It falls to the right and loses its mesentery on becoming pressed against the dorsal body wall. The remainder of the small intestine, thrown into loops since its elongation while in the umbilical cord, is given the somewhat arbitrary names of *jejunum* and *ileum.* The original cæcal bulge grows and makes a definite, blind sac which extends the large intestine beyond its junction with the ileum (Fig. 213 *A*). The distal end of this sac lengthens rapidly for a time (*B*), but it fails to keep pace in thickness with the rest of the sac (*C*). As a result, the characteristic *vermiform process* of the higher apes and man becomes distinct from the *cæcum* (*D*).

As the liver decreases in relative size and accordingly 'retreats' craniad, a kink appears in the originally oblique proximal limb of the colon and becomes increasingly sharper. This *right colic* (or hepatic) *flexure* progressively demarcates an ascending from a transverse colic limb (Fig. 212,*).[7]

The *ascending colon,* beginning to elongate as such in the middle of fetal life, is not completed until early childhood (Fig. 242, *A, B*). The *transverse colon* necessarily courses in front of (*i.e.,* ventral to) the duodenum, which never left the abdomen, and hence most of the colon continues to be suspended by its mesocolon (*B*). Quite different is the *descending colon,* which, like the ascending colonic limb, is applied against the body wall; each then loses its free mesentery and becomes anchored (*C, D*) in a way to be explained on p. 280). A relatively early and pronounced elongation of the most caudal portion of the colon is retained as the *sigmoid colon.* The terminal portion of the intestine, or *rectum,* is derived from the subdivision of the cloaca; Figures 275, 276 illustrate the process of separation, which is described in full on p. 309. After the anal membrane ruptures at the end of the eighth week, a short ectodermal pit, or *proctodeum,* is added to the entodermal digestive tube. This *anal canal* results from the encircling growth of several anal hillocks (Figs. 274, 299 *C, D*).[8] The junction between ectoderm and entoderm is somewhat below the so-called anal valves.

Between the fifth week and birth the intestine increases its length 1000 times, while the small intestine becomes six times the length of the large intestine. The small intestine is originally thicker than the large intestine; it is not until the fifth month that the large intestine becomes greater in diameter. Proliferation of the epithelial lining of the duodenum leads to its occlusion in the sixth and seventh weeks (Fig. 214 *A, B*), but vacuolation soon restores a continuous lumen (*C, D*). All of the small and large intestine shows a similar phenomenon, but in lesser degree; the entire intestinal tract is finally clothed with a single-layered epithelium. *Villi* begin to appear at eight weeks as independent, rounded elevations of the lining membrane (*E*). *Intestinal glands* (of Lieberkühn) arise as tubular ingrowths of the epithelium about the bases of the villi (Fig. 205 *D*). They first appear toward the end of the third month and are closely followed by the compound *duodenal glands* (of Brunner). Both villi and glands increase greatly in number during childhood. *Lymph nodules* and *Peyer's patches* are present at five months. The colon bears temporary villi in the middle third of fetal life; the *teniæ* are linear thickenings of the longitudinal muscle layer. At ten weeks the cæcum makes a sharp bend with the colon proper, and this flexure is responsible for the production of the *colic valve.* Peri-

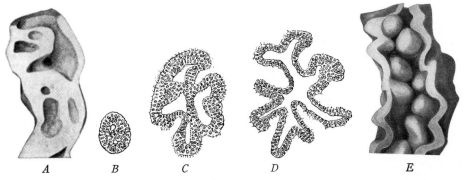

A	*B*	*C*	*D*	*E*

Fig. 214. Differentiation of the epithelium of the human small intestine (after Johnson). × 70. *A,* Duodenal epithelium, reconstructed and cut lengthwise to display the temporary occlusion and vacuolation at eight weeks. *B–D,* Transverse sections of the duodenal epithelium at six, eight and nine weeks. *E,* Jejunal epithelium, reconstructed and cut lengthwise to display the villi at eight weeks.

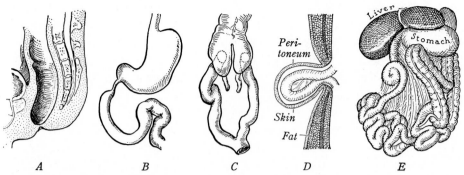

Fig. 215. Anomalies of the human intestine. *A,* Imperforate anus; *B,* atresia of the duodenum; *C,* duplication of the lower ileum and cæcum; *D,* congenital umbilical hernia, in section; *E,* non-rotation of the intestine.

stalsis of the intestine has been observed at 11 weeks. Intestinal glands secrete enzymes by the middle of fetal life.

Meconium begins to collect in the intestine after the third month. This mass is a pasty mixture of mucus, bile and cast-off epithelial cells, to which are added lanugo hairs, desquamated epidermal cells and sebaceous secretion swallowed with amniotic fluid. It is green in color and is wholly voided by the third or fourth day after birth—a fact sometimes of medicolegal value in proving how long an infant has lived. At birth the intestine and its contents are perfectly sterile, but a bacterial flora is acquired promptly.

Causal Relations. Transplantation experiments on amphibians show that the fore-gut has been irreversibly determined at the end of the neurula stage, not only as a whole but also regionally (mouth; pharynx; esophagus; stomach; intestine). Other experiments prove that the factors related to the rotation of the stomach and intestine, and the asymmetrical displacement of these organs and the liver, are already present in the gastrula stage (and probably earlier). This conclusion follows since reversal of the archenteron-roof leads to the subsequent transposition of the stomach, liver and intestine as in a mirror-image.

Entoderm, like ectoderm, is dependent on associated mesoderm in its differentiation. It is said that the differentiation of smooth muscle in the wall of all hollow organs depends on inductive influences from their epithelial lining. The invagination of ectoderm to form the proctodeum is induced by the presence in that region of ventral-lip mesoderm of the amphibian gastrula. But it is the entoderm of the hind-gut, thus brought into contact with the floor of this pit, that is responsible for the perforation of the ectoderm which establishes the anus.

Anomalies. The failure of the anal portion of the cloacal membrane to rupture results in an *imperforate anus* (Fig. 215 *A*); it may be combined with atresia of the rectum. More or less of a permanent cloaca follows the incomplete separation of rectum from urogenital sinus (Fig. 278 *A, B*). The intestine may undergo stenosis or atresia; this occurs most often in the duodenum (Fig. 215 *B*). Some of these specimens represent a partial or complete retention of the temporary fetal occlusion. Others are attributed to a local interruption in a rapidly elongating tube or, possibly, to an interference with the blood supply. Two per cent of all adults show a persistence of the proximal end of the yolk stalk which forms a pouch, known as Meckel's *diverticulum of the ileum* (Fig. 84 *A*). This may extend even to the umbilicus (*B*), and when patent constitutes a fecal *umbilical fistula* (*B, C*).

Duplication of the digestive tube occurs at all levels. This may result from longitudinal subdivision and take the form of parallel, communicating tubes (Fig. 215 *C*).

Another duplicating method is sacculation, in which persistent diverticula or even separated sacs range in shape from the commoner spheroid to elongate, blind tubes.

Umbilical hernia designates a secondary protrusion of peritoneum and intestinal wall through a poorly closed umbilical ring; it causes the overlying skin to bulge. Other hernias of the bowel are explained on pp. 293, 335.

Rarely there is *non-rotation;* the returning jejuno-ileum then lies on the right side and the colon on the left (Fig. 215 *E*). Transposition of the digestive tract right for left, as in a mirror-image, is one feature of the more general condition known as *situs inversus* (p. 271); the participation of the intestine in this process is characterized by a complete reversal of the normal course of rotation (Fig. 232 *B*). Twisting of the bowel (*volvulus*) and other conditions resulting from an unanchored intestine are explained on p. 283).

THE LIVER

The liver is a ventral outgrowth from the gut-entoderm in the region of the anterior intestinal portal. In embryos with 17 somites (2.5 mm.) its shallow primordium lies between the pericardial cavity and the attaching yolk stalk (Fig. 172 *A*). Here is the floor of the future duodenum which presently gives rise to a definite sacculation named the *hepatic diverticulum* (Fig. 216 *A*). This consists of a cranial portion, which will differentiate into the glandular tissue and its bile ducts, and a caudal portion which becomes the gall bladder and cystic duct (*C*). The hepatic diverticulum forces its way ventrad into a mass of splanchnic mesoderm that is the basis of much of the future diaphragm; at this stage this tissue-mass is named the *septum transversum* (*A, B*). A little later, the region of the septum occupied by the liver

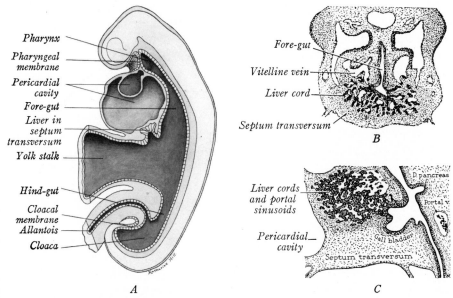

Fig. 216. Origin and relations of the human hepatic diverticulum. *A*, At 3 mm., in hemisection (Prentiss, after His; × 25). *B*, At 3.5 mm., in transverse section (× 60). *C*, At 5 mm., in sagittal section (× 60).

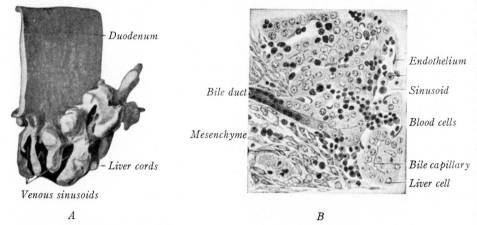

Duodenum

Bile duct

Mesenchyme

Liver cords

Venous sinusoids

Endothelium

Sinusoid

Blood cells

Bile capillary

Liver cell

A   *B*

Fig. 217.   Differentiation of the human liver. *A,* Model of half of the duodenal wall and
liver, at 4 mm. (Bremer; × 100). *B,* Section, at 16 mm. (Bloom; × 350).

becomes drawn out as the ventral mesentery, and the final relation of the
liver is then more intimately related to this mesentery than to the diaphragm proper (Fig. 234).

Straightway after its appearance, the cranial portion of the hepatic diverticulum buds-off epithelial cords. These invade the septum transversum
farther and continue to proliferate there into a rapidly expanding spongework (Figs. 216, 217). From the first the diverticulum lies close to the paired
vitelline veins which flank the gut, and these veins send branches into the
region of proliferation (Fig. 216 *B*). The result is a mutual, intimate intergrowth of tortuous liver cords and sinusoidal channels (*C*). Perhaps it is because of its rich blood supply that the hepatic mass enlarges so rapidly (Fig.
229). In any event, the liver of a 5 mm. embryo is a large crescentic mass
with a wing extending upward on each side of the gut (Fig. 208 *A*). While
these changes have progressed, the stem portion of the original diverticulum
is elongating and differentiating into the duct system (Fig. 221).

Glandular Tissue and Blood Vessels.   The early epithelial strands are
the forerunners of the definitive *liver cords* (actually thin plates[9]) against
which the endothelium of the broad sinusoids becomes closely applied (Fig.
217). In its early growth upward around the gut, the wings of the liver come
to enclose and interrupt the nearby vitelline veins. After this occurs, only
delicate *sinusoids* interconnect the supplying (portal) and draining (hepatic)
vessels (Fig. 323). At first relatively far apart, these two venous trees grow
steadily as the liver expands and thus progressively 'approach' each other in
an alternating (or dovetailing) manner (Fig. 218 *A-C*). The regularity of
the system of branching that is employed is responsible for the creation of
the characteristic *hepatic lobules* from the epithelial tissue and sinusoids.

From the second to the seventh month of fetal life blood cells are actively
differentiating between the hepatic cords and the sinusoidal lining, but
only small foci remain at birth (Fig. 217 *B*). This potential generative abil-

ity remains latent throughout life, and blood formation can be resumed in the liver whenever the need for replacement is sufficiently urgent. The sinusoidal lining partially transforms into large macrophages which become the so-called *Kupffer cells*. Typical bile is secreted by the hepatic cells in fetuses five months old and colors the meconium.

In a 4 mm. embryo the whole liver is a single, complete lobule; at 7.5 mm. it is bilobed and has two primary lobules; at 11 mm. there are six lobules, while a late fetus has many thousands. Each lobule is surrounded by several terminal branches of the portal vein and is drained by a single hepatic vessel (Fig. 218 *A, B*). Toward the end of the fetal period, but mostly after birth, these primary lobules subdivide into smaller, secondary units. Each central (hepatic) vein takes the initiative and bifurcates or gives off a side branch. New lobules then arise by the simple splitting (*i.e.,* through connective-tissue invasion) of such a lobule which has thus acquired two central veins (*C, D*).[10] The portal veins at the periphery branch correspondingly as they push in between the new lobules to keep the vascular relationship unchanged. A clear demarcation of the definitive lobules, some 500,000 in number, is not seen until early childhood.

The Ducts and Gall Bladder. The main portion of the hepatic diverticulum elongates into the *ductus choledochus* (common bile duct) and its direct continuation to the liver, the *hepatic duct* (Fig. 221 *A, B*). During the seventh week the original ventral site of origin of the common duct shifts, through torsion of the duodenum, until its attachment becomes permanently dorsal (*B, C*). The *bile ducts* within the liver, which are tributary to the hepatic duct, arise in a secondary manner beginning at eight weeks. Wherever the liver cords come under the influence of connective tissue that grows in with the branching portal vein, they transform into *interlobular ducts* (Fig. 217 *B*).[11, 12] The liver cords are said to be microscopically hollow at their earliest appearance and hence the tiny *bile capillaries* of the permanent cords are primary lumina and not secondary acquisitions.[11]

The *gall bladder* constitutes a separate, caudal region of the originally shallow hepatic diverticulum (Fig. 216 *C*). In a 5 mm. embryo it is a solid,

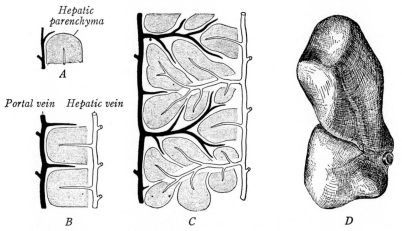

Hepatic parenchyma

A

Portal vein   Hepatic vein

B   *C*   *D*

Fig. 218. Method of vessel branching and the origin of hepatic lobules. *A–C,* Diagrams of successive stages of growth and subdivision (after Mall). *D,* Bifurcating lobule of the postnatal pig's liver (after Johnson; × 35).

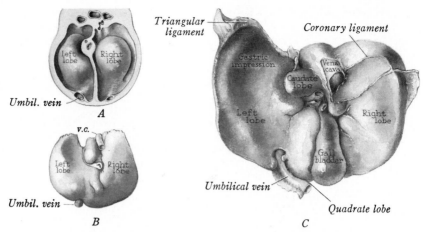

Fig. 219. External form of the human liver, viewed from the dorso-caudal surface. *A,* At 8 mm. (× 10); *B,* at 16 mm. (× 5); *C,* at birth (× ⅔).

epithelial cylinder which is carried away from the duodenum by the elongating common duct (Fig. 221 *A*). A distinct stem, or *cystic duct,* is then recognizable (*B, C*), and in the seventh week a lumen has been established throughout most of the tract which then appears like an offshoot from the main biliary passage.

Accessory Tissues. The growing liver expands greatly that particular portion of the septum transversum (soon ventral mesentery) within which it lies (Fig. 235 *B*). Superficially the mesentery becomes an *hepatic capsule* about the liver. More internally the mesenchyme differentiates into the connective-tissue framework (*Glisson's capsule*) about the lobules and into the muscular walls of the larger ducts and gall bladder.

The Liver as a Whole. The primary attachment of the liver to the septum transversum causes it to 'descend' (*cf.* p. 248; Fig. 250) with the latter organ from a cervical level of origin. The liver soon outgrows its original location within the septum transversum, and at four weeks bulges caudad into the abdominal cavity (Fig. 237). The continued progressive separation of liver from septum occurs at the time when the gut is also drawing away from the septum, and thus producing a definite ventral mesentery. This is the reason why the later liver is intimately associated with both the septum and ventral mesentery (Fig. 234). Such relations and the development of the *hepatic ligaments* will be described on p. 283. The history of the vitelline and umbilical veins with respect to the liver may be found on pp. 361–364. The subdivision of the liver into its characteristic *lobes* is largely the result of unequal growth, while its final shape is a passive response to adjacent pressures (Fig. 219).

The primary gross swellings are the paired *right* and *left lobes* (Fig. 219 *A*). Originally these are of equal size, but the right lobe becomes larger after the third month. In part this asymmetry is due to intrinsic growth factors, although the greater available space on that side plays a practical rôle; the vitelline and umbilical veins are usually credited with some influence as well. Parts of the early right lobe become set off as subordinate lobes. Thus, at six weeks the *caudate lobe* is recognizable, bounded by the ventral mesentery and inferior vena cava (Fig. 219 *B*). The *quadrate lobe* originates later with the atrophy of the liver tissue overlying the intrahepatic portion of the umbilical vein; it lies between that vein and the gall bladder (*C*).

The developing liver is spongy and highly plastic, so that it tends to adapt itself to the available space not used by firmer, neighboring organs. This passive molding accounts largely for its general final shape. In certain regions the hepatic tissue undergoes degeneration (due to pressure atrophy?) and especially is this true peripherally in the left lobe. For a time the liver grows at a faster rate than other organs or the body as a whole. Its maximum relative size is attained at nine weeks when it constitutes 10 per cent of the body volume and occupies most of the belly cavity.

Causal Relations. Suitable transplants of the ventral yolk-mass of an amphibian neurula show that the liver and pancreas, although structurally indistinguishable from the future gut-entoderm of this period, are already irreversibly determined as to their respective fates. Even further, the more dorsal portion of the actual liver bud is a region that is already determined as gall bladder. These results prove that both glands are not secondary specializations, budding out from a previously established gut-entoderm, as sections seem to imply, but that their primordia merely occupy for a time, along with gut-entoderm, a common medial strip on the floor of the archenteron. At a still earlier stage every cell of the gut is presumably pluripotent in these respects. This is supported by abnormal human development, since nests of stomach or pancreas cells can occur in a Meckel's diverticulum or elsewhere in the digestive tube. In the chick it appears that contact of the cardiac primordium with the gut floor is a prerequisite to liver formation.

Anomalies. The liver rarely develops elsewhere than in its normal site. A reduction or an increase in the external lobation of the liver is a rare occurrence (Fig. 220 *A, C*). An increase sometimes results in lobation resembling that in lower mammals. The main ducts and the gall bladder are subject to partial or complete duplication (*B*). Absence of the gall bladder (as occurs normally in the horse and elephant) is well known (*C*). A congenitally narrowed, solid or interrupted condition of the gall bladder or of the chief ducts is related to the temporary embryonic occlusion (*D*).

THE PANCREAS

Two outpocketings from the entodermal lining of the gut represent the earliest indications of the future pancreas. These buds arise on opposite sides of the duodenum in embryos of 3 to 4 mm. (Fig. 221 *A*). One pushes

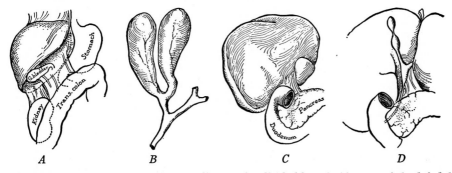

A	*B*	*C*	*D*

Fig. 220. Anomalies of the human liver and gall bladder. *A*, Absence of the left lobe of the liver. *B*, Double gall bladder. *C*, Absence of the gall bladder, and the consequent lack of a quadrate lobe. *D*, Stenosis of the gall bladder and common duct.

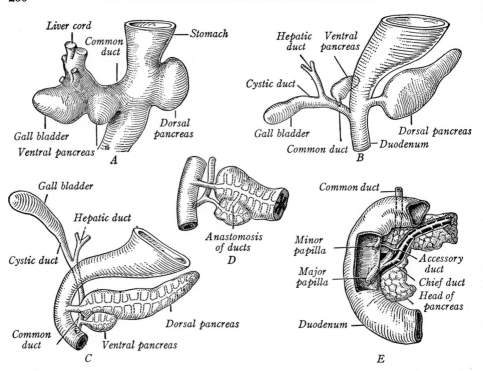

Fig. 221. Development of the human pancreas, shown by models viewed from the left side. *A,* At 6 mm. (× 83); *B, C, D,* at 8, 12 and 16 mm., respectively (× 42); *E,* at birth (× 1).

out from the dorsal wall, just craniad of the level of the hepatic diverticulum; it is the *dorsal pancreas.* The other, probably originally paired,[13] appears ventrally in the caudal angle between gut and hepatic diverticulum and consequently is designated the *ventral pancreas.* These two primordia meet and unite, thus producing a joint organ (*B-E*).

Of the two pancreatic primordia, the dorsal one grows more rapidly (Fig. 221 *B*). In the sixth week it is an elongate, nodular structure, which extends into the dorsal mesentery (Fig. 623). Since it arises near the mouth of the developing omental bursa, it continues its growth within the dorsal layer of that mesenterial sac (Fig. 239 *A*). The ventral pancreatic bud remains smaller; it is carried away from the duodenum by the lengthening common bile duct and then arises directly from the latter (Fig. 221 *B*). Unequal growth of the duodenal wall shifts the bile duct dorsad and thereby brings the ventral pancreas into the dorsal mesentery, near the stem of the dorsal pancreas (*B, C*). During the seventh week the two primordia interlock intimately (*D*). Grossly the dorsal pancreas forms all of the mature gland except most of the head and the uncinate process, which arise from the ventral primordium (*E*).[14]

The Ducts. Both pancreatic buds have an axial duct. The dorsal duct arises directly from the duodenal wall, but the base of the ventral duct is carried upward onto the elongating common bile duct and shares a common stem with it (Fig. 221 *A, B*). When duodenal torsion brings the two pancreatic primordia side-by-side, the short ventral duct taps the dorsal duct (*C, D*). Thereafter the long distal segment of the dorsal duct plus the entire ventral duct will serve as the chief line of drainage (*E*). This combined tube is known in adult anatomy as the *pancreatic duct* (of Wirsung). The proximal, stem segment of the dorsal duct constitutes the so-called *accessory duct* (of Santorini). It becomes tributary to the main duct, but commonly retains its duodenal outlet as well (*E*).

The occurrence of a permanent common outlet into the duodenum for bile and pancreatic juice is a direct consequence of the close relationship between the bile and ventral pancreatic ducts (Fig. 221 *A, B*). The region of the common outlet is the *ampulla* (of Vater) which opens at the major *duodenal papilla* (Fig. 221 *E*). This joint duct gains a circular sheath of smooth muscle (*sphincter of Oddi*) in the seventh week. A final arrangement of ducts similar to that in man occurs in sheep, while the hog and ox reverse the relation and use the dorsal duct as the chief stem; less specialization occurs in the horse and dog, which retain both ducts as functional outlets into the intestine.

The Glandular Tissue. Secretory *acini* begin to appear in the third month as terminal and side buds from the primitive ducts (Fig. 222 *A*). *Pancreatic islands* (of Langerhans) also are differentiating from the ducts at about the same time (*A*).[15] They are composed of distinctive cells which take the form of single sprouts, but later through growth (and, it is claimed, through union)[16] become complex island masses (*B*). In all about a million islets are formed, some of which retain their original (but soon impervious) connections with the parent ducts.

No histological distinction exists between the acini of the dorsal and ventral pancreatic masses, but probably the dorsal pancreas alone differentiates islands. The alpha and beta cells specialize relatively early.

Trypsin has been detected at five months and insulin seems to be present still earlier. The mesenchymal bed, in which the gland develops, furnishes a connective-tissue *capsule* and subdivides the organ into *lobes* and *lobules*.

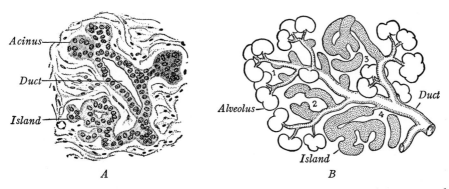

A   *B*

Fig. 222. Differentiation of the human pancreas. *A*, Section, at fourteen weeks, demonstrating the origin of acini and islands from ducts (after Lewis; × 350). *B*, Diagram showing four progressive stages (1–4) in the organization of islands.

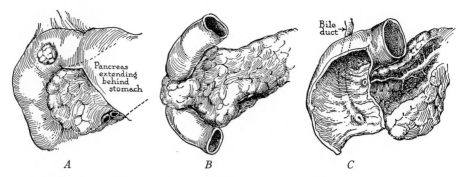

Fig. 223. Anomalies of the human pancreas. *A,* Accessory pancreas on the wall of the duodenum. *B,* Annular pancreas, encircling the duodenum. *C,* Absence of the stem of the ventral (chief) duct, the 'accessory duct' serving as the sole outlet.

Causal Relations. A pancreas-forming potency is irreversibly determined before the primordium can be recognized as an entity in the gut-lining of amphibians (p. 259). Moreover, experiments can prove the existence at this time (neurula stage) of separate dorsal and ventral primordia; only the former is able to differentiate islands. Pancreatic primordia, like liver and gut, will self-differentiate in culture medium even to the extent of producing secretion.

Anomalies. Accessory pancreases are relatively common (Fig. 223 *A*). Many of these lie within the wall of the intestine and stomach; others are associated with the spleen and omentum. The development of supernumerary primordia and the displacement of parts of the diffuse, early pancreas are responsible for these several conditions. An annular pancreas encircling and constricting the duodenum, and less frequently the bile duct or portal vein, sometimes occurs (*B*). It is perhaps related to the development of the normally suppressed member of the two ventral pancreatic buds. The ventral pancreas, and accordingly the main duct of the adult gland, may arise directly from the duodenum. Absence of that part of the gland derived from either primordium, failure of union between the dorsal and ventral pancreatic components, completely independent ducts, and a single duct (*C*) are all well known.

REFERENCES CITED

1. Pernkopf, E. 1922–26. Z'ts. Anat. u. Entw., *64,* 96–275; *73,* 1–144; *77,* 1–143; *85,* 1–130.
2. Kanagasuntheram, R. 1957. J. Anat., *91,* 188–206.
3. Botha, G. S. M. 1959. Anat. Rec., *133,* 219–239.
4. Enbom, G. 1939. Anat. Rec., *75,* 409–414.
5. Dott, N. M. 1923. Brit. J. Surg., *11,* 251–286.
6. Frazer, J. E. & R. H. Robbins. 1916. J. Anat. & Physiol., *50,* 76–110.
7. Fitzgerald, M. J. T. *et al.* 1971. J. Anat., *109,* 71–74.
8. Politzer, G. 1931. Z'ts. Anat. u. Entw., *95,* 734–768.
9. Lipp, W. 1952. Z'ts. f. mikr.-anat. Forsch., *58,* 289–318; *59,* 161–186.
10. Johnson, F. P. 1919. Am. J. Anat., *25,* 299–331.
11. Bloom, W. 1926. Am. J. Anat., *36,* 451–465.
12. Horstmann, E. 1939. Arch. Entw.-mech. d. Organ., *139,* 363–392.
13. Odgers, P. N. B. 1930. J. Anat., *65,* 1–7.
14. Russu, I. G. & A. Vaida. 1959. Acta Anat., *38,* 114–125.
15. Conklin, J. L. 1962. Am. J. Anat., *111,* 181–193.
16. Neubert, K. 1927. Arch. Entw.-mech. d. Organ. *111,* 29–118.

Chapter XV. The Respiratory System

The nose and naso-pharynx belong to the respiratory apparatus, but since this relation is a secondary adaptation their development is described in other chapters (pp. 243, 524). As with all hollow viscera, the larynx, trachea and smaller respiratory passages are lined with an epithelium (in this instance, entoderm) which is strengthened and supported by other layers differentiated from the surrounding mesenchyme. In addition, the lungs expand into the cœlom (pleural cavities) and in so doing gain a covering of splanchnopleure (visceral pleura), whose free surface is mesothelium.

The earliest indication of the future respiratory tree is in embryos of 3 mm., with 20 somites. It constitutes a groove that runs lengthwise in the floor of the gut, just caudal to the pharyngeal pouches.[1] In surface view the entoderm projects as a ventral *laryngo-tracheal ridge* (Figs. 206, 224 *A*). A lateral furrow appears on each side, along the line of junction between ridge and esophagus (Fig. 224*C*). Becoming progressively deeper and extending craniad, the furrows join, thereby splitting off a laryngo-tracheal tube. At the upper end, the laryngeal portion of the tube advances slightly rostrad until it lies between the fourth branchial arches. Meanwhile the caudal end of the original ridge has become rounded, and is commonly called the *lung bud* (*B, C*). At the 4 mm. stage the bud begins to bifurcate (*D*) and the respiratory organs are then represented by: (1) a *laryngeal region,* opening off the pharynx; (2) the tubular *trachea;* and (3) two *primary bronchi.* The latter are potentially more than bronchi, since by growth and branching they will ultimately produce all the subdivisions of the respiratory tree (bronchial branches, bronchioles and air sacs).

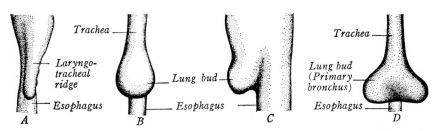

Fig. 224. Early stages of the human respiratory primordium (after Grosser and Heiss). × 75. *A,* At 2.5 mm., in ventral view; *B, C,* at 3 mm., in ventral and lateral views; *D,* at 4 mm., in ventral view. (See Fig. 195 *A* for total relation.)

263

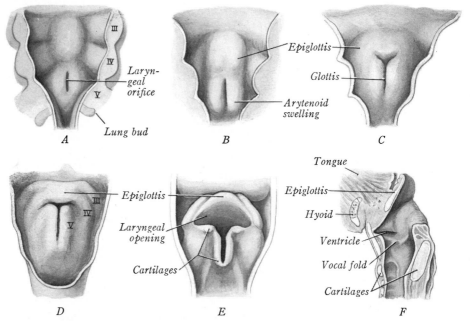

Fig. 225. Development of the human larynx. *A*, At 5 mm.; *B*, at 9 mm.; *C*, at 12 mm.; *D*, at 16 mm.; *E*, at 40 mm. (× 7); *F*, sagittal hemisection, at birth (× 1.5).

THE LARYNX

This organ develops somewhat differently in its upper and lower halves. The lower portion organizes around the cranial end of the trachea, whereas the part above the vocal folds rises out of the pharyngeal floor, in the region of the laryngeal orifice, as a sort of vestibule (Fig. 225 *A*).

The *epiglottis* is peculiar to mammals. Embryos of 5 mm. show a rounded prominence that elevates midventrally from the bases of the third and fourth arches (Fig. 225*A*). This soon alters its shape (*B-D*) and consolidates into the transverse flap that guards the entrance to the larynx during swallowing (*E*). It becomes concave on its laryngeal surface and in the middle of fetal life differentiates cartilage internally (*F*).

The slit that opens from the floor of the pharynx into the trachea is the primitive *glottis* (Fig. 225 *A*). Presently it is bounded on each side by a rounded eminence, of fourth and fifth arch origin, known as an *arytenoid swelling* (*B*). These two swellings straightway begin to grow in a tongueward direction. On meeting the primordium of the epiglottis they arch upward and forward against its caudal surface (*C*). In the seventh week this results in the original, sagittal slit adding a transverse groove to its upper end, so that the laryngeal orifice becomes **T**-shaped (*D*). Nevertheless, the entrance to the larynx ends blindly for some time because fusion of the epithelium in the upper larynx has obliterated the lumen there. When the epithelial union is dissolved (10 weeks) the entrance becomes more oval in contour (*E*) and a pair of lateral recesses (*laryngeal ventricles*) is evident in the restored cavity. Each of these is bounded cranially and caudally by a projecting, lateral shelf. The caudal pair of shelves, lying at the same level as the primitive laryngeal slit, are the *vocal folds* (*F*); they appear at eight weeks, and later differentiate elastic tissue.

The epithelial lining of the larynx is supported by dense mesenchyme derived from the fourth and fifth branchial arches (Fig. 615). Early in the seventh week this mass shows localized condensations that foretell the *laryngeal cartilages* (Fig. 225 *D-F*). These belong to the skeleton and will be treated more in detail in a later chapter (p. 419). The *laryngeal muscles* also originate from the same branchial arches and consequently continue to be innervated by the vagus nerves which supply those arches.

THE TRACHEA AND PRIMARY BRONCHI

Soon after its appearance, the 'lung bud' bifurcates (at 4 mm.) into the two *primary bronchi,* and the tubular system then becomes λ-shaped. The right bronchus extends more directly caudad than does the left bronchus (Fig. 226 *A*). This difference is maintained throughout life and accounts for the more frequent aspiration of foreign bodies into the right bronchial

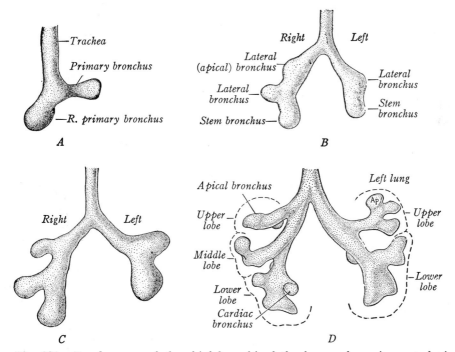

Fig. 226. Development of the chief bronchi of the human lung, in ventral view (after Heiss and Merkel). ×50. *A,* At 5 mm.; *B,* at 7 mm.; *C,* at 8.5 mm.; *D,* at 10 mm. *Ap.,* Apical bronchus (homologue) of left lung.

tube (Fig. 222 *B*). The tracheal tube elongates rapidly for a time, and the point of bifurcation ultimately 'descends' (like the lungs) a distance of eight body segments.

The *epithelial lining* changes from its early columnar form to the final pseudostratified ciliated type. Smooth-muscle fibers and incomplete cartilaginous 'rings' and *plates* are differentiating from the surrounding condensed mesenchyme at the end of the seventh week. The *glands* develop as ingrowths from the epithelium after the fourth month.

THE LUNGS

In an embryo 7 mm. long the right primary bronchus gives rise to two side buds, or *lateral bronchi,* while the left bronchus forms but one; the end of each main tube constitutes the so-called *stem bronchus* (Fig. 226 *B*).

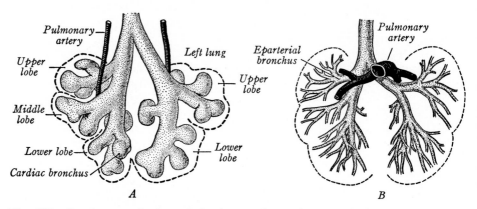

Fig. 227. Developmental plan of the human lungs, in ventral view. *A,* At 14 mm. (after Ask; × 33); *B,* at birth (adapted; × 1).

Even at this early stage the plan of the future lobed lung is begun, as slightly older embryos make clear (*C, D*). On the right side the upper bud is small and is called the *apical bronchus,* since it presages the *upper lobe;* the other lateral bud is the axis for the *middle lobe,* whereas the stem bronchus will form the *lower lobe.* On the left side the single lateral bud identifies the future *upper lobe* and the stem bronchus is the forerunner of the *lower lobe.*

The development of the lungs proceeds much like the branching of a compound gland into a bush-like set of tubules. Their prenatal development is in three phases which occupy consecutive periods, as follows:[2]

1. Establishment of the larger conducting tubes—*bronchi* and *bronchioles* (five weeks to nearly four months).

2. Laying down of the *respiratory bronchioles* (four to six months).

3. Extension into a system of *alveolar ducts* and the differentiation of early *alveoli* (six months to term). The lung loses its glandular appearance and becomes highly vascular.

Epithelial Growth and Differentiation. As the bronchial buds continue to grow and branch, the tubular system in each pulmonary lobe becomes increasingly bush-like with dorsal, ventral, lateral and medial rami (Fig. 227). In the fifth month the epithelial lining of the terminal buds of the respiratory tree is cuboidal (Fig. 228 *A*). Early in the sixth month the adjoining capillaries begin to push against the epithelium which seems to disappear so that the capillaries lie exposed (*B*).[3] Actually, the epithelial lining merely flattens to extreme thinness and a complete layer of this sort continues throughout life.[4] At birth all branching is finished and clusters of *alveoli* (*i.e., alveolar sacs*) come off the smallest twig-like ducts.[5] Within two months after birth *alveolar ducts* make their appearance by alveoli budding out of the walls of previous respiratory bronchioles; in a similar manner terminal bronchioles produce alveoli and transform into *respiratory bronchioles.*[5]

Even in the seventh week, ten bronchi of the third order of branching have arisen for the right lung and eight for the left lung (*cf.* Fig. 227 *A*). These are the trunk bronchi that will supply the clinically important *broncho-pulmonary segments* that become separated from each other by connective-tissue septa. Postnatal increase in lung size is brought about by increases in length and caliber of all respiratory passages peripheral to the definitive terminal bronchioles.[6]

The early apical bronchus of the right upper lobe (Fig. 226 *D*) comes to be called the *eparterial bronchus* because it alone lies upon the pulmonary artery. Originally this bronchus is dorsal to the artery, as the name implies (Fig. 227 *A*). Later, when the heart descends, it is cranial with respect to this vessel (*B*); yet, on the basis of upright posture, the designated meaning 'upon the artery' is again wholly appropriate. It is commonly stated that this bronchus was anciently a secondary branch in what was then the upper lobe of the lung; in the course of evolutionary advance it is supposed to have migrated upward onto the main stem and induced the formation of a new lobe about it.[7] Others view this bronchus as an entirely independent replacing outgrowth, at a higher level, that became selected as the basis for a new lobe.[8]

The left upper lobe seems to contain a bronchial branch that is the equivalent of the entire apical bud on the right side (Fig. 226 *D, Ap*). Since, however, this branch remains

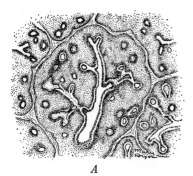
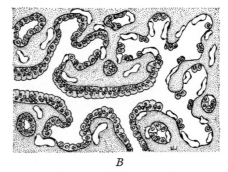

A   *B*

Fig. 228. Sections of the human lung. *A,* Developing lobules, at four months (× 75). *B,* Seeming loss of epithelium in the terminal air buds, at eight months (× 125).

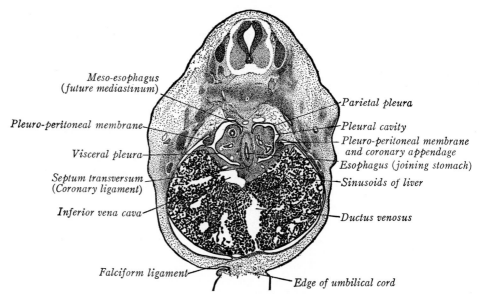

Meso-esophagus (future mediastinum)

Pleuro-peritoneal membrane

Visceral pleura

Septum transversum (Coronary ligament)

Inferior vena cava

Falciform ligament

Parietal pleura

Pleural cavity

Pleuro-peritoneal membrane and coronary appendage

Esophagus (joining stomach)

Sinusoids of liver

Ductus venosus

Edge of umbilical cord

Fig. 229.   Mediastinum, lungs and pleural cavities of a 10 mm. human embryo, in transverse section (Prentiss). × 23.

small and fails to induce the formation of a separate lobe about it, the upper lobe of the left lung is homologous to both the upper and middle lobes of the right side.[8] The suppression of the upper left lobe has been interpreted as an adaptation to facilitate the normal 'caudal recession' of the aortic arch (p. 354).[7] An alternative explanation stresses the lessened opportunity for pulmonary expansion on the left side consequent on the results of the rotation of heart and esophagus in opposite directions.[8] As a contributory factor, the more caudal position of the left common cardinal vein is perhaps significant.[1, 9] Also on the left side an important branch is suppressed in the lower lobe, owing to the position of the heart and pulmonary vein; this, however, affords opportunity for an excessive development of the corresponding right ramus which then projects into the space between the heart and diaphragm as the *cardiac bronchus* (Fig. 226 *D*).

Relations to Mesenchyme and Cœlom.   The entodermal lining of the early respiratory primordium develops within a median mass of mesenchyme, located dorsal and cranial to the main peritoneal cavity. This tissue resembles a broad mesentery; it is later named the *mediastinum* (Fig. 230). The original right and left bronchial buds grow out laterally into their respective pleural cavities, carrying before them dome-shaped investments of mesenchyme surfaced with mesothelium (Fig. 229). The subsequent branching of the bronchial buds takes place within these simultaneously growing tissue-masses. The mesenchyme adapts itself to the shape of each bronchial tree (Fig. 230), and gradually the external lobation of the two lungs takes form (Fig. 231), including the subdivision of *lobes* into *bronchopulmonary segments*. Internally each lobe becomes subdivided into *lobules* (Fig. 228 *A*). The mesenchyme actually encasing a bronchial tree ultimately differentiates into the muscle, connective tissue and cartilage plates of the walls of the air tubes and the supporting tissue of the alveolar sacs. Into the connective tissue grow blood vessels and nerve fibers.

As the lungs enlarge, they make room at the expense of the spongy tissue of the adjacent body wall (Fig. 258). This burrowing advance splits off an increasingly extensive *pericardium* from the thoracic wall and allows the lungs more and more to flank the heart on each side (Fig. 230). When the pleural cavities are completed, the mesothelial and connective-tissue covering of the lungs becomes the permanent *visceral pleura*. The facing layers, lining the thoracic wall, constitute the *parietal pleura*. These two pleural layers are derived respectively from the visceral (splanchnic) and parietal (somatic) mesoderm of the embryo.

Birth Changes. Respiratory-like movements of the chest, which tend to aspirate amniotic fluid into the lungs, sometimes occur in fetuses;[10] they are probably due to oxygen want.[11] Nevertheless, until normal breathing distends the lungs with air, these organs are relatively small; in particular they leave vacant the ventral and caudal portions of the pleural cavities (Fig. 231 *B*). With the onset of respiration after birth, the lungs gradually expand and occupy better the space allotted them. The pulmonary tissue, which was previously compact and resembled a gland in structure (Fig. 231 *C*), slowly becomes spongy owing to a great increase in the size of its alveoli and blood vessels (*D*). For weeks the alveoli remain small and shallow, but eventually the alveolar sacs interlock and press against each other until their arrangement is both intricate and intimate. Evidence of such expansion (the lungs float in water) has medicolegal value in determining whether respiration ever occurred. When gross inflation has been completed and the amniotic fluid in the lungs has been absorbed (at three days after birth), the lungs are considerably larger in every diameter and have more

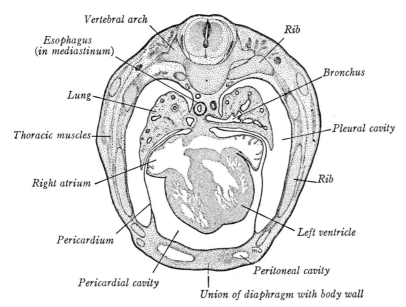

Fig. 230. Growth of the human lungs and pleural cavities (and the consequent extension of the pericardium), shown in a transverse section at seven weeks. × 12.

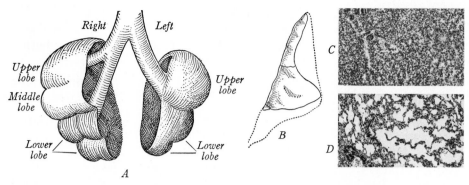

Fig. 231. The lungs as a whole, and their expansion. *A,* Human lungs, at 13 mm., in ventral view (after Blisnianskaja; × 19). *B,* Human right lung, at birth, in ventral view (× ½); broken lines indicate the unfilled extent of the pleural cavity before breathing begins. *C, D,* Sections demonstrating the appearance of the lungs of guinea pigs, just before breathing began and after eleven minutes of breathing (× 45).

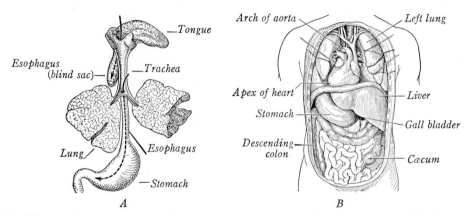

Fig. 232. *A,* Atresia in the esophagus of a newborn, with a fistulous opening of its lower segment into the trachea (× ½). *B,* Complete transposition of the adult viscera.

rounded margins. Because of the greater amount of blood admitted to the lungs after birth, their absolute weight increases considerably. Premature babies, born before the alveolar sacs have developed properly, appear to accomplish their respiratory exchanges through the terminal duct system, which functions well enough to sustain life.

Causal Relations. A grafted lung bud fails to undergo branching and characteristic differentiation unless its normal investment of mesoderm is retained. At least, a mechanical rôle for mesoderm is suggested; a true induction is claimed.

Anomalies. A highly reduced lung, or complete absence, occurs rarely, as does also an accessory lung. Variations occur in the size and the number of the major lobes. Rarely there is an eparterial bronchus, or even a third lobe, on the left side. The right eparterial bronchus at times arises directly from the trachea, as in the sheep, pig and ox; contrariwise, it may imitate the relations on the left side. The presence of a distinct *cardiac lobe* of the lung, though infrequent, is interesting since it occurs regularly in some mammals, including certain primates. Congenital *cysts* arise as derivatives of the bronchioles.

A serious anomaly is presented when there is a fistulous connection between the trachea and esophagus (Fig. 232 *A*); the esophagus usually is atretic and divided transversely, the trachea opening into its lower segment while the upper portion ends as a blind sac. The cause seems to lie in an incomplete separation of the early laryngo-tracheal groove from the gut.[12, 13]

Situs Inversus. This is a striking malformation of the viscera, in which the various organs are transposed in position, right for left and left for right, as in a mirror image. This reversal may affect all the internal organs (Fig. 232 *B*). On the other hand, an independent transposition of the thoracic or abdominal viscera alone may occur, or a single organ or part alone may be involved (Fig. 352 *B*). Positive knowledge of the cause is lacking, but this reversal is merely a part of the larger problem of how bilateral symmetry and asymmetry are established normally. Experiments on amphibian gastrulæ suggest that the organizer is asymmetrical, since damage to the mesoderm on the left side produces transpositions. Transposition of the viscera is an extreme type of symmetry reversal; left-handedness and counter-clockwise hair whorl on the crown are other more familiar but milder expressions of the same tendency. The relation of these reversals to identical twinning is discussed on p. 196.

REFERENCES CITED

1. Heiss, R. 1919. Arch. Anat. u. Physiol., Anat. Abt., Jahrg., 1–129.
2. Loosli, C. G. & E. L. Potter. 1951. Anat. Rec., *109,* 320–321.
3. Short, R. H. D. 1950. Trans. Roy. Soc. London (Ser. B), *235,* 35–86.
4. Onishi, I., *et al.* 1962. Ann. Paed. Jap., *8,* 72–84.
5. Boyden, E. A. & D. H. Tompsett, 1965. Acta Anat., *61,* 164–193.
6. Loosli, C. G. & E. L. Potter. 1959. Am. Rev. Resp. Dis., *80,* 5–23.
7. Flint, J. M. 1906. Am. J. Anat., *6,* 1–138.
8. Huntington, G. S. 1920. Am. J. Anat., *27,* 99–201.
9. Ekehorn, G. 1921. Z'ts. Anat. u. Entw., *62,* 271–351.
10. Davis, M. E. & E. L. Potter. 1946. J. Am. Med. Assn., *131,* 1194–1201.
11. Windle, W. F., *et al.* 1939; '47. Surg., Gyn. & Obst., *69,* 705–712; J. Am. Med. Assn., *133,* 125.
12. Gruenwald, P. 1940. Anat. Rec., *78,* 293–302.
13. Smith, E. I. 1957. Carnegie Contr. Embr., *36,* 41–57.

Chapter XVI. The Mesenteries and Cœlom

To a degree the middle germ layer resembles entoderm and ectoderm by differentiating some of its organs directly from solid layers of tissue (in this case, *mesoderm*). Such mesodermal derivatives, in the strict sense of that term, are the following: skeletal muscle (from the myotome plates of somites); kidneys and their ducts (from the nephrotome plates); and serous membranes, cardiac muscle, spleen and suprarenal cortex (from the somatic and splanchnic plates of lateral mesoderm). Another group of derivatives arises from the primitive filling-tissue that is known as *mesenchyme*. This tissue consists of cells that gave up their compact epithelioid arrangement in the original mesoderm and became loosely arranged, star-shaped elements (Fig. 10 *B*). Its derivatives are: connective tissue; cartilage; bone; blood; smooth muscle; and endothelium.

THE MESENTERIES

The Primitive Mesentery.   The gut arises when the splanchnopleure is folded into a tube (Fig. 233 *A, B*). The splanchnic mesoderm, which is associated with the entoderm, then takes the form of a double-layered partition, extending momentarily all the way from the roof of the cœlom to the midventral body wall (*C*). This median partition is the *primitive mesentery;* it divides the cœlom into halves and contains the gut between its component sheets. The early, straight gut naturally subdivides the mesentery into two parts. The portion above the gut is an important membrane named the *dorsal mesentery*. The portion below the gut is a transitory *ventral mesentery,* whose permanent representative is a secondary development limited to the region of the stomach, duodenum and liver (p. 282).

A mesentery is a double layer, fused back to back, of the *serous membrane*

272

that bounds the body cavity. Each layer of the combined membrane differentiates into connective tissue overlaid by an epithelium (mesothelium). The mesentery (and its continuation enclosing the gut and liver) comprises the visceral layer of the peritoneum (Fig. 233 *C*). Besides enclosing the gut and serving as its outer coat (*serosa*) the splanchnic mesoderm also differentiates into the muscle and connective tissue of the gut-wall.

In addition to the mesenteries of the digestive tube and its associated organs, there is a temporary mesentery of the heart and special supports for the genital organs. These will be described in later chapters.

Specializations of the Dorsal Mesentery. At first the gut is broadly attached dorsally, but presently this region becomes relatively narrower and the gut is then suspended throughout most of its length by a definite *dorsal mesentery* (Fig. 233). This extends like a curtain in the midplane and supplies the pathway through which blood vessels and nerves reach the gut. Only the pharynx and upper esophagus lack a mesentery, since they lie craniad in regions where there is no permanent cœlom. To the continuous primitive mesentery of the rest of the digestive canal are given distinctive names at its successive, divisional levels (Fig. 234). Thus there are the *meso-esophagus,* the *dorsal mesogastrium* (or *greater omentum*) of the stomach, the *mesoduodenum,* the *mesentery proper* of the jejunum and ileum, the *mesocolon* and the *mesorectum.*

As development advances, parts of the primitive dorsal mesentery become specialized; other regions, following the displacements of the gut, depart from the original midline position and gain secondary attachments, while still other regions are lost by obliteration. Yet most of this mesenterial system persists permanently in some form or other.

THE MESO-ESOPHAGUS. A middle stretch of the esophagus courses in a mesentery that never thins into a membrane but becomes the dorsal portion

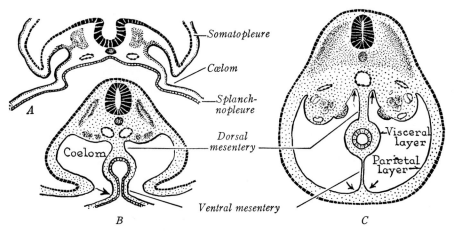

Fig. 233. Stages illustrating the formation of the primitive mesentery in human embryos. *A,* At 2 mm. (× 100); *B,* at 4 mm. (× 50); *C,* at 8 mm. (× 40). Four arrows in *C* indicate the junctions of somatic and splanchnic mesoderm.

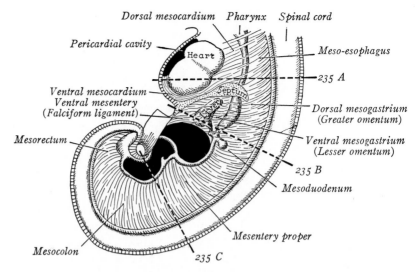

Fig. 234. The primitive human mesenteries, shown as a diagram viewed from the left side (after Prentiss). Broken lines indicate the levels of Fig. 235 *A–C*.

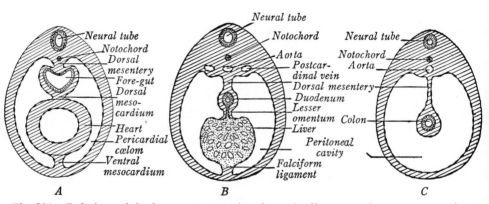

Fig. 235. Relations of the human mesenteries, shown in diagrammatic transverse sections through the levels (*A–C*) indicated on Fig. 234 (after Prentiss).

of a thick and specialized median partition known as the *mediastinum* (Fig. 230). This partition encloses all the thoracic viscera except the lungs, which expand progressively beyond its confines (Fig. 258). The dorsal meso-esophagus at these levels is short and the esophagus is broadly attached. But below the diaphragm, near its junction with the stomach, the meso-esophagus becomes a typical mesentery. Figure 229 shows an intermediate condition between the two extremes.

THE MESOGASTRIUM. The history of the dorsal mesogastrium is chiefly concerned with the development of a huge, secondary sacculation, known as the *omental bursa* or *lesser peritoneal sac*. It is so important that its origin and final relations must be traced in some detail.

Although the bursa is often described as a folding of the omentum, brought about by the rotation of the stomach, it actually arises as an independent invagination into the interior of the originally thick mesentery, starting even before rotation begins.[1, 2] The earliest indication of the bursa is in 4 mm. embryos, when a shallow pocket appears on the right surface of the dorsal mesogastrium and straightway proceeds to burrow deeper into the substance of the mesentery. One subdivision of this recess extends craniad between the esophagus and the right lung bud (Fig. 236 *A*). Such an extended passage is permanent in reptiles, but in human embryos it is soon interrupted by the developing diaphragm; the pinched-off apex then constitutes a small sac (the *infracardiac bursa*) that frequently persists in the adult (*B*).[3] The other subdivision of the original recess is located more caudally. It enlarges toward the left, behind the stomach (*A*), and thus creates a blind pocket extending into the interior of the mesogastrium (*C*). This is the beginning of the *omental bursa*.[1, 2] After the stomach has rotated, the bursa lies dorsal to the stomach; in a sense, the stomach is then carried on the ventral bursal wall (*D, E*).

The bursa proper is a progressively growing sac whose expanding walls become thinner as it pushes to the left of the general, medially located mesogastrium (Fig. 238 *A*). Subsequent sagging of the stomach to a partially transverse position changes the direction of growth of the sac so that it extends caudad (*B*). This flattened, saccular portion of the bursa then over-

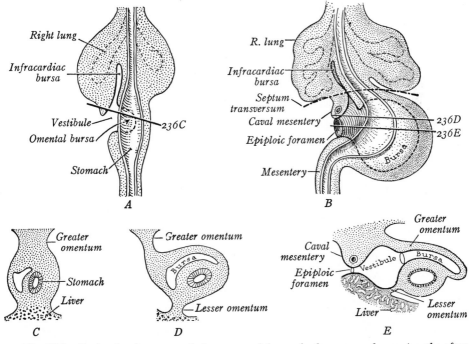

Fig. 236. Early development of the omental bursa in human embryos (partly after Frazer). *A, B,* Ventral views, at four and six weeks. *C–E,* Transverse sections, at the levels indicated on *A, B.*

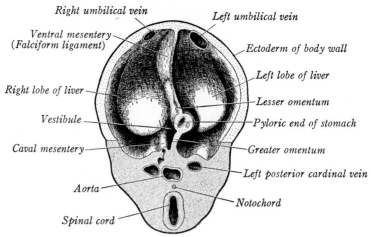

Right umbilical vein

Left umbilical vein

Ventral mesentery
(Falciform ligament)

Ectoderm of body wall

Left lobe of liver

Right lobe of liver

Lesser omentum

Vestibule

Pyloric end of stomach

Caval mesentery

Greater omentum

Left posterior cardinal vein

Aorta

Notochord

Spinal cord

Fig. 237. Early relations of the human vestibule, shown in a simplified model at five weeks. (Prentiss). × 30. The observer looks craniad.

lies the intestines and looks like an apron hanging from the greater curvature of the stomach (Fig. 239 *A*).

The narrowed mouth of the bursal sac opens into a common *vestibule* which also receives the proximal remnant of the recess that extended lungward (Fig. 236 *A, B*). The vestibule, in turn, communicates through an aperture (*epiploic foramen*) with the general peritoneal cavity (Fig. 238). In fact, the vestibule is nothing more than a recess off that cavity. It is necessary to emphasize that the epiploic foramen is a wholly different thing from the aperture into the omental bursa proper. Also, sections passing through both foramina give the false appearance of a long mesogastrium folded simply upon itself (Fig. 236 *E*); the true nature of a small-mouthed sacculation and its two divisions (vestibule and bursa proper), connected by a narrower passage, is not revealed by such a section.

Although the progressive invasion of the bursal sac into the mesogastrium is commonly described as an invagination, the actual process is one in which separate clefts arise in the mesenchyme of the mesogastrium and rapidly coalesce.[4] The method is identical with that utilized in the development of the cœlom in general (Fig. 246). It is the extension of these clefts into the growing mesentery that produces the expansive dorsal wall of the sac and its final attachment to the greater curvature of the rotated stomach (Fig. 236).[4]

The *vestibule,* already mentioned, is a peritoneum-lined space captured from the general peritoneal cavity. It is bounded cranially and laterally by a lip-like fold of the dorsal mesentery that continues caudad along the dorsal body wall into the right mesonephric fold; this is the *caval mesentery* in which the upper segment of the inferior vena cava develops (Fig. 238). Moreover, as the liver comes to locate within the ventral mesentery, its primitive right lobe both enters into relation with the caval mesentery and grows caudad (Fig. 237). In this manner the cavity of the vestibule is extended caudad, to the level of the pyloric stomach, while the caval mesentery and right hepatic lobe form its lateral wall on the right side (Fig. 243). The left wall of the vestibule is furnished by the stomach and dorsal mesogastrium. Dorsally the vestibule is limited by the dorsal body wall. As the stomach rotates so that its midventral line becomes the lesser curvature and lies at the right, the position of the lesser omentum (*i.e.*, the ventral mesogastrium between stomach and liver) is necessarily shifted from a sagittal to a frontal plane (Fig. 239). This mesentery then makes a ventral floor to the vestibule.

When the changes outlined in the preceding paragraph have been completed, the *epiploic foramen* presents a slit-like opening leading from the peritoneal cavity into the vestibule of the omental bursa (Fig. 240). The foramen is bounded ventrally by the free border (originally caudal edge) of the lesser omentum, dorsally by the inferior vena cava, cranially by the caudate process of the liver and caudally by the wall of the upper (transversely directed) duodenum. All these parts are, of course, surfaced with peritoneum. The communication between the vestibule and the bursal sac proper is an orifice that is bounded permanently by sickle-shaped *gastro-pancreatic folds.*[5]

In the third and fourth months the omental bursa makes secondary at-

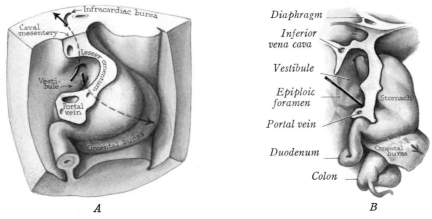

Fig. 238. The early vestibule and omental bursa of human embryos, in ventral view. *A*, At six weeks (after Frazer; × 25); *B*, at eight weeks (after Braus; × 5). In these reconstructions the liver has been cut away.

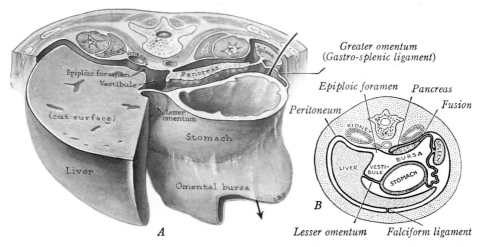

Fig. 239. Relations of the human omenta and general peritoneum, at about four months. *A*, Model, cut transversely; *B*, transverse section.

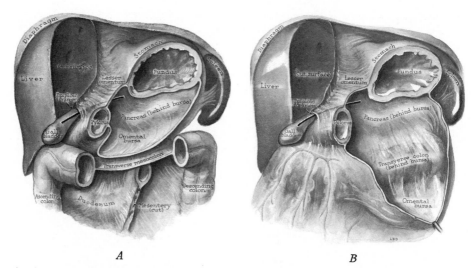

A   *B*

Fig. 240.   Dissections, showing the relations of the human vestibule and omental bursa. *A,* Before union of the bursa and transverse mesocolon; *B,* after union.

tachments. Its flat, dorsal lamella, into which the pancreas has extended, fuses with the dorsal body wall, thereby fixing the tail of the pancreas and covering the left suprarenal gland and part of the left kidney (Fig. 239). This results in the mesogastrium acquiring a new line of origin at the left of the midplane (Fig. 244). Where the dorsal lamella of the bursa lies upon the transverse mesocolon and colon, it likewise adheres and fuses (Figs. 240, 241). This results in the transverse mesocolon becoming fundamentally a double mesenterial structure (Fig. 244), but, as in all similar fusions, any evidence of compounding soon vanishes. The omental connection between stomach and colon is henceforth designated as the *gastro-colic ligament.* Caudal to this colonic attachment, the walls of the omental bursa proper unite after birth and obliterate its cavity (Fig. 241 *C*). The cavity of the adult omental bursa thus may be limited chiefly to a space between the stomach and the dorsal lamella of the greater omentum, which latter layer is largely fused to the peritoneum of the dorsal body wall (Fig. 240 *B*). The spleen develops in the cranial portion of the greater omentum; that stretch of the omentum between stomach and spleen is known as the *gastro-splenic ligament,* while its continuation beyond the spleen to the left kidney and diaphragm is the *phrenico-splenic ligament* (Fig. 239 *B*).

THE INTESTINAL MESENTERY.   As long as the gut remains a straight tube, the dorsal mesentery is a simple sheet whose two attached edges are equal in length. But when the intestine begins to elongate faster than the body wall, the intestinal border of the mesentery grows correspondingly (Fig. 208 *B, C*). The result is an elongate, somewhat fan-shaped mesentery, and in this state it is carried out into the umbilical cord between loops of the gut. On the return of the now highly coiled intestine into the abdomen, the characteristic rotation, already begun at the time of herniation into the

cord, is completed. It will be remembered that in this process the cæcal end
of the colon is carried over to the right, whereby the future transverse colon
crosses ventral to the duodenum (Fig. 242 *A*) and the small intestine lies
at the left of the cæcum and future ascending colon. There is thus accomplished a torsion of the mesentery (about the origin of the superior mesenteric artery as an axis), and this rotation is accentuated even more as the
limb of the ascending colon elongates and its flexure beneath the liver gains
prominence (Fig. 212). From a focal point at the root of the artery the
continuous mesentery of the entire intestine spreads out like a funnel
(Fig. 245 *B*).

Previous to the fourth month the entire intestine is freely movable within
the scope of its restraining mesentery, while the latter still retains its primitive line of origin along the mid-dorsal abdominal wall (Fig. 242 *A*). At
this period, however, secondary fusions begin which affix certain portions
of the gut and thereby produce new lines of attachment. Rotation of the
stomach and the enlarging head of the pancreas cause the duodenum to
become curved and laid to the right of the midplane against the body wall
(Fig. 238 *B*). A short segment of the duodenum nearest the stomach retains
its dorsal and ventral mesentery, but the rest of the *mesoduodenum* fuses
with the peritoneum of the dorsal body wall at the end of the third month
and this portion of the small intestine then becomes permanently fixed
(Figs. 242 *B*, 244). The pancreas, growing dorsad into the mesoduodenum
and into the dorsal lamella of the greater omentum, necessarily shares the
fates of these anchoring mesenteries, and hence assumes a retroperitoneal
position (Figs. 239, 241). The *mesentery proper* of the jejuno-ileum is
thrown into numerous folds, corresponding to the loops of the intestine,
but normally remains entirely free (Fig. 242 *A, B*). It does, however, acquire
a secondary line of origin where it continues into the presently fixed me·
sentery of the ascending colon (*B*).

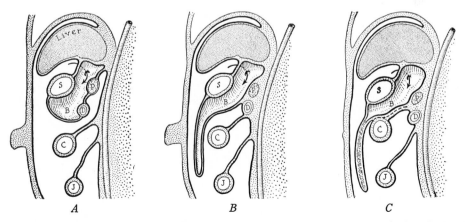

Fig. 241. Secondary fusions of the omental bursa, shown by schematic longitudinal
sections of the body (after Kollmann). *A,* At two months; *B,* at four months; *C,* adult.
B, Omental bursa; *C,* transverse colon; *D,* duodenum; *J,* jejunum; *P,* pancreas;
S, stomach.

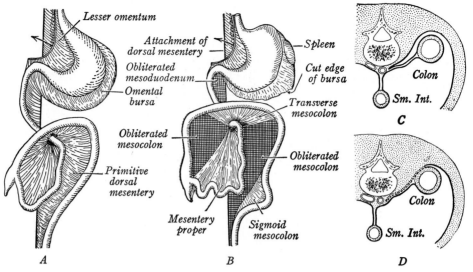

Fig. 242. Secondary fusions of the duodenum and mesocolon. *A*, At three months, before fusions begin. *B*, Later stage; fused surfaces indicated by cross hatching. *C, D*, Method of obliteration, shown in transverse section.

The large intestine suffers extensive mesenterial loss. The *ascending* and *descending mesocolons* grow rapidly, carrying the corresponding segments of the colon to the right and left, respectively, far laterad in the abdomen. The mesocolons themselves become pressed against the dorsal body wall and their flat surfaces progressively fuse (in a lateromedian direction) with the adjacent peritoneum (Fig. 242 *B-D*).[4] In this manner these two limbs of the colon become permanently anchored by the end of the fifth month, and they themselves attach broadly to the general peritoneum (Fig. 244). The *transverse mesocolon* remains largely free (Fig. 242 *B*), although it does fuse with and cover the duodenum where the colon crosses ventral to it (Fig. 244); this makes that portion of the duodenum become, for the second time, secondarily retroperitoneal in position. The line of junction of the free, transverse mesocolon with the neighboring, obliterated, mesocolic sheets gives a new (and transverse) line of origin to the free mesocolon (Fig. 242 *B*). The fusion between omental bursa and transverse colon has been described in an earlier paragraph (p. 278). The *sigmoid mesocolon* remains largely free (Fig. 242 *B*), but the primitive *mesorectum* obliterates as the rectum comes to lie against the sacrum (Fig. 244). The free and obliterated portions of the mesentery of the large intestine are shown in Fig. 244.

Obliteration of considerable portions of the primitive dorsal mesentery and the resulting fixation of the gut and pancreas are apparently related to upright posture, since these processes especially characterize the anthropoid apes and man. Such mesenterial adhesions, a part of the normal developmental plan, have much in common with those occurring pathologically after inflammation of the peritoneum. It is interesting that the original left side of the dorsal mesentery alone effects fusions with the body wall. This is true

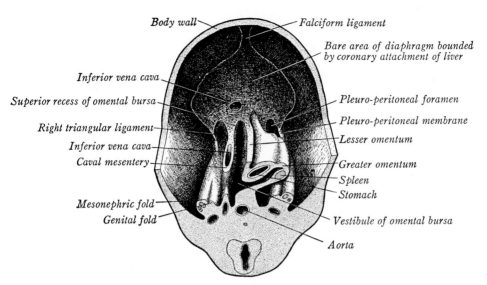

Body wall — *Falciform ligament*

Bare area of diaphragm bounded by coronary attachment of liver

Inferior vena cava

Superior recess of omental bursa — *Pleuro-peritoneal foramen*

Pleuro-peritoneal membrane

Right triangular ligament — *Lesser omentum*

Inferior vena cava

Caval mesentery — *Greater omentum*

Spleen

Stomach

Mesonephric fold

Genital fold — *Vestibule of omental bursa*

Aorta

Fig. 243. Mesenterial relations in the region of the diaphragm, shown in a simplified model of a 14 mm. human embryo (Prentiss). × 15. The liver has been cut away from its attachments; the observer looks craniad toward the septum transversum.

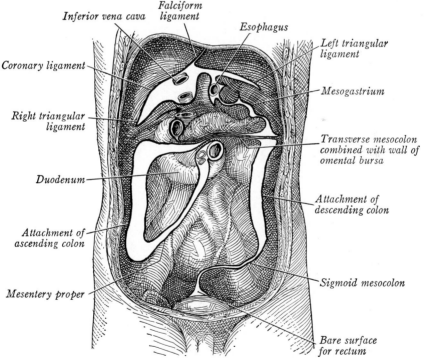

Inferior vena cava *Falciform ligament* *Esophagus*

Left triangular ligament

Coronary ligament

Mesogastrium

Right triangular ligament

Transverse mesocolon combined with wall of omental bursa

Duodenum

Attachment of descending colon

Attachment of ascending colon

Mesentery proper

Sigmoid mesocolon

Bare surface for rectum

Fig. 244. The final lines of attachment of the human mesenteries (cut away) to the dorsal abdominal wall and diaphragm.

of both the omentum and the mesentery proper, except for the short mesoduodenum which by its rotation presents a special case. The fusion of apposed peritoneal surfaces is not merely the result of nearness and pressure; it represents the carrying out of a definite hereditary pattern.

Specializations of the Ventral Mesentery.   The two layers of splanchnic mesoderm that produce a dorsal mesentery, or mediastinum, in the thoracic region come into relation, ventral to the esophagus, with the lungs and heart. Each layer serves as a mesentery to a corresponding lung (*i.e.,* its pulmonary root, or *mesopulmonon*), and this tissue, carried outward by the enlarging lungs, encloses them as the *visceral pleura* (Fig. 258). The splanchnopleure is also associated intimately with the heart, which becomes an elongate, single tube by the progressive union of paired blood vessels, each coursing in a corresponding fold of splanchnic mesoderm (Fig. 336). Hence, through its very manner of formation, the heart lies beneath the fore-gut and is fashioned within a specialized region of the primitive ventral mesentery (Figs. 234, 235 *A*). The dorsal portion of this mesentery constitutes the *dorsal mesocardium*. For a brief period it suspends the heart (Fig. 337 *B*), but soon disappears. Thereafter the heart lacks any mesenterial support, yet its wall is made of the same substance as the mesentery (Fig. 336 *C*). The ventral portion, or *ventral mesocardium,* is at best transitory and in mammals is said to have no real existence as such.[6] Secondary shiftings and re-arrangements cause the heart and pericardium to occupy the ventral region of the permanent mediastinum (Fig. 258).

In the region of the lower esophagus, stomach and upper duodenum a permanent ventral mesentery arises secondarily, subsequent to the growth of the liver bud into a mass of splanchnic mesoderm (the septum transversum) that represents the primitive diaphragm (Fig. 216). At first these divisions of the fore-gut directly overlie the septum (Fig. 248 *A*). When, however, the fore-gut soon draws away, its region of attachment with the septum stretches and thins into the definitive ventral mesentery (mostly *ventral mesogastrium*) of this region. At the same period the rapidly enlarging liver begins to project caudad from the surface of the septum and the relations come to be as in Figs. 234 and 235 *B*. Henceforth the liver can be said to lie within split halves of the ventral mesentery. Caudal to the upper duodenum, no definite ventral mesentery persists after the gut finishes its folding-off process.

LIGAMENTS OF THE LIVER.   Since the ventral mesentery encloses the liver, it gives rise to its fibrous capsule and mesenterial supports; the latter are specially designated as *ligaments*. Except where the liver impinges on the diaphragm, the enveloping hepatic capsule is covered by mesothelium that is continuous with the general peritoneum (Fig. 235 *B*). Along its mid-dorsal and midventral lines the liver maintains permanent connections with the ventral mesentery. The portion of the mesentery that extends from the stomach and duodenum to the liver is the *lesser omentum* (Fig. 240).

For convenience it is more specifically subdivided and given two regional designations: the more cranial part is the *hepato-gastric ligament,* while the more caudal portion is the *hepato-duodenal ligament.* The mesenterial attachment of the liver to the ventral body wall is named the *falciform ligament* (Fig. 239 *B*) because it extends caudad, from diaphragm to umbilicus, in a sickle-shaped fold (Fig. 237).

The peritoneum does not invade the area of contact where the liver abuts against the septum transversum (later, the diaphragm). Instead it reflects from the diaphragm to the otherwise exposed surfaces of the liver, leaving a '*bare area*' on the liver and diaphragm. This area is continued dorsolaterad by prolongations of the lateral liver lobes known as the *coronary append-ages* (Fig. 229). The attachment of the liver to the septum transversum then has the outline of a crown (Fig. 243) whose name, the *coronary liga-ment,* is more appropriate at an early stage than later (Fig. 244). As these illustrations show, the dorsoventral extent of the coronary ligament is relatively reduced during later development and the shape becomes more crescentic. Nevertheless, the coronary ligament is extended caudad some-what by an attachment established between the right lobe of the liver and the ridge ('caval mesentery') in which the inferior vena cava of this level is developing (Fig. 243). The later extensions of the coronary appendages upon the diaphragm give rise to a *triangular ligament* on each side (Fig. 219 *C*).

In general, the several displacements and secondary fusions of the primi-tive mesentery, already recorded, cause its line of attachment with the body wall and diaphragm to depart markedly from the original midsagittal posi-tion. The final condition is illustrated in Fig. 244.

The change in position of the lesser omentum to a frontal plane and its participation in creating the vestibule of the omental bursa have already been described (p. 277). Its caudal free margin, bordering the epiploic foramen, contains the bile duct and the vessels supplying the liver. The original ventral, later right, margin of the omentum attaches to the hilus of the liver and to the groove in which courses the ductus venosus (after birth called the ligamentum venosum). For a time it enclosed the gall bladder and conducted the vitelline veins from yolk sac to duodenum (Fig. 622); later regression of the omentum exposed the caudal surface of the gall bladder and freed the fused, common vein. The falciform ligament remains in the midplane and carries the umbilical vein (after birth called the ligamentum teres) embedded in its free border. The liga-mentum venosum and ligamentum teres are not mesenteries but obliterated blood channels (p. 394).

Anomalies. The mesenteries show frequent variations of form and relations. These are commonly due to the retention of simpler conditions that mark the normal develop-mental course of the intestinal canal. In about one-fourth of all cases the ascending or descending mesocolon is more or less free, owing to faulty fusion with the dorsal peri-toneum. At times the ascending colon, unanchored as yet, follows the retreating liver and becomes fixed at a high position; the condition is then improperly termed '*unde-scended cæcum*' (Fig. 245 *A*). Fixation of the entire intestine may fail completely (*B*); in this instance the bowel may twist about the root of its fan-shaped mesentery (*volvulus*) and give rise to obstruction (*C*). The primitive cavity of the omental bursa sometimes falls short of its normal degree of obliteration; the cavity may then extend even to the

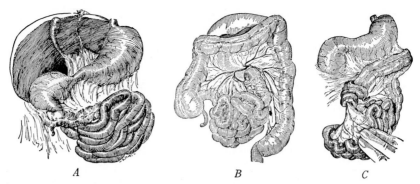

A   *B*   *C*

Fig. 245. Anomalies of the human mesenteries (Ladd). *A,* The cæcum is fixed in a high position ('undescended cæcum'); this condition has obstructed the duodenum. *B,* Free mesenteries, through failure to make any fusions. *C,* Twisting (volvulus) of the unfixed intestine, producing obstruction.

bottom of the sac, as is normal in many mammals. Less common than underfixation are conditions based on changes that exceed normality. For example, reduction in the length and breadth of the sigmoid mesocolon lessens the mobility of that portion of the bowel.

THE CŒLOM

The Primitive Cœlom. Originally the cœlom of animals was used as a temporary reservoir for excretory wastes, but this function has been superseded in vertebrates so that it now serves as a large bursa to permit frictionless movement of the heart, lungs and abdominal viscera. From the standpoint of development, the cœlom permits the visceral organs to grow and shift position without hindrance. The heart, lungs and abdominal organs of mammals occupy separate cœlomic compartments, whose respective linings are named *pericardium, pleura* and *peritoneum.* The line of junction between somatic and splanchnic mesoderm marks a subdivision of each lining into a parietal and a visceral layer (Fig. 233).

The first occurrence of a body cavity in early human stages is in the extra-embryonic mesoderm which lies between the embryo proper and the primitive chorionic capsule (Fig. 67 *A*). The earlier appearance here than in the embryo proper is presumably correlated with the precocious development of the extra-embryonic membranes. A cleft appears toward the end of the second week of development; it divides the extra-embryonic mesoderm into a *somatic layer,* which lines the chorion, and a *splanchnic layer* which invests the yolk sac. The space itself is the *cœlom,* while the cuboidal mesodermal cells that bound it soon flatten into a limiting membrane called *mesothelium.*

About one week later (at the beginning of somite formation) numerous horizontal clefts appear also in the unsegmented mesoderm belonging to the embryo itself; these lie lateral to the midline and begin to split the solid mesodermal sheet of each side into a somatic and a splanchnic layer (Fig. 246). Such isolated, cœlomic spaces coalesce first in the folding-off head region where they form a canal on each side. The cranial ends of the two cœlomic channels are continuous with a similar space located ahead of the embryo, which is destined to be the cardiac region (Fig. 247 *A, B*). After this beginning new spaces continue to appear, lateral to the somites, as fast as differentiation of the embryo in a tailward direction permits. These then link up progressively to extend the cœlomic cavities caudad. In the region where the lungs will develop just caudal to the heart, the cœlom remains as two separate canals (Fig. 247 *C*). At the heart-lung level the head-end of the embryo is separating from the underlying blastoderm, and the body cavity does not connect laterally with the extra-embryonic cœlom. Caudal to each prospective lung-region the embryonic cœlom communicates freely with the extra-embryonic cœlom. Thus, in an embryo about 2.5 mm. long, the cœlom of the embryo comprises a ∩-shaped system; the thick bend of the ∩ corresponds to the *pericardial cavity*, whereas the right and left limbs may be called *pleural canals* at this stage. The future *peritoneal cavity* communicates broadly with the extra-embryonic cœlom of each side.

The earliest cœlom within the embryo occupies a flat, horizontal plane (Fig. 247 *A, B*), but the forward growth of the head-end of the embryo and the accompanying reversal of the cardiac region presently swing the peri-

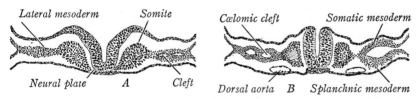

Fig. 246. Origin of the human intra-embryonic cœlom, shown by transverse sections. × 65. *A*, At two somites; *B*, at seven somites. The right half of each section is somewhat more advanced than the left.

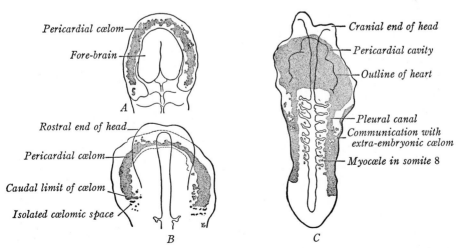

Fig. 247. Early cœlom of human embryos, in dorsal view (adapted). × 25. *A*, At one somite; *B*, at two somites; *C*, at nine somites.

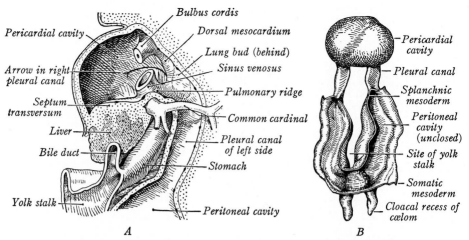

Fig. 248. Reconstructions of the body cavities in human embryos of 3 mm. *A*, Section, cut at left of the midplane, viewed from the left side (after Kollmann; × 40). *B*, The cœlomic system, isolated, in ventral view (adapted after Davis; × 24).

cardial cavity to a more ventral position beneath the embryo (Fig. 73). This single, pericardial chamber still communicates at a right-angled bend with the paired pleural canals, which now lie more dorsally (Fig. 248). Farther caudad the pleural canals, in turn, connect with the *peritoneal cavity.* As the gut and abdominal wall fold off, the primitive ventral mesentery (Fig. 228 *C*) is lost and the right and left temporary cavities become a single, common chamber. At the end of this early period the cœlomic system thus consists of a single pericardial cavity and a single peritoneal cavity, interconnected by a pair of pleural canals. As the embryo continues its folding and elongation, the peritoneal chamber is separated progressively from the extra-embryonic cœlom; the last region of closure is at the site of the developing umbilical cord.

Three specialized portions of the intra-embryonic cœlom will not be considered in the account that follows: One, the *myocœles* or tiny cavities of the mesodermal segments, disappears early and has only an historical significance (Figs. 71, 247 *C*). A second region, the temporary *umbilical cœlom* within the umbilical cord at the time of intestinal herniation, has already been discussed (p. 250). The third portion, the saccular *vaginal processes,* extends from the inguinal region of the abdominal cavity into the scrotum (Fig. 297); their development will be described on p. 332.

The division of the continuous, primitive cœlom into separate, permanent cavities is accomplished through the development of the three sets of partitions. They are: (1) the unpaired *septum transversum,* which serves as an early, partial diaphragm; (2) the paired *pleuro-pericardial membranes,* which soon join the septum and complete the division between pericardial and pleural cavities; and (3) the paired *pleuro-peritoneal membranes,* which also unite with the septum and complete the partition between each pleural cavity and the peritoneal cavity.

The Septum Transversum. When the pericardial region undergoes the reversal of position that brings it beneath the embryo proper, the original cranial margin of the pericardium becomes its definitive caudal wall (Fig. 73). This unsplit mass of mesoderm then constitutes a transverse partition occupying the space between the gut, yolk stalk and ventral body wall (Fig. 250 *A*). Standing thus between the pericardial and abdominal cavities, it is called the *septum transversum*. It is, however, an imperfect septum since the paired pleural canals, which connect the pericardial and abdominal portions of the general cœlom, course dorsally above the septum on each side (Figs. 248 *A*, 249). Such permanent communications between the pleural and abdominal cavities characterize amphibians, reptiles and birds. Sharply contrasted are the mammals, which supplement this partial septum with additional membranes; these complete the isolation of the

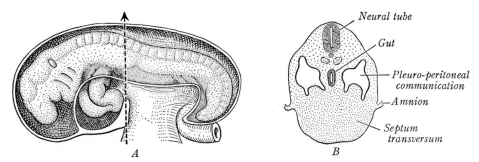

Fig. 249. Relation of the septum transversum to the cœlom of a human embryo of four weeks. × 15. *A*, Opened cœlom, viewed from the left side. *B*, Transverse section, in the plane indicated by the arrow in *A*, showing the position and relations of the septum.

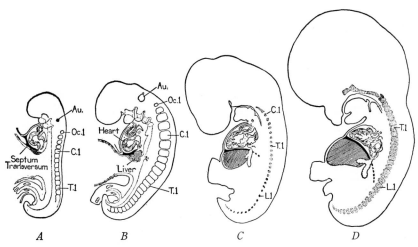

Fig. 250. Stages illustrating the 'caudal migration' of the human septum transversum to its final position at two months (Patten). *A*, At 2 mm. (× 17); *B*, at 3.6 mm. (× 10); *C*, at 11 mm. (× 4); *D*, at 25 mm. (× 2.5).

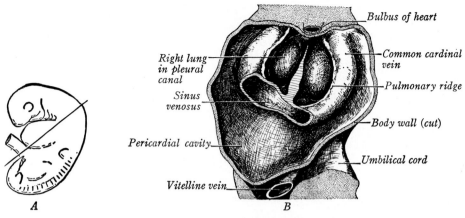

Fig. 251. Model of the human pleuro-pericardial cavity, at 5 mm., opened ventrally (after Frazer). × 32. The plane of section in *B* is indicated on *A*.

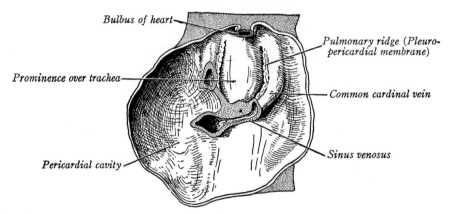

Fig. 252. Model of the human pericardial cavity, at 10 mm., opened ventrally (after Frazer). × 20. The plane of section is indicated on Fig. 251 *A*.

pericardial, pleural and peritoneal cavities and, in so doing, produce a true *diaphragm*.

Only the cranial part of the original septum transversum continues in its rôle as an actual partition (Fig. 248 *A*). The liver bud penetrates the more caudal portion of the septum and, as the liver increases in size, this caudal mass draws away, thus producing the ventral mesentery (containing the liver) as already described (p. 255). Since both the primitive heart and liver abut against the septum, the stems of all the great veins (vitelline, umbilical and cardinal) course through its substance as they join the heart (Fig. 310).

The septum transversum of a 2 mm. embryo occupies a position opposite the highest occipital somite. It then enters upon what is usually described as an extensive caudal migration (Fig. 250). This displacement is, however, largely relative and is caused chiefly by a faster forward growth of the

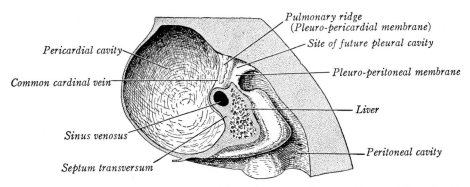

Fig. 253. Model of a right portion of the human cœlom, at 5 mm. (adapted after Frazer). × 35. The cut surface represents a longitudinal section, near the midplane; the heart and lung have been removed.

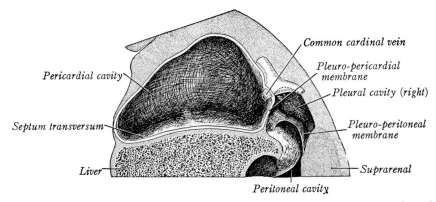

Fig. 254. Model of a right portion of the human cœlom at 13 mm. (adapted after Frazer). × 15. The model is cut longitudinally near the midplane.

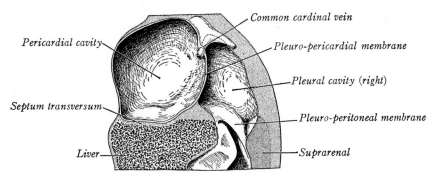

Fig. 255. Model of a right portion of the human cœlom, at 16 mm. (adapted after Frazer). × 10. The model is cut longitudinally near the midplane.

dorsal region of the body, which leaves the more ventral structures behind (in relation to the somites as 'fixed' points of reference). When opposite the fourth cervical segment the septum receives the phrenic nerve, by way of the pleuro-pericardial membrane, and carries it along (Fig. 257 *A*). The final location of the septum dorsally is at the level of the first lumbar seg-

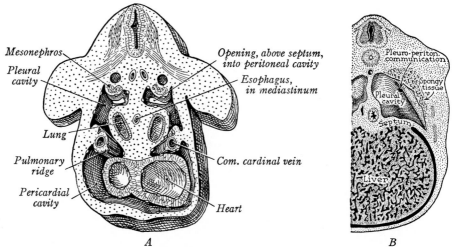

Fig. 256. Models of human embryos, cut across the pleural cavities and viewed in a caudal direction. *A,* At 7.5 mm., cut cephalad of the septum (after Kollmann; × 17); *B,* at 16 mm., cut just caudal to the septum (after Frazer; × 12).

ment; this position is attained in an embryo of two months (Figs. 250 *D,* 257 *B*).

The Pleuro-pericardial Membranes. In a 4 mm. embryo the lungs begin to develop within the medial mass of mesenchyme that separates the two pleural canals, and soon bulge into them (Figs. 248 *A,* 256 *A*). The canals thereby become potential *pleural cavities,* and will be so termed hereafter. At this period the common cardinal veins (ducts of Cuvier), on their way to the heart, curve around the pleural cavities laterally in the somatic body wall (Fig. 248 *A*). Each vein courses in a mesodermal ridge that projects medially into the adjacent pleural canal (Fig. 256 *A*). This major elevation ends in a projecting, irregular edge known as the *pulmonary ridge* (of Mall). As the common cardinal veins shift toward the midplane they draw out a mesentery-like fold on each side which bears the name of *pleuro-pericardial membrane.*[7, 8] By the end of the sixth week (12 mm.) the membrane of each side comes into contact with the median mass of tissue (primitive mediastinum) and fuses with it. The separation of pericardial and pleural cavities is then complete. Stages in this process can be traced in Figs. 251–255.

The two stages shown as Figs. 251, 252 represent models of the pericardial and pleural cavities after the front half of the body wall had been removed along the plane indicated in Fig. 251 *A.* They illustrate the way in which the pulmonary ridges, containing the common cardinals, come into relation with the median mass of mediastinal tissue, thereby closing the communication between the pleural and pericardial cavities and masking the lungs from view. In Fig. 252 a mere slit still exists, somewhat like the permanent condition in sharks, but this aperture closes in a stage immediately following. Other views of the pleuro-pericardial membrane are displayed in Figs. 253–255. These illustrations represent models of the body cavities on the right side of the body; they have been exposed by sections cutting a little to the right of the median plane, and the heart and lungs have been removed. The free border of the pulmonary ridge is apparent in the early stage shown in Fig. 253, but in an embryo 12 mm. long it joins the median mass of mediastinal tissue. Hence Figs. 254, 255 show the completed (and greatly expanded) pleuro-pericardial membrane.

The Pleuro-peritoneal Membranes.   This pair of membranes is largely produced when the lungs can find room for lateral expansion only by invading the adjoining body wall. Representative stages are shown in Figs. 253–255. At first there is merely a shallow and narrow space between the pulmonary ridge, located more cranially, and a quite separate fold now appearing caudally (Fig. 253). The latter represents a dorsolateral extension of the caudalmost portion of the septum transversum. Soon, however, growth of the lung and shiftings of the liver and common cardinal vein create more room between these pleural boundaries. In such manner a definite pleuro-peritoneal membrane is brought into existence (Fig. 254).[8, 9] Continued expansion of the pleural cavity progressively increases the area of this membrane, and of the pleuro-pericardial membrane as well (Fig. 255). The openings between pleural and peritoneal cavities become reduced during the seventh week (Fig. 256 B) and close shortly after the end of that week (17 mm.). Figure 257 illustrates the relations of the body cavities, at two important stages in their history, as are revealed by lateral dissections.

The Pericardium.   The primitive pleural cavities are small (Fig. 258 A). In order to accommodate the rapidly expanding lungs huge extensions are added, so that a major portion of each definitive pleural sac is a new formation brought into existence in the following manner. Since enlargement of the lungs is limited medially by the mediastinal contents,

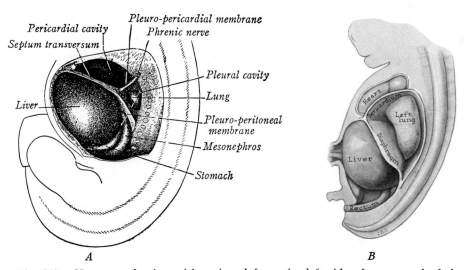

A   *B*

Fig. 257.   Human cœlomic cavities, viewed from the left side after removal of the lateral body wall. *A*, At 11 mm., with incompletely partitioned cavities indicated by arrow (Prentiss, after Mall; × 8). *B*, At 28 mm., after partitioning by the pericardium and diaphragm is complete (after Frazer; × 3).

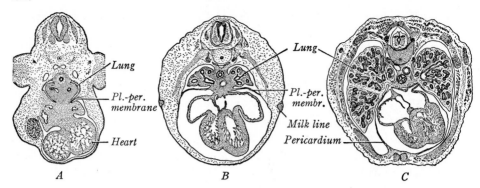

Fig. 258. Formation of the definitive human pericardium, illustrated by transverse sections. *A*, At 8.5 mm. (× 12); *B*, at 16 mm. (× 8); *C*, at 35 mm. (× 4).

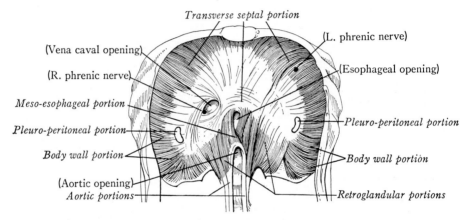

Fig. 259. The components of the definitive diaphragm (after Wells). Only the extent of portions contributed by the pleuro-peritoneal membranes can be estimated successfully.

they have to grow in other directions. Room is made for the lungs at the expense of the adjacent body wall by the obliteration of its loose, spongy mesenchyme, some of which is indicated in Fig. 256 *B*. In this growth the cavities expand, especially in lateral and ventral directions, splitting off additions to the pleuro-pericardial membranes as they advance (Fig. 258 *B, C*). Thus the lungs more and more come to flank the heart. The membrane then separating heart from lungs represents not only the original pleuro-pericardial membranes but also the additions to them captured from the body wall. The final, partitioning membrane encloses the heart like a sac and is named the *pericardium* (Fig. 351 *B*).

The Diaphragm. The complete separation of the pleural cavities from the abdomen by a diaphragm is a distinctive mammalian characteristic. It increases greatly the power of inspiration and, in its capacity as a septum, restricts to the thorax the negative pressure produced during inspiration.

The liver grows enormously during the second month, and on both sides some of the adjacent body wall is taken up into the primitive diaphragm.

When completed the *diaphragm* is derived from six sources, but the limits of these several contributions cannot be set exactly. The individual components are as follows (Fig. 259):[8, 9] (1) a large ventral portion from the septum transversum; (2) paired lateral portions from the mesenchyme of the costal body walls; (3) small intermediate portions from the paired pleuro-peritoneal membranes; (4) a small dorsomedian portion from mesenchyme of the meso-esophagus; (5) dorsolateral portions from mesenchyme behind paired glandular groups, each consisting of the suprarenal, gonad and mesonephros; and (6) paired, elongate portions from mesenchyme bordering the sides of the aorta. In addition to these components there is the striated muscle of the diaphragm, whose origin is customarily attributed to tissue supposedly derived from the cervical myotomes when the septum transversum still stood at a high level.

Actually the exact source of the muscular component of the diaphragm is not surely established. The early passage of cervical nerve fibers (the phrenic nerve) to the primordial diaphragm is only circumstantial evidence of a concomitant migration of myotomic, premuscle masses and the retention of their original nerve supply (Fig. 257 *A*). On the contrary, the muscle may be derived from the body wall when the burrowing lungs and expanding liver strip off tissue that is added to the periphery of the diaphragm.[9] The parts of the diaphragm derived from the pleuro-peritoneal membranes are not so expansive in man as in some other mammals. It seems probable that the closure of the pleuro-peritoneal communication, as well as that between the pleural and pericardial cavities, is not owing primarily to the activity of the membranes themselves, but is passive and caused by the growth of adjacent organs.[9]

Anomalies. The diaphragm is subject to developmental defects that permit herniation of abdominal viscera into the chest. The commonest *diaphragmatic hernia* of congenital origin is related to imperfect development of a pleuro-peritoneal membrane, and usually (5:1) the left one is at fault. The great majority of these specimens show an actual dorsolateral aperture, resulting from imperfect closure by the pleuro-peritoneal membrane (Fig. 260 *B*). Any of the abdominal viscera, except the stomach and descend-

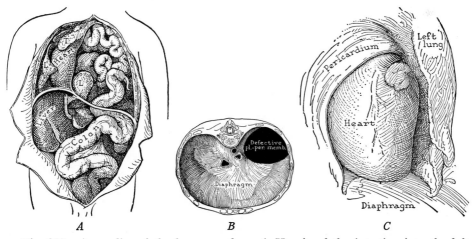

A   *B*   *C*

Fig. 260. Anomalies of the human cœlom. *A,* Hernia of the intestine into the left pleural cavity. *B,* Cranial surface of a diaphragm (with a defect like that in *A*). *C,* Incomplete pericardium, with the heart and left lung occupying a common cavity.

L, Lung; *T,* thymus.

ing colon, may then project into the corresponding pleural cavity (*A*). In 10 per cent of specimens the herniating viscera occupy a double-layered sac (pleura and peritoneum); the peritoneal component represents a pleuro-peritoneal membrane devoid of muscular content. Much rarer is the faulty development of a pleuro-pericardial membrane, which results in the heart and lung occupying a common chamber (*C*). *Cysts* outside the pericardium probably originate during the splitting away of the pericardium from the chest wall.

Another type of hernia is related to an abnormally large aperture where the esophagus passes through the diaphragm. In this type, more or less of the stomach protrudes into the mediastinum, carrying with it a membranous portion of the diaphragm. The condition is unlike that of a thoracic stomach whose descent is halted by a short esophagus (p. 249). There may be a retention of the temporary herniation of the intestine into the cœlom of the umbilical cord (Fig. 215 *D*), while weakness of the body wall at an imperfectly constructed umbilical region predisposes to acquired herniation there. The sacs of peritoneum that invade the scrotum may retain their connecting stalks and admit a loop of the bowel (Fig. 298 *C*).

REFERENCES CITED

1. Broman, I. 1905. Ergeb. Anat. u. Entw., *15*, 332–409.
2. Pernkopf, E. 1922. Z'ts. Anat. u. Entw., *64*, 96–275.
3. Viikari, S. J. 1950. Ann. Chir. et Gyn. Fenn., *39* (Suppl.), 3–97.
4. Kanagasuntheran, R. 1957. J. Anat., *91*, 188–206.
5. Crymble, P. T. 1913. J. Anat. & Physiol., *47*, 207–224.
6. Davis, C. L. 1927. Carnegie Contr. Embr., *19*, 245–284.
7. Elliott, R. 1933. Am. J. Anat., *48*, 355–390.
8. Bremer, J. L. 1943. Arch. Path., *36*, 539–549.
9. Wells, L. J. 1954. Carnegie Contr. Embr., *35*, 107–134.

Chapter XVII. The Urinary System

The urinary and reproductive systems are intimately associated in origin, development and certain final relations. Both arise in mesoderm that initially takes the form of a common urogenital ridge, located on each side of the median plane (Fig. 274). Both systems continue to develop in close approximation; they drain into a common cloaca and, slightly later, into the urogenital sinus which is a subdivision of the cloaca. The history of neither system is simple and direct. Some organs result from the association of structures that were originally quite separate. Other parts arise, only to disappear after a transitory existence during which they may never have functioned. Still other structures, designed for one kind of purpose, abandon their original course and are turned to a wholly new use. Certain common primordia transform differently, as will be appropriate to the emerging male or female. Yet because of the interwoven nature of this story, it is far simpler to pursue separate narratives for the urinary and genital systems than to attempt a synchronized, single description of what is often called the *urogenital system*.

Vertebrates have made three distinct experiments in the production of kidneys. Each was an improvement over the preceding type. As might be anticipated, the embryos of the higher vertebrates indicate this progress by repeating the same kidney sequence during development; nowhere can be found a better illustration of the principle of recapitulation. The earliest and simplest excretory organ was the *pronephros,* functional today only in cyclostomes and a few fishes. The pronephros, nevertheless, does serve as a provisional kidney in larval fishes and amphibians, but it is replaced by the *mesonephros* which remains as the permanent kidney of these animals. The embryos of reptiles, birds and mammals develop first a rudimentary and functionless pronephros and then a mesonephros (functional during a part of fetal life), whereas the final kidney is a new organ, the *metanephros*. These three kidneys develop overlappingly, one caudad of the other, in the order indicated by their names (Fig. 261). Some prefer to emphasize the essential identity of these kidneys and their similar origins, and hence consider all three as a unit, parts of which develop more or less separately in time and space.[1,2]

All three kidney types are organs composed of units known as *uriniferous*

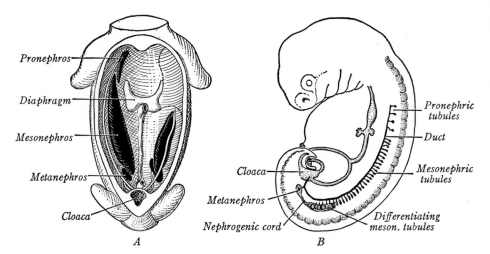

Fig. 261. Locations and relations of the three kidney-types in mammals (semi-diagrammatic). *A,* Ventral dissection, the left side showing a later stage than the right. *B,* Lateral view, with the nephric system and gut showing through.

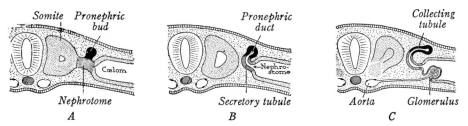

Fig. 262. Development of a functional pronephric tubule, illustrated by transverse sections of stages in a lower vertebrate (semidiagrammatic). × about 140.

tubules, which have a common source of origin and exhibit somewhat the same structural plan. They arise from the mesoderm of the *nephrotome;* this plate lies just lateral to the somites and connects the latter with the somatic and splanchnic layers of mesoderm which enclose the cœlom (Figs. 262 *A,* 263 *A*). In close relation with all three types of secretory tubules (*nephrons*) there is a vascular tuft (*glomerulus*), specialized for separating urinary wastes from out the blood (Fig. 267 *A*). The collected waste products are then conducted by *collecting tubules* to a common *excretory duct* which discharges them from the body. All three kinds of kidneys, as well as both kinds of gonads, differ from other exocrine glands in that their secretory tissue differentiates directly from an unorganized, cellular blastema. Union with the duct system is secondary.

THE PRONEPHROS

The functional pronephros of lower vertebrates consists of paired *pronephric tubules,* arranged segmentally. They arise as buds off the nephrotome plate (Fig. 262 *A*); their ends unite, thus producing a longitudinal *pronephric duct (B, C;* Fig. 263). One end of each tortuous tubule opens into the cœlom, the other into the duct which is excretory in function and drains into the cloaca. The ciliated, funnel-shaped communication with the body cavity is the *nephrostome.* Near by, but entirely separate from each tubule, an arterial tuft projects into the cœlom (Fig. 262 C). These external *glomeruli,* covered only by thin epithelium, filter wastes from the blood into the cœlom. The mixture of urine and cœlomic fluid is then taken up by the tubules and carried by means of ciliary currents into the main excretory duct. As implied by its name, the pronephros is located well craniad in the body (Fig. 261); for this reason it has often been called the 'head kidney.'

Although the human pronephros is vestigial, it is as well developed as that of other amniote embryos. It consists of several pairs of rudimentary pronephric 'tubules,' arising as dorsolateral sprouts from the longitudinally fused nephrotomes (the *nephrogenic cord*) of each side. Even preceding the appearance of tubule primordia at 10 somites, a cellular strand has split away from the corresponding cord.[2] This hollows as the primitive *excretory duct,* but the pronephric tubules do not join it. The total extent of the pronephros is in the region of junction between the future neck and thorax. The exact lower limit of the series of pronephric tubules is uncertain because they grade into the mesonephric type.[2] The degeneration of pronephric tubules is complete at about the 5 mm. stage, but the excretory duct persists. Caudal to somite 14 the duct no longer arises by delamina-

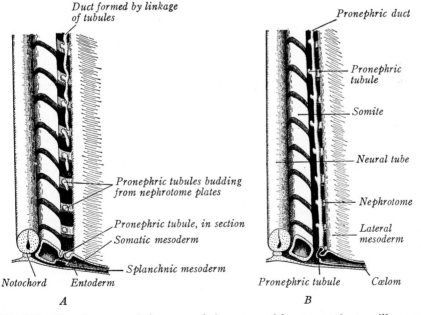

Fig. 263. Development of the pronephric system of lower vertebrates, illustrated by models (after Felix). *A,* Younger stage, with tubules still forming and linking together. *B,* Older stage, with tubules and duct completed.

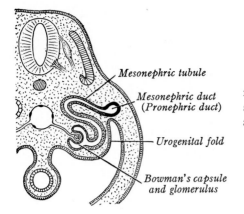

Mesonephric tubule

Mesonephric duct
(Pronephric duct)

Urogenital fold

Bowman's capsule
and glomerulus

Fig. 264. Form and relation of a human mesonephric tubule, shown in a transverse section at 5 mm. (semidiagrammatic). × about 75.

tion. Instead, its caudal blind end grows freely toward the cloaca, which it reaches and perforates in embryos about 4 mm. long.

THE MESONEPHROS

The mesonephros, or Wolffian body of vertebrates, is larger than the pronephros; not only does it contain more tubules, but also these are longer and more complicated. It is located farther caudad and is appropriately named the 'middle kidney' (Figs. 261, 265 *A*). Whereas the pronephros is entirely functionless in higher vertebrates, the mesonephros serves these embryos as a temporary excretory organ that overlaps the initial activity of the permanent kidney. In most, but not all, mammals function is attained; even in man, whose mesonephros is not large, this is apparently true until the tenth week.[3]

The mesonephros of each side, like the pronephros, consists of a series of tubules, each of which at one end becomes associated with a knot of blood vessels and at the other end opens into the excretory duct (Fig. 264). But the mesonephric tubule differs in two important respects: (1) the glomerulus is internal (*i.e.*, it indents the blind end of the tubule, and excreta from the blood pass directly into the lumen of the tubule); and (2) the nephrostome is at best transitory and never serves as a functional cœlomic mouth to the tubule proper. Mesonephric tubules drain into the same excretory duct that began its development in relation to the pronephros. It is henceforth known as the *mesonephric* (or Wolffian) *duct* (Fig. 265 *B*). As a whole, the mesonephric tubules bear no significant relation to body segmentation, and several commonly develop alongside a single somite.

Differentiation of the Mesonephros. Mesonephric tubules arise from the same general source as did the pronephric tubules. This is the *nephrogenic cord*, which extends as far caudad as the twenty-eighth somite. In embryos of 17 somites tubules begin to take origin from spherical masses of cells, brought into existence by subdivision of the nephrogenic cord

(Fig. 266 *A*). They appear progressively, adding to the series chiefly in a caudal direction. Each spherical mass of mesonephrogenic tissue hollows and the vesicle, so formed, sends out a solid extension which unites with the mesonephric duct near by (*B*). To complete a mesonephric tubule there is further canalization, growth with S-shaped bending, and association with a glomerulus (*C, D*). The free end of the tubule enlarges and becomes thin-walled when a knot of blood vessels (the *glomerulus*) indents one side. The double-walled cup, thus formed, is the *glomerular* (or *Bowman's*) *capsule*. Capsule and glomerulus together comprise a unit known as the *mesonephric corpuscle* (Fig. 267 *A*). Next in order comes a thicker, lighter-staining *secretory segment* of the tubule and then a thinner, darker-staining *collecting segment* which, in turn, opens into the mesonephric duct (*B*).[4]

When the developing tubules begin to enlarge, there is not room for them in the body wall and accordingly they bulge ventrolaterally into the cœlom. On each side of the dorsal mesentery there is thus produced a longi-

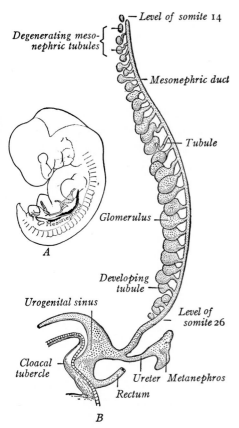

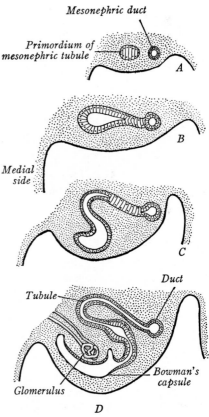

Fig. 265. Location and composition of the human mesonephros. *A*, At 8 mm. (after Shikinami; × 4.5). *B*, At 10 mm., showing the mesonephric region reconstructed in greater detail (after Felix; × 35).

Fig. 266. Stages in the differentiation of a human mesonephric tubule, shown in simplified sections (after Felix). × about 100.

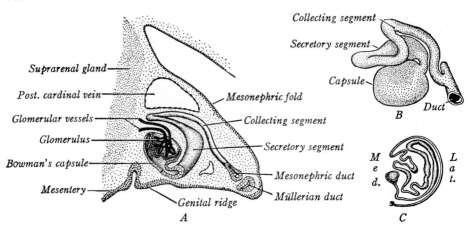

Fig. 267. Models of mature mesonephric tubules. *A,* Human tubule, at 10 mm., partly opened and superimposed on a section of the left mesonephric ridge (× 95). *B,* Human tubule, at 13 mm. (× 75). *C,* Pig's tubule, at 80 mm. (after McCallum; × 3).

tudinal *urogenital ridge,* which attains its greatest relative length at about five weeks (7 mm.) by extending through a distance of some 15 somites (Fig. 279 *A*). Soon after its formation this common fold subdivides longitudinally into a more lateral *mesonephric ridge* and a medial *genital ridge* (*B, C*). After the fourth week progressive degeneration of the more cranial tubules and continued new formation at the caudal end of the mesonephric ridge effect a wave-like settling of the gland caudad. As a result of this, the upper five-sixths of its extent is lost by the end of the second month. The cranial remnant is reduced to a band known as the *diaphragmatic ligament* of the mesonephros; it presently will serve as a *suspensory ligament* of the gonad (Fig. 294). In the remaining one-sixth, some new tubules arise, but at the end of ten weeks none remains as an unbroken tubule.

How the male genital system salvages the mesonephric duct and remnants of surviving tubules, and utilizes them for new purposes, will be traced in the following chapter. Yet with regard to the permanent kidneys it is necessary first to describe how their entire system of drainage ducts traces origin from a bud off each mesonephric duct, and how these ducts unite secondarily with a new set of secretory tubules.

The formation of mesonephric primordia commences at about somite 11 in human embryos with some 18 somites. In a 7 mm. embryo the caudal limit is reached at the twenty-sixth somite (fourth lumbar) (Fig. 265). Embryos four to nine weeks old have a rather constant number of about 30 tubules in each mesonephros, and within these time limits the gland reaches the height of its development.[5, 6] In all, a maximum number of about 40 pairs of tubules is possible, of which some 17 pairs still persist at nine weeks. Half of these are already non-functional, while within another week all become discontinuous; yet the maximum degeneration attained is not complete until the end of the fourth month.

The glomeruli occupy a median column in the gland; the duct is lateral and the tubules are intermediate and dorsal in position (Fig. 267 *A*). Lateral branches from the aorta supply the glomeruli, while the posterior cardinal veins, dorsally placed, break up into a network of sinusoids about the tubules; these latter channels are continuous in turn with the subcardinal veins and constitute a true renal-portal system, as in lower vertebrates. The mesonephros of man, along with that of the cat and guinea pig, is somewhat small (*B*). By comparison, the mouse and rat have a tiny, non-functional gland, while that of the sheep is medium-sized and the pig and rabbit have extremely large mesonephroi with more completely coiled tubules (*C*). These varying degrees of size and differentiation are correlated with the size of the allantois, but stand in inverse relation to the permeability of the placenta.[5]

THE METANEPHROS

The permanent kidney of amniotes (reptiles, birds and mammals) arises far caudad in the body (Fig. 261). As in the case of the mesonephros, the final kidney consists of an aggregate of tubules which drain into a common duct. Also like the mesonephros, the metanephros is of double origin; but in this instance the boundary between the two components lies midway of the uriniferous tubules themselves. Thus the inert system of drainage ducts (ureter; pelvis; calyces; papillary ducts; and straight collecting tubules) is derived from a bud growing off the mesonephric duct (Fig. 268). On the other hand, each secretory unit, or *nephron* (Bowman's capsule, both convoluted tubules and Henle's loop), differentiates from the substance of the caudal end of the nephrogenic cord; it thus has an origin similar to that of the entire mesonephric tubule. A collecting and secretory tubule then unite secondarily to complete a continuous *uriniferous tubule* (Fig. 270). Yet in structure and function these two components remain as different as was their origin.

The mesonephric duct makes a sharp bend just before joining the cloaca. It is at this angle (level of the twenty-eighth somite, or the future first sacral vertebra) that the so-called *ureteric bud* soon arises, dorsal and somewhat medial in position (Fig. 268 *A*). Each primordium appears in embryos of four weeks (5 mm.), taking the form of a hollow bud which first grows dorsad and then turns cephalad. The proximal, rapidly elongating stalk of this diverticulum is the future *ureter*, while the distal, blind end dilates at once into the primitive *renal pelvis* (*B*). Shortly after its first appearance the ureteric bud pushes into a mass of condensed tissue which is the caudalmost portion of the nephrogenic cord (*A*). This *metanephrogenic mass* promptly separates from the more cranial mesonephrogenic tissue and then surrounds the pelvic dilatation like a cap (*B*). Such are the early primordia that jointly give rise to the permanent kidney; compare models, Fig. 275 *B, D*.

Differentiation of the Ureteric Bud. The ureteric stem elongates and, toward the end of the sixth week, the primitive renal pelvis both flattens from side to side and bifurcates into the two future *major calyces* (Fig. 268 *B, C*). From these calyces secondary branches bud out, which in turn give rise to tertiary branches (*D*). The branching process is repeated still farther in radial directions until, at five months, some twelve generations of collecting tubules have been developed (*E, F*); this amount of branching

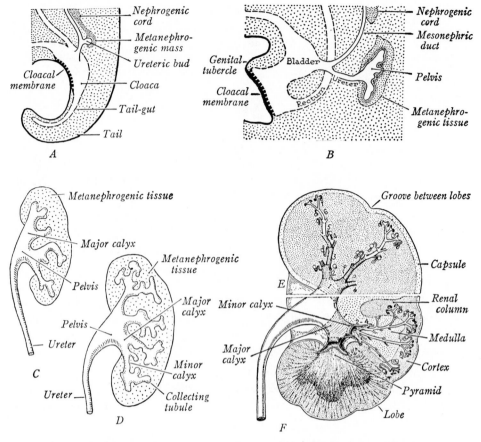

Fig. 268. Origin of the human metanephros and the development of its duct system, as shown by reconstructions. *A, B,* Origin and early relations (*A,* at 5 mm., × 35; *B,* at 11 mm., × 25). *C–F,* Further stages in the development of the ureteric bud into the duct system (*C,* at 15 mm., × 45; *D,* at 20 mm., × 40); *E,* at nine weeks, × 40; *F,* at birth, × 1.5).

increases but little thereafter. The several secondary tubules soon enlarge and are said to absorb into their walls the branches of the third and fourth order; each of the approximately nine composite receptacles so formed is a *minor calyx* (*E, F*). As a result of the absorption just mentioned, the tubules of the fifth order, up to 25 in number, then open into each minor calyx as *papillary ducts*. The remaining higher orders of diverging tubules constitute the straight *collecting tubules*. These are abundant in the *medulla* of the definitive kidney and they also project into the *cortex* where they course in the rays (*pars radiata*).

The aggregate of all the tubular 'trees' (whose trunks are papillary ducts) that drain into any one, minor calyx comprises a renal unit known as a *pyramid*. Its base adjoins the overlying cortex and its apex, or *papilla*, projects into the calyx (*cf.* Fig. 268 *E*). Later, each primary pyramid (but not its papilla) subdivides into two or more secondary pyramids in such a way

that each papilla comes to serve as a common outlet for the several definitive pyramids (*F*). The human kidney, with about nine papillæ, thus differs
from the general primate plan of a single great pyramid and papilla.

The simple epithelium of the collecting tubules and papillary ducts elevates to a distinctive, pale columnar type. By contrast the calyces, pelvis and
ureter differentiate into a stratified, so-called transitional epithelium; these
parts of the urinary tract become invested with coats of smooth muscle and
connective tissue, produced by specialization of the surrounding mesenchyme.

Differentiation of the Metanephrogenic Tissue.   The early cap of tissue
that covers the dilated, pelvic portion of the ureteric bud shows two layers
of different density (Fig. 269 *A*). The internal, denser layer differentiates
into the secretory tubules, or *nephrons,* whereas the external, looser layer
is destined to become the interstitial connective tissue of the kidney and
its enveloping *capsule* (*B, C*).

When the primitive renal pelvis starts on its program of branching, the
internal layer of the metanephrogenic tissue subdivides into a corresponding
number of masses. One such lump covers the end of each pelvic subdivision
(Fig. 269 *A, B*). As new orders of collecting tubules arise progressively, each
mass of metanephrogenic tissue not only increases steadily in amount but
also subdivides in the same rhythm (*C*). One small lump is left behind in
association with each terminal collecting tubule while other parts of the
mass are lifted to higher and higher levels as the stem tubules advance. The
shell of tissue that thus comes to overlie the bases of the pyramids is the
cortex of the kidney; some of its substance fills in the spaces between individual pyramids and is there designated as *renal columns* (Fig. 268 *F*).

The renal cortex consists of two kinds of territories. The *pars radiata,* or
cortical rays, result from the massive extension of radial bundles of collecting tubules, as already explained. The *pars convoluta,* or labyrinth, consists
of the aggregate of secretory tubules differentiated from out the metanephrogenic tissue. Each tiny ball of this formative substance is the forerunner of
a *secretory tubule,* or nephron, whose developmental course can be followed

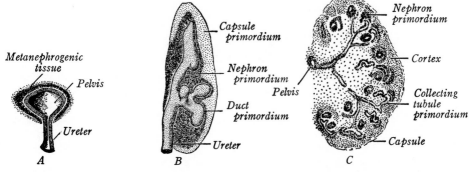

Fig. 269.   The early differentiation of human metanephrogenic tissue. *A,* Halved model,
at 8 mm. (× 80); *B,* half model, at 12 mm. (× 65); *C,* section, at 20 mm. (× 40).

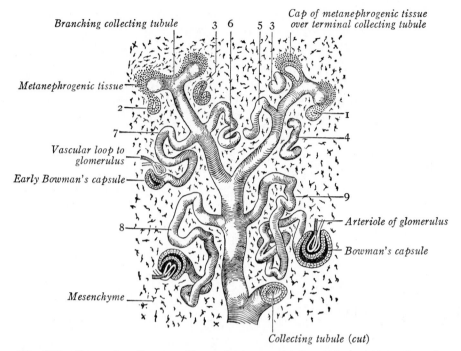

Branching collecting tubule   3 6   5 3   *Cap of metanephrogenic tissue over terminal collecting tubule*

Metanephrogenic tissue

Vascular loop to glomerulus

Early Bowman's capsule

Arteriole of glomerulus

Bowman's capsule

Mesenchyme

Collecting tubule (cut)

Fig. 270. Composite diagram, illustrating stages in the differentiation of nephrons and their linkage with branching collecting tubules (after Corning). Progressive stages of development are numbered serially, 1–9.

in Fig. 270. A ball hollows into a vesicle which elongates and becomes tortuous. At one end a Bowman's capsule and glomerulus differentiate, while the other end establishes communication with the nearby collecting tubule. Since the newer generations of tubules develop progressively in a centrifugal direction from out the self-perpetuating metanephrogenic tissue, it follows that the oldest tubules are those nearest the medulla. The differentiation of new tubules terminates about one month before birth in the zone just beneath the external capsule;[7] at this time more than one million have been produced in each kidney. All later increase in renal size results from the enlargement of tubules already present. The steps in tubule-differentiation can be followed precisely in Fig. 271.

The details of the tubule-differentiation are as follows: During the seventh week some of the nephrogenic tissue about the ends of the collecting tubules condenses into spherical masses; these hang down in the angles between the end-buds of collecting tubules and their parent stems (Fig. 271 *A*). One such metanephric sphere is the forerunner of each secretory tubule. The formation of new spheres and their transformation into tubules continue at progressively higher levels as the cortex thickens and the stem tubules continue to branch. The stage of a solid sphere is soon converted into a vesicle with an eccentrically placed cavity (*A, B*). The vesicle then elongates, thereby producing an S-shaped *secretory tubule* (*C*) which unites at one end with the adjacent terminal *collecting tubule* (*D*). The thinner-walled, blind end of the tubule becomes the *glomerular capsule* (Bowman's) of a renal corpuscle (*D, E*). The stage of the S-shaped tubule is followed by marked elongation and twisting (*F, G*).

The fully formed uriniferous tubule is arranged in a definite and orderly manner (Fig. 271 *G*). Beginning with *Bowman's capsule* each tubule consists of a *proximal convoluted* portion, a U-shaped *loop* (of Henle) with descending and ascending limbs, a *connecting piece* which lies close to the renal corpuscle and a *distal convoluted* portion continuous over into the collecting tubule. These parts are derived from the S-shaped primordium in a manner more easily traced by the differential markings in Fig. 271 *E–G* than through a written description. It should, however, be noted that the primitive loop (of Stoerck) includes not only the definitive Henle's loop but a portion of the proximal convoluted tubule as well. Into the concavity of Bowman's capsule first grows the afferent arteriole, then the capillary loops of the *glomerulus* differentiate; from these loops the efferent arteriole finally buds outward.[8] The concavity of the capsule is at first shallow (*E*). Later, the walls of the capsule grow about and enclose the vascular knot except at the point where the arterioles enter and emerge (*F*). The first few generations of secretory tubules are temporary, or provisional, and ultimately degenerate (Fig. 268 *E*).[9]

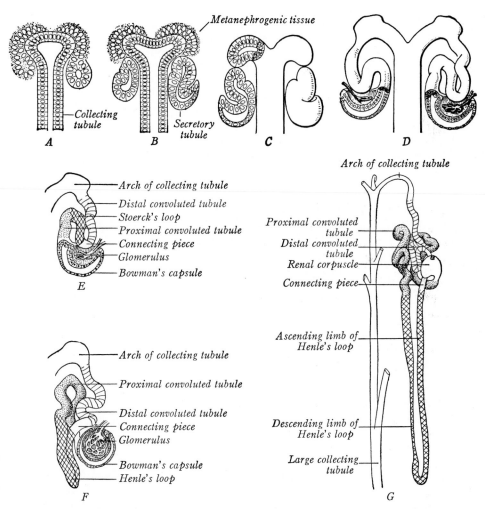

Fig. 271. The differentiation of human uriniferous tubules (after Huber). *A–D*, Stages so arranged that the left tubule in each drawing illustrates an earlier condition than the right tubule. *E–F*, Reconstructions, differentially marked to show the changing relations during the growth and specialization of a tubule.

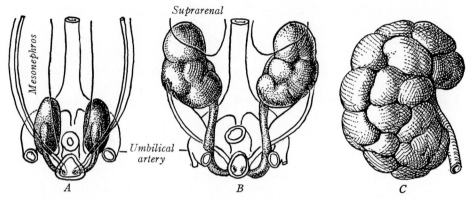

Fig. 272. Ascent and lobation of the human kidney. *A, B,* Shift in position between six and nine weeks, and early lobation (after Kelley; × 25). *C,* Lobation at birth (× 1).

There is considerable regional specialization of the original epithelial lining of the secretory tubules. This produces the characteristic modifications of the simple layer that are encountered at the various levels of a functional tubule.

Other Features. The kidneys are unusual among the organs of the body by coming to lie at a higher level than their site of origin (Fig. 272 *A, B*). In this shift the ureters do elongate, but it is perhaps a straightening of the body curvature at this time, together with a marked growth of the lumbosacral region, that chiefly brings about a displacement in which the kidney primordium seems actually to move craniad.[10] This displacement is through the distance of four somites, and at eight weeks the kidneys have met the large, caudally migrating suprarenal glands. Then the center of each kidney is already at its permanent level (opposite the future second lumbar vertebra). As the kidneys rise out of the pelvis they undergo a rotation of 90 degrees, whereby the original dorsal border becomes the convex, lateral border and the hilus faces medially rather than ventrad (Fig. 272 *A, B*). At all times the kidney is retroperitoneal in position.

As the cortex organizes, the region over a primary pyramid becomes demarcated from similar territories by a deep groove (Fig. 268 *E*). Later, as the pyramids subdivide, there is a corresponding increase in the number of cortical areas (*F*). These *lobes* may reach a maximum of twenty in the newborn (Fig. 272 *C*), but such external expression disappears progressively in infancy and early childhood as the grooves fill in. Within the lobes are smaller functional units, the *lobules*. Each contains all of the uriniferous tubules whose collecting ducts course in a particular cortical ray.

The human kidney is capable of secretion early in the third fetal month.[2, 11] Even though excretion is adequately performed by the placenta during fetal life and the physiological conditions are unfavorable to efficient renal activity, urine is produced slowly. Not only does the bladder fill in the early months, but also some urine voids into the amniotic sac; this, in turn, is drunk along with the amniotic fluid proper (p. 109).

Causal Relations. The pronephros seems to originate through an inductive influ-

SUMMARY CONCERNING HUMAN EXCRETORY ORGANS

ORGAN	SOURCE OF SECRETORY TUBULES	SOURCE OF COLLECTING TUBULES	ORIGIN OF EXCRETORY DUCT	SOMITE LEVEL OF ORIGIN	STAGE OF EARLIEST APPEARANCE	STAGE OF MAXIMUM DEVELOPMENT
Pronephros	Nephrotomes. (Segmental.)	Nephrotomes. (Segmental.)	None.	9-12	13 somites.	23 somites. (3 mm.)
Mesonephros	Nephrogenic cord.	Nephrogenic cord.	Delamination from cord; terminal growth.	9–26	18 somites. (2.5 mm.)	4–9 weeks.
Metanephros	Nephrogenic cord.	Branches from ureteric bud.	Bud from mesonephric duct.	26–28 (Ureter, 28)	5 mm. (4 weeks.)	After birth.

ORGAN	PERIOD OF DEGENERATION	TUBULES IN EACH KIDNEY	TUBULE CHARACTERISTICS	URINARY FUNCTION	PERMANENT FEATURES	FUNCTIONAL DERIVATIVES
Pronephros	25–40 somites. (3.5-5 mm.)	4 ±	Segmental; short. Nephrostome. Ext. glomerulus.	None.	None.	None.
Mesonephros	7–110 mm. CR. (5–16 weeks.)	38 ±	Larger; nonsegmen. Int. glomerulus. Renal-portal blood system.	Transitory, in embryo and early fetus.	Some tubules and duct retained.	Efferent ductules. Ductus deferens.
Metanephros	7–20 weeks. (Early orders of tubules.)	2,000,000 ±	Complex; long. Orderly arrangement.	Prenatal (meager). Postnatal.	Entire organ and duct.	Permanent excretory system.

ence emanating from the trunk organizer (chorda-mesoderm). To a considerable degree, at least, the differentiation of mesonephric tubules is induced by the nearby mesonephric duct. In the absence of a definite ureteric bud, which normally grows out of the mesonephric duct and into the nearby metanephrogenic tissue, the latter is unable to differentiate tubules—and hence no kidney is formed.

Anomalies. One or both kidneys may be lacking. A primary cause of such agenesis lies in the failure of the mesonephric ducts to develop or to give off ureteric buds. Less severe as an anomaly is a kidney that appears but remains stunted and poorly differentiated. Another kind of failure is *non-ascent,* or the retention of the primary pelvic position of the organ. This may involve one (Fig. 273 *A*) or both kidneys; when both remain they commonly fuse into a 'cake' (*B*). A joined, common kidney may lie on one side of the body and drain into one normal and one crossed-over ureter. Most commonly joined kidneys unite only by their lower ends, forming a *horse-shoe kidney,* and

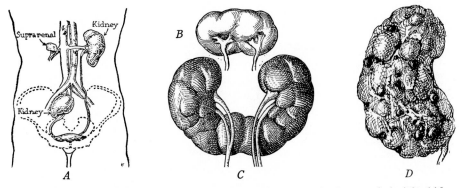

Fig. 273. Anomalies of the human kidney and ureter. *A,* Unascended right kidney. *B,* Fused, unascended kidneys forming a 'cake' (× ¼). *C,* Horse-shoe kidney, with fetal lobation retained (× ⅓); the right ureter is cleft, the left double. *D,* Congenital cystic kidney (× ¼).

these may or may not ascend properly (*C*). Such union presumably results from early fusion due to a converging course taken by the ureteric buds. An arrest by which the external lobation persists (*C*) duplicates the normal adult condition in reptiles, birds and some mammals (*e.g.,* whale; bear; ox).

Opposite in nature is developmental excess. Wholly separate supernumerary kidneys are rare. They arise when two ureteric buds grow out of the same mesonephric duct, and a separate metanephrogenic primordium then organizes about each one. More frequent is a double (merged) kidney whose two pelves may be drained by two ureters or by a Y-shaped ureter (*cf.* Fig. 273 *C*). Congenital *cystic kidney* is usually bilateral. It is characterized by the presence of blind secretory tubules that become dilated cysts through the retention of fluid (*D*). The cause has been attributed commonly to the primary non-union of secretory and collecting tubules, but no single explanation[12, 13] accounts satisfactorily for all conditions encountered.

THE CLOACA

The Primitive Cloaca. Vertebrates below the placental mammals retain a common entodermal chamber into which fecal, urinary and reproductive products all pass, and from which they are expelled to the exterior. Higher mammals have subdivided this *cloaca* into a dorsal rectum and a ventral region, consisting chiefly of a bladder and urogenital sinus. In such manner two separate outlets were gained for fecal and urogenital discharge. These changes were consequent on the evolution of an external penis in higher mammals. Cloacal subdivision also brought into existence that portion of the *perineum* separating the rectal orifice from the urogenital vent. The developmental course of the human cloaca, before complete division is attained, recapitulates several stages permanent in lower mammals.

In human embryos with six somites the future cloaca is merely a blind, caudal expansion of the hind-gut which already stands in contact ventrally with the ectoderm. This area of union between ectoderm and entoderm presently becomes the temporary *cloacal membrane* (Fig. 274 *A*), a region

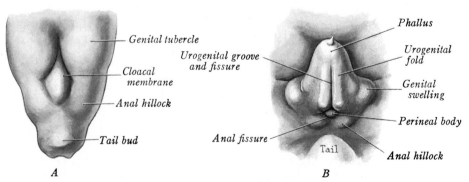

Fig. 274.   Region of the human cloacal membrane, in ventral view. *A,* At 3 mm. (after Keibel; × 60); *B,* at 21 mm. (after Otis; × 16).

originally caudal to the primitive streak that has been turned under by the tail fold (Fig. 73).[14] At first the cloacal membrane extends from the end bud to the body stalk (Fig. 268 *A*), but later this expanse is diminished relatively by the ingrowth of mesoderm to produce the infra-umbilical belly wall (*B*).[15] At its cranial end the early cloaca gives off the ventrally directed *allantoic stalk;* laterally the cloaca receives the *mesonephric ducts,* while it is prolonged caudad as the transitory *tail-gut* (Fig. 275 *A, B*).

Subdivision of the Cloaca. The facing walls of the hind-gut and allantois meet in a saddle-shaped notch, or fold, whose apex points caudad (Fig. 275 *A*, B**). The wedge of mesenchyme filling this interval is the so-called *cloacal septum;* it is also known as the *urorectal septum.* This mesenchymal mass pushes caudad as the fold advances, thereby dividing the cloaca into a dorsal *rectum* and a ventral *bladder* and *urogenital sinus.*[16] Division is completed during the seventh week (Fig. 276 *C*).

Even at the end of the sixth week (11 mm.), before the cloacal division is wholly finished, certain future regions can be recognized in the ventral

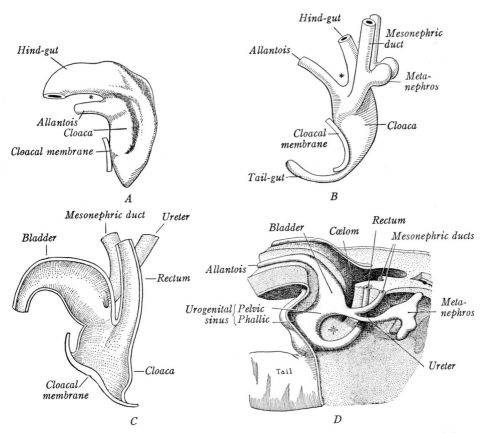

Fig. 275. Partial division of the human cloaca, illustrated by models viewed from the left side. *A, B,* At 3.5 mm. and 4 mm., respectively (after Pohlman; × 50); *C,* at 8 mm. (× 50); *D,* at 11 mm. (after Keibel; × 25). An asterisk indicates the position of the cloacal septum in *A, B*.

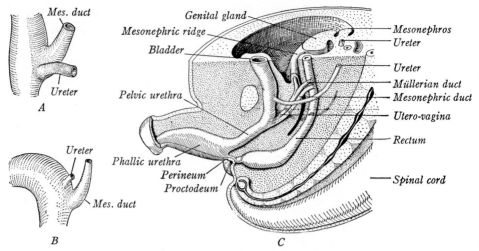

Fig. 276. Completed division of the human cloaca, and associated changes, shown by models viewed from the left side. *A, B,* At six weeks ($\times$ 50) and seven weeks ($\times$ 35), respectively, illustrating the absorption of the left ureter and mesonephric duct into the wall of the bladder. *C,* At nine weeks (after Keibel; $\times$ 15).

half (Fig. 275 *D*). The primitive *bladder* is continuous with the allantois and, at its caudal end, receives the two common stems of the paired mesonephric ducts and ureters. These stems also mark the approximate upper end of the primitive urogenital sinus, which shows two emerging regions. Nearest the bladder there is a *pelvic portion;* more distad is the slit-like *phallic portion,* which extends as a solid plate into the terminal part of the genital tubercle. Presently these two parts become clearly defined in both sexes (Fig. 276 *C*); their fates will be explained in a subsequent paragraph.

The Perineum.   When the cloacal septum has extended to the level of the cloacal membrane, rupture of this entodermal-ectodermal plate follows promptly (7 weeks). For this reason, separate *anal* and *urogenital membranes* have no significant existence as such. The rupture exposes the caudal edge of the septum, which is surfaced, necessarily, with the entoderm of the advancing fold (Fig. 291).[17] This projecting wedge, interposed between anus and phallus, is the *perineal body* (Fig. 274 *B*). The external fissure, resulting from the disappearance of the cloacal membrane, is closed again in its middle region by the merger of the perineal body with lateral folds flanking the fissure. The area so produced, covered finally by encroaching ectoderm and marked by a median raphé, is the primary *perineum* (Fig. 299 *D, F*). Hillocks, first located behind the anus (Fig. 274 *B*), encircle the anal orifice and create a definite proctodeum or *anal canal* (Fig. 299 *C, D*).[18] This canal is lined with ectoderm as far as the level of the now ruptured anal portion of the cloacal membrane.

The Bladder.   At the time of its emergence as a separate entity, the *bladder* still receives on each side the common stem of a mesonephric duct and ureter (Fig. 275 *D*). Growth processes quickly lead to the absorption of these stems, so that the four ducts acquire individual openings (Fig. 276 *A*). A somewhat complicated shifting then displaces the mesonephric ducts farther caudad (*B, C*).[16, 19] The two ureters come to lie well apart from each other, but the mesonephric ducts open close together at an elevation in the future urethra. This important landmark is known as *Müller's tubercle* (Fig. 277 *A*). The triangular area on the dorsal wall of the bladder, and its continuation along the dorsal wall of the urethra to Müller's tubercle, as marked off by these four ducts, is the *trigone* of adult anatomy. Temporarily, at least, it is an island of mesodermal epithelium amid the general entodermal lining of the bladder, because of the process of absorption already described. However, this original mesodermal island is said to be replaced by encroaching sinus epithelium, so that the trigone is finally entodermal also.[16, 19]

After the second month the bladder proper expands into an epithelial sac whose apex tapers into an elongate tube; this latter portion is named the *urachus* (Fig. 296 *B*). The early urachus, in turn, is continuous at the umbilicus with the proximal remnant of the allantoic stalk, but it is believed that the latter contributes nothing to either urachus or bladder.[16, 20] The bladder and urachus elongate proportionately as the infraumbilical body wall is progressively brought into existence. The general organization of the bladder, both as regards its stratified epithelial lining and muscular wall, is attained during the third month. The urachus persists throughout life as a cord which may retain some of its epithelial canal and muscle.[21] After birth it is known as the *middle umbilical ligament*.

The Urethra and Urogenital Sinus. The more caudal region of the

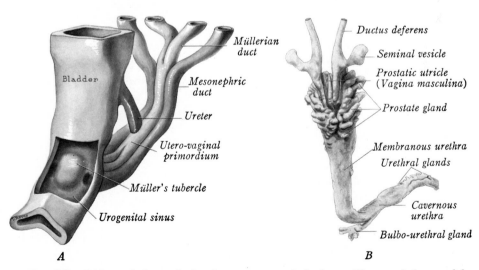

A   *B*

Fig. 277. Differentiation of the human urogenital sinus, illustrated by models. *A,* Female fetus of nine weeks, in left front view (after Keibel; × 120). *B,* Male fetus of four months, in right rear view (after Broman; × 13).

cloaca that separates away from the rectum becomes the primitive *urogenital sinus;* its pelvic and phallic segments have already been mentioned (Fig. 276 *C*).

THE FEMALE.   The originally short neck connecting the bladder and the urogenital sinus elongates into the permanent *urethra* (Fig. 291).[22] The pelvic and phallic portions of the sinus merge to create the shallow slit-like *vestibule* into which the urinary and genital tracts open separately. Hence the vestibule serves as a permanent urogenital sinus. The female urethra does not extend into the clitoris, which otherwise is homologous to the penis of the male (Fig. 13).

THE MALE.   The male urethra is longer and has a more complicated origin than that of the female. The counterpart of the entire female urethra is a short tube extending from the bladder to Müller's tubercle, which presently becomes the permanent seminal colliculus (Fig. 277 *A*). This canal comprises most of the *prostatic urethra.* Next beyond the level of Müller's tubercle is the pelvic portion of the urogenital sinus; this becomes the rest of the *prostatic* and all of the *membranous urethra* (*B*). Finally the phallic portion of the sinus adds the *cavernous urethra,* which extends through the penis (*cf.* p. 337). Since the mesonephric ducts are utilized by the male as the chief genital ducts, all of the permanent urethra distal to their outlets on Müller's tubercle serves as a true, canal-like urogenital sinus. This includes both the membranous and the cavernous urethra (*B*).

Accessory Genital Glands.   Several glands, associated with the genital system, develop from the tissue of the urogenital sinus and can logically be included in the present chapter.

THE PROSTATE GLAND.   This organ develops as multiple outgrowths of the entodermal urethral epithelium, both above and below the entrance of the male ducts (Fig. 277 *B*). The tubules arise at eleven weeks in five distinct groups and total an average number of 63. Each group corresponds to a future lobe of the final gland. The surrounding mesenchyme differentiates both connective-tissue and smooth-muscle fibers, into which the prostatic buds grow. The prostate of the new-born shows some evidence of temporary activation by maternal hormones, but it remains relatively undeveloped until puberty.[23]

In the female the presumptive homologues are certain *urethral glands,* a group of which on each side is drained by *para-urethral ducts* (of Skene).[24]

THE BULBO-URETHRAL GLANDS.   These glands (of Cowper) arise in male embryos of nine weeks as a pair of solid buds from the entodermal epithelium of the urogenital sinus at the beginning of the future cavernous urethra (Fig. 277 *B*). The outgrowths extend backward, almost paralleling the sinus, and penetrate through the investing mesenchyme of the primitive corpus spongiosum. At four months the epithelium becomes glandular.

The *major vestibular glands* (of Bartholin) are the female homologues. They appear at the same age as the male glands, grow through puberty and involute after the menopause. These glands open into the vestibule, near the hymen.

THE URETHRAL GLANDS. The cavernous urethra begins to bud off numerous small glands (of Littré) at three months (Fig. 277 *B*). Homologous structures, the *minor vestibular glands,* open into the vestibule of the female.

THE SEMINAL VESICLES. Although not of sinus origin, these saccular glands belong functionally in the present group. They are exclusively male organs which outpouch from the lower ends of the mesonephric (now deferent) ducts in fetuses of thirteen weeks and gain a muscular wall from the adjacent mesenchyme (Fig. 277 *B*). By the seventh month the seminal vesicles have attained their adult form; until puberty they grow but slowly.

Anomalies. *Imperforate anus* results from a retention of the anal portion of the cloacal membrane; the blind anal canal (proctodeum) is shallow (Fig. 210 *A*). A conspicuous malformation is a *persistent cloaca,* as occurs normally in most vertebrates. The usual outcome is a *recto-vestibular fistula* in the female (Fig. 278 *A*), or a *recto-vesical* or *recto-urethral fistula* in the male (*B*). All such conditions are due to the failure of the rectum and urogenital sinus to separate completely.

Only rarely is the bladder absent, duplicated or subdivided into two chambers. It sometimes opens and everts broadly onto the ventral body wall; this condition is known as *exstrophy of the bladder* (Fig. 278 *C*). Such severe eversion is accompanied by spread pubic bones and either a grooved or bifid penis or clitoris. Failure of mesoderm to invade the infra-umbilical region and reduce the expansive cranial extent of the early cloacal membrane would predispose to this condition (*cf.* Fig. 268 *A*); rupture of this membrane would then expose the bladder. Because of the primary relation of the mesonephric ducts to the ureters, and their normal absorption into the differentiating cloaca, variations in the ureteric openings occur. They may terminate in the seminal vesicles, urethra, rectum, uterus or vagina. Rarely the urachus remains patent even to the umbilicus, and establishes a *urachal fistula* there through which urine escapes (Fig. 278 *D*).

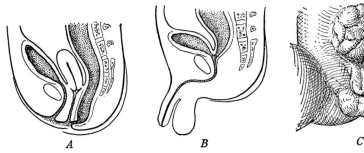

A   *B*   *C*

Fig. 278. Anomalies resulting from faulty differentiation of the human cloaca. *A,* Persistent cloaca (recto-vestibular fistula) in the female, shown in sagittal section. *B,* Similar condition (recto-urethral fistula) in the male. *C,* Exstrophy of the bladder in a newborn, combined with epispadias of the penis and undescended testes. *D,* Patent urachus in a female, shown in sagittal section; the urachus opens on the abdomen as a urachal fistula (Cullen).

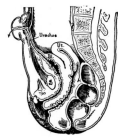

D

Less complete remnants of the urachus are blind sinuses leading from the bladder, and isolated epithelial remnants. Anomalies of the urethra and accessory genital glands are not common.

REFERENCES CITED

1. Frazer, E. A. 1935. Biol. Rev., *25,* 159–187.
2. Torrey, T. W. 1954. Carnegie Contr. Embr., *35,* 170–197.
3. Gersch, I. 1937. Carnegie Contr. Embr., *26,* 33–58.
4. Kozlik, F. & B. Erben. 1935. Z'ts. f. mikr.-anat. Forsch., *38,* 483–502.
5. Bremer, J. L. 1916. Am. J. Anat., *19,* 179–210.
6. Altschuler, M. D. 1930. Anat. Rec., *46,* 81–91.
7. Potter, E. L. & S. Thierstein, 1943. J. Pediat., *22,* 695–706.
8. Edwards, J. G. 1951. Anat. Rec., *169,* 495–501.
9. Kampmeier, O. F. 1926. Anat. Rec., *33,* 115–120.
10. Gruenwald, P. 1943. Anat. Rec., *85,* 163–176.
11. Chambers, R. & G. Cameron. 1938. Am. J. Physiol., *123,* 482–485.
12. Kampmeier, O. F. 1928. Surg., Gyn. & Obst., *36,* 208–216.
13. Potter, E. L. 1961. Pathology of the Fetus and Newborn (Year Book Publishers).
14. Florian, J. 1933. J. Anat., *67,* 263–276.
15. Wyburn, G. 1937. J. Anat., *71,* 201–231.
16. Chwalla, R. 1927. Z'ts. Anat. u. Entw., *83,* 615–733.
17. Politzer, G. 1931; 1932. Z'ts. Anat. u. Entw., *95,* 743–768; *97,* 622–660.
18. Tench, E. M. 1936. Am. J. Anat., *59,* 333–343.
19. Gyllensten, L. 1949. Acta Anat., *7,* 305–344.
20. Felix, W. 1912. Chapter 19 in Keibel & Mall: Human Embryology (Lippincott).
21. Hammond, G., *et al.* 1941. Anat. Rec., *80,* 271–287.
22. Politzer, G. 1952. Z'ts. f. mikr.-anat. Forsch., *58,* 6–28.
23. Swyer, G. I. M. 1944. J. Anat., *78,* 130–145.
24. Huffman, J. W. 1948. Am. J. Obst. & Gyn., *55,* 86–101.

Chapter XVIII. The Genital System

The intimate relations of the urinary and genital organs as components of the developing urogenital system have already been mentioned (p. 295). As with other hollow viscera, the epithelial constituents of all these organs are primarily important and it is their development that naturally receives major attention. The accessory, investing coats of muscle and connective tissue organize in both systems during the third month from condensed, neighboring mesenchyme.

THE INDIFFERENT STAGE

During the fifth and sixth weeks (5–12 mm.) the genital system makes its appearance. This has been named the 'indifferent period' because the sex of the embryo cannot be determined at that time, either by gross inspection or by sections of the gonads. In addition to such a pair of generalized sex glands, all vertebrate embryos are equipped at an early stage with a double set of sex ducts (male and female). Both are held in readiness for the time when sexuality is declared, but only the one appropriate duct system will then advance significantly beyond its primitive state; the complementary set suffers regression.

To be sure, the sex-determining mechanism is present from the moment of fertilization (p. 59), and the diagnosis of future sex on the basis of the presence or absence of sex chromatin (p. 60) in resting nuclei of somatic cells can be made as early as three weeks.[1] Yet it is not until the eighth week, at the earliest, that sex recognition becomes fairly practicable by inspection of the external genitalia.

The Gonads. As long as the prospective testis and ovary are structurally indistinguishable they are given the noncommittal name, *gonad*. The primitive sex gland makes its appearance within a localized region of the thickening that has already been described as the *urogenital ridge* (p. 295); this folded ridge is appropriately named since it contains both the nephric and genital primordia (Fig. 279).

On the ventromedial surface of the urogenital ridge the peritoneal epithelium begins to thicken (6 mm. embryos) and rapidly becomes several layers thick. Proliferation soon causes this region to bulge into the cœlom as the *genital ridge* (Fig. 279). This thickened strip extends longitudinally and thus parallels the mesonephric ridge, but lies medial to it. At six weeks

315

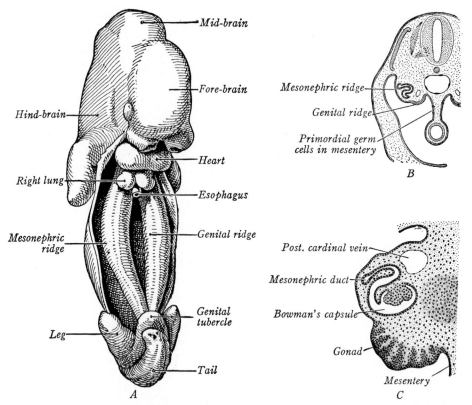

Fig. 279. Urogenital ridge of the human embryo. *A,* Dissection, at 9 mm., in ventral view (Kollman; × 11). *B, C,* Transverse sections, at 7 mm. (× 35) and 10 mm. (× 75).

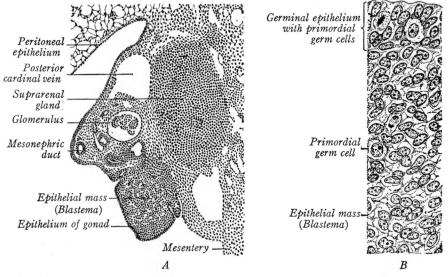

Fig. 280. Indifferent stage of the human gonad, illustrated by transverse sections. *A,* At 12 mm., including associated regions (after Prentiss; × 82). *B,* At 12 mm., showing structural details in a sample area (× 550).

the resulting, 'sexless' gonad consists of a superficial *germinal epithelium* and an internal *blastema,* somewhat loosely arranged (Fig. 280). The blastemal mass is derived, partly at least, by direct, proliferative ingrowth from the epithelium which, at about this time, loses its basement membrane.[2] Longitudinal furrows separate the indifferent sex gland from the mesonephros laterally, and from the mesentery of the gut, medially. During the seventh week the gonad begins to assume characteristics that identify it as a testis, but distinctive ovarian features do not appear until several weeks later.

Even in presomite embryos certain large, distinctive cells can be recognized caudal to the embryonic disc in the yolk-sac entoderm. Soon afterward they lie in the cloacal entoderm, and in 4 mm. embryos they are proliferating and migrating craniad, by way of the entodermal gut and dorsal mesentery, into the region of the future genital ridge (Fig. 279 B).[3, 4] Such cells are called *primordial germ cells* (Fig. 280 B); in a 4 mm. embryo nearly 1400 were counted. The best evidence favors the view that all definitive sex cells of the genital glands are descended from them.[4, 5] A minority opinion holds that all functional sex cells originate locally from the germinal epithelium, as needed in later life.[6]

The Primitive Genital Ducts. The male does not elaborate any ducts intended primarily for the service of the testis. Instead, with the degeneration of the mesonephros, it merely appropriates the abandoned mesonephric ducts and some of the mesonephric tubules and converts them into genital canals. The origin and early history of these parts have been adequately described in previous paragraphs (pp. 297–301). The duct is complete at four weeks and new tubules cease differentiating at five weeks. Use will also be made of the urethra as a terminal sexual passage.

Both sexes develop somewhat more tardily a pair of female ducts (of

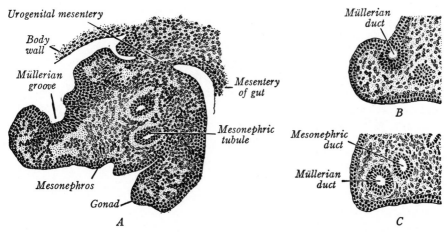

Fig. 281. Origin of the human Müllerian duct, illustrated by transverse sections of the urogenital ridge at 12 mm. × 165. *A,* Through open groove. *B,* Slightly lower level, showing closure. *C,* Still lower level, showing free tube.

Müller), also named *paramesonephric ducts*. Embryos of nearly six weeks (10 mm.) first indicate the future *Müllerian ducts* by a groove in the thickened epithelium of each urogenital ridge (Fig. 281 *A*). This furrow is located laterally on the mesonephros, near its cranial pole. The extreme cranial end of the groove remains open, while more caudally the lips of the groove close and create a funnel-like tube (*B, C*). Starting thus as an epithelial inrolling, the Müllerian duct continues to advance in a caudal direction, seemingly by the progressive growth of its solid, blind end. On the other hand, the female duct courses just lateral to the mesonephric (male) duct, with which its tip is intimately related (*cf.* Fig. 638).[7] Hence it is possible that the mesonephric duct contributes directly to the growth of the Müllerian duct.

Near the cloaca the two urogenital ridges swing toward the midplane and fuse into the so-called *genital cord* (Fig. 282 *A*). In this maneuver the

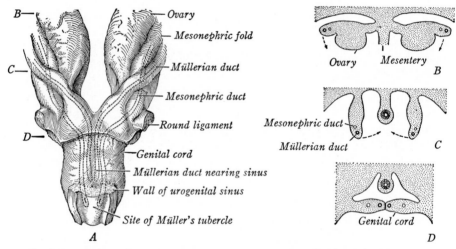

Fig. 282. Course of the human urogenital ducts, and formation of the genital cord. *A*, Model, at two months (× 25). *B–D*, Transverse sections, at the three levels indicated alongside *A*.

progressively elongating Müllerian ducts, originally lateral in position, necessarily are brought side-by-side in the midplane, whereas the mesonephric ducts assume a more lateral position (*B-D*). In embryos of nine weeks the portions of the Müllerian ducts coursing within the genital cord have fused and end blindly at *Müller's tubercle* (Fig. 277 *A*). This tubercle is a median protuberance projecting from the dorsal wall of the urogenital sinus. It was originally brought into existence by the earlier arrival of the mesonephric ducts. The fused, common Müllerian tube is the first indication of a *uterus* and *vagina*, whereas the more cranial portions of the ducts remain separate and will serve as the *uterine tubes*.

The double set of ducts, male and female, is similar in all embryos throughout the second month. After the sex of the embryo is well established, the provisional ducts of the opposite sex regress and largely disappear (Fig. 288).

The External Genitalia. Embryos at the end of the fifth week (8 mm.) show a conical *genital tubercle* in the midline of the ventral body, between the umbilical cord and tail (Fig. 279 *A*). Its caudal slope bears a shallow *urogenital groove,* whose roof is the thin *urogenital membrane* and whose side walls are slightly elevated *urogenital folds.* During the seventh week the genital tubercle elongates into a somewhat cylindrical *phallus,* with its tip foreshadowing the knob-like *glans* (Fig. 274 *B*). Lateral to the base of the phallus, a rounded ridge also makes its appearance on each side; they are the *genital swellings,* presently to be designated labial or scrotal swellings. Rupture of the urogenital membrane in a region near the base of the phallus provides an external opening for the urogenital sinus at seven weeks. From this generalized set of primordia, the external genital organs of the male or female will be modeled in an appropriate and distinctive manner during the ensuing weeks.

Causal Relations. The formation of a genital ridge is conditioned by local influences, such as entoderm and mesonephros, and its development is wholly independent of the presence of germ cells. The determination of a genital ridge is established early, before it is visible as such. Hence a nephrotome region (prior to its development into a germinal ridge and before the arrival of germ cells), when transplanted into the body wall of a larval amphibian, will undergo development, but it becomes a sterile gonad. The appearance of Müllerian funnels in these animals seems to depend on the presence of the pronephroi. The growth caudad of the Müllerian ducts of amphibians and birds can be proved to be dependent on the presence of mesonephric ducts alongside.

INTERNAL SEXUAL TRANSFORMATIONS

THE GONADS

Differentiation of the Testis. As the future male genital glands increase in size, they shorten relatively into more compact organs located farther caudad (*cf.* Fig. 293 *A*). At the same time the originally broad attachment to the mesonephros is converted into a gonadial mesentery known as the *mesorchium* (Fig. 284 *A*). In embryos about 15 mm. long, destined to be males, the gonads begin to show two characteristics that mark them as testes (Fig. 283): (1) the appearance of prominent, branched and anastomosing strands of cells, the *testis cords;* and (2) the occurrence, between the covering (germinal) epithelium and the centrally located testis cords, of a layer of tissue that foreshadows the *tunica albuginea,* or fibrous capsule of the gland.

The testis cords of human embryos, containing scattered primordial germ cells, organize suddenly within the blastemal mass (Figs. 280, 283). The radially arranged testis cords converge toward the mesorchium where a dense portion of the blastemal mass is also emerging as the primordium of the *rete testis.* Soon the cell clusters of the rete primordium become a

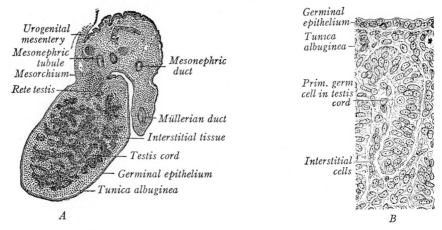

Fig. 283. Early differentiation of the human testis, illustrated by transverse sections at nearly eight weeks. *A*, General organization and relations of the urogenital ridge (after Prentiss; × 70). *B*, Structural details of the testis (× 300).

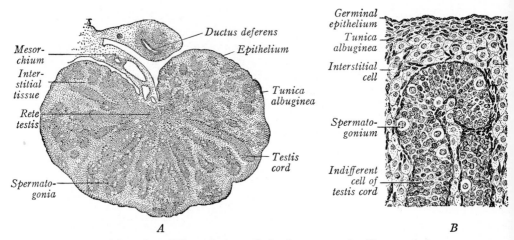

Fig. 284. Advancing differentiation of the human testis, illustrated by transverse sections at fourteen weeks. *A*, General plan and relations (Prentiss; × 44). *B*, Structural details in a sample area (× 150).

network of strands which are continuous with the testis cords (Fig. 284 *A*). Each of the latter splits into three to four daughter cords—the forerunners of the *seminiferous tubules*. Their peripheral ends join in looping arches (*B*), while the main portions of the cords soon elongate into twisted *tubuli contorti*. Nearer the rete testis, however, they remain straight, as the *tubuli recti* (Fig. 285 *A*). The rete testis unites the tubuli recti with the mesonephric components of the duct system in a manner to be described presently (p. 324). Actually the testis cords do not canalize completely until the time of puberty (*B, C*).[8] Their central cavities then unite with the cavities of the rete cords which were completed before birth. Thus the originally solid cords of both kinds end their development as a continuous system of tubules, lined with epithelium.

The early testis cords are composed chiefly of so-called indifferent cells, which transform ultimately into *sustentacular cells* (of Sertoli). The primordial germ cells, enclosed within the cords, remain few in number until the onset of puberty when they begin to proliferate and produce many *spermatogonia* (Fig. 284). The full course of development of spermatogonia into spermatozoa has been described in an earlier chapter (pp. 41–43).

The general bed of mesenchymal tissue, in which the tubules of the testis lie, organizes into the connective-tissue framework of the organ. Thus the 250 *lobules* of the testis, each containing three or four seminiferous tubules derived from a primitive testis cord, become isolated by partitions (Fig. 285 *A*). In one direction these *septula* converge to the *mediastinum testis* (where the rete tubules lie); in the opposite (peripheral) direction they extend to the *tunica albuginea* which, after the second month, becomes a definite encapsulating layer (Figs. 283, 284). Coincidentally with the differentiation of a tunica, the germinal epithelium reverts to the ordinary type of peritoneal mesothelium; persisting, it will accompany the testis on its scrotal journey. Certain cells of the mesenchymal bed specialize into large, pale elements which lie in the unspecialized connective tissue between the seminiferous tubules and hence are designated *interstitial cells*. They are abundant in fetuses of the fifth and sixth months (Fig. 284 *B*), whereupon they decline suddenly and remain inconspicuous in infancy and childhood. In the puberal years they again increase in number. Interstitial cells are held to be transformed fibroblasts, responsible for the endocrine secretion (testosterone) of the testis, both in fetal and postnatal life.[9]

Differentiation of the Ovary. This gland does not exhibit any distinctive ovarian features until several weeks after the gonad of the male has

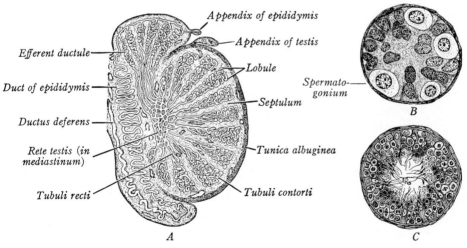

Fig. 285. Late differentiation of the human testis. *A*, Hemisection, showing the plan of organization of the testis and its ducts in a newborn ($\times$ 4). *B*, Section of a seminiferous tubule of a newborn ($\times$ 400). *C*, Section of a seminiferous tubule of an adolescent ($\times$ 110).

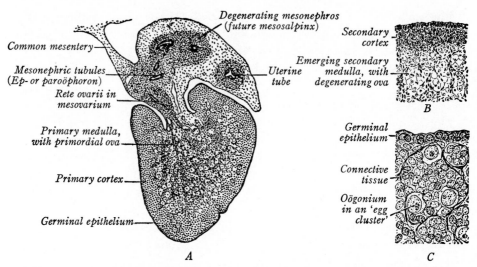

Fig. 286. Early differentiation of the human ovary, illustrated by transverse sections. *A*, General organization and relations, at three months (after Prentiss; × 44). *B*, Structural details, at fourteen weeks (× 90). *C*, Secondary cortex, at four months (× 370).

declared itself as a testis. However, gonads that do not differentiate epithelial cords internally during the seventh week can be diagnosed negatively as ovaries. In the eighth week the blastemal mass of the indifferent period begins to show clusters composed of small, indifferent cells and one or more primordial germ cells. Soon there may be distinguished a denser *primary cortex* beneath the germinal epithelium and a looser *primary medulla* internally. In addition, a compact cellular mass bulges from the medulla into the mesovarium and establishes there the primitive *rete ovarii*, or homologue of the rete testis (Fig. 286 *A*). Neither epithelial cords nor tunica albuginea are developed at this stage, as in the testis. In the third month the ovary, like the testis, gains a mesentery (*mesovarium*) and settles to a more caudal position (Figs. 286 *A*, 294).

In fetuses about four months old two important changes are taking place (Fig. 286): (1) The ovary enlarges rapidly, due to the deposition of a definitive *cortex* upon the original blastemal mass (*B, C*). This secondary cortex arises by the division of primordial germ cells and other cells of the superficial blastema, already present, and also through a renewal of proliferation by the germinal epithelium.[2] (2) Ingrowth of connective tissue (accompanied by blood vessels) from the region of the rete ovarii produces supporting structures similar to the mediastinum and septula of the testis. The septula now isolate the cortical substance into cord-like masses, often called *Pflüger's tubules* (Fig. 286 *C*). At the periphery of the ovary the septula spread out during the seventh month, forming a connective-tissue layer known as the *tunica albuginea* (Fig. 287 *A*).

Coincidental with the addition of new cells (secondary cortex) at the periphery of the ovary goes the decline of the earlier germ cells that were growing in the primary medulla and cortex (Fig. 286 *B*). Such clusters of germ cells, separated by invading connective tissue, regress and are replaced by a vascular, fibrous stroma; in this way the permanent *medulla* comes into being (Fig. 287 *B*). In the secondary cortex the epithelial cords fragment, and egg clusters and single egg cells are similary isolated by connective tissue, but they do not succumb (Fig. 286 *C*). Instead, neighboring cells (of germinal epithelium origin) surround the young cortical oögonia in the later fetal months, and even after birth, and thereby produce the *primary follicles* (Fig. 287 *B*). Although some of these advance further during fetal life and after birth, the development of *vesicular* (Graafian) *follicles* is particularly characteristic of the active sexual years (Fig. 17). The later history of growing oögonia and follicles has been described in Chapters III and IV.

Bipotentiality of the Gonads. The gonads of vertebrates show a definite tendency toward bisexual organization: A testicular homologue in the human ovary is represented by the primary cortex and medulla. Similarly, the functional cortex of the ovary is represented in normal testicular development by a rudimentary cortical zone.[2, 10] The differentiation of a primordial germ cell into sperm or egg is not self-contained, but depends on whether it comes to lie in a competent cortical or medullary environment.[11] There is an antagonism between the inductive power of the cortex and medulla, and normally the superiority of one or the other is hereditarily fixed. Yet the latent potentiality of the presumptive subordinate part is shown in instances of sex reversal in a gonad or by production of a combined sex gland (ovotestis).

Causal Relations. Experiments on the newt and chick indicate that only sterile

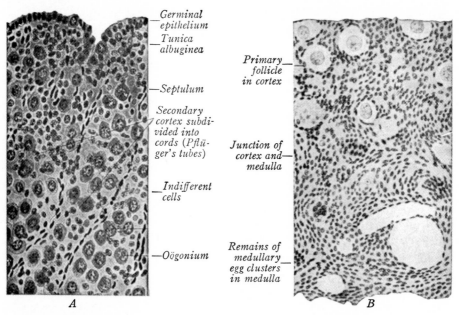

Fig. 287. Later differentiation of the human ovary, shown in vertical sections. × 185. *A,* Cortex, at six months (after DeLee). *B,* Junction of cortex and medulla, at eight months (after Felix).

gonads develop when the normal migration of primordial germ cells into the gonad is prevented. This argues against the potency of the intact 'germinal' epithelium as a source of sex cells. The favored interpretation is that the primordial cells are, indeed, the only source of definitive sex cells, and not merely inductors of a secondary crop of different origin. As stated in the previous paragraph, the early germ cells cannot self-differentiate completely into the male or female type. Rather, they depend on the kind of tissue in which they come to lie (*i.e.,* cortex or medulla) for the specific influences that bring about their specialization.

Anomalies. Congenital absence or duplication of the testes or ovaries is very rare. Fused testes and lobed ovaries have been recorded. A combined *ovotestis* is a common accompaniment of the rare, true hermaphroditism (p. 338). The tumor-like growths, known as *dermoid cysts* and *teratomas,* which among other sites occur in the ovary and testis, have been discussed in a previous chapter (p. 180).

TRANSFORMATION OF THE MESONEPHROI

The mesonephric system of male amphibians performs a double function. Some of the more cranial tubules unite with the testis, while the caudal ones continue to excrete urine. Hence the mesonephric duct conveys both urine and spermatozoa to the cloaca. In higher vertebrates the same potential arrangement is laid down, but the replacement of the mesonephros by the permanent kidney makes available individual ducts for the sexual and urinary products of the male. In the female the two kinds of ducts are separate from the start.

The growth of the gonad soon surpasses that of the mesonephros, which thereafter appears as an adjunct alongside (Fig. 286 *A*). Nevertheless, both in male and female embryos of nine weeks there still remain some 17 mesonephric tubules; of these, half are intact and the rest more or less fragmented. All tubules that escape complete degeneration can be divided into a cranial and a caudal group on the basis of their subsequent history (Fig. 288). The cranial group soon consists of about five tubules; these project against the adjacent primordium of the rete testis or rete ovarii, as the case may be. Union of the rete cords and mesonephric tubules begins in fetuses of three or more months. The point of union may be at any level where a break occurs, or even at Bowman's capsule.[12] The caudal group of tubules does not make such unions, yet it tends to remain as a vestige.

The Male. Portions of the paired, mesonephric organs are salvaged by the male and used as functional sexual ducts. The fates of these parts can be followed in Fig. 288 *A*. The lumina of the rete tubules and the cranial group of mesonephric tubules become continuous by the end of the sixth month, whereupon the mesonephric tubules are given a new name—the *efferent ductules* of the epididymis. Each coiled ductule makes a conical mass known as a *lobule of the epididymis.*

The efferent ductules are destined to convey spermatozoa from the rete testis into the mesonephric duct. The latter, accordingly, undergoes certain regional specializations which transform it into the chief genital duct (Fig. 285 *A*). In completing these changes the cranial end of the mesonephric

duct becomes highly convoluted and is named the *duct of the epididymis.*
The caudal portion remains straight and, as the *ductus deferens* and terminal *ejaculatory duct*, extends from epididymis to urethra; it opens into
the urethra at Müller's tubercle (Fig. 282 *A*). The lower end of the ductus
deferens will dilate as the *ampulla;* in this region the saccular *seminal vesicle* evaginates in fetuses of 13 weeks (Figs. 14, 277 *B*).

VESTIGES.   One of the cranial group of mesonephric tubules ends blindly; it is the
cranial aberrant ductule (Fig. 288 *A*). The entire caudal group of tubules is functionless,
yet it persists as the *paradidymis* and the blindly ending *caudal aberrant ductule*. The
cranial tip of the mesonephric duct is generally believed to become the cystic *appendage
of the epididymis* (Fig. 285 *A*). (Some have claimed that this appendage is formed by
certain cranial mesonephric tubules and others attribute its origin to an accessory
Müllerian funnel.[13])

The Female.   The entire mesonephric system undergoes loss or atrophy in the
female. Fates of the several parts are illustrated in Fig. 288 *B*. The *rete ovarii* is vestigial, though retained in the adult.[14] Some time before birth it canalizes and often
unites with the persisting cranial group of mesonephric collecting tubules, thus duplicating the functional connections in the male. Nevertheless, the cranial group of
mesonephric tubules always remains a functionless vestige, located within the broad
ligament. Most of its components are tiny, blind canals attached to a short, persistent
segment of the mesonephric duct; the whole complex is the *epoöphoron*. A few of the
most cranial tubules of the cranial group may become cystic *aberrant ductules*. The
caudal group of mesonephric tubules constitutes the smaller *paroöphoron;* it usually
disappears before adult life is attained.

The greater part of each mesonephric duct atrophies and disappears in the female,
the process beginning early in the third month. A cranial portion persists as the *duct of
the epoöphoron* and its tip becomes the cystic *vesicular appendage* (Fig. 288 *B*). More
caudal portions, known as *Gartner's duct*, may occur as vestigial structures at any level
between the epoöphoron and hymen (Fig. 302 *B*). Representatives are to be found in
about one-fourth of all adult females; usually they are located in the broad ligament or
in the wall of the uterus or vagina, and sometimes they give rise to cysts.

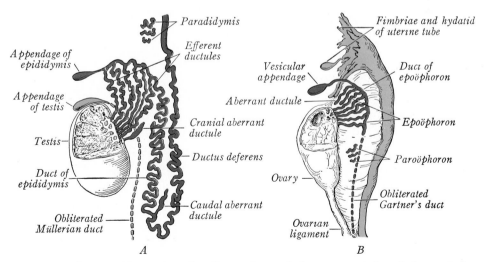

Fig. 288.   Diagrams illustrating the diverse fates of the mesonephric tubules and the
mesonephric and Müllerian ducts in the two sexes. *A,* Male; *B,* female.

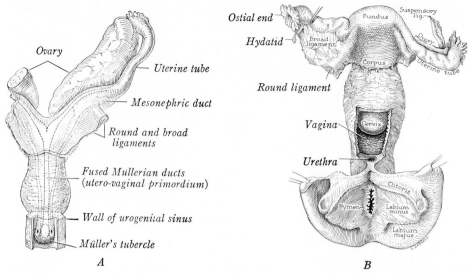

Fig. 289. Genital tract of the human female, in ventral view. *A,* At ten weeks ($\times$ 15); *B,* at birth ($\times$ 1).

TRANSFORMATION OF THE MÜLLERIAN DUCTS

All female vertebrates below marsupials acquire female ducts that remain separate and open independently into the permanent cloaca. In placental mammals, on the other hand, there is fusion to varying degrees at the caudal ends (*cf.* Fig. 302). In primates, complete caudal union of these ducts produces a common uterus as well as a common vagina, but only in the highest primates does the union reach its fullest expression.

A previous page has described how the female ducts develop in the urogenital ridges, enter the genital cord, fuse there, and end at Müller's tubercle (Fig. 277 *A*). When the urogenital ridges are crowded laterad by the enlarging suprarenal glands and permanent kidneys, the Müllerian ducts naturally participate in this displacement (Fig. 289 *A*). As a result, each duct in its course makes two bends which roughly establish three regions, different in future potentialities: (1) a cranial, longitudinal portion (uterine tube); (2) a middle, more or less transverse portion (uterine fundus and corpus); and (3) a caudal, longitudinal portion which fuses with its fellow to produce a common tube (uterine cervix and, at least, a provisional vagina).

The Female. The cranial segment of each Müllerian duct retains its separate existence and is presently called a *uterine tube*. Its open, upper end early gains a fringe, or fimbriæ (Fig. 289). Originally the future fundus-corpus portion of the *uterus* is represented by the transverse limbs of the Müllerian ducts. After a time the cranial walls of both these tubes bulge in a cranial direction, so that their original angular junction becomes a convex dome (Figs. 289, 290). In this manner a considerable extent is added to the uterus; it comprises the definitive *fundus* and *corpus* of man and apes. The uterine *cervix* is usually said to arise from the more cranial portion of the primary fusion between the Müllerian ducts. On the other hand, some believe that its original epithelium is replaced, as in the vagina, by invading entoderm from the urogenital sinus.[15, 16] The more caudal portion of the fused ducts represents the primitive *vagina*. Its permanent lining was formerly believed to be the original epithelium, but it is now claimed that entodermal epithelium of the urogenital sinus invades this level of the genital cord and replaces the Müllerian epithelium.[15, 17] The *hymen* arises at the site of Müller's tubercle (Fig. 291 *A*) as a ring-shaped fold between the future vagina and the urogenital sinus.[18] When the vagina acquires a lumen, the hymen serves as a perforate membrane guarding the entrance to the vagina (Fig. 289 *B*).

The uterine tube and uterus are lined with a simple epithelium. Only the uterus develops *glands*; these invaginate by the seventh month, yet remain small until puberty. A distinction between uterus and vagina is not evident until the middle of the fourth month when the *fornices* appear (Figs. 290 *C*, 291 *B*). For a time the vaginal epithelium is a solid column; when the lumen reappears as a central cleft in fetuses of about five months, the epithelium continues to be stratified. The muscular wall of the entire genital tract is foreshadowed at three months by mesenchyme condensing about the epithelial lining. This investment is especially thick in the genital cord where the uterus develops (Fig. 290). The uterus grows rapidly in the last fetal months, and the cervix becomes the longest segment by far. Shortly after birth the uterus loses one half its length, most of which is at the expense of the cervix,[19] and does not recoup this loss until just before puberty. This strong prenatal and pubertal growth is directed by the female hormone, estrogen, supplied first by the mother and later by the maturing daughter. The vagina, including its epithelial lining, shares in this interrupted response.[20]

The young uterine tubes fail to match the elongation of the trunk as a whole, and their trumpet-like upper ends finally lie opposite the fourth lumbar vertebra, thirteen segments below their level of origin (Fig. 289). The vagina is originally some distance above the outlet of the urogenital sinus, similar to the permanent condition in many mammals (Fig. 291 *A*). The intervening stretch of sinus thereafter undergoes a great relative shortening to become the shallow vaginal *vestibule* into which both urethra and vagina open independently (*B*). From the standpoint of specialization of the primitive cloaca, this arrangement is an advance over the condition found in the male since a common urogenital sinus has been practically eliminated.

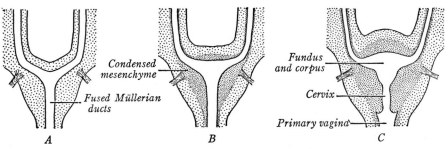

Fig. 290. Diagrams illustrating the later history of the transverse limbs of the Müllerian ducts and the fused portions of the ducts within the genital cord.

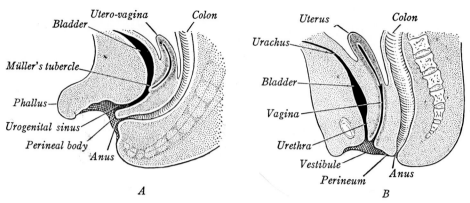

Fig. 291. Sagittal sections of female fetuses, demonstrating the relative shortening through which the tubular urogenital sinus becomes a shallow vestibule. *A*, At ten weeks ($\times$ 4); *B*, at five months ($\times$ 1).

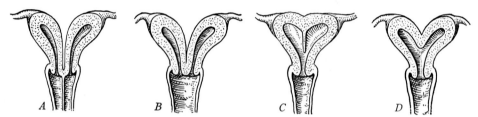

Fig. 292. Anomalies of the human uterus. *A*, Double uterus and vagina; *B*, double uterus; *C*, bipartite uterus; *D*, bicornuate uterus.

The Male. Similar primordia also develop in the male, but remain rudimentary (Fig. 302 *C*). Degeneration of the Müllerian ducts occurs in the third month and only the extreme cranial ends are finally spared; each vestige is called an *appendix testis*. The vaginal primordium persists as a tiny pouch on the dorsal wall of the urethra to which has been given the name *prostatic utricle*. Like the vagina of the female, its original Müllerian epithelium is replaced by invading epithelium from the urogenital sinus.[21] The Müllerian tubercle is represented by the elevated *seminal colliculus,* from whose summit leads off the prostatic utricle. Unlike the female, the male retains its primitive urogenital sinus as a tube (the permanent membranous and cavernous urethra) (Fig. 14).

The results of the transformation of the Müllerian ducts in both sexes are summarized in Fig. 302.

Hormonal Influences. The administration of androgens or estrogens to developing amphibians and birds swerve the gonad to that of the opposite sex by inhibiting either its cortical or medullary component. In mammals the manner of selection of appropriate genital ducts from the double set differs in the two sexes. The fetal ovary does not secrete a hormone in sufficient amounts to have any influence, and the development of the female duct system, until the puberal period, is self-differentiating and independent of such control. On the contrary, the male duct system and the indifferent external genitalia are directly responsive to hormonal substances elaborated by the developing testes, and this dependence is absolute. At the same time, the male hormone suppresses the development of female ducts. Hence, in the absence of fetal testes, the female set of ducts prevails. Stated differently, the embryo develops in the female direction whether ovaries are present or absent; it is the dominant testes that prevent genetic males from differentiating as females.[22]

Anomalies. The more common anomalous conditions of the uterus are based on the failure of the Müllerian ducts to approximate, fuse or form a fundus (Fig. 292). These conditions are: (1) Duplication of the uterus and vagina (as in monotremes and lower marsupials; *A*). (2) Duplication of the uterus, but not the vagina (as in many rodents and bats; *B*). (3) Bipartite uterus, more or less subdivided by a median septum (as in carnivores, some bats and the sow; *C*). (4) Bicornuate uterus, in which the uterine horns fail to elevate into a domed fundus (as in ungulates, cetaceans and most bats; *D*).

Retention of the fetal or infantile type of uterus results from inadequate puberal estrogen, or from failure to respond normally to this hormone. Congenital absence of one or both uterine tubes, of one uterine horn, or of the uterus or vagina occurs rarely, but it may also be associated with hermaphroditic conditions. The vagina sometimes remains solid. The hymen may retain its primary imperforate condition which, at puberty, prevents the discharge of menstrual products. An accessory Müllerian funnel may give rise to an extra tube, a branched tube or an accessory tubal mouth; an occasionally seen cyst (hydatid) has a similar origin (Figs. 288 *B*, 289 *B*).

THE GENITAL LIGAMENTS AND DESCENSUS

THE GENITAL LIGAMENTS

At six weeks the urogenital ridge is attached broadly near the root of the intestinal mesentery (Fig. 280 *A*), but soon a common urogenital mesentery suspends the gonadic and mesonephric regions (Fig. 281 *A*). Toward the end of the second month there develop additional mesenterial and ligamentous supports for the internal genitalia. These are comparable in both sexes, but only in the female do they become structures of permanent importance.

The Female. The ovary is primarily suspended by a short mesentery, named the *mesovarium,* which comes into prominence as the gonad outgrows the mesonephros (Fig. 286 *A*). The remains of the atrophic urogenital ridge at more cranial levels connects with the cranial pole of the ovary and persists as the *suspensory ligament* (Figs. 293, 294). Similarly, the terminal portion of the genital ridge unites the caudal end of the ovary first to the transverse bend of the urogenital ridge and then to the uterus which develops in it. This connection becomes fibromuscular and is known as the *proper ligament of the ovary* (Fig. 294).

With the degeneration of the mesonephric system, the uterine tube lies in a mesenterial fold, the *mesosalpinx* (Figs. 286 A, 294). Somewhat earlier the mutual fusion of the caudal portions of the two urogenital ridges has produced the *genital cord* (Fig. 282 *B-D*). This a mesenchymal shelf that bridges in the frontal plane between the two lateral body walls and contains the uterus in its center (Fig. 293). The shelf itself persists as the sheetlike *broad ligaments* on each side of the uterus. The pelvic cœlom is thereby subdivided into two blind bays, the *recto-uterine pouch* (of Douglas) and the *vesico-uterine pouch.* After the ovary and uterine tube 'descend' (*cf.* p. 332) to a lower position, the mesovarium and mesosalpinx are in obvious direct connection with the broad ligament (Fig. 289 *B*). Owing to

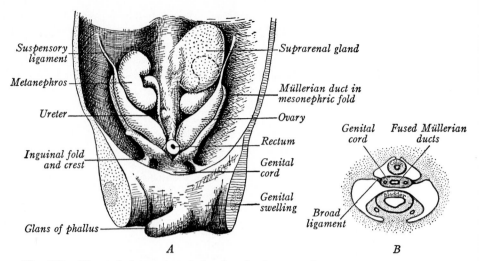

Suspensory ligament
Metanephros
Ureter
Inguinal fold and crest
Glans of phallus
Suprarenal gland
Müllerian duct in mesonephric fold
Ovary
Rectum
Genital cord
Genital swelling
Genital cord
Fused Müllerian ducts
Broad ligament

A   B

Fig. 293. Urogenital organs of the female fetus, and especially the early genital ligaments. *A*, Ventral dissection, at two months (Prentiss; × 8). *B*, Transverse section through the lower trunk, at the level of the genital cord, at three months.

the manner of formation of the genital cord and broad ligament (Fig. 282), the ovary will be suspended from the dorsal surface of the broad ligament.

During the seventh week a pair of uterine ligaments of another sort is begun. At the level where each urogenital ridge bends horizontally toward the midplane in forming the genital cord, a continuation of the common urogenital mesentery (*inguinal fold*) bridges across to a prominence (*inguinal crest*) on the adjoining abdominal wall (Fig. 293 *A*). Within these parts the mesenchyme condenses as a primitive ligament. A direct continuation of this band extends to the subcutaneous tissue of the genital (future labial) swelling of the same side. Thus, by the beginning of the third month, a continuous cord of dense mesenchyme extends from the region of the uterus to the labium majus (Fig. 294). This cord soon becomes fibromuscular and persists as the *round ligament* of the uterus (Fig. 289). Where the ligament passes through the abdominal wall, muscles organize about it and constitute the slanting, tubular *inguinal canal* (*cf.* Fig. 295 *B*).

The Male. The primitive mesentery of each testis is named the *mesorchium* (Figs. 283 *A*, 284 *A*). It loses prominence in later fetal life and is represented in the adult merely by the fold between the testis and epididymis. A *suspensory ligament*, equivalent to that of the ovary, for a time extends down to the cranial pole of the testis, but later it disappears. The *ligamentum testis*, comparable to the proper ligament of the ovary, develops in a caudal continuation of the genital ridge; it extends from the caudal pole of the testis to the transverse bend in the urogenital ridge. From the caudal surface of this bend a primitive ligament soon bridges across to the adjacent body wall and continues into the scrotal swelling of the same side of the body, similar to the formation of the round ligament in the female (*cf.* Figs. 293 *A*, 294). Since a uterus does not develop in the male, at the beginning of the third month there thus exists a continuous mesenchymal cable extending from the caudal pole of each testis through the inguinal canal to the scrotal swelling below (Fig. 295 *B*). This cable is named the *gubernaculum testis*. It is composed of three regional segments (Fig. 295 *A*): (1) the ligamentum testis; (2) a connecting cord in the urogenital ridge (region of the regressive mesonephros and uterine pri-

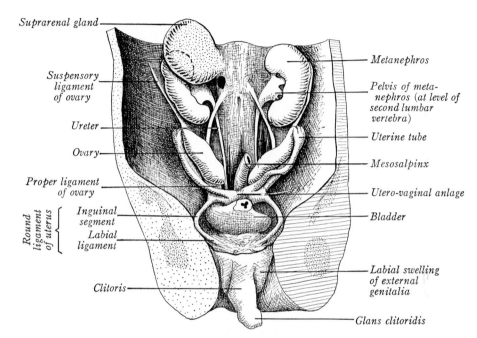

Fig. 294.   Urogenital organs of a female fetus, at ten weeks (Prentiss). × 5. The ventral dissection displays especially the genital ligaments.

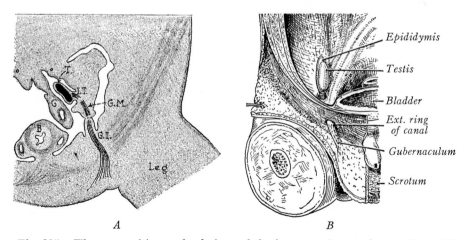

A   *B*

Fig. 295.   The composition and relations of the human gubernaculum testis. *A*, The several components, demonstrated by a schematic section through a male embryo of two months (after Felix; × 10). *B*, Relations at the start of descent (seventh month; × 1).

B, Bladder; *G*, gut; *T*, testis; *L.T.*, ligamentum testis; *G.M.*, mesonephric segment of gubernaculum; *G.I.*, inguinal and scrotal segment of gubernaculum.

mordium); and (3) a ligament extending from this ridge into the scrotal swellings. In terms of female counterparts it is the homologue of the proper ligament of the ovary plus the round ligament of the uterus, between which the uterus intervenes. As in the female, the early gubernaculum antedates the appearance of muscles, and hence an *inguinal canal* arises in the abdominal wall.

DESCENT OF THE GONADS

The Male. The original positions of the testis and ovary change during development. At first they are slender structures, extending caudad from the diaphragm (Fig. 279 *A*). A faster elongation of the trunk craniad, in contrast to the slower growing gonad, produces a relative shift of the latter in a caudal direction until the sex gland lies ten segments below its level of origin (Fig. 294).[23] When this process of growth and shifting is complete (10 weeks), the caudal end of the gonad lies at the boundary between abdomen and pelvis, and close to the groin.

In addition to its early 'migration' caudad (internal descent), just mentioned, the testis later leaves the abdominal cavity and descends bodily into the scrotum (external descent). At the beginning of the third month, sac-like pockets of the peritoneum protrude into each side of the ventral abdominal wall. These are the beginnings of the so-called *vaginal processes,* and until the end of the sixth fetal month the lower poles of the testes continue to lie near them (at the site of the future internal abdominal ring of the inguinal canal) without change of position (Figs. 296 *A*, 297 *A*). Each process, or sac, evaginates (herniates) through the ventral abdominal wall, following the path of the gubernaculum in the slanting *inguinal canal.* The final passage of the sac over the pubis and its extension into the scrotum occurs when the testis, to which the local pelvic peritoneum is attached, migrates into the scrotum, carrying with it some of the loosely folded peritoneum of that region.[24]

Also during the seventh to ninth months the testes descend along the same path (Figs. 296, 297). The mechanics of this process, including the rôle of the gubernaculum testis, is poorly understood. The actual time spent in passing through the inguinal canal is very short, whereas the continued journey into the scrotum proper is leisurely. The testes are usually found in the scrotum early in the ninth lunar month, or at least before birth.

It should be realized that the testis and gubernaculum are covered by peritoneum before the descent begins. Consequently, in its journey the testis must always remain behind (*i.e.,* dorsal to) the peritoneal tube (Fig. 297 *A, B*). Hence on reaching the scrotum, the testis is covered by a reflected fold of the processus vaginalis and merely bulges into its cavity (*C*). Even before birth the narrow peritoneal canal, which connects the main sac of the processus vaginalis with the abdominal cavity, sometimes begins to obliterate, whereupon its epithelium disappears.[25] The vaginal sac, now isolated, represents the *tunica vaginalis* (Fig. 297 *C*). Its visceral layer is closely wrapped about the protruding testis and epididymis, whereas the parietal layer forms a lining to the scrotal sac.

Thus the scrotum proves to be a specialized pouch of skin into which an extension of the muscular and fascial body wall accompanies the evaginating sac of peritoneum. The ductus deferens and the spermatic vessels and nerves are carried down into the scrotum along with the testis and epididymis. They are embedded in connective tissue and constitute the *spermatic cord*. Owing to the path taken by the testis in the scrotal migration, the ductus deferens loops over the ureter (Fig. 14). Fascia and muscle from the internal oblique of the abdominal wall are commonly said to be

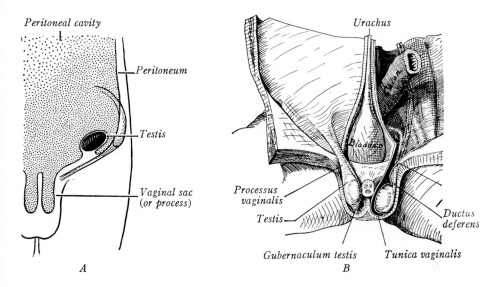

Fig. 296. Peritoneal relations before and after the descent of the human testis. *A*, Diagrammatic phantom at the onset of external descent (seventh month). *B*, Dissection of a newborn; the left vaginal process has been opened and the testis rotated 90° (partly after Corning; × ½).

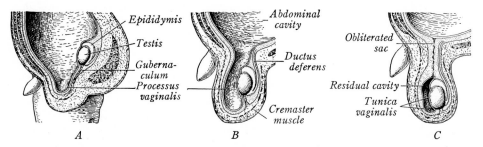

Fig. 297. Descent of the human testis and its subsequent relations, shown in diagrammatic hemisections.

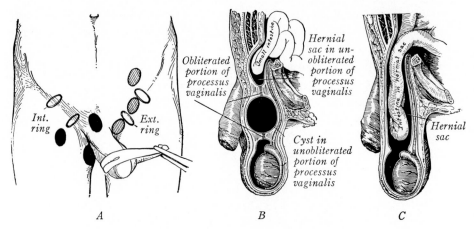

Fig. 298. Anomalies of testicular descent. *A,* Diagram showing final positions of the testis in arrested descent (hatched black) and in ectopic descent (solid black). *B, C,* Dissections of partial and total inguinal hernia, correlated with different degrees of failure of the vaginal process to obliterate (Callander).

added about the spermatic cord and carried down with it. Yet, in the pig at least this *cremaster muscle* arises independently.[26]

During the seventh month the gubernaculum not only ceases to grow, but also shortens by one-half. Yet this shortening, both relative and actual, cannot draw the testis into the scrotum since the lower end of the gubernaculum is not attached to the scrotal skin.[24, 26] On the contrary, the gubernaculum converts into soft mucoid tissue and becomes at least as broad as the testis and epididymis. Through this change the inguinal canal is dilated and the testis perches on the crater-like upper end of the gubernaculum.[26, 27] There is a marked loss in the sustaining power of the soft, jelly-like tissue. Aside from these relations the rôle of the gubernaculum in testis descent remains obscure. Equally unproved are the rôles of endocrines and other suggested influences. The gubernaculum of a newborn infant is only one-fourth of its length when descensus began; after birth it atrophies almost beyond recognition.

A permanent scrotal location of the testes is normal for primates and various other mammals. This is advantageous because of a lower temperature there, since spermatogenesis does not occur at the higher temperature of the abdomen. In a few mammals an abdominal position of the testes is permanent, but the body-temperature in this group is well below that found in mammals with scrotal testes. Still other mammals maintain open inguinal canals; their testes remain in the abdomen except during the mating season when they descend to the cooler scrotum and enter upon spermatogenesis. Some hibernating animals have a periodic descent of the testes that follows the sharp rise in temperature on awakening.

The Female. The internal descent, already mentioned (p. 332), brings the ovaries to the pelvic brim during the third month. Here they still lie in a newborn female. Afterwards the ovary and uterus attain their final positions in the minor (true) pelvis as the result of marked growth by the pelvis. Each ovary rotates into a transverse position and also revolves about the uterine tube until it comes to rest dorsal to the tube (Fig. 289 *B*).

Shallow peritoneal pockets, frequently persistent as the *diverticula of Nuck,* correspond to the primary vaginal processes of the male. There is no external descent of the ovaries; one reason lies in the mechanical block presented by the uterus which is interposed between the ovarian and round ligaments (Fig. 294).

Anomalies. Testicular descent may be arrested at any point along its normal path, but most often at the external inguinal ring (Fig. 298 *A*, on right). This condition, known as *cryptorchism* (*i.e.*, concealed testis), is due to developmental anomalies, mechanical obstruction or, perhaps, to hormone deficiency (Fig. 278 *C*). At birth, three per cent of boys still have imperfectly descended testes. In permanent cryptorchism of any significant degree the testis fails to produce competent sperms or makes none at all. After transversing the inguinal canal, descent into an abnormal location, such as the pubes, thigh or perineum, results in an *ectopic testis* (Fig. 298 *A*, on left). Rarely, in instances of faulty development of the internal genitalia, more or less complete descent of the ovary into the labium majus occurs.

When the stalk of a vaginal process does not obliterate, wholly or in part, conditions are favorable for congenital *inguinal hernia* of the intestine into the inguinal canal or even into the scrotum (Fig. 298 *B*, *C*). Structural weakness of the body wall in this region predisposes to other types of hernia.

THE EXTERNAL GENITALIA

The acquisition of an external penis, with a penile urethra, in the male of higher mammals has paralleled the evolution of a vagina, uterus and the intra-uterine development of the young in the female. The relation of the external penis to the subdivision of the cloaca and the production of a perineum has already been mentioned (p. 308). Progressive stages in these several changes are illustrated in reptiles, monotremes, marsupials and placental mammals.

The indifferent state of the external genitalia was described on p. 319. For a week or more after the first indication of these primordia, their sexless appearance continues (Fig. 299 *A*, *B*). During the eighth week the sex of an embryo begins to be indicated grossly through certain external characteristics, chief among which are the erectness of the phallus, the length of the urethral groove and the relations of the urogenital folds to the genital swellings (*C*, *E*).[28, 29] Yet for a time these criteria are only fairly reliable; especially is there liability to error in diagnosing retarded males for females.[30] At three months the progressive modeling of the external genitalia has attained characteristics that are recognizable as distinctively male or female.

The Male. Fetuses of ten weeks are at the beginning of the definitive stage. The simplest change occurs in the genital swellings, which shift caudad and are then known as *scrotal swellings* (Fig. 299 *C*). Each becomes a half of the *scrotum,* separated from its mate by the *scrotal septum* whose position is marked superficially by the scrotal raphé (*D*).[31] The phallus becomes the *penis.* An early indication of this is when the edges of the urogenital groove fold together and unite progressively in the direction of the glans. This fusion and restoration of the roof of the groove (where the

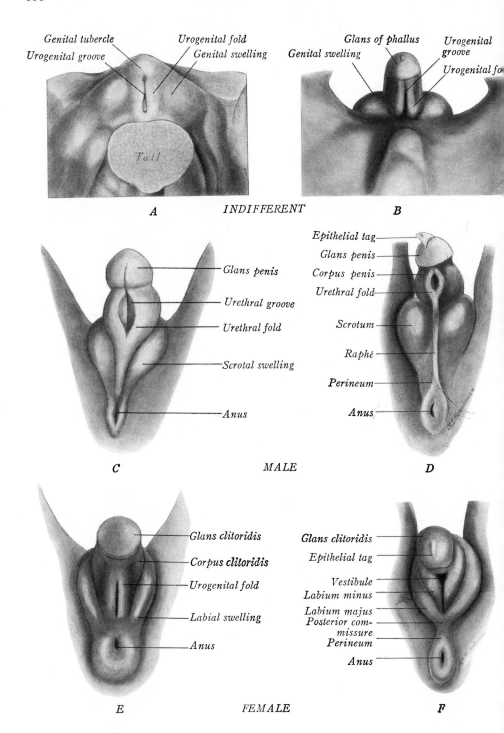

Fig. 299. Differentiation of the human external genitalia (after Spaulding). *A, B,* The indifferent period, at nearly seven and nearly eight weeks (× 15). *C, D,* The male, at ten and twelve weeks (× 8). *E, F,* The female, at ten and twelve weeks (× 8).

urogenital membrane originally ruptured; p. 319) transform a slit-like urogenital sinus, bordered by what may now be called *urethral folds,* into the tubular *cavernous urethra* within the shaft of the penis. The fused edges continue the raphé from the scrotal region (Fig. 299, *C, D*). By the fourteenth week the urethra has closed as far as the glans (Fig. 300 *A*). Here a solid *urethral plate,* outgrown from the early urogenital sinus, partitions the glans incompletely into right and left halves. A combination of trough formation externally and hollowing of the urethral plate produces a tube that continues the urethra to an opening at the tip of the glans (*B*).[32]

During the third month a fold of skin at the base of the glans begins growing toward the tip, and two months later it surrounds the spheroidal glans.[33] This is the tubular *prepuce,* or fore-skin (Fig. 300 *A*). Fusion occurs between the epithelial lining of the prepuce and the covering of the glans (*B*), but clefts appear later in this combined membrane and free the prepuce once more; this separation, however, is still incomplete at birth. A region of incomplete prepuce-formation on the under surface of the glans produces the fold known as the *frenulum.* The *corpora cavernosa* are indicated in the seventh week as paired mesenchymal columns within the shaft of the penis. The unpaired *corpus spongiosum* results from the linking of similar mesenchymal masses, one in the glans and the other in the shaft (Figs. 14, 300). The development of accessory glands (prostate; bulbo-urethral; urethral), arising as buds from the urethra, is described on p. 312, as are the seminal vesicles from the ductus deferens on p. 313.

The Female. Changes in the female are less profound, yet slower (Fig. 299 *E, F*). The phallus lags in development and becomes the *clitoris,* with its homologous *glans clitoridis* and *prepuce.* The urogenital groove is shorter than in the male, never invading the clitoris. It opens permanently to the exterior, since the floor of the groove (the urogenital membrane) ruptures and fails to reclose. After this happens, the combined, shallow cavity is called the *vestibule;* into it open both the urethra and vagina. The urogenital folds, which flank the original groove, constitute the *labia minora;* the junction of ectoderm and entoderm is assumed to be at the

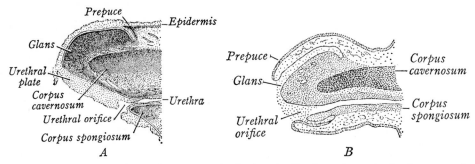

Fig. 300. The tip of the human penis, in longitudinal section. *A,* At four months (× 18); *B,* at five months (after Hunter; × 12).

margin of the lip (Fig. 13). The primitive genital swellings grow caudad and fuse in front of the anus as the *posterior commissure,* while the swellings as a whole enlarge into the *labia majora;* these parts now form a horseshoe shaped rim, open toward the umbilicus (Fig. 299 *E, F*). The cranially located *mons pubis* arises later to complete the gap in the ring; it develops independently of the primitive swellings.[28]

Anomalies. Absence or doubling of the penis or clitoris is very rare; doubling, by non-union of the paired primordia, is usually associated with severe exstrophy of the bladder. The penis may remain rudimentary (Fig. 301 *C*), or the clitoris may hypertrophy; both conditions are common in hermaphroditism. If the lips of the urogenital groove on the under surface of the penis fail to close anywhere along their extent, *hypospadias* results (*B*); a type involving the scrotal region is a common occurrence in false hermaphroditism that simulates the female type (*C*). Far more rarely the urethra opens as a gutter of variable length on the upper surface of the penis, an abnormality known as *epispadias* (*D*). This defect commonly accompanies severe exstrophy of the bladder, just above (Fig. 278 *C*); it is perhaps consequent on a displacement of the paired penile primordia (*cf.* Fig. 274 *A*) caudad of the cloacal membrane. The scrotum may be infantile in size, especially when the testes do not descend. It may be cleft (local hypospadias). A position cranial to the penis is extremely rare.

HERMAPHRODITISM

The name *hermaphroditism (i.e.,* Hermes plus Aphrodite) has been given to the conditions that actually or apparently combine both sexes in one person. *True hermaphroditism* is a condition of intersex in which testis and ovary are both present in the same individual. It occurs rarely in birds and mammals, is not uncommon in the lower vertebrates, and is the normal condition in hag fishes and many invertebrates (worms; molluscs). Numerous human cases have been described that can be accepted as authentic. These include persons with separate ovary and testis (Fig. 301 *A*), with a pair of combined ovotestes, and with an ovary or testis paired with an ovotestis. No such inter-

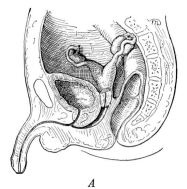

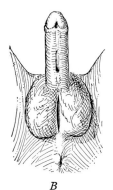

A	*B*	*C*

Fig. 301. Anomalies of the human genitalia. *A,* True adult hermaphrodite, in sagittal section; the external genitalia are typically male, except for the empty scrotum; the internal genitalia are female and include a bicornuate uterus; an imperfect testis and ovary occur on each side. *B,* Hypospadias, showing in one drawing a composite of the different locations. *C,* Hypospadias, of a severe degree, in a false male hermaphrodite. *D,* Epispadias.

D

sex is known to have been fertile either as a male or as a female. The internal genitalia are faultily bisexual, although double ducts and gonads may be represented on one or both sides. When a female gonad and duct occur on one side and a male gonad and duct on the other, this combination approximates the condition of *gynandromorphism*. The external genitalia of hermaphrodites often show features intermediate between male and female characteristics. The secondary sexual characters (beard; breasts; voice; etc.) are usually mixed, tending now one way, now the other.

Far more common is *false hermaphroditism*. This condition is characterized by the presence of the genital glands of one sex in an individual whose secondary sexual characters and external genitalia tend to be indeterminate or to approximate those of the opposite sex. The correct diagnosis of sex is often not established until puberty, and even then it may require tests such as gonad sampling, hormone assay or sex-chromatin determination (p. 61). The internal sexual tract can be that of either sex, or it may be double or mixed; it is commonly atrophic in some of its parts. In *masculine hermaphroditism* an individual possesses testes, often undescended, but the external genitals (by retarded development and severe hypospadias) and secondary sexual characters are like those of the female (Fig. 301 C). In the rarer *feminine hermaphroditism* ovaries are present and sometimes descended, but the other sexual characters, such as enlarged clitoris or perfectly fused labiæ, tend to simulate the male.

No one theory accounts satisfactorily for all hermaphroditic conditions. In general,

TABULATION OF UROGENITAL HOMOLOGIES

MALE	INDIFFERENT STAGE	FEMALE
Testis (1) .... (2) Seminiferous tubules (3) Rete testis	Gonad	Ovary (1) Cortex (2) Medulla (primary) (3) *Rete ovarii*
(1) *Mesorchium* (2) .... (3) *Ligamentum testis* (4) *Gubernaculum testis* (caudal part) (5) *Gubernaculum testis* (as a whole) (6)	Genital ligaments	(1) Mesovarium (2) Suspensory ligament of ovary (3) Proper ligament of ovary (4) Round ligament of uterus (5) .... (6) Broad ligament of uterus
(1) Efferent ductules; cranial *aberrant ductule* (2) *Paradidymis;* caudal *aberrant ductule*	Mesonephric collecting tubules (1) Cranial group (2) Caudal group	(1) *Epoophoron; aberrant ductules* (2) *Paroöphoron*
(1) *Appendix epididymidis* (2) Ductus epididymidis (3) Ductus deferens; seminal vesicle (4) Ejaculatory duct (5) Ureter, pelvis, etc.	Mesonephric (Wolffian) duct	(1) *Vesicular appendage* (2) *Duct of the epoöphoron* (3, 4) *Gartner's duct* (5) Ureter, pelvis, etc.
(1) *Appendix testis* (2) .... (3)	Müllerian duct	(1) Uterine tube (2) Uterus (3) Vagina (upper part?)
Seminal colliculus	Müller's tubercle	Hymen (site of)
(1) Bladder (2) Upper prostatic urethra	Vesico-urethral primordium	(1) Bladder (2) Urethra

(1) Lower prostatic urethra
  (a) *Prostatic utricle* (*vagina masculina*)
  (b) Prostate gland
(2) Membranous urethra
(3) Cavernous urethra
  (a) Bulbo-urethral glands
  (b) Urethral glands (of Littré)
	Urogenital sinus
	(1, 2) Pelvic portion
	
	(3) Phallic portion
		(1) Vestibule (nearest vagina)
		  (a) Vagina (lower part, at least)
		  (b) *Para-urethral ducts;* urethral glands
		(2) Vestibule (middle part)
		(3) Vestibule (between labia minora)
		  (a) Vestibular glands (of Bartholin)
		  (b) Lesser vestibular glands

(1) Penis
  (a) Glans penis
  (b) Urethral surface of penis
  (c) Corpora cavernosa penis
  (d) Corpus cavernosum urethræ
(2) Scrotum
(3) Scrotal raphé
 	
(4) ....
	(1) Phallus
	  (a) Glans
	  (b) Lips of urogen. groove
	  (c, d) Shaft
	
	(2, 3) Genital swellings
	
	(4) Median cranial swelling
		(1) Clitoris
		  (a) Glans clitoridis
		  (b) Labia minora
		  (c) Corpora cavernosa clitoridis
		  (d) Vestibular bulbs
		(2) Labia majora
		(3) Posterior commissure
		
		(4) Mons pubis

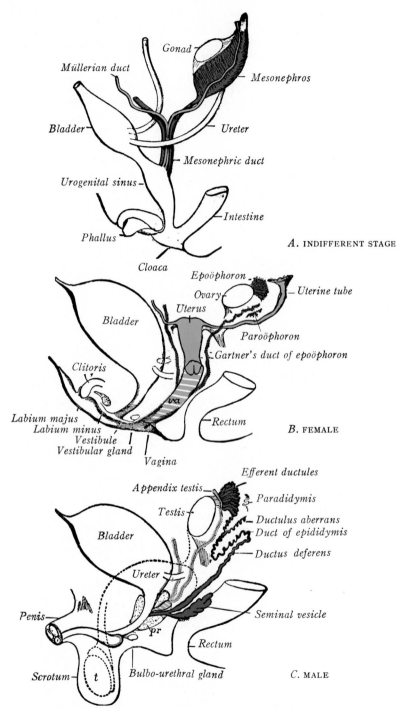

Fig. 302. Diagrams illustrating the transformation of an indifferent, primitive genital system into the definitive male and female types (Thompson).

an abnormal genic balance, controlling the relative activity of the cortical (female) and medullary (male) components of the primitive bisexual gonad, is seemingly responsible for some of the conditions observed (p. 323). This would offer a possible explanation for the separate testis and ovary or the combined ovotestis of a true hermaphrodite. The masculine trend in genetically female pseudohermaphrodites may result either from an excessive production of male hormone by an enlarged cortex of the suprarenal gland, or from a tumor with testicular structure in the ovarian medulla. On the other hand, the feminization of a congenital male pseudohermaphrodite can result from an inadequate suppressive influence by the testicular hormone during the fetal period (p. 328), or by sterile testes that produce much androgen.

HOMOLOGIES OF THE UROGENITAL SYSTEM

In the table on p. 339 are summarized the equivalent permanent derivatives of the indifferent reproductive system; persisting vestigial parts are printed in italics. Figure 302 sets forth the same facts in pictorial form.

REFERENCES CITED

1. Park, W. W. 1957. J. Anat., *91*, 369–373.
2. Gillman, J. 1948. Carnegie Contr. Embr., *32*, 81–131.
3. Witschi, E. 1948. Carnegie Contr. Embr., *32*, 67–80.
4. Pinkerton, J. H. M., *et al.* 1961. Obst. & Gyn., *18*, 152–181.
5. Mintz, B. 1960. J. Cell. & Comp. Physiol., *56* (suppl.), 31–47.
6. Evans, H. M. & O. Swezy. 1931. Mem. Univ. Calif., *9*, 119–224.
7. Burkl, W. & G. Politzer. 1953. Z'ts. Anat. u. Entw., *116*, 552–572.
8. Stieve, H. 1930. Bd. 7, Teil 2 in Möllendorf: Handbuch (Springer).
9. Fawcett, D. W. & M. H. Burgos. 1956. Ciba Found. Coll. on Ageing, *2*, 86–98.
10. Grünwald, P. 1942. Am. J. Anat., *70*, 359–397.
11. Witschi, E. 1956. Development of Vertebrates, Chapter 10 (Saunders).
12. Bremer, J. L. 1916. Am. J. Anat., *19*, 179–210.
13. Zuckerman, S. 1937. Phil. Trans. Roy. Soc. London (Sect. B.), *228*, 147–172.
14. Wilkerson, W. V. 1923. Anat. Rec., *26*, 75–78.
15. Matêika, M. 1959. Anat. Anz., *106*, 20–37.
16. Fluhman, C. F. 1960. Obst. & Gyn., *15*, 62–69.
17. Bulmer, D. 1957. J. Anat., *91*, 490–509.
18. Vilas, E. 1933. Z'ts. Anat. u. Entw., *101*, 752–767.
19. Hunter, R. H. 1930. Carnegie Contr. Embr., *22*, 91–108.
20. Fränkel, L. & G. N. Papanicolau. 1938. Am. J. Anat., *62*, 427–451.
21. Vilas, E. 1933. Z'ts. Anat. u. Entw., *99*, 599–621.
22. Jost, A. 1960. Mem. Soc. Endocrin., No. 7, 49–62.
23. Higuchi, K. 1932. Arch. f. Gynäk., *149*, 144–172.
24. Hunter, R. H. 1926. Brit. J. Surg., *14*, 125–130.
25. Mitchell, G. A. G. 1939. J. Anat., *73*, 658–661.
26. Backhouse, K. M. & H. Butler. 1960. J. Anat., *94*, 107–120
27. Forssner, H. 1928. Acta Obst. et Gyn. Scand., *7*, 379–406.
28. Spaulding, M. H. 1921. Carnegie Contr. Embr., *13*, 67–88.
29. Szenes, A. 1925. Morph. Jahrb., *54*, 65–135.
30. Wilson, K. M. 1926. Carnegie Contr. Embr., *18*, 23–30.
31. Politzer, G. 1932. Z'ts. Anat. u. Entw., *97*, 622–660.
32. Glenister, T. W. 1954. J. Anat., *288*, 413–425.
33. Hunter, R. H. 1935. J. Anat., *70*, 68–75.

Chapter XIX. The Vascular System

ANGIOGENESIS

Both the blood cells and blood vessels arise from mesenchyme. The earliest formative tissue of this kind has long been called *angioblast*, while the process of vessel development is known as *angiogenesis*. It is claimed that the earliest origin of angioblast can be traced to distinctive cells that separate away from the primary trophoblast which constitutes the wall of the blastocyst. This occurs at the same time that the more generalized extra-embryonic mesoderm is similarly delaminating there.[1] Slightly later, at the stage of the early primitive streak, angioblast also appears (by spreading?) in the body stalk and in the wall of the yolk sac. In the latter location the angioblast takes the form of isolated masses and cords, termed *blood islands,* in the splanchnic mesoderm (Fig. 303 *A, B*). Originally solid, they soon hollow out (*C, D*). In this process the peripheral cells become arranged as a flattened endothelium; the more central cells are the earliest *blood cells* and these float in the primitive *blood plasma.* The latter makes its appearance as a clear fluid, apparently secreted by the cells of the blood island. The plasma first occupies discrete intercellular clefts, but these spaces soon coalesce and produce a common lumen. For a time, a cluster of primitive blood cells may adhere to the side of such an endothelial space; these cell groups are sometimes termed blood islands also. Such primitive blood cells soon separate, differentiate mostly into red blood cells, and are swept into the general circulation. However, the majority of blood cells, both red and white, do not trace origin to angioblastic elements on the yolk sac; they arise progressively from the mesenchyme of the embryo proper in a way to be described under the following topic (p. 344).

By growth and union the originally isolated vascular spaces, derived from solid angioblast, are converted into plexuses of blood vessels. These are present on the yolk sac, body stalk and chorion of human embryos at the late head-process stage. In the wall of the yolk sac this network comprises the *area vasculosa* which eventually envelops the entire sac (Fig. 310). The first vessels within the embryo itself appear at the time when the earliest somites are being laid down. In general, these vessels are not direct extensions of extra-embryonic vessels that progressively invade the embryo. The primary source of intra-embryonic vessels is mesenchyme that differentiates

342

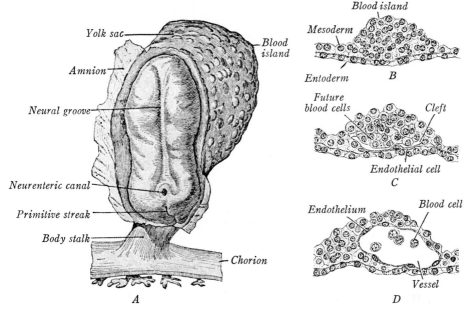

Fig. 303.   The differentiation of blood vessels from blood islands on the human yolk sac. *A,* Presomite embryo (Spee; × 23). *B–D,* Sections from an embryo of six somites, showing three progressive stages of angiogenesis (× 325).

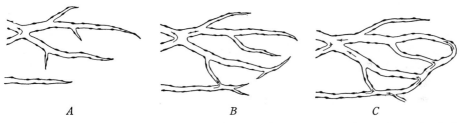

Fig. 304.   Stages in the development of a capillary network, observed in the living rabbit. × 110.

locally into cleft-like, vascular primordia whenever and wherever the need arises in early development.[2, 3, 4]

Proliferative growth of the endothelium, thus primarily established, links the simple vascular spaces into continuous channels; the latter further expand their primitive network by independent sprouting (Fig. 304). After a system of closed vessels (and a primitive blood circulation within them) has been established, new vessels arise only as outgrowths of pre-existing vessels.[5] The causative stimulus that induces budding is unknown. The several steps, already described, in the creation of vascular spaces and in their linkage and subsequent growth into networks are common to all vertebrates.

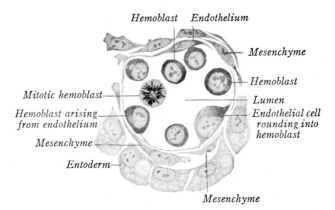

Fig. 305. Section of a blood vessel on the yolk sac of a rabbit embryo, at five somites, showing early blood-cell formation from endothelium (Maximow). × 500.

HEMOPOIESIS

The development of blood cells (*hemopoiesis*) is similar in all embryonic sites. The parent tissue is the versatile mesenchyme which has other lines of specialization as well (Fig. 11). At any local site where hemopoiesis is to begin, certain mesenchymal cells round up and detach; they become free, basophilic elements that are the progenitors of all types of blood cells.

As a preliminary to hemopoiesis the primitive mesenchyme of the embryo begins differentiating into three sorts of tissues: (1) blood islands on the yolk sac; (2) endothelium (elsewhere than on the yolk sac); and (3) fixed mesenchyme cells. Of these, the primitive blood cells of the yolk sac form mostly into early generations of large, nucleated red blood cells; these serve the embryo for a time and then die out.[6] The endothelium of early stages has capacities identical with the mesenchyme which gave it origin, but these powers are soon lost (Fig. 305). The mesenchyme is the chief blood-forming tissue of the embryo while its successor, the fixed connective-tissue cells, serves the same function in the adult. In all the locations about to be mentioned, hemopoiesis is made possible by the freeing of mesenchymal cells which then serve as proliferative stem cells.

During the prenatal period several locations are utilized successively for the formation of blood cells (and especially red elements).[7] These regions tend to have rich meshworks of slender channels through which blood flows slowly. Their sequence and the time of first appearance of hemopoietic centers in them are as follows: (1) yolk sac (fourth week); (2) body mesenchyme and blood vessels (fifth week); (3) liver (sixth week); (4) spleen, thymus and lymph glands (second to fourth month); and (5) bone marrow (fourth month). There is considerable overlap in the activities of these foci. For example, the yolk sac abandons hemopoiesis in the second month; by contrast, the liver is the most active site until the middle of fetal life, when its activity decreases slowly and ceases at birth. One by one, these organs give up total blood-cell formation. The red marrow becomes dominant at five months, and eventually serves as the single source from which red blood corpuscles and granular leucocytes are recruited normally during post-natal life. A few lymphocytes and monocytes may continue to be formed in the mature marrow, but these nongranular leucocytes are supplied mostly by the lymphoid organs (and fixed connective-tissue cells other than fibroblasts?). All marrow is of the red type throughout infancy and much of childhood.

Two sharply contrasted views are held as to the exact mode of origin of the various blood elements. According to the *monophyletic theory,* a common mother cell gives rise to all types of blood elements, both red and white. As the term indicates, they are all of 'one family' (Fig. 11). The *polyphyletic theory,* although variously presented, most commonly asserts that the erythrocytes and granular leucocytes are derived from one mother cell, while the nongranular leucocytes trace their ancestries to a separate stem cell. The total evidence seems to favor the monophyletic view, and the descriptions that follow will be based upon it.[8] Yet it should be recognized at the outset that hemopoiesis is a difficult and baffling study concerning which other opinions, divergent in certain respects from those set forth here, have been advanced.

The generalized mother cell from which the various blood elements are thought to differentiate is called the *hemocytoblast* (Fig. 11). It has the typical appearance of a large lymphocyte, and accordingly is an ameboid cell with a large, open-structured nucleus and a relatively small amount of finely granular, basophilic cytoplasm. From such parent cells, according to the monophyletic view, all blood elements arise. Specialization proceeds in divergent directions, but the determining factors responsible for such diverse differentiations are not easy to assign.

Differentiation of Red Cells.   A generic name for the differentiating red cell is *erythroblast.* Springing from the totipotent 'hemocytoblasts of the blood islands, body mesenchyme, liver, lymphoid tissue and bone marrow, it undergoes in each location an identical transformation whereby the cytoplasm gains hemoglobin and the nucleus condenses and is lost. In this metamorphosis there are recognized three principal stages (Fig. 11):

1. *Erythroblasts* or *rubricytes* (once termed ichthyoid blood cells because of their resemblance to the typical red blood cells of fishes). They are characterized by checkered nuclei and the presence of some hemoglobin in the cytoplasm. From the third through the sixth week of development the erythroblast is the only red blood cell found. For a time, it multiplies within the blood vessels, but after the third month practically disappears from the blood stream.

2. *Normoblasts* or *metarubricytes* (once termed sauroid blood cells because they resemble the red blood cells of adult reptiles and birds). This stage first transforms from erythroblasts in the liver, and is predominant in embryos of two months. Normoblasts are distinguished by their small, dense nuclei and richer hemoglobin, but in spite of this specialization they

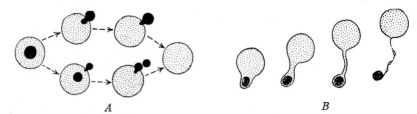

Fig. 306. Methods by which the nucleus may be lost from mammalian normoblasts. *A*, Loss by extrusion, either as a whole (above) or in fragments (below). *B*, Loss by cytoplasmic constriction, as followed in blood cultures during a thirty-minute period (after Emmel).

still undergo mitosis and so continue to aggregate in clusters. At seven weeks almost every red cell in the circulating blood is nucleated; at 11 weeks only a few with nuclei remain.[9]

3. *Erythrocytes* (red blood corpuscles). These elements, characteristic of mammals, originate from normoblasts through the loss of their nuclei. The way in which the nucleus disappears is disputed. It has been said to be extruded as a whole or in fragments (Fig. 306 *A*), or that it is completely absorbed. It can be shown that the cytoplasm buds away from the nucleated remnant (*B*).[10] The earliest red blood corpuscles are spherical elements; they are first formed during the second month, chiefly in the liver. Early in the third month the enucleated corpuscles first predominate in the blood stream, and by the middle of the month they have nearly supplanted those with nuclei.[9] Fetal hemoglobin differs somewhat from the definitive, adult type which begins to replace it in the sixth month of fetal life.

Differentiation of Granulocytes. In the locations already enumerated, the hemocytoblasts also give origin to differentiating granular leucocytes named *myelocytes* (Fig. 11). The young granulocytes, thus produced, elaborate within their cytoplasm specific kinds of granules. While still immature, these cells continue to proliferate, yet they cannot transform into any cell type other than the one already begun. Ultimately the ability to divide is lost and differentiation into mature granulocytes proceeds to an end. These elements first appear in the blood stream at three months. Three types may be recognized:

1. *Neutrophils* have a finely granular and neutrally staining cytoplasm. The nucleus changes through crescentic to complex, lobate shapes.

2. *Eosinophils* develop coarse granulations and a bilobed nucleus. The granules stain intensely with acid dyes.

3. *Basophils* acquire an irregularly shaped nucleus and differentiate coarse cytoplasmic granules that stain heavily with basic dyes. These blood elements are a type entirely distinct from the tissue basophils, or mast cells.

Differentiation of Nongranular Leucocytes. There is no essential difference between the hemocytoblast and the definitive larger lymphocytes. In fact, the latter must be considered as retaining their original totipotent potentialities which may be exercised when needed.[8] The larger *lympho-*

Fig. 307. The origin of blood platelets (Wright). × 1000. A megakaryocyte extends processes into a blood vessel (*V*) and detaches platelets (*bp*).

cytes regularly give rise to the small type, and the latter in turn grow into larger ones. Lymphocytes arise in relation to lymph vessels and become the dominant component of all emerging lymphoid organs. They first appear in the blood stream at two months. From the primitive embryonic mesenchyme are differentiated the *fixed macrophages (reticulo-endothelium)* of lymphoid, hepatic and marrow sinuses; both the fixed macrophages and the hemocytoblasts produce freely wandering, highly phagocytic elements (*free macrophages*) of close affinity. The hemocytoblast mother-cells (and possibly the fixed macrophages) give rise to *monocytes* with a characteristic, kidney-shaped nucleus, and to connective-tissue *mast cells* with basophilic cytoplasmic granules. Finally, some hemocytoblasts specialize into *megakaryocytes,* which are the giant cells found typically in bone marrow. From them *blood platelets* arise by fragmentation of the cytoplasm (Fig. 307).[11]

Causal Relations. The determination of cells as angioblast takes place in diverse regions and times—the earliest probably by the end of gastrulation. In amphibians the entoderm is said to act as the inductor of blood. The blood-island material of a frog embryo occupies an elongate strip in the splanchnic mesoderm, between the heart and anus. It is already determined as the former of red blood cells at the neurula stage and, if extirpated, no erythrocytes are formed. If isolated in culture, it self-differentiates into these elements. Moreover, bits of the yolk-sac wall of young rat embryos, when implanted into a neutral environment, such as the anterior chamber of the eye, prove that the first free cells of the blood islands are capable of differentiating both primitive erythrocytes and also definitive red blood corpuscles and granular leucocytes.

THE PRIMITIVE VASCULAR SYSTEM

The differentiation of blood vessels occurs first in the chorionic region of very young embryos of higher mammals. This precocity is correlated with the absence of a significant amount of nutritive yolk and the consequent early need for vessels that will extract nourishment and oxygen from the maternal circulation and distribute them to the tissues of the embryo. For a while the arteries and veins are not distinguishable structurally, yet even in young embryos they are named in anticipation of the particular vessels that are destined to arise from them.

Reconstructions demonstrate the linkage of separate primordia and delicate injections show that diffuse, capillary plexuses precede the formation of definite arterial and venous trunks in any region (Fig. 308 *B*). It is only through the selection, enlargement and differentiation of appropriate paths

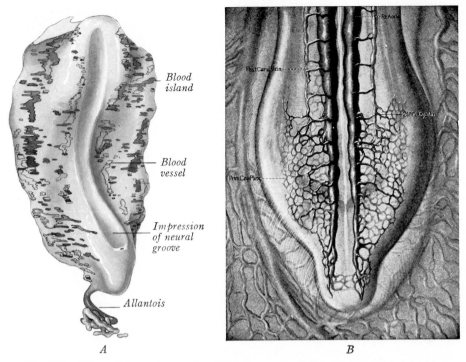

Blood
island

Blood
vessel

Impression
of neural
groove

Allantois

A   B

Fig. 308. The origin and growth of blood vessels. *A*, Model of a human embryo of two somites, in which ectoderm and all but blood-island mesoderm have been omitted (after Ludwig; × 40); dark masses indicate still solid islands; light areas outline hollowed vessel-primordia. *B*, Primitive vascular plexus in the caudal end of a sixty-hour chick embryo (Evans; × 35); the sciatic artery will differentiate from the primary capillary plexus of each limb bud; aortæ are already consolidating from the medial margins of the plexuses.

in such networks that the definitive vessels arise; those capillaries from which the flow has been diverted, atrophy.[12] The selection of appropriate channels from the diffuse capillary bed results both from the action of inherited patterns and from the hemodynamic factors incident to the blood flow. What determines the differentiation of some channels into arteries and others into veins is not fully known; presumably the mechanical conditions of the blood flow (speed, pressure and pulse) play a rôle.[12, 13]

The first indications of paired vessels are found in human embryos that stand at the beginning of somite formation (Fig. 308 *A*). These are solid cell clusters and actual mesenchymal clefts which are aligned longitudinally in four incompletely linked courses. A medial pair of alignments represent the beginnings of *dorsal aortæ*. They are continued interruptedly through what will become the first pair of *aortic arches* to the *heart primordia* beneath the folded head. From the heart a second pair of alignments extend caudad along the borders of the embryonic disc. These contain representatives of both the organizing *vitelline* and *umbilical veins,* but the latter have not connected with the better organized umbilical vessels already present in the body stalk.

Embryos with about 12 somites are characterized by a single heart which now communicates by a first pair of *aortic arches* with the paired *dorsal aortæ* (Fig. 309). Further advances include a vitelline circuit (paired *vitelline arteries* and *veins*) to the yolk sac and a separate placental circuit (paired *umbilical arteries* and *veins*) which uses the body stalk to reach the chorion; the latter is already participating in the organization of a placenta. By this time the heart begins to twitch and then to propel the blood in an actual circulation through the vitelline and umbilical circuits. Because of the early decline of the yolk sac, a complete vitelline circulation, as such, will last but a short time. By contrast, the umbilical circuit remains functional until birth. New vessels, designed primarily for the body of the embryo, are the *dorsal intersegmental arteries*, arising as offshoots from the aorta, and a pair of *precardinal veins* which have arisen to return blood from the head end of the embryo to the heart.

The primitive vascular system is well represented in embryos four weeks old (5 mm. long and now possessing the full number of 42 somites). In Fig. 310 four pairs of aortic arches have appeared, while the paired *precardinal* (or anterior cardinal) and *postcardinal veins* drain into the heart by a *common cardinal* stem on each side. The dorsal aortæ have fused throughout much of their lengths into the *descending aorta* and this vessel bears numerous dorsal, lateral and ventral branches. Of the ventral (primitively vitelline) series three branches are gaining prominence and will continue permanently as important *arteries* (Fig. 319): (1) the *cœliac* in the stomach-pancreas region; (2) the *superior mesenteric* in the small-intestine region; and (3) the *inferior mesenteric* in the large-intestine region.

The stage of paired, symmetrical vessels represents the primitive vascular plan. This embryonic state is profoundly altered and made somewhat asymmetrical by fusions, differential enlargement, atrophy and the emergence of certain new vessels and routings. These alterations represent adjustments

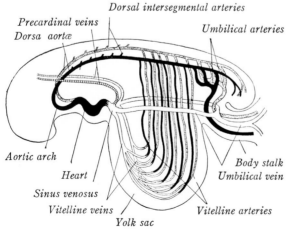

Fig. 309. Diagram of the arrangement of blood vessels, viewed from the left side, in a human embryo of three weeks (Prentiss).

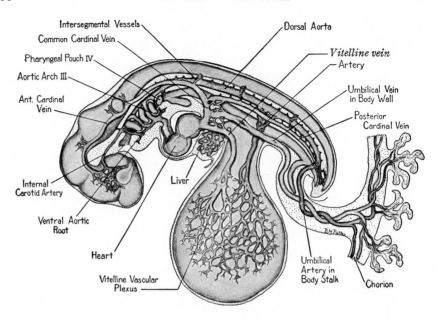

Fig. 310. The vascular plan, viewed from the left side, in a human embryo of four
weeks (Patten). × 20.

to changing form and pattern within the developing organ-systems of the
body. The ensuing descriptions of the development of arteries and veins
will explain the essential changes that supervene between the primary,
symmetrical set and the definitive arrangement. As the primitive blood
vessels advance, the neighboring mesenchyme adds accessory coats about
the endothelium, which is the primary tissue of the vascular system. These
are: (1) the *tunica intima* (endothelial and fibrous); (2) the *tunica media*
(muscular and fibrous); and (3) the *tunica externa* or *adventitia* (fibrous).
Through folding, the tunica intima of veins gives rise to pocket-like *valves*.

DEVELOPMENT OF THE ARTERIES

In the second month the primitively paired set of arteries gives way to
the partly unpaired system that characterizes later stages. The *descending
aorta* has already been consolidated into a common trunk by fusion of the
paired dorsal aortæ in all but their most cranial portions. It bears serially
repeated dorsal, lateral and ventral branches. The caudal termination of
the aorta is the so-called *middle sacral artery* whose final dorsal position,
as an apparent aortic branch, is the result of secondary shifting through
growth. There is little in a human embryo that could be called ventral
aortæ. Almost from the start the short truncus of the heart continues into
an enlargement which has received the name *aortic sac* (Fig. 312 *B*). There
is no elongate ventral aorta, as in lower vertebrates.[14] From this sac the
several *aortic arches* radiate and curve upward around the pharynx to

reach the dorsal aortæ (Fig. 311 *A*); each courses like a central axis in a corresponding branchial arch (Fig. 192 *A*).

The chief changes leading toward the definitive arterial system are: (1) the transformation of the aortic arches; (2) the specialization of certain branches of the aorta; and (3) the development of arteries in the extremities.

Transformation of the Aortic Arches.   The aortic arches of the human embryo have great significance when viewed comparatively. Five or more pairs of arches are provided in connection with the functional gills of fishes, and either three or four pairs serve the same purpose in tailed amphibians. In higher vertebrates there is both a reduction in number and an extensive transformation into vessels more appropriate to air-breathing animals. Birds

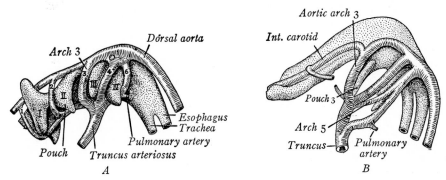

Fig. 311.   Reconstructions of the pharyngeal pouches (Roman numerals) and aortic arches (Arabic numerals), viewed from the left side (after Tandler). $\times$ 38. *A*, Composite at 3–5 mm.; *B*, at 9 mm.

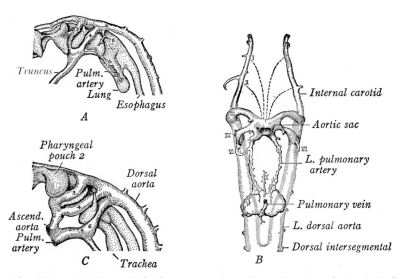

Fig. 312.   Reconstructions of the human aortic arches (after Congdon). *A, B,* Lateral and ventral views, at 5 mm. ($\times$ 23); *C*, lateral view, at 11 mm. ($\times$ 17). The development of the pulmonary artery and arch are shown especially well.

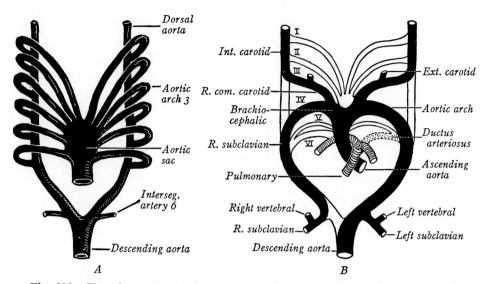

Fig. 313. Transformation of the human aortic arches, illustrated in ventral views. *A,* Model of the full set of aortic arches, before transformation begins. *B,* Scheme, with the dorsal vessels spread laterad, to show persisting vessels (black), discontinued vessels (white) and atrophic vessels (stipple).

use the right arch of the fourth pair, while mammals use the left arch as the route to the descending aorta.

In the embryos of man and other mammals six pairs of aortic arches develop but all are not present at any one time; Figs. 311 *A,* 313 *A* are inaccurate in this respect. This total includes a rudimentary and inconstant pair (number five of the series) whose status as true arches has not escaped challenge (Fig. 311 *B*).[14, 15] It is largely for this reason that some prefer to call the last arch not the sixth in the series, but merely the pulmonary arch since its history is bound up with the formation of arteries to the lungs. The period of development of the aortic arches extends throughout the fourth week, and their transformation mainly occupies the fifth, sixth and seventh weeks. Some characteristic changes in this set of vessels are brought about by the loss or interruption of some arches and portions of the aortæ, correlated with a local reduction or stagnation of the blood flow. Innovations involve the enlargement of certain vessels and the new formation of still others.

The *first* and *second* pairs of aortic arches involute early and contribute only slightly to the permanent arterial plan (Figs. 312, 313 *B*).[16] The dorsal aortæ at the level of these arches persist, but between the third and fourth arches both aortæ atrophy (Fig. 313 *B*). The outcome on each side is a compound vessel, beginning with the *third arch* and continued by way of the dorsal aorta to the head region. These vessels are the primitive *internal carotid arteries,* which not only branch in the head to supply the brain, eyes and ears but also connect with the vertebral-basilar arterial supply in a way to be described presently (Fig. 316). The *external carotid arteries* are new, direct outgrowths of the aortic sac which move their bases up onto the third arches and for a time supply merely the territory of the first and second branchial arches (Fig. 316 *A*). Henceforth the common stem of each third aortic arch, proximal to the origin of the external carotid, is known as the *common carotid* (Fig. 313 *B*). Eventually the external carotids are carried onto the surface of the entire head by migrating muscles that originate in the first, and especially the second, branchial arches.

Both *fourth arches* persist, but their histories differ (Fig. 313 *B*). On the left side the arch (with some annexations of the adjacent ascending aorta and left dorsal aorta) becomes the permanent *arch of the aorta*.[17] The remainder of the left dorsal aorta is added to the two merged dorsal aortæ (farther caudad), thereby completing the upper segment of the descending aorta. On the right side a right portion of the aortic sac elongates into the *brachio-cephalic* (or *innominate*) *artery*, which then serves as the main stem for both the common carotid and subclavian vessels of that side. The *right subclavian* itself begins with the right fourth arch and then continues caudad by annexing the right dorsal aorta down to a level where this vessel has lost connection with its mate. The continuation of the right subclavian farther into the arm bud is by a branch off this aortic remnant, and this part alone corresponds to the entire *left subclavian*.

The so-called *fifth aortic arches* have been mentioned. They are inconstant, incomplete and transitory. Shortly after the 7 mm. stage they disappear without trace.

The *pulmonary* (*sixth*) *arches* come into being when a sprout from each dorsal aorta bridges across to the primitive *pulmonary arteries* which are already growing caudad from the aortic sac to the lung buds (Fig. 312 *A, B*). The distal portion of each pulmonary vessel, so tapped, then appears as a mere offshoot set at right angles to an arch of composite origin (*C*). On the right side this pulmonary arch loses connection with the right dorsal aorta, but on the left the corresponding distal segment remains as an important channel, the *ductus arteriosus* (of Botallo), until birth (Figs. 313 *B*, 314). Meanwhile, when the arches are transforming, the aortic sac

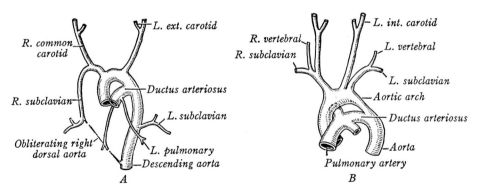

Fig. 314. Changing relations, especially of the human subclavian arteries, shown in ventral views. *A*, At 17 mm. (× 20); *B*, at birth (× ½).

and the primitive truncus have been splitting into aortic and pulmonary stems. This division proceeds in such a manner that the aortic trunk is continuous with the third and fourth arches, while the pulmonary trunk opens into the left sixth arch (Fig. 312 *B*).

The final relations of the heart, aorta and aortic-arch derivatives result from so-called caudal displacements and readjustments. Nevertheless, any 'caudal migration' is relative rather than actual; it is due to a failure of some parts to keep pace with the growth craniad of adjacent structures like the neural tube and pharynx. The brachio-cephalic and common carotid arteries elongate in step with this upward growth (and the appearance of a neck), while the root of the left subclavian shifts considerably higher on the permanent aortic arch (Fig. 314).

The roundabout courses and the slightly different relations taken by the recurrent laryngeal nerves find an explanation in various facts already cited. Originally each vagus nerve gives off a branch that extends directly to the larynx, then located at the same level; in doing this the two rami pass caudal to the sixth aortic arches. But when the aortic arches are left behind in the upward growth of the head and neck, then both nerves necessarily become looped around the sixth arches. Hence, after the arch-transformations have been completed, the left recurrent nerve remains hooked around the ligamentum arteriosum (former sixth arch), while the right nerve, released by the degeneration of the fifth and sixth arches on that side, bears a similar looped relation to the right subclavian (*cf.* Figs. 314, 477).

Branches of the Dorsal Aorta. Previous to the fusion of the dorsal aortæ during the fourth week, each vessel bears dorsal, lateral and ventral branches. These are repeated serially and each set is arranged in a longitudinal row. After aortic fusion has occurred, the relations at a typical level are as shown in Fig. 315. It is with the transformation of these paired arteries into more specialized vessels that the following paragraphs deal; the period involved is mainly that of the fifth through the seventh weeks of embryonic life.

1. DORSAL BRANCHES.   These total some thirty pairs. Since they arise

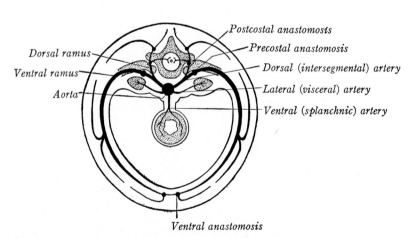

Fig. 315.   Arrangement of the primitive aortic branches, shown in a schematic transverse section of the trunk. Longitudinal anastomoses are drawn as bead-like enlargements; vertebra, ribs, suprarenals and gut are stippled.

in a regular dorsolateral series and pass between successive somites, it is appropriate to call them *dorsal intersegmental arteries* (Fig. 312 *B*). At first each vessel extends to the spinal cord only, but this primary distribution is soon overshadowed by a secondary branch that supplies the somites and body wall. As a result, the original arteries come to look like minor off-shoots from larger vessels (Fig. 315). It is convenient to designate the two divisions of the original stem as *dorsal* and *ventral rami*.

a. The *dorsal rami* supply the region of the cord, including dorsal muscles and skin. The *vertebral arteries* are an important pair of vessels, which arise as secondary developments from two series of dorsal rami belonging to the neck. These rami undergo longitudinal linkage just dorsal to the ribs (Fig. 315; 'postcostal anastomosis'). All of their original stalks then atrophy except the most caudal one in the series (Fig. 316 *A*). The resulting longitudinal vessel is a *vertebral artery;* it takes origin, along with the subclavian, from the sixth cervical intersegmental artery (Fig. 317 *A*).[18] The vertebral arteries establish functional communications in the head with certain branches of the internal carotids. The intermediary vessel, responsible for this linkage, is the *basilar artery*. It arises quite independently of the others through the consolidation of two longitudinal channels beneath the brain.[14, 19] Anastomotic unions on the part of the basilar and internal carotids beneath the brain produce the *circulus arteriosus* of Willis (Fig. 316 *B*). At its caudal end the basilar similarly joins the two vertebrals. By these separate routes (carotid and vertebral-basilar), the brain is doubly guaranteed its arterial blood supply.

b. The *ventral rami* of the dorsal intersegmental arteries become especially prominent in the thoracic and lumbar body wall where they persist as the serially arranged *intercostal* and *lumbar arteries* (Fig. 317 *A*, *C*). The common stem of the sixth intersegmental continues into an enlarged ventral ramus that becomes the important *subclavian* (all of the left and the distal segment of the right; p. 353) and its continuation as the *axillary artery*.[16, 19]

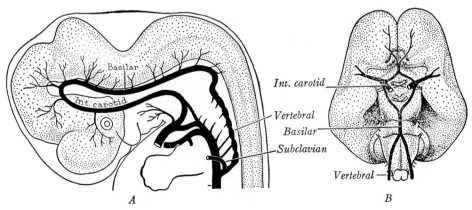

Fig. 316. Human arteries of the head and neck regions. *A*, Phantom, at six weeks, viewed from left side ($\times$ 6). *B*, Arteries of the brain, at fourteen weeks, in ventral view ($\times$ 1).

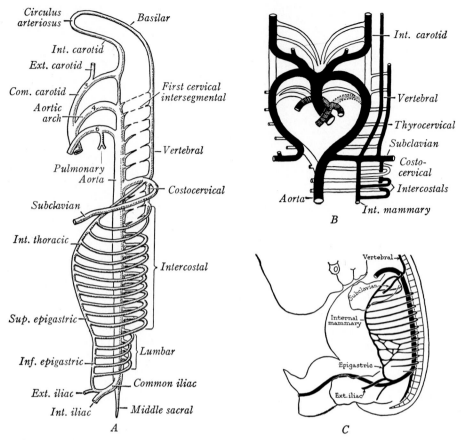

Fig. 317. Derivatives of the dorsal branches (intersegmental arteries) of the human aorta. *A,* Diagram, viewed from the left side. *B,* Scheme, in ventral view, explaining origins in the vicinity of the subclavian artery. *C,* Relations of the internal thoracic (or mammary) and epigastric arteries, at 16 mm., viewed from the left side (after Mall).

A secondary branch off the subclavian provides an origin for an interesting vessel. Longitudinal ventral anastomoses between the tips of the ventral rami (Fig. 315) produce a vascular chain known as the *internal thoracic* and the *superior* and *inferior epigastrics* (Fig. 317 *A, C*). Also connected with the subclavian are the *thyro-cervical* and *costo-cervical* trunks. The former comes from longitudinal precostal anastomoses (*i.e.,* ventral to the ribs; Fig. 315) of those ventral rami that are located just craniad of the subclavian; their stems then drop out. The costo-cervical trunk arises in a similar fashion from precostal anastomoses that link the three ventral rami next caudal to the subclavian. In this instance the distal portions of the second and third of these vessels survive as intercostal arteries (Fig. 317 *A, B*).

2. LATERAL BRANCHES. These sprouts from the sides of the descending aorta are at best irregularly segmental (Fig 315). They supply structures arising from the nephrotome region (mesonephros, sex glands, metanephros and suprarenal glands) (Fig. 318 *A*). The original series of vessels was developed primarily in relation to the mesonephros. As this organ involutes,

the total number of lateral branches is reduced; from the vessels that escape involution emerge the *inferior phrenic, suprarenal, renal,* and *internal spermatic* or *ovarian arteries (B).*

3. VENTRAL BRANCHES. This series is imperfectly segmental. Primitively they constitute the paired *vitelline arteries* to the yolk sac and allantois (Fig. 309). As the dorsal aortæ combine, single ventral vessels appear— apparently by fusion (Fig. 315). The total number persisting is progressively reduced until, at 8 mm., they occur mainly at four levels. Three of these remaining vessels pass by way of the mesentery to the gut where they are converted into (Fig. 319 *A, C*): the *cœliac artery* (to the stomach, duodenum, pancreas, liver and spleen); the *superior mesenteric* (to the small and part of the large intestine); and the *inferior mesenteric* (to the descending colon, sigmoid colon and rectum). Minor twigs supply the esophagus and bronchi.

The fourth set of ventral branches is established in very young embryos as the arteries that accompany the allantois and then continue through the body stalk into the chorion. They remain paired and are known as *umbilical arteries* because they traverse the umbilical cord and supply the placenta (Fig. 309). By the end of the fourth week each artery acquires a secondary connection with a dorsal intersegmental branch of the aorta (Fig. 319 *B*), and the earlier ventral stem promptly disappears (*A*). The new replacing stem (from the aorta to the level of the *external iliac* which buds from this new trunk) becomes known after birth as the *common iliac (B).* The remainder of the original umbilical trunk (located distad, but annexed by the common iliac) will be the *internal iliac artery (B, C).*

When the placental circulation ceases at birth, the distal portions of both internal iliac arteries, from bladder to umbilicus, collapse. They reduce to the solid cords which persist as the *lateral umbilical ligaments* of adult anatomy (Fig. 354 *A*). The more proximal portions of the internal iliacs do not obliterate, and so continue to supply the walls and viscera of the pelvic region.

All of the ventral aortic vessels undergo caudal displacement from the levels where

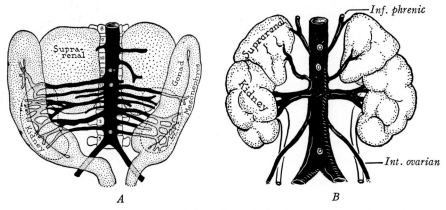

Fig. 318. Derivatives of the lateral branches of the human aorta, shown in ventral views. *A,* At seven weeks (after Felix; × 20); *B,* at birth (× ½).

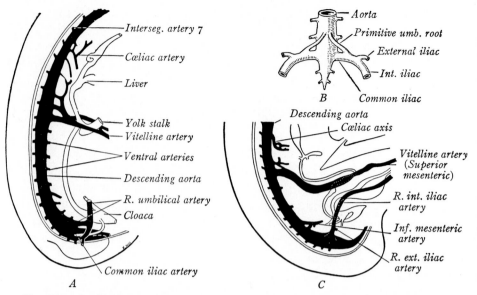

Fig. 319. Derivatives of the ventral branches of the human aorta. *A, C,* At 5 mm. (× 23) and 9 mm. (× 12), respectively, viewed from the right side (after Tandler). *B,* Terminal aorta at 5 mm., in ventral view, showing the replacement of the original umbilical roots by the common iliacs.

they first appear; in this descent the cœliac wanders 13 segments, the superior mesenteric 11 and the inferior mesenteric 3. To explain this migration two views have been advanced:[20] one emphasizes the attachment to new, caudal roots and the simultaneous atrophy of old, higher roots; the other attributes the cause to unequal growth on the part of the dorsal and ventral walls of the aorta.

THE GLOMUS COCCYGEUM. The *coccygeal body* is an arterio-venous anastomosis developed in connection with the middle sacral artery. Appearing in the third month, it becomes a channelled mass whose polyhedral cells have been interpreted either as highly modified, smooth-muscle elements[21] or as postembryonal angioblasts.[22]

Arteries of the Limbs. A limb bud arises as a simple swelling of the trunk, and extends over several body segments. As it enlarges it acquires a capillary plexus that is fed by several adjacent dorsal intersegmental arteries (Fig. 308 *B*). An axial vessel soon arises by enlargement and consolidation within the limb plexus, and this vascular axis then appears as if it always had been a direct extension from the single remaining aortic stem (Fig. 329).[23] The definitive vessels are largely additional and replacing vessels (Figs. 320, 321).

THE ARM. In human embryos of 5 mm. it is the *subclavian-axillary artery,* whose origin has already been described (p. 353), that becomes the sole arterial stem. In the future free arm it is continuous with the differentiating *brachial artery* of the upper arm and the *interosseous artery* of the forearm (Fig. 320 *A*). The *median artery* soon branches off the brachial and annexes the vessels of the hand (*A, B*). Following this, first the *ulnar* (*B*) and then the *radial* (*C*) also arise as brachial branches. They become the most prominent vessels of the forearm and take over the vessels of the hand (*D*). Before the end of the second month these rearrangements are complete, and the early primacy of the interosseous and median arteries is no longer apparent (*E*).[23]

THE LEG. A branch known as the axial, or *sciatic artery* is given off from the umbilical (future internal iliac) artery, and in embryos of 9 mm. it is the chief arterial stem of the lower extremity (Fig. 321 *A*). A little later the sciatic is being superseded by the *femoral* which is a continuation of the *external iliac* (*A, B*) whose origin from the new common iliac was described on p. 357. The femoral artery annexes the sciatic and its branches distal to the middle of the thigh (*C*). Eventually the sciatic persists proximally only as the *inferior gluteal artery;* its original, distal course is marked by the *popliteal* and *peroneal* vessels (*D*). The *anterior tibial artery* is a branch from the popliteal (*B, C*). The *posterior tibial* arises by union of the lower femoral with the popliteal (*C*). These two tibial vessels take over the arteries of the foot (*D*). All these alterations are completed in the third month.[24]

Anomalies. Anomalous blood vessels are of common occurrence. They may be due: (1) to the choice of unusual paths in the primitive vascular plexuses; (2) to the persistence of vessels normally obliterated; (3) to the disappearance of vessels normally retained; (4) to incomplete development; and (5) to fusions and absorptions of parts usually distinct. Outstanding anomalies specific to arteries include the following conditions:

Transposition of the ascending aorta and pulmonary artery, and the stenosis (*i.e.,* narrowing) of these vessels, will be discussed in the following chapter (p. 390). The *aortic arch* may turn to the right, as in birds; in this instance it is usually a part of the larger picture of transposition which includes also reversed relations between the ascending aorta and pulmonary artery, and a right-sided ductus arteriosus (Fig. 232 *B*). The aortic arch may be double, as is normal for reptiles, and then enclose both esophagus and trachea in a vascular collar (Fig. 322 *A*). Variations in the origins, positions and relations of the *carotids, subclavians* and *vertebrals* are common (*B*); some combinations,

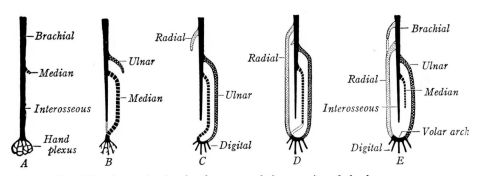

Fig. 320. Stages in the development of the arteries of the human arm.

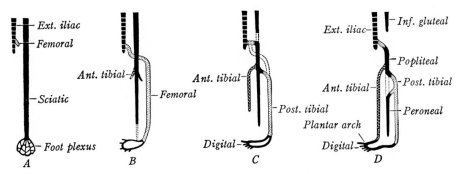

Fig. 321. Stages in the development of the arteries of the human leg.

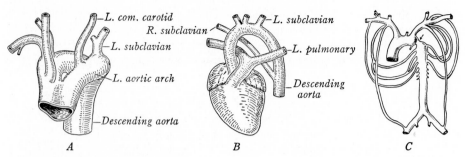

L. com. carotid
R. subclavian
L. subclavian
L. aortic arch
Descending aorta

L. subclavian
L. pulmonary
Descending aorta

A   B   C

Fig. 322.   Arterial anomalies, in ventral view. *A,* Double aortic arch. *B,* Anomalous right subclavian artery. *C,* Coarctation of the aorta (at arrow), compensated by the institution of a by-pass system of vessels.

atypical for man, occur regularly in lower mammals. Almost all of these irregularities result from the variable selection of definitive vessels out of the primitive pattern of the aortic arches and dorsal aortic branches. Persistence of an unclosed *ductus arteriosus* in an otherwise normal vascular system brings on a shunt from the aorta to the pulmonary artery where pressure is lower and results in both ventricles having to work harder than normal. When a patent ductus accompanies a stenosis of the aorta or pulmonary artery, the otherwise limited flow in either vessel is increased and the cyanosis is alleviated. A patent ductus also confers some benefit when it occurs in conjunction with coarctation of the aorta, as is explained in the next paragraph.

The completeness of fusion of the *dorsal aortæ* varies, and even complete doubleness has been recorded. *Coarctation of the aorta* is a constriction occurring most commonly at the level of the ductus arteriosus. When the constriction is above that level, it resembles the fetal condition (Fig. 314 *B*). The restricted flow is sometimes aided by a patent ductus arteriosus. Constriction, more or less opposite the normally fibrosed ductus, is commonly ascribed to traction by that ligamentous cord. Compensation for restricted blood flow through the aorta is commonly established by extensive collateral anastomoses which by-pass the constriction (Fig. 322 *C*). Variations in the branches of the descending aorta are frequent. Accessory renal arteries are referable to a retention of additional vessels belonging to the early series of lateral (mesonephric) branches *(cf.* Fig. 318 *A*).

DEVELOPMENT OF THE VEINS

Three systems of paired veins are present in embryos with 20 somites and about 3 mm. long (Fig. 310): (1) the *umbilical veins* from the chorion; (2) the *vitelline veins* from the yolk sac; and (3) the *cardinal veins* from the body of the embryo itself. The latter are really in two sets: the precardinals which drain blood from the head region, and the postcardinals which return blood from levels caudal to the heart; both pairs unite at the heart into short *common cardinal veins* (ducts of Cuvier). At this stage, therefore, it is characteristic for three venous stems to open into the right horn of the sinus venosus of the heart and three into the left (Fig. 308 *B*). Somewhat later two other pairs of veins, the *subcardinals* and the *supracardinals,* successively replace and supplement the postcardinals.

The subsequent history of venous development is a recital of the changes that these primitive, symmetrical vessels undergo. Such alterations are more extensive than those occurring among arteries. The factors responsible for the final, asymmetrical venous plan are: (1) shifts of position and direction of flow; (2) anastomoses; (3) local transformations and size readjustments; (4) loss by atrophy; and (5) new formations, including secondary replacements.

Transformation of the Vitelline Veins. The developing liver exerts a profound influence in modifying the primitive vitelline and umbilical vessels. The paired *vitelline veins* follow the yolk stalk into the body. They then turn craniad, continue alongside the short fore-gut to the septum transversum, and enter the sinus venosus (Fig. 310). Also into the septum transversum grows the liver bud, already proliferating into cords. It will be remembered that there is a mutual intergrowth between hepatic cords and vitelline endothelium (Fig. 216). As a result, the vitelline vessels at the level of the liver resolve during the fourth week into networks of sinusoids which are incorporated into the expanding right and left hepatic lobes (Fig. 323).

Each vein, thus interrupted by a sinusoidal labyrinth, is effectively divided into a distal segment, which follows the gut from yolk sac to liver, and a short, proximal segment, which returns blood to the corresponding horn of the sinus venosus (Fig. 323). In anticipation of the descriptions to

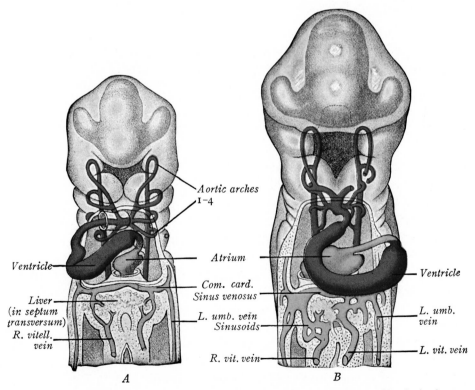

Fig. 323. Veins in the vicinity of the liver, in ventral view (His). × 30. *A,* At 3 mm.; *B,* at 4 mm.

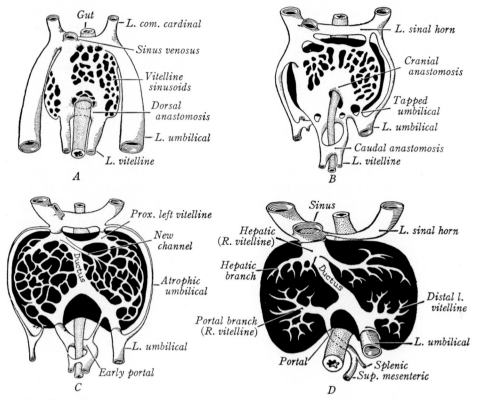

Fig. 324. Transformation of the human vitelline and umbilical veins in the region of the liver, seen in ventral view (adapted). *A*, At 4.5 mm.; *B*, at 5 mm.; *C*, at 6 mm.; *D*, at 9 mm. The liver substance is shown in black.

follow it can be stated here that the distal portions are converted into the *portal vein,* the intermediate sinusoids mostly remain as such but in part expand into the *ductus venosus,* while the right proximal stem represents the *hepatic vein* (Fig. 324 *C, D*).

The early symmetrical relations, still essentially present in Fig. 324 *A,* are promptly disturbed by further changes that lead quickly to the final arrangement of vessels. Even at the 5 mm. stage the distal segments of the paired vitelline veins communicate by three cross anastomoses, of which the middle one lies dorsal to the duodenum (*A, B*). In this manner, two venous rings are created, through which the gut weaves. When the stomach and duodenum rotate to their final positions, the blood flow favors a more direct course through this system of vessels. Disuse and atrophy quickly lead to the disappearance of all components except the right limb of the cranial ring and the middle anastomosis common to both rings (*C, D*). The composite, **S**-shaped vessel surviving these changes is the *portal vein.* Its caudal limit is established by the union of the splenic vein with the superior mesenteric vein (*D*), just caudal to the transverse anastomosis. Actually the *superior mesenteric* is not the distal portion of the left vitelline, as might be supposed, but a new, replacing vessel arising in the dorsal mesentery of the intestinal loop; it supersedes the vitelline veins in this region, as the latter disappear in company with the decline of the yolk sac. Within the liver some distal remnants of the right and left vitelline veins persist as portal outlets (*D*). On the right there is a branch, directly continuous with the portal vein. A corresponding set of vitelline vessels on the left is related to the umbilical vein, but after that channel becomes functionless at birth, the tributaries are taken over by the portal.

The proximal segments of the early vitelline veins drain blood from the sinusoids into the respective horns of the sinus venosus (Fig. 324 *A, B*). When the right horn soon is favored as the drainage outlet, the left vitelline and left horn cannot compete and both decline and disappear (*C, D*). The blood from the sinusoids on the left side is then rerouted across to the right sinal horn, and a new drainage channel arises to take care of this territory (*C, D*).[25] The surviving stems of the right vitelline vein are the *hepatic veins*, which subsequently become tributaries of the later-formed inferior vena cava.

Transformation of the Umbilical Veins. Accompanying the vitelline alterations go important, related changes in the *umbilical veins*. These early vessels return blood from the placenta by way of the umbilical cord and body wall; they drain into the sinus venosus (Figs. 310, 323). As the primitive right and left lobes of the liver expand laterally, they soon come in contact with the umbilical veins coursing nearby (Fig. 324 *A*). Both of these vessels are then tapped and their blood, so diverted, finds a more direct route to the heart by way of the hepatic sinusoids (*B*). When all of the umbilical blood enters the liver, as happens in embryos of 6 mm., the entire right umbilical and the proximal segment of the left atrophy (*C*) and soon disappear (*D*). At 7 mm. the distal remainder of the left umbilical is already large; it continues to maintain itself throughout fetal life, shifting to the midplane and occupying the free edge of the falciform ligament (Fig. 325).

Since the early channel of the right vitelline vein within the liver is larger than the left, the blood from the tapped left umbilical vein first takes that route to reach the right horn of the sinus venosus (Fig. 324 *A, B*). But with the progressive growth of the right lobe of the liver, this pathway becomes increasingly circuitous. A more direct course is, therefore, created through the enlargement of a diagonal passage out of the hepatic sinusoids (*C*).

Fig. 325. Relations of veins to the human liver, at the time of birth. × ½.

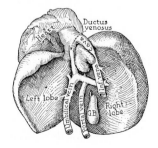

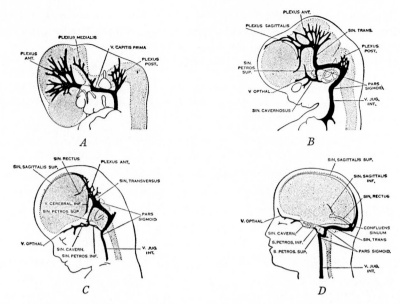

Fig. 326. Transformation of the primary head vein into dural sinuses (Streeter). *A,* At six weeks; *B,* at eight weeks; *C,* at eleven weeks; *D,* at birth.

This is the *ductus venosus* (*D*). It continues in direct line with the left umbilical vein and empties into the common hepatic vein (which later is incorporated into the inferior vena cava). Consequently, the purer blood from the placenta avoids the general system of hepatic sinusoids to some extent; yet the umbilical vein gives branches to the liver, receives branches within the liver and makes an anastomotic connection with the portal vein (Fig. 324 *D*). A ductus sphincter regulates blood admitted to the liver.

Transformation of the Precardinal Veins. Each precardinal vein (also called anterior cardinal) consists of two parts (Fig. 310): (1) the *primary head vein,* which courses ventrolateral to the brain wall throughout all but the caudal end of the head; and (2) the true *precardinal,* located laterally in the segmented portion of the head and in the neck, and emptying into the common cardinal vein.

The primary head veins drain three pairs of tributary plexuses that extend dorsad over the brain (Fig. 326 *A*). This system of vessels becomes separated into two parts by the development of the membranes about the brain. The deeper layer of vessels converts into the *cerebral veins;* the more superficial layer aids the primary head vein itself in producing the various *dural sinuses* (*B-D*). The steps in the development of these sinuses are rather complicated and the enormous growth of the two cerebral hemispheres is chiefly responsible for their definitive positions and relations.[26]

Coincidental with the growth of the internal ear, the segment of the old head vein just ventral to it disappears and a new channel, connecting the middle and posterior plexuses, develops dorsal to the ear (Fig. 326 *A, B*). The rostral portion of the head vein is spared; it receives the ophthalmic vein and constitutes the *cavernous sinus* (*B-D*). The original stem of the middle plexus, retained as the *superior petrosal sinus,* interconnects the cavernous sinus with the new dorsal channel (*B*). The *transverse sinus* is a main line of drainage. It arises from portions of the middle and posterior plexuses that are linked by the new dorsal vessel; part of its extent is the *sigmoid sinus* (*B, C*). The *inferior petrosal sinus* results from the re-establishment of a channel along the course of the degenerated segment of the head vein (*C*). The *superior sagittal sinus* develops in the midplane from portions of the anterior plexus (*B, C*). The *inferior sagittal* and *straight sinuses* arise from a part of the plexus that extends downward between the cerebral hemispheres (*C, D*).

The true precardinals begin near the base of the head and run caudad into the heart (Fig. 326 *A, B*). They communicate during the eighth week by an oblique cross-channel which shunts blood from the left vein across to the right one (Fig. 327 *A, B*). As a result of this diversion, the stem portion of the left precardinal, just caudad, soon loses its communication with the common cardinal on the same side and survives merely as part of the highest intercostal vein (*C*). The left common cardinal comprises most of the inconstant *oblique vein* of the left atrium (*cf.* p. 382). The right common cardinal and the right precardinal, as far up as the intercardinal anastomosis, become the *superior vena cava*. The anastomosis itself forms the *left brachio-cephalic vein,* while the portion of the right precardinal next craniad (between the anastomosis and the right subclavian vein) is known as the *right brachio-cephalic.* Still more craniad, the precardinals continue as *internal jugular veins.* The *external jugular* and *subclavian veins* are both extraneous vessels that develop independently and attach secondarily (p. 369).

Transformation of the Post-, Sub- and Supracardinal Veins. Caudal to the heart there appear in succession three pairs of veins, whose histories both overlap and interweave. These vessels care for the venous drainage from the lower limbs, body wall and viscera (Fig. 328). The first set to be seen (colored blue) is the *postcardinals,* which are developed primarily as the vessels of the mesonephroi (*A*); they run dorsal to the mesonephroi and also receive tributaries from the lower limbs and body wall. Next to appear are the *subcardinal veins* (colored red), which lie ventromedial to the meso-

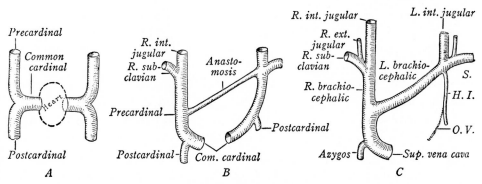

Fig. 327. Transformation of the cardinal veins of human embryos, shown by diagrams in ventral view. *A,* At six weeks; *B,* at eight weeks; *C,* adult.
H.I., Highest intercostal; *O.V.,* oblique vein of the left atrium; *S.,* left subclavian.

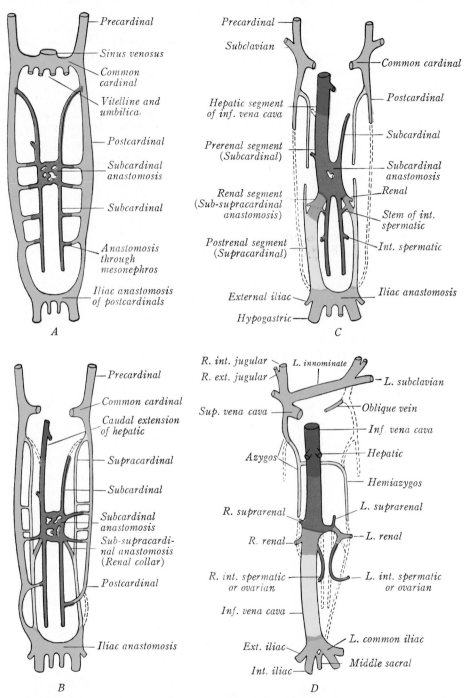

Fig. 328. Transformation of the primitive veins of the trunk of the human embryo, shown by diagrams in ventral view (adapted after McClure and Butler). *A*, At six weeks; *B*, at seven weeks; *C*, at eight weeks; *D*, adult.

nephroi; they connect not only with each other but also, through the meso-nephric sinusoids, with the postcardinals (*A*). Finally the *supracardinals* (colored yellow) make their appearance (*B*); they course dorsomedial to the postcardinals and in a sense replace them. The transformations, leading to the final venous plan, are largely accomplished in the sixth, seventh and eighth weeks.[27] These changes are complex and subject to some differences of interpretation, especially as regards the rôle of the supracardinals.[28, 29]

The postcardinals appear with the mesonephroi and disappear as that organ wanes. An early anastomosis where the caudal and leg veins are received unites the two postcardinal trunks (Fig. 328 *A*); this connection is destined to persist as the longer and more oblique left common iliac vein. The only permanent representatives of the postcardinal system are (*D*): (1) the root of the *azygos,* where it joins the superior vena cava on the right; and (2) the *common iliacs* (which are later annexed by the inferior vena cava; the veins of the legs are tributary to them).

The subcardinals follow closely on the appearance of the postcardinals. Their original cranial connections, permitting drainage into the respective postcardinals, are soon lost (Fig. 328 *A, B*). Anastomoses with the latter veins also disappear, but a prominent set of connections between the two subcardinals, midway in their course, is destined to serve as the stem unit-ing the *left renal vein* to the vena cava (*A-D*). The only further subcardinal contributions to the permanent system of veins are the paired *suprarenal* and *sex veins,* and a segment of the *inferior vena cava* formed from the right subcardinal (*D*).

The supracardinals become broken in the region of the kidneys (Fig. 328 *C*). Above this level they unite by a cross anastomosis and become the *azygos* and *hemiazygos veins* (*D*). Below the kidneys the right supracardinal alone is taken over as the caudalmost section of the *inferior vena cava* (*D*), whereas the left vein drops out without trace.

THE INFERIOR VENA CAVA. The unpaired inferior vena cava is complex and requires some additional description. Its history involves the several changes through which blood, returning from the lower body, is progres-sively shifted from the left side to the right. This is accomplished by anas-tomoses between some vessels, by the enlargement and consolidation of others, and by the regression and replacement of still others. The outcome is a new, compound path to the heart which rapidly straightens and consoli-dates into what looks thereafter like a simple vein of unitary origin.

The inferior vena cava is composed, in cranio-caudal order, of the follow-ing four portions: (1) An *hepatic segment* is derived from the hepatic vein (proximal right vitelline) and hepatic sinusoids. It connects with the right subcardinal through a vein in the caval mesentery (Fig. 328 *B*; colored purple). The latter structure is a ridge which extends caudad from the attachment of the right lobe of the liver to the dorsal body wall (Fig. 237). Capillaries invade this mesentery from the liver and meet and fuse with similar capillaries growing craniad from the right subcardinal. (2) A *pre-*

renal segment begins at the junction of the hepatic segment with the right subcardinal and continues down this vessel to the level of the kidneys (Fig. 328 *B*). (3) A *renal segment* represents an important anastomosis ('renal collar') that unites the right sub- and supracardinal veins (*B, C*; colored green). (4) A *postrenal segment* is the lumbar portion of the right supra- cardinal down to the level of the iliacs (*C, D*). These latter vessels are annexed when the degeneration of the postcardinals would otherwise leave them without central connections.

CAVAL TRIBUTARIES. Accompanying these complex changes there emerge tributary vessels to the vena cava, some of whose origins and rela- tions are correspondingly intricate. As the permanent kidneys assume their final positions, a *renal vein* appears on each side and drains into the anas- tomosis developed between the sub- and supracardinals (Fig. 328 *C*). Since on the right this anastomosis is incorporated into the vena cava as its renal segment, the corresponding renal vein empties directly into the vena cava (*D*). On the left the situation is more complicated because the primitive renal vein opens into the sub-supracardinal anastomosis which, in turn, must find its way to the vena cava through the great anastomosis between the subcardinals. For this reason the adult left renal vein is longer and more complex than its mate (*D*).

The two suprarenal veins, likewise, are not wholly homologous vessels. The *right suprarenal* is a simple tributary of the subcardinal at a level where this vessel becomes a part of the inferior vena cava (Fig. 328 *C, D*). On the other hand, the *left suprarenal* is a prerenal portion of the subcardi- nal itself; it corresponds to the right subcardinal contribution to the infe- rior vena cava (*C, D*). The prospective left suprarenal first communicates with the right subcardinal through the great anastomosis, but after this anastomosis becomes the stem of the left renal, it serves as a tributary to that vessel.

The *spermatic* or *ovarian veins* are the remnants of the paired subcardi- nals below the level of the kidneys (Fig. 328 *C*). The right opens into that portion of the right subcardinal that is incorporated into the inferior vena cava. The left early drains into the left caudal border of the great sub- cardinal anastomosis which, as already described, becomes the stem of the left renal vein (*A-C*). Soon secondary roots arise from new anastomoses (*C*) and both sex veins shift their origins onto the sub-supracardinal anasto- moses (*D*). The final attachments are still to the inferior vena cava and left renal vein, respectively, but to segments of different origin than originally was the case.

The *posterior intercostal* and *lumbar veins* are at first tributaries of the postcardinals. As the latter vessels degenerate, these veins connect second- arily with the replacing supracardinal veins. Later they of necessity drain into the azygos veins and inferior vena cava, respectively.

The Pulmonary Veins.   Each primitive vessel buds from the left atrium and joins a pulmonary plexus (Fig. 312 *B*).[30] Its trunk is absorbed progressively into the left atrium until the stems of four main branches become separate vessels (Fig. 346).

Veins of the Limbs. The primitive capillary plexus of the flattened limb buds gives rise to a peripheral *border vein* which serves as an early drainage channel to blood brought in by the axial arterial vessels (Fig. 329). Along the cranial margin of the limb bud this vein is smaller and mostly disappears, but on the caudal margin it transforms into permanent vessels. The border vein appears in the arm and leg in the sixth and eighth weeks, respectively; the general venous plan becomes outlined within the next two weeks.

THE ARM. The radial extension of the border vein atrophies but the ulnar portion persists, forming at different levels the *subclavian, axillary* and *basilic veins.* The border vein originally opens into the postcardinal but, as the heart shifts caudad, the subclavian ultimately transfers into the precardinal (internal jugular) vein. The *cephalic vein* develops secondarily in connection with the radial border vein; later it anastomoses with the external jugular, but finally opens into the axillary vein, as in the adult.

THE LEG. Homologous with conditions in the arm, the tibial continuation of the primitive border vein disappears, while the fibular part persists to a large degree. The *great saphenous vein* arises separately from the postcardinal, gives off the *femoral* and *posterior tibial veins,* and then annexes the fibular border vein at the level of the knee. Distal to this junction, the border vein develops into the *anterior tibial* and, probably, the *small saphenous;* proximally, it becomes greatly reduced, forming the *inferior gluteal.*

The Fetal Circulation. The course of blood in the fetus differs from that after birth because of the presence of a placenta, an incompletely partitioned heart and nonfunctional lungs. An explanation of this provisional circulation and the changes that occur at birth will be postponed until after the development of the heart has been described (p. 392).

Anomalies. Among the anomalous veins produced through the operation of the general factors already cited (p. 359) are found such conspicuous specimens as the fol-

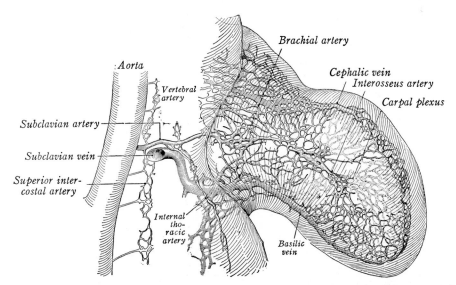

Fig. 329. Primitive blood vessels in the left fore-limb of a 12 mm. pig embryo, seen in ventral view (Woollard). × 30.

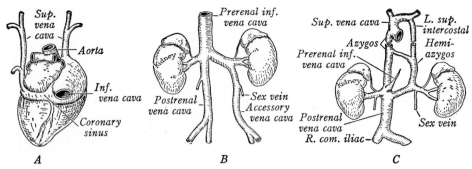

Fig. 330. Anomalous veins of man. *A,* Double superior vena cava. *B,* Accessory postrenal vena cava on left side. *C,* Rudimentary prerenal vena cava and compensatory development of the azygos system of vessels.

lowing (Fig. 330): paired superior (*A*) or inferior venæ cavæ (*B*); unpaired left superior vena cava opening into the left common cardinal; unpaired left inferior vena cava opening into the coronary sinus; azygos vessels serving as a main venous pathway (*C*); accessory hemiazygos vein opening into the coronary sinus (*i.e.,* by way of the left common cardinal); retention of the original, single trunk of the pulmonary vein (Fig. 346 *A*), or the ending of one of the pulmonary stems in the superior vena cava, left innominate or azygos vein.

THE LYMPHATIC SYSTEM

The lymphatics develop quite independently of blood vessels and any temporary venous connections which they may show are acquired secondarily. They originate as discrete spaces in the mesenchyme; the mesenchymal cells bordering each space flatten into an endothelial lining.[31] By progressive fusion such locally formed clefts link into continuous channels which also grow, branch and extend the system further (Fig. 331).[5] Through the combination of both processes the lymphatics attain their final form as a one-way system of closed vessels for the return of tissue fluid.

Fig. 331. Stages in the growth of a lymphatic capillary, observed in a living rabbit (after Clark). × 90.

Vessels. The first plexus of lymphatic capillaries is distributed along the primitive, main venous trunks. The dilatation and coalescence of these networks at definite regions give rise to six lymph sacs (Fig. 332 *A*): (1, 2) Paired *jugular sacs* appear at six weeks, lateral to the internal jugular veins. (3) In embryos of two months the unpaired *retroperitoneal sac* is developing at the root of the mesentery, adjacent to the suprarenal glands; (4) at this stage the *cisterna chyli* also differentiates nearby. (5, 6) Likewise at the end of the second month paired *posterior* (or sciatic) *sacs* arise in relation to the sciatic veins.

Downgrowths from the jugular sacs unite with each other and meet upgrowths from the cisterna to produce the longitudinally coursing *thoracic duct* which links the several sacs into a common system (Fig. 332 *B*). With relation to the lymph sacs as centers the peripheral lymphatics develop rapidly. Thus lymphatic vessels grow to the head, neck and arm from the jugular sacs (including its subsidiary subclavian sacs); to the hip, back and leg from the posterior sacs; and to the mesentery from the retroperitoneal sac. The jugular sacs are the only ones to acquire permanent connections with the venous system. They drain into the internal jugular veins by valve-guarded openings; the left one of these is utilized later by the thoracic duct (Fig. 332 *B*). The various sacs themselves eventually break up into networks and are largely replaced by chains of lymph nodes. Lymphatic vessels differentiate like thin-walled veins, and at intervals develop valves. These appear even earlier than those formed in veins.

Anomalies. Even the largest lymphatic vessels are highly variable. The commonest pattern of origin of the thoracic duct from a plexus is shown in Fig. 332 *B*. But the temporary double channel may be retained, as is normal for lower vertebrates; or a single duct may take any course and ending made possible by selection from the initial

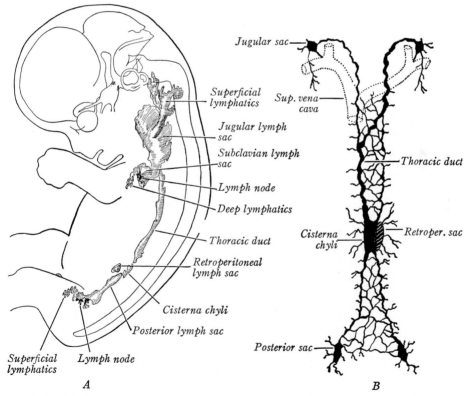

Fig. 332. Development of the primary lymphatic vessels in man. *A*, Profile reconstruction of the primitive lymphatic system in an embryo of nine weeks (after Sabin; × 3). *B*, Diagram, in ventral view, of the definitive thoracic duct emerging from a lymphatic plexus.

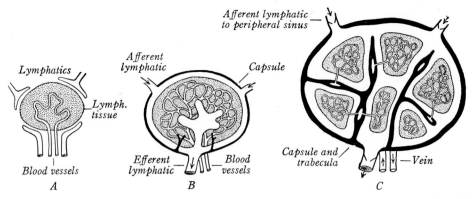

Fig. 333. Diagrams of a developing lymph node (adapted after Braus). Lymph sinuses are shown as broad, open channels and the smaller blood vessels as fine plexuses connecting with trabecular vessels.

plexiform arrangement shown. The cisterna chyli varies in size and position; frequently it is lacking.

Lymph Nodes. The earliest, or primary lymph nodes appear during the third month as the lymph sacs begin to break down into plexuses of lymphatic vessels (Fig. 332 *A*). Secondary lymph nodes develop later along the course of the peripheral lymphatics which spread from these centers.[32] The first stage of development is marked by a lymphatic plexus which lies in association with strands of mesenchymal tissue. Continued proliferation and differentiation enlarge these strands into nodular lymphoid masses, and the vessels are crowded to the periphery where the *peripheral sinus* then makes its appearance (Fig. 333 *A, B*). About the whole a connective-tissue *capsule* condenses, while the *trabeculæ* and *central sinuses* spread inward from the *hilus* (*B, C*).[33] Blood vessels, which even at an early period supply the lymphoid masses, also enter and leave at the hilus. *Medullary cords* differentiate from the common lymphoid primordium, but a true *cortex* appears much later since definite *cortical nodules*, with germinal centers, are completed mostly after birth.

Hemal (Hemolymph) Nodes. The development of a hemal node is much like that of lymph nodes, but its origin is traced to a condensation of mesenchyme that develops in relation to blood vessels, not lymphatics.[34] The peripheral sinus arises independently; its vascular connections are secondary. Hemal nodes occur in some mammals, but their presence in man is doubtful.

The Spleen. Embryos of 8 mm. exhibit a swelling on the left side of the dorsal mesogastrium (Fig. 334 *A*). This bulge is caused by an accumulation of mesenchymal cells located just beneath the surface epithelium. This covering layer of peritoneal mesothelium possibly adds to the mass by proliferation, since it is merely a more superficial layer of the splanchnic mesoderm.[35] As the mass increases in size, it projects above the omental surface, usually as several hillocks which slowly merge (*A, B*). The region

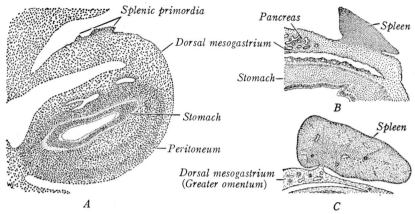

Fig. 334. Development of the spleen, shown by transverse sections through human embryos. *A,* At six weeks ($\times$ 65); *B,* at two months ($\times$ 40); *C,* at four months ($\times$ 18).

of union of the spleen with the dorsal mesogastrium, or greater omentum (*B*), fails to keep pace with the general enlargement and is reduced to a narrow band (*C*). This portion of the mesentery, attaching at the *hilus* of the spleen, is the *gastro-splenic ligament* (Fig. 240 *A*). At three months the spleen acquires its characteristic form.

The mass of splenic mesenchyme is well vascularized, and from it differentiate the *capsule, trabeculæ* and *pulp cords.* The specialization into red and white pulp seems to be dependent upon the development and distribution of the vascular channels, but there is no agreement as to which type is primary. Lymphoid tissue appears early, but it is not until six months that the *splenic corpuscles* form ovoid nodules about arteries. From the fifth through the eighth month of fetal life the formation of white blood cells is supplemented by red corpuscles which develop actively within the splenic mass. From the start the blood vessels are said to terminate by opening freely into the splenic tissue. The peculiar *splenic sinuses* originate as separate cavities by the middle of fetal life.[36] They come to drain into pulp veins, but whether or not arterial terminations ever open into the sinuses is still in dispute.

Causal Relations. The proper development of the spleen and lymph nodes depends on the presence of a fetal thymus. Emigrating thymic cells, or a thymic secretion, stimulate development and the acquisition of immunological competence.

Anomalies. The spleen is sometimes partially subdivided (Fig. 335), or even multiple. Smaller, accessory spleens are also common (Fig. 335). These types result either

Fig. 335. Spleen of a newborn, showing partial subdivision and an accessory spleen. $\times$ 1.

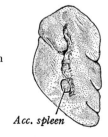

Acc. spleen

from the continuance of the early, multiple hillocks or from an exaggeration of temporary incisures that appear in the third and fourth months.

The Tonsils and Thymus.   For their development see pp. 238–240.

REFERENCES CITED

1. Hertig, A. T. 1935. Carnegie Contr. Embr., *25*, 37–82.
2. McClure, C. F. W. 1921. Anat. Rec., *22*, 219–237.
3. Reagan, F. R. 1915. Anat. Rec., *9*, 329–341.
4. Jolly, J. 1940. Arch. d'Anat. Micr., *35*, 295–361.
5. Clark, E. R. & E. L. Clark. 1909; 1932. Anat. Rec., *3*, 183–198; Am. J. Anat., *51*, 49–87.
6. Bloom, W. & G. W. Bartelmez. 1940. Am. J. Anat., *67*, 21–53.
7. Gilmour, J. R. 1941. J. Path. & Bact., *52*, 25–45.
8. Bloom, W. 1938. Sect. 13 in Downey: Handbook of Hematology (Hoeber).
9. Thompson, E. L. 1951. Anat. Rec., *111*, 317–325.
10. Bessis, M. & M. Bricka. 1952. Rev. d'hémat., 7, 407–435.
11. Yamada, E. 1957. Acta Anat., *29*, 267–290.
12. Arey, L. B. 1963. Chapter 1 in Orbison: The Peripheral Blood Vessels (Williams & Wilkins).
13. Hughes, A. F. W. 1943. J. Anat., *77*, 266–287.
14. Congdon, E. D. 1922. Carnegie Contr. Embr., *14*, 47–110.
15. Golub, D. M. 1929. Z'ts. Anat. u. Entw., *90*, 690–693.
16. Padget, D. H. 1948. Carnegie Contr. Embr., *32*, 205–261.
17. Barry, A. 1951. Anat. Rec., *111*, 221–238.
18. Padget, D. H. 1954. Anat. Rec., *119*, 349–356.
19. Schmeidel, G. 1932. Morph. Jahrb., *71*, 315–435.
20. Pernkopf, E. 1922. Z'ts. Anat. u. Entw., *64*, 96–275.
21. Schumaker, S. 1907. Arch. mikr. Anat. u. Entw., *71*, 58–115.
22. Krompecher, S. 1940. Z'ts. Anat. u. Entw., *110*, 423–442.
23. Woollard, H. H. 1922. Carnegie Contr. Embr., *22*, 139–154.
24. Senior, H. D. 1919; 1920. Am. J. Anat., *25*, 55–95; Anat. Rec., *17*, 271–279.
25. Schneider, H. 1937. Z'ts. Anat. u. Entw., *107*, 326–352.
26. Padget, D. H. 1957. Carnegie Contr. Embr., *36*, 79–140.
27. McClure, C. F. W. & E. G. Butler. 1925. Am. J. Anat., *35*, 331–383.
28. Reagan, F. P. 1927. Anat. Rec., *35*, 129–148.
29. Grünwald, P. 1938. Z'ts. mikr.-anat. Forsch., *43*, 275–331.
30. Neill, C. A. 1956. Pediat., *18*, 880–887.
31. Kampmeier, O. F. 1960. Am. J. Anat., *107*, 153–176.
32. Hellman, T. 1930. Bd. 6, Teil 1 in Möllendorff: Handbuch (Springer).
33. Heudorfer, K. 1921. Z'ts. Anat. u. Entw., *61*, 365–401.
34. Meyer, A. W. 1917. Am. J. Anat., *21*, 375–495.
35. Holyoke, E. A. 1936. Anat. Rec., *65*, 333–349.
36. Thiele, G. A. & H. Downey. 1921. Am. J. Anat., *28*, 279–339.

Chapter XX.   The Heart and Circulation Changes

As a mammalian embryo advances through the stages characterized by cleavage, morula, blastocyst and germ layers, it satisfies all its metabolic needs by simple, diffusive interchanges with the fluid medium in which it is immersed. But as the embryo continues to gain size and begins to take form, a functioning circulatory system becomes necessary in order to make use of the required food and oxygen obtainable from the mother's blood. Hence it is that the heart and blood vessels are the first organ system to reach a functional state. This is when the embryo possesses about 12 somites.[1] The heart of any animal is a specialized blood vessel with a large lumen and extremely thick, muscular walls.

In lower fishes and in amphibians the heart seemingly develops in a simple, direct manner. A tubular cavity appears within the mesodermal cells that lie just beneath the fore-gut; about this cavity the mesoderm straightway differentiates into the lining and wall of the heart. Nevertheless, the heart material is originally furnished by the merged border-regions of the paired plates of lateral mesoderm.

In bony fishes, reptiles and birds, the early stages of cardiac development are more complicated. This is the result of a flattened blastoderm, due to excessive yolk, and the consequent necessity for the heart to develop as two obviously lateral halves. At first well separated, these halves secondarily swing together and fuse in the midplane (Figs. 537–539). Most mammals similarly merge two cardiac tubes that are originally separate and bilateral.

THE PRIMITIVE TUBULAR HEART

The Cardiac Primordium. The human heart does not arise wholly by the simple fusion, side-by-side, of bilateral halves.[2] A *cardiogenic plate* lies beyond the head region—in the splanchnic mesoderm, beneath the single pericardial chamber located there (Fig. 336 A). With the forward growth of the head (chiefly neural plate) there is a reversal of this portion of the blastoderm, and the region thus turned under becomes the floor of the fore-gut (B). In this process the heart primordium is necessarily reversed end-for-end, with respect to its original orientation. Also it then lies above, instead of below, the pericardial cœlom, and in the splanchnic mesoderm that is situated beneath the fore-gut. The now caudal end of the heart is continuous with the mass of mesoderm, just craniad of the anterior intes-

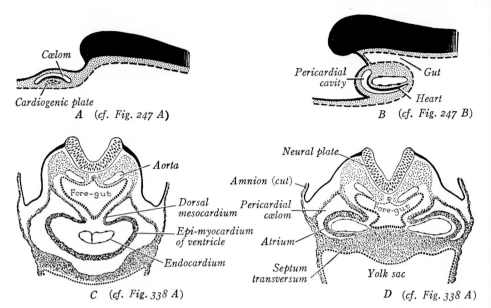

Fig. 336. The origin and relations of the early human heart. *A, B,* Diagrammatic sagittal sections, at a presomite stage and at seven somites, respectively, showing the reversal of the heart and pericardial cœlom. *C, D,* Transverse sections through the ventricle and paired atria, at seven somites ($\times$ 75).

tinal portal, that forms the septum transversum. Here it receives the several veins that enter the septum to drain blood into the heart (Fig. 310).

The Cardiac Tube. The earliest identifiable cardiac primordia are aggregates of splanchnic mesodermal cells that appear in the cardiogenic plate, beneath the cœlom. They arrange themselves side-by-side as two longitudinal strands, each of which gains a cavity just as primordial vessels do elsewhere in the embryo. These thin-walled, endothelial tubes lie within corresponding longitudinal folds of the splanchnic mesoderm (*cf.* Fig. 336 *D*). In the cranial half of the future heart the two endothelial vessels quickly fuse into a single tube and their individual mesodermal folds merge into a single, trough-like fold which encloses them (Figs. 336 *C*, 338). Traced caudad from the short, common pericardial cavity, where these events have occurred, separate tubes and folds are still seen. This is because they necessarily follow the course of the two lateral cœlomic canals which continue the common chamber of the pericardial cavity caudad (Fig. 336 *D*). As the anterior intestinal portal retreats in a caudal direction, thereby elongating the fore-gut, opportunity is offered for these paired cardiac primordia to join the median, unpaired portion, already formed. This they do in pace with the enlarging pericardial cavity which progressively incorporates the lateral cœlomic canals (Fig. 247). Thus the paired cardiac primordia merge progressively until the entire heart is a single organ (Figs. 338–340).

In embryos with but few somites the heart is a simple tube within a tube (Fig. 336 *C*). The internal, endothelial component is destined to become the actual lining layer of the *endocardium*; the external, thick covering gives rise to the *myocardium* and *epicardium*. Such a heart is suspended by a mesenterial attachment where the lateral margins of the mesodermal folds are reflected upon the ventrolateral sides of the fore-gut. This mesentery, named the *dorsal mesocardium,* is only temporary (Fig. 337 *B*); it is lost before the heart has advanced greatly. A peculiarity of the mammalian heart, in contrast to other vertebrates, is that there is no ventral mesocardium (Figs. 336 *C*, 337 *B*).[2, 3] This is because the cœlom arises very early from the coalescence of separate spaces and forms a complete cavity in the region of the heart before the head fold and heart, as such, begin to differentiate (Fig. 336 *A, B*).

The Cardiac Regions. Even while the bilateral cardiac halves are merging, they each bear alternate dilatations and constrictions which indicate the future atrium, ventricle and bulbus (Fig. 337 *A*). The union of the bulbar and ventricular halves is complete in embryos with some nine somites, but the atria are still paired sacs. Such a heart shows at first three divisions (Fig. 339 *A*): (1) the paired *atria,* which receive blood from the primitive veins; (2) the *ventricle,* or chief pumping region; and (3) the *bulbus cordis,* which in lower vertebrates helps maintain a continuous blood supply to the gills. Within the next day or two a fourth division, the *sinus venosus,* becomes recognizable as a region distinct from the caudal end of the atrium (Figs. 339 *B,* 340). At this period it lies within the septum transversum and is a center of confluence for the veins draining the body (cardinals), placenta (umbilicals) and yolk sac (vitelline). As in fishes, it serves as a reservoir during atrial contraction. At the extreme cranial end of the heart it is convenient to distinguish a fifth division, the *truncus arteriosus,* where the bulbus continues as a canal that conducts blood into the aortic sac (Figs. 339 *A,* 341 *A*). Internally, a pair of *sinus valves* (right and left) guards the entrance into the atrium and prevents backflow during

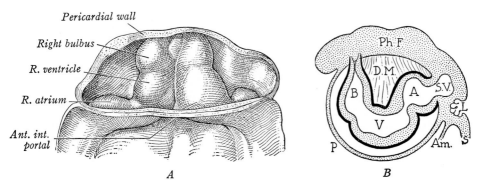

Fig. 337. Human hearts in the simple, tubular stage. *A,* Paired cardiac tubes, at six somites, within the ventrally opened pericardial cavity (after Davis; × 75). *B,* Sagittal section to show schematically the heart and its relations (after Frazer).

A, Atrium; *Am.,* amnion; *B,* bulbus; *D.M.,* dorsal mesocardium; *L.,* liver; *P.,* pericardium; *Ph.F.,* pharyngeal floor; *S.,* septum transversum; *S.V.,* sinus venosus; *V.,* ventricle.

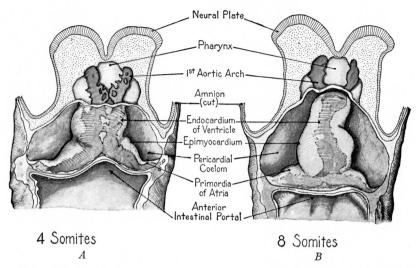

4 Somites
A

8 Somites
B

Fig. 338. The human heart, at the stage of fusion of the paired primordia, exposed as a transparent organ in ventral view (Patten). × 65. *A,* At four somites; *B,* at eight somites.

atrial contraction. Swollen *endocardial cushions* (dorsal and ventral) also narrow the heart locally into an *atrio-ventricular canal,* while elongate *bulbar ridges* (dorsal and ventral) course in the bulbus and truncus (Fig. 339 *B*).

THE ESTABLISHMENT OF EXTERNAL FORM

Between the stages of 7 and 16 somites the dorsal mesocardium has arisen and disappeared, thereby leaving the heart unattached except at its two ends; here it is anchored by the venous roots and the truncus (Fig. 338). Extending through and beyond this same period the cardiac tube grows faster than the pericardial chamber in which it lies, and as a result the heart is compelled to bend. The method of asymmetrical growth is such that the entire tube is thrown first into a simple bend and then into a spiralled **S**; the chief primary flexure is to the right, and by means of it the bulbus and ventricle become a **U**-shaped loop (Fig. 339). A continuation of this growth-process drops the bulbo-ventricular loop still farther caudad and ventrad (Fig. 340 *A*). At the same time the sinus venosus is drawn out of the septum transversum and its horns partially merge (*B*). Both the atrium and sinus shift in a cranial direction until both lie dorsal and cranial to the rest of the heart (Fig. 341). This shift is caused by a more oblique position taken by the septum transversum.

These changes thus result in an essential reversal of the original cranio-caudal relations of the primitive parts of the heart; in addition, the venous and arterial ends are brought close together as in the adult (Fig. 346). The growing atrium is now constricted dorsally by the sinus venosus and ventrally by the bulbus-truncus. For this reason it can enlarge rapidly only in lateral directions, and in so doing forms a sacculation on each side which foreshadows the future right or left atrium, respectively (Fig. 341); the location of the internal partition separating the two is marked superficially by a groove in which the truncus courses. Meanwhile the right horn of the sinus venosus enlarges more rapidly than the left (Fig. 340 B), owing to an important shift in the blood flow from the left side of the body across the liver, and the sinus itself comes to open into the right side of the common atrium (Fig. 342 B).

As the bulbo-ventricular loop increases in size, the duplication of the wall between its two limbs lags in development (perhaps hastened by actual atrophy)[4] and disappears during the fifth week (Fig. 341). The result is the merging of the two into a single chamber, the primitive ventricle (Fig. 347), which is separated from the atria by a deep *coronary sulcus*. At about the same time, the ventricle develops a median longitudinal groove that indicates the position of an internal septum already partitioning the unpaired

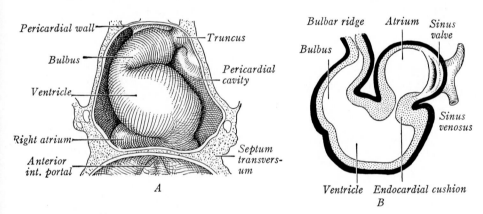

Fig. 339. Human hearts in early flexion. *A,* Ventral view, at eleven somites, *in situ* (after Davis; × 75). *B,* Diagrammatic sagittal section of the heart (after Frazer).

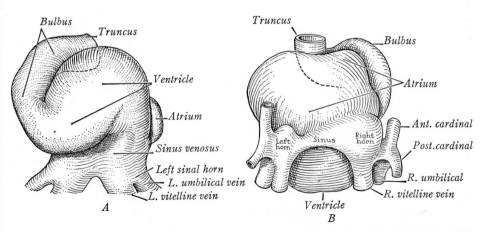

Fig. 340. Human hearts in advanced flexion. *A,* Ventral view, at sixteen somites (× 60). *B,* Dorsal view, at twenty-two somites (× 45).

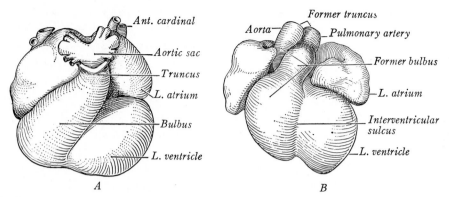

Fig. 341. Human hearts, in ventral view, progressing toward the definitive external form. *A,* At 5 mm. (partly after Ingalls; × 33); *B,* at 12 mm. (after Wirtinger; × 15).

chamber into two chambers; this external groove is the *interventricular sulcus* (Fig. 341 *B*). Thus in an embryo of six weeks (about 12 mm. long) the heart exhibits the general external shape and markings that characterize it permanently.

The fetal heart has the task of maintaining a placental circuit in addition to a circulation within the body of the embryo itself. For this reason it is not astonishing to find that the size of the heart at six weeks, relative to the embryo as a whole, is about nine times that in the adult. At first the heart lies high in the cervical region, but lengthening of the pharynx and the structures dorsal to it causes a relative recession toward the definitive position in the thorax. This caudal 'migration' through the distance of about 15 somites is attested permanently by the downwardly displaced courses of the recurrent and cardiac nerves (Fig. 477). After the diaphragm reaches its final location, the heart rotates so that the ventricles, which previously were ventral to the atria, henceforth become more caudal.

INTERNAL SPECIALIZATIONS

In an embryo of 5 mm., the heart contains three as yet undivided chambers: (1) the *sinus venosus,* opening dorsally into the right dilatation of the atrium; (2) the bilaterally dilated *atrium,* communicating, in turn, by a common atrio-ventricular canal with (3) the primitive *ventricle,* which is beginning to incorporate the bulbus into itself. This is the type of heart found in adult fishes, where it needs only to force venous blood to the gills for oxygenation. But the acquisition of lungs by higher vertebrates introduced the need for a heart that acts as a double pump, each with its own supply and exhaust. Hence the heart came to be partitioned into a venous half, which sends blood to the lungs, and an arterial half which provides for the rest of the body. Birds and mammals have a four-chambered heart, completely partitioned by septa that arise independently in the atrium, ventricle and bulbus; in it venous blood circulates on the right side and arterial blood on the left. Amphibians and reptiles have intermediate types, with partially separated atria and ventricles.

Important changes, chiefly concerned with the elaboration of septa and valves, next follow; they lead to the formation of the four-chambered human heart. These developments include: (1) the partitioning of the common atrium into separate right and left chambers; (2) the division of the atrio-ventricular canal into two canals; (3) the absorption of the sinus venosus into the wall of the right atrium and of the pulmonary veins into the left atrium; (4) the merging of the bulbus into the definitive right ventricle, and the longitudinal division of the bulbus and truncus into aorta and pulmonary artery; (5) the partitioning of the single ventricle into right and left chambers; and (6) the histogenetic differentiation of the cardiac wall, including the development of valves. Although most of these processes go on simultaneously it is more convenient to describe them separately; this will be done, in the order listed, in the topics that follow. When practically completed, as happens in an embryo of two months, the fetal heart has attained the general structural features that will characterize it permanently.

Partitioning of the Atrium and Atrio-ventricular Canal. In human embryos of 6 mm. a thin, sickle-shaped membrane grows down from the mid-dorsal wall of the atrium (Fig. 342 *A*). This is called the *septum primum* (or septum I), for it grows toward the ventricle as a first partition whose free edge soon fuses with the so-called *endocardial cushions*, thereby obliterating the previous free communication between the right and left halves of the atrium. The two cushions, just mentioned, are local endocardial thickenings; one bulges from the dorsal, the other from the ventral wall of the common canal that originally connects atrium with ventricle (Fig. 342). By the time the septum primum arrives, these thickenings have already fused midway, figure-of-eight fashion, and so divide the single canal into a right and a left *atrio-ventricular canal* (Fig. 349). It is on the merged tissue between these canals that the septum primum attaches (Fig. 343 *A*). Until this occurs, the lower border of the growing septum bounds a temporary *foramen primum*. Meanwhile the septum primum has thinned and

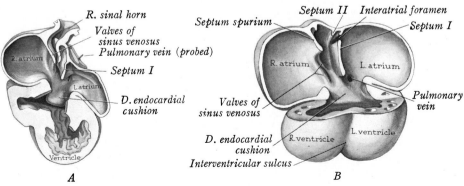

Fig. 342. Human hearts, cut away to show the internal structure and dorsal wall (after Tandler). *A,* At 6.5 mm. ($\times$ 40); *B,* at 9 mm. ($\times$ 50).

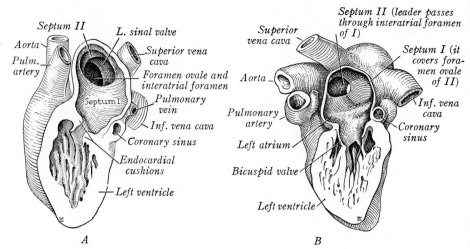

Fig. 343. Human hearts, opened from the left side (after Prentiss). *A,* At six weeks (× 40); *B,* at three months (× 5).

become perforate dorsocranially in a previously intact region, thereby form-ing secondarily a *foramen secundum;* it is commonly called the *interatrial foramen* (Fig. 342 *B*). So it is, that (except for this foramen) there is already at the end of the sixth week (12 mm.) a separate right and left atrial cham-ber; each connects through its respective atrio-ventricular canal with the right or left ventricle, also incompletely partitioned at this time.

In the seventh week the thicker *septum secundum* (or septum II) makes an appearance just at the right of the septum primum (Fig. 342 *B*). Its origin seems to be in relation to the left sinus valve and the fused region of the two endocardial cushions.[5] Like the septum primum, this second par-tition is an incomplete membrane. It retains a prominent defect, near the inferior vena cava, known as the *foramen ovale* (Fig. 343). The growth of the two atrial septa proceeds in such a manner that each persists as a sep-arate, perforate partition; but each is so shaped that it covers the defect in the other. The main expanse of the septum primum overlaps the fora-men ovale and serves as a flap (*valve of the foramen ovale*) which permits blood to pass from the right to the left atrium, but not in the reverse di-rection (Fig. 345). After birth the two septa unite and so produce the per-manent *atrial septum* (p. 394).

Atrial Absorptions. The primitive atria are enlarged and modified by the incorporation into their walls of much of the sinus venosus and the stem portions of the pulmonary veins.

THE SINUS VENOSUS. As in fishes, the sinus originally serves as a res-ervoir for the purpose of collecting blood during atrial contraction. The left horn of the sinus venosus begins to lose functional importance in the fifth week when the returning blood is diverted to the right horn (Fig. 324). It continues to dwindle and persists merely as the stem of the *oblique vein of the left atrium* (Fig. 346). The transverse portion of the sinus receives this oblique vein and other cardiac veins from the heart walls. It opens into the dorsal wall of the right atrium and becomes known as the *coronary sinus.*

In embryos of about seven weeks the superior vena cava has been formed to return blood from the head end of the embryo, and the inferior vena cava to serve similarly for lower levels of the body. Both vessels drain into the right horn of the sinus venosus (Fig. 346 *A*). Between the sixth and eighth week the atria increase rapidly in size and the right horn of the sinus venosus, failing to keep pace, is taken up into the wall of the right atrium. By this absorption the superior vena cava of necessity drains directly into the cranial wall of the atrium, while the inferior vena cava opens into its caudal wall (*B*). The boundary between the original atrium and the absorbed sinus is marked externally by a permanent groove, the *terminal sulcus.* Internally the absorbed sinal wall represents the region between the right sinal valve and the atrial septum (Figs. 344, 345); it is known as the *sinus venarum* and is characterized permanently by a smooth lining. The remainder of the atrium, including all of the *right auricle,* gains a thick and uneven muscular wall.

The opening of the early sinus venosus into the dorsal wall of the right atrium is guarded on each side (right and left) by a valvular fold (Fig. 342 *A*). Along the dorsal and cranial wall of the atrium these two folds unite into the so-called *septum spurium,* which has no significance beyond that of keeping the two valves tense (*B*). Caudally the valves flatten out on the

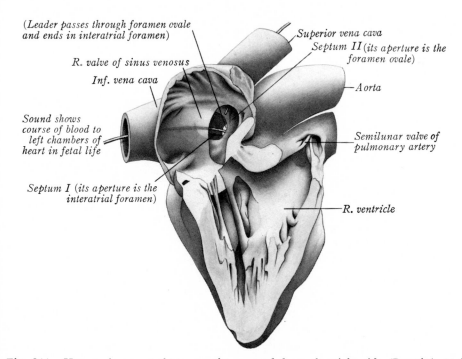

Fig. 344. Human heart, at three months, opened from the right side (Prentiss). × 8.

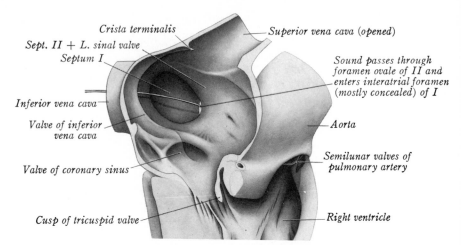

Crista terminalis
Sept. II + L. sinal valve
Septum I
Superior vena cava (opened)
Sound passes through foramen ovale of II and enters interatrial foramen (mostly concealed) of I
Inferior vena cava
Valve of inferior vena cava
Aorta
Valve of coronary sinus
Semilunar valves of pulmonary artery
Cusp of tricuspid valve
Right ventricle

Fig. 345. Human heart, at four months, opened from the right side (Prentiss). × 5.5.

floor of the atrium. Through the slower growth of the intervening space, the *left valve* of the sinus venosus approaches and fuses with the limbus region of the septum secundum. The *right valve* of the sinus venosus is expansive until the end of the third month and nearly divides the atrium into two chambers (Fig. 344), but later it diminishes greatly in relative size. Its cranial portion becomes a rudimentary crest on the wall of the right atrium; it is known as the *crista terminalis* (Fig. 345). The remainder of the valve is subdivided by a ridge into two parts; of these, the larger, cranial division persists as the *valve of the inferior vena cava* (Eustachian valve), located at the right of the opening of that vein, while the smaller, caudal portion becomes the *valve of the coronary sinus* (Thebesian valve).

THE PULMONARY VEINS. The final relation of the pulmonary veins to the heart is also the result of an absorptive process through which the left atrium is markedly enlarged. In embryos of about 6 mm., a single pulmonary vein drains into the caudal wall of the left atrium at the left of the septum primum (Fig. 346 *A*). This vessel bifurcates into right and left veins which in turn divide again, so that two terminal branches drain each lung. As the atrium grows, these pulmonary vessels are progressively drawn into the atrial wall. As a result, at first one, then two, and finally four pulmonary veins open separately into the left atrium (*B*). The absorbed stems of the veins are permanently recognizable as a smooth portion of the atrial wall. The primitive atrium thereby becomes restricted to the definitive region that is known as the *left auricle;* its wall is ridged internally by the *pectinate muscles.*

Fate of the Bulbus and Truncus. In the description of the external development of the heart, mention was made of the incorporation of the proximal part of the bulbus into the right ventricle. This absorption occurs through the laggard growth (and, perhaps, atrophy)[4] of the bulbo-ventricular fold (Fig. 347). As a result, the bulbus loses its separate identity during the fifth week and the definitive right ventricle comes into being (Fig. 341). Nevertheless, the former bulbar region of this ventricle is permanently designated as the *conus arteriosus*.

In embryos of 5 mm. there arise in the bulbus and truncus two prominent longitudinal thickenings of the endocardial lining (Fig. 349 *A*).[6] These so-called *bulbar ridges* presently meet and fuse, thereby creating a septum that divides the truncus and unabsorbed portion of the bulbus into an aortic and a pulmonary trunk (Fig. 348 *A-C*). Distally these vessels connect with fourth and sixth aortic arches, respectively (Fig. 312 *C*). Proximally the two thickenings so pursue spiral courses that the separated *ascending aorta* and *pulmonary artery* slightly intertwine, the latter crossing ventral to the aorta (Fig. 349 *B*). Still more proximally the spiral division of the bulbus is continued toward the ventricular septum in such a way

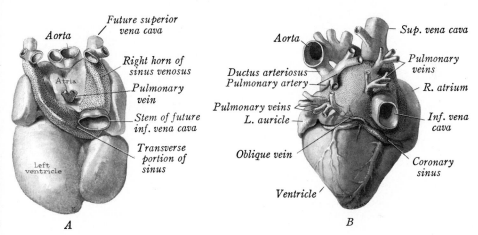

Fig. 346. Human hearts, in dorsal view, showing the absorption of the sinus venosus (dark stippling) and pulmonary veins (pale stippling). *A,* At 7 mm. (after Braus; × 28); *B,* in newborn (× ⅔).

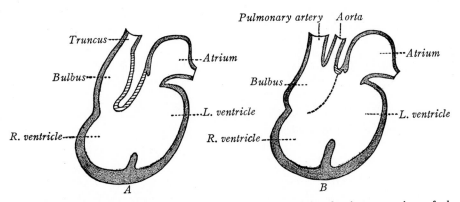

Fig. 347. Diagrams of the mammalian heart to explain the incorporation of the bulbus into the right ventricle through the slower growth and atrophy of the bulboventricular fold (hatched). Stage *B* is older and should be drawn much larger than *A;* the broken line marks the former extent of the fold (modified after Keith).

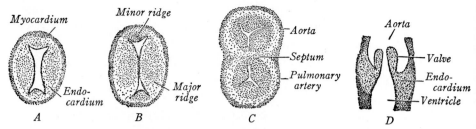

Fig. 348. Subdivision of the human bulbus and truncus, and the origin of the semi-lunar valves. *A–C,* Transverse sections, at five to seven weeks (× 27). *D,* Longitudinal section of the aortic-ventricular junction, at seven weeks (× 45).

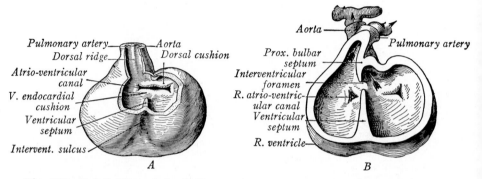

Fig. 349. Subdivision of the bulbus, atrio-ventricular canal and common ventricle, shown in schematic ventral views (after Kollmann). *A,* At 5 mm.; *B,* at 8 mm. The middle portion of the ventricular septum has been cut away. In *A* its caudal horn is the one labeled; the cranial horn is a small, stippled protuberance lying just above the dorsal endocardial cushion.

that the base of the pulmonary trunk (now to the right and somewhat ventral) opens off the right ventricle, while the base of the aorta (now lying on the left and somewhat dorsal) opens off the left ventricle.

In addition to the longitudinal thickenings of the endocardium that split the bulbus and truncus lengthwise, there are two minor thickenings (Fig. 348 *A*). After the division of the truncus occurs, both the aorta and the pulmonary artery contain one of the smaller ridges and a half of each of the larger ridges (*B, C*). Distally, the three plump thickenings, then present in each vessel, disappear. Proximally, at the level of the aortic and pulmonary roots, they enlarge and hollow out on their distal surfaces (*D*). Each set of three thin-walled pockets, formed in this manner, henceforth serves as *semilunar valves* (Fig. 344). Just distal to the aortic valves arises a pair of *coronary arteries* which supply the heart wall.

Partitioning of the Ventricle. At the end of the fourth week (5 mm.) a median partition begins to project inward from the base of the common ventricle (Fig. 349 *A*). This *ventricular septum* is brought into existence by the enlargement of the future halves of the ventricle on each side of it. Increasing in height as the ventricular sacs become deeper, it grows as a crescentic plate whose two horns respectively join the dorsal and ventral endocardial cushions. For a short time the septum makes an incomplete partition which partially divides the ventricle into right and left chambers; throughout this stage the communication between the two ventricles is known as the *interventricular foramen* (*B*). This foramen is bounded by: (1) the ventricular septum; (2) the proximal bulbar septum, continued downward from the longitudinally dividing bulbus; and (3) the fused, middle portions of the endocardial cushions. By the end of the seventh week the foramen is being closed by tissue proliferated from these several sources,[6] but especially from the endocardial cushions.[7] The resulting thin (and permanently nonmuscular) membrane, which completes the partition, is the *septum membranaceum,* whereas all the rest of the partition becomes the *septum musculare.*

DIFFERENTIATION OF THE HEART WALL

Identical types of tissue differentiation and spatial organization occur throughout the whole heart, but they attain highest expression in the ventricles which become thick and especially rich in muscle. The internal endothelial tube of the primitive heart continues as the principal constituent of the *endocardium.* The investing folds of splanchnic mesoderm transform into both the massive *myocardium,* with its specialized type of muscle, and the serous coat known as the *epicardium.*

At first the endothelial cardiac tube is widely separated from the thick, outer coat, not yet differentiated beyond the stage of a common epi-myocardium. The intervening space is filled with a fluid jelly which serves the early, pumping heart in a valvular capacity. Later it is invaded by cells and comes to resemble mucous tissue (Fig. 350 *A*);[8] the space is finally reduced as the jelly transforms into the connective tissue of the *endocardium* (*B*). The endocardial cushions and the bulbar thickenings are prominent because of an exaggeration and longer retention of this otherwise temporary condition.

The *myocardial coat* differentiates into a thin, cortical layer of dense muscle and a thick, spongy layer whose loosely arranged trabeculæ project into the heart cavity

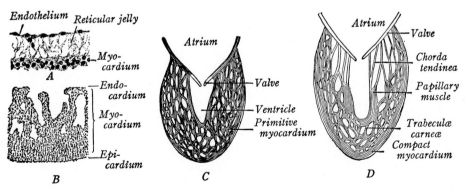

Fig. 350. Differentiation of the human ventricular wall and the atrio-ventricular valves. *A*, *B*, Vertical sections, at 2 mm. ($\times$ 115) and 7 mm. ($\times$ 55), respectively. *C*, *D*, Diagrammatic longitudinal sections (after Gegenbauer).

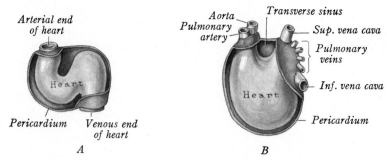

Arterial end
of heart

Heart

Pericardium   Venous end
of heart

A

Aorta
Pulmonary
artery

Transverse sinus

Sup. vena cava

Pulmonary
veins

Inf. vena cava

Heart

Pericardium

B

Fig. 351.   Diagrams of the changing relations of the heart and pericardium (after Braus). *A,* At the period of early flexion; *B,* at the definitive stage.

(Fig. 350 *B, C*). As the muscular trabeculæ increase, the originally simple sac of endocardium dips into their interspaces and wraps around them. Before long there is a condensation of the spongy myocardial tissue, and especially is this marked at the periphery. As a result, the superficial musculature becomes increasingly compact, whereas the trabeculæ nearer the lumen retain an open arrangement for a longer period (*D*). Such a condition is permanent for lower vertebrates, but in mammals the entire cardiac wall finally becomes compact. The irregular muscle bundles that persist next the ventricular cavities make up the *trabeculæ carneæ*. The musculature of the ventricles is far better developed than that of the atria. However, the thicker wall of the left ventricle is largely acquired after birth as the result of the harder work that is entailed by the permanent plan of circulation.

The myocardium, at first continuous throughout the entire heart, becomes divided by connective tissue at the atrio-ventricular canal and retains only a small bridge there. This connecting strand of modified muscle is located behind the dorsal endocardial cushion and is a part of the *atrio-ventricular bundle* (of His).[9] It is specialized to conduct the cardiac impulse from atria to ventricles. The first heart beats are spasmodic twitchings that soon gain in force and regularity. By analogy with other vertebrates it is supposed that the human heart begins to set the blood in motion during the part of the fourth week when the embryo has 7 to 17 somites. At that time the rhythmic contractions are purely muscular phenomena, since the nerves first invade the heart several weeks later. The rate of the heart beat is 65 pulsations a minute in 15 mm. embryos, but this is increased to 130 to 145 in older fetuses. Electrocardiograms can be recorded first in fetuses of 11 weeks.[10]

The surface of the original epi-myocardial coat flattens into mesothelium; combined with a substratum of connective tissue it constitutes the *epicardium* (Fig. 350 *B*).

The Cardiac Valves.   The blood is kept in its proper course through the heart by means of valves which prevent backflow. A previous paragraph (p. 384) has described how the right valve of the sinus venosus adapts itself as the *valve of the inferior vena cava* and *valve of the coronary sinus*. The origin of the *semilunar valves* of the aorta and of the pulmonary artery has likewise been discussed (p. 386).

An important valve occurs between each atrial and ventricular chamber.[11] Their development is bound up with that of the endocardial cushions (p. 381) which by fusion, figure-of-eight fashion, convert the single atrio-ventricular canal into two canals (Fig. 349). Elevated folds of the endocardium appear at the margins of these canals, and each set of thickenings becomes both invaded by muscle and attached to the muscular trabeculæ of the ventricular wall. Three such flaps, or valvular cusps, are formed about the right atrio-ventricular canal, two around the left. The size of the primitive cusps is presently increased by an undermining process in which the attached muscular cords, beneath, become less numerous and more widely spaced (Fig. 350 *C, D*). Degeneration ensues both in the muscle tissue of the valves and in that of the subjacent muscular cords. As a result, the valvular cusps turn fibrous and connect with guy-like *chordæ tendineæ* similarly transformed from the muscular cords; the latter, in turn, continue into unaffected *papillary muscles*. Thus there are developed the three cusps of the *tricuspid valve* between the right chambers of the heart, and the two flaps of the *bicuspid (mitral) valve* between the left chambers (Fig. 343 *B*).

THE PERICARDIUM

The pericardial cavity is bounded peripherally by the *pericardium,* which is composed of somatic mesoderm. This serous sac was originally in complete continuity with the epicardium of the heart (splanchnic mesoderm) through the presence of a dorsal mesocardium (Figs. 336 *C,* 337 *B*). Since this mesentery disappears promptly and the ventral mesocardium is lacking from the first, it soon happens that the only region of continuity is at the two ends where veins enter and arteries leave (Fig. 351 *A*). Flexion of the tubular heart brings these ends close together, so that the regions of continuity are separated from each other only by a space, the *transverse sinus* of the pericardium (*B*). The *oblique sinus* is brought into being by the changed direction of the lines of reflection of the pericardium onto the pulmonary veins and venæ cavæ; this change is consequent on the absorption of those vessels into the atria.

Causal Relations. In amphibians heart-development is induced by the entoderm. Experiments show that as early as the neurula stage the fate of the cardiac primordium

FATES OF MAIN CARDIAC PRIMORDIA

PRIMARY DIVISION	PERMANENT REPRESENTATIVE	PRIMITIVE SEPTA	FATE OF PRIMITIVE SEPTA	PRIMITIVE VALVES	FATE OF PRIMITIVE VALVES
Right sinal horn	Smooth portion of right atrium			Right sinal	Crista terminalis Valve of inf. v. cava Valve of cor. sinus
Trans. portion	Coronary sinus				
Left sinal horn	Oblique vein (stem)			Left sinal	Contributes to atrial septum
Atrium	Auricles and some of main atria	Septum I } Septum II	Atrial septum		
				Endocardial cushions	Bicuspid valve Tricuspid valve Adds to septum II Adds to septum membranaceum
Ventricle	Right ventricle (except conus) Left ventricle	Ventricular	Ventricular septum (muscular part)		
Bulbus	Absorbed into right ventricle (conus)	Prox. bulbar	Ventricular septum (membranous part)		
Truncus	Aorta and pulmonary artery	Dist. bulbar (or truncal)	Splits truncus (see column 2)	Bulbar ridges	Semilunar valve

is already determined sufficiently so that it will develop into beating heart-tissue when transplanted. But not until the tail-bud stage is it wholly free of entodermal, guiding influence. Early half-primordia of the heart, prevented from joining, will develop into two complete hearts. Moreover, if an entire primordium is inserted between the split primordium of a host animal, the total complex becomes a single, normal heart. Continued development of a heart depends on the presence of blood flow and on its amount.

Anomalies. Absence of the heart (*acardia*) may occur in twins sharing a common placenta. The so-called acardiac monster is either a separate or conjoined member of a twin pair (Figs. 152, 154). Among the rare anomalies of the heart is a tendency toward doubling; the apex is bifid, as occurs permanently in some aquatic mammals (Fig. 352 *A*).

Some mishaps affect the position or external relations of the heart. It may occupy a high position through failure to complete its descent into the thorax. Rarely it may be exposed on the surface of the chest (*ectopia cordis*) and this tends to be associated with an open pericardium and widely parted sternal halves (Fig. 352 *A*). *Dextrocardia* denotes a condition of transposition in which the heart and its vessels are reversed, left-to-right, as in a mirror image; it takes origin from a primary reversal of the primitive cardiac loop. Although dextrocardia may occur by itself, it is usually but one feature in the larger picture of thoracic or total transposition of viscera (Fig. 232 *B*). Yet the ascending aorta and pulmonary artery may be *transposed* in the absence of total dextrocardia, owing to the spiral septum failing to accomplish the usual crossed relation of these vessels (Fig. 352 *B*). In this instance the two great vessels connect with the wrong ventricles so that two separate circuits are set up. Even when there is a retention of function by the foramen ovale, ductus arteriosus or interventricular foramen, postnatal life is limited to a year or less.

Septal defects are common and one or both septa may even be lacking. An incomplete ventricular septum is usually attributable to faulty development of the complexly formed septum membranaceum (Fig. 352 *C*). Although it can exist alone, this defect more often occurs in conjunction with irregularities in the subdivision of the truncus (*E*). Most frequent of all cardiac anomalies is the persistence of an interatrial communication, due to imperfect obliteration of the foramen ovale after birth. The valve of the foramen ovale may fuse incompletely with the septum secundum or it may even be unable to cover completely the foramen ovale (*D*). Incomplete anatomical closure occurs in nearly one out of four individuals. In spite of this frequency the defect, by itself, is significant physiologically only when an actual aperture exists during atrial contraction; even then the flow is from left to right, and the loss of oxygenated blood to the right atrium merely tends to overwork the right side of the heart. Rarely the failure of the septum primum to fuse with the endocardial cushions (persistent foramen primum) creates a similar condition (*cf.* Fig. 342 *B*).

Valvular anomalies occur both in the atrio-ventricular region and in the semilunar valves of the aorta and pulmonary artery. Such deviations often involve the size or number of cusps that constitute a valve (Fig. 352 *E*); the patency of the passage that is guarded may range from valvular incompetence to stenosis and complete atresia. Such embryonic survivals as an unpartitioned truncus and a common atrio-ventricular canal necessitate special valvular adaptations.

The vessels connecting with the heart sometimes show abnormal relations. A spiral septum may fail to subdivide the truncus into ascending aorta and pulmonary artery (Fig. 352 *E*), or subdivide them unequally. The superior and inferior venæ cavæ may open into the left atrium; some or all of the pulmonary veins at times drain into the right atrium or superior vena cava; one or both of the coronary arteries may arise from the pulmonary artery; the pulmonary veins or a left superior vena cava may drain into the coronary sinus. *Stenosis* of the aorta or pulmonary artery is a localized stunting that can entail serious consequences through the ensuing difficulty in distributing blood properly (*F*).

Chief among the hazards introduced by cardiac anomalies is the distribution of inadequately oxygenated blood to the body as a whole. Many abnormal conditions are compatible with fetal life and a placental circulation, because of the presence of shunts through the foramen ovale and the ductus arteriosus. But after birth these abnormalities interfere with pulmonary aeration and are incapacitating to some degree, even if not incompatible with postnatal life. The lack of oxygen in the peripheral circulation is made apparent by the bluish tinge of the skin and mucous membranes (*cyanosis*). Typical is the condition known as the *tetralogy of Fallot,* for which 'blue-baby operations' have been devised. Its four features are (Fig. 352 *F*): (1) stenosis of the pulmonary artery, so that only part of the blood received by the right ventricle can be pumped to the lungs; (2) over-riding of the ventricular septum by the mouth of the aorta, so that this vessel receives blood from both ventricles; (3) presence of a permanent interventricular foramen, which makes possible such over-riding and a double blood supply to the aorta; (4) hypertrophy of the right ventricle, caused by its receiving and pumping more than its ordinary share of blood. These conditions result in little blood reaching the lungs and this amount being degraded in the aorta by venous blood also entering that vessel.

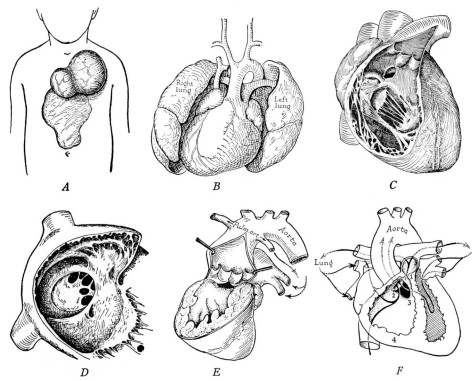

Fig. 352. Anomalies of the human heart. *A,* Ectopia of a heart, which also had a bifid ventricle; more caudad the body wall is closed by a membranous layer, lacking muscle. *B,* Dextrocardia, accompanied by transposition of both of its great vessels; not of the lungs. *C,* Opened right ventricle, displaying an incomplete ventricular septum; the persistent foramen has a high location, near the origin of the pulmonary artery. *D,* Persistent foramen ovale, guarded by a defective (multiperforate) valve, as viewed from the right atrium. *E,* Persistent truncus arteriosus, displayed by incision and by removal of part of the right ventricle; pulmonary arteries arise from the common trunk; unfused bulbar ridges become a four-cusped valve, beneath which is an unclosed interventricular foramen. *F,* The tetralogy of Fallot, displaying four defects numbered as in the text.

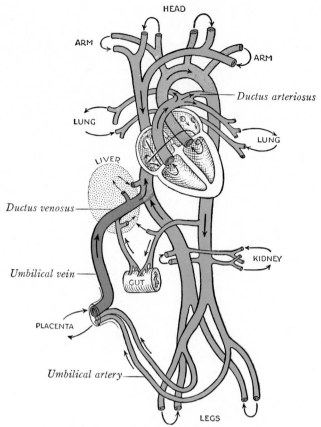

Fig. 353. Plan of the human circulation before birth (partly after Dodds). Colors show the quality of the blood and arrows indicate its direction of flow.

FETAL CIRCULATION AND BIRTH CHANGES

The Fetal Circulation. During fetal life oxygenated blood, returning from the placenta, enters the embryo by way of the large umbilical vein and is conveyed to the liver (Fig. 353; in red). Thence it flows to the inferior vena cava, in part directly through the ductus venosus but to a definitely larger extent indirectly through the liver sinusoids and hepatic veins.[12] Nevertheless, the impure blood of the portal vein and inferior vena cava contaminates only partially the large volume of placental blood. Accordingly, the mixture entering the right atrium from the lower body is relatively well oxygenated (in purple). By contrast the superior vena cava carries oxygen-poor blood, returning from the upper body (in blue). It also enters the right atrium.

The course of blood through the fetal heart has been recorded in motion pictures after injecting radiopaque material into the blood stream (Fig. 353).[13] The less pure (superior caval) blood is directed into the right ventricle, whence it leaves the heart through the pulmonary artery. Some of this blood reaches the nonfunctioning lungs, while the greater part continues through the ductus arteriosus to the descending aorta. Here it is distributed to the trunk, abdominal viscera, limbs and placenta. On the other hand, the purer (inferior caval) blood takes a double course through the heart. A smaller amount goes directly to the right ventricle and then follows the course already described. The main volume, however, crosses through the foramen ovale into the left atrium, where the pressure is lower because of the small pulmonary return, and thus reaches the left ventricle. From here it is pumped into the ascending aorta and thence much of it passes to the coronary arteries, head and arms. In this manner the heart and brain, in particular, are given preferential treatment with respect to oxygenated blood from the placenta. The remainder of the aortic blood joins and freshens somewhat the flow received by the descending aorta from the ductus arteriosus. It supplies the general body and placenta.

Changes at Birth. When the lungs become functional, the placental circulation ceases quickly. This transfer of the seat of oxygenation not only changes the character of the blood in many vessels but throws some important fetal vessels and passages into disuse (Fig. 354 *A*). As a group these latter channels seem to abandon their functional rôles suddenly and completely,[14] although there may be some variation among mammals in the promptness and completeness of this response.[15] By contrast, anatomical

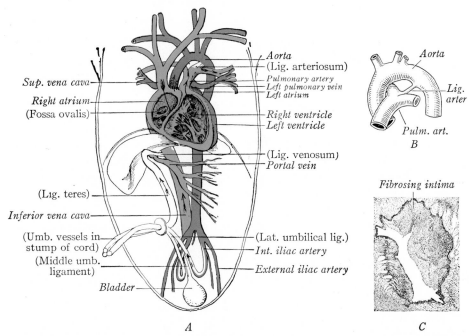

Fig. 354. Changes in the human circulation after birth. *A*, Plan of circulation, in ventral view; obliterated fetal passages are designated by Roman type within parentheses (Heisler). *B*, Ligamentum arteriosum, at three months (*cf.* Fig. 314). *C*, Transverse section of the interior of the obliterating ductus arteriosus, at one month after birth (after Schaeffer).

obliteration is a gradual process of fibrosis which proceeds slowly in the postnatal months.[16]

Since the pressure is equal or higher in the right fetal atrium than in the left atrium, it is seemingly not true that during atrial contraction the two septa are pressed together to prevent back flow into the right atrium.[17] After birth, however, the pressure declines in the right atrium and rises in the left atrium because of the ending of a placental circulation and the sharp increase in the volume of blood returning from the lungs. This re-adjustment permits the two septa to lie in constant apposition, whereupon the septum primum, which serves as the valve of the foramen ovale, fuses gradually with the margin of the foramen ovale. This union is completed after about one year, but more than 20 per cent of all individuals never obtain perfect closure. The site of the previous aperture in the septum secundum is marked permanently by the depressed *fossa ovalis* (Fig. 354 *A*). The thick rim of septum-secundum tissue, bounding the fossa, constitutes its border, or *limbus*.

When breathing begins, muscular contraction closes the ductus arteriosus; perhaps this is in response to increased blood-oxygen.[18] Its anatomical obliteration is caused by the internal coat proliferating pads of fibrous tissue into the lumen (Fig. 354 *C*).[19] After one month it has usually become impervious, at least in part of its length, and the resulting cord is the *ligamentum arteriosum*.

Shortly after birth the empty umbilical vessels contract, first the arteries and then the vein. They gradually lose their lumina by fibrous invasion; this extends through the first two or three months of postnatal life. Distally the arteries obliterate into the *lateral umbilical ligaments;* proximally they continue as functional internal iliac arteries (Fig. 354 *A*). The vein becomes cord-like and persists as the unpaired *ligamentum teres* of the liver; it courses in the margin of the falciform ligament from umbilicus to liver. The distal ends of the umbilical vessels join the depressed scar that is the *umbilicus*. The ductus venosus likewise atrophies; within two months it has transformed into the fibrous *ligamentum venosum,* superficially embedded in the wall of the liver (*cf.* Fig. 325).

REFERENCES CITED

1. Gilmour, J. R. 1941. J. Path. & Bact., *52*, 25–45.
2. Davis, C. L. 1927. Carnegie Contr. Embr., *19*, 245–284.
3. Yoshinaga, T. 1921. Anat. Rec., *21*, 239–308.
4. Frazer, J. E. 1917. J. Anat. & Physiol., *51*, 19–29.
5. Odgers, P. N. B. 1935. J. Anat., *69*, 412–422.
6. Kramer, T. C. 1942. Am. J. Anat., *71*, 343–370.
7. Odgers, P. N. B. 1938. J. Anat., *72*, 247–259.
8. Barry, A. 1948. Anat. Rec., *102*, 289–298.
9. Walls, E. W. 1947. J. Anat., *81*, 93–110.
10. Larks, S. D. 1961. Fetal Electrocardiography (Thomas).
11. Odgers, P. N. B. 1939. J. Anat., *73*, 643–657.

12. Franklin, S. J., *et al.* 1940. J. Anat., *75*, 75–87.

13. Barclay, A. E., *et al.* 1941. Am. J. Anat., *69*, 383–406.

14. Barron, D. H. 1944. Physiol. Rev., *24*, 277–295.

15. Everett, N. B. & R. J. Johnson. 1951. Anat. Rec., *110*, 103–111.

16. Scammon, R. E. & E. H. Norris. 1918. Anat. Rec., *15*, 165–180.

17. Hamilton, W. F., *et al.* 1937. Am. J. Physiol., *119*, 206–212.

18. Reynolds, S. R. M. 1961. Chapter 2 in Luisada: The Cardiovascular System (McGraw-Hill).

19. Jager, B. V. & O. T. Wollenmann. 1942. Am. J. Path., *18*, 595–613.

Chapter XXI. The Skeletal System

HISTOGENESIS OF THE SUPPORTING TISSUES

Connective tissue, cartilage and bone all differentiate from that type of diffuse mesoderm known as *mesenchyme*. It arises chiefly from mesodermal somites and the lateral layers of somatic and splanchnic mesoderm (Fig. 364). Mesenchyme is a spongy meshwork composed of branching ameboid cells whose processes perhaps merely touch rather than actually anastomose (Fig. 357).[1] Between the cells occur open, labyrinthine spaces which are filled with a *ground-substance;* this material is a structureless, semifluid jelly. In early embryos the mesenchyme serves as an unspecialized packing material between the emerging organs and parts of the embryo, but it soon enters on various lines of differentiation (Fig. 11).

Of the numerous derivatives of mesenchyme, the supporting tissues, whose function is predominantly mechanical, are peculiar in an important regard. This is because the ground-substance becomes bulkier than the cellular elements themselves and acquires characteristic fibers. This ground-substance may remain jelly-like (in fibrous tissues); it may stiffen greatly (in cartilage) or even attain maximum firmness through the deposition of calcium salts in it (in bone). The composite, consisting of ground-substance, fibers and any impregnating materials, is known as *matrix*. In each type of supporting tissue the matrix is considered by most authorities to represent a nonliving substance, laid down in the interspaces between cells and organized under their guiding influence. In each division of the general group of supporting tissues it is nonliving, formed substance, rather than the cells themselves, that performs the mechanical function characteristic of that particular tissue. The maintenance of the matrix in a normal, functioning state, nevertheless, is dependent on the presence of associated, specialized cells in active condition.

CONNECTIVE TISSUE

The fibrils of connective tissue develop in a gelatinous ground-substance that embeds the distinctive cells named *fibroblasts*. The main problem in fibrilogenesis concerns the site of origin of the fibrils and their relation to the ground-substance. It is now generally believed that the ground-substance is lifeless material, secreted by the cells. In it the fibrils once seemed to appear by a process akin to crystallization. More recently, presumptive prefibrillar material has been seen leaving the fibroblasts,[2] and even proto-fibrils have been described within the cytoplasm, which, on gaining the exterior, are believed to fuse into the definitive fibrils.[3]

Reticular Tissue. Except for the jelly-like mucous tissue of the umbilical cord, *reticular tissue* departs least from the embryonal type (Fig. 355, at top). Its stellate cells, little different from those of mesenchyme, are usually described as maintaining a clasping relation, wrapped about the reticular fibrils. The latter are fine filaments, staining electively with silver. They can develop into collagenous fibers and are, in this sense, an immature stage of these elements.[4] Their different appearance and staining qualities are apparently due to physical fineness rather than to chemical constitution.

Fibro-elastic Tissue. The predominant arrangement of connective tissue in the body is a mixture of *collagenous fibers* and *elastic fibers*. The early mesenchymal cells of these tissues specialize into *fibroblasts,* which are specifically related to the genesis and maintenance of fibers,[2] and into other free types such as *macrophages* and *mast cells* (Fig. 11).

The differentiation of collagenous fibers can be divided into two phases: first there is a preliminary stage marked by the appearance of thin fibrils, resembling those of reticular tissue (Fig. 355; at top); next, in the third month, the fibrils assume parallel positions and aggregate into more or less wavy bundles which acquire the physical and chemical characteristics of the mature tissue (Fig. 355; at middle). Chemically they consist of a substance known as *collagen.* Elastic fibers differentiate later (fifth month) than the collagenous fibers, but largely in the same general manner (Fig. 356).[5] Typically they remain as solitary, coarse fibers that both branch and anastomose. They consist of *elastin* which is wholly different from collagen both chemically and in physical properties.

The collagenous and elastic fibers occur loosely arranged or in compact concentration; their relative proportions, in each instance, meet the mechancial requirements of that particular location. In *areolar tissue* and fibro-elastic sheets of denser weave, such as *fascia* and *periosteum,* the col-

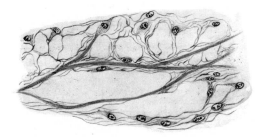 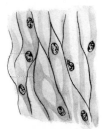

Fig. 355. The differentiation of collagenous fibers in the skin of a 5 cm. pig fetus (after Mall). × 270. At top, reticular tissue.

Fig. 356. The differentiation of elastic fibers in the umbilical cord of a 7 cm. pig fetus (after Mall). × 270.

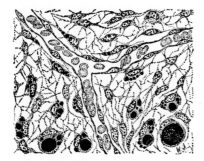

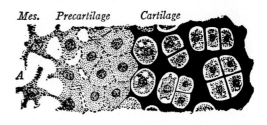

Fig. 357. -Fat cells differentiating in the fourth month from the mesenchyme near a capillary. × 250.

Fig. 358. Stages in the development of cartilage (Lewis).

lagenous fibers predominate in a fabric of network pattern. In *tendons* and ordinary *ligaments* elastic fibers are virtually lacking, whereas the white collagenous fibers are arranged in compact, parallel groups (fascicles). In certain locations elastic fibers merge into dense plates or they collect into compact *elastic ligaments*.

ADIPOSE TISSUE

Certain of the mesenchymal cells give rise to *lipoblasts,* which resemble fibroblasts and are the forerunners of fat cells (Fig. 11).[6] A lipoblast takes up fat in a soluble form and stores it within its cytoplasm as droplets which increase in size and become confluent (Fig. 357). As the amount of fat increases, the cell body becomes rounded. When at last a huge globule distends the cell, the nucleus is pressed to the periphery. Fat cells arise in close association with developing blood vessels in areolar tissue; they appear first during the fourth month. The primitive clusters are foci for the later *lobules,* which do not become prominent until near the end of fetal life. It has been urged by some that these lobules qualify for recognition as specific organs.[7]

It is possible that fibroblasts, though differentiated as such, on demand can transform into fat cells and thus add to the adipose tissue of the body. On the contrary, it seems more probable that new fat cells always come from the store of undifferentiated mesenchymal cells which the body maintains as a reserve.

At various locations in the fetus there are groups of distinctive, granular lipoblasts that are set apart as so-called *adipose glands.* Fat cells derived from them are peculiar in containing multiple fat droplets which for a time do not coalesce; yet in early infancy they become indistinguishable from the ordinary type.[8] In the cat this specific tissue continues until maturity, and in some rodents it is a permanent feature. Pigment gives the tissue a brownish color. It is a type not only histogenetically distinct from ordinary fat, but also it is more active functionally.[6, 9]

Causal Relations. Certain cells are already determined as lipoblasts while they are still indistinguishable from the ordinary cells of embryonic connective tissue. If tissue from a known adipose site in a rat embryo is implanted into an adult rat it becomes typical fat; transplants of similar looking, prospective connective tissue do not yield fat.

CARTILAGE

The earliest indication of future cartilage can be detected in the fifth week. This preliminary stage begins when the mesenchymal cells of a local region proliferate, enlarge and differentiate into a compact, cellular *pre-cartilage* that assumes the same shape as the cartilage to be formed (Fig. 399). The large, rounded cells become definitive *cartilage cells.* Between them *cartilage matrix* makes its appearance, and as it increases in amount the cells are forced farther and farther apart (Fig. 358). The firm matrix of *hyaline cartilage* contains a masked feltwork of fine collagenous fibers. In *fibrocartilage* heavier collagenous fibers predominate throughout most of the matrix; in *elastic cartilage,* elastic fibers similarly differentiate.

Cartilage grows internally and also at its periphery. Internal development, or *interstitial growth,* results both from the division of cartilage cells and from the production of new matrix by them. It continues as long as the matrix remains plastic enough to permit such expansile growth. Peripheral development, or *appositional growth,* starts somewhat later. It takes place through the mitotic activity of an enveloping connective-tissue sheath, the *perichondrium.* Its inner cells become specialized elements, known as *chondroblasts.* They transform into young cartilage cells, which deposit matrix and become buried by their own activities. This sequence repeats as long as the cartilage increases in size.

Causal Relations. The development of the cartilaginous, branchial-arch skeleton of amphibians and birds can be proved to be dependent on the migration of cells of the neural crest into that region (p. 419). The exercise of an inductive power by these cells is shown on their implantation, at the neurula stage, into epidermis; this tissue then differentiates into cartilage. The development of the cartilage of the axial skeleton depends on an influence exerted by the neural tube or notochord.

BONE

It is customary and convenient to recognize two types of bones: (1) the *membrane bones* of the face and cranial roof, which develop directly within blastemal (*i.e.,* mesenchymal) sheets; and (2) *cartilage bones,* which replace a provisional cartilaginous skeleton and comprise the deeper, remaining bones of the body. The actual method of bone formation, however, is identical in each instance; there is one, and only one, mode of histogenesis. Bone matrix begins to be deposited in the eighth week. It is laid down through the activity of specialized mesenchymal cells, named *osteoblasts* (*i.e.,* bone-formers). A soft preosseus tissue, made up of fibrillæ and a structureless ground-substance, first differentiates, but this so-called *osteoid tissue* becomes impregnated with lime salts almost as fast as it appears (Fig. 359 *A*). As this *bone matrix* is progressively laid down, some osteoblasts be-

Osteoid tissue

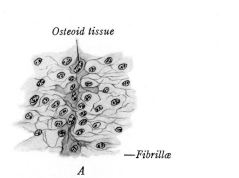

—*Fibrillæ*

A

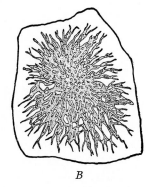

B

Fig. 359.   The development of human membrane bone. *A,* Growing tip of a spicule, at two months (after Mall; × 270). *B,* Parietal bone, at three months, in surface view to illustrate the spread of ossification (× 270).

come trapped and remain imprisoned as *osteocytes,* or *bone cells (C);* these are lodged in spaces termed *lacunæ* (Fig. 360). The total process of calcium deposition is not understood.

Development of Membrane Bones.   The flat bones of the face and cranial vault are preceded by a dense, blastemal membrane. At one or more well-vascularized points in this primitive connective tissue, *intramembranous ossification* begins. Such centers of ossification are characterized by the appearance of osteoblasts which promptly deposit bone matrix in the form of needle-like *spicules.* Since the osteoblasts are arranged in an epithelioid layer upon the surface of a spicule, the latter grows both in thickness and at its tip (Fig. 360). The expanding spicules unite into a meshwork of *trabeculæ* (*i.e.,* little beams) that spread radially in all directions (Fig. 359 *B*).

After these primary internal centers are well under way, the entire primordium becomes enclosed within a periosteum (Fig. 360 *A*). This is a fibrous membrane, condensed from the local mesenchyme. Osteoblasts differentiate on its inner surface and deposit first spongy bone, and later parallel plates (*lamellæ*) of compact bone. This process is known as *periosteal ossification.* In such manner are developed the dense inner and outer *tables* of the cranium. The mass of spongy bone joining the two tables is the *diploë.*

Much of the matrix that is first formed in any bone is provisional, and so is resorbed and replaced in varying degrees as the bone grows and assumes its final modeling. During resorption multinucleate cells, formed by the fusion of osteoblastic cell types,[10, 11] usually appear upon the surface of the bone matrix, and to these giant cells the name *osteoclast* (bone destroyer) has been given (Fig. 360 *B*). It is, however, by no means certain that the osteoclasts or any other cells are directly responsible for actual bone dissolution. Although the shape of a bone and its architectural design are hereditary, the final arrangement of the plates and trabeculæ of any bone is modified in conformity with the stresses actually encountered in postnatal life; in all instances maximum strength is gained from a minimum of material.

The open spaces of spongy bone are filled with cellular and fibrous derivatives of the mesenchyme. This at first constitutes the *primary marrow.* Later it differentiates into the reticular tissue, fat cells, sinusoids and developing blood cells that characterize the *red bone marrow.* Until the middle of fetal life the marrow resembles lymphoid tissue. It then differentiates erythroblasts, and at birth all marrow is of the red variety. Later much of the red marrow is replaced by fat cells, thereby producing what is known as *yellow bone marrow.*

Development of Cartilage Bones. Most bones of the body are preceded by a temporary cartilaginous model of the same shape as the definitive bone. This intermediate state is advantageous because cartilage is the only skeletal tissue that can grow rapidly enough to match the period of fastest growth in the fetus when a skeleton of some kind becomes necessary. The chief peculiarity of this method of bone formation is the preliminary destruction of the cartilage, which is provisional and must be got rid of before ossification can proceed. For this reason these skeletal elements are sometimes called *replacement,* or *substitution bones.* When the cartilage is once

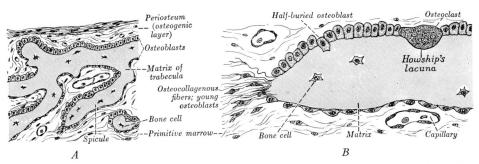

Fig. 360. The development of membrane bone. *A,* Trabeculæ near the surface of a fetal bone (× 100). *B,* Detail of the spicule tip shown in *A* (× 300).

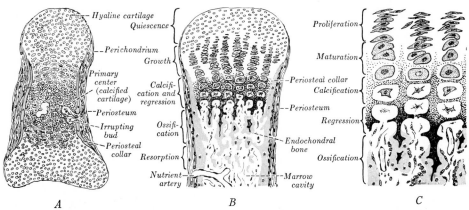

Fig. 361. Development of a cartilage bone of the finger. *A,* Stage of the primary center (× 50). *B,* Later advance of ossification (× 60). *C,* Detail of region midway of *B* (× 150).

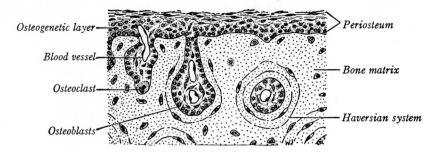

Osteogenetic layer
Blood vessel
Osteoclast
Osteoblasts
Periosteum
Bone matrix
Haversian system

Fig. 362. Three stages illustrating the origin of Haversian systems. × 250.

removed from a local region, the course of events is essentially as in the development of a membrane bone. Ossification occurs both within the eroded cartilage and peripherally beneath its perichondrium. In the first case the process is *endochondral* (or intracartilaginous); in the second instance, *perichondral* or, better, *periosteal*.

ENDOCHONDRAL BONE FORMATION.   In the center of the hyaline cartilage the cells multiply (aligning themselves in characteristic radial rows) and enlarge; at this time some lime is deposited in their matrix. The cartilage cells and part of the calcified matrix then disintegrate and disappear, thereby bringing into existence primordial marrow cavities. This destruction is paralleled by the appearance of vascular *primary marrow tissue* which simultaneously invades the cartilage and occupies the cavities. The early marrow tissue arises from the inner, cellular layer of the perichondrium and burrows into the cartilage in bud-like masses (Fig. 361 *A*). Such *irruptive tissue* gives rise both to osteoblasts and to the vascular marrow which occupies the early marrow cavities. The osteoblasts deposit matrix at many points, and at first they utilize spicules of cartilage that have escaped destruction as convenient surfaces on which to aggregate and work (*B, C*). For this reason endochondral bone is characteristically spongy.

In a progressive manner the hitherto intact regions of cartilage also undergo similar invasion, destruction and replacement until eventually the main cartilage-mass is superseded by spongy (cancellous) bone. But this bone is subject to erosion and reworking, and in older regions of long bones a central *marrow cavity* arises. Successive stages in the growth of cartilage, its replacement by bone and the creation of a marrow cavity are shown in Fig. 361 *B, C*.

PERIOSTEAL BONE FORMATION.   While the foregoing changes are occurring within the cartilage, first spongy- and later compact bone develops around it. This process is identical with the formation of the tables of the flat bones, and results from a corresponding activity of the inner osteogenetic layer of the perichondrium which henceforth is called more appropriately the *periosteum* (Fig. 361 *A, B*). During the waves of destruction that accompany the postnatal remodeling of compact bone wherever found, grooves and tubular channels become hollowed out. Penetrating buds of osteogenetic tissue then lay down secondary deposits of bone as concentric cylinders whose central axis is a tube containing blood vessels; the whole is known as an *Haversian system* (Fig. 362).

Growth of Bones. Flat membrane bones increase in lateral extent by continued marginal ossification from osteoblast-rich connective tissue at the site of the later sutures. Both cartilage and membrane bones grow in thickness through the further deposition of periosteally-formed matrix at their peripheral surfaces. In a long bone this superficial accretion is accompanied by a central resorption that destroys not only the endochondral osseous tissue but also the earlier periosteal layers (Fig. 363 *E-G*). As a result, the main shaft becomes a hollow cylinder, whereas spongy bone persists only at the ends. Red bone marrow fills all these cavities; its replacement by *yellow bone marrow* begins before puberty and is completed at about the twenty-fifth year.[12]

Many cartilage bones (especially long bones and vertebræ) increase in length by an interesting method. While still in the fetal condition, the cartilage at each end of such a bone continues to grow rapidly and ossify in the same manner that was used from the start (Fig. 363 *A-C*). However, at some time between birth and puberty, or even later, osteogenetic tissue invades these terminal cartilages and one or more secondary ossification centers, the *epiphyses,* are established there (*D*). The distal surface of the cartilaginous *epiphyseal plate,* left between the original bone and its epiphysis, continues to develop new cartilage as long as the bone lengthens. Replace-

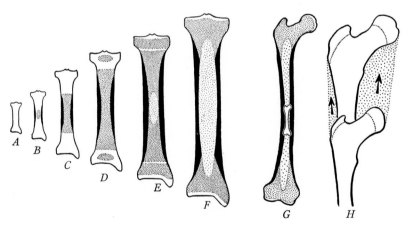

Fig. 363. Ossification and growth in a long bone. *A*, Cartilaginous stage. *B, C,* Deposit of spongy, endochondral bone (stipple) and compact, perichondral bone (black). *D*, Appearance of an epiphysis at each end. *E*, Appearance of the marrow cavity (sparse stipple) by resorption of endochondral bone. *F*, Union of epiphyses, leaving a plate of articular cartilage at free ends; enlargement of marrow cavity by the resorption of periosteal bone centrally, as deposition continues peripherally. *G*, Superposition of a human femur at birth in the marrow cavity of an adult femur, to show their relative sizes and the amount of deposition and internal resorption during growth. *H*, Two periods in the elongation of a femur, illustrating the shape (in stipple) that would prevail were the steady deposition not corrected by resorption and remodeling.

ment of cartilage by bone proceeds in equal steps at the proximal surface
of the plate (*E*).

Finally, when the adult length is attained, the cartilage plate ceases proliferation and submits to ossification throughout. As a result, the epiphyses
become firmly united to the rest of the bone (Fig. 363 *F*); the adult *epiphyseal lines* mark this union. Such union, and the cessation of bone-growth in
general, apparently occurs when the sex hormones reach a critical level in
the blood of adolescents. The omission of constant remodeling during
elongation would make many bones unnecessarily thick, heavy and un-
wieldy (*H*). If the free end of an epiphysis is to become a movable articulation, it remains cartilaginous permanently. Short tubular bones, such as
those of the fingers, have an epiphysis at one end only. Small bones, such
as those of the wrist, have a single center of ossification and no epiphyses.

Epiphyses are of three sorts: (1) pressure epiphyses, developed at the ends
of long bones; (2) traction epiphyses, affording processes for the attachment
of muscles (*e.g.*, the trochanters of the femur); and (3) atavistic epiphyses,
representing a formerly separate bone (*e.g.*, the coracoid process of the
scapula).

Most bones have more than one center of ossification (*cf.* Fig. 363 *D*). In
all there are over 800 such centers, but half of them do not arise until after
birth. On the average, therefore, there are four centers for each mature
bone. All of these appear earlier in females than do the centers of corresponding bones in males. The epiphyses of females also unite sooner with
the diaphyses, so that growth in length ceases earlier by some three years.
But even in the male most of the fusions are ending at about the twentieth
year.

Causal Relations.   The course of bone development and the assumption of characteristic form by different bones are inherent and determined through gene action. Bones
are self-differentiating organs in which the histogenesis and morphogenesis of the
cartilaginous model and its early replacing bone are not dependent on mechanical or
environmental influences. A graft of a suitable piece of an early limb bud of a chick
embryo onto the chorio-allantoic membrane, or a limb bud or primordial femur grown
in culture media, differentiates into a femur; this both develops and maintains its typical
form in cartilage, and also starts ossifying. Yet extrinsic influences are of real importance
in furnishing the conditions necessary to normal development and, as mechanical forces,
in producing the final perfections of form that are required of a functioning skeleton.

MORPHOGENESIS OF THE SKELETON

The skeleton includes both the axial skeleton (skull, vertebræ, ribs and
sternum) and the appendicular skeleton (pectoral and pelvic girdles, and
the limb bones). The plate-like bones of the face and cranial vault are
dermal elements that develop directly in a connective-tissue membrane.
This style of development is especially correlated with the acquisition of
such newer features as large cerebral hemispheres, and a prominent nose
and palate. The remaining, deeper bones of the skeleton exhibit first a

blastemal (*i.e.,* mesenchymal or membranous) stage, next a cartilaginous phase and finally a permanent, osseous replacement. It has been customary to interpret the appearance of cartilage as a recapitulation of the condition permanent in the most primitive vertebrates. On the other hand, it has been urged that the fossil evidence indicates that the earliest vertebrates had a bony skeleton, whereas cartilage came in later as a secondary, embryonic adaptation, retained as an embryonic survival in cartilaginous fishes.[13] Cartilage also occurs in the development of higher vertebrates, since it is an expansile, fast-growing tissue, of practical use as a temporary skeleton, but destined to be replaced largely by slow-differentiating bone.

The human body contains about 270 bones at birth (Fig. 387 *B*). Fusion of some of these in infancy reduces this number slightly, but from then until puberty there is a steady increase, owing to the appearance of the epiphyses and the bones of the carpus and tarsus. At puberty there are 350 separate bony masses, and this number is increased still further during adolescence. Thereafter, fusions again bring about a reduction to the final quota of 206, yet this reduction often is not completed until middle life. Age is determinable from inspection (by x-rays or otherwise) of the progress of ossification.[14, 15]

The full developmental history of the skeleton is long and detailed. Only enough basic facts will be presented in this chapter to enable the reader to appreciate general trends, to hint at the actual complexities, and to explain how the more important final results are achieved.

THE AXIAL SKELETON

The primitive axial support of all vertebrates is the *notochord,* or chorda dorsalis, the origin of which has been traced on pp. 78–81. The cellular notochord constitutes the only skeleton of Amphioxus and nearly the entire skeleton of cyclostomes, but in higher animals it is increasingly replaced by a stiffer axial skeleton composed of a skull and a jointed vertebral column. Among mammals a notochordal rod serves as a transient axial support; it persists only at the intervertebral discs, within which it swells into the mucoid *nuclei pulposi* (Fig. 366 *B*).

The axial skeleton differentiates from mesenchyme, most of which traces origin to the serially-arranged pairs of mesodermal somites. During the fourth week the ventromedial wall of each somite breaks down into a mass of diffuse cells (Fig. 364 *A*). This aggregate of mesenchyme, designated a *sclerotome,* migrates toward the notochord and surrounds it (*B*). The sclerotomes are destined to form vertebræ and ribs.

The Vertebræ. The segmental masses of sclerotomic mesenchyme are separated from similar masses before and behind by the intersegmental arteries. In embryos of about 5 mm. each sclerotome proliferates in its caudal half so that this region is denser (Fig. 365 *A*). A fissure next separates these parts and the component halves of adjacent sclerotomes reunite

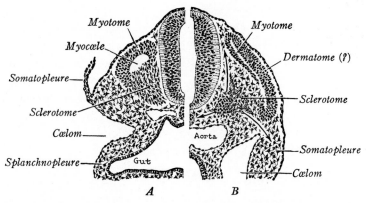

Fig. 364. Growth and separation of the human sclerotome, shown in transverse sections. *A,* Beginning migration toward notochord (at seventeen somites; × 140). *B,* Separate sclerotomic mass; arrows indicate the directions of further spread which will produce the body, arch and costal process of a vertebra (at twenty-five somites; × 115).

in new combinations.[16] That is, the denser, caudal part of each original sclerotome joins the looser, cranial half of the sclerotome next caudad (*B*). These recombinations, and not the primitive sclerotomes, are the primordia of the definitive vertebræ.

From each lateral half of a pair of primordia, growth takes place in three principal directions (Fig. 364 *B*): (1) mesad, to surround the notochord and establish the *vertebral body;* (2) dorsad, flanking the neural tube, to constitute the neural, or *vertebral arch;* and (3) ventrolaterad, in the interspaces between myotomes, to provide the *costal processes,* or primordia of the ribs. The denser portion is the dominant part of the vertebral primordium. It shares in the formation of the centrum and, almost unaided, gives rise to the processes that become the vertebral arch and ribs.[17] At this stage the mesenchymal vertebræ acquire the proportions modeled in Fig. 366 *A.*

The recombination of sclerotomic masses, already mentioned, establishes intervertebral fissures between the organizing vertebræ (Fig. 365 *B*). Mesenchymal tissue, derived from the denser portion of the vertebral primordium, condenses into an *intervertebral disc* within each such interspace.[16, 18] It is at these intervals, opposite the middle of nearby myotomes, that remnants of the notochord become incorporated in the discs and persist as the pulpy nuclei (Fig. 366 *B*).[18] Since a vertebra develops from parts of two adjacent sclerotomes, it is also evident from Figs. 365 and 366 *B* that the originally intersegmental arteries come to pass midway across the body of a vertebra, that the segmental nerves of the myotomes lie at the level of the intervertebral discs, and that the myotomes and definitive vertebræ alternate in position. This alternation is a fundamentally necessary arrangement in order that the myotomic muscles may move the spine.

Following this blastemal stage, centers of *chondrification* begin to appear in the seventh week in the highest vertebræ, and quickly follow at successively lower levels. There are two centers in the vertebral body, or *centrum,* and one in each half of the still incomplete *vertebral arch* (Fig. 367 *A*). The four centers enlarge and merge into a solid, cartilaginous vertebra. The vertebral arches do not unite and enclose the spinal cord until well into the third month. From them the transverse and spinous processes grow out between the myotomes; they will serve as levers on which the spinal muscles can act (Fig. 230).

Finally, at nine weeks, the stage of *ossification* makes a beginning (Fig. 387 *A*). Each lateral half of the vertebral arch has a single center. Whereas the vertebral body commonly appears to have but one, there are often transient indications of doubleness and even separate centers. In the fifth month centers are present in all but the sacral and coccygeal vertebræ. The full union of these primary bony components is not completed until several years after birth. At about the sixteenth year secondary centers arise in the cartilage still covering the cranial and caudal ends of the vertebral body and resolve it into disc-like bony *epiphyses* (Fig. 367 *B*). These plates, peculiar to mammals, unite with the rest of the vertebræ by the twenty-fifth year at the latest. Still other secondary centers appear during adolescence and fuse equally late (*C*). The various *ligaments* of the vertebral column differentiate from mesenchyme in proximity to the vertebræ.

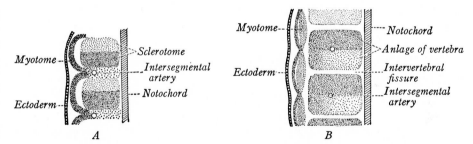

Fig. 365.   Early stages in the differentiation of human vertebrae, illustrated by frontal sections through the left somites. × about 75. *A,* At about 4 mm., showing the differentiation of each sclerotome into a less dense and denser region. *B,* At about 5 mm., illustrating the new combination of the halves of successive sclerotomes into definitive vertebral primordia.

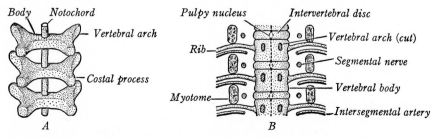

Fig. 366.   The form and relations of early human vertebræ, seen in ventral view. *A,* Models of mesenchymal vertebræ, at 7 mm. (after Bardeen; × 30). *B,* Diagram of vertebral relations, at 12 mm.

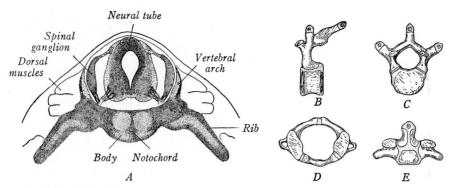

Fig. 367. The later development of human vertebræ. *A,* Chondrification centers, shown in a transverse section at 13 mm. (after Bardeen; × 18). *B, C,* Ossification of a thoracic vertebra (side and front views); composites up to seventeen years. *D,* Atlas, at three years, in cranial view. *E,* Axis, in third year, in dorsal view.

While the foregoing account holds for vertebræ in general, a few marked deviations occur. When the *atlas* is forming, its body differentiates typically but is soon taken over by the body of the *axis.* It thereafter serves as the peg-like extension, or *dens,* of the latter (Fig. 367 *E*). The definitive atlas, therefore, is merely the vertebral arch which, closing-in ventrally, takes the shape of a ring (*D*). The sacral and coccygeal vertebræ represent types with reduced vertebral arches. Between puberty and about the twenty-fifth year the sacral vertebræ unite progressively into a single bony mass; a similar fusion occurs between the rudimentary coccygeal vertebræ (Fig. 369 *B*).

The vertebral column and its associated muscles served primitively as a flexible locomotor apparatus for propelling vertebrate animals through the water. Terrestrial life introduced many functional changes, and man has altered conditions still further by adopting an erect posture; this position, and the modified locomotion that accompanies it, has made necessary certain peculiar adaptations. A narrowing of the spine occurs both in the upper thoracic region and toward its lower end. The former is correlated with the presence of ribs and sternum, which help relieve the spine in its function as a support. Similarly, the transference of weight to the pelvis relieves the lower spine. The curves in the vertebral column, which appear when the child learns to walk, have been mentioned in an earlier chapter (p. 208).

Causal Relations. The development of the cartilaginous axis of an amphibian or chick embryo depends on the spinal cord which exerts an influence over the modeling and proportions of its arches and over the jointing of its vertebral segments. Under the influence of the neural tissue even the presumptive myotome substance can be converted into skeletogenous cells. The notochord is not a positive factor in these activities.

Anomalies. Remnants of the notochord may persist within the vertebral column and give rise to tumors (*chordomas*). With the exception of the cervical region, numerical variations below (not above) the normal number of vertebræ are not infrequent (Fig. 369 *B*). The change to a secondary segmentation that results in a vertebral column may be irregularly carried out, so that a half vertebra is missing or unjoined to its mate;

successive vertebræ may fuse asymmetrically and the relations to ribs are similarly irregular (Fig. 368 *A*). Most vertebral defects are due either to the absence of certain cartilages or bony centers, or to the imperfect fusion of otherwise well-formed components. The non-union of the paired vertebral arches is *rachischisis*, or cleft spine, also known as *spina bifida*. Extensive involvement of the spine (*B*) is much less common than are relatively localized defects which tend to favor lumbar or sacral locations (*C*). The abnormality may be exposed to view (*B*) or concealed (*spina bifida occulta*) beneath intact skin (*C*).

The Ribs. The history of the ribs begins with the *costal processes* which grow out from the primitive vertebral mass and extend in the clefts between myotomes as already mentioned (p. 406). Yet only in the thoracic region do they become long bars, following the curvature of the body wall (Fig. 369 *A*). The mesenchymal rib-tissue acquires a single center of chondrifi-

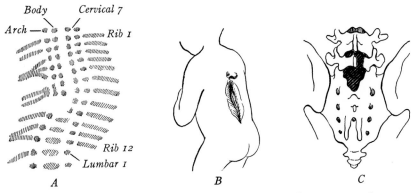

Fig. 368. Anomalies of the vertebral column. *A*, Irregular segmentation, produced by suppressed and fused centers in the vertebræ and ribs, as revealed by x-ray in a fetus of five months (after Harris). *B*, Rachischisis, or cleft spine, exposing a flat spinal cord; above is an adventitious tuft of hair and a separate opening. *C*, Spina bifida occulta, affecting the last lumbar and first sacral vertebræ.

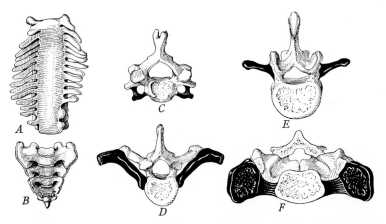

Fig. 369. Human ribs and their relations to vertebræ. *A*, Growing thoracic ribs, in ventral view, at 13 mm. (after Müller). *B–F*, Types of vertebræ, and ribs (*C–F*, in black); *B*, anomalous sacrum composed of four (instead of five) vertebræ; *C*, cervical; *D*, thoracic; *E*, lumbar; *F*, sacral.

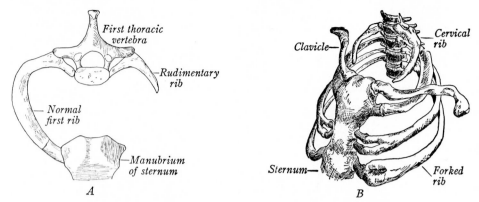

First thoracic vertebra

Rudimentary rib

Normal first rib

Manubrium of sternum

A

Clavicle—

Cervical rib

Sternum—

Forked rib

B

Fig. 370.   Cervical ribs and bifurcated ribs in an adult.

cation in the seventh week and transforms into cartilage (Fig. 367 *A*). The original union of costal process with vertebra is replaced by a joint in which a concavity on the vertebra receives the *head* of the rib; at the same time, a prominent *transverse process* of the vertebra extends outward and makes an articulation with a growing *tubercle* of the rib (Fig. 369 *D*).

A center of ossification appears in the ninth week, near the future angle of each rib, even before any centers occur in the corresponding vertebra. The cartilaginous rib progressively converts into bone, but the distal ends of the thoracic ribs always remain cartilaginous. At the fifteenth year two epiphyseal centers appear in the tubercle and one in the head. The highest development of ribs is realized in the thoracic region where they maintain movable articulations with the vertebræ and follow the curving body wall to join the sternum in the midventral line (Fig. 370 *B*). In the neck they are tiny, and unite with the cervical vertebræ; their tubercles fuse with the transverse processes and their heads with the vertebral bodies, thus leaving an interval, the *transverse foramen,* through which the vertebral arteries course (Fig. 369 *C*). In the lumbar region the ribs are short and fused to the transverse processes (*E*). The modified ribs of the sacral vertebræ are represented by prominent, flat plates which unite on each side to produce most of a *pars lateralis* of the common *sacrum* (*B, F*). Only in the first of the coccygeal vertebræ do traces of ribs remain (*B*).

Causal Relations.   Rib development in amphibians is dependent on the presence of well-formed rib-bearers on the vertebræ. In the chick the influence of the axial skeleton is less complete. Ribs, in turn, control the production and arrangement of the jointed segments (sternebræ) of the sternum.

Anomalies.   Underdevelopment produces a rudimentary rib (Fig. 370 *A*). Overdevelopment of costal processes may lead to a supernumerary rib in connection with the lowest cervical (*B*), or highest lumbar vertebra (Fig. 368 *A*). The former is important practically, since through pressure it may injure the brachial plexus or subclavian artery nearby. Forking of ribs sometimes occurs at their ventral ends (Fig. 370 *B*). The ribs tend to participate in the rare irregularities of axial metamerism (Fig. 368 *A*).

The Sternum.   The earliest indication of a breast bone, or *sternum,* is seen in a pair of mesenchymal bands that can be identified in human

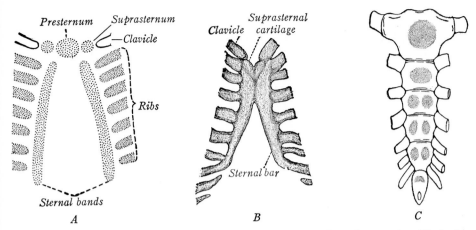

Fig. 371. Development of the human sternum. *A*, Mesenchymal stage (modified after Kingsley). *B*, Cartilaginous stage, at nine weeks. *C*, Ossification centers in a child.

embryos of six weeks. These lie ventrolaterally in the body wall and at first have no connection either with the ribs or with each other (Fig. 371 *A*).[19] Following the prompt attachment of the ribs to the sternal bars, these paired bands unite progressively in a craniocaudal direction, at the same time incorporating cranially a smaller medial mass which corresponds to the presternum of lower animals, and two variable suprasternal elements (*A, B*).[20] At nine weeks the union of the cartilaginous bars is complete. The cranial end of the developing sternum bears two imperfectly separated suprasternal cartilages with which the clavicles articulate (*B*). They usually join the manubrium of the sternum and lose their identity.

Ossification begins at about five months, but all the centers are not present until childhood. Variations in the ossification centers are not uncommon, although a bilateral tendency is evident (Fig. 371 *C*). The segmentation of the sternum into *sternebræ* is acquired secondarily.

Causal Relations. The sternal primordia of the chick and mouse, when removed and cultured, will develop a sternum in the absence of ribs, clavicles or coracoids. Explanted sternal bands move together, fuse, undergo segmental hypertrophy and ossify as in the embryo. The morphogenesis of a sternum is largely controlled by factors intrinsic in the primordia. Extrinsic factors are probably important in providing optimal conditions for the expression of potencies inherent in the primordia and in maintaining normal form. Union of the two sternal bars is due, primarily, not to their expansion or to the growth of the attached ribs, but to neighboring cellular shifts toward the midline. Segmentation into sternebræ results from an influence exerted by the ribs, and the arrangement can be altered experimentally.

Anomalies. Specimens of cleft sternum, perforated sternum and notched xiphoid process all reflect the bilateral origin of this organ (Fig. 372). 'Suprasternal bones' are merely ossified suprasternal cartilages that fail to attach to the manubrium sterni (*cf.* Fig. 371 *B*); they perhaps correspond to the epicoracoids of lower vertebrates.

The Skull. The head-skeleton includes three primary components: (1) the brain case; (2) capsular investments of the sense organs; and (3) a

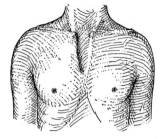

Fig. 372. Adult with a markedly cleft sternum.

branchial-arch skeleton, derived from the embryonic counterparts of the gill arches that support the mouth and pharynx of adult fishes and tailed amphibians (p. 200; Fig. 77). Apart from exceptions in the third group, these several elements unite intimately into a composite mammalian skull. The branchial-arch components originally subserved the functions of respiration and mastication, and this morphological relationship has been largely maintained.

The early notochord extends into the head as far as the pharyngeal membrane, but its termination is identified later by the caudal border of the fossa for the hypophysis located in the sphenoid bone. In replacing the notochord of the head region, early vertebrates evolved a cartilaginous cranium that is still used by sharks and allied fishes, but is represented in mammals only by the floor of the cartilaginous, and later osseous, cranium. The chordal part of this ancient cranium exists in shark embryos as two *parachordal cartilages* which accompany the notochord into the head (Fig. 373 A). In man they are united as a single *basal plate* from their first appearance; this is the forerunner of the occipital bone. Farther rostrad the prechordal part of a shark's cranium is represented by two *trabecular cartilages* (the future sphenoid bone), which flank the pituitary gland, and their fused extensions (the nasal septum of the ethmoid). A slight trace of this doubleness can be seen in human embryos.

Alongside the parachordals, cartilaginous capsules were primitively built around the otocysts, only to fuse later with the parachordals (Fig. 373). Their counterparts in mammals give rise to the petrous and mastoid portions of the temporal bone. The eyes have remained as independent, movable bulbs—although most vertebrates below mammals have incorporated cartilage or bone into their optic capsules. On the other hand, the capsules surrounding the olfactory sacs joined with the trabecular cartilages and thus contributed to the ethmoids.

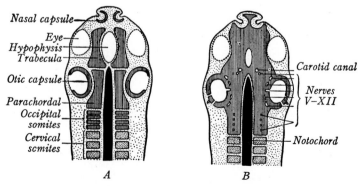

Fig. 373. Development of the chondrocranium, somewhat schematized (after Clara).

The hind- and mid-brain were primitively supported by the parachordal cartilages. However, the functional relation of the ears (both auditory and equilibratory) to the hind-brain led to the development of a prominent cerebellum, and this became housed by newer bones. These bones originated in the skin covering the brain of fishes, and they continue to differentiate similarly as membrane bones in man. The prechordal part of the skull was a primitive support for the fore-brain which expanded greatly as the result of its relation with the eyes and nose. In consequence of this, a capacious dome of dermal bone was added secondarily, and mammals similarly build a cranial vault of intramembranous origin.

Typical somites do not occur in the head. Accordingly, except in the base of the skull where some four somites are incorporated into the occipital bone (Fig. 373),[21] this part of the skeleton lacks any direct evidence of segmentation (cf. p. 436) The lines of union of the several cartilaginous components of the primitive skull are indicated by various foramina through which nerves and blood vessels find their outlets (Fig. 373 B).

The Desmocranium.   The earliest indication of the skull is a mass of dense mesenchyme which, during the fifth and sixth weeks, envelops the cranial end of the notochord and extends craniad into the nasal region (Fig. 374 A). Laterally it expands into wings which are continuous with the general head mesoderm that houses the brain. Ventrally it communicates with the mesenchymal cores of the branchial arches. The mesenchyme of the branchial-arch region is serially segmental, but this is a special kind of metamerism that reflects similar, repeated relations to the series of pharyngeal pouches. The blastemal stage of the skull is the *desmocranium.*

The Chondrocranium.   During the seventh week chondrification begins medially in the future occipital and sphenoidal regions of the blastemal skull. From here it spreads laterad and to a slight extent dorsad, and also extends into the nose (Fig. 375 A). At the same time, the internal ears

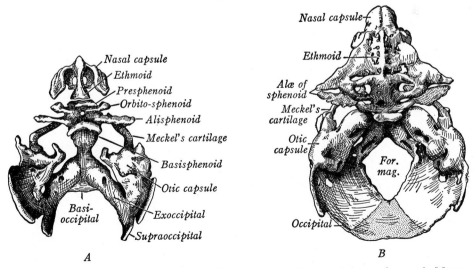

Fig. 374.   Developmental stages of the human skull, viewed from above. *A,* Mesenchymal and cartilaginous skull, at about seven weeks (adapted; × 5). *B,* Cartilaginous skull, at three months (after Hertwig; × 3).

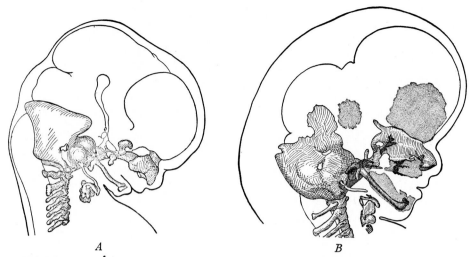

A   *B*

Fig. 375. Chondrocranium of the human fetus. *A,* During the eighth week (adapted after Lewis; × 5). *B,* At ten weeks (after Macklin; × 3). Cartilage bones are drawn in line; membrane bones are in stipple.

become invested with cartilaginous otic capsules which eventually unite with the occipital and sphenoidal cartilages. The *chondrocranium,* as it is termed, is thus confined chiefly to the base of the skull, whereas the rest of the sides and the roof is at this period a connective-tissue capsule in which membrane bones are destined to appear. Chondrification also occurs more or less extensively in the branchial arches. The process as a whole is at its height by the middle of the third month and the chondrocranium is then a unified cartilaginous mass without clear boundaries that indicate the limits of future bones (Figs. 374 *B,* 375 *B*). Some of the cartilaginous skull traces origin to migrant cells of the neural crest.[22, 23]

The Osteocranium.   In the period of ossification, which now ensues, it becomes evident that most bones develop from two or more formative centers. To a large extent these multiple origins betray ancestral histories, since it is obvious that various bones, separate in lower animals, have thus combined to produce certain compound bones of the human skull. As such individual components may have arisen either in membrane or in cartilage, the mixed nature of some compound, adult bones is also explained.

The basal part of the skull is modeled in cartilage (Fig. 375). The bones that develop in it by endochondral ossification grow in the typical manner, at the expense of proliferating cartilage. This growth is determined by the needs of not only the brain case but also the sense organs and those visceral organs that stand in relation to the branchial-arch system. By contrast, the sides and roof of the skull develop directly in membrane. A striking feature of the fetal skull is the great relative size of the neural portion (Fig. 155); the ratio of cranial to facial (or visceral) volume decreases from 8:1 at birth to 2.5:1 in the adult. Ossification of the chondrocranium begins early in the third month, but some membrane bones are even more precocious (Fig.

387 *A*). The union of the several components of compound bones is not completed until after birth; in certain ones many years elapse before final fusion is accomplished.

THE OCCIPITAL BONE. Four centers appear in the cartilage about the foramen magnum (Fig. 376). From the ventral center comes the basilar part (*basioccipital*) of the future bone; from the lateral centers, the lateral portions (*exoccipitals*) which bear the condyles; from the dorsal center, the squamous part (*supraoccipital*) below the superior nuchal line. The squamous area (*interparietal*) above that line is an addition of intramembranous origin which develops from two centers.

THE SPHENOID BONE. Ten principal centers arise in the cartilage that corresponds to this bone (Fig. 377): (1, 2) in each ala parva (*orbitosphenoid*); (3, 4) in each ala magna (*alisphenoid*); (5, 6) in the corpus between the alæ parvæ (*presphenoid*); (7, 8) in the corpus between the alæ magnæ (*basisphenoid*); and (9, 10) in each *lingula*. Intramembranous bone also enters into its composition, one center forming the orbital and temporal portion of each ala magna and another center the medial lamina of each *pterygoid process* (except the hamulus).

THE ETHMOID BONE. The ethmoidal cartilage consists both of a medial mass, which extends from the sphenoid to the tip of the nasal process, and

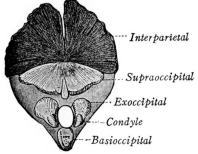

Fig. 376. Human occipital bone, at four months. × 1.5. Unossified cartilage is shown as a homogeneous background.

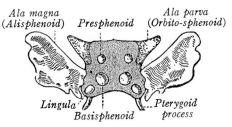

Fig. 377. Human sphenoid bone, at nearly four months. × 2. Parts still cartilaginous are represented in stipple.

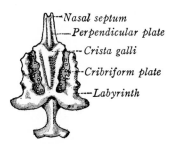

Fig. 378. Human ethmoid bone, at four months. × 1.5.

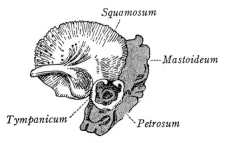

Fig. 379. Human left temporal bone, at birth. × 1. The portion of intracartilaginous origin is represented in stipple.

of a pair of masses (developed in the lateral nasal processes) lateral to the olfactory sacs (Fig. 378). The terminal part of the medial mass persists as the cartilaginous *nasal septum,* but ossification of the upper portion produces the *perpendicular plate* and the *crista galli* which complete the septum. The lateral masses ossify at first into the spongy bone of the *ethmoidal labyrinths.* From these, the definitive honeycomb structure (*ethmoidal cells*) and the *conchæ* differentiate through invaginations of the nasal mucous membrane and the simultaneous resorption of bone. (In other regions of the cranium a similar invasion of the mucous membrane and a corresponding dissolution of bone produce the frontal, sphenoidal and maxillary *sinuses;* p. 527). Fibers of each olfactory nerve at first pass between the unjoined medial mass and its respective lateral mass. Later on, cartilaginous trabeculæ surround these bundles of nerve fibers and interconnect the three masses. Upon ossifying, the perforated parts of the completed ethmoid are designated *cribriform plates.*

THE TEMPORAL BONE. Multiple centers of ossification in the cartilage of the otic capsule produce a composite, bony shell about the inner ear. This constitutes the *petrous portion* of the temporal bone (Fig. 379). A definite *mastoid process* first develops after birth by an outward bulging of the petrous bone. Its internal cavities, the *mastoid cells,* result from a postnatal invagination by the epithelial lining of the middle ear which first induces erosions and then lines the spaces thus excavated. The *squamosal* and *tympanic portions* of the temporal bone are of intramembranous origin, while the *styloid process* originates from the dorsal end of the second (hyoid) branchial arch.

MEMBRANE BONES OF THE SKULL. From the preceding account it is evident that, although the bones of the base of the skull arise primarily in cartilage, they receive substantial contributions from membrane. The remainder of the sides and the entire roof of the brain case are wholly of intramembranous origin, each of the *parietals* developing from a single (originally double) center and the *frontal* from a pair of centers (Fig. 380). To accommodate the rapid expansion of the brain, these bones grow in breadth at their margins. Progressively decreasing curvatures are accomplished by the deposit of matrix unequally on the external surface, as corresponding resorption proceeds on their internal surface. At the incomplete angles between the parietals and their adjacent bones, union is delayed for months after birth. These membrane-covered spaces constitute the *fontanelles,* or 'soft spots' (Fig. 381). Inconstant *sutural* (Wormian) *bones* appear frequently in such locations.

The *vomer* forms from two centers in the connective tissue that flanks the lower border of the perpendicular plate of the ethmoid. The cartilage of the ethmoid, thus invested, undergoes resorption. Single centers of ossification in the mesenchyme of the nasal region give rise to the *nasal* and *lacrimal bones.* The membrane bones of the maxillary and mandibular regions are described in the next paragraphs.

BRANCHIAL-ARCH DERIVATIVES. The *first branchial arch* on each side forks into a more rostral *maxillary* and a more caudal *mandibular process* (Fig. 78). Each maxillary process becomes an upper-jaw region by undergoing ossification directly within its mesenchymal tissue. In this way arise the *premaxilla* (incisive bone), *maxilla, palate, zygomatic* and *squamous* (temporal, in part) *bones* (Fig. 380).

Besides the maxilla proper there arises a separate element, the *premaxilla*, which bears the incisor teeth and fuses with the rest of the maxilla. The premaxilla is a derivative of the medial nasal processes of the embryonic face, whose extent in relation to the maxillæ is disputed (Fig. 187 D).[24] Each half of the maxilla and palate bones develops from a single center, as does also the zygomatic and squamous component of the temporal bone. The premaxilla has a major and minor center in each half.[25]

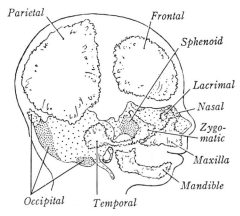

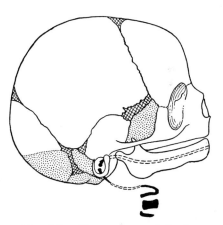

Fig. 380. Components of the fetal skull at three months. × 2. Pale areas with irregular margins represent membrane bones; densely stippled areas, cartilage bones; sparsely stippled areas, unossified chondrocranium.

Fig. 381. Components of the human skull, at birth. × ½. White areas represent membrane bone; stippled areas, cartilage bones; black areas, branchial-arch derivatives; cross-hatched areas, fontanelles.

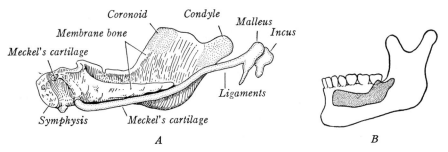

Fig. 382. Ossification and growth of the human mandible. *A*, Relation of Meckel's cartilage to the mandible at two months (after Low; × 8); the right half of the mandible is viewed from the medial side. *B*, Mandible of the newborn (stippled) superimposed on the adult mandible to show their relative size and shape (× ⅓).

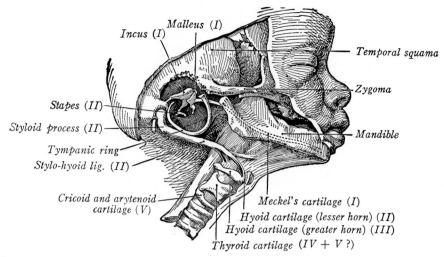

Fig. 383. Derivatives of the human branchial arches, demonstrated in a lateral dissection of a fetal head (after Kollmann).

The mandibular process and succeeding branchial arches also contribute, either through persistent cartilage or by replacement bone, to the composition of the skull. These contributions from the cartilaginous cores of the branchial arches are shown in solid black in Fig. 381. They also are illustrated more in detail in Fig. 383 and are tabulated on p. 203.

The mesenchymal core of the mandibular process transforms into a cartilaginous bar, *Meckel's cartilage,* which extends dorsad into the tympanic cavity of the ear where it becomes surrounded by the temporal bone (Fig. 375). The dorsal end of each cartilege is replaced by two auditory ossicles; these are the *incus* and *malleus,* except for minor contributions from the second arch (Figs. 382 *A*, 383).[26] Next ventrad the sheath of this cartilage converts into the *anterior malleolar* and *spheno-mandibular ligaments.*

Despite the presence of cartilaginous material, which constitutes the whole lower jaw of sharks, the main mass does not convert directly into the definitive *mandible.* Instead, membrane bone has been substituted, originally for the purpose of strengthening the primitive mandible and providing supports for the teeth. Such replacing bone develops ventrally in the body of the future lower jaw and encloses both Meckel's cartilage and the inferior alveolar nerve, whereas more dorsally, in the ramus, it takes the form of a plate that merely lies lateral to these structures (Fig. 382 *A*). This relation explains the position of the adult mandibular foramen, where the nerve enters the jaw bone. The portion of Meckel's cartilage invested by bone disappears and contributes nothing to the permanent jaw except at the tip (chin).[27] The total mandible represents the fusion of the two substitute halves, each with a single center. It changes greatly in shape as the result of growth and the acquisition of permanent teeth (Fig. 382 *B*).

Each *second branchial arch* (Reichert's cartilage) also enters into relation dorsally with the otic capsule (Fig. 375). This dorsal segment is replaced by another auditory ossicle, the *stapes* (except its footplate), and the *styloid process* which combines with the temporal bone (Figs. 383, 515).[28] The succeeding portion of the sheath of this cartilage is converted into the *stylo-hyoid ligament*. This ligament connects the styloid process with the ventral end of the cartilage, which undergoes endochondral ossification to form a lesser horn of the *hyoid bone* and the cranial part of the hyoid body (Fig. 383).

Cartilage occurs only in the ventral portions of the *third branchial arches*. These ossify and refashion into the paired greater horns of the *hyoid bone*, while the extreme ventral ends unite to constitute the caudal portion of its body (Fig. 383).[29]

The *fourth branchial arches* differentiate in their more ventral regions into the *cuneiform* and *thyroid* cartilages (Fig. 383).[30]

The *fifth branchial arches* transform similarly into the *corniculate, arytenoid* and *cricoid* cartilages (Fig. 383).[31] It is possible that a contribution is made to the thyroid cartilage, as well.

Causal Relations. The appearance of branchial arches depends on the presence of the stomodeum and pharyngeal pouches. In amphibians and birds, at least, their chondrification is accomplished by cells that emigrate from the neural crest under the influence of the pharyngeal-pouch entoderm. Parts of the trabecular cartilages also are of neural-crest origin, and the same may be true for all of the nasal capsule. Since the crest cells are ectodermal, the derived mesenchyme is sometimes called *mesectoderm*. Each epithelial otocyst induces the surrounding mesenchyme to form a cartilaginous otic capsule; a similar influence is probably exerted by the epithelium of the nasal pits on the nasal capsules.

Anomalies. An unclosed roof to the skull is designated *cranioschisis*. Severe examples are given the name *acrania;* in such specimens the head is set on the shoulders, without a neck, while the brain is virtually missing (Fig. 384 *A*). Sometimes combined with these defects is a similar open condition of the vertebral column, *cranio-rachischisis* (*B*). Premature closure of certain sutures, while growth continues along other margins, can result in distortions known as *scaphocephaly* (wedgeshaped cranium; *C*), *acrocephaly* (Fig. 156 *D*), or *plagiocephaly* (asymmetrical or twisted skull). Cleft palate (p. 230), cleft lip and allied conditions (p. 207) have been discussed previously. An undersized cranium (*microcephaly*) and an oversized cranium (*macrocephaly*) are illustrated in Fig. 156 *B, C*. In the former, premature ossification of the sutures accompanies arrested growth of the

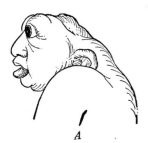

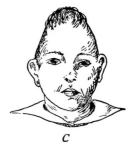

A   *B*   *C*

Fig. 384. Anomalies of the human axial skeleton. *A*, Cranioschisis, or acrania, in a newborn. *B*, Cranio-rachischisis in a newborn. *C*, Scaphocephaly.

brain. In the latter, the membranous capsule of the brain case expands rapidly to match the distending brain, but ossification lags and the expansive fontanelles are closed by secondary, sutural (or Wormian) bones.

THE APPENDICULAR SKELETON

The appendicular skeleton consists of shoulder- and pelvic supports, or *girdles,* and the skeleton of the free appendages attached to them. Fundamentally the two sets of girdles and limbs are comparable, but especially in the highest vertebrates and man has specialization complicated some of the existing homologies. Torsion in opposite directions also adds to the superficial differences between the limbs (p. 210).

The limb buds arise as elevations of the somatopleure of the body wall, at some distance from the somites (Fig. 77), and the bones of the limbs also seem to be derived directly from the unsegmented somatic mesenchyme and not from the sclerotomes.[33] In embryos of five weeks mesenchymal condensations have formed definite blastemal masses both at the sites of the future pectoral and pelvic girdles and within the primitive limb buds (Figs. 385, 396 *A*). Following this condition, the various primordia pass into a cartilaginous stage at seven weeks (Fig. 399) and then the largest of them begin to transform into bone in the eighth week (Fig. 387 *A*). Differentiation proceeds in a proximo-distal direction in the limb. The arm develops somewhat in advance of the leg, and it is not until the second year of life that the lower limb becomes longer. The appendicular skeleton of the newborn is incompletely ossified; some elements, like those of the wrist, are still wholly cartilaginous (*B*). Secondary centers organize as epiphyses between birth and the twentieth year; fusions with the main bony mass occur mostly in late adolescence.

The Upper Limb. The human pectoral girdle is drastically modified from the generalized type in land vertebrates and primitive mammals. The *sternum* (p. 410) is actually a part of the pectoral girdle; custom wrongly includes it with the axial skeleton.[32] The *clavicle* is highly developed in man and some other mammals that climb or clasp powerfully (Fig. 386).

Fig. 385. Blastemal primordium of the arm bones in the mesenchyme of a 12 mm. human embryo. × 30.

It is the first bone of the skeleton to ossify, two primary centers appearing for the shaft in embryos 15 mm. long. The style of ossification is peculiar, showing first membranous and then atypical cartilaginous features.[34]

The *scapula* arises as a single plate (Fig. 386) with two chief centers of ossification and several later epiphyseal centers. An early primary center forms the *body,* including the *spine* and *acromion.* The other center, after birth, gives rise to the rudimentary *coracoid process.* In lower forms the coracoid is a separate bone extending from scapula to sternum, but in man and other mammals that enjoy great mobility of the fore limb it unites with the body of the scapula and persists merely as a small projection.

The *humerus, radius* and *ulna* chondrify early in the seventh week (Fig. 386), and each ossifies from a single primary center in the diaphysis and an epiphyseal center at each end (Fig. 363 D-F). Additional epiphyseal centers are typical for the humerus and may occur on the radius and ulna.

In the cartilaginous *carpus* there is a proximal row of three, and a distal row of four elements (Fig. 386 B). Other inconstant or uncertain cartilages (like the centrale and pisiforme) may appear and subsequently disappear, or they may become incorporated into the carpal bones. The carpal cartilages delay their ossification until after birth (Fig. 388); each has a single center. Each *metacarpal* and *phalanx* likewise develops from a single primary center, but there is also an epiphyseal center at one end.

The Lower Limb. The human pelvic girdle follows the general plan evolved in land vertebrates and, as such, is much less modified than the scapula. The early cartilaginous plate of the future coxal, or *hip bone,* comes into relation with the first three sacral vertebræ (Fig. 389 A). A retention of the blastemal condition in the lower half of this plate accounts for the *obturator membrane* which permanently fills the foramen of the same name (C). Three main centers of ossification appear and gradually shape into the primitively dorsal *ilium,* cranioventral *pubis* and ventrocaudal *ischium* (D). Where the three elements join there is a cup-shaped depression, the *acetabulum,* which receives the head of the femur. The two pubic bones unite in the *symphysis pubis* along their midventral lengths, while the two ilia articulate with the sacrum.

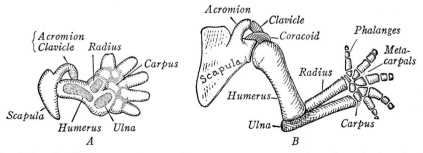

Fig. 386. Cartilaginous skeleton of the human upper limb (adapted after Lewis). × 9. *A,* At 11 mm., with chondrification (in stipple) beginning. *B,* At 20 mm., with advanced chondrification.

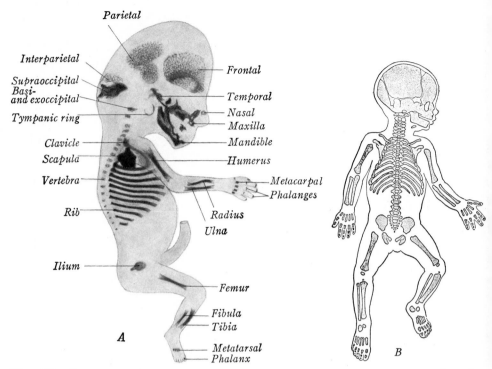

Fig. 387. Stages of ossification in human fetuses. *A,* At eleven weeks in a cleared specimen (after Broman; × 1.5). *B,* At birth (Scammon; × ⅙).

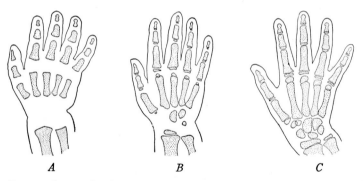

Fig. 388. Postnatal contributions to the ossification of the human wrist and hand, as shown by x-ray. *A,* At birth; *B,* at three years; *C,* at seven years.

The general development of the *femur, tibia, fibula, tarsus, metatarsus* and *phalanges* is similar to that of the corresponding bones of the upper extremity (Fig. 389 *A, B*). The *patella* is regarded as a sesamoid bone which develops within the tendon of the quadriceps femoris muscle.

Casual Relations. See pp. 211, 404.

Anomalies. Congenital absence (wholly or in part) of the clavicles occurs infrequently; it permits the shoulders to be approximated (Fig. 390). The radius or fibula is occasionally lacking and is sometimes accompanied by the absence of the thumb or great

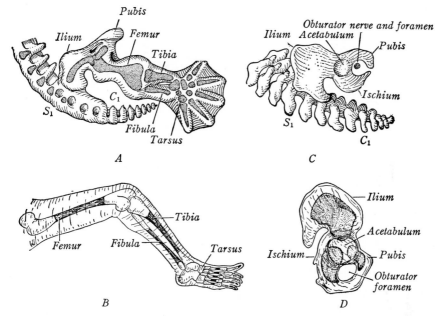

Fig. 389. Developmental stages of the human lower limb. *A,* Early chondrification, in stipple, at 14 mm. (after Bardeen; × 20). *B,* Ossification in cleared leg, at three months (× 2). *C,* Cartilaginous primordium of hip bone, at two months (after Bardeen; × 15). *D,* Ossification in cleared hip bone, at birth (× ½).

toe. External malformations of the appendages have been described on p. 211 and some conditions related to gigantism and dwarfism on p. 21.

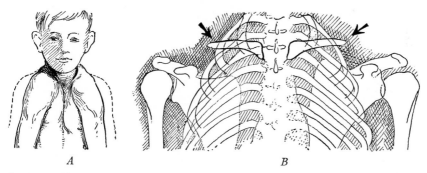

Fig. 390. Absence or reduction of the clavicle. *A,* Demonstration of how the shoulders can then be approximated. *B,* Clavicles (at arrows) represented only by their sternal ends, as revealed by x-ray.

THE ARTICULATIONS

The joints, or *articulations,* occur at regions where bones meet. These include two general groups: (1) *synarthroses,* in which little or no movement is allowed; and (2) *diarthroses,* or freely movable joints.

In joints of the synarthrodial type, the mesenchyme that intervenes

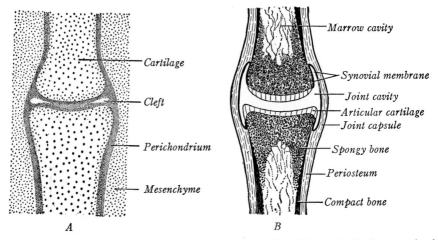

Fig. 391. Stages in the development of a diarthrodial joint. *A*, Early organization; *B*, completion.

between developing bones differentiates into a uniting layer. This union may be by connective tissue (*syndesmosis; e.g.,* cranial suture), cartilage (*synchondrosis; e.g.,* pubic symphysis) or bone (*synostosis; e.g.,* epiphyseal line).

Diarthrodial joints are characterized both by a prominent joint cavity, between the movable skeletal parts, and by a fibrous capsule at the periphery. The joint cavity arises in the third month from clefts in the dense mesenchyme, located circumferentially between the prospective bones (Fig. 391 *A*). These clefts progress in a central direction until they meet, after which the ends of the apposed cartilages retain for a time a thin cellular covering.[35] The sleeve-like *capsule* is derived from the denser peripheral tissue that borders the cavity and its fibers are continuous with the perichondrium (and, later, with the periosteum). The cells of the inner layers of the capsule become a lubricative sheet called the *synovial membrane;* it reflects onto the periphery of the apposed cartilages (or bones) but does not extend onto the articular cartilages (*B*). When there is an *articular disc,* the joint cavity develops as separate compartments while the intervening tissue differentiates into a partitioning plate of fibrocartilage. An *articular meniscus* develops similarly; since it projects from the capsule as a crescentic disc, it subdivides the joint cavity but partially.

Ligaments are fibrous bands that interconnect articulating bones and hold them in proper relation to each other. They arise from the mesenchyme associated with regions where joints are developing. Some differentiate within the joint capsule as local condensations. Others organize within the joint cavity; still others develop outside the capsule.

Sesamoid bones (e.g., patella; digitals) develop in relation both to tendons and to joints; most of them arise in the substance of the primitive joint capsule and sometimes exhibit a cartilaginous stage. *Bursæ (i.e.,* fluid-filled sacs at regions of frictional play) arise in the later fetal months as clefts in the connective tissue that stand in relation to the skin, muscles, fascias and tendons. Union with joints is secondary and late in life.

Causal Relations. A joint primordium, isolated from its surroundings and grown in culture medium, has only a limited capacity for self-differentiation; complete joint-formation cannot take place unless a minimal amount of the neighboring bones is included in the explant. Excision of the joint region from a precartilaginous blastema does not prevent joint-formation by the remaining tissue. If, however, the blastema is already partly chondrified, such excision results in fusion and lack of a joint. The characteristic shape of a joint is intrinsic in the limb mosaic. The joint cavity is produced through resistance of the more sluggish peripheral tissues to the axial expansion of the faster growing cartilaginous center. A mass-streaming of cells or a nervous, muscular or vascular influence does not enter as a factor.

Anomalies. Congenital dislocation of the hip is discussed on p. 212.

REFERENCES CITED

1. Lewis, W. H. 1922. Anat. Rec., *23,* 177–184.
2. Stearns, M. L. 1940. Am. J. Anat., *67,* 55–97.
3. Wassermann, F. 1954. Am. J. Anat., *94,* 399–400.
4. McKinney, R. L. 1929. Arch. f. exp. Zellforsch., *9,* 14–35.
5. Takagi, K. & O. Kamase. 1967. J. Elect. Micr., *16,* 330–339.
6. Hoffmann, A. 1951. Z'ts. f. mikr.-anat. Forsch., *56,* 415–449.
7. Wassermann, F. 1926. Z'ts. Zellforsch. u. mikr. Anat., *3,* 235–328.
8. Shaw, H. B. 1901. J. Anat. & Physiol., *36,* 1–13.
9. Fawcett, D. W. 1952. J. Morph., *90,* 363–405.
10. Arey, L. B. 1920. Am. J. Anat., *26,* 315–345.
11. Bloom, M. A., *et al.* 1958. Am. J. Anat., *102,* 411–453.
12. Piney, A. 1922. Brit. Med. J., *2,* 792–795.
13. Romer, A. S. 1942. Am. Nat., *76,* 394–404.
14. Noback, C. R. 1944; 1951. Anat. Rec., *88,* 91–125; Am. J. Anat., *89,* 1–28.
15. Flecker, H., *et al.* 1942. Am. J. Roent. & Rad. Ther., *47,* 95–159.
16. Prader, A. 1947. Acta Anat., *3,* 68–83.
17. Sensenig, E. C. 1949. Carnegie Contr. Embr., *33,* 20–42.
18. Peacock, A. 1951; 1952. J. Anat., *85,* 260–274; *86,* 162–179.
19. Gladstone, R. J. & C. P. G. Wakely. 1932. J. Anat., *66,* 508–564.
20. Cobb, W. M. 1937. J. Anat., *71,* 245–291.
21. Sensenig, E. C. 1957. Carnegie Contr. Embr., *36,* 141–151.
22. Stone, L. S. 1929. Arch. Entw.-mech. d. Organ., *118,* 40–77.
23. Yntema, C. L. 1944. J. Comp. Neur., *81,* 147–167.
24. Krause, B. S. & J. D. Decker. 1960. Acta Anat., *40,* 278–294.
25. Woo, J. K. 1949. Anat. Rec., *105,* 737–761.
26. Hanson, J. R. & B. J. Anson. 1962. Arch. Otolaryn., *76,* 200–215.
27. Richany, S. F. *et al.* 1956. Q. Bull. Northwest. Univ. Med. Sch., *30,* 331–355.
28. Cauldwell, E. W. & B. J. Anson. 1942. Arch. Otolaryn., *36,* 891–925.
29. Smith, S. 1925. J. Anat., *54,* 388–389.
30. Frazer, J. E. 1910. J. Anat. & Physiol., *44,* 156–191.
31. Jenkinson, J. W. 1911. J. Anat. & Physiol., *45,* 305–318.
32. Seno, T. 1961. Anat. Anz., *110,* 97–101.
33. Saunders, J. W., Jr., 1948. Anat. Rec. *100,* 756.
34. Zawisch, C. 1952. Z'ts. f. mikr.-anat. Forsch., *59,* 187–226.
35. Haines, R. W. 1947. J. Anat., *81,* 33–55.

Chapter XXII. The Muscular System

The muscular system is composed of specialized cells, called *muscle fibers,* whose specific contractile elements are the component *myofibrils.* These cells constitute a distinctive tissue in which contractility has become the predominant function. The fibers are of three sorts: (1) *smooth,* found principally in the walls of the hollow viscera, glandular ducts and blood vessels; (2) *cardiac,* localized in the myocardium of the heart; and (3) *skeletal,* chiefly attached to the skeleton. Of these types, cardiac and skeletal muscle fibers are banded with cross stripes, but only skeletal fibers are under voluntary control. All three differentiate from formative *myoblasts,* originating in the middle germ layer; the only exceptions are the smooth muscles of the iris and those of sweat- and mammary glands, which are ectodermal.

Terminal, naked branches of nerve fibers end in intimate contact with muscle fibers. In smooth and cardiac muscle the endings are simply knobbed branches. In skeletal muscle, flattened terminal networks differentiate in the fourth month; they overlie a specialized region of muscle protoplasm. This type of ending is called a *motor end plate* (Fig. 395 *E*).

THE HISTOGENESIS OF MUSCLE

All cells that are destined to become *myoblasts* of any kind, by virtue of particular locations in the embryo, begin their program of differentiation by elongating. Their general course of *myogenesis* also shows certain features in common. At an early stage the myoblasts seem to interconnect in a syncytial manner. It may be, however, that the alleged cytoplasmic bridges represent touching processes rather than actual anastomoses.[1, 2] Thread-like myofibrillæ, coursing lengthwise in the elongating myoblasts, soon make an appearance in the general cytoplasm (*sarcoplasm*) through the linear arrangement and union of cytoplasmic granules (Fig. 393).[3, 4] Such fibrils are not artefacts but real structures that can be observed in living cultures of embryonic muscle.[5] The most primitive, thin *myofilaments* join into compound *myofibrils* and both of these are said to multiply by splitting.[4, 6] The fibrils of smooth muscle remain homogeneous threads; those of cardiac and skeletal muscle become more specialized by acquiring alternate *dark-* and *light bands,* the former being denser regions (Fig. 395 *C*). These bands and the *Z-lines* are not seen in the ground sarcoplasm and the Z-lines are not continuous with the cell membrane, or *sarcolemma.* After a time all

426

of the fibrils become aligned in such a manner that the dark and light bands occur at the same levels, and so the entire fiber appears to be cross-striped.

Smooth Muscle. As early as the fifth week cells of the general mesenchyme aggregate about the epithelial tube of the esophagus, and this advance is soon duplicated whenever smooth muscle is to develop. These cells enlarge and elongate, and in them myofibrils make an appearance (Fig. 392). The cell nucleus retains a central position and also elongates to adapt itself better to the spindle shape of the muscle fiber.

Some of the primitive fibrils of the myoblasts cluster together, making thick bundles (Fig. 392 B). Perhaps these correspond to the coarser *border fibrils* which sometimes have been interpreted as inert, supporting elements.[7] Part of the reticular fibrils that surround muscle fibers have been claimed to be produced by the myoblasts themselves.[8] In older fetuses new muscle elements arise not only by the transformation of interstitial cells located between fibers and the specialization of mesenchyme at the surface of the muscle mass, but also by the mitotic division of fibers already present.

Cardiac Muscle. The cardiac type of involuntary muscle develops from the splanchnic mesoderm that invests the primitive heart tubes as the primitive myocardium (Fig. 336 C). Myofibrils arise first at the periphery of cells and soon seem to extend long distances because cell limits are not plain (Fig. 393). The nuclei remain centrally placed; after birth, cells may contain more than one nucleus. Z-lines are the first markings to appear. The characteristic *intercalated discs* are earliest recognizable (with the light microscope) in later fetal stages.[9] As specializations about cell junctions, they are seen with increasing frequency as age advances.

The former belief that cardiac muscle remains permanently syncytial has been disproved by electron microscopy.[10] The *Purkinje fibers* of the impulse-conducting system, located directly under the endocardium, take a different line of specialization from ordinary cardiac fibers.[11] They are thick elements, swollen about the nuclei; their few myofibrils are located peripherally in the fiber. The sinu-atrial node appears at birth.

Skeletal Muscle. Striated voluntary muscle is derived either from the paired somites (muscles of the neck, trunk and, possibly, limbs) or from mesenchyme of the branchial arches (muscles of the head and, in part, of the neck). The portion of a somite that is left after the emigration of the sclerotome mass to form a vertebra is the *myotome,* or muscle plate (Fig. 394). This plate thickens and its cells differentiate into myoblasts. These spindle-shaped elements arrange themselves parallel with the long axis of

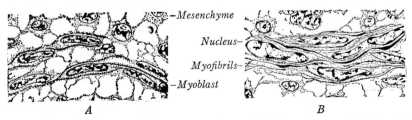

Mesenchyme

Nucleus—

Myofibrils—

—Myoblast

A   B

Fig. 392. Stages in the histogenesis of smooth muscle (adapted after McGill). *A,* At 13 mm. (× 550); *B,* at 27 mm. (× 850).

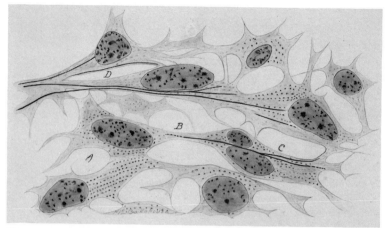

Fig. 393. Stages in the histogenesis of cardiac muscle from a 9 mm. rabbit embryo (adapted after Godlewski). *A*, Linear arrangement of granules; *B*, coalescence of granules into a fibril; *C*, fibril splitting; *D*, fibrils extending through apparent syncytium.

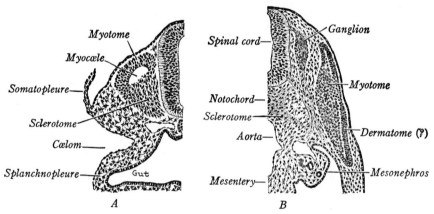

A   *B*

Fig. 394. Growth and differentiation of the human myotome, shown in transverse half-sections. *A*, At 3 mm., with the myotome becoming a distinguishable entity (× 140). *B*, At 7 mm., with the myotome separate from the sclerotome and undergoing histogenetic differentiation (× 48).

the body and transform into skeletal muscle fibers. Certain mesenchymal cells of the branchial arches undergo a similar metamorphosis into additional skeletal muscle that is histologically indistinguishable.

At the beginning of differentiation the nucleus lies centrally within a fiber, surrounded by a granular cytoplasm (Fig. 395 *A*). The latter gives rise to the myofibrils, at first located solely at the periphery of the myotube (*B*, *D*). In the third month the light and dark markings on the fibrils first come to coincide in alignment and thereafter appear as continuous bands across the fiber (*C*). During the later fetal months numerous nuclei within a fiber are crowded to the surface by the uneven distribution of newly forming fibrils. Until the middle of fetal life, at least, new muscle fibers arise by

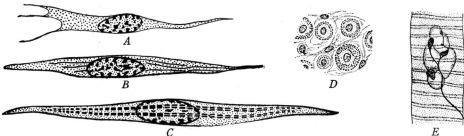

Fig. 395. The histogenesis of skeletal muscle. *A–C,* Myoblast stages, showing fibril formation, splitting and striation (after Godlewski). *D,* Transverse sections of fibers, at different stages of differentiation ($\times$ 400). *E,* Motor end plate, at birth (after Tello; $\times$ 1000).

the continued differentiation of myoblasts and, it has been claimed, by the longitudinal splitting of fibers already present. When this proliferation ceases, all further enlargement of a muscle is by the relatively enormous increase in the size of individual fibers.

There is a possibility that the lateral wall of the somite is not myotome material but becomes a *dermatome,* which furnishes connective tissue for the skin (Fig. 394 *B*). The most controverted topic of muscle histogenesis is how the early fibers become long and multinucleate.[12] One interpretation views a completed fiber as a greatly drawn-out myoblast whose nucleus has undergone repeated division. Other investigators, however, maintain that the same end is accomplished by the union of separate myoblasts into a composite fiber. The latter claim is now well proven.[31] Sensory nerve endings (*muscle spindles*) differentiate at three months (Fig. 484 *D*) and the motor nerve endings (*end-plates*) in the fourth month (Fig. 395 *E*). Both cardiac and skeletal muscle fibers become contractile before myofibrils differentiate or nerves make endings on them.[13, 14] During myogenesis there is a variable destruction of partly formed fibers to make room for blood vessels and connective-tissue stroma; some believe that this degenerative phase is less widespread than has commonly been credited.

MORPHOGENESIS OF THE MUSCLES

The muscles of the body are distributed in two systems; these comprise the visceral musculature and the skeletal musculature.

The Visceral Musculature. This group, largely of splanchnic meso-dermal origin, is associated chiefly with the hollow viscera and is under the involuntary control of the autonomic nervous system. Except for the striated cardiac muscle in the wall of the heart, the visceral muscles are smooth. The individual smooth-muscle fibers are bound together by reticular fibrils; groups of fibers, assembled as interlacing bundles or orderly sheets, are united by coarser fibrous tissue. In tubular organs, the fibers of a layer are usually oriented either in a longitudinal or circular direction.

The Skeletal Musculature. As the name indicates, these striated voluntary muscles are attached primarily to the skeleton. Except for the branchial arch muscles of the head and neck, and probably those of the limbs, the skeletal muscles originate from myotomes. Mesodermal somites first appear in the future occipital region of embryos about 1.5 mm. long (Fig. 74) and the full number of about forty is acquired at 8 mm. (Fig. 76). Early in the fifth week, the older myotomes enter upon the differentiation of muscles; within the remarkably short space of the next three weeks the definitive muscles of the fetus become well fashioned and begin to be capable of correlated movements (Fig. 396 B).[15] In this process of morphogenesis, the muscle fibers aggregate in groups that constitute the individual muscles. These are true organs, supported and enclosed by connective tissue differentiating from out the local mesenchyme. *Tendons* are fibrous cords that arise independently and attach muscles to bones.[16]

Since there are more than 650 named muscles in the human body, the treatment of the musculature as organs must be limited to essential generalities. First it will be instructive to examine the basic methods that muscle primordia employ in attaining their final state. Then the history of the more important regional groups of muscles will be traced in outline.

FUNDAMENTAL PROCESSES. With only a few exceptions, muscles retain their original innervation throughout life.[17] This is true whether a muscle is of myotomic origin and innervated by a segmental spinal nerve (or nerves, if compound) or is of branchial-arch origin and innervated by a cranial nerve. Similarly if a nerve supplies more than one muscle it can be assumed that these muscles are subdivisions of an original myotome. For this reason the histories of adult muscles formed by early fusion, splitting, migration or other modifications have been reconstructed with considerable certainty. A nerve enters its muscle mass at or near the midpoint.

An analysis of how muscles develop grossly shows that several basic principles are operating, and that these are utilized again and again by different muscles throughout the body. Six important factors of this sort are the following:

1. A *change in direction* of muscle fibers from their original craniocaudal orientation in the myotome. The fibers of but few muscles retain their initial orientation parallel to the long axis of the body.

2. A *migration* of muscle primordia, wholly or in part, to more or less remote regions. Thus the latissimus dorsi originates from cervical myotomes, but finally attaches to the lower thoracic and lumbar vertebræ and to the crest of the ilium. A shift in the opposite direction is shown by the facial musculature of expression, which takes origin in the second branchial arches. The muscles of the ventral trunk illustrate ventral growth from the dorsally placed myotomes.

3. A *fusion* of portions of successive myotomes into a composite muscle. Both the rectus abdominis and the sacro-spinalis illustrate this process. In fact, few muscles are derivatives of single myotomes.

4. A *longitudinal splitting* of myotomes or branchial-arch muscle primordia into subdivisions. One example is found in the sterno- and omohyoid, another in the trapezius and sterno-mastoid.

5. A *tangential splitting* into two or more layers. The oblique and the transverse muscles of the abdomen are formed in this common way.

6. A *degeneration* of myotomes or parts of myotomes. By this method aponeuroses and some ligaments are produced.

MUSCLES OF THE NECK AND TRUNK. The previous chapter has explained how and why the myotomes come to alternate with the permanent verte-bræ (Figs. 365, 366 *B*). From them arise not only the dorsal musculature, but also the lateral and ventral muscles of the neck (except branchial-arch derivatives) and trunk. In embryos 8 mm long all the myotomes are fusing superficially, while ventral extensions are growing down from myotomes of the cervical and thoracic regions (Fig. 396 *A*). In this manner the body wall becomes temporarily segmental, through invasion, as a secondary phe-nomenon. Fishes retain such original myotomic segments throughout the trunk, as shows well in their easily separable layers of muscle. Man retains a similar muscular metamerism in his intercostal muscles, whose individual fibers stretch from rib to rib.

At five weeks the fusing and ventrally spreading myotomic tissue on each side makes a start toward becoming subdivided into a dorsal or *epaxial*

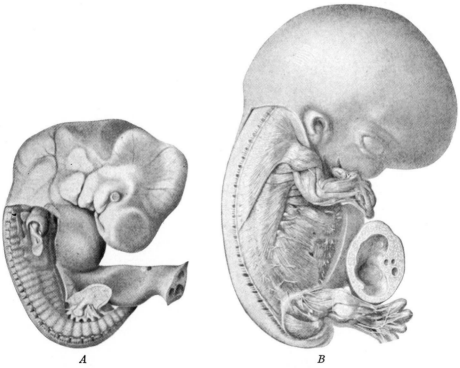

A   B

Fig. 396. Muscles of the trunk and limbs in human embryos (after Bardeen and Lewis). *A,* At 9 mm., showing the partially fused myotomes and the premuscle masses of the limbs (× 7); distally, in the upper limb, the radius, ulna and hand plate are disclosed; in the lower limb the primordial hip bone and the border vein show. *B,* At 20 mm., with definitive muscles in an advanced state of organization (× 4.5).

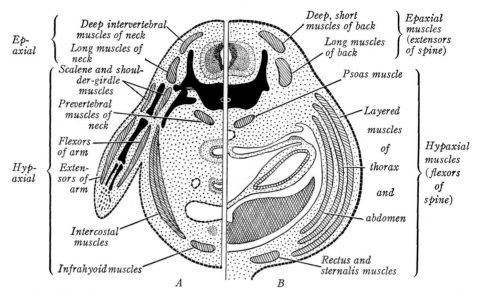

Fig. 397. The arrangement of primitive muscle-masses, shown diagrammatically in a human embryo of seven weeks. *A,* Half-section through the neck and shoulders. *B,* Half-section through the thorax and abdomen.

column and a more ventral or *hypaxial series*. These portions are inner-vated, respectively, by dorsal and ventral rami of the serially repeated spinal nerves (Fig. 470).

The epaxial column subdivides further. A more superficial, fused muscle-portion gives rise, by longitudinal and tangential splitting, to the various *long muscles* of the back of the neck and trunk; these serve as extensors of the vertebral column (Figs. 396 *B*, 397). A deeper muscle-portion retains better the primitive segmental arrangement and gives rise mainly to short *intervertebral muscles*. The spinal musculature was primitively a powerful sculling mechanism—the first locomotor apparatus of vertebrates. The de-veloping transverse processes of vertebræ are agents that help separate the dorsal and ventral sets of muscle primordia. As the rib elements develop, this separation is intensified and the segmental arrangement of the hypaxial set in the thoracic region becomes emphasized (Figs. 230, 398 *A*). The hypaxial masses invade the region ventral to the vertebræ and become the *prevertebral muscles* (Fig. 397 *A*); they also extend into the lateral and ventral body wall where they form the layered *muscles of the chest and abdomen* (*B*).[18, 19] These muscles are primarily flexors of the vertebral col-umn. Such specializations are well advanced in embryos of seven weeks (Fig. 396 *B*).

The *cervical muscles* are strongly developed dorsally as epaxial extensors, both superficial and deep. The lateral and ventral regions are highly modi-fied by the presence of branchial arches and fore-limbs, and by the caudal recession of the cœlom from these levels (Fig. 397 *A*). Nevertheless, the hypaxial masses are represented laterally by certain prevertebral muscles, the scalenes and parts of other muscles; ventrally, by the infrahyoid muscles. The *muscles of the diaphragm,* which organ in early stages lies in the future neck region, perhaps differentiate from ventral extensions of cervical myotomes. It is at this level that the phrenic nerves arise, enter the septum transversum and participate in the caudal migration of the diaphragm. But this evidence is wholly circumstantial (p. 293).[20]

The *thoraco-abdominal muscles* can be regarded as type forms, illustrating the muscle-morphogenesis of the trunk. The dorsal (epaxial) musculature, both superficial and deep, is well formed at all levels. In the thoracic region the lateral and ventrolateral musculature (hypaxial) are represented by the *intercostal muscles* which split tangentially into three layers (Fig. 230). Ventrally there is occasionally an ancestral muscle, the *sternalis.* Similarly in the abdominal region the expansive lateral portion of the fused muscular sheet on each side splits tangentially into the *transverse-* and two *oblique muscles* (Figs. 396 *B,* 397 *B*). Ventrally the invading muscle takes the form of paired longitudinal bands, the *rectus muscles,* whose segmental appearance is a secondary acquisition. Only the first pair of the lumbar myotomes participates at all typically in the formation of the lateral and ventral walls. An important lumbar feature is contributed by dorsomedial representatives of the hypaxial myotomic tissue; these form the *psoas muscles,* which correspond to the prevertebral muscles of the neck.

The primitive body wall of an early embryo is merely somatopleure and, as such, is extremely thin and transparent. Dorsally and laterally it begins to thicken as the myotomes expand and send muscular sheets toward the ventral midline (Figs. 229, 230). At seven weeks a large, diamond-shaped area surrounding the attachment of the umbilical cord still remains uninvaded. When the advancing thick margins, right and left, finally meet and fuse, the rectus muscles lie side-by-side and the *linea alba* (or white line

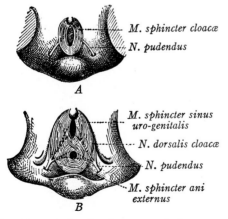

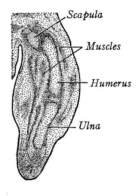

Fig. 398. Early development of the human perineal muscles (Popowsky). *A,* At two months; *B,* at three months.

Fig. 399. Muscle primordia of the human arm, at 16 mm. (× 18).

of junction) is created, thereby closing the true body wall midventrally at three months.

The *pelvic musculature* is highly modified from type. The dorsal division, in the lower sacral and all coccygeal somites, transforms into ligaments. If the ventral extensions of the second lumbar to second sacral myotomes persist at all, it is possible that they contribute to the formation of the lower limbs. The third sacral to first coccygeal myotomes, however, do form the muscles of the pelvic diaphragm and, probably, the other *perineal muscles* as well.[21] This region develops somewhat tardily. Temporarily there is a common cloacal sphincter (Fig. 398 *A*), but this undergoes subdivision in conformity with the partitioning of the cloaca into rectal and urogenital canals (*B*). The more dorsal of the secondary sphincters persists as the external anal sphincter, while the ventral sphincter differentiates into the muscles related to the urogenital sinus of both sexes.

FATES OF THE MYOTOMES OF THE NECK AND TRUNK

MYOTOME PORTION	CERVICAL MUSCLES	THORACIC MUSCLES	ABDOMINAL MUSCLES
Epaxial:			
Superficial......	Long muscles of back of neck and trunk, not associated with limbs		
Deep..........	Deep, short intervertebral muscles of back of neck and trunk		
Hypaxial:			
Dorsomedial....	Prevertebral group	....	Psoas
Lateral........	Scalenes Trapezius (in part) Sterno-mastoid (in part)	External intercostals	External oblique Quadratus lumborum (dors. layer)
		Internal intercostals	Internal oblique Quadratus lumborum (vent. layer)
		Transversus thoracis	Transversus abdominis
Ventral........	Infrahyoid group	Sternalis (inconstant)	Rectus abdominis

MUSCLES OF THE LIMBS. The early limb buds are ectodermal sacs, stuffed with mesenchyme of seemingly local origin (Fig. 619). At about the 9 mm. stage this mesenchyme condenses into premuscle masses (Fig. 396 *A*). From them the girdle- and limb muscles differentiate; the proximal ones are the first to appear and, at any level, the extensors appear sooner than the flexors (Fig. 399). The progressive modeling of distinct muscles reaches the level of the hand and foot in embryos of seven weeks (Fig. 396 *B*). The upper limbs naturally gain an initial advance over the lower limbs.

The developing muscles tend to become arranged into a dorsal group of *extensors* and a ventral group of *flexors* (Fig. 397 *A*). These are innervated by nerves that divide into a dorsal division (*e.g.,* radial nerve), supplying extensors, and a ventral division (*e.g.,* ulnar and median nerves), supplying flexors. Because of the opposite rotations of the arms and legs in reaching their definitive positions (p. 211), the muscles on the inner side of one set of limbs are homologous to those on the outer side of the other.

In sharks it is plain that buds from the myotomes grow into the embryonic fins and there break down into mesenchyme that is the source of the fin muscles. With higher vertebrates this is not so clearly the case, and in birds and mammals a direct myotomic origin of the muscles of the appendages is usually denied. In this connection it should be emphasized that the segmental nerve supply of the limb muscles of higher animals is merely suggestive, not proof, of a myotomic origin. Although a diffuse migration of cells from the ventral edges of human cervical myotomes has been claimed[22] and even tiny myotomic extensions toward the limbs are on record, there is still doubt that the myotomes contribute significantly to the musculature of the limbs.[19, 23, 30]

MUSCLES OF THE HEAD. With minor exceptions, to be mentioned presently, the muscles of the head differ from all other skeletal muscles in that they arise from the general head mesoderm, located rostral to the somites and body cavities; more specifically this parent tissue is mesenchyme within the paired series of branchial arches. Such premuscle masses in the arches are innervated by nerves (visceral), quite different in origin and functional relations from those (somatic) that supply myotomic muscles (p. 510). Assuming that all the muscles derived from the several arches retain their primitive branchial-arch innervation, it is easy to establish origins and relationships (Fig. 400 *A*; table, p. 203). In practice, the rapidly decreasing size of arches caudal to the second, makes their derivatives difficult to trace with certainty, so that some conclusions have been reached as logical deductions rather than through actual demonstration. All of these transformations occupy the sixth to eighth weeks of development.

It is established that the mesenchyme of the *first branchial arches* gives

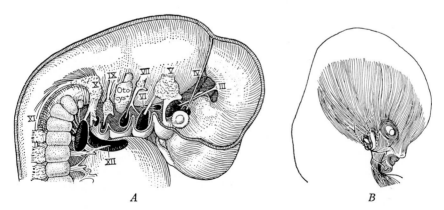

A   *B*

Fig. 400. Development of the muscles of the human head and neck. *A*, Premuscle masses, at 8 mm. (× 10). The following muscles (in black) are identified by their numbered cranial nerves: ocular (III, IV, VI); masticatory (V); facial (VII); pharyngeal (IX); laryngeal and palatine (X); sterno-mastoid and trapezius (XI); lingual (XII). *B*, Superficial muscles of the head, migrated from the second arch at seven weeks (after Futamura; × 4); the distribution of the facial nerve is also shown.

rise to the muscles of mastication and a few others, all of which are innervated by the trigeminal (fifth) nerves which supply a premuscle mass in each
arch (Fig. 400 *A*). Similarly the facial muscles of expression, and those
others supplied by the facial (seventh) nerves, originate from the *second
arches* (*A, B*). The *third arches* sink below the surface and are the source
of the stylo-pharyngeal muscles (and perhaps the upper pharyngeal con-
strictors as well), which receive branches of the glossopharyngeal (ninth)
nerves. The *fourth* and *fifth arches* share the vagus (tenth) nerves; they
innervate muscles, presumably derivatives of these arches, such as the laryngeal muscles and most of the pharyngeal and palate group. The accessory
(eleventh) nerves, each really a motor portion of a vagus complex, innervate the sterno-mastoid and trapezius muscles, which are regarded as of
caudal branchial-arch origin in part.[24]

The *muscles of the tongue* are innervated by the hypoglossal (twelfth)
nerves. Since this pair of nerves stands in relation to the early occipital
somites, it is held that the lingual muscles are derived from myotomes of
the occipital region, even though the tissue of direct origin is mesenchyme
(Fig. 400 *A*).

The extrinsic *ocular muscles* trace origin from three mesenchymal masses,
located over each optic cup. These condensations differentiate into the
several muscles, which are activated by the third, fourth and sixth cranial
nerves (Fig. 400 *A*). There is some reason for suspecting that the premuscle
masses are descendants of ancestral myotomes.

Definite somites do not occur in the head region, except at the base (Fig. 400 *A*).
Here the occipital somites are situated, as are also their associated nerves. The hypoglossal pair of nerves belonged originally to a member of the spinal series and each
rootlet is still wholly comparable to the ventral (motor) root of a spinal nerve. That
the lingual muscles are historical derivatives of the occipital myotomes is undoubtedly
true. Occipital somites of the chick embryo, thymidine labeled, yield surely marked
tongue muscle.[25] In the mammal a mesenchymal mass has been followed from the occipital myotomes to the tongue region.[26]

An interpretation of the ocular premuscle masses as altered myotomes finds support
in comparative embryology. Lower fishes and reptiles have three specialized myotomes,
rostral to the otocyst, from which the extrinsic ocular muscles develop. The three cellular
masses observable in the embryos of birds and mammals seem to be comparable.[27, 28]
That these premuscle masses are, indeed, homologous to myotomes is enhanced by the
fact that the eye-muscle nerves are somatic motor in nature. They thus are of the same
type (in morphological and functional relations) as the ventral roots of spinal nerves,
which innervate muscles derived from typical myotomes. Such somatic motor nerves are
quite different from those supplying the head muscles in general, but they are in the
same category as the hypoglossus where the relation to myotomes is similar.

Segmentation of the Vertebrate Head. The vertebrate head consists of fused segments. This was suggested to the earlier workers by the serial arrangement of the
branchial arches (*branchiomerism*), by the presence of supposedly significant '*neuromeres*'
in the brain wall (p. 478), and by the discovery of the specialized somites mentioned in
a preceding paragraph.

The only somites to persist are related to the eyeball and tongue. All the remaining
muscles of the head are derived from branchiomeres. Even assuming that the branchio-

meres represent portions of the primary head somites—and there are sufficient observations which tend to disprove this—their segmentation still is not comparable to that of the trunk; this is because the branchial arches originate through the serial division of lateral mesoderm, tissue which in the trunk never segments. The branchial arches, therefore, represent a different sort of metamerism.[29] From what has been said it is evident that one cannot compare the relation of the cranial nerves to the branchiomeric muscles with the relation of a spinal nerve to its myotomic muscles. Because of this, the cranial nerves furnish unreliable evidence as to the primitive number of cephalic segments. Using various methods of attack, different investigators have set this hypothetical number between eight and nineteen.

Causal Relations. Somites arise through the inductive influence of the notochord, but muscles, differentiating after the neurula stage, do so independently of this stimulus. Although the establishment of myoblast proliferation-centers is then governed by the neural tube, muscle development becomes a self-differentiating phenomenon even at an early stage. Muscles will develop normally in the absence of nerve, as is shown in certain types of monstrous development and by experiments in which nerves have been removed from a region or limb buds have been transplanted to a nerve-free site. Hence nerves do not supply a necessary organizing stimulus, and muscle-morphogenesis will proceed to completion without function ever having entered. On the other hand, a developing muscle normally provides some attractive forces that serve to guide a nerve to it. When a muscle has attained full histogenetic development, it is subject to some kind of trophic influence exerted by its nerve and, if denervated, the muscle undergoes atrophy.

Anomalies. Sometimes a whole muscle or a part of a compound muscle is lacking because of agenesis. In general, muscles are notoriously variable. Some variations are progressive (such as producing a separate deep flexor of the fingers). Others are regressive (such as the palmaris longus, whose fleshy belly may be insignificant or lacking). Some muscles, not normally encountered, are occasionally represented, while other constant components may have abnormal relations or attachments; since both conditions simulate features found regularly in lower primates, these occurrences are viewed as an expression of atavism. Certain muscles are represented regularly (*e.g.,* ear; scalp), but are vestigial when compared to their presence and use in other mammals. Others appear only occasionally and are interpreted as atavisms (*e.g.,* elevator of clavicle; sternalis; tail muscles). Variations in the form, position and attachments of the muscles are common. Most muscular anomalies are referable to an over- or underexpression of particular developmental factors, as listed on pp. 430–431.

FASCIÆ

The variously located *fasciæ* consist of connective tissue that may take the form of dense membranes or loose fibrous tissue. Although often described as if they were separate sheets, they actually constitute a continuous system or spongework of sheaths in which are contained muscles, bones, vessels, nerves and glands. Fasciæ can best be understood on the basis of development. When the various primordia are differentiating into muscles, vessels and other organs in any region, some of the mesenchymal building material is left over. This indifferent tissue, which embeds the specific primordial organs, then differentiates into connective tissue and so creates the fascial system of that region. Tendinous fasciæ are called *aponeuroses.*

REFERENCES CITED

1. Lewis, W. H. 1926. Carnegie Contr. Embr., *18*, 1–21.
2. Hogue, M. J. 1937. Anat. Rec., *67*, 521–535.
3. Weed, I. G. 1937. Z'ts. Zellforsch. u. mikr. Anat., *25*, 516–540.
4. Challice, C. E. & G. A. Edwards. 1961. pp. 44–73 in Pres: Specialized Tissues of the Heart (Elsevier).
5. De Renyi, G. S. & M. J. Hogue. 1938. Anat. Rec., *70*, 441–449.
6. Van Breeman, V. L. 1952. Anat. Rec., *113*, 179–195.
7. McGill, C. 1907. Internat. Monats. Anat. u. Physiol., *24*, 209–245.
8. Haggquist, G. 1931. Bd. 2, Teil 3 in Möllendorff: Handbuch (Springer).
9. Zschiesche, E. S. & E. F. Stillwell. 1934. Anat. Rec., *60*, 477–486.
10. Muir, A. R. 1957. J. Biophys. & Biochem. Cytol., *3*, 193–201.
11. Walls, E. W. 1947. J. Anat., *81*, 93–110.
12. Konigsberg, I. R. 1965. pp. 337–358 in DeHaan and Ursprung: Organogenesis (Holt).
13. Goss, C. M. 1940. Anat. Rec., *76*, 19–27.
14. Strauss, W. L., Jr. & G. Weddell. 1940. J. Neurophysiol., *3*, 358–369.
15. Fitzgerald, J. E. & W. F. Windle. 1942. J. Comp. Neur., *76*, 159–167.
16. Long, M. E. 1947. Am. J. Anat., *81*, 159–197.
17. Haines, R. W. 1935. J. Anat., *70*, 33–55.
18. Seno, T. 1961. Acta Anat., *45*, 60–82.
19. Theiler, K. 1957. Acta Anat., *30*, 842–864.
20. Wells, L. J. 1954. Carnegie Contr. Embr., *35*, 107–134.
21. Power, R. M. H. 1948. Am. J. Obst. & Gyn., *55*, 367–381.
22. Zechel, G. 1924. Z'ts. Anat. u. Entw., *74*, 593–607.
23. Glücksmann, A. 1934. Z'ts. Anat. u. Entw., *102*, 498–520.
24. McKenzie, J. 1955. J. Anat., *89*, 526–531.
25. Hazelton, R. D. 1970. J. Embr. & Exp. Morph., *24*, 455–466.
26. Bates, M. N. 1948. Am. J. Anat., *83*, 329–355.
27. Adelmann, H. B. 1929. J. Morph., *44*, 29–87.
28. Gilbert, P. W. 1957. Carnegie Contr. Embr., *36*, 59–78.
29. Kingsbury, B. F. 1926. J. Morph. & Physiol., *42*, 83–109.
30. Saunders, J. W., Jr. 1948. Anat. Rec. *100*, 756.
31. Bassleer, R. 1962. Zts. Anat. u. Entw., *123*, 184–205.

C. ECTODERMAL DERIVATIVES

Chapter XXIII. The Integumentary System

The contributions of ectoderm to the development of the teeth, tongue, palate, salivary glands, hypophysis and anal canal are described in earlier chapters. Here will be presented the histogenesis of the skin and the development of its specialized derivatives, all of which make up what is known as the *integument.*

THE SKIN

The skin is an organ of double origin. Its superficial component is a stratified epithelium, called the *epidermis,* that specializes from the general ectoderm other than that involved in making the nervous system. The epidermis lies upon a basement membrane of noncellular, polysaccharide nature. This separates the epidermis from a fibrous *dermis,* or *corium,* of mesenchymal origin. Beneath these two layers of the skin is the loosely fibrous *subcutaneous layer,* containing much fat.

The Epidermis. The embryonic ectoderm is originally a single sheet of cuboidal cells (Fig. 401 *A*), but in the fifth week it begins to add a second layer (*B*). The outer cells make up a distinct, transitory layer named the *periderm.* Its cells flatten, cornify and eventually spread to several times the diameter of the deeper cells. The basal cells, cuboidal in shape, are the reproducing elements that presently give rise to new layers above them. During the third and fourth months the epidermis is typically three-layered, an intermediate stratum being gradually interposed between the basal and periderm cells (*C, D*).

After the fourth month the epidermis becomes increasingly stratified and specialized (Fig. 401 *E, F*). The lower layers consist of living cells, whereas the upper layers constitute 'dead skin.' The deepest stratum (basal cells)

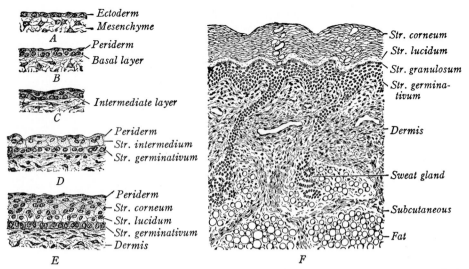

Fig. 401. Development of the human skin, shown in vertical sections. × 160. *A,* At 4 mm.; *B,* at 12 mm.; *C,* at two months; *D,* at three months; *E,* at five months; *F,* at birth (× 100). Epidermis of the palm and sole (*F*) is more highly differentiated than that of the body surface in general.

and its immediate descendants in the layers next above (prickle cells) constitute the definitive *stratum germinativum*. All of its cells are the actively dividing elements of the epidermis. Daughter cells are crowded upward by still newer ones and eventually reach the free surface. As a cell attains higher and still higher levels it undergoes progressive changes, culminating in cornification. Thus, directly above the germinative cells is the thin *stratum granulosum,* containing keratohyalin granules. Next higher lies the thin and clear *stratum lucidum* whose content is a fluid *eleidin* that replaces the granules. Still nearer the surface, the epidermal cells flatten steadily and comprise the many-layered *stratum corneum.* The peripheral cytoplasm of cells in this layer becomes cornified through a transformation of the eleidin, and the epidermis thereby loses its transparency.

It is important to understand that only in the thickened epidermis of the palm and sole are all the layers, just mentioned, distinguishable; over the general body surface the granular and lucid strata are not clearly represented. In a few regions, like the red margin of the lip and the anus, cornification is slight. *Pigment granules* appear in the cells of the stratum germinativum of all races. Such granules are obtained by transfer from the processes of *melanocytes.* These cells migrate from the primitive neural-crest tissue, invade the basal layer of the epidermis and specialize in pigment formation (Fig. 402).[1] Pigment development is incomplete at birth, as shown by the marked darkening of Negro skin during the next six to eight weeks.

When the hairs emerge, at about the sixth fetal month, they do not penetrate the toughened periderm of the epidermis but loosen or break it. Hence in mammals this layer, which is also characteristic of land vertebrates in general, is known also by another name, the *epitrichium* (*i.e.*, upon the hair). Desquamated epitrichial and epidermal cells mingle with cast-off lanugo hairs and sebaceous secretions to form the pasty *vernix caseosa* which smears the fetal skin. This material is said to protect the epidermis against a macerating influence that otherwise would be exerted by the amniotic fluid. As a lubricant, it also prevents chafing-injuries from the amnion as the growing fetus becomes progressively confined in its fluid-filled sac.

The plane of union between epidermis and dermis is smooth until early in the fourth month when epidermal thickenings grow down into the dermis of the palm and sole. About two months later corresponding elevations first appear on the skin surface (Fig. 401 F). These epidermal ridges complete their permanent, individual patterns (*e.g.*, finger prints) in the second half of fetal life.

The Dermis. The fibrous layer of the integument is customarily traced to cells proliferated from the lateral walls of the paired somites. In consequence, this region of a somite has received the name *dermatome*, or cutis plate (Fig. 394 B). Evidence in support of this claim is not plain in mammals and it has even been urged that the so-called dermatome really belongs to the myotome.[2] In any event, the dermatome could not be expected to supply connective tissue far beyond the vicinity of the somites. Much of the dermis must differentiate from nonspecific mesenchyme subjacent to the epidermis, most of which comes from the lateral sheets of somatic mesoderm.[3]

Collagenous fibers appear in the third month and *elastic fibers* in the sixth month; their manner of differentiation has been described previously (p. 397). Only gradually does a distinction between the compact dermis and the looser, subcutaneous tissue become recognizable (Fig. 401 F). Columnar papillæ project upward from the dermis into the germinative stratum; the dermal papillæ are of two kinds, depending on whether they contain blood vessels or nerve endings. Some of the dermal cells acquire pigment granules. In the lower sacral region deep-lying pigment tends to give local areas a bluish to brownish color. They occur more commonly in infants and young children of the dark-skinned races and are known as 'Mongolian spots.' Fat develops in the *subcutaneous layer*, but does not become abundant until the later months of fetal life.

Melanocyte mostly in dermis

Melanocyte in epidermis

Fig. 402. Early melanocytes entering the human epidermis during the fourth month.
× 410.

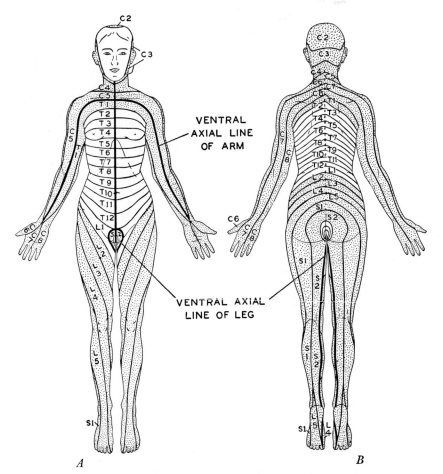

Fig. 403. Distribution of the segmental spinal nerves to 'dermatomic' cutaneous areas (Keegan and Garrett). See also Fig. 168.

The skin is innervated by segmental spinal nerves that supply successive zones of the integument known as *dermatomic areas* (Fig. 403). The implication, however, that all of the dermis derives from portions of paired somites and that each girdling dermatome carries with it a sensory nerve, just as myotomes maintain their original motor innervation, is not at all secure. It will be noted in Fig. 403 that the radial (outer) surface of the arm and tibial (inner) surface of the leg receive nerves from higher levels than do the respective inner and outer surfaces. This is in agreement with the primitive cranial and caudal surfaces of the limbs before rotation occurred in opposite directions (p. 210).

Causal Relations. Presumptive epidermis is for a time a highly plastic tissue. When pieces from an amphibian gastrula are transplanted into the flank of a neurula, they will develop into various things depending on the location. Among such products are spinal cord, cartilages, myotomes and pronephros. After determination occurs, skin can self-differentiate in culture media. The normal growth and functioning of epidermis depends on early contact with a basement membrane and on the proximity of mesenchyme.

Anomalies. Skin may fail to complete normal differentiation, thereby retaining its fetal character. Rarely it is pendulous or astonishingly elastic. The deposition of pig-

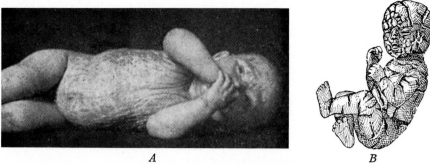

A   *B*

Fig. 404. Ichythyosis. *A,* Moderately severe condition in an infant (Andrews). *B,* Severe condition (harlequin fetus; alligator skin) in a stillborn.

ment in the epidermis and elsewhere sometimes fails (*albinism*) or is over-abundant (*melanism*). Such atypical pigmentation may affect local areas only. *Nævus* is a name given either to a pigmented spot ('mole') or to a red to purple discoloration caused by a cavernous, vascular plexus in the dermis ('birthmark'). *Ichthyosis* designates a rough, scaly skin due to abnormal cornification of the superficial layers (Fig. 404 *A*); in extreme cases the epidermis shows thick plates, separated by cracks ('alligator skin'; *B*).

THE NAILS

Nails are modifications of the epidermis that correspond to the claws and hoofs of lower mammals. The first indication of a nail is foreshadowed at ten weeks by a thickened area of epidermis (*nail field*) on the dorsum of each digit (Fig. 405 *A*). The adjoining territory, on each side and at the base, tends to overgrow this field, thereby giving rise to shallow *lateral nail folds* which continue into a much deeper *proximal nail fold* that extends nearly to the proximal end of the terminal phalanx (*B*).

Although the primitive nail field undergoes some local cornification ('false nail'; Fig. 405 *B*), the material of the true nail is developed within the under layer of the proximal nail fold. This layer is accordingly named the *matrix* (*C*). During the fifth month specialized keratin fibrils differentiate in the matrix layer, without having passed through a keratohyalin or eleidin stage as in the ordinary method of cornification.[4] The keratinized cells flatten and consolidate into the compact tissue of which the *nail plate* is composed. In this manner the nail substance differentiates in the proximal nail fold as far distad as the outer edge of the *lunula,* which is the whitish crescent at the base of the exposed nail. Beyond the lunula, the nail plate merely shifts progressively over the *nail bed* and reaches the tip of the finger one month before birth. As might be expected, the nails of the toes are begun and completed slightly later than the finger nails. The dermis, beneath the nail, is thrown into parallel longitudinal folds which are said to produce the characteristic ridging and grooves.

The stratum corneum and periderm of the epidermis for a time cover completely the free nail and are jointly termed the *eponychium* (*i.e.,* upon

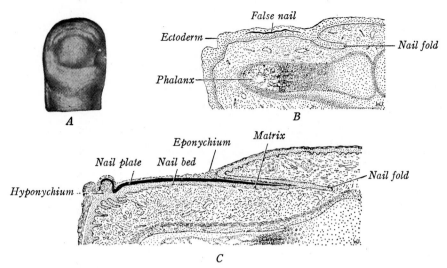

Fig. 405. Development of the human nail. *A,* Dorsum of finger, at ten weeks (Kollmann; × 10). *B,* Longitudinal section, at fourteen weeks (× 33). *C,* Longitudinal section, at birth (× 15).

the nail; Fig. 405 *C*). In late fetuses this layer is lost, except for horny portions that continue to adhere to the nail plate along the curved rim of the nail fold. Underneath the free end of the nail the epidermal cells also accumulate to constitute a piled-up epidermal mass known as the *hyponychium,* or substance beneath the nail; this region is much more important in a claw, and still more so in a hoof where it forms the 'sole.' The opacity of the lunula has been interpreted variously.[5]

Anomalies. Misshapen nails occur. Absence of nails (*anonychia*) may accompany other failures or defects among epidermal derivatives.

THE HAIR

Hairs are specialized epidermal threads that rest upon a sunken papilla of the dermis. They are produced only by mammals and are a distinctive characteristic of that vertebrate group. The comparative hairlessness of man today is a feature acquired within relatively recent times. Since it is similar to the condition found in late fetuses of anthropoid apes, this reduced hairiness is regarded as an example of arrested development. The primary insulating and heat-conserving function of hairs is compensated for in man by a heavy deposit of fat beneath the skin. Hairs of a fetus tend to be grouped in threes or fives, with the central one larger, and also to be arranged in lines. These relations are interpreted as the survival of a primitive mammalian condition in which the hairs stood in definite relation to scales which covered the skin, after the manner still seen in certain living forms (Fig. 406).

Hairs begin to develop early in the third month on the eyebrows, upper lip and chin; those of the general integument make a start one month later. The first evidence of a future hair is the crowding and elongation of a cluster of germinative cells in the epidermis (Fig. 407 *A*). Their bases sink root-like into the dermis, and active proliferation soon produces a cylindrical, epithelial peg (*B*). At this stage the hair follicle consists of an outer wall of columnar cells, continuous with the basal layer of the epidermis, and an internal mass of polyhedral cells. About the whole is a mesenchymal investment (the later *connective-tissue sheath*), and at the clubbed base the mesenchyme condenses into a mound-like *papilla*.

As development proceeds and the hair peg pushes deeper into the dermis, its base enlarges into the *bulb* which becomes molded like an inverted cup over the papilla (Fig. 407 *C*). The actual hair substance is a proliferation from the basal epidermal cells lying next the papilla (Fig. 408). These cells give rise to an axial core, destined to become the *inner epithelial sheath* and *shaft;* the latter grows progressively upward toward the surface. Quite distinct are the peripheral cells on the sides of the original downgrowth, which comprise the *outer epithelial sheath.*

The young 'hair cone' grows by the steady addition of new cells in the

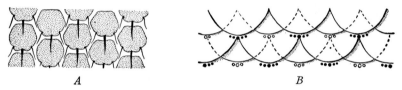

A   B

Fig. 406. The relation of hair groups to scales. *A*, Arrangement on the oppossum's tail (after Danforth and de Meijere). *B*, Arrangement in the human fetus, with hypothetical dermal scales drawn in (after Stöhr).

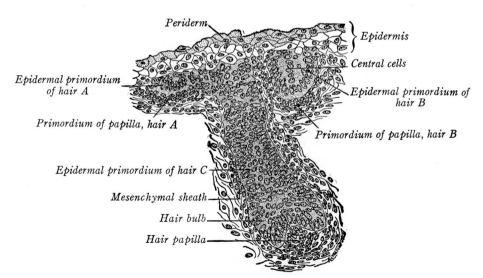

Fig. 407. Human hair follicles, at three months, shown in longitudinal section (Prentiss). × 330. Three stages (*A, B* and *C*) are included.

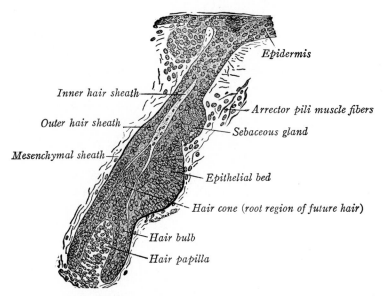

Fig. 408. Human hair follicle, at six months, shown in longitudinal section (after Stöhr). × 220.

bulb. In this manner it is pushed up through the central cells of the solid, primordial follicle and is molded into shape by the organizing inner sheath. The *shaft* finally reaches the general epidermis, follows up a *hair canal* produced in this layer (Fig. 410 *A*), and erupts at the surface. Above the level of the bulb, the cells of the hair shaft cornify and differentiate into an outer *cuticle,* middle *cortex,* and central (inconstant) *medulla.* Two swellings of the outer epidermal sheath appear on the lower side of the obliquely directed follicle (Fig. 408). The upper of these becomes the *sebaceous gland,* which will remain permanently associated with the hair; the deeper swelling is the *epithelial bed,* a region of rapid mitosis that also contributes to the growth of the periodically regenerating hair follicle. Mesenchymal tissue near the epithelial bed transforms into the smooth fibers of the *arrector pili* muscle, which attaches to the side of the follicle. *Pigment granules* appear midway of fetal life in the basal cells of the beginning hair; they are acquired from migratory *melanocytes* that originated in the neural crest.[1, 6] Such cells are carried upward along with other hair cells and cause the characteristic coloration.

The first generation of fetal hairs is a downy coat termed *lanugo.* It constitutes a dense covering to the body, especially on the back and limbs, prominent by the fifth month. Lanugo hairs are short-lived, all being cast off either before birth or soon afterward. The replacing hairs develop, at least in part, from new follicles; they are likewise fine and constitute the so-called *vellus* (*i.e.,* fleece) of the prepuberal years. Hair is shed and formed anew periodically throughout life. At the termination of any growth cycle the hair is carried upward by its shortening, regressive follicle. After a time the follicle reorganizes and begins to elaborate a new hair in the same general manner as already described (Fig. 410 *B*).

Some hairs remain permanently of the vellus type. In the female such occur on the face, neck and trunk; in the male they remain on the face (except beard), the flexor surface of the upper arms and various regions of the trunk. The replacing hairs of the brows, eyelashes and scalp of children are progressively larger and coarser than the first set. Under the influence of hormones, and especially those of the gonads, coarser and darker hairs appear at puberty on the pubis and axillæ of both sexes and on the beard-area and trunk of the male. The hair coat shows definite, directional patterns (streams; whorls). These are established by similar angular slants of the hair follicles at their first development in any local region.

Causal Relations. Mouse ectoderm of the tenth day, isolated and grown in culture medium, develops into epidermis and normal hairs. Probably, as with feathers, it is the presence of the dermal component of a hair papilla that induces the differentiation of the epithelial component. Ectoderm and hair follicles have no inherent power to form pigment. Only those cultures containing presumptive neural crest can differentiate melanoblasts—and hence pigment.

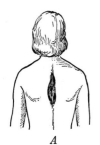  

A   *B*   *C*

Fig. 409. Hypertrichosis, or excessive hairiness. *A,* Local tuft; *B,* local area, normally 'downy'; *C,* general hairiness.

Anomalies. *Hypertrichosis* refers to excessive hairiness which may be localized (Fig. 409 *A, B*) or general, as in exhibited 'hairy monsters' (*C*). It is undecided whether this is due to an augmented development of the later hair follicles or to a persistent overgrowth of the lanugo-vellus set. In the rare *hypotrichosis* the congenital deficiency of hair may be complete (*atrichia*); the latter is usually associated with defective teeth and nails.

THE SEBACEOUS GLANDS

Most of the sebaceous glands accompany hairs. However, some independent ones, such as those on the genitalia, anus, nostrils and upper eyelids, develop from the general epidermis. Many of these do not organize until after birth, and some (*e.g.,* upper lip) not until puberty.

A gland primordium appears first in the fifth month as a swelling on the outer epithelial sheath of the hair follicle (Fig. 408). The swelling becomes a lobulated, flask-shaped sac whose lumen arises by the fatty degeneration of the central cells (Fig. 410). The resultant oily secretion is an important con-

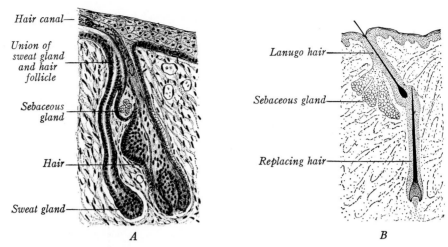

Fig. 410. Human cutaneous glands. *A*, Early stage in the development of a sebaceous gland and an apocrine sweat gland, both in association with a hair follicle (after Pinkus; × 125). *B*, Later sebaceous gland, connected to the follicle of a lanugo hair which is being replaced by a coarser, vellus hair (× 80).

stituent of vernix caseosa (p. 441); it is usually credited with helping to preserve the fetal skin from maceration. Cells in the basal layer of a sac are the reproducing elements.[7] Throughout life they supply new cells, which are forced into the lumen and disintegrate there during the process of oil elaboration. Such a gland is *holocrine* (*i.e.*, the secretion consists of altered, disintegrated gland cells themselves).

THE SWEAT GLANDS

Sudoriferous glands first begin to develop at four months from the deep epidermal ridges of the finger tips, palms of the hands, and soles of the feet. They are formed as solid, cylindrical ingrowths, but differ from hair primordia in being more compact and in lacking mesenchymal papillæ at their bases (Fig. 411 *A, B*). During the sixth month the deep ends of the simple cords twist, and at seven months internal clefts, which arose earlier by hollowing, now unite into a continuous lumen (*C, D*). The complete layer of cells about the lumen constitutes the gland cells. By contrast, certain peripheral cells transform into flattened elements that are usually considered to be smooth muscle fibers; this interpretation is of special interest, since such muscular elements would then be ectodermal. The duct portion of the gland at first ends blindly at the epidermis, but later, as cells are replaced during the course of growth in the stratified epithelium, a canal is left there which continues the duct lumen to the surface. Glands appear to be capable of functioning at seven months, but probably are not needed until after birth.

In certain regions of the body, destined to be supplied with coarse hairs

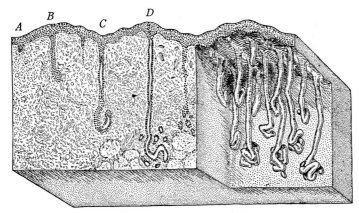

Fig. 411. The development of human sweat glands, shown in a model of the skin. × 40. Left half, gland stages (*A–D;* four to seven months) in longitudinal section. Right half, epidermis and glands at similar stages isolated from the dermis.

(pubis; axilla; areola; eyelids), there develop large, specialized sweat glands. Many of these originate on the sides of the hair follicles, and then move upward until they acquire separate openings on the epidermis (Fig. 410 *A*).[8] (An association with hair follicles is characteristic for sweat glands in general in most mammals, but this is not true of the ordinary kind in man.) Human glands of the axillary type are *apocrine* (*i.e.,* the tops of their secretory cells break away along with the secretion). They have a thicker secretion than that formed by the ordinary type.

Anomalies. Skin glands (sebaceous or sweat glands) may be overdeveloped in local regions. Contrariwise they may be underdeveloped and even lacking, either regionally or generally.

THE MAMMARY GLANDS

Mammary glands are peculiar to mammals. It is remarkable that they appear so early in development, not only since they are of use to adults alone but also because they are a late acquisition among vertebrate organs. Early in the sixth week of human development an ectodermal thickening extends on each side as a longitudinal band between the bases of the limb buds. At about 9 mm. it makes a distinct linear elevation that has been called the *mammary ridge,* or *milk line* (Fig. 412 *A*). In man this usually is inconspicuous except in the pectoral region, and in any event all but the cranial third normally vanishes quickly.[9] By contrast, lower mammals with serially repeated glands, like the sow, have a prominent milk line extending from axilla to groin (Fig. 598).

Each human mammary gland begins as a localized thickening on its appropriate epidermal ridge, in the region of the future breast. At first lens-shaped (Fig. 412 *B*), the primordium gradually becomes globular (*C*), and

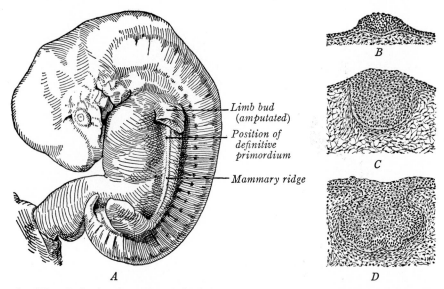

Fig. 412. Early development of the human mammary gland. *A,* Unusually prominent mammary ridge at 13 mm. (after Kollmann; × 5). *B–D,* Vertical sections of gland primordia, at six weeks, nine weeks and four months, respectively (× 80).

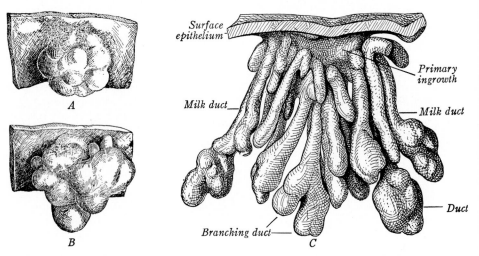

Fig. 413. Intermediate development of the human mammary gland, shown by models (after Broman and Lustig). *A,* At fourteen weeks (× 45); *B,* at five months (× 45); *C,* at six months (× 30).

then bulbous and lobed (*D*). During the fifth month from 15 to 20 solid cords begin to bud inward, pushing aside the dermal connective tissue as they advance (Fig. 413 *A, B*). These primary *milk ducts* continue to grow and branch throughout fetal life (*C*). The duct system also slowly acquires *lumina* by hollowing. The originally elevated free surface of the primordium flattens and cornifies (Fig. 414 *A*); hollowing produces a pit into which the ducts open (*B*). About the time of birth, or even considerably later, this sunken area elevates into the *nipple*. The *areola* is first recognizable as a circular area, free of hair primordia but acquiring branched *areolar glands* (of Montgomery) in the fifth month (*B*).

In the female the areolar region becomes elevated in late childhood (Fig. 415 *A–D*). During the puberal years and adolescence there is rapid enlargement of the breast, owing to further branching of the duct system and the deposition of fat. As a result, the breast becomes a hemisphere, bearing the areola and nipple at the apex (*D–F*). The gland-complex is further augmented during pregnancy, when *acini* first become recognizable as swollen end-pieces and the system of epithelial ducts and acini advance greatly both in bulk and structural differentiation.

Two or three days after parturition the glands become functionally active (Fig. 415 *G*). This culmination of development is a response to hormonal stimulation. In this process, estrogen (and progesterone?) excite the preliminary changes, whereas prolactin of the hypophysis is the final activator responsible for actual lactation (p. 165). The early milk of the mother is a

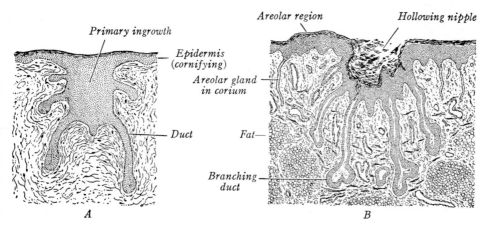

Fig. 414. Later prenatal development of the mammary gland, in vertical section. *A*, At five months (× 50); *B*, at eight months (× 23). These stages are in sequence with Fig. 412 *B–D*.

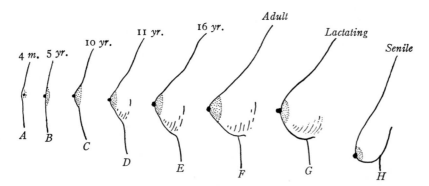

Fig. 415. Profiles of the female breast throughout postnatal life.

thin fluid called *colostrum*. The mammary glands of the newborn of both sexes also yield a little secretion (*'witch milk'*) within a few days after birth. Their activity at this time depends upon changing hormonal relations similar to those that bring about lactation in the mother.[10]

The glands of both sexes show equal development throughout fetal life and until late childhood.[11] The male gland reaches its full development at twenty years; at this time it resembles the female gland at an early stage of puberty.[12] After the menopause the mammary glands of the female undergo a regression that parallels a similar decline of the internal genitalia (Fig. 415 *H*). All of these parts were built up and maintained by ovarian hormones; when the ovary declines, they correspondingly suffer regression.

The mammary glands are regarded by most authorities as modified and specialized sweat glands of the apocrine type. This homology is made because their development is similar and because in the lowest mammals their structure is the same. Moreover, rudimentary mammary glands (the areolar glands), which also resemble sweat glands, occur about the nipple.

In many mammals several pairs of mammary glands are developed along the mammary ridge (hog; dog). Some have a single pair, occupying the pectoral region (primates; elephant) or even the axilla (fruit bat; flying lemur); in others they are confined to the inguinal region (sheep; cow; horse), or even occur near the genitalia (some whales). The human gland on each side develops from one of several local thickenings along the ridge; this multiple appearance of potential mammary sites possibly represents an expression of atavism.

Anomalies. Absence of one or both mammary glands (*amastia*), retention of the prepuberal condition (*micromastia*) and the attainment of abnormal size (*macromastia*) are all well known (Fig. 416 *A, B*). In some instances the male develops a breast, more or less of the female type (*gynecomastia; C*). This condition is dependent on a disturbed androgen-estrogen balance and is frequently present in hermaphrodites. Two examples of actual milk secretion by an adult male have been recorded.[13] Supernumerary mammary glands (*hypermastia*) are quite rare, but accessory nipples (*hyperthelia*) are fairly common in both sexes (*D*). It is said that at least 1 per cent of large populations may show traces of them.[14] They occur chiefly between the axilla and groin, but have been

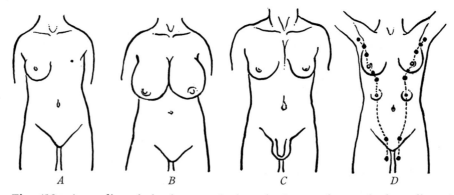

Fig. 416. Anomalies of the breasts. *A*, Amastia (except for a nipple rudiment); *B*, macromastia; *C*, gynecomastia; *D*, accessory pair of glands on abdomen. The commoner sites of accessory glands (or nipples) are indicated in *D* by dots along the course of the former mammary ridges (shown as broken lines).

recorded on the arm, back and thigh. Such structures represent independent differentiations, usually along the primitive milk line, such as occur normally in lower animals.

REFERENCES CITED

1. Billingham, R. E. & W. K. Silvers. 1960. Quart. Rev. Biol., *35,* 1–40.
2. Bardeen, C. R. 1900. Johns Hopkins Hosp. Rep., *9,* 367–400.
3. Murray, P. D. F. 1928. Austr. J. Exp. Biol., *5,* 237–256.
4. Hoepke, H. 1927. Bd. 3, Teil I in Möllendorff: Handbuch (Springer).
5. Burrows, M. T. 1917. Anat. Rec., *12,* 161–166.
6. Hardy, M. H. 1949. J. Anat., *83,* 364–384.
7. Montagna, W. 1956. The Structure and Function of Skin (Academic Press).
8. Steiner, K. 1926. Z'ts. Anat. u. Entw., *78,* 83–97.
9. Lustig, H. 1915. Arch. f. mikr. Anat., *87,* 38–59.
10. Lyons, W. R. 1937. Proc. Soc. Exp. Biol. & Med., *37,* 207–209.
11. Thölen, H. 1949. Acta Anat., *8,* 201–235.
12. Pfaltz, C. R. 1949. Acta Anat., *8,* 293–328.
13. Haenel, H. 1928. Münch. med. W'sch., *75,* 261–263.
14. Speert, H. 1942. Quart. Rev. Biol., *17,* 59–68.

Chapter XXIV. The Histogenesis
of Nervous Tissues

Both the nervous system and the sensory epithelia of the nose, eye and ear are derived from portions of the primitive integument. The material basis of the brain and spinal cord is a thickened area of ectoderm (*neural plate*) stretching along the mid-dorsal line of the embryo (Fig. 426). This band lies ahead of the primitive groove and knot and is added to as the retreating knot (and, later, the 'end bud') pays out material for the lower trunk (p. 94; Fig. 72). At first the neural plate is flat and single-layered, but it rapidly becomes thick and stratified. The plate folds into a *neural groove* by the time somites are appearing; the groove itself is bounded on each side by an elevated *neural fold* (Fig. 417 *A–C*). The groove continues to narrow and the thickened neural folds presently meet and fuse above it, thereby rolling the original plate into a *neural tube* (*C, D*). At the completion of this process the tube lies below the surface of the ectoderm and has detached from it. The enclosed cavity is the *neural canal*.

The substance of the neural tube gives rise to all nervous elements whose cell bodies lie within the brain and spinal cord; it also furnishes the non-nervous neuroglial cells of those organs. The lateral margins of the neural plate are not incorporated into the neural tube. Instead, each becomes a band of cells known as the *neural crest* (Fig. 422). From it differentiate all

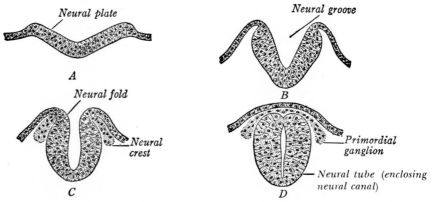

Fig. 417. Origin of the neural tube and neural crest, illustrated by transverse sections from early human embryos. × 125.

ganglion cells (cranial; spinal; autonomic), as well as various other derivatives (p. 458).

DIFFERENTIATION OF THE NEURAL TUBE

The Stem-Cells.   At the beginning of its development, the neural plate is composed of undifferentiated, proliferative epithelium. Soon its daughter cells enter into two lines of specialization (Fig. 418). One path leads toward *nerve cells,* in which irritability and conductivity have become predominant functions; the other course is toward *ependymal* and *neuroglial cells,* which constitute the distinctive supporting tissue of the nervous system.

The embryonic nerve cell is a *neuroblast;* it passes through a bipolar stage, with a process at each end, before reaching a multipolar stage, or immediate precursor of the typical *neuron* of the central nervous system. The *spongioblast* is the forerunner both of the *ependymal cells* and of neuroglial cells known as *astrocytes.* Some spongioblasts are migratory in nature; these differentiate into *oligodendroglia* and into additional astrocytes, as well. It has been claimed that some migratory spongioblasts convert into neuroblasts, but this interpretation is open to doubt.[1]

The Layers.   The neural plate is originally a single layer of columnar cells (Figs. 419 *B*), but it rapidly becomes pseudostratified. The bases of all cells are anchored on a basement membrane at the surface bordering the central canal of the tube (*C*). At this period the component cells seemingly constitute a syncytium (*D*), but actually they are always distinct.[2] In the sixth week the neural wall is bounded on the outer and inner surfaces by an *external* and *internal limiting membrane,* respectively, while the cellular elements of the wall are arranged radially (*D*). At this stage the neural tube is sufficiently organized so that three concentric zones may be distinguished: (1) an inner *ependymal layer,* with its cell bodies

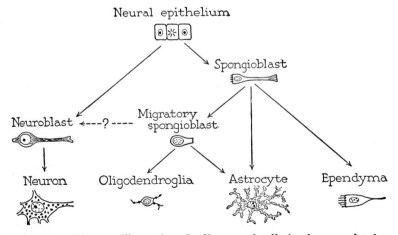

Fig. 418.   Diagram illustrating the lineage of cells in the neural tube.

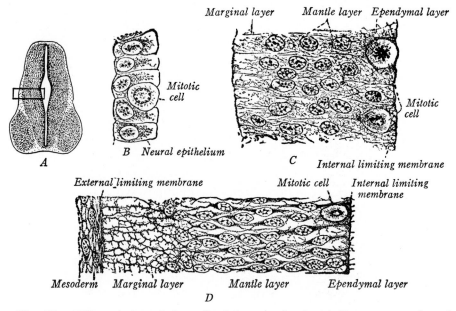

Fig. 419. Differentiation of the wall of the neural tube. *A,* Transverse section of a spinal cord; a rectangle explains the relations of the detail-sections, and especially *D. B–D,* At stage of the neural plate, 5 mm. and 10 mm. (after Hardesty; × 690).

abutting on the internal limiting membrane (next the neural canal) and their processes extending peripherally; (2) a middle, nucleated *mantle layer,* whose main cell-bodies lie at an intermediate level and represent the future gray substance; and (3) an outer, non-nucleated *marginal layer,* into which many nerve processes (nerve fibers) grow, the white substance.

The ependymal layer, originally the uppermost stratum of the neural-plate stage, not only contains the nucleated bodies of inertly-supporting ependymal cells but it also harbors mitotic cells. These are not special stem-cells, peculiar to that layer. Rather, it is now known that whenever cells of the mantle zone are about to divide they retract and round up, so that they come to lie temporarily in the ependymal zone.[3, 4] The mantle layer makes up the future *gray substance* of the central nervous system; it is predominantly cellular in structure and contains the cell bodies of the neurons and many neuroglial cells. The marginal layer is a 'fibrous' mesh which lacks nuclei in the early months, but later gains neuroglia. It provides an enveloping zone into which the processes of nerve cells grow and reach their destinations; thereby distant neurons become linked with other neurons, and center with center. The marginal layer becomes the *white substance* of both the brain and spinal cord. The details of the transformation of neuroblasts into neurons and spongioblasts into ependyma and neuroglia demand further treatment and will occupy the descriptions that follow.

The Differentiation of Neuroblasts. The process by which neural epithelium specializes into neurons is named *neurogenesis.* In accomplishing this end each neuroblast (and its descendants) sooner or later loses the

power of cell division, develops cell processes and converts into a definitive
neuron. A neuron is a discrete structural and functional unit of nervous
tissue; it consists of a nerve cell and all of its processes. The only relation
of one neuron to another is that of contact between processes, or between
a process and a cell body; this area of touching constitutes a functional
contact known as a *synapse*. Mitosis among neuroblasts ceases during the
first year of postnatal life. Thereafter the nervous system matures and enlarges, but the ability to produce new neurons is forever lost. The total
number of neurons developed for the use of the human nervous system is
remarkably constant, and this is true regardless of the size of the individual.

For a time neuroblasts are *apolar*—that is, rounded and without processes
(Fig. 420 *A*). As differentiation proceeds, a fiber-like process grows out from
the opposite ends of each cell, thus producing *bipolar neuroblasts* (*B*). This
intermediate stage becomes pear shaped, and from the slender end of each
cell the thin, growing extension differentiates into the chief process, or
axon. It is destined to conduct efferent (motor) impulses away from the cell
body. The main cell-mass soon becomes multipolar by the development of

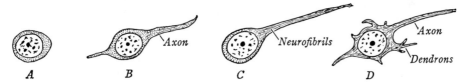

Fig. 420. Stages in the development of a multipolar neuron (× 750). *A*, Apolar; *B*,
bipolar; *C–D*, multipolar.

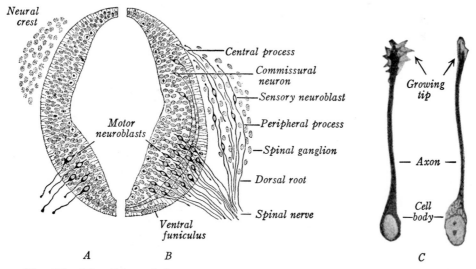

Fig. 421. The differentiation and growth of human neuroblasts. *A*, Spinal cord, in
transverse hemisection, at 4 mm. (× 225). *B*, Spinal cord, in transverse hemisection, at
5 mm. (× 140). *C*, Two neuroblasts, demonstrating neurofibrils and the enlarged, growing tip (Cajal; × 500).

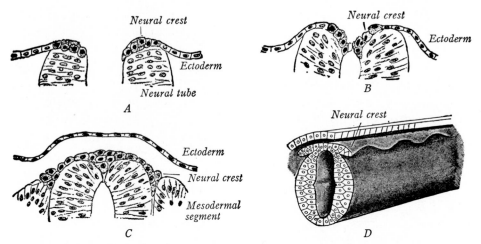

Fig. 422. Development of the human neural crest. *A–C,* Successive stages, at 2.5 mm., in transverse section (Lenhossék; × 250). *D,* Model of the early spinal cord and its beaded neural crest (Kingsley).

branched, secondary processes, named *dendrons* (or *dendrites*) (*C, D*). Within the cytoplasm (*neuroplasm*) of even young nerve cells and their processes, fine *neurofibrils* can be demonstrated in life;[5] also by refined fixation and staining methods (Fig. 421 *C*),[6] and by the electron microscope. Considerably later, *chromophil substance,* known also as *Nissl bodies,* differentiates in the neuroplasm. The sprouting and growth of axonal processes can be followed in neuroblasts that have been isolated and cultivated in clotted lymph, while the growth of axons day by day has been observed directly in living tadpoles.

Toward the end of the first embryonic month, neuroblasts are arising in the ventrolateral walls of the neural tube. These produce axons that are motor, or efferent, in function. Many such nerve fibers penetrate the marginal layer and pierce the external limiting membrane. On emerging at segmental levels, they combine as *ventral roots* of *spinal nerves* (Fig. 421 *A, B*). The fibers of each root grow into association with a myotome of that level. Similar associations of efferent axons participate in the formation of cranial nerves. Still other neuroblasts remain wholly within the neural tube and become interconnecting *association neurons.*

DIFFERENTIATION OF THE NEURAL CREST

The longitudinal band of cells that appears on each side where the ectoderm and neural folds join (Fig. 422 *A*), possesses many potentialities. In addition to differentiating neuroblasts and the sheath cells of nerve fibers, some of the crest cells produce such dissimilar materials as cartilage (p. 399), epidermal pigment (p. 440) and suprarenal medulla (p. 519).

When the neural folds become a tube and the ectoderm detaches from it, the crest-substance separates into right and left linear halves, distinct from the neural tube (Fig. 422 *B–D*). These portions settle to a position between the tube and the myotomes. On its arriving in this location, each half of the original crest-substance is a cellular band extending the full length of the spinal cord and far rostrad along the brain wall. As long as the neural tube grows in a caudal direction, new crest-tissue is added progressively to these bands.

At regular intervals, agreeing with the position of somites, the proliferating cells of the crest give rise to bead-like enlargements, the *spinal ganglia*. The serially repeated ganglia of each side are interconnected for a short time by parts of the originally continuous crest-substance (Fig. 469), but these bridges soon disappear. In the hind-brain region the ganglia of cranial nerves V to X develop also from the local crest. They are spaced in relation to the branchial arches, which they supply.

The Cranio-spinal Ganglia. The cells of the primordial ganglia differentiate into *ganglion cells* and *supporting cells*. While still in a formative stage these groups are comparable to the neuroblasts and spongioblasts of the neural tube. The neuroblastic forerunners of ganglion cells differentiate slightly later than do the multipolar cells, already described (Fig. 421 *A*). Each elongates into a spindle-shaped element and, by developing a primary process at both ends, transforms into a neuron of the bipolar type (*B*). The central processes from each ganglionic mass grow toward the neural tube, converge into a distinct bundle, and thus constitute a *dorsal root* (Fig. 423 *A*).

Each component fiber of the root penetrates the dorsolateral wall of the

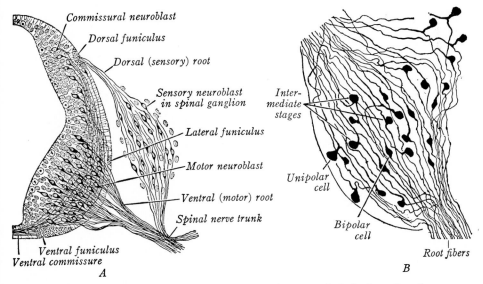

Fig. 423. Development of a human peripheral nerve. *A*, Spinal cord and nerve, at 7 mm., in transverse section (× 120). *B*, Stages in the transformation of bipolar into unipolar nerve cells, shown in a longitudinal section of the spinal ganglion at ten weeks (Cajal).

neural tube, courses in the marginal layer of the spinal cord and, by means of end-twigs, comes in contact with a neuron of the mantle layer. The peripheral processes of the bipolar ganglion cells complete a dorsal spinal root by passing outward and joining the corresponding ventral root; the common bundle, thus formed, constitutes the trunk of a *spinal nerve* (Fig. 423 *A*). The peripheral processes, sensory (or afferent) in nature but structurally like an axon, end chiefly in the skin. Although bipolar at first, ganglion cells become unipolar in a way not surely understood. Presumably a part of the cell body draws out into a common stem that bears the two processes at its tip.[7] In Fig. 423 *B* there can be traced intermediate stages between typical bipolar and unipolar cells.

The Autonomic Ganglia.   Certain neuroblasts of crest origin migrate ventrad and differentiate into cells of the autonomic ganglia. The source of these neuroblasts was long disputed, but modern evidence favors an origin from the neural crest (Fig. 480 *A, B*).[8, 9, 10] The course of differentiation of autonomic ganglion cells differs from that in the cranio-spinal series inasmuch as the final product is multipolar cells whose chief process is an axon. Functionally these ganglion cells are efferent.

THE DIFFERENTIATION OF SUPPORTING ELEMENTS

Both the primitive neural tube and the early ganglia furnish cells that become non-nervous elements. These are permanent constituents of the brain and spinal cord and of the peripheral nervous system, respectively.

Elements of the Neural Tube.   The brain and spinal cord are given stability and metabolic support by ectodermal, interstitial tissue in the form of *ependymal cells,* which bound the spinal canal and extend outward toward the periphery, and by *neuroglia cells* which are more irregularly distributed. A preceding paragraph has described how the spongioblasts originate from the undifferentiated cells of the neural-plate tissue and become more or less altered. The degree and direction of this specialization determine whether they result in ependyma or neuroglia.

For a while the spongioblastic elements are radially arranged, like columnar epithelium. One end, which also contains the nucleus, borders the cavity of the neural canal and projects cilia into it; in the other direction the slender cell extends even to the periphery of the neural tube (Fig. 424 *A, B*). Those spongioblasts that retain their primitive shape and bordering relation to the neural canal are known as *ependymal cells* (*C*, 'Floor plate'). The majority of spongioblasts, however, differentiate further. These elements lose their relation with the neural canal, migrate outward and convert into *neuroglia cells* (*C*); some cells, so displaced, retain a peripheral attachment, but most abandon both central and distal connections. It is of interest to note that the several developmental stages encountered in mammals recapitulate the progressive neuroglial conditions found within the chordate group.

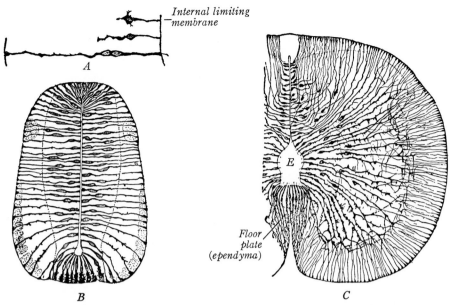

Fig. 424. The differentiation of supporting tissue, demonstrated in transverse sections of the spinal cord (Cajal). *A,* Growth stages of spongioblasts in a chick embryo of one day (× 1000). *B,* Spongioblasts in a chick embryo of three days (× 240). *C,* Human embryo of ten weeks (*E*, ependyma, adjoining the central canal; midway in the wall are detached spongioblasts, destined to become neuroglia). (× 45).

In its final state the ependymal tissue consists of elements whose nuclei lie next the cavity of the brain or spinal cord, and whose cell bodies radiate outward like columnar epithelium. Primitive ependymal relations are clearly retained only at the midplane of the spinal cord and medulla (Fig. 424 *C*); in other regions the peripheral processes of ependymal cells extend merely for varying distances beyond the cell body. Elsewhere in the brain and spinal cord the supporting elements are neuroglia cells, distributed throughout the mantle and marginal layers. They are of two morphological types: (1) *astrocytes,* stellate in shape and with long processes (Fig. 425 *A, B*); and (2) *oligodendroglia,* with a smaller cell body and fewer, finer processes (*C*). A third type, *microglia* (*D*), should be mentioned although apparently they do not belong developmentally, structurally or functionally with the true neuroglia. These elements, which are potential ameboid phagocytes, appear late in the neural tube and seem to originate from invading mesenchymal cells.[11] They could, therefore, be appropriately named *mesoglia.*

The astrocytes are derived from full-length primitive spongioblasts, from spongioblasts that (because of the thickness of the tube) never connect with the periphery, and from wandering spongioblasts (Fig. 418). Astrocytes appear first in the third month. Those occupying the gray substance are named *protoplasmic astrocytes* (Fig. 425 *A*); another type, *fibrous astrocytes,* develop fibrils within their cytoplasm and are typical of the white substance (*B*). The oligodendroglia, derived solely from migratory spongioblasts, arise at a later period than astrocytes (*C*).

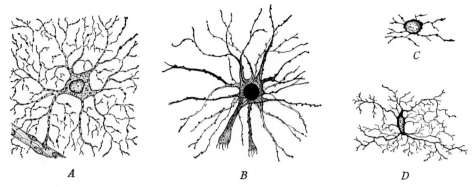

Fig. 425.   Types of neuroglial cells (after Penfield). × 650. *A*, Protoplasmic astrocyte and capillary; *B*, fibrous astrocyte; *C*, oligodendroglial cell; *D*, microglial cell.

Elements of the Ganglia.   The supporting cells of the ganglia at first make up an apparent syncytium, in the meshes of which are found the neuroblasts. The interstitial elements differentiate both into flattened *satellite cells,* which invest the ganglion cells as a capsule, and into *sheath cells,* which migrate peripherally along the developing nerve fibers and envelop the axons (Fig. 480 *A, B*).[12] Still other migratory non-nervous cells, of early neural-crest origin, mingle with the general mesenchyme. Their subsequent history and diverse specializations (p. 458) have been explained, at least in part, by properly designed experiments.

The Neurolemma Sheath.   Peripheral nerve fibers are enclosed within a cellular sheath (of Schwann), called the *neurolemma.* These migratory cells, although indistinguishable from one. another, have two sources of origin. As already mentioned, some differentiate from the tissue of the early neural crest. Slightly later others emerge from the neural tube by way of the ventral roots.[10, 13] The young sheath cells are spindle-shaped and wrap about bundles of nerve fibers. Multiplication on the part of the sheath cells then produces a series of cell links, keeping pace with the elongating nerve fibers. Some fiber-bundles retain this shared relation with a common series of neurolemma cells. Such fibers do not develop a myelin sheath. By contrast, further proliferation of sheath cells separates other bundles into single fibers, each with its own spirally wrapped neurolemma.[14] Such sheaths are destined to produce an investment of myelin about the fiber.

Direct observation in living tadpoles permits the advance of neurolemma cells along a nerve fiber to be traced. Somewhat later, myelin appears at the level of neurolemma nuclei and spreads in both directions to the extremities of each cell.

The Myelin Sheath.   Between the fourth month of fetal life and the third month following birth a fatty *myelin sheath,* also called a medullary sheath, begins to appear about many nerve fibers. It surrounds the chief process (axon) and is a product of the neurolemma sheath. It is deposited in layers corresponding to the spirally wrapped units of the neurolemma.[19] In the central nervous system there is no typical neurolemma sheath investing the nerve fibers, yet many acquire a spiral myelin sheath. Here rows of neuroglia cells (oligodendroglia?) are said to substitute for the neurolemma of peripheral nerves.[15]

Myelination. The *myelinated fibers* (*i.e.,* those with a myelin sheath) have a glistening, white appearance which gives the characteristic color to the *white substance* of the brain and spinal cord and to most peripheral nerves. Many of the fibers of the central nervous system remain *unmyelinated;* this is true of certain fibers coursing in the white substance, while the portions of all fibers lying within the *gray substance* never acquire myelin. Many fibers in the peripheral nerves are also unmyelinated.

Myelin is deposited first near the cell body of a neuron and then spreads progressively along the fiber. The myelin sheath also appears at widely different times in the various fiber-systems. The oldest tracts historically, which are also the earliest to attain function, are myelinated soonest. In general, tracts acquire myelin at about the time they become capable of functioning.[16] Since myelin is deposited in the various fiber tracts at different developmental periods, this condition has been of great help in tracing the origin, course and extent of the various fiber-groups within the nervous system.

The development of myelin in the spinal cord begins at the middle of fetal life and is not completed in some fibers until adolescence. It occurs first in the cervical cord and then extends progressively to lower levels. Fibers of the ventral roots acquire myelin before those of the dorsal roots. Tardiest of all are certain descending motor tracts (cortico-spinal; tecto-spinal) which myelinate largely during the first and second postnatal years. The brain begins to myelinate in the sixth fetal month, but progress is slow and only the fibers of the basal ganglia and those that continue the structure of the spinal cord upward possess myelin sheaths at the time of birth. In fact, the brain of a newborn is still largely unmyelinated, so that deposition goes on chiefly from birth through puberty. First to acquire sheaths in this period are the primary sensory-motor fields—that is, the olfactory, optic and auditory cortical fields and the motor cortex. The projectional and commissural fibers myelinate last.

THE MENINGES

These membranes serve as closed coverings to the brain and spinal cord. All three coats arise as condensations of the neighboring mesenchyme (Fig. 419 D),[17] although migrant cells from the neural crest may contribute slightly to the delicate *pia-arachnoid* which lies next to the neural tube.[18] More externally is the tough *dura mater* which is a distinct membrane at eight weeks.

Causal Relations. The ectoderm that overlies the chorda-mesoderm of the gastrula stage is subject to induction by that layer. By the end of the stage, the ectoderm loses its former adaptive potentialities and becomes irreversibly determined as neural plate (and neural crest). The specific factors that then determine and direct the further divergent courses of neuroblasts and spongioblasts are not known.

Nerve fibers reach their appropriate destinations, even when their normal pathway
is blocked or when the target-organ has been displaced to a strange location. The nerve
fibers act as if they were attracted, yet this guiding principle is nonspecific since a nasal
placode or an optic vesicle, substituted for a myotome or limb bud, will influence the
same nerves in a similar manner. The directing force is apparently neither chemical nor
electrical, as was once believed. Nerve fibers can grow forward only when in contact with
a solid medium, and the mechanical structure of the ground substance of embryonic
tissue supplies the requisite substrate for growth. An organ such as a limb bud is a
region of high chemical activity and it is credited with so altering the density and
arrangement of the surrounding ground substance that structural pathways converge
toward the active center. These oriented ultrastructures then serve to guide the growing
nerve fibers to their destination.

REFERENCES CITED

1. Penfield, W. 1928. Sect. 30 in Cowdry: Special Cytology (Hoeber).
2. Sauer, F. C. 1935. J. Comp. Neur., *63,* 13–23.
3. Watterson, R. L. 1956. Anat. Rec., *124,* 379.
4. Sauer, M. E. & A. C. Chittenden. 1959. Exp. Cell Res., *16,* 1–6.
5. Weiss, P. & H. Wang. 1936. Anat. Rec., *67,* 105–117.
6. Hoerr, N. L. 1936. Anat. Rec., *66,* 81–90.
7. Truex, R. C. 1939. J. Comp. Neur., *71,* 473–486.
8. Hammond, W. S. 1949. J. Comp. Neur., *91,* 67–85.
9. Brizzee, K. R. & A. Kuntz. 1950. J. Neuropath. & Exp. Neur., *9,* 164–171.
10. Weston, J. A. 1963. Devel. Biol., *6,* 279–310.
11. Kershman, J. 1939. Arch. Neur. & Psychiat., *41,* 24–50.
12. Brizzee, K. R. 1949. J. Comp. Neur., *91,* 129–146.
13. Jones, D. S. 1939. Anat. Rec., *73,* 343–357.
14. Peters, A. & A. R. Muir. 1959. Q. J. Exp. Physiol., *44,* 117–130.
15. Bunge, M. B., *et al.* 1961. J. Biophys. & Biochem. Cytol., *10* (Suppl.), 67–94.
16. Langworthy, O. R. 1933. Carnegie Contr. Embr., *24,* 1–137.
17. Sensenig, E. C. 1951. Carnegie Contr. Embr., *34,* 145–157.
18. Spofford, W. R. 1945. J. Exp. Zoöl., *99,* 35–52.
19. Geren, B. B. 1970. Exp. Cell Res., *7,* 558–662.

Chapter XXV. The Central Nervous System

The primitive neural tube is produced by the folding of the neural plate into an epithelial tube, as described in the previous chapter. The neural groove begins to close about midway of its length in embryos with six somites, and the closure advances progressively in both directions (Fig. 426). With continued growth of the embryo caudad, the neural groove extends steadily in that direction; at first an open trough, it folds into a tube as fast as is mechanically possible. The open caudal end of the neural tube is called the *posterior neuropore;* it closes off at about the 25-somite stage. Below this level the remainder of the neural tube cannot be added by folding. Instead, it differentiates progressively, along with the rest of the caudal trunk, out of the formative cell-mass that constitutes the end bud' (p. 95).[1]

In the meantime continued fusion at the rostral end of the groove has extended the neural tube into the future brain region. In embryos with 15 somites the tube is complete as far forward as the fore-brain, and shortly afterward (20 somites) the terminal aperture, known as the *anterior neuropore,* seals off (Figs. 426 *E,* 427). This point of closure is not located at the original rostral end of the neural plate, since ventral fusions have advanced somewhat to meet the main closing folds. The exact site in the brain of the end of the primitive neural plate is not surely known, but is thought to be at the optic recess (Fig. 452).[2] Even in early stages of neural-tube formation, and before closure involves the future brain region, the rostral half of the neural tube has enlarged locally to indicate the three *primary brain vesicles* (Fig. 426). The rest of the neural tube, which remains smaller in diameter, is the *spinal cord.* Its further elongation waits on the development of the caudal end of the body; in Fig. 426 *E* closure has progressed only to a low thoracic level.

The neural tube, in regions corresponding to both the brain and spinal cord, enters early into generalized organizational activities. These advances are preliminary to the acquisition of the specialized structure that eventually characterizes each organ. As a result of such differentiation the entire neural tube at an early period can be analyzed both into concentric layers and into longitudinal strips. The concentric layering is the outcome of

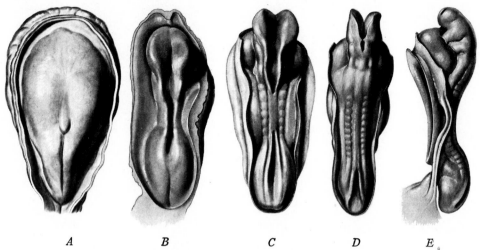

A   B   C   D   E

Fig. 426. Developmental stages of the human neural groove and tube (Streeter). All but *E* are in dorsal view. *A,* Presomite embryo, with neural plate and primitive streak (× 40). *B,* At three somites, with deep neural groove (× 37). *C,* At seven somites, with closure beginning midway (× 31). *D,* At ten somites, with closure extending into brain region (× 31). *E,* At nineteen somites, with closure complete except for neuropores at each end (× 20).

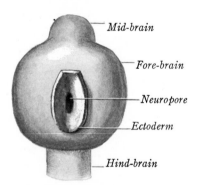

Mid-brain

Fore-brain

Neuropore

Ectoderm

Hind-brain

Fig. 427. Anterior neuropore, shown in a front view of the brain of an 18-somite human embryo (after Sternberg). × 64.

the histogenetic differentiation already described (p. 455). Viewed as a whole, the neural tube really consists of three concentric cylinders, which are in order (Fig. 431): (1) an inner *ependymal layer,* bounding the central canal; (2) a middle, cellular *mantle layer;* and (3) an outer, fibrous *marginal layer.* The neural tube also can be subdivided longitudinally into six strips or bands (Fig. 431). The primitive dorsal and ventral walls are primarily ependymal in structure and do not participate in the marked thickening that characterizes the lateral walls; these dorsal and ventral walls are named, respectively, (1) the *roof plate* and (2) the *floor plate.* Midway on the inner surface of each lateral wall is a groove, the *sulcus limitans,* which marks the subdivision of the wall into (3, 4) a more dorsal *alar plate* (sensory and co-ordinating) and (5, 6) a more ventral and thicker *basal plate* (motor). A clear understanding of the concentric and longitudinal organization of the neural tube is fundamental to an appreciation of later specializations.

The central nervous system is relatively large throughout the fetal period. Even at birth the brain constitutes 10 per cent of the body weight, whereas in the adult it is but 2 per cent. The spinal cord relatively outgrows the brain during the postnatal years, increasing from 0.9 per cent of the brain weight to 2 per cent. After the cessation of cell division in the nervous system and, accordingly, the ending of new neuron production, an important factor in the further increase in size of the spinal cord and brain is the thickening of myelin sheaths. Some of these investing sheaths are already present at birth, but all are still thin (p. 463).

The remainder of the present chapter will be devoted to descriptions of how the spinal cord and the brain organize both in external form and internal structure.

THE SPINAL CORD

External Form. There is no special boundary between the brain and spinal cord; the latter can be considered as beginning at the level of the first pair of spinal nerves. For a time the spinal cord is a thick-walled tube which tapers gradually to a caudal ending. In the fourth month it enlarges at the levels of the nerve plexuses that supply the upper and lower limbs. This is because of the additional sensory and motor neurons at these levels, and the presence here of shorter segments between successive spinal nerves. The *cervical enlargement* is located at the level of origin of the nerves of the brachial plexus to the arm; the *lumbo-sacral enlargement* is opposite the origins of the nerves of the lumbo-sacral plexus to the leg. (Fig. 428).

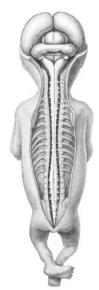

Fig. 428. Form and extent of the human spinal cord, at three months, exposed by a dorsal dissection. × 1.

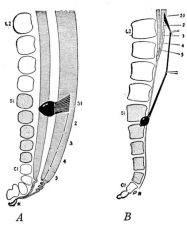

A   *B*

Fig. 429. Diagram of the 'recession' of the human spinal cord in relation to the faster growing vertebral column (Streeter). The formation of the filum terminale and the drawing-out of a sample nerve are illustrated. An asterisk indicates the coccygeal vestige. *A,* At nine weeks (× 6); *B,* at six months (× ⅘).

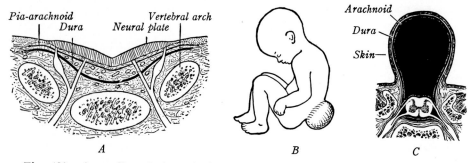

Fig. 430. Anomalies of the spinal cord. *A,* Section across a cleft spine and unclosed spinal cord (*cf.* Fig. 368 *B*). *B, C,* Meningocele shown in external view and in vertical section.

After the third month the vertebral column grows faster than the spinal cord. Since the cord is anchored by the brain, the vertebræ of necessity shift caudad along the spinal cord; this movement drags down inside the vertebral canal the roots that still have to find exits between vertebræ that were originally located directly opposite (Fig. 429). For this reason the spinal cord appears to recede up the vertebral canal, until at birth it ends at the level of the third lumbar vertebra, and in the adult opposite the first lumbar vertebra. Thus the roots of the lumbar to coccygeal nerves leave the spinal cord at a fairly high level; continuing downward within the vertebral canal, the nerves emerge between the vertebrae at lower levels. As might be expected, the thoracic nerves are displaced to a less degree, while the cervical nerves incline but little in a caudal direction.

The tip of the neural tube retains its terminal connections during this period of unequal growth; it becomes stretched and dedifferentiated into the slender, fibrous strand known as the *filum terminale* (Fig. 429).[3] The obliquely coursing spinal nerves, surrounding the filum terminale, constitute the *cauda equina* which was so named from its fancied resemblance to a horse's tail. Traces of the original saccular termination of the neural tube in the integument are recognizable at birth (Fig. 429). It constitutes the *coccygeal vestige,* located near the tip of the coccyx; the site is frequently marked superficially by a dimple or pit in the skin (Fig. 166 *B*).[4, 5]

Anomalies. Some striking malformations of the spinal cord and its investing membranes often accompany a cleft spine (*spina bifida*). The spinal cord may be widely exposed, like an unclosed neural plate (*myeloschisis;* Figs. 368 *B,* 430 *A*), or practically absent (*amyelus*). Herniation of the membranes (with or without participation of the cord) may accompany an unclosed spine. This is most frequent in the lumbo-sacral region where the skin-covered sac may become the size of an infant's head (Figs. 368 *C,* 430 *B*); yet in some instances the swelling is so small as not to be visible externally. When the protruding, fluid-filled sac is formed solely from the membranes about the spinal cord, the condition is called a *meningocele* (Fig. 430 *C*). When the cord also herniates locally into the membranous sac, it is *meningo-myelocele.* Duplication of the central canal, especially toward its caudal end, sometimes occurs. A *pilonidal sinus* is interpreted as an infected coccygeal fovea, or a similar dimple at a higher level (Fig. 166 *B*).[6]

Internal Organization. The wall of the spinal portion of the neural tube thickens so quickly that in the fourth week the typical three layers have already made their appearance (Fig. 431). Coincidental with this growth comes a relative narrowing of the internal cavity. For a time the neural canal is somewhat diamond-shaped in transverse section, its lateral angle on each side (the sulcus limitans) subdividing the side walls into plainly seen alar- and basal plates. The roof- and floor plates are relatively thin and poor in cells.

THE EPENDYMAL LAYER. This innermost stratum is a prominent component of the roof plate, dorsally, and the floor plate, ventrally (Fig. 431). As the alar plates thicken, the roof plate is obliterated as such and the facing ependymal layers of this region unite progressively into a median seam, the *dorsal median septum* (Figs. 432–434). This fusion steadily reduces the extent of the *central canal;* at three months it has become limited to the most ventral portion of the original cavity. The cells lining the final canal are definitive, ciliated *ependymal cells.*

As the proliferative activity in the spinal cord reaches its height, the rapidly thickening walls of the basal plates overlap the laggard floor plate. Nevertheless, the basal plates do not meet and fuse; instead, there is produced the longitudinal furrow known as the *ventral median fissure* of the spinal cord (Figs. 431–434). The ependymal cells of the thin floor plate

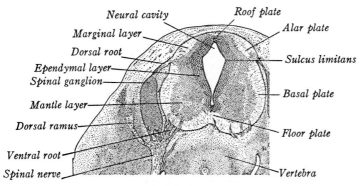

Fig. 431. Human spinal cord, at six weeks, in transverse section (after Prentiss) × 30.

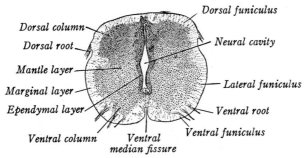

Fig. 432. Human spinal cord, at nearly eight weeks, in transverse section (after Prentiss). × 30.

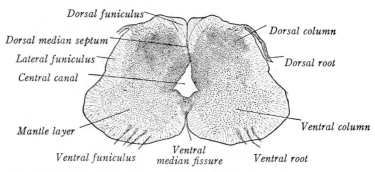

Fig. 433. Human spinal cord, at nine weeks, in transverse section (after Prentiss). × 30.

retain their original radial orientation and extend to the surface of the cord (Fig. 424 *C*).

THE MANTLE LAYER. The origin of this stratum from cells proliferating in the ependymal layer has been described previously (p. 456). Neuroblasts are the important constituent of this tissue which lacks myelinated fibers and hence receives the name, *gray substance*. Increase in mass of the gray substance depends upon cell division, and when this ceases shortly after birth this tissue has reached the end of its growth. On each side of the midplane the gray substance forms bulky masses named *dorsal, lateral* and *ventral columns* (sometimes also called horns) (Fig. 433).

With respect to the functional specialization of neuroblasts it can be said that those concerned with the primary reception and transmission of sensory messages have their cell bodies situated in the spinal ganglia; those that have to do with motor impulses are located in the ventral and lateral columns; all other neurons are concerned in linking up the sensory and motor systems.

In embryos of 10 mm. a thickening of the mantle layer on each side becomes prominent ventrolaterally (Fig. 431). It constitutes the *ventral gray column* which, in considerably later stages, supplies migrant cells that organize also a *lateral gray column*. Both are derivatives of the basal plate. In embryos of 20 mm. tardier dorsolateral thickenings of the mantle layer are likewise seen, the cells of which represent the *dorsal gray columns* (Figs. 432–434). The dorsal root fibers from spinal ganglion cells end in synapses about these neuroblasts which, as derivatives of the alar plate, become sensory relay neurons. Above and below the central canal, the mantle layer remains thin, constituting the *dorsal* and *ventral gray commissures* (Fig. 434). Fetuses in the fourth month have their gray substance arranged in what is essentially the permanent form.[7]

THE MARGINAL LAYER. This stratum is composed primarily of a framework made up of the processes from ependymal and neuroglial cells (Fig. 431). Into this mesh grow the axons of nerve cells, so that the significant thickening of the marginal layer is due entirely to nerve fibers contributed by cord neuroblasts and ganglion cells located elsewhere. The development of myelin about many of the fibers in the marginal zone is responsible for the appearance of a definite peripheral layer of *white substance* in the spinal cord; continued deposition of myelin accounts for the thickening of this layer even into adolescence. The white substance is subdivided by the dorsal and ventral nerve roots into *funiculi* (Fig. 433). These, in turn, contain various bundles, or *tracts;* each is composed of fibers of the same functional character.

The dorsal root fibers from the spinal ganglion cells, entering the cord dorsolaterally, subdivide the white substance in this region into *dorsal* and *lateral funiculi* (Fig. 433). In similar manner, the lateral funiculus is marked off by the ventral root fibers from the *ventral funiculus*. In the ventral floor plate, nerve fibers cross over from both sides of the cord as the *ventral white commissure*. The white substance as a whole is arranged in *tracts*, whose general relations and proportions are attained at the middle of the fetal period. The dorsal funiculus is formed chiefly by the dorsal root fibers of the ganglion cells, which enter and course both craniad and caudad in the marginal layer. It is subdivided into two distinct bundles, the *fasciculus gracilis,* median in position, and the *fasciculus cuneatus,* lateral (Fig. 434). The lateral and ventral funiculi are composed: (1) of *fasciculi proprii,* or ground bundles, originating in the spinal cord and interconnecting adjacent regions; (2) of *ascending fiber tracts* from the cord to the brain; and (3) of *descending tracts* from the brain.

The activities of higher vertebrates are dominated by the brain to a much greater degree than is the case in lower forms. Stated differently, the nerve centers in the spinal cord of lower vertebrates are far more independent and autonomous. With this in mind it is only to be expected that the earliest tracts of nerve fibers to appear in the marginal zone of man differentiate (early in the second month) for the purposes of distributing the dorsal root fibers within the cord and of linking together the nerve centers of the spinal cord itself. In the third month long association tracts of two kinds come into existence. Some begin with cell bodies in the cord and ascend to the brain; these serve to relay to the fore-, mid- and hind-brains the sensory impulses that are arriving in the cord from without. Others originate in the mid-brain and hind-brain and descend, thereby making possible an influence of higher centers over lower ones. Finally, in the fifth month, the cortico-spinal (pyramidal) tracts begin growing downward from the motor cells of the cerebral cortex; it is through these

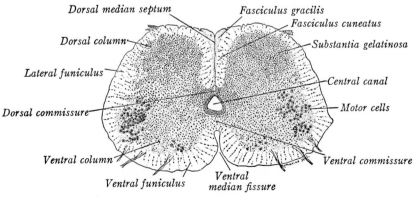

Fig. 434. Human spinal cord, at three months, in transverse section (after Prentiss). × 30.

neurons that the brain controls the motor cells of the spinal cord. The pyramidal tracts of man are not only the largest in any animal but they also contain both crossed and uncrossed fiber bundles. The latter (direct tracts) are peculiar to man and anthropoid apes, and as late acquisitions are extremely variable in size.

THE BRAIN

FUNDAMENTAL CONSIDERATIONS

The brain becomes increasingly specialized in proportion to the distance its several divisions lie above the spinal cord. In a corresponding manner the form and structure of the brain at higher levels also depart markedly from the simple plan that is laid down in the embryonic neural tube and is still retained in its essentials by the spinal cord. It is helpful at the outset of a study of brain development to take note of the more important of these fundamental and distinctive characteristics.

Primary Divisions.   The neural axis in embryos 2 mm. long (and with somites just appearing) is still nearly straight, but its rostral end is enlarging into the primitive brain (Fig. 426 *B*). Even before this region of the neural groove begins to close, three points of expansion, separated by two retarded zones of relative constriction, subdivide the brain into three parts (*C, D*): (1) the fore-brain (*prosencephalon*); (2) the mid-brain (*mesencephalon*); and, (3) the hind-brain (*rhombencephalon*). When the brain becomes a closed tube, these divisions are referred to as the *primary brain vesicles* (Fig. 435 *A*). The human brain at this stage is shown in Fig. 436 *A,* but the three divisions are not so clearly demarcated as they are, for example, in the chick (Fig. 543).

Both the fore- and the hind-brain vesicle promptly give rise to two secondary vesicles, whereas the mid-brain remains permanently undivided (Fig. 435 *B*). In embryos of about 3 mm. (early fourth week) the fore-brain shows indication dorsally of a groove that subdivides it into the *telenceph-*

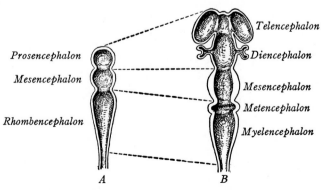

Fig. 435.   Subdivisions of the brain. *A*, Three-vesicle stage; *B*, five-vesicle stage.

alon, with its primitive cerebral hemispheres, and the *diencephalon* which bears the optic vesicles (Fig. 436 *B*). The mid-brain retains its original designation, the *mesencephalon*. Somewhat later the hind-brain specializes into the *metencephalon,* or future region of the cerebellum and pons, and the *myelencephalon,* or medulla oblongata (Fig. 437 *D*). A constricted region, the *isthmus,* lies at the junction of mesencephalon with metencephalon. The progressive separation and growth of these five brain vesicles can be followed easily in Fig. 437.

Flexures. While the several divisions of the brain are differentiating, certain flexures appear in its roof and floor, due largely to unequal growth processes. In part these correspond to those external bendings seen in the head and neck regions of young embryos. The first, or *cephalic flexure* occurs in the mid-brain region of embryos 3 to 4 mm. long, where the end of the primitive head takes a sharp bend ventrad (Fig. 436). Soon the angle is so acute that the long axes of the fore- and hind-brains make an acute angle (Fig. 438 *A*). At about the same time a *cervical flexure* appears at the junction of the brain and spinal cord. It is produced by the entire head flexing ventrad at the level of junction with the future neck. The *pontine flexure* begins to gain prominence, at the 10 mm. stage, at the junction of the met- and myelencephalon. It bends in a direction opposite to the others and is limited to the brain wall (Fig. 438 *B*). Eventually these two caudal flexures straighten and practically disappear, but the diencephalon and hemispheres become set permanently at an angle with the rest of the brain axis (Figs. 437 *E,* 453).

Cavities. The lumen of the primitive, tubular brain undergoes change that corresponds to the regional specialization of the walls (Figs. 435, 439). The cavity of the telencephalon extends into the paired cerebral hemispheres as the *lateral ventricles.* That of the diencephalon (and the median portion of the telencephalon) is designated the *third ventricle.* The narrow

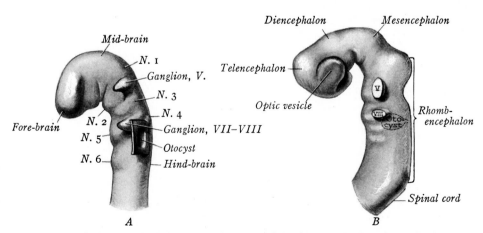

Fig. 436. Early human brains, viewed from the left side. *A,* At 3 mm., with eighteen somites (after Politzer; × 55); *B,* at 4 mm. (after Hochstetter; × 17). N.1–6, Neuromeres.

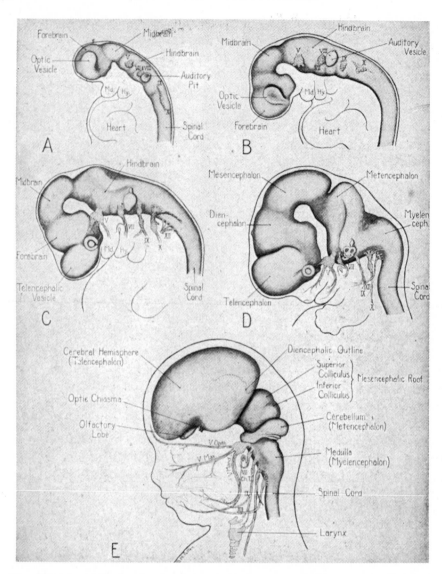

Fig. 437. Stages in the development of the human brain (Patten). *A*, At 3 mm.; *B*, at 4 mm.; *C*, at 8 mm.; *D*, at seven weeks; *E*, at three months. In *B* and *C*, the parts labelled 'forebrain' are telencephalon and diencephalon, respectively. In *C*, the 'hind-brain' is differentiating into metencephalon and myelencephalon.

canal of the mesencephalon becomes the *cerebral aqueduct*. The lumen of the metencephalon and myelencephalon is the *fourth ventricle*. The latter is demarcated rostrally by a constricted *isthmus* and is continuous caudally with the *central canal* of the spinal cord. A cast of these cavities from the brain of a newborn shows how the lateral ventricles finally overshadow the other three of the series (Fig. 461 *B*).

Derivatives. The primary divisions of the early neural tube and the parts eventually derived from them are summarized in the appended table:

DERIVATIVES OF THE NEURAL TUBE

PRIMARY DIVISIONS	SUBDIVISIONS	CONSTITUENT PARTS	CAVITIES
Prosencephalon	Telencephalon	Rhinencephalon Corpora striata Cerebral cortex	Lateral ventricles Rostral portion of the third ventricle
	Diencephalon	Epithalamus Thalamus (including Metathalamus) Hypothalamus	Most of the third ventricle
Mesencephalon	Mesencephalon	Colliculi Tegmentum Crura cerebri	Cerebral aqueduct
Rhombencephalon	Metencephalon	Cerebellum Pons	Fourth ventricle
	Myelencephalon	Medulla oblongata	
Spinal cord	Spinal cord	Spinal cord	Central canal

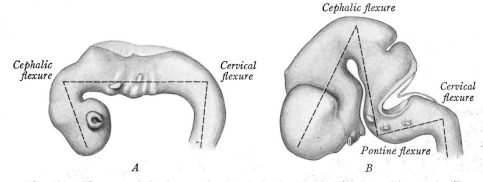

Fig. 438. Flexures of the human brain. *A,* At 6 mm. ($\times$ 13); *B,* at 14 mm. ($\times$ 7).

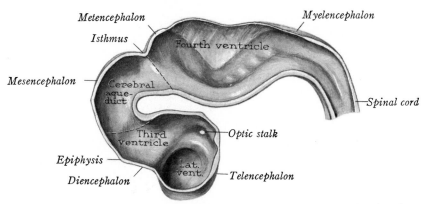

Fig. 439. Cavities of the human brain, at 11 mm., shown in a hemisection. $\times$ 10.

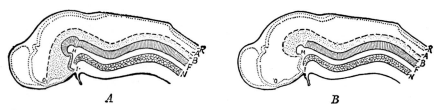

Fig. 440. Diagrams of the vertebrate brain, in sagittal section, illustrating the forward extent of the roof-, alar-, basal-, and floor plates (after Kingsbury). *A,* According to His; *B,* according to Kingsbury.

A, Alar plate; *B,* basal plate; *F,* floor plate; *I,* infundibulum; *M,* mammillary recess; *N,* notochord; *O,* optic recess; *R,* roof plate. Broken line is the sulcus limitans.

Composition of the Wall. The general plan of the spinal cord is continued into the brain. Yet the roof plate, and especially the alar and basal plates, undergo such a high degree of specialization that much of the brain, on casual inspection, appears to have little in common with the cord. Three striking peculiarities in structure may be singled out for mention. First, the roof plate in certain regions of the brain expands into a broad, thin sheet that tends to become associated with a rich network of blood vessels (Figs. 443, 444). Some portions of it protrude into the cavity of the brain as tufted extensions; these constitute the *chorioid plexuses* (Fig. 456). Second, contrary to earlier teaching, the floor plate of the brain is now said to extend only as far as the caudal boundary of the mid-brain, while the basal plate terminates at its rostral limit (Fig. 440).[2] The corollary of this conclusion is that both diencephalon and telencephalon are developed almost entirely out of alar-plate material.[8]

A third peculiarity is modifications in the structural arrangements within the alar and basal plates. The proliferation of neuroblasts in localized regions of the mantle layer leads to aggregations of cell bodies that are functionally alike. These masses of gray substance are called *nuclei* (Fig. 444). They may be subdivided by ingrowing nerve fibers into several parts, or they may invade the white substance and assume new locations there (Fig. 449 *C*). Such massing of nerve cells and fibers leads to regional thickenings of the brain wall and is one of the chief agencies through which the brain takes form and acquires its characteristic internal organization. Especially notable is the histological structure of certain regions of the brain, best illustrated by the cerebral and cerebellar hemispheres. Here the positions of the gray and white substances are largely reversed, and these layers are renamed as the *cortex* (gray) and *medulla* (mostly white) (Fig. 464 *B*). This new relation is brought about by many neuroblasts migrating through the mantle layer into the marginal layer and there giving rise to a superficial gray stratum that becomes strongly folded (*A*). Some neuroblasts do not participate in this migration and constitute scattered, deep nuclei in the mantle layer, now the medulla (Fig. 457).

Cell Shifts. The gray substance of the brain stem suffers marked displacements in comparison to the regular alignments of the gray columns in the spinal cord. Moreover, in the various vertebrate groups the nuclei of the brain occupy quite different positions depending on the particular trends of brain specialization that have been followed. Such changing relations through mass migration to new locations not only can be noted from group to group within the ascending vertebrate series, but also they are demonstrable in the development of individual embryos.[9] It is claimed that the shift is accomplished by the cell bodies moving closer to the source from which they receive most of their messages, that is to say, a shift against the flow of the nervous impulse (Fig. 441). Such a presumptively directed and oriented response has been named a *neurobiotaxis,* but the responsible factors are obscure. Examples of neurobiotactic shifting are furnished by the migration of the visceral motor nuclei of the cranial nerves to a lateral position (Figs. 443, 447).

Causal Relations. The induction that sets the fate of the neural plate also establishes a cranio-caudal polarity that results in a brain and spinal cord. This regional difference is said to result from a gradient in concentrations brought about by the activities of head- and trunk organizing substances (p. 179). The outcome is a self-differentiating neural plate that is a mosaic of potentially specific organ-regions (such as eye and mid-brain regions of the brain, and dorsal sensory and ventral motor strips in the spinal cord). Later, after the closure of the neural tube, more specific histogenetic fates become assigned to the primitive cells of the tube. Some, like the large motor cells of the spinal cord, occur in particular locations only, but the factors responsible for these specializations remain unknown.

During the establishment of a neural plate, its prospective cells move in from both sides toward the organizing dorsal midline. In part they are pushed passively by an expanding movement taking place in the adjacent non-neural ectoderm. On the other hand, folding of the neural plate into a tube is essentially an active process, but the exact mechanics of the act is not wholly understood. The normal presence and positions of the notochord and somites, with respect to the early spinal cord, are responsible for its characteristic shape and bilateral symmetry (thick lateral walls; radially arranged cells; compressed canal; median fissure). Conversely, the absence of myotomes and the presence of sense organs in the head region are correlated with the development of a differently shaped brain, bearing distinctive local peculiarities. The hollow brain molds its shape as the result of differential growth rates; the local rate is influenced by the ingrowth of sensory nerve fibers which brings on cell division and differentiation in the centers penetrated. Moreover, the presence and abundance of nerve cells in various centers and levels influence quantitatively the development of neurons in other regions. Perhaps the molding of the brain is aided by the turgor of the distending fluid, actively secreted by the

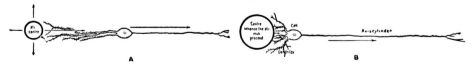

Fig. 441. Diagrams illustrating the principle of neurobiotaxis (Kappers). Axons grow in the direction of the nervous current (indicated by arrow), whereas first the dendrites and then the cell body grow against the current toward the source of stimulation.

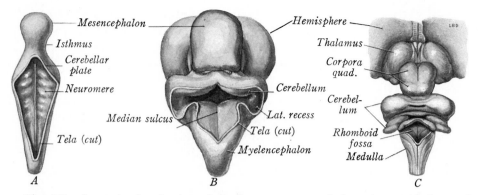

Fig. 442. Stages in the shaping of the human mesencephalon, metencephalon and myelenchephalon, in dorsal view (after Hochstetter). *A,* At five weeks ($\times$ 7); *B,* at nine weeks ($\times$ 4.5); *C,* at fifteen weeks ($\times$ 1.3).

ependymal layer of the neural tube, which bulges the thinner regions while the more resistant parts remain as relative constrictions.

After an experimental interchange of positions, portions of the still differentiating neural tube are capable of adjusting themselves to the quantitative neuronal requirements of the new locations. The spinal cord is more easily influenced by external environmental factors than is the brain, in which the intrinsic factors apparently play a larger rôle. The spinal cord also is more subject to the influence of higher levels of the nervous system than is the brain.

REGIONAL SPECIALIZATIONS

The Myelencephalon. The most caudal part of the brain, commonly called the *medulla oblongata,* is bounded rostrally by the early pontine flexure; this level is identified later by the caudal border of the pons (Fig. 445). The caudal limit of the medulla is the first cervical nerve.

The myelencephalon is transitional in structure between the spinal cord and the more highly specialized parts rostrad of it. Among other functions it is notable for serving as a great pathway linking brain and cord into a functional whole. All the typical features of the cord are continued into the medulla oblongata where they are gradually displaced to new positions and relations, are altered to a greater or less degree, and in most instances receive new names. Other elements, not represented at lower levels, also appear and enter into association with these basic structures. Hence as one progresses rostrad through the myelencephalon the familiar picture of the cord becomes more and more confused.

Among the more obvious differences from the spinal cord may be mentioned several features: (1) First is the loss of the serially segmental repetitions of the cord. To be sure, in the fifth and sixth weeks the floor of the hind-brain (rhombencephalon) is furrowed transversely by pairs of *rhombic grooves,* seven in number,[10] whose external bulgings are the so-called *rhombomeres,* or *neuromeres* (Figs. 436 *A,* 442 *A*). Some view these as evidence of a fundamental segmented condition of the head (p. 436). It seems more probable, however, that their serial arrangement is the expression of a growth by which they stand in rather regular relation to the nerves supplying the branchial arches.[11, 12] (2) Another difference is the addition, on each side, of a lateral row of nerves, intermediate in position between the dorsal sensory and the ventral motor series. These lateral nerves of the hind-brain are by number: *V, VII, IX, X* and *XI.* They are associated primarily with the branchial arches—rather than with the segmental trunk and its appendages, as were the spinal nerves (Fig. 437). (3) Still another difference is the disappearance of a sharp demarcation between gray and white substance. Nerve fibers, crossing in every direction, largely break up the gray substance into a mixture of gray and white known as the *reticular formation;* nevertheless, some is spared to form definite but isolated nuclear masses (Figs. 443, 444).

The wall of the developing myelencephalon shows the typical organization into longitudidal *plates* and the *sulcus limitans* (Figs. 443, 444). All but the roof plate are fairly comparable to their homologues in the spinal cord; by contrast, this local region is stretched into a broad, thin layer of ependymal tissue. Vascular mesenchyme (the pia mater; here also called the *tela chorioidea*) comes to lie on the ependymal roof. Rostrally the combined membrane infolds as vascular tufts that project into the cavity of the myelencephalon. This arrangement constitutes the *chorioid plexus* of the fourth ventricle (Fig. 451). As the roof plate thus expands into a non-nervous cover, the alar and basal plates are spread laterally like an opened book whose hinge is the floor plate (Figs. 443, 444). Both the alar and basal

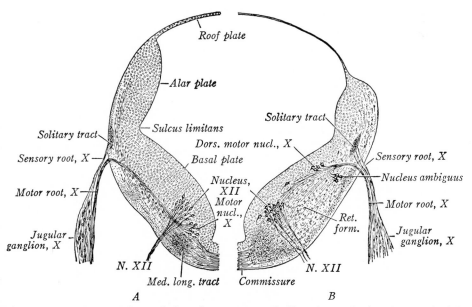

Fig. 443. Human myelencephalon, in transverse half-sections. *A,* At 10 mm. ($\times$ 75); *B,* at 12 mm. ($\times$ 45).

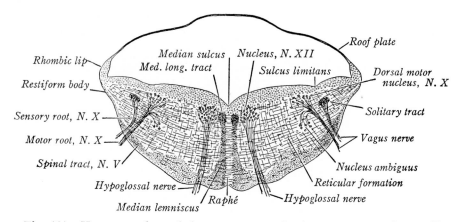

Fig. 444. Human myelencephalon, at two months, in transverse section. × 18

plates are at first represented by distinct ependymal, mantle and marginal layers, but the rapid proliferation of neuroblasts, the complex course of fibers extending from them, and the invasion of fibers from without, all tend soon to mask the primitive layering. The floor plate remains ependymal but becomes a thin seam, the *raphé,* located in the midplane. Some gray substance, not involved in the reticular formation, organizes into definite but isolated nuclear masses (Figs. 443, 444). These comprise: (1) *sensory relay nuclei,* about which the ascending sensory fibers of spinal nerves and sensory fibers of cranial nerves *IX* and *X* end; (2) *motor nuclei,* which give origin to the efferent fibers of nerves *IX* to *XII;* and (3) still other nuclei, whose fibers connect with the brain stem, cerebellum and fore-brain.

It is instructive to indicate some of the more important connections that the processes of neuroblasts in the alar and basal plates (and similar ones of the hind-brain as a whole) make in the myelencephalon. The sensory relay nuclei of this part of the brain effect four general types of communication: (1) correlating connections with the motor nuclei of the myelencephalon by means of the reticular formation; (2) descending connections with the motor centers of the spinal cord; (3) connections with the cerebellum; and (4) connections with the diencephalon which, in turn, are relayed to the cerebral cortex. Other important tracts are found in the medulla oblongata, but they are merely passing through to terminate at higher or lower levels.

The ependymal *roof plate* becomes a broad and flattened layer in a passive manner. Coincidental with the formation of a marked pontine flexure, at about the middle of the second month, the alar plates bulge laterally and the thinner roof plate is widened. Especially is this true in the rostral portion of the myelencephalon (Fig. 442 *B*). The cavity of the rhombencephalon (fourth ventricle) is thereby spread out from side to side and flattened dorsoventrally, a change most marked rostrally where *lateral recesses* of the fourth ventricle occur. Through local resorptions of the roof plate, paired lateral apertures (*foramina of Luschka*) and a medial aperture (*foramen of Magendie*) appear rostrally; they permit communication with the subarachnoid space.[13] The ridge where the roof plate joins the alar plate is known as the *rhombic lip* (Fig. 448 *A, B*).

Sensory nerve fibers, entering the *alar plates* from cranial nerves *VII, IX* and *X,* collect into the *solitary tract* located in the marginal layer (Figs. 443, 444, 476). Alar-plate neuroblasts migrate into the primitive marginal layer and partly surround this terminal tract (which corresponds to the dorsal root fibers of a spinal nerve); here they organize into the receptive *secondary sensory nuclei* of nerves *IX* and *X.* More caudally the *gracile nucleus* and *cuneate nucleus* are similarly developed as secondary relay stations for sensory fibers which ascend through the spinal cord from its nerves. Still other nuclei of alar-plate origin include the conspicuous *olivary nuclei,* which migrate far ventrad.[14]

The *basal plates* of the myelencephalon differentiate a little earlier than the alar plates. In embryos of the sixth week their neuroblasts give rise to the *motor nuclei* of origin for several cranial nerves; these nuclei occupy distinctive positions (Figs. 443, 444). Laterally, nearer the sulcus limitans, is located the nucleus (*nucleus ambiguus*) from which nerves *IX, X* and *XI* acquire the fibers (special visceral motor) that pass to the musculature derived from the third and fourth branchial arches. More medially (primitively, ventrad) lies the nucleus of somatic motor nerve *XII.* Later descriptions will make plain that these two sites of origin are also characteristic for other nerves of similar functional qualities that arise at higher levels in the brain. Among the motor nerve fibers coursing in the marginal layer are those descending from the motor cortex which produce the prominent *pyramids* (Fig. 445).

The ependymal cells of the *floor plate* elongate to keep pace with the thickening of the ventral wall of the medulla oblongata. Their processes extend from lumen to surface as the septum-like *raphé* (Fig. 444). On the floor of the medulla this raphé lies at the bottom of the *median sulcus* (Fig. 442 *B*).

The Metencephalon. This division of the brain extends from the isthmus to the pontine flexure at the caudal border of the pons. The metencephalon continues the general structure of the myelencephalon upward. It, however, adds two specialized and historically more recent parts. These are the *cerebellum* dorsally and the *pons* ventrally (Fig. 445). The former is

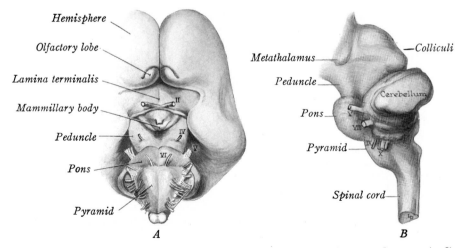

Fig. 445. Human brain, at fourteen weeks (after Hochstetter). *A,* Ventral aspect ($\times$ 2). *B,* Left lateral aspect, caudal to the diencephalon ($\times$ 2.5).

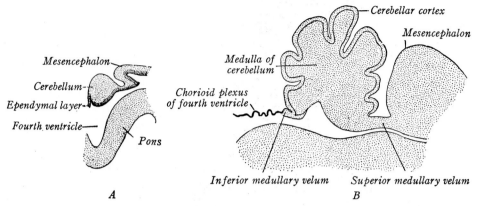

Fig. 446. Human metencephalon, in sagittal section (after Prentiss). *A,* At two months (× 5); *B,* at five months (× 8).

the chief co-ordination center of unconscious stimuli related to body position and movement; the latter is the principal conduction pathway between the cerebral cortex and cerebellar cortex.

The early metencephalon is made up of the six typical plates of the neural tube, but the primitive relations become modified profoundly, exceeding those in the myelencephalon by the differentiation of a cerebellum and pons. The *roof plate* transforms chiefly into a thin layer of white substance known as the *medullary velum* (Fig. 446). The *alar plates* supply receptive *sensory relay nuclei* for cranial nerve *V* and, in part, for *VII* and *VIII*. They also elaborate the *cerebellum* and contribute to the *cerebellar peduncles* which, as three pairs of stalks, connect the cerebellum with the rest of the brain. The *basal plates* supply the *motor nuclei* of origin for cranial nerves *V, VI* and *VII*. The *floor plate* forms a *raphé*, as in the medulla. Upward continuations of most of the structures of the medulla comprise the *tegmental portion of the pons;* it makes a thick floor to the cavity (fourth ventricle) of the metencephalon and is underlaid by the more conspicuous *basilar portion of the pons,* which is a purely mammalian, secondary acquisition. Surprisingly, the *pontine nuclei,* about which the basilar portion is developed, are migrants from the alar plates.

Since in early embryos the hind-brain lies directly above the pharynx, fore-gut and heart (Fig. 172), it is natural that the centers concerned with the regulation of chewing, tasting, swallowing, digestion, respiration and circulation remain located in the hind-brain, even though the organs innervated become considerably dislocated in position.

The *roof plate* transforms into a thin sheet of white substance rostrad of the cerebellum; caudad it is ependymal. These regions constitute, respectively, the *superior* and the *inferior medullary velum;* the rest of the roof plate is lost in the substance of the cerebellum (Fig. 446). The *alar plates* feature prominently. They differentiate the *secondary sensory nuclei* for the sensory components of cranial nerve *V* and, in part, for *VII* and *VIII*. The so-called *sensory nucleus of the fifth nerve* is found in the tegmental portion of the pons, the caudal region of it also extending into the medulla oblongata in company with the descending *spinal tract of the fifth nerve* (Figs. 444, 447, 476). The relation of the sensory fibers of the seventh nerve to the *solitary tract* and its nucleus has been mentioned already, since cranial nerves *IX* and *X* have similar connections (p. 481). Other cells of the tegmentum aggregate as the *cochlear* and *vestibular nuclei* of the eighth nerve; these originate through the proliferation of neuroblasts from the margin of the rhombic lip, and the cochlear nuclei are pushed ventrad. Other nuclei differentiating from the rhombic lip, but displaced to even a greater degree, are the numerous *pontine nuclei;* these (by way of the transverse fibers of the pons and those of the middle peduncle) relay to the cerebellum the afferent impulses descending from the motor cortex.

The *basal plates* are responsible for the *reticular formation* and the *motor nuclei* of cranial nerves *V, VI* and *VII* (Fig. 447). Many fibers from higher and lower levels pass uninterruptedly through the pons toward their destinations. Most imposing are the *pyramidal tracts.* The ependymal cells of the *floor plate* elongate to keep pace with the thickening of the ventral wall of the medulla oblongata. Their processes extend from lumen to surface as the septum-like *raphé* (cf. Fig. 444). On the floor of the hind-brain this seam lies at the bottom of the *median sulcus* (Fig. 442).

THE CEREBELLUM. The alar plates of the metencephalon are bent out laterally by the pontine flexure, and during the second month their rhombic lips thicken and bulge into the fourth ventricle (Fig. 448 *A*). Near the midline paired swellings indicate the future *vermis*, while the more lateral portions are destined to become *cerebellar hemispheres (B)*. During the third month the cerebellar mass everts and forms on each side a convex cerebellar hemisphere (*C*). In the meantime the paired primordia of the vermis have fused in the midline, thereby producing a single structure. The rhombic lip of this region gives rise to those most ancient parts of the cerebellum known as the *flocculus* and *nodulus (D)*. Between the third and fifth months the cerebellar cortex grows faster than the deeper layers, and in this way the principal lobes and fissures are produced (*C, D*). The hemispheres are the last to undergo such specialization; their fissures do not appear until the fifth month, but in fetuses of seven months the cerebellum has attained its final configuration.

The cerebellum shows at first a differentiation into the same three layers that typify the neural tube as a whole. During the second and third months proliferating cells from the rhombic lip, and perhaps from the mantle layer

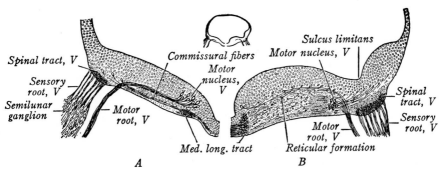

Fig. 447. Human metencephalon, in transverse partial section. *A*, At 6 mm. (× 66); *B*, at 11 mm. (× 57). Above is an orientation drawing of the total half-sections.

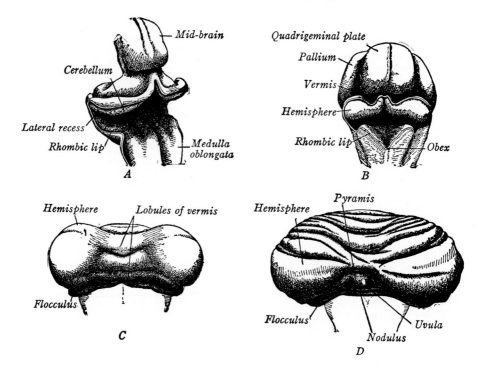

Fig. 448. Human cerebellum, in dorsal view (*A*, His; *B–D*, Prentiss). *A*, At six weeks (× 8); *B*, at two months (× 4); *C*, at four months (× 3); *D*, at five months (× 2.8). In *A*, *B* the roof plate has been removed.

of the cerebellum as well, migrate into the marginal layer; here they organize the *cerebellar cortex,* with its characteristic molecular and granular layers (Fig. 446 *B*). The final differentiation of this gray cortex is not completed until after birth. The axons of the conspicuous Purkinje cells and entering afferent fibers comprise the deep-lying *medulla* of the cerebellum.

The white substance of the medulla is continuous with three pairs of *cerebellar peduncles* which connect the cerebellum with the mid-brain, pons, medulla and spinal cord. Many cells of the primitive mantle layer take no part in the development of the cerebellar cortex, but give rise to neuroglial tissue and to the internal nuclei. Of these latter, the largest is the *dentate nucleus* which is not seen until the end of the third month.

The cerebellum and pons are not especially conspicuous except in animals with finely adjusted equilibrium and well-developed muscular co-ordination. The metencephalon reaches its highest expression in primates, but is also large in flying and swimming vertebrates. The cerebellum and pons have evolved through association with the adjacent otocysts, which are organs not only for hearing but also for equilibration; afferent systems arising from the vestibular mechanism, muscles, bones and joints are the prime factors, in so far as the evolution of the cerebellum and pons is concerned. Besides its intimate connection with all the sensory centers concerned with body equilibrium and the maintenance of muscle tone, the cerebellum also receives fibers from the cerebral cortex and gives off fibers from its nuclei to the motor centers of the brain stem.

The Mesencephalon. A plane passing just caudal to the posterior commissure dorsally, and the mammillary bodies ventrally, defines the rostral limits of the mesencephalon; its caudal limit is the isthmus (Fig. 453).

In general form the mid-brain is least modified from the primitive neural tube (Fig. 454). After the third month it is soon overshadowed and concealed by the much bulkier fore- and hind-brains and then serves, as a thick-walled tube, to interconnect them (Fig. 453). The roof-, alar- and basal plates are all represented, but the floor plate is now said to terminate with the metencephalon (Fig. 440 B).[2] For a time the roof plate constitutes a seam uniting the alar plates, yet even this loses its separate identity at two months. Conspicuous derivatives of the alar plates are the *superior* and *inferior colliculi,* related to sight and hearing (Fig. 442 C). The basal plates produce the local components of the equally prominent *cerebral peduncles* and the *motor nuclei* of origin of nerves *III* and *IV* (Fig. 449). The primitive neural cavity is reduced to the slender *cerebral aqueduct,* which after the third month narrows both relatively and absolutely (Fig. 453).

The mesencephalon is primarily associated with reflexes of the eyes and head in response to visual stimuli. The rostral pair of colliculi receive fibers from the retina, and from deep motor nuclei is derived the chief nerve supply of the muscles of the eyeball. The caudal pair are reflex acoustic centers. The mid-brain also becomes the main highway for motor fibers that unite the fore-brain with the nuclei of lower levels, and for sensory paths that connect in the reverse direction.

As at other levels, the *alar plates* develop more tardily than the basal plates. Here they give rise to the roof, or *tectum,* of the mid-brain which bears the *colliculi* (Figs. 442, 445 B). These latter are two pairs of rounded eminences appearing in the fourth month to serve as centers for visual and auditory correlation (Fig. 449 C). The rostral pair (the *superior colliculi*) are nuclei which, among other connections, receive fibers from the optic tracts; the caudal pair (the *inferior colliculi*) are nuclei which receive fibers from nuclei associated with the cochlear nerve. Neuroblasts migrate toward the surface of the colliculi and there organize stratified ganglionic layers, which are com-

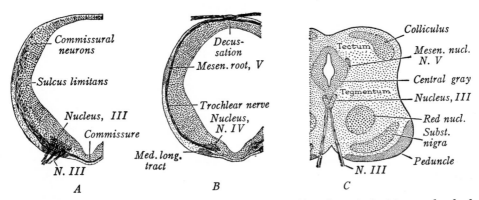

Fig. 449. Human mesencephalon, in transverse half-section. *A,* At 10 mm., level of oculomotor nucleus (× 48). *B,* At 10 mm., level of trochlear nucleus and decussation (× 48). *C,* Later stage (semidiagrammatic).

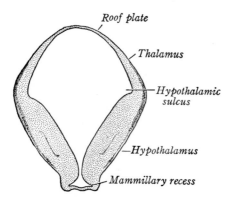

Roof plate

Thalamus

Hypothalamic sulcus

Hypothalamus

Mammillary recess

Fig. 450. Human diencephalon, at 12 mm., in transverse section. × 33.

parable to the cortical layers of the cerebellum; the deeper cell masses correspond to the cerebellar nuclei. The continuation rostrad to this level of the sensory nucleus (here named the *mesencephalic nucleus*) of the fifth nerve is worthy of note; it furnishes the only instance in which sensory fibers of a peripheral nerve have their cell bodies implanted in the wall of the neural tube (Fig. 449 C).[15]

Early in the second month neuroblasts of the *basal plate* condense into the *motor nuclei* of the third and fourth cranial nerves (Figs. 449, A, B). The *tegmentum* is similar to and continuous with the reticular formation of lower levels; it contains *tegmental nuclei (C)*. The *red nucleus* and the pigmented *substantia nigra* presumably differentiate from the basal plate,[16] although the former has commonly been said to originate from migratory, alar-plate neuroblasts. Important descending and ascending tracts lie dorsal and ventral to the substantia nigra. They course in two rounded masses, known as the *cerebral peduncles,* which bulge conspicuously on the ventral surface of the brain (Fig. 445).

The Diencephalon.   The rostral extent of the diencephalon is established by folds that set caudal limits to the cerebral hemispheres and corpora striata; on the floor of the brain this boundary passes just rostral to the optic chiasma (Fig. 452 A). The caudal limit includes the posterior commissure dorsally and the mammillary bodies ventrally (B).

Although prominent during the second month (Figs. 451 A, 452 A), the diencephalon becomes largely concealed by the greater expansion of adjoining parts of the brain (Figs. 451 B, 461). It is almost wholly given over to various kinds of correlations, and through it pass all the nervous impulses that reach the cerebral cortex with the single exception of those from the olfactory organs. The wall of the diencephalon differentiates a dorsal roof plate and paired alar plates, the latter including both the sides and the floor of the tube (Fig. 450). It seems probable that neither the basal plate nor the floor plate of lower levels extends this far rostrad (Fig. 440 B).[2] Except for this difference, and the absence of typical nerves, the diencephalon appears in early stages not unlike the primitive spinal cord. It remains preponderatingly composed of gray substance, derived from the mantle layer, and aggregated as *nuclei*. The cavity of the diencephalon is the *third ventricle;* for a time it is relatively broad (Fig. 442 A), but the strongly thickening lateral walls later compress it to a narrow, median cleft (Fig. 453).

The roof plate remains as a thin, ependymal lining (Fig. 450). Its folds, in association with the vascularized pia mater (*tela chorioidea*), forms a *chorioid plexus* which invaginates into the third ventricle during the second month (Figs. 453, 465). Far caudad the *epiphysis,* or pineal body, evaginates during the seventh week; in man it becomes conical, solid and possibly glandular (Figs. 452–454).[17]

The rest of the diencephalon is of alar-plate origin. It soon shows three main regions (Fig. 452): the *epithalamus,* dorsally; the *thalamus,* laterally; and the *hypothalamus,* ventrally.[18] The epithalamus and hypothalamus are the more primitive in character, and their differentiation precedes that of the thalamus which is best developed in higher vertebrates. The *epithalamus* lies at the junction-zone of the roof plate and alar plate and includes some of the latter, especially caudally. Each thickened alar plate proper is divided by the *hypothalamic sulcus* into thalamus (above), and hypothalamus (below). The *thalamus* rapidly outgrows the epithalamus, and nerve fibers separate the massive gray substance of its wall into numerous

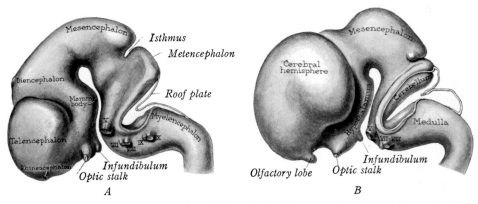

Fig. 451. Human brain, in left lateral view (after Hochstetter). *A,* At 14 mm. ($\times$ 7.5); *B,* at 27 mm. ($\times$ 5).

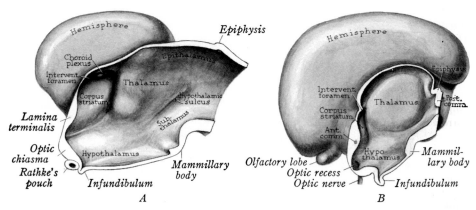

Fig. 452. Human telencephalon and diencephalon, hemisected and viewed from the left side (after Hochstetter). *A,* At seven weeks ($\times$ 10); *B,* at ten weeks ($\times$ 4.5).

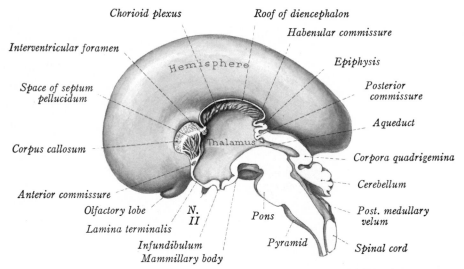

Fig. 453. Human brain, hemisected, at fourteen weeks (after Hochstetter). × 2.5.

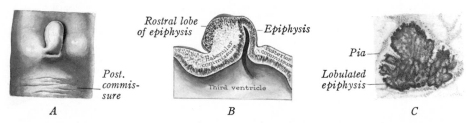

Fig. 454. Stages in the development of the human epiphysis. *A,* Surface view, at eight weeks (after Hochstetter; × 30). *B,* Longitudinal hemisection, at ten weeks (after Hochstetter; × 27). *C,* Transverse section, at five months (after Gladstone; × 8).

thalamic nuclei. The two thalami grow into close approximation and usually unite by a bridge across the median plane, thereby forming the *massa intermedia.* The walls of the *hypothalamus* contain various *hypothalamic nuclei* and in the floor are the *optic chiasm,* formed by the region of crossing of the optic nerves, and the *infundibulum* which specializes into the stalk and neural lobe of the hypophysis (p. 230).

The *epithalamus* is a synaptic region for the correlation of olfactory impulses. The *epiphysis* lies between the *habenular* and *posterior commissures,* which interconnect corresponding nuclei of the two sides (Figs. 453, 454). The epiphysis of mammals, birds and some reptiles is interpreted as an organ that has differentiated in quite a different direction than does the median parietal eye of certain fishes, amphibians and reptiles. In this sense it is not a vestigial or degenerated parietal eye; the latter even develops from an entirely separate primordium.[19] The *thalamus* consists of a more ancient part, which can act independently of the cerebral cortex in effecting reflexes having to do with pleasurable and painful sensations, and a newer part which is the larger by far in man. The latter portion is the main corridor through which impulses of cutaneous, visual and auditory sensibility are relayed by other neurons to the cerebral cortex. A special region of the newer thalamus, named the *metathalamus,* contains the *geniculate bodies* which are concerned with the transmission of impulses of visual and auditory sensibility. In addition to the stalked optic cups and the infundibulum, already mentioned, the *hypothalamus* also develops the *tuber cinereum* and *mammillary bodies* (Fig. 453). The hypothalamus is the co-ordinating headquarters of the autonomic system; it regulates such visceral functions as digestion, sleep, heat regulation and emotional behavior. A transition region, interposed between the thalamus, hypothalamus and the tegmentum of the mid-brain, is often recognized as a distinct subdivision, the *subthalamus.*

The Telencephalon. The caudal boundary of this most rostral subdivision of the brain has already been defined (p. 486). The telencephalon consists of a median portion and of two lateral outpouchings from this region. The median portion is continuous with the diencephalon and encloses a rostral extension of the third ventricle. The outpouchings are the paired *cerebral hemispheres,* each of which contains a *lateral ventricle* (Fig. 455). Like the diencephalon, the telencephalon is mostly a product of greatly expanded alar plates. Basal and floor plates are lacking, and the roof plate is chiefly concerned with the formation of a chorioid plexus.

The roof of the original telencephalon is inconsiderable in comparison to the evaginated hemispheres and does not take part in their extensive development (Fig. 455). The cerebral hemispheres begin to be prominent during the sixth week and expand rapidly until, at the middle of fetal life, they overgrow the diencephalon and mesencephalon and overlie the cerebellum somewhat (Figs. 451, 461). During this period of enlargement the original rostral end of the neural tube remains a medial band, relatively unchanged in position; for this reason it is named the *lamina terminalis*

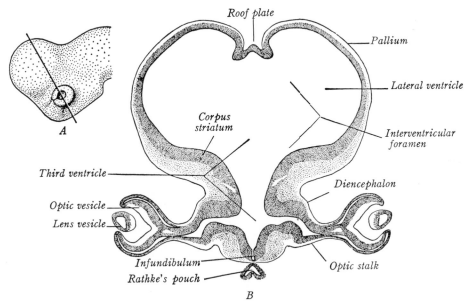

Fig. 455. Human telencephalon, at 10 mm. *A,* Left lateral view (× 10). *B,* Transverse section, through the level indicated on *A* (after Prentiss; × 30).

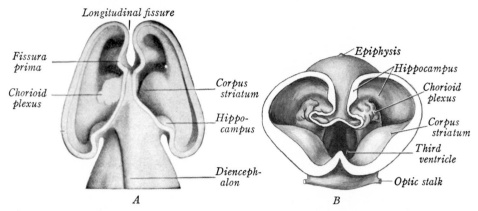

Fig. 456. Human fore-brain, with a portion of the wall removed. *A,* At six weeks, in dorsal view (His; × 13). *B,* At seven weeks, in front view (after Hochstetter; × 9).

(Figs. 452, 453). Since the two hemispheres grow forward on each side of the lamina, this plate becomes buried at the bottom of the resulting *longitudinal fissure* that separates the hemispheres (Fig. 456 *A*). The *lateral ventricles,* or cavities of the hemispheres, at first communicate broadly with the third ventricle through the paired *interventricular foramina* of Monro (Figs. 455 *B*). Later each foramen is narrowed to a slit, not by constriction but because its boundaries grow more slowly than the rest of the telencephalon (Fig. 452).

The telencephalon is also divisible into three portions that differ in functional nature. One is the *corpus striatum,* directly continuous with the thalamus; like the latter it is a reflex and reinforcing center, but of a higher order (Figs. 452, 454). The second division is the *rhinencephalon,* including what is known as the *paleopallium* and *archipallium.* The remainder, far bulkier than the other two, constitutes the *neopallium* (Fig. 453). The neopallium represents almost all of the externally visible hemispheres, whose wall as a whole may be called the *pallium.*

In higher vertebrates the telencephalon becomes the most specialized and complex region of the brain. Practically all of the nervous mechanisms of lower levels are concerned with rigid responses involving reflex and instinctive activities. In lower vertebrates the telencephalon is still of this nature, but in mammals the characteristically variable types of response (acquired and mostly consciously performed) are mediated through the gray covering of the hemispheres known as the *cerebral cortex;* accordingly, this substance gains increasing prominence until its elaboration reaches a climax in man. The rhinencephalon is the olfactory part of the brain; in fishes it represents almost the entire cerebral hemispheres, but in higher forms it is progressively subordinated as smell declines as the dominant sense. The neopallium, or non-olfactory cortex, advances in importance in reptiles and birds, becomes very large in mammals and constitutes almost all of the exposed portions of the human cerebrum.

THE CORPUS STRIATUM. The floor of each primitive hemisphere produces a thickening, which at six weeks bulges prominently into the lateral
ventricle (Fig. 456). The *corpus striatum,* so formed, is in line caudally
with the thalamus of the diencephalon and is closely related to it both
developmentally and functionally. The thalamus and corpus striatum are
separated by a deep groove until the end of the third month (Fig. 452).
As the two structures enlarge, the groove between them disappears and
they then seem like one continuous mass. (In Fig. 457 the corpus striatum
is labeled as caudate and lentiform nuclei.) Since the corpus striatum is
thus anchored, it does not share greatly in the displacements experienced
by the rest of the expanding hemisphere. On the contrary, it serves as a
fixed area from which the cerebral expansion is produced.

The *corpus striatum* thickens, owing to an active proliferation that gives rise to a
prominent mass of mantle-layer cells. Nerve fibers, passing in both directions between
the thalamus and the cerebral cortex, course through the corpus striatum; here they are
arranged in a lamina which takes the form of a wide **V**, open laterally. This band-like
tract of white fibers is the *internal capsule.* Its rostral limb partly divides the corpus
striatum into the *caudate* and *lentiform nuclei;* the caudal limb of the capsule extends
between the lentiform nucleus and the thalamus (Fig. 457). The corpus striatum elongates in company with the cerebral hemispheres, its caudal portion curving around to
the tip of the inferior horn of the lateral ventricle and forming the slender tail of the

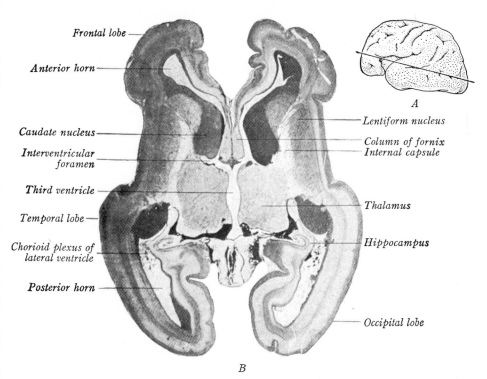

Fig. 457. Human telencephalon and diencephalon, at five months. *A,* Left lateral view.
B, Horizontal section, through the level indicated on *A* (His; × 2).

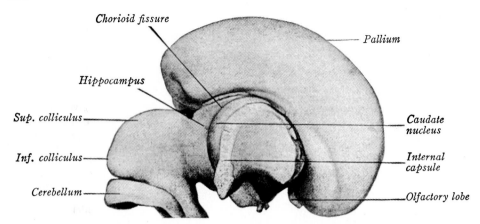

Fig. 458. Human brain, at nearly three months, viewed from the right side after removal of most of the right pallium (His). × 5.

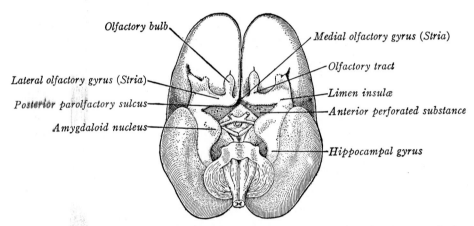

Fig. 459. Human brain, at eighteen weeks, showing the rhinencephalon in ventral view (Hardesty after Retzius). × 1.5.

caudate nucleus (Fig. 458). By the middle of fetal life the corpus striatum has attained its adult shape and relations.

THE RHINENCEPHALON AND PRIMITIVE PALLIUM.   During the sixth week a swelling appears on the ventral surface of each cerebral hemisphere (Fig. 451). These enlarge into distinct *olfactory lobes* which, however, remain small in man (Fig. 458). Each lobe is arbitrarily divided into a rostral and caudal division (Fig. 459). The former (or pars anterior) represents the *olfactory bulb* and *tract;* the tract receives the backward-growing olfactory fibers, while bulb and tract alike lose their original lumen. The caudal division (pars posterior) is a thickening of the brain wall, which matures into the *anterior perforated substance* and the *parolfactory area* medial and dorsal to it. The olfactory apparatus includes two pallial regions (Figs. 459, 460). The *paleopallium* is composed of proximal portions of the olfactory lobes; the *archipallium* comprises the *hippocampal formation,* developed along the medial margin of each hemisphere.

THE NEOPALLIUM. The non-olfactory cortex becomes predominant in mammals, while in man it so outgrows the fore-brain as a whole that other parts are largely hidden from surface view. It comprises almost all of the hemispheres except the hippocampal formation.

EXTERNAL FORM OF THE HEMISPHERES. The rapid progress of the hemispheres can be followed in Figs. 437, 461, 462. The telencephalon expands in such a fashion that four lobes can be distinguished in each hemisphere (Fig. 461 *A*). These have no close correlation with functional specializations, but are convenient for descriptive purposes. They are: (1) a rostral *frontal lobe;* (2) a dorsal *parietal lobe;* (3) a caudal *occipital lobe;* and (4) a ventrolateral *temporal lobe,* produced when a part of the primitive occipital lobe turns ventrad and rostrad.

The surface expanse of the cerebral wall, the gray *cortex,* increases far more rapidly than the white medullary layer which underlies it. As a result, the cortex of each hemisphere is folded into *gyri,* or *convolutions,* between which are prominent furrows. The main, deep furrows are called *fissures;* the lesser ones receive the name of *sulci* (Fig. 462). The earliest fissures appear during the fourth month, yet throughout the first half of fetal life the exposed surface of the brain remains quite smooth. In the later months, however, other fissures arise and complete the series. Developing at the same time as this last group of fissures are the numerous but shallower sulci (Figs. 460, 462). The secondary and tertiary sulci, peculiar to the human brain, are developments of the final fetal months. Previous to their appearance the brain resembles that of an adult monkey. Although the gyri and sulci have a definite and regular arrangement, they correspond inexactly either to localized cortical structure or to function.

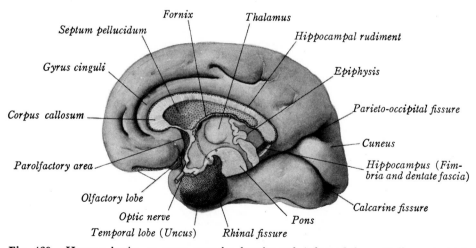

Fig. 460. Human brain, at seven months, hemisected (adapted from Kollmann). × 1. The rhinencephalon is designated by stipple.

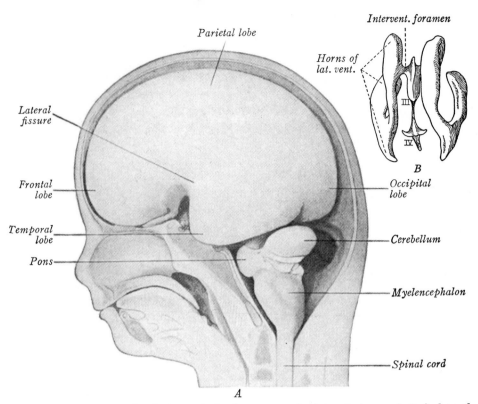

Fig. 461. Form and relations of the human cerebral hemisphere. *A,* Left lateral view of the brain, at fourteen weeks, *in situ* (His; × 2.5). *B,* Cast of the **ventricular** system in a newborn, viewed from above (after Welker; × ½).

The first two fissures to appear are the *rhinal* (Fig. 460) and *hippocampal fissures,* which develop during the fourth month in association with the rhinencephalon. The hippocampal fissure represents a curved infolding along the medial wall of the temporal lobe; the corresponding elevation on the internal surface of the pallium is the hippo- campus itself (Figs. 457, 458). At about the same time, the *lateral fissure* (of Sylvius) makes its appearance but is not completed until after birth (Fig. 461). Its development is due to the fact that the cortex overlying the corpus striatum expands more slowly than do the surrounding areas; this region is consequently overgrown by opercular (covering) folds of the frontal, parietal and temporal lobes. The area thus incompletely enclosed is the *insula* (island of Reil), and the depression so formed is the lateral fissure (Fig. 462). These opercula do not close-in over the insula and come into contact until after birth. The *chorioid fissure* results from the early ingrowth of the chorioid plexus (Fig. 458). Since the temporal lobe carries with it both the chorioid plexus and the fissure through which it entered, the chorioid fissure is transferred to the under aspect of the hemisphere.

Among the neopallial depressions that appear at six to seven months are several that become important landmarks in cerebral topography. They are: (1) the *central sulcus,* or fissure of Rolando, which forms the dorsolateral boundary between the frontal and parietal lobes (Fig. 462); (2) the *parieto-occipital fissure* which, on the median wall of the hemisphere, is the line of separation between the occipital and parietal lobes (Fig. 460); (3) the *calcarine fissure,* which marks the position of the visual area of the cere- brum (Fig. 460) and internally causes that convexity termed the calcar avis; and (4) the *collateral fissure,* on the ventral surface of the temporal lobe, which produces the **inward** bulging on the floor of the posterior horn of the ventricle known as the collateral eminence. In this same period, and continuing until birth, arise the various sulci.

COMMISSURES. For the purpose of securing co-ordination, the reflex centers of the two sides of the neural tube are connected by bundles of crossing fibers called *commissures*. They occur both in the brain and in the spinal cord. Besides the *optic chiasma* (Fig. 452 *A*) and the *habenular* and *posterior commissures* of the diencephalon (Fig. 453), already mentioned (p. 488), there are three in the telencephalon (Fig. 463). The *hippocampal* and *anterior commissures* are the more ancient cross-connections for the archipallium, while the larger *corpus callosum* is the great transverse bridge of the neopallium. These three commissures develop in relation to the lamina terminalis, since this is the natural, direct path from one cerebral vesicle to the other (Fig. 452). They cross partly in the lamina and partly in the fused adjacent portions of the median pallial walls. Owing to the union of the pallial walls dorsal and rostral to the lamina, the latter thickens rapidly during the fourth and fifth months. It is at this time that the significant development of the commissures occurs.

In the rostral portion of the lamina terminalis, fibers crossing the midplane unite the two hippocampi and produce the *hippocampal commissure* (Fig. 463 *A*); with the later growth of the corpus callosum this commissure shifts farther caudad (*B*). The hippocampal commissure is closely associated with the *fornix* which is made up of paired, symmetrical fiber tracts that pursue arching courses to connect the hippocampi with the hypothalamus. The fibers of the *anterior commissure* cross in the lamina terminalis, ventral to the primitive hippocampal commissure (Fig. 463 *A, B*). They arise in paired rostral and caudal divisions which unite into a common bundle near the midplane. The rostral part interconnects the olfactory bulbs in a horse-shoe bow of fibers. The caudal division passes ventrally between the corpora striata and the cerebral cortex and may be derived from one or both of these regions.

The *corpus callosum* develops in the roof-region of the thickened lamina terminalis, located both rostral and dorsal to the primitive hippocampal commissure (Fig. 463 *A*).

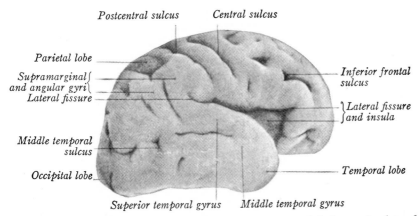

Fig. 462. Right cerebral hemisphere, from a seven-months' fetus, in lateral view (Kollmann). × 1.

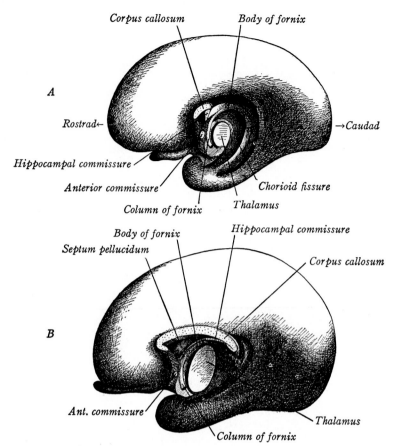

Fig. 463. Development of the commissures, shown in hemisected human brains, viewed from the left side (adapted by Prentiss). × about 2. *A,* At three months; *B,* at four months.

Within a short time it has extended particularly in the caudal direction, and thereafter constitutes a conspicuous landmark of the telencephalon (*B*). Through its fibers, which grow out from neuroblasts in the wall of the neopallium, nearly all regions of one hemisphere are associated eventually with corresponding regions of the other. In fetuses of five months this great commissure has attained the structure and shape that is characteristic of the adult (Fig. 460).

The triangular interval between the fornix and corpus callosum contains a thin partition that separates the two lateral ventricles (Fig. 463 *B*). This *septum pellucidum* is a membranous portion of the lamina terminalis and really consists of thinned, median pallial wall (Fig. 460). As a result of stretching, caused by the growth of the corpus callosum, the septum sometimes splits and contains a cleft-like cavity bounded by distinct laminæ. This cleft is designated the *space of the septum pellucidum,* or often, inappropriately, the fifth ventricle (Fig. 463).

HISTOGENESIS OF THE PALLIUM.   In the wall of the pallium are differentiated the ependymal, mantle and marginal layers typical of the neural tube in general. During the first two months the pallium remains thin and differentiation is slow. Early in the third month neuroblasts are migrating from the mantle zone into the marginal zone and there they soon lay down a *primordial cortex,* close to the surface (Fig. 464). Stratification within this mass then proceeds and at six months the six concentric layers of pyramidal and other cells that characterize the *cerebral cortex* are demarcated. The final differentiation of the outer layers, however, is not complete until middle childhood. These outer three layers are more highly developed in man than in lower mammals.

Beginning with the fourth month the pallial wall thickens rapidly, owing both to the intrusion of fibers from the thalamus and to fibers derived from the neuroblasts of the cortex itself. These fibers as a whole are arranged in a thick internal layer, which is white in color and surrounded by the much thinner gray cortex. This *medulla* occupies the position of the mantle zone of the spinal cord and the ordinary relations of the main white and gray substances are, of course, reversed. As the cerebral wall increases in thickness, the size of the lateral ventricles diminishes relatively; especially is this true of their lateral diameters.

THE VENTRICLES AND THEIR CHORIOID PLEXUS. The original third ventricle of the telencephalon expands into *lateral ventricles,* which follow the development of the hemispheres and extend into the four lobes belonging to each. The *body* of each lateral ventricle occupies a corresponding parietal lobe, while the *anterior, posterior* and *inferior horns* extend, respectively, into the frontal, occipital and temporal lobes (Fig. 461 *B*). Just as the chorioid plexus of the third ventricle develops in the folded roof plate of the diencephalon, so the thin median wall of the pallium (originally dorsal, and largely roof plate) at its junction with the wall of the diencephalon is folded into each lateral ventricle along the chorioid fissure (Figs. 456, 458). A vascular plexus, continuous with that of the third ventricle, grows into this fold and projects into the corresponding lateral ventricle (Fig. 465). The entire plexus system is a paired structure which, with the plexus of the third ventricle, makes a Y-shaped figure; the stem of the Y overlies the third ventricle, and its curved arms project into the lateral ventricles just caudal to the interventricular foramen. Later, as the pallium expands, the chorioid plexus elongates (along with the chorioid fissure) in a caudal direction (Fig. 458). It eventually extends far into the occipital and tem-

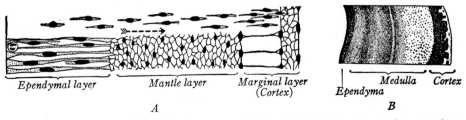

| *Ependymal layer* | *Mantle layer* | *Marginal layer* (*Cortex*) | *Ependyma* | *Medulla* | *Cortex* |

A   *B*

Fig. 464. Histogenesis of the human pallial wall. *A,* Schematic section, at three months (after His). Below is the spongioblastic framework only; above are wandering neuroblasts migrating into the cortical layer. *B,* Vertical section of the pallium, at four months, with its cortex thickening rapidly (× 15).

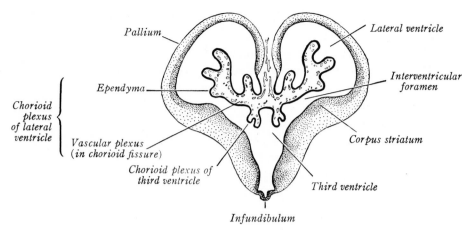

Fig. 465. Invagination of the human chorioid plexuses into the lateral and third ventricles, shown by a transverse section through a fetal brain of the third month.

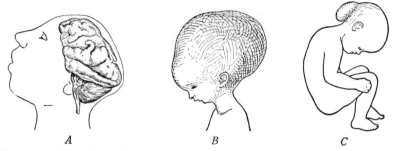

A   *B*   *C*

Fig. 466. Anomalies of the human brain. *A*, Micrencephaly, associated with a microcephalic head. *B*, Hydrocephaly, producing a macrocephalic head. *C*, Encephalocele.

poral lobes, where it protrudes into the posterior and inferior horns of the lateral ventricles (Fig. 457).

Anomalies. The brain is subject to great variation in size. Marked smallness and underdevelopment is *micrencephaly;* the mentality of the individual shown in Fig. 466 *A* was slight. Since in all such instances the brain case is correspondingly small, the condition is also known as *microcephaly* (Fig. 156 *B*). Opposite in nature is excessive largeness of the cranium (*macrocephaly*). This oversize is due to a distention of the brain and the still unfused cranium through interference with the drainage of cerebral fluid; hence the anomaly is also designated as *hydrencephaly* or *hydrocephaly* (Fig. 466 *B*). Accompanying a roofless skull (*acrania,* Fig. 379 *A*) goes a severe arrest of brain development, and even a degenerative regression (*anencephaly*). The cerebral hemispheres always suffer such damage, and other regions of the brain usually do also. There may be a local herniation of the membranes (*meningocele*) through a defective portion of the skull. More commonly this protrusion involves both the brain and its membranes (*meningoencephalocele*); this occurs most frequently in the occipital region and the sac, enveloped with skin, may attain the size of the head (Fig. 466 *C*). Faulty development or differentiation of the cerebral cortex is correlated with conditions of congenital idiocy. If the motor region of the cortex is involved in this manner, a spastic paralysis results.

REFERENCES CITED

1. Gaertner, R. A. 1949. J. Exp. Zoöl., *111*, 157–174.
2. Kingsbury, B. F. 1922. J. Comp. Neur., *34*, 461–491.
3. Streeter, G. L. 1919. Am. J. Anat., *25*, 1–11.
4. Kunitomo, K. 1918. Carnegie Contr. Embr., *8*, 167–198.
5. Alegais & Peyron. 1920. Compt. rend. Soc. biol., *83*, 230–232.
6. Willis, R. A. 1962. The Borderland of Embryology, pp. 303–304 (Butterworth).
7. Romanes, G. J. 1944. J. Anat., *75*, 145–152.
8. von Schulte, H. W. & F. Tilney. 1915. Ann. N. Y. Acad. Sci., *24*, 319–346.
9. Windle, W. F. 1933. J. Comp. Neur., *58*, 643–733.
10. Nilsson, F. 1926. Z'ts. f. mikr.-anat. Forsch., *7*, 191–230.
11. Neal, H. V. 1918. J. Morph., *31*, 293–315.
12. Adelmann, H. B. 1925. J. Comp. Neur., *39*, 10–171.
13. Wilson, J. T. 1937. J. Anat., *71*, 423–428.
14. Kooy, F. H. 1917. Folia Neur.-biol., *10*, 205–369.
15. Pearson, A. A. 1949. J. Comp. Neur., *90*, 1–46.
16. Shaner, R. F. 1932. J. Comp. Neur., *55*, 493–511.
17. Gladstone, R. J. & C. P. G. Wakeley. 1935. J. Anat., *69*, 427–454.
18. Gilbert, M. S. 1935. J. Comp. Neur., *62*, 81–115.
19. Tilney, F. & L. F. Warren. 1919. Am. Anat. Memoirs, No. 9, Pt. 1, 257 pp.

Chapter XXVI. The Peripheral Nervous System

The peripheral nervous system consists of bundles of myelinated and unmyelinated *nerve fibers,* and aggregations of nerve cells known as *ganglia.* The fibers are of two types: *afferent fibers,* which carry sensory impulses to the central nervous system, and *efferent fibers,* which dispatch motor impulses away from the nervous centers (Fig. 470). The afferent fibers develop from neuroblasts located in the neural crest alongside the neural

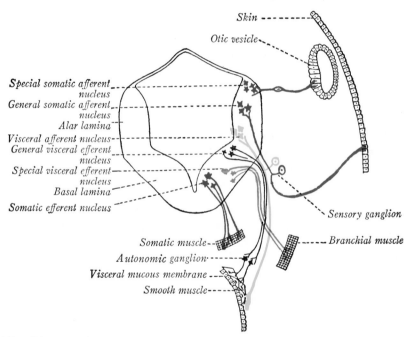

Fig. 467. Diagram showing the arrangement of the functional cell columns of the brain and the functional components of the cranial nerves (Ranson).

tube (p. 459). From each cell one fiber extends outward, while another fiber passes back into the neural tube. The efferent fibers originate from neuroblasts in the basal plate of the tube, from whence they grow to the outside. Fibers of one or both of these categories converge and join into distinct cables called *nerves;* these are arranged in pairs which innervate corresponding regions of the bilaterally symmetrical body. The nerves belong to two main systems: the *cranio-spinal* series and the *autonomic* division.

Functional Classification of Fibers. The early observation that sensory impulses travel in the dorsal root fibers and motor impulses in ventral root fibers has been supplemented by a more complete analysis (Fig. 467). All neurons fall within four chief functional groups, which are further subdivided as indicated in the following list. No single nerve contains representatives of all fiber types; those components designated 'special' are peculiar to the cranial nerves alone.

A. *Afferent (or sensory)*.
    1. *Somatic afferent*. (Fibers ending on skeletal muscles.)
        (a) *General* (fibers ending chiefly in the integument).
        (b) *Special* (fibers from the sensory epithelia of the eye and **ear**).
    2. *Visceral afferent*.
        (a) *General* (sensory fibers from the viscera).
        (b) *Special* (fibers of smell and taste).

B. *Efferent (or motor)*.
    1. *Somatic efferent*. (Fibers ending on skeletal muscle.)
    2. *Visceral efferent*.
        (a) *General* (fibers ending about autonomic ganglion cells which, in turn, control smooth muscle, cardiac muscle and glandular tissue).
        (b) *Special* (cranial nerve fibers terminating on the striated musculature derived from branchial arches).

THE SPINAL NERVES

The spinal nerves become arranged segmentally, in agreement with the myotomes they supply. Each is attached to the spinal cord by two roots. One root is dorsal (posterior) in position and bears a spinal ganglion along its course; the other is ventral, or anterior (Fig. 470). Toward the end of the fourth week (4 mm.) the *ventral root fibers* can be seen growing out from the ventrolateral wall of the spinal cord (Fig. 468 B). At this time the spinal ganglia are represented by local enlargements along the continuous ganglion crest. Slightly later (5 mm.) the cells of the primitive *ganglia*

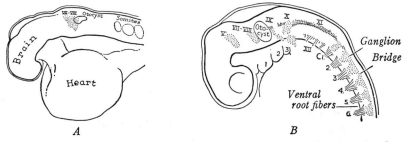

A   *B*

Fig. 468. Early human brains, viewed from the left side, showing the developing cranio-spinal nerves. *A*, At 2 mm., with ten somites ($\times$ 30); *B*, at 3.5 mm., with twenty-five somites ($\times$ 14).

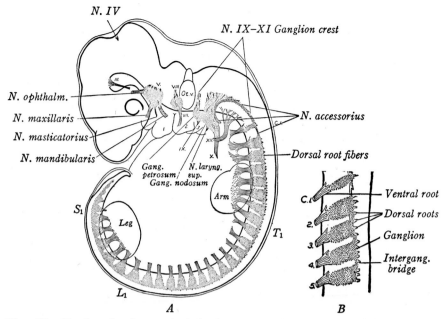

Fig. 469. Further development of the human cranio-spinal nerves (Streeter). *A,* At 7 mm. (× 12); the first nerve of each spinal group is numbered. *B,* At 10 mm., showing cervical spinal nerves (× 17).

begin to develop centrally directed processes, which enter the marginal zone of the cord as *dorsal root fibers,* while their peripheral processes join the fibers of the ventral root, thereby completing the serially repeated nerve trunks (Fig. 469). Successive stages in the differentiation of a spinal nerve, from a segmental mass of neural-crest tissue and from a corresponding group of neuroblasts located in the basal plate, are illustrated in Figs. 421 *A, B;* 423 *A.*

At the 10 mm. stage the cellular bridges of the neural crest, which previously interconnected the spinal ganglia, have begun to disappear (Fig. 469 *B*), and the several parts of a typical spinal nerve become evident (Fig. 470). In this differentiation the nerves more cranial in position maintain a slight advance over those at lower levels. Just beyond the union of the dorsal and ventral roots, the trunk of a nerve gives off laterally the *dorsal ramus.* Its motor fibers supply the dorsal muscles, while its sensory fibers make endings in the integument (Fig. 470). The stouter *ventral ramus,* continuing distad, branches off medially the *ramus communicans* to an autonomic ganglion and then divides into the lateral and ventral *terminal divisions.* The motor fibers of the terminal divisions supply the muscles of the lateral and ventral body wall, while the sensory fibers end in the integument of the same general regions.

Nerve Plexuses. At the points where the ventral and lateral terminal divisions arise, connecting loops may extend from one spinal nerve to adjacent nerves in the series, thereby forming distinct *nerve plexuses.* Especially favorable for this condition are those regions where the muscles of the limbs superimpose themselves upon the ordinary regularity of the trunk musculature (Fig. 469 *A*). The nerves supplying the arm and leg thus unite and are clearly indicated in embryos of five and six weeks, respectively. The plexus related to the arm is the *brachial plexus;* the one to the leg is the *lumbo-sacral plexus* (Fig. 636). Both of these divide into dorsal and ventral divisions, whose branches are distributed respectively to the dorsal and ventral surfaces of the limbs. The dorsally located nerves innervate the extensor muscles of the dorsal side, the ventral nerves the flexor muscles of the ventral side. The cutaneous innervation of these regions is shown in Fig. 403.

The innervation of the limbs is supplied by nerves that correspond to the early extent of the limb buds (Fig. 469 *A*). Hence it is the trunks of the last four cervical nerves, together with the first thoracic, that unite into the flattened plate that represents the primitive *brachial plexus*. From this plate nerve cords extend into the intermuscular spaces and end in the premuscle masses. The developing skeleton of the shoulder splits the brachial plexus into dorsal and ventral laminæ. From the dorsal lamina arise the axillary and radial nerves; from the ventral lamina the musculocutaneous, median and ulnar nerves. Similarly the lumbar and first three sacral nerves to the leg associate in a plate-like mass that differentiates into the *lumbo-sacral plexus*. This plate is divided by the skeletal elements of the pelvis and femur into two lateral (primitively dorsal) and two medial (primitively ventral) trunks. Of the cranial pair, the lateral component becomes the femoral nerve; the medial, the obturator nerve. The caudal pair constitutes the primitive sciatic nerve; its lateral trunk will be the peroneal nerve, the medial trunk the tibial nerve.

The limb nerves, though large in comparison to the size of the limb buds, do not interlace their fibers in plexuses because of simple crowding. On the contrary, it is a

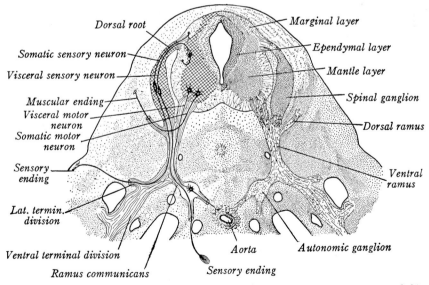

Fig. 470. Typical spinal nerve (on right) and its functional components (on left), shown in a transverse section of a 10 mm. human embryo. × 30.

functional adaptation related to several conditions, but more especially to the necessity of linking the sensory fibers from a muscle to the spinal centers of all its helpers and antagonists.

Causal Relations. The neural tube and neural crest are not inherently metameric. The segmental arrangement of the spinal nerves is a secondary acquisition that is dependent on the presence of a primary metamerism established by the myotomes. Experimental proof is afforded by the early removal of somites from a local region or the substitution there of a larger number of small somites. The size of ganglia and the number of ganglion cells differentiated in them depend on the size of the skin area supplied, as may be proved by shifting the position of an early limb bud to a strange location. Similarly, but less precisely, the expected number of motor neurons at a particular level of the spinal cord can be diminished by excising adjacent myotomes or a limb bud, and can be increased by grafting an additional limb bud near a normal one.

Nerve fibers reach their terminal fields as the result of guidance. This factor acts at an early stage when the path of growth to the end organ is short and direct. It may be a kind of chemical selectivity, but also involves the organization of the intervening ground substance into a field of oriented ultrastructure. Thus a limb bud moved to a new site acquires a nerve supply arranged in the pattern characteristic for that limb. But since growing nerves will also converge on a nasal placode, eye cup or tail bud that is transplanted into the flank, it follows that the 'attraction' is unspecific; the ingrowth is apparently a response to the presence of a localized region of intensive growth. In a similar way sensory fibers enter the brain and spinal cord at points corresponding to local regions of mitotic activity; a relocated nasal placode sends its fibers to that part of the brain which was at that particular time at peak activity.

THE CRANIAL NERVES

Twelve pairs of cranial nerves appear during the fifth and sixth weeks (Fig. 471). They are not arranged segmentally, and attempts to interpret all of them satisfactorily as serial homologues of spinal nerves have failed. In addition to the general sensory and motor components of spinal nerves, the cranial group contains special fibers distributed to the major sense organs and to muscles derived from branchial arches. The several motor nuclei and the nuclei about which the sensory nerves end are arranged in definite masses and columns within the basal or alar plate, respectively (Fig. 467). Unlike the nerves of the spinal series, which are fundamentally alike, the several cranial nerves vary widely in functional composition. Those in the first two groups of the subjoined list have but a single kind of fiber. Quite different is the third group, all of whose representatives are mixed; notable are the seventh, ninth and tenth nerves, each of which contains five different fiber types.

SPECIAL SENSORY	SOMATIC MOTOR	VISCERAL SENSORY AND MOTOR
I. Olfactory.	III. Oculomotor.	V. Trigeminal.
II. Optic.	IV. Trochlear.	VII. Facial.
VIII. Acoustic.	VI. Abducent.	IX. Glossopharyngeal.
	XII. Hypoglossal.	X. Vagus complex (including XI, Accessory).

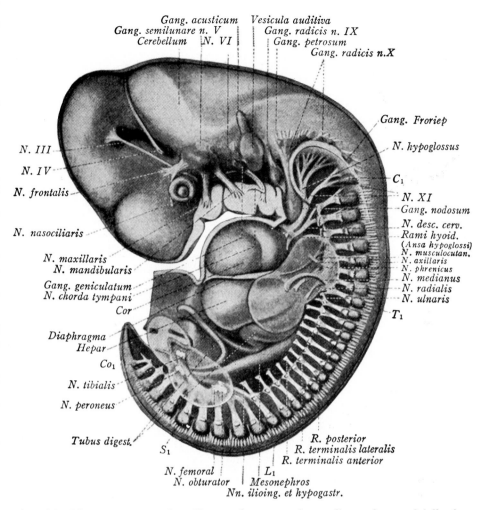

Gang. acusticum   Vesicula auditiva
Gang. semilunare n. V   Gang. radicis n. IX
Cerebellum   N. VI   Gang. petrosum
Gang. radicis n.X

Gang. Froriep
N. hypoglossus

N. III
N. IV
N. frontalis

C₁

N. XI
Gang. nodosum
N. desc. cerv.
Rami hyoid.
(Ansa hypoglossi)
N. musculocutan.
N. axillaris
N. phrenicus
N. medianus
N. radialis
N. ulnaris

N. nasociliaris

N. maxillaris
N. mandibularis
Gang. geniculatum
N. chorda tympani
Cor

T₁

Diaphragma
Hepar
Co₁

N. tibialis
N. peroneus

Tubus digest.
S₁

R. posterior
R. terminalis lateralis
R. terminalis anterior

N. femoral
N. obturator   L₁
Mesonephros
Nn. ilioing. et hypogastr.

Fig. 471. Nervous system of a 10 mm. human embryo, dissected superficially from the left side (Streeter). × 12.

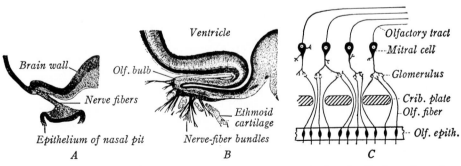

Ventricle
Olfactory tract
Mitral cell
Brain wall   Olf. bulb
Glomerulus
Nerve fibers
Crib. plate
Ethmoid   Olf. fiber
cartilage
Epithelium of nasal pit   Nerve-fiber bundles   Olf. epith.
A   B   C

Fig. 472. Development of the olfactory nerve, shown in longitudinal sections from human embryos. *A*, At seven weeks (Pearson; × 17); *B*, at thirteen weeks (Pearson; × 7); *C*, diagram of neuron relations.

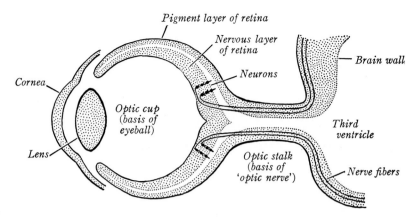

Fig. 473. Diagram showing the retina as an extended portion of the brain, and the course of optic nerve fibers through a connecting stalk (so-called optic nerve) to reach the brain-wall proper.

THE SPECIAL SENSORY NERVES

I. The **Olfactory Nerve,** though purely sensory (special visceral), has no ganglion. Its neurons differentiate from cells that occur in the epithelial lining of the primitive olfactory sac (Fig. 472 *A*). These cells become bipolar elements. Short peripheral processes end in receptive 'hairs' that protrude beyond the surface of the olfactory epithelium. Proximal processes grow brainward during the fifth week and later separate into some 15 to 20 strands, around which the cribriform plate of the ethmoid bone develops (Fig. 472 *A, B*). These unmyelinated fibers end in the *glomeruli* of the olfactory bulb, where they make contact with dendrites of the *mitral cells,* or olfactory neurons of the second order (*C*).

In close association with the olfactory nerve are two minor bundles. The *terminal nerve* courses along the medial side of the olfactory nerve. It is apparently rudimentary in man, although ganglion cells occur along its extent, and its unmyelinated fibers end in the epithelium of the nasal septum.[1] The *vomero-nasal nerve* is usually a temporary fetal structure which passes back from the transient vomero-nasal organ (Fig. 491 *B*). It courses in company with the olfactory nerve, of which it is a specialized part.

II. The **Optic Nerve** is formed by more than a million fibers that grow brainward from neuroblasts located in the nervous layer of the primitive retina (Fig. 473). Since the retina differentiates from the evaginated wall of the fore-brain, the optic nerve is not a true peripheral nerve but belongs rather to the system of cerebral nerve tracts; its fibers, nevertheless, are customarily classified as special somatic sensory. The neuroblasts from which the optic nerve fibers develop constitute the ganglion-cell layer of the retina (Fig. 474 *A*). During the sixth and seventh weeks these cells give rise to central processes which become sensory fibers that converge toward the optic stalk and grow through the tissue of its wall back to the brain (Figs. 473, 500). In the floor of the diencephalon, at its boundary with the telencephalon, the two optic nerves unite at about the end of the second

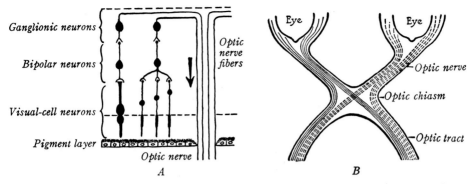

Fig. 474. Relations of the optic nerve fibers, shown in diagrams. *A,* Neuron links of the retina, and the origin and course of the fibers that constitute the optic nerve. *B,* The different courses taken by optic nerve fibers of mammals through the region of crossing (optic chiasm).

month and produce the *optic chiasm.* This is a region where there is a partial crossing of optic-nerve fibers, whereupon the reorganized nerves continue into the brain as the *optic tracts* (Fig. 474 *B*). Optic nerve fibers are myelinated but lack neurolemmal sheaths;[2] a few of these may originate from neuroblasts located in the brain.[3] The sum total is more than one-third of all the fibers found in all cranial and spinal nerves.[2]

VIII. The **Acoustic Nerve** is composed of sensory fibers (special somatic) which grow out of a ganglion located just rostral to the otocyst. The early ganglion-mass is common to the facial and acoustic nerves (Fig. 468), but a more caudal portion becomes recognizable at 7 mm. as the *acoustic ganglion* (Fig. 469). This mass subdivides into the *spiral ganglion* of the cochlear duct and the *vestibular ganglion* of the utricle, saccule and semicircular ducts (Fig. 475). The nerve cells remain bipolar throughout life—central processes uniting the ganglia to the acoustic tubercle of the myelencephalon, and peripheral fibers connecting it to the specialized sensory areas of the developing otocyst.

The primitive *facial-acoustic ganglion* arises from neural-crest substance which secondarily separates away from the tissue of the brain wall.[4] This is the first sensory primordium that can be identified in an embryo. The portion that becomes the primitive *acoustic ganglion* derives from placodal cells, part of the wall of the otocyst.[5] It differentiates into two separate, definitive ganglia in the following manner (Fig. 475). The original ganglionic mass elongates and is subdivided into superior and inferior portions in 7 mm. embryos (*A, B*). The superior part and some of the inferior portion co-operate in innervating the utriculus, sacculus and the semicircular ducts; the combined ganglionic mass becomes known as the *vestibular ganglion* (*C, D*). Most of the pars inferior, however, differentiates into the *spiral ganglion,* the peripheral fibers of which innervate the auditory hair cells of the spiral organ (of Corti) in the cochlea. The spiral ganglion is recognizable in 9 mm. embryos and conforms to the spiral turns of the cochlea—hence its name. Its centrally directed nerve fibers produce the *cochlear division* of the acoustic nerve, auditory in function. This is distinctly separated from the corresponding fibers of the vestibular ganglion, which constitute the *vestibular division* of the acoustic nerve, equilibratory in function. In spite of this, the component

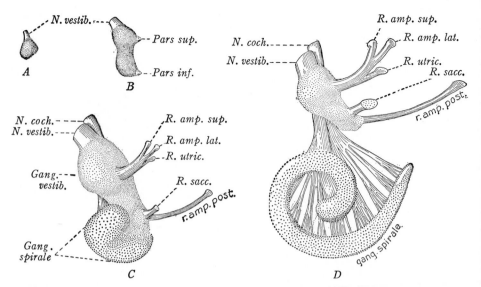

Fig. 475. Differentiation of the primitive left acoustic ganglion and nerve of man (Streeter). × 30. The vestibular ganglion is finely stippled, the spiral ganglion coarsely stippled. *A,* At 4 mm.; *B,* at 7mm.; *C,* at 20 mm.; *D,* at 30 mm.

of the vestibular ganglion derived from the pars inferior does become closely associated with the n. cochlearis, and thus in the adult it appears as though the sacculus and posterior ampulla were supplied by the cochlear nerve.

THE SOMATIC MOTOR NERVES

This group, consisting of the three nerves to the eye muscles and the hypoglossal nerve to the musculature of the tongue, is somatic motor; their nuclei of origin within the brain are colored red in Fig. 476. They are regarded as homologues of the ventral motor roots of the spinal cord that have lost their segmental arrangement and are otherwise modified. The fibers of these nerves originate from neuroblasts located in the basal plate, near the floor plate (Figs. 443, 444). In embryonic stages some sensory nerve fibers may become included with these nerves.[5]

III. The **Oculomotor Nerve** develops from neuroblasts aggregated as a nucleus in the basal plate of the mesencephalon (Figs. 449 *A, C;* 476). The fibers emerge ventrally as small fascicles, located in the concavity under the mid-brain that was brought about by the cephalic flexure (Fig. 471). The fascicles collect into a nerve trunk and end in a premuscle mass alongside the eye (Fig. 400 *A*). The nerve eventually supplies all of the extrinsic eye muscles, save the superior oblique and external rectus.

IV. The **Trochlear Nerve** fibers arise from a group of neuroblasts located just caudal to the nucleus of the oculomotor nerve (Fig. 476). Strangely enough, they are directed dorsad and curve around the cerebral aqueduct within the neural wall; crossing in the roof of the mesencephalon, the two nerves leave the brain at the isthmus (Figs. 449 *B,* 469 *A*). From such a dorsal, superficial origin each passes ventrad as a slender nerve that connects with the primordial superior oblique muscle of the eye, on the side opposite to its nucleus of origin (Fig. 471).

VI. The **Abducent Nerve** takes origin from a motor nucleus located in the pontine region of the metencephalon (Fig. 476). The fibers pass out ventrally at a point caudal to the pons and, as a single trunk, course rostrad to end in the external rectus muscle of the eye (Fig. 471).

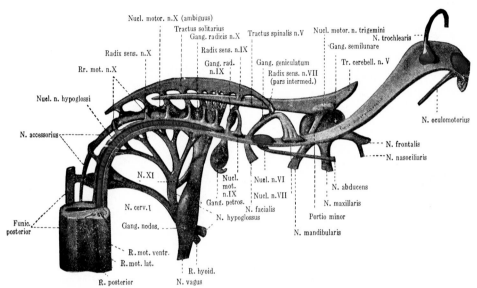

Fig. 476. Reconstruction of the nuclei of origin and termination of the cranial nerves in a 10 mm. human embryo (Streeter). × 30. The somatic motor nuclei are colored red.

XII. The **Hypoglossal Nerve** has resulted from the fusion of the ventral root fibers of three or four occipital nerves located just rostral to those commonly recognized as belonging to the cervical series (Fig. 468 *B*). Sensory roots, corresponding to the dorsal roots of spinal nerves, have dropped out. The somatic motor fibers originate from a nucleus continuous with the ventral gray column of the spinal cord (Figs. 443. 444). These fibers leave the ventral wall of the myelencephalon in several groups which converge ventrally to form the common trunk of the nerve (Fig. 469). Later they grow rostrad and eventually end in the muscles of the tongue.

Additional Details. The minor, sensory contributions to the oculomotor and trochlear nerves are fibers arising from cells of the *mesencephalic nucleus* buried in the substance of the mid-brain (p. 486).[6] Rudimentary *accessory ganglia* may be present in the region of the hypoglossal nerve (Fig. 478); in some mammals these (and especially the caudalmost in the series, *Froriep's ganglion*) send dorsal rootlets to the hypoglossus (Fig. 600). Also some sensory-type cells occur along the hypoglossal nerve roots within the myelencephalon of the embryo[7] and on the free nerve.[8]

The final position of the motor nuclei of the trochlear nerves and the crossing of their nerves appear to be the consequence of secondary shifts within the vertebrate

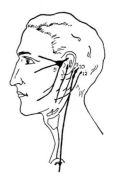

Fig. 477. Position of certain cranial nerves with respect to the head.

series.[9] Transient vestigial rootlets of the abducent and hypoglossal nerves tend to fill in the gap between these two nerves.[10] The hypoglossus historically innervates occipital somites whose myotomes become lingual muscles and whose sclerotomes have been assimilated into the base of the skull (Fig. 373).

THE VISCERAL MIXED NERVES

This group of similar nerves is primarily related to the branchial arches, as Fig. 400 *A* shows; the later relations to the head and neck are indicated in Fig. 477. The motor roots of these nerves arise in a lateral series, distinct from the roots of other nerves already discussed. This position is the result of an early migration of motor neuroblasts from an original, strictly ventral location to a more lateral position (Figs. 443, 444, 447). The sensory elements are derivatives of a neural crest in direct line with that of the spinal cord (Fig. 468 *B*). Additions are said to come from epibranchial placodes.[12, 15]

V. The **Trigeminal Nerve** is the nerve of the first branchial arch (Fig. 400 *A*). It is chiefly sensory, its large *semilunar ganglion* lying near the rostral end of the hind-brain (Fig. 469 *A*). Centrally directed processes from the ganglion form the large sensory root that enters the wall of the metencephalon at the level of the pontine flexure (Fig. 447). The processes peripheral to the ganglion separate into three large divisions (the *ophthalmic, maxillary* and *mandibular nerves;* Fig. 471); their sensory fibers (general somatic) supply the integument of the face as well as the lining of the nose and mouth.

The motor fibers of the trigeminal nerve grow out from neuroblasts in the pons (Fig. 447). They enter the mandibular division of the nerve and supply the premuscle masses derived from the first branchial arch (Fig. 400 *A*). This mass transforms into the muscles of mastication.

VII. The **Facial Nerve** supplies the second (hyoid) branchial arch (Fig. 400 *A*). It is composed for the most part of efferent fibers that leave the myelencephalon just medial to the acoustic ganglion (Fig. 469 *A*); continuing ventrad, they become lost in the tissue of the second arch. Because of this primary relationship they innervate the muscles of facial expression and all other muscular derivatives of the second arch (Fig. 400 *B*).

The sensory fibers of the facial nerve grow from the cells of the *geniculate ganglion,* which develops from a more rostral portion of an originally common facial-acoustic ganglionic primordium (Figs. 468, 471). The proximal processes from the cells enter the pons. The more important peripheral fibers pass by way of the *chorda tympani* branch to taste buds on the body of the tongue.

IX. The **Glossopharyngeal Nerve** is the nerve of the third branchial arch (Fig. 400 *A*). It takes its superficial origin from the myelencephalon by several rootlets that emerge just caudal to the otocyst (Figs. 468 *B*, 469 *A*). The sensory fibers arise from a subdivided ganglion mass. The *superior ganglion* occurs in relation to the nerve rootlets, while the *petrosal ganglion* lies farther distad on the nerve trunk (Fig. 478). These fibers comprise the greater part of the nerve; distal to the ganglia they separate into rami that pass to the second and third branchial arches and eventually innervate most of the root of the tongue and pharynx. The motor fibers are few; some come to innervate such muscles of the pharynx as are derived from the third arch; other efferent fibers activate the parotid gland.

X, XI. The **Vagus** and **Accessory Nerves** represent a composite through the union of nerves that supply the fourth to sixth branchial arches of aquatic vertebrates (Fig. 400 *A*). As might be expected, the vagus has several ganglia which arise as local enlargements along the course of the more caudal neural crest of the brain region (Fig. 478). The most dorsal and rostral of these is the *jugular ganglion;* the other dorsal ones, termed *accessory ganglia,* are vestigial structures which are not segmentally arranged. In addition to such root ganglia there is the *nodose ganglion* located farther distad on the main trunk. Beyond this level the trunk is largely sen-

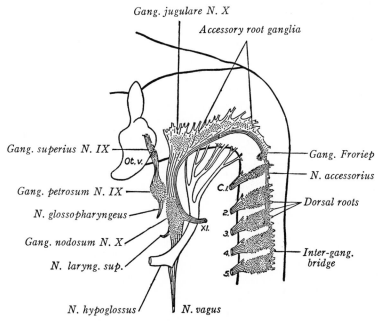

Fig. 478. Peripheral nerves in the occipital region of a 10 mm. human embryo (Streeter). × 17.

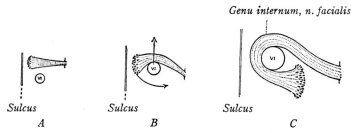

Fig. 479. Diagram illustrating three stages in the development of the genu of the human facial nerve (Streeter). *A,* At 10 mm.; *B,* at 13 mm.; *C,* at 20 mm.

sory, the peripheral processes of the ganglion cells distributing to the ear region and various viscera.

The motor fibers of the vagus proper leave the myelencephalon by several rootlets that converge into the vagal trunk. They innervate muscles of the pharynx and larynx (derivatives of the fourth and fifth arches) and serve as initial links in the autonomic innervation of the thorax and abdomen. The more caudal rootlets of the vagus become predominantly motor and form a distinct bundle which has acquired additional rootlets by spreading down into the territory of the spinal cord. This bundle is described as a separate entity (*accessory nerve*) in adult anatomy. It ascends, joins the trunk of the vagus proper and quickly becomes separate again (Fig. 478). Its motor fibers pass to premuscle masses of branchial-arch origin that form the sterno-mastoid and trapezius muscles of the arm girdle.

Additional Details. *V.* The sensory fibers (general somatic) of the *trigeminal nerve* arise from cells in the semilunar ganglion which lies opposite the second neuromere. They make connections with the sensory relay nuclei of the pons and medulla, some turning caudad to constitute the *spinal tract* of the trigeminal nerve (Fig. 476). The sensory fibers of the mandibular division arise from unipolar cells of the *mesencephalic nucleus* of the mid-brain; this presence of ganglion cells within the substance of the brain is unique (Fig. 449 C).[11] The ophthalmic division supplies the apical region of the head and thus is not related to the first branchial arch, as are the other two divisions of the trigeminus. The motor fibers (special visceral) take origin from a *motor nucleus* that lies within the pons at the same level as the semilunar ganglion (Figs. 447, 476). These fibers leave the brain wall as a separate *masticatory nerve* (Fig. 469), but later they become incorporated into the mandibular nerve.

VII. Sensory fibers of the *facial nerve* enter the alar plate of the pons by way of the intermediate nerve (of Wrisberg) and reach the *solitary tract* and its nucleus (Fig. 476). The peripheral processes (general and special visceral) distribute through all branches of the nerve. The motor fibers (special visceral) of the facial nerve arise from a cluster of neuroblasts that comprise its *motor nucleus;* this is located in the pons beneath the third neuromere of the rhombencephalon (Fig. 476). These motor fibers at first grow straight laterad from the motor nucleus, passing rostral to the nucleus of the abducent (Fig. 479 A). The nuclei of the two nerves later shift their positions, that of the facial nerve moving caudad and laterad while the nucleus of the abducent nerve shifts rostrad (B). As a result, the motor root of the facial nerve bends around the nucleus of the abducent (C), producing the *genu,* or knee, of the former. Besides the efferent fibers already discussed, still others (general visceral) originate in the *superior salivatory nucleus* which lies close to the main motor nucleus. These fibers emerge through a small root (which becomes the *intermediate nerve* of Wrisberg) and are distributed to the lacrimal, submandibular and sublingual glands.

IX. The *glossopharyngeal nerve* leaves the brain at the level of the sixth neuromere. The sensory fibers that grow from cells in the superior ganglion are general somatic, while those originating in the petrosal ganglion are general and special visceral. Processes growing centralward from these ganglion cells enter the alar plate of the myelencephalon and mostly join similar fibers of the facial nerve coursing caudad in the *solitary tract* (Fig. 476). Some motor fibers (special visceral) arise from neuroblasts beneath the fifth neuromere-groove; they help form the *nucleus ambiguus* (Fig. 476). Other motor fibers (general visceral) develop from the *inferior salivatory nucleus.*

X–XI. The jugular ganglion of the *vagus nerve* is located at the level of the seventh neuromere. Some of its sensory fibers (general somatic) have central processes that end in the *spinal nucleus* of the trigeminus and peripheral processes that pass to the skin of the ear-region. The nodose ganglion sends central processes (special visceral) to the *solitary tract* and its nucleus (Fig. 476) and distal processes to taste buds on the epiglottis and pharynx. Other nodose fibers (general visceral) take a similar central course and extend also to the *dorsal sensory nucleus* of the vagus; peripherally they supply the digestive and respiratory tracts and the heart. The trunk ganglia of both the vagus and glossopharyngeal nerves may be derived from epibranchial placodes (see beyond) rather than from the migration of the neural crest to more ventral positions.[12] Some motor fibers (special visceral) of the vagus proper spring from neuroblasts of the *nucleus ambiguus* of the myelencephalon; others (general visceral) arise from the *dorsal motor nucleus* nearby (Figs. 443, 444). The former innervate striated muscles of the pharynx and larynx; the latter extend to autonomic ganglia of the viscera. The motor fibers of the *accessory nerve* that come from the medulla are similar in origin to the special and general visceral components just described for the vagus. The spinal contribution arises from cells in the lateral part of the anterior gray column of the cervical spinal cord; these cells are directly continuous with those of the nucleus ambiguus. Some embryos show ganglion cells along the spinal course of this nerve, implying a mixed sensory-motor composition.[13]

Placodes. In lower vertebrates two series of ectodermal thickenings occur in connection with certain cranial nerves. Their plate-like nature has suggested the designation 'placode' for them. The *dorsolateral placode* is developed in relation to the auditory placode as a focal point; spreading rostrad and caudad, .it is responsible for the sense organs and nerves of the acoustic and lateral-line systems. *Epibranchial placodes* originate at the dorsal ends of the branchial clefts, and in some lower vertebrates cellular proliferation from them clearly adds to the neighboring ganglia. In higher animals, including man, the relations of all the placodes are less plain. The dorsolateral system is represented by the sense organs of the internal ear and perhaps by additions to the acoustic nerve.[12, 14] Indications of contributions from the epibranchial placodes to the ganglia of nerves *V, VII, IX* and *X* have also been reported.[12, 15]

Anomalies. There are many variations in the arrangement and distribution of the peripheral nerves. The more striking anomalies are usually accompanied by correlated disturbances of the central nervous system and axial skeleton.

THE AUTONOMIC NERVOUS SYSTEM

The autonomic nervous system is composed of neurons (visceral motor) arranged in a two-link chain. The first neuron extends from the central nervous system to an autonomic ganglion. The second neuron begins with a ganglion cell in that ganglion; its axon leaves the ganglion and extends

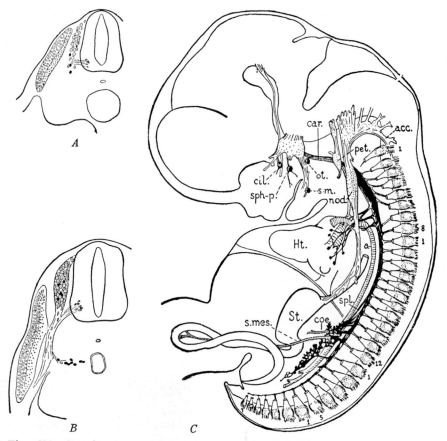

Fig. 480. Development of the human autonomic system (after Streeter). *A, B,* Schematic sections through the lumbar and thoracic levels of a 7 mm. embryo (× 40). *C,* Autonomic system of a 16 mm. human embryo (× 7); the ganglionated trunk is heavily shaded.

cil., Ciliary ganglion; *cœ.,* cœliac artery and plexus; *Ht.,* heart and cardiac plexus; *ot.,* otic ganglion; *pet.,* petrosal ganglion; *s-m.,* submandibular ganglion; *sph-p.,* spheno-palatine ganglion. In *A* and *B,* spinal ganglion cells are represented by dotted circles, autonomic cells by black ovals, and sheath cells by white rings.

to (and activates) gland cells or the involuntary muscle of the viscera and blood vessels. The exact source of origin of the ganglion cells is not wholly agreed upon, but the best evidence favors the neural crest[16] rather than the neural tube.[17] In the fifth week some of the crest cells are migrating down the dorsal roots of the spinal nerves (Fig. 480 *A*). Leaving the nerve trunk, they take position in paired masses dorsolateral to the aorta (*B*). Growth quickly merges them into continuous longitudinal strands, whose segmental enlargements represent primordial *autonomic ganglia,* each containing an aggregation of neuroblasts (Fig. 481). These soon differentiate into multipolar *ganglion cells,* encapsulated by satellite cells also of crest origin.

There are three groups of ganglia, based on the distance of cell migration: (1) The first set to appear is forming in embryos of 6 mm. in the vicinity of the aorta throughout most of the extent of the trunk (Fig. 483 *A, D*). The prospective ganglion cells consolidate into well-demarcated, segmental masses during the seventh week, whereupon the paired ganglia become linked, chain-fashion, by longitudinal nerve cords. The resulting ganglionated cords are the *sympathetic trunks* which extend along the vertebral column, on each side (Fig. 480 *C*). In the neck region such primary

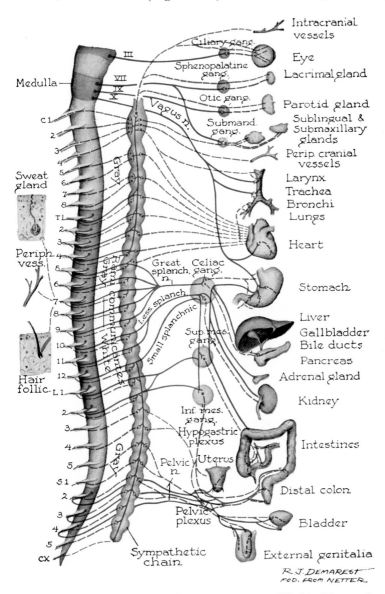

Fig. 481. Diagram of the autonomic nervous system (Clark). Thoraco-lumbar outflow is in red; cranio-sacral outflow in blue; preganglionic fibers in solid lines; post ganglionic fibers in broken lines.

ganglia develop in the lower cervical segments only; growth craniad then carries the trunks to higher cervical levels. (2) Other cells migrate farther away and form *collateral ganglia* arranged about the aorta, such as the cœliac and mesenteric ganglia (Figs. 480 C, 481). They are conspicuous about a week after the trunk ganglia appear. (3) At about this same time, still other cells migrate to organs that they will innervate and form *terminal ganglia,* located near or even within these organs. Such ganglia are found in the head, heart, lungs and pelvic viscera, and in the myenteric and submucous plexuses of the digestive tube (Fig. 481). The cranial terminal ganglia, related to the brain, are at no time segmental. They appear slightly later than those of the trunk, and are derived chiefly from cells that migrate from the primitive semilunar ganglion, although the geniculate and petrosal ganglia also contribute (Fig. 480 C).[18] The cells of the remaining terminal ganglia advance along the vagus nerves, except those of the gastro-intestinal region which come from collateral ganglia.[19]

All autonomic ganglion cells are stimulated into action by *preganglionic fibers* whose cell bodies lie in the brain or spinal cord. The ganglion cells themselves send out *postganglionic fibers* to the involuntary muscle and gland cells that they, in turn, activate (Figs. 470, 481). The ganglia of the sympathetic trunks and the collateral ganglia belong to the *thoraco-lumbar autonomic* (or sympathetic) *system.* The terminal ganglia belong to the *cranio-sacral autonomic* (or parasympathetic) *system.* These two divisions are different, and in many instances antagonistic, in their functional effects (Fig. 481).

Besides the pre- and postganglionic fibers of the autonomic chain, there are afferent fibers that merely pass through the autonomic ganglia and join the cranio-spinal nerves. These bring visceral sensory impulses directly from the viscera to the cranial and spinal ganglia and thence to the brain and spinal cord; they are largely incidental components that utilize these routes as convenient pathways (Fig. 470). Both of the fiber-types just mentioned acquire myelin sheaths and, in the thoraco-lumbar region, constitute the *white communicating rami.* On the other hand, all axons extending from the autonomic ganglion cells are unmyelinated and efferent in function (*i.e., postganglionic*). Some of them grow back into the spinal nerves by way of separate *gray communicating rami* and are then distributed through these nerves to hairs, blood vessels and sweat glands (Fig. 481).

Anomalies. Occasionally accessory clusters of ganglion cells, called *intermediate ganglia,* occur along spinal nerves and communicating rami. Terminal autonomic ganglion cells may fail to invade the wall of the lower colon and rectum. In this instance the muscular coat does not exhibit peristalsis, and the gut cannot empty itself completely. The resulting condition is known as *megacolon* or *congenital dilatation of the colon.*

THE CHROMAFFIN SYSTEM

Not all of the cells of the primitive autonomic ganglia transform into neurons. Some differentiate into satellite and neurolemmal cells, associated with the neurons.[20] Still others become distinctive endocrine elements that stain brown with chrome salts and hence are designated *chromaffin cells.* This reaction is due to the presence within the cells of the hormone, *epinephrine.* Cells of this type give rise to a *chromaffin system,* the most conspicuous member of which is the medulla of the suprarenal gland (Fig. 482 *A*).

The Paraganglia. Some chromaffin cells collect in rounded masses in close relation to autonomic ganglia and plexuses. Because of this association they have received the appropriate name *paraganglia* (Fig. 482 *C*). They are becoming organized at two months and before birth attain a diameter of a millimeter or more.[21] Other chromaffin masses, similar in nature, occur along the course of the aorta. The largest, found on the abdominal sympathetic plexus, is the pair of *aortic chromaffin bodies* (of Zuckerkandl). These are first recognizable toward the end of the second month about the root of the inferior mesenteric artery (Fig. 482 *A, B*). At birth they are about 1 cm. long. All representatives of this group are composed of cords of chromaffin cells intermingled with strands of connective tissue; the whole mass is surrounded by a connective-tissue capsule. After birth the chromaffin bodies decline but do not disappear entirely until puberty.[21]

The Aortic Bodies. Associated with the vessels derived from the aortic arches are several small masses, one of which belongs to the chromaffin category. Best known are the *carotid bodies* which organize in the seventh week from a mesodermal condensation on the wall of each internal carotid artery.[22] They are invaded by nonchromaffin cells from a nearby cervical autonomic ganglion and receive afferent fibers chiefly from the glossopharyngeal nerves; both nerve and artery belong to the third branchial arch.

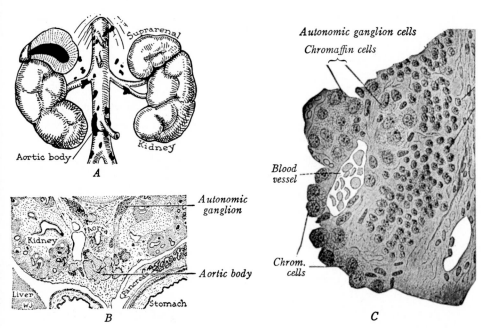

Fig. 482. Human chromaffin bodies. *A,* Distribution of chromaffin tissue (in black), at six months (× 1.5). *B,* Aortic body, at eight weeks, shown in a transverse section (× 22). *C,* Sectioned paraganglion, at ten weeks (after Kohn; × 450).

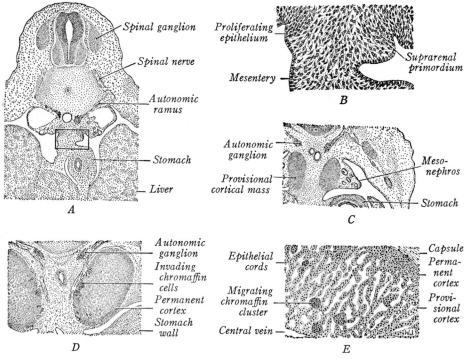

Fig. 483. Development of the human suprarenal gland, shown in transverse sections. *A,* At 8 mm. (× 24); *B,* detail of area marked by a rectangle in *A* (× 75); *C,* at 12 mm. (× 20); *D,* at 16 mm. (× 24); *E,* sample detail-area at four months (× 70).

The invading neuroblasts presumably become the chemoreceptive cells of the mature organ which serves as a reflex system in the regulation of blood pressure.

The Suprarenal Gland.   Each gland has a double origin and in reality is two distinct glands secondarily combined within a common capsule. The *cortex* is derived from mesoderm, the *medulla* from ectodermal chromaffin tissue. In fishes the cortex and medulla occur normally as separate organs; in higher animals there is an increasingly intimate association between the two parts, until a climax is reached in mammals by the cortex fully encapsulating the medulla.

The Cortex.   In an embryo of 8 mm. the material of the cortical portion of each gland begins to gather beneath the peritoneal epithelium (*i.e.,* mesothelium) at the attached base of the dorsal mesentery near the cranial pole of the mesonephros (Fig. 483 *A, B*).[23] Its origin from the proliferating epithelium is similar to that of other mesenchyme bordering the serous cavities. Rapid growth of the two cortical primordia produces a pair of prominent mesenchymal condensations (*C, D*). These then enter on a course of specialization leading to the differentiation of distinctive, well-vascularized organs. The enlarging suprarenals soon project from the dorsal wall of the cœlom, between the urogenital organs and mesentery. Here they become relatively huge, encapsulated organs (Fig. 293 *A*); even at birth each is one-third the weight of a kidney, in contrast to the final ratio of 1:28. The caudally migrating glands meet the ascending kidneys at eight weeks, and thereafter perch upon them.

No sooner has the original cortical primordium established itself as a cellular mass (Fig. 483 *C*) than it begins to become enveloped by a second proliferation of cells that are more compactly arranged and of another type ('permanent cortex' in *D, E*).[23] The original mass, now internal in position, is a *provisional cortex* that is especially characteristic of the fetus. It constitutes the chief bulk of the organ at birth, but enters on a rapid decline at this time and its involution is largely completed in the first few weeks of postnatal life.[24] Simultaneously with the regression of the provisional cortex, whose significance to fetal life is unknown, the secondary *permanent cortex* (characteristic of primates) comes into rapidly increasing prominence. This cellular layer differentiates as it advances in a central direction. The glomerular zone, next the capsule, is already present at birth, and there is the beginning of a fasciculate zone as well. The fasciculate and the reticular zones become well defined within a few months.[25]

THE MEDULLA. The chromaffin cells of the medulla are descended from migratory cells that leave the primitive ganglia of the cœliac plexus of the autonomic system. In embryos of the seventh week, when the cortex is already prominent, masses of these cells begin to touch and invade the medial side of the cortical primordium (Fig. 483 *D*). The continued migration of these cell clusters brings some of them to a central position in the gland by the middle of the fetal period (*E*). Such immigration ceases at the end of fetal life and the chromaffin tissue becomes grouped in cords and masses. Like ductless glands in general, the suprarenal tissue is permeated with a profuse network of sinusoidal capillaries.

Anomalies. Multiple primordia or secondarily separated portions of the parent gland frequently form *accessory suprarenals*. As a rule, such accessory glands are composed of cortical substance only. They may be located in the region of the suprarenals and kidneys, in the retroperitoneal area below the kidneys, or in relation to the gonads.

REFERENCES CITED

1. Pearson, A. A. 1941. J. Comp. Neur., *75*, 39–66.
2. Bruesch, S. R. & L. B. Arey. 1942. J. Comp. Neur., *77*, 631–665.
3. Arey, L. B. 1916. J. Comp. Neur., *26*, 213–245.
4. Bartelmez, G. W. & H. M. Evans. 1926. Carnegie Contr. Embr., *17*, 1–67.
5. Batten, E. H. 1958. J. Embr. & Exp. Morph. *6*, 597–615.
6. Pearson, A. A. 1943; 1944. J. Comp. Neur., *78*, 29–43; *80*, 47–63.
7. Pearson, A. A. 1939. J. Comp. Neur., *71*, 21–39.
8. Tarkham, A. A. & S. A. El-Malek. 1950. J. Comp. Neur., *93*, 219–228.
9. Larsell, O. 1947. J. Comp. Neur., *86*, 447–466.
10. Bremer, J. L. 1908. J. Comp. Neur. & Psychol., *18*, 619–639.
11. Pearson, A. A. 1949. J. Comp. Neur., *90*, 1–46.
12. Yntema, C. L. 1944. J. Comp. Neur., *81*, 147–167.
13. Pearson, A. A. 1964. Am. J. Anat., *114*, 371–392.

14. Van Campenhout, E. 1935. Arch. de Biol., *46*, 273–286.

15. Van Campenhout, E. 1948. Arch. de Biol., *59*, 251–266.

16. Nawar, G. 1956. Am. J. Anat., *99*, 473–505.

17. Brizzee, K. R. & A. Kuntz, 1950. J. Neuropath. & Exp. Neur., *9*, 164–171.

18. Cowgill, E. J. & W. F. Windle. 1942. J. Comp. Neur., *77*, 619–630.

19. Van Campenhout, E. 1932. Physiol. Zoöl., *5*, 333–353.

20. Brizzee, K. R. 1949. J. Comp. Neur., *91*, 129–146.

21. Coupeland, R. E. 1952. J. Anat., *86*, 357–372.

22. Boyd, J. D. 1937. Carnegie Contr. Embr., *26*, 1–31.

23. Crowder, R. E. 1957. Carnegie Contr. Embr., *36*, 193–210.

24. Benner, M. C. 1940. Am. J. Path., *16*, 787–798.

25. Keene, M. F. L. & E. E. Hewer. 1927. J. Anat., *61*, 302–324.

Chapter XXVII. The Sense Organs

The cell-body of sensory neurons in primitive animals, such as worms, is ectodermal in origin, superficial in position, and generalized in its receptive capacities. Only the sensory cells of the vertebrate olfactory organ retain this primitive, ectodermal location. During evolutionary history the cell-bodies of all other primary sensory neurons are believed to have migrated inward and become dorsal ganglia. As a result of such centralization the peripheral processes of sensory neurons have assumed new relations. That is, they end freely in the epithelium and connective tissue, become enclosed within connective-tissue capsules, or appropriate new epithelial cells that serve as sensory receptors (taste; hearing).

Among the sense organs are receptive elements of general sensibility which belong to the integument, muscles, tendons and viscera; these mediate such general sensations as touch, pressure, muscle and tendon sensibility, temperature, and pain. Other organs, of a special sensory nature, are responsible for the sensations of taste, smell, vision and hearing. Each is attuned to a specific and exclusive kind of stimulus. The organs of smell, vision and hearing are *distance receptors;* they stand in contrast to all others that collect information from the organism itself, and especially from its integument. The apparatus for smell and taste consists of little more than the special sensory cells and fibers alone; at the other extreme are the eye and ear which possess elaborate accessory mechanisms for receiving the external stimulus and converting it into a form suitable to affect the sensory cells proper.

THE GENERAL SENSORY ORGANS

Free nerve terminations (for pain and common chemical sense) are by far the commonest of all the general sensory organs. They are also the simplest since the sensory nerve fibers merely push in among the cells of the epithelium or into the connective tissue and branch there (Fig. 484 *A*). Free nerve endings begin to invade the epidermis at the end of the third month, while Merkel's tactile discs organize one month later.[1] More specialized are the plexuses ending in hair sheaths.

Lamellated corpuscles (for deep pressure) include several variant types (Pacinian; Golgi-Mazzoni), but all are fundamentally alike; best known is that of Pacini. Their differentiation begins in fetuses four months old and

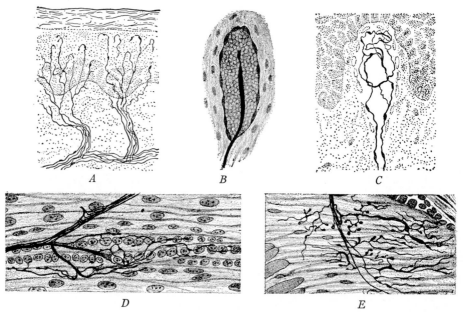

Fig. 484. Differentiation of sensory nerve endings (partly after Tello and Szymono-
wicz). *A,* Free nerve endings in the epidermis, at eight months (× 150). *B,* Lamellated
(Pacinian) corpuscle of chick embryo (× 275). *C,* Tactile corpuscle at seven months
(× 500). *D,* Neuromuscular spindle of chick embryo (× 500). *E,* Neurotendinous spindle,
at six months (× 400).

is completed at eight months.[1] A corpuscle starts as a mass of mesenchymal
cells clustered around a nerve termination (Fig. 484 *B*). These cells multiply, flatten and give rise to concentric fibrous lamellæ which increase in
number from without inward.[2] In the cat, at least, lamellar corpuscles
increase in number by budding.

A *tactile corpuscle* (of Meissner) originates as a looping plexus of terminal nerve fibers located just beneath the epidermis in certain regions (Fig.
484 *C*). This plexus becomes encapsulated along with a cluster of mesenchymal cells. Differentiation begins at four months, but it is not completed
until a year after birth.[1] The history of certain specialized variants, such as
Ruffini's terminal cylinders (heat), Krause's end bulbs (cold) and the genital
corpuscles is less well known. Temperature differences are appreciated in
the early postnatal weeks, but the sense of touch and pain are less acute
at birth than later.

Neuromuscular spindles probably begin their differentiation during the
third month.[3] A plexus of nerve fibers first comes into relation with a group
of myoblasts. The latter take the form of a tapering bundle and the whole
is encased in a connective-tissue sheath (Fig. 484 *D*).

Neurotendinous end organs develop concurrently with muscle spindles.[2]
Their branching nerve fibers end on an ensheathed bundle of tendon
fibers (Fig. 484 *E*). These organs, along with muscle spindles, give informa-

tion concerning muscle tension and the relative positions of the various parts of the body.

THE GUSTATORY ORGAN

In fetuses of two months local thickenings of the lingual epithelium presage the first *taste buds*.[4] In general, the parent tissue is entoderm, yet some buds are located in presumable ectodermal territory. The basal cells of such a thickened spot lengthen and extend toward the surface of the epithelium (Fig. 485 *A*). This produces an epithelial cluster which, in later fetal months, differentiates further. Some of the elements specialize into slender *taste cells,* ending in hair-like receptive tips, while others become columnar 'supporting' cells, supposedly nonsensory (*B*). Taste buds are supplied by nerve fibers of the seventh, ninth or tenth cranial nerves; the fibers branch and end about the periphery of the taste cells (*C*). However, the functional relationship between nerve and epithelium is more intimate than one might assume, since the nerve seems to exert an organizing influence on the development of taste buds.[5] Moreover, a taste bud degenerates when its nerve is cut and does not reform until the nerve regenerates.

Between the fifth and seventh fetal months taste buds are more widely distributed in the mouth and pharynx than in the adult. It is possible that this represents a transitory recapitulation of the more widespread distribution occurring in lower vertebrates. In late fetuses and after birth many taste buds degenerate; those that survive are to be found on the vallate and foliate papillæ, on a few fungiform papillæ and on the soft palate and laryngeal surface of the epiglottis. Their location at the brink of the pharynx, just before swallowing becomes an involuntary act, is advantageous. Reflex responses to taste are present in a premature infant of seven months.

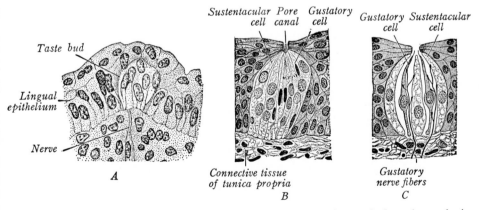

Fig. 485. Development of the human taste bud, shown in vertical sections. *A,* At two months (× 300). *B, C,* At birth, in true section and interpretative diagram (Arey in Morris; × 340).

The development of the various lingual papillæ has been described in an earlier chapter (p. 235).

THE NOSE

The development of the nose is bound up with the changes that produce the face and palate, already considered (Figs. 160, 187). In primitive verte- brates the olfactory organ is a blindly ending sac, used solely for the sense of smell. But when air-breathing was adopted, these chambers opened into the mouth and then came to subserve respiration as well.

The Nasal Cavities. The first indication of the olfactory organ is an oval area of thickened ectoderm occurring on each ventrolateral surface of the head in embryos about 4 mm. long (Fig. 486 *A*). Each is an *olfactory placode,* which straightway becomes an *olfactory pit* bounded by an ele- vated margin (*B*). Actually the early pit is more like a groove, since it is deficient ventrally and communicates with the oral cavity, as in sharks. At this period it is convenient to designate the peripheral wall, on either side of the pit, as a median and a *lateral nasal process.* Figure 487 shows similar stages as they appear in sections. Close at hand laterally are the *maxillary processes* of the first branchial arches, while in the midplane is the tissue that represents the future bridge of the nose and *nasal septum* (Fig. 486 *B*).

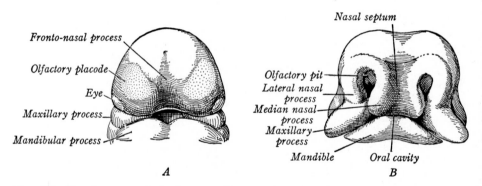

Fig. 486. Development of the human olfactory pits (after Peter). *A,* At 5 mm. (× 24); *B,* at 11 mm. (× 12).

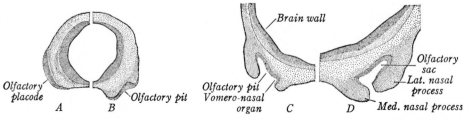

Fig. 487. Development of the human olfactory pits, shown in transverse half-sec- tions through the head. *A,* At 5 mm. (× 25); *B,* **at** 7 mm. (× 17). *C,* At 9 mm. (× 21); *D,* at 12 mm. (× 21).

How these parts combine to produce the external features of the nose is shown clearly in Fig. 160.

When fusion of the maxillary and nasal process on each side presently converts the nasal grooves into well-delimited blind sacs, the opening into each can be called a *nostril* (Fig. 488 *A*). At its deep end the olfactory sac is separated from the mouth cavity by an epithelial plate (the oro-nasal membrane) which thins and ruptures during the seventh week (*B*). Each nasal cavity then opens to the outside through a nostril and communicates internally, by way of its *primitive choana*, with the oral cavity as in amphibians. A thickened region beneath and between the two early olfactory sacs constitutes the primitive palate (Fig. 488). This will differentiate both into a median part of the upper lip and into the so-called *premaxillary palate* (Fig. 187).

Parallelling these changes comes a broadening of the head, so that the olfactory pits shift from lateral locations to a more ventral position and seem to approach the midplane (Fig. 160). In accomplishing this, the deep frontonasal region between the two is relatively compressed; it becomes the *nasal septum* (Fig. 489). Additions to the original nasal sacs are gained when the palate halves unite and separate a dorsal portion of the primitive mouth cavity away from the rest (Fig. 488 *B*). The nasal septum extends ventrad and caudad correspondingly; it presently fuses with both the primary (premaxillary) and secondary palate, thus completing the separation of the two nasal cavities (Figs. 490 *A*, 491 *A*). The permanent nasal passages then consist of the original nasal sacs plus a portion of the primitive mouth cavity that has been captured secondarily by the development of the palate (Figs. 491 *B*, 493 *A*). Their internal openings into the pharynx is by secondary, permanent *choanæ*. From the second to the sixth month the nostrils are closed by epithelial plugs.

The Epithelial Lining. The entire nasal passage, both that derived from the primitive olfactory sacs and that captured from the primitive mouth, is

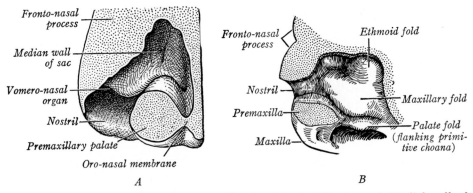

Fig. 488. Early nasal cavity of man, illustrated by hemisections. *A,* Medial wall of left cavity, at 12 mm. (after Schaeffer; × 55). *B,* Lateral wall of right cavity, at seven weeks (after Frazer).

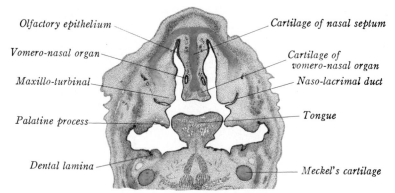

Fig. 489. Human nasal cavities, before the completion of the palate, shown in a frontal section at seven weeks (Prentiss). × 20.

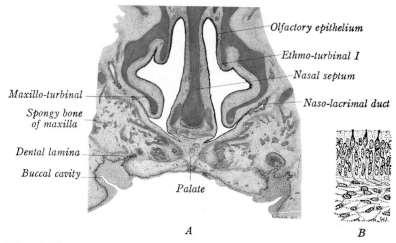

Fig. 490. *A,* Human nasal cavities, after the union of the septum and palate, shown in a frontal section at three months (after Prentiss; × 10). *B,* Olfactory epithelium, at seven weeks, showing three neurons in vertical section (× 200).

ectodermal. The lining of the upper part of each cavity quickly transforms into a specialized, pseudostratified *olfactory epithelium* (Figs. 490 *A,* 491 *B*). Many of its components become elongate, bipolar sensory elements which are really the cell bodies of *sensory neurons* (Fig. 490 *B*). At the bulbous free end of each cell a diplosome divides to produce six to eight basal bodies, each of which sends forth a fine bristle that serves as a sensory receptor. The basal end of the cell tapers into an *olfactory nerve fiber* which joins others and grows brainward (Fig. 472). Interspersed between the olfactory cells are inert, columnar and basal elements, supporting in function. Olfactory perception is present at eight months, but is not well established as yet. The rest of the nasal epithelium is ciliated and glandular in structure. It covers most of the septum and conchæ and is in contact with the flow of respiratory air (Fig. 490 *A*).

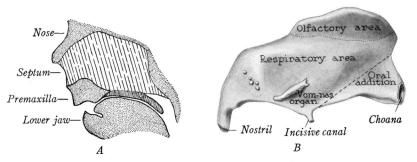

Fig. 491. *A*, Relation of the human nasal septum and palate, shown in a sagittal section of a fetal head (after Frazer). *B*, Right nasal cavity, in median view, shown by a cast at 11 weeks (after Broman; × 12); compare with Fig. 493 *A* for the extent of the oral addition in the adult.

The *vomero-nasal organs* (of Jacobson) are rudimentary diverticula which first appear in 8 mm. embryos as a pair of grooves, one on the median wall of each nasal cavity (Figs. 487 *C, D;* 488 *A*). The grooves deepen and close caudally to produce blind, tubular sacs which open toward the front of the nasal septum (Fig. 491 *B*). Nerve fibers, arising from the epithelial cells of the lining, join the olfactory nerve. Special cartilages are developed for the support of the vomero-nasal organ (Fig. 489), and during the sixth month it attains a length of 4 mm. (Fig. 492 *B*). In late fetal stages the vomero-nasal organ often degenerates, but it may persist in the adult. This organ is not functional in man, yet in many tetrapods it evidently constitutes a supplementary olfactory apparatus, perhaps useful in picking up odors from food in the mouth cavity.

The Conchæ. Mammals have increased the surface area of the nasal chambers by creating local folds that sometimes take the form of extensive spiral scrolls. By contrast, the human *conchæ* are relatively simple. In the fetus they arise as a series of elevated folds on the lateral wall of each nasal passage; these folds secondarily become supported first by cartilage and then by bone (Fig. 492 *A*). After birth their number is reduced to three permanent conchæ.

The *maxillo-turbinal* develops first and is followed by five *ethmo-turbinals* arranged in a series of decreasing size (Figs. 490 *A*, 492 *A*). The ethmo-turbinals arise, at least in part, on the median walls and by a process of unequal growth are transferred secondarily to the lateral walls.[6] The *naso-turbinal* is very rudimentary and appears merely as a slight elevation near the rostral end of the maxillo-turbinal.

Important growth changes and consolidations continue even into childhood; the final condition, resulting from these modifications, is shown in Fig. 493 *A*. The definitive *inferior concha* represents the little-changed maxillo-turbinal of Fig. 492 *A*; similarly, the *middle concha* comes from the first (or lowest) ethmo-turbinal. On the contrary, the *superior concha* is the combined second and third ethmo-turbinals; when a *supreme concha* exists it is the representative of the highest ethmo-turbinals. The naso-turbinal becomes the inconspicuous *agger nasi* of man.

The Paranasal Sinuses. Lodged within the adjoining bones, and in communication with the nasal cavity, are several irregular chambers known collectively as the *paranasal sinuses* (Fig. 493 *A*). They are lined with exten-

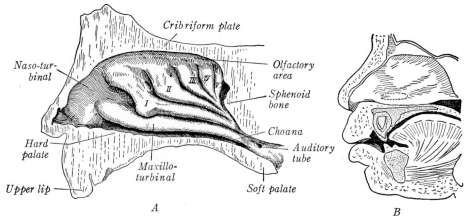

Fig. 492. Nasal walls in human fetuses near term. *A,* Reconstruction of the lateral wall of the right nasal passage (after Killian); *I–V,* ethmo-turbinals. *B,* Left surface of the nasal septum, with the position of the vomero-nasal organ indicated in oval outline (after Corning).

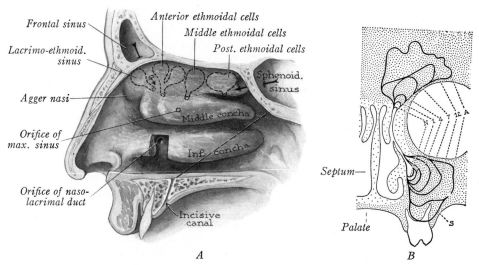

Fig. 493. Later relations of the human nasal wall. *A,* Conchæ and paranasal sinuses, shown in the right half of a longitudinally sectioned head; the broken, straight line indicates the position of the primitive choana. *B,* Stages of postnatal growth of the frontal and maxillary sinuses, indicated in outline on a frontal section (after Torrigiani). *A,* Adult; *N,* newborn; *S,* old age; 1–12, age in years.

sions of the nasal epithelium. All are first indicated at about the fourth month, but most of their expansion occurs after birth (*B*).[7] Destruction of the bone that neighbors the nasal cavities, in order to make room for the expanding sinuses, proceeds apparently under the influence of the lining epithelium. This nasal epithelium advances at equal pace with the destruction, and its invaginated sacculations serve thereafter as a lining to the sinuses.

The *ethmoidal cells* develop in the grooves between the primitive ethmo-turbinals

Fig. 494. Cleft nose, accompanied by median hare lip and an abnormally large mouth.

and are fairly well differentiated in the later fetal months (Fig. 493 *A*). The *maxillary sinus* invaginates from the groove between the maxillo-turbinal and first ethmo-turbinal and is of appreciable size in the newborn. The superior portion of the same furrow gives rise to the *frontal sinus* which undergoes most of its development after birth. The caudal end of each nasal fossa is set aside as a *sphenoidal sinus,* but actual invasion of the sphenoid bone does not occur until the third year of childhood.

Causal Relations. The production of an olfactory organ in amphibians is induced in sequence by the mesoderm and fore-brain. Such a fate is irrevocably fixed before the stage of the neural plate. Deletion experiments show that the ectodermal area with olfactory-sac potency is larger than that which normally becomes the olfactory placode. Still larger is the area that can form a nostril but no sac. The development of a choana depends on the presence of an olfactory placode and the fusion of its sac with oral epithelium; even a rudimentary sac suffices to accomplish this end.

Anomalies. Atresia or stenosis of the nostril results from the retention and organization of the normally temporary epithelial plug. Similarly, the more frequent atresia of the choana is said to be due to the organization and persistence of the primitive oro-nasal membrane which is carried back with the secondary extension of the nasal chamber and septum. An incomplete septum likewise represents the retention of the condition preceding normal attachment to the palate. The rare failure of the region between the nasal sacs to consolidate into a typical septum leads to a doubling of the nose; this ranges from mere apical bifurcation and fully cleft nose (Fig. 494) to complete duplication. The most striking anomaly is associated with cyclopia; in such cases the nose is a tubular proboscis attached above the single, median eye (Fig. 507 *A*). Other malformations may be introduced by cleft lip and palate, as already described (pp. 207; 230).

THE EYE

Comparative anatomy fails to give any clue to the evolution of the vertebrate eye, since it is highly organized even in the lowest groups. Its materials come from three sources: (1) the *optic nerve* and *retina* are derivatives of the fore-brain; (2) the *lens* arises from the surface ectoderm of the head; and (3) the *accessory tunics,* which provide support, nutrition and accommodation, differentiate from the adjacent mesenchyme.

The Optic Primordia. In embryos with eight somites a vaguely expressed optic field can be identified on each of the widely spread halves of the future fore-brain (*cf.* Fig. 426 *C*). A little later a shallow pit occurs within this region (Fig. 495 *A, B*). Presently the evagination (*optic vesicle*) becomes more extensive as the union of the neural folds into a tubular brain advances toward completion (*C, D*). Embryos 4 mm. long have pro-

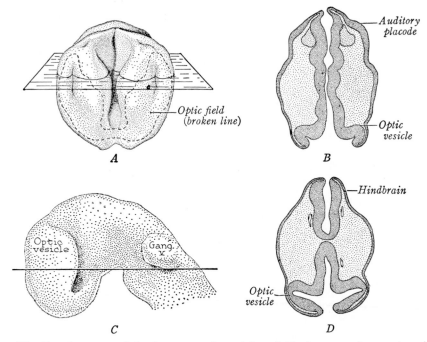

Fig. 495. Development of the human optic vesicles. *A,* Brain, at twelve somites, in front view (after Bartelmez; × 50). *B,* Section, at fourteen somites, in the plane indicated on *A* (× 50). *C,* Brain at fifteen somites, in lateral view (after Wen; × 66). *D,* Section, at sixteen somites, in the plane indicated on *C* (× 50).

gressed to the extent that the swollen optic vesicles are attached to the brain wall by somewhat slenderer *optic stalks* (Fig. 436 *B*). This condition is followed quickly by the stage of the *optic cup,* which is characterized by an indenting of the distal wall of the vesicle brought about by rapid, marginal growth. The result is a double-layered cup, connected to the diencephalon by a tubular optic stalk (Fig. 496).

The optic cup is destined to become the retina or the essential sensory lining of the eye; its stalk is the tissue through which fibers of the *optic nerve* grow back from the retina to the brain. When the optic vesicle begins to invaginate (4.5 mm.), the local area of ectoderm directly overlying it thickens into a *lens placode* (Fig. 496 *A*). This plate straightway pockets inward to produce the *lens vesicle,* or lens primordium, which then occupies the concavity of the optic cup (*B, C*). Slightly later, when the fundamental parts, already mentioned, are differentiating further, the *accessory coats* (vascular and fibrous) of the eyeball organize from the surrounding mesenchyme (Fig. 503). The axes of the primitive eyes in an embryo not yet six weeks old are set at an angle of 180 degrees to each other; between three months and the time of birth, broadening of the head has reduced this divergence to practically the final condition (Fig. 160). Such convergence makes possible the binocular vision of primates.

The eye, like the brain, grows precociously and at birth is three-fourths its final diameter.

The Optic Cup and Stalk. From its beginning, the optic cup is imperfect because of a notch in its double wall. This defect is brought about by the original invagination involving also the under side of the cup and then continuing as a groove that extends along the optic stalk (Fig. 496 *B, C*). The complete defect comprises the *optic fissure* (long badly named as the *chorioid fissure*). As a necessary result of this type of invagination, the internal layer of the optic cup is continuous with a corresponding infolding of the stalk (Fig. 497). During the seventh week the lips of the fissure close, so that the double-walled cup is complete, while the stalk becomes a tube within a tube (Fig. 500). This continuity of the inner tube with the inner layer of the optic cup creates a direct path along which optic nerve fibers, originating in the inner layer, pass to the brain (Fig. 474 *A*). Otherwise they would have to follow a route around the rim of the cup. The same arrangement likewise furnishes a tunnel which the hyaloid artery utilizes as a short-cut in reaching the interior of the eyeball without piercing its layers (Fig. 505 *A*).

Returning to the stage of the optic vesicle, it is obvious that its continued deepening as an optic cup brings the invaginating layer progressively closer to the external layer. Soon the two come to lie essentially in apposition, thereby obliterating the primitive lumen of the cup (Fig. 497 *A*). These facing layers now transform into an epithelial *retina,* while the rim of the

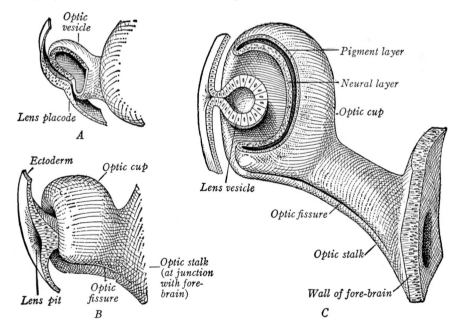

Fig. 496. Human optic primordia, shown as models in side view (after Mann). × 100. The lens is sectioned and the optic cup in *A* and *C* has been partly cut away. *A,* At 4.5 mm.; *B,* at 5.5 mm.; *C,* at 7.5 mm.

Mesenchyme   Lens vesicle   Vitreous body   Optic stalk   Optic recess of brain

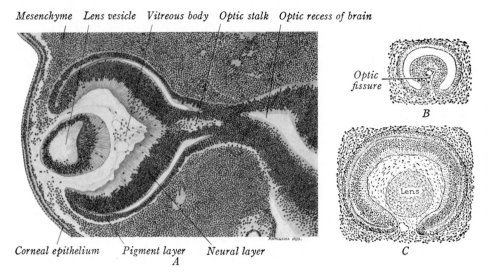

Optic
fissure

B

Lens

C

Corneal epithelium   Pigment layer   Neural layer
                              A

Fig. 497. Optic primordia of a 10 mm. human embryo. *A*, Longitudinal section (Prentiss; × 80). *B, C,* Optic stalk and optic cup, respectively, sectioned transversely to show the optic fissure (× 65).

cup represents the border of the future *iris*. The circular opening into the cup is the primitive *pupil*.

Differentiation of the Retina. The combined layers of the optic cup, or retina, soon show two zonal regions: (1) the thicker *pars optica,* a truly nervous portion that lines most of the cup; (2) the thinner *pars cæca,* an insensitive zone bordering the rim of the cup. The line of demarcation between these two regions makes a wavy circle, the *ora serrata.*

The Pigment Layer. The outer, thinner component of the optic cup becomes an epithelium known as the *pigment layer* of the retina. In embryos of 7 mm. pigment granules, elaborated from the cytoplasm,[8] appear in its cells and the pigmentation is soon dense (Fig. 505 *A*). This layer necessarily extends even to the pupillary margin of the optic cup (Fig. 504). Presently it becomes only one cell thick, but in the iris it also gives rise to the *pupillary muscles* which are unusual because of their ectodermal origin.[9]

The Neural Layer. The internal, thicker layer of the optic cup differentiates mostly into photoreceptive and impulse-transmitting neurons. In the *pars optica,* or nervous portion of the retina, this differentiation begins near the optic stalk, from which center the process extends progressively peripherad. An outer, nucleated layer (next the pigmented epithelium) and an inner, clear layer (next the cavity of the cup) can be distinguished in 12 mm. embryos (Fig. 497 *A*). These correspond, respectively, to the cellular layers (ependymal and mantle) and marginal layer of the neural tube. At two months the retina shows three strata—neuroblasts (including early ganglion cells) in the meantime having migrated inward from the outer neuroblastic layer (Fig. 498). In a fetus of six months all the layers

Cone cell
Rod cell
Rod cell
Fiber of Müller
Amacrine cell
Ganglion cell
Optic fibers

External limiting
membrane
Outer neuroblastic layer
Inner neuroblastic layer
Fibrous layer
Internal limiting
membrane

Fig. 498. Early differentiation of the neural layer of the human retina, shown in a vertical section at three months (Prentiss). × 440. At left, Cajal's analysis of the component elements after silver impregnation; at right, the appearance with ordinary stains.

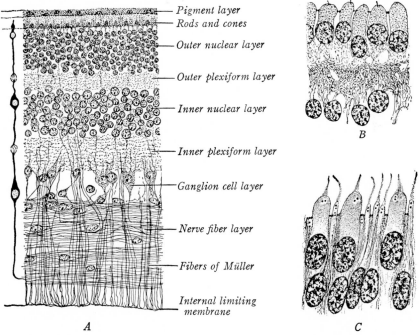

Pigment layer
Rods and cones
Outer nuclear layer
Outer plexiform layer
Inner nuclear layer
Inner plexiform layer
Ganglion cell layer
Nerve fiber layer
Fibers of Müller
Internal limiting
membrane

A   B   C

Fig. 499. Later differentiation of the human retina. *A,* At seven months, in vertical section (after Prentiss). × 440; at left, the chief neurons shown by silver technique; at right, appearance with ordinary stains. *B,* Early cone cells during the fifth month (after Magitot; × about 750). *C,* Rods and cones during the seventh month (after Seefelder; × 750).

of the adult retina can be recognized, including the developing, photo-receptive *rods* and *cones* (Fig. 499 *A*).

As in the wall of the neural tube, both supporting and nervous elements differentiate (Fig. 498). The supporting elements, or *fibers of Müller,* resemble ependymal cells and are arranged vertically. Their terminations produce the *internal* and *external limiting membranes* which bound the neural layer; the actual membranes are described as a formed product, elaborated by these cells.[10] The neuroblasts of the retina nearest the pigment layer transform into *visual cells,* which are of two kinds. These are *rod-* and *cone cells,* both of which are at first unipolar. In fetuses of seven months specialized processes protruding from the visual cells through the external limiting membrane have differentiated into the visual *rods* and *cones* (Fig. 499 *B, C*). These are the actual photoreceptive elements of the retina. Next deeper in position comes an intermediate layer of cells, composed mostly of *bipolar neurons;* they make connections both with the visual cells above and with ganglionic neurons below. The innermost stratum of nerve cells is the *ganglion cell layer.* Its multipolar cells give off axons that extend into the innermost (marginal) zone, and there comprise the *nerve fiber layer.*

How the retina came to be inverted, so that light has to pass through it before encountering the rods and cones, is a matter of speculation.[11] The site of keenest vision in the retina is a small area, known as the *macula lutea,* that differentiates late in fetal life and in early infancy. The macula lies in the direct visual axis; this spot is particularly characterized by a thinner, highly specialized center which lies at the bottom of a shallow pit, or *fovea centralis.* These structures and the partial crossing of optic fibers at the chiasm (Fig. 474 *B*) are associated with binocular vision and the fusion of images in higher primates. The human eye is sensitive to light in the seventh month, but form perception and color discrimination are not acquired until the first and second years, respectively.

In the *pars cæca,* or blind portion of the retina located peripheral to the ora serrata, the appearance of radial folds foreshadows the organization of *ciliary processes,* while beyond this wavy region is the plane territory of the future *iris.* This recognizable difference makes it possible to distinguish the two subdivisions of the retina known as the *pars ciliaris* and *pars iridica* (Fig. 504). The pigmented layer and a simplified neural layer are both represented in these regions. In the ciliary portion of the retina they cover the definitive ciliary processes. The iridic portion, bordering the pupil, owes its existence to the continued growth of these layers at the margin of the original optic cup.[9] In both regions the continuation of the neural epithelium, elsewhere stratified, reduces to a simple, insensitive layer. Where it covers the ciliary processes it becomes a secretory epithelium (like the ependyma of the chorioid plexuses of the brain); in the iris it becomes pigmented.

Differentiation of the Optic Nerve.   Fibers that arise from ganglion cells converge radially to a point where the optic stalk leaves the cup. In embryos of the seventh week they are growing back in the tissue of its inner tube toward the brain (Fig. 500 *A*). Eventually more than one million nerve fibers take this route.[12] The cells of the optic stalk presently convert into a scaffolding of neuroglial supporting tissue, and the canal in the stalk is rapidly obliterated (Fig. 500). The optic stalk is thus transformed into the so-called *optic nerve* (p. 506), containing a central artery and vein which originally coursed along the open groove of its optic fissure. Their branches will vascularize the neural retina.

Differentiation of the Lens. For a short time the saccular primordium of the lens is still attached to the parent ectoderm and nearly fills the cavity of the optic cup (Fig. 496). In embryos of 8 mm. it has detached and lies free as the *lens vesicle;* at this stage it is a hollow spheroid whose back wall is already thicker than the front one, next the surface epithelium (Fig. 497 *A*). The cells of the front wall retain a low columnar shape and constitute the permanent *lens epithelium.* The cells of the back wall are also single-layered, but they increase rapidly in height and at about seven weeks practically obliterate the original cavity (Fig. 501 *A-C*). These slender cells transform into transparent *lens fibers.* Toward the end of the second month such primary fibers attain a length of 0.3 mm., whereupon they cease dividing into new fibers and their nuclei degenerate. This mass forms the core of the lens. All additional fibers arise from proliferating cells located in an equatorial zone where the less specialized lens epithelium joins the lens-fiber mass (Fig. 501·C). The tips of these new fibers grow toward the front and back poles of the lens and such fibers are added, layer on layer, about the core.

As new fibers are progressively superposed on older ones, they become longer and longer, but still do not extend the whole distance from the back pole to the front pole of the lens. Characteristic *linear sutures* make their appearance on the front and back surfaces when such newer fibers fail to complete the full interpolar distance (Fig. 501 *D*). As the lens enlarges, these simple sutures expand into complex *lens stars* containing three, and finally six or even nine rays (*E*). Lens fibers continue to be added until adult life and the diameter of the lens does not increase much in the adult years. The total number of lens fibers formed is fairly constant (2250). The structureless *capsule* enclosing the lens is apparently derived from the cells of the lens vesicle. The lens of the early fetal months is spherical and relatively large (Fig. 504). At birth it has attained its final

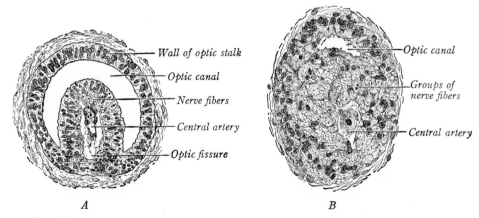

A   *B*

Fig. 500. Transformation of the human optic stalk into the optic nerve, shown in transverse sections (after Bach and Seefelder). *A*, At 14.5 mm. (× 275); *B*, at 19 mm. (× 350).

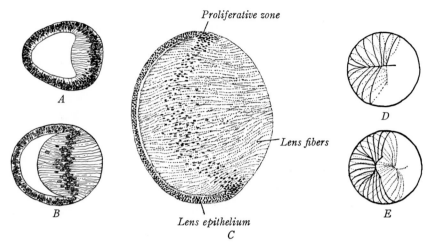

Fig. 501. Differentiation of the human lens. *A–C,* Sections through the lens of embryos at 10 mm., 13 mm. and 20 mm. ($\times$ 100). *D, E.,* Diagrams of the formation of sutures (Mann); *D* represents a stage when sutures are linear, like the adult dogfish lens (front suture, horizontal; back suture, vertical); *E* shows a four-pointed star of the early adult.

thickness and two-thirds of its final diameter. The increasing inequality of these measures results from the progressive failure of lens fibers to extend from pole to pole.

The lens tissue itself is at all times wholly nonvascular. For a time, however, blood vessels do spread over its surface. Some of these come from the *hyaloid artery.* This is a continuation of the *central artery* which in 6 mm. embryos courses along the gutter-like groove of the optic stalk, enters the back of the optic cup through the optic fissure, and (renamed the *hyaloid artery*) extends to the back surface of the lens (Fig. 502). Here it ramifies in a mesenchymal coat that soon encloses the lens and is known as the *vascular tunic of the lens* (Fig. 504). Other vessels, spreading from the iris, supply the front of the lens in a portion of this tunic called the *pupillary membrane* (Fig. 503). The vascular tunic flourishes during the period of rapid growth of the lens and attains its highest development in the fifth month; before birth the tunic has normally disappeared. The hyaloid artery also degenerates completely, the only permanent trace being the lymph path through the jelly of the vitreous body that is called the *hyaloid canal.*

Accessory Coats. During the seventh week the mesenchyme surrounding the optic cup begins to specialize into two accessory coats (Fig. 503). The outer one is more compact and becomes a definitely fibrous tunic, the *sclera* and *cornea.* The inner, looser covering organizes into the vascular *chorioid;* it also contributes to the ciliary body and iris.

THE FIBROUS COAT. The mesenchymal *sclera* transforms into dense fibrous tissue. It covers the base and sides of the eyeball and receives the insertions of the ocular muscles. This sclerotic coat corresponds to the dura mater of the brain, with which it is continuous by way of the sheath of the optic nerve. Toward the front of the eyeball the fibrous coat is named the *cornea* (Fig. 504). It consists of transparent connective tissue, surfaced externally with ectodermal *corneal epithelium* and lined internally with the flat *endothelium of the anterior chamber.* The former is modified epidermis. The latter represents the first mesenchymal cells that grow in from the sides, whereas the main substance of the cornea traces origin to mesenchyme that fills in secondarily between epidermis and endothelium.[13]

THE VASCULAR COAT. The *chorioid* is the inner of the two primary, mesenchymal tunics of the eyeball (Fig. 503). It is located between the sclerotic coat and the pigment layer of the retina. The chorioid primordium acquires a high vascularity in embryos as young as six weeks; moreover, its cells become stellate and pigmented, so that the tissue is loose and spongy. This vascular layer, in which course the chief vessels of the eye, corresponds to the pia-arachnoid of the brain. The lustrous inner layers, or *tapetum,* which reflect light in various mammals, are not represented in the chorioid of man. Distal to the level of the ora serrata, the primitive vascular coat differentiates into: (1) the vascular connective tissue of the *ciliary body;* (2) the unstriped fibers of the *ciliary muscle;* and (3) the connective-tissue stroma of the *iris.* The presence of pigment in both layers of the retina in the region of the iris, and the differentiation of the pupillary muscles from the external (original pigment) layer of the optic cup, have been mentioned previously (pp. 532, 534).

The Ocular Chambers. The hollow interior of the eyeball consists of chambers that contain either a watery fluid (*aqueous humor*) or a jelly (*vitreous body*). The lens separates these two different components.

THE AQUEOUS CHAMBERS. The *anterior chamber* is not a simple cleft occurring in the mesenchyme between ectoderm and lens (Figs. 503, 504). Rather, the cornea differentiates first, whereas the mesodermal layer (*pupillary membrane*) overlying the lens is an independent and secondary ingrowth from all sides.[14] The intervening space is the anterior chamber; its continued peripheral extension is responsible for the separation of a definite *iris* from the cornea. Close to the margin of the anterior chamber, at the

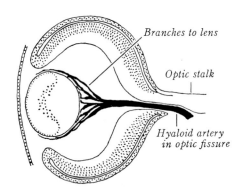

Fig. 502. Relation of the hyaloid artery to the lens, shown in a section of a 17 mm. human embryo (after Streeter). × 45.

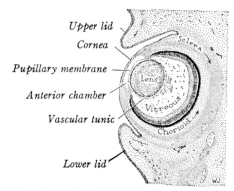

Fig. 503. Human eyeball and eyelids, at two months, in longitudinal section. × 15.

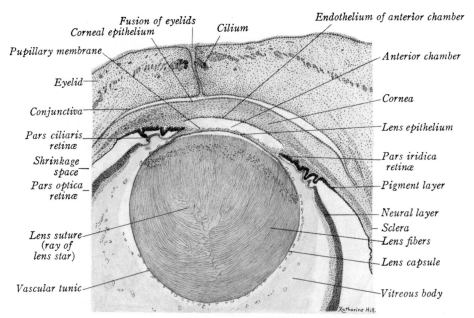

Pupillary membrane

Corneal epithelium Fusion of eyelids Cilium Endothelium of anterior chamber

Eyelid

Anterior chamber

Conjunctiva

Cornea

Pars ciliaris retinæ

Lens epithelium

Shrinkage space

Pars iridica retinæ

Pars optica retinæ

Pigment layer

Neural layer

Sclera

Lens suture (ray of lens star)

Lens fibers

Lens capsule

Vascular tunic

Vitreous body

Fig. 504. Human eyeball and eyelids, at three months, in longitudinal section (after Prentiss). × 27. The neural and pigment layers of the retina have separated artificially.

junction of cornea and sclera, there is an important ring-shaped drainage space, the *scleral venous sinus* (canal of Schlemm). The *posterior chamber,* between the iris, lens and vitreous body, makes a relatively tardy appearance since it is brought into existence by the growth forward of the iris. It communicates with the anterior chamber when the pupillary membrane disappears.

THE VITREOUS BODY. The crescentic cavity between the lens and the optic cup becomes filled during the second month with a transparent, fibrillar jelly which comes to be known as the *vitreous body* (Fig. 505 *A*). Modern investigations agree that this substance is primarily an epithelial product.[15] Its early 'secretion' from lens tissue soon ceases, but the vitreous substance is progressively increased by fibrillar processes that project from the surface of the retina; they probably grow out from the basement membrane of the supporting cells of Müller (*B*). Those stouter fibers laid down by the pars ciliaris retinæ seemingly become the *zonula ciliaris,* or suspensory ligament of the lens.

Only when the primitive vitreous humor is partly developed does mesenchyme first appear within the optic cup. Some of it enters through the optic fissure with the hyaloid artery (Fig. 505 *A*). Still other mesenchymal cells gain entrance around the edge of the cup in association with the lens. The fate of all this invading mesenchyme—whether it contributes significantly as a secondary vitreous or whether it degenerates—is not yet decided beyond question.

Associated Organs. The *Eyelids* develop as folds of the integument adjacent to the eyeball (Fig. 503). These folds are indicated at the end of the seventh week, and two weeks later their edges have met and fused (Fig. 504). This epidermal union begins to break down in fetuses five months old, but the eyes do not reopen until the seventh or eighth month; in some mammals this is delayed until after birth. A third, rudimentary eyelid, perhaps incorrectly homologized with the functional nictitating membrane of lower vertebrates,[16] is represented by the adult *plica semilunaris* at the inner angle of the eye. The ectoderm of the outside of the lid differentiates into epidermis. Contrasted is the continuation of ectoderm on the internal surface of the lid and its reflection over the front half of the sclera and all of the cornea; this is a mucous membrane named the *conjunctiva* (Fig. 504). The *cilia*, or eyelashes, develop like ordinary hairs at the edges of the lids; they are provided both with sebaceous glands (of Zeis) and with modified sweat glands (of Moll). About thirty *tarsal glands* also arise along the edge of each lid (Fig. 506); these glands (of Meibom) are sebaceous in nature. The hair follicles for the eyelashes begin developing during the tenth week. They are followed closely by the several glands associated with the cilia and eyelids. These start budding inward early in the fourth month while the eyelids are still fused.

The *Lacrimal Glands* appear during the ninth week as approximately six knobbed outgrowths of the conjunctival epithelium (Fig. 506). They lie dorsally near the external angle of the eye. At first the primordia are solid epithelial cords, but they soon branch and acquire lumina. Other buds appear in the third month.

Each *Naso-lacrimal Sac and Duct* arises in 12 mm. embryos as a ridge-

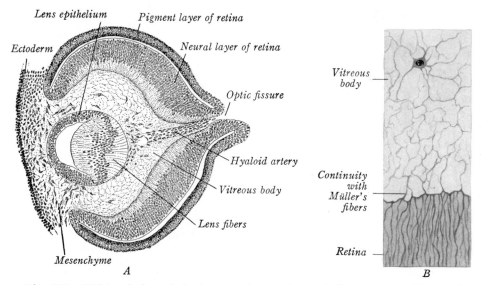

Fig. 505. Differentiation of the human vitreous body. *A,* Optic cup, at 12 mm., in longitudinal section (after Prentiss; × 100). *B,* Detail of the vitreous humor and its relation to the retina, at two months.

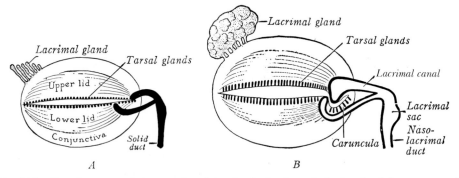

Fig. 506. Development of the tarsal and lacrimal glands and the lacrimal-duct system of the human eye, shown by diagrams (adapted after Ask).

like thickening of the ectoderm, parallel to the naso-lacrimal groove (Fig. 396 *A*).[17] It extends from the inner angle of the eye to the primitive olfactory sac and separates the maxillary and lateral nasal processes of its respective side. The thickening becomes cut off and, as a solid cord, sinks into the underlying mesenchyme. A secondary sprout grows out to each eyelid to comprise the *lacrimal canals* (Fig. 506), while an extension in the opposite direction connects with the nose (Fig. 490 *A*). The nasal end of the lumen is not completed until birth. The *caruncula* (the reddish elevated mass at the inner angle of the eye) is a part of the lower lid, secondarily upraised by the lacrimal duct (Fig. 506 *B*).[18]

Causal Relations.   The optic field is at first a diffusely localized area which becomes established under the influence of the rostral portion of the roof of the archenteron (Fig. 495 *A*). Within this field the same continuing influence of the ento-mesoderm establishes two centers of eye formation which become irreversibly determined as eye-cup primordia in the neural plate stage of amphibians. The invagination of the optic vesicle to produce a cup is a self-governed process, wholly independent of the lens primordium, which is developing at the same time. Moreover, the eye cup will self-differentiate successfully when cultured or grafted into an indifferent region. On the contrary, lens formation in man and various other animals is dependent on induction of the overlying ectoderm (epidermis) by the optic vesicle (p. 177). The ability of epidermis as a whole to respond to contact with an optic vesicle varies. In many amphibians the ectoderm in general can so react; in others, only head ectoderm; in some, only the ectoderm directly over the optic vesicle is competent to form a lens. Also dependent is the differentiation of the cornea which is induced to specialize from the primitive skin, chiefly by the lens. The chorioid and sclera differentiate only in the presence of the pigmented epithelium.

Anomalies.   Undersize of the eye (*microphthalmia*) and even its absence (*anophthalmia*) are known, as is the virtual absence of the lens (*aphakia*). In *cyclopia* a single, median eye replaces the usual paired condition (Fig. 507 *A*). All intergrades exist from closely approximated eyes to perfect unity. The cyclopic condition is the result of the faulty organization of paired optic centers in a primitive, common 'eye field.' More specifically it is attributable to the failure of a wedge-shaped area of neural plate and head mesoderm to differentiate in this region. In cases of cyclopia the nose is usually a cylindrical proboscis, situated at the base of the forehead above the median eye (*A*).

Opacities of the lens (congenital *cataract*) and cornea are acquired aberrations, since there is no normal stage of development when these tissues are not clear.[19] Cataract is

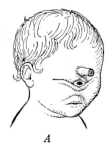
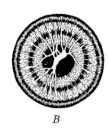
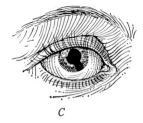

A B C

Fig. 507. Anomalies of the human eye. *A,* Cyclopia of a newborn, with a single eyeball but partial doubling of the lids; above the eye is the proboscis-like nose. *B,* Persistent pupillary membrane in an adult. *C,* Coloboma of the iris.

common in infants whose mothers had German measles during the second month of pregnancy. Retained portions of the pupillary membrane may cross the pupil and so interfere with vision (Fig. 507 *B*); a similar obstruction in the visual path is presented by a variably persistent hyaloid artery. Lack of pigment in the retina and iris is usually associated with general *albinism*. Congenital *glaucoma* results when the canal of Schlemm or its channels of supply fail to develop properly and hence do not provide for the normal drainage of aqueous humor. The absence of a sector (or any local area) of the iris, ciliary body, retina or chorioid tunic produces a defect known as *coloboma* (Fig. 507 *C*). It results from a disturbance in the mechanism of closure of the optic (chorioid) fissure. Of eyelid defects, the best known is a cleft or split in the upper lid.

THE EAR

The human ear consists of a sound-conducting apparatus and a receptive sense organ (Fig. 514 *B*). The reception and transmission of sound waves is the function of the *external* and *middle ears.* The end organ proper is the *internal ear,* with auditory sensibility residing in its cochlear duct. The remainder of the internal ear (semicircular ducts; utriculus; sacculus) serves as an organ of equilibration; this apparatus constitutes the entire ear of fishes.

The Internal Ear. The epithelium of the internal ear is derived from the ectoderm. It first takes the form of a thickened ectodermal plate, the *auditory placode,* located midway alongside the hind-brain. Such a pair of placodes is prominent in embryos with nine somites and this stage is quickly followed by the appearance of distinct *auditory pits* (Fig. 508 *A, B*). In embryos of about 24 somites (nearly 4 mm.) the cup-like pits close into *otocysts* which soon lose their union with the ectoderm and become detached as ovoid sacs (*C*). The otocyst, or auditory vesicle, lies opposite the fifth neuromere and is in contact rostrally with the facial-acoustic ganglionic mass (Fig. 436).

Approximately at the point where the otocyst joined the ectoderm, a tubular recess, the *endolymph duct,* straightway pushes out as a new growth and then shifts to a medial position (Fig. 509). This secondary origin makes the endolymph duct somewhat different from that of selachian fishes which, as the retained stalk of the otocyst, opens permanently to the

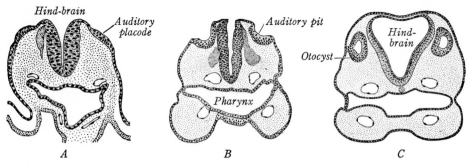

Fig. 508. Development of the human otocyst, illustrated in transverse sections. *A,* At nine somites ($\times$ 80); *B,* at sixteen somites ($\times$ 60); *C,* at about 4 mm. ($\times$ 40).

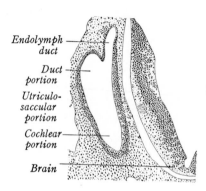

Fig. 509. Human otocyst, at 7 mm., in longitudinal section (His). $\times$ 50.

exterior. In higher vertebrates the blind extremity of the duct dilates into the *endolymph sac* (Fig. 510). In the fifth week the ovoid otocyst elongates still further in the dorsoventral axis (Figs. 509, 510 *A*). Its slenderer, ventral part represents the future *cochlear duct.* The most dorsal portion of the otocyst already shows indications of the developing *semicircular ducts,* while an intermediate region is destined soon to subdivide into *utricle* and *saccule.* These progressive changes are illustrated in Fig. 510.

The *semicircular ducts* are well outlined at six weeks as two flattened pouches—the superior and posterior ducts arising from a single pouch at the dorsal border of the otocyst, the lateral duct from a horizontal outpocketing; the cochlear region has elongated and become J-shaped (Fig. 510 *A, B*). The seventh week is occupied with the rough modeling of the otocyst into an approximation of the definitive system of sacs and ducts, the *membranous labyrinth* (*C*). Centrally the walls of the two secondary pouches, just mentioned, flatten and fuse into epithelial plates; in this process canals are left at the periphery, and these communicate with the remainder of the otic vesicle. Soon the solid, central portions of the epithelial plates break down and set the semicircular ducts free, except at their ends.

Early in the eighth week the endolymph duct and the three semicircular ducts are well represented; at the same time the main sac is dividing into *utricle* and *saccule,* and the *cochlear duct* has begun to coil like a snail's shell (Fig. 510 *C*). It will be noticed that the superior and posterior ducts have a common limb (or crus) opening dorsally into the utricle; their opposite ends and the rostal end of the lateral duct are dilated to form *ampullæ.* Constriction separates the utriculo-saccular region into a dorsal portion, the *utriculus,* to which are attached the semicircular ducts, and a ventral portion, the *sacculus,* connected with the cochlear duct (*C*). Early in the third month the general adult form of the internal ear is nearly attained (*D*). At this time the sacculus and utriculus are less broadly connected; the semicircular ducts are relatively longer, their ampullæ more prominent, and the cochlear duct is coiled to its final extent of two and one-half turns. In the final condition the utriculus and sacculus are completely separated from each other, but each remains attached to the endolymph duct by a slender canal. Similarly, the cochlear duct is further constricted from the sacculus; the basal end of the former becomes a blind process, while a narrow canal, the *ductus reuniens,* is the sole connection between the two.

The totally differentiated otocyst, with all its subdivisions and their fibrous support, is called the *membranous labyrinth.* The utriculus and sacculus alone correspond to the entire 'ear' of various invertebrates, in which the organ functions merely for equilibration and not for hearing. The semicircular ducts (of vertebrates in general) and the cochlear duct

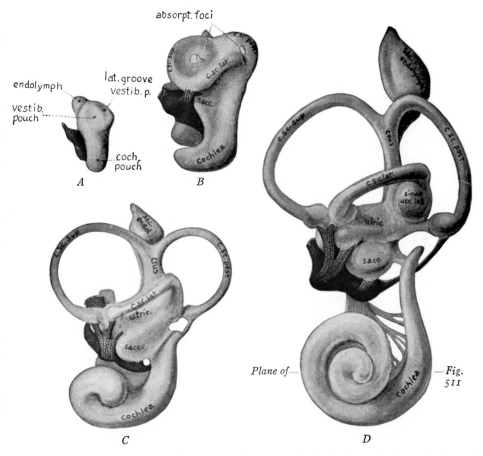

Fig. 510. Development of the left membranous labyrinth, shown in lateral views of models (Streeter). × 25. *A*, At 6.6 mm.; *B*, at 13 mm.; *C*, at 20 mm.; *D*, at 30 mm. The colors, yellow and red, indicate respectively the cochlear and vestibular divisions of the acoustic nerve and their accompanying ganglia.

absorpt. foci, Area where absorption is complete; *crus,* crus commune; *c. sc. lat., c. sc. post., c. sc. sup.,* lateral, posterior and superior semicircular ducts; *endolymph,* endolymph duct; *sacc.,* sacculus; *sac. endol.,* endolymph sac; *utric.,* utriculus.

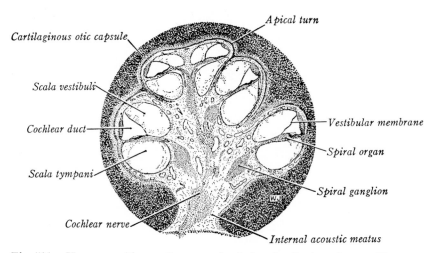

Fig. 511. Human cochlea, at four months, in longitudinal section. × 12.

(of mammals) historically are secondary outgrowths from this older part. In fishes and amphibians the cochlear portion is rudimentary, while in reptiles, birds and monotremes it is a straight tube (the lagena); only in true mammals does a coiled canal differentiate. The epithelium of the membranous labyrinth is composed at first of a single layer of low columnar cells. At an early stage, fibers from the acoustic nerve grow between the epithelial cells in certain regions, and these areas are then thickened and modified into special sense organs. Such end organs are the *cristæ ampullares* in the ampullæ of the semicircular ducts, the *maculæ acusticæ* in the utriculus and sacculus, and the *spiral organ* (of Corti) in the cochlear duct (Figs. 511–513).

The cristæ and maculæ are sense organs for maintaining equilibrium and giving information concerning the direction and extent of body movements. They differentiate during the seventh week. In each ampulla, transverse to the long axis of the duct, the epithelium and underlying tissue form a curved ridge, the *crista* (Fig. 513 *B*). The cells of the epithelium differentiate both into sense cells, which bear delicate 'hairs' at their ends, and into supporting cells. The latter elements secrete a jelly-like substance (the *cupula*) upon the free surface; into it the sensory hairs project. The *macula* of the utriculus or sacculus resembles the cristæ in its development, save that larger areas of the epithelium specialize into cushion-like end organs. The free surface becomes covered with a gelatinous *otolithic membrane* which bears superficial calcareous deposits, the *otoconia*.

The true organ of hearing, the *spiral organ*, develops slowly in the epithelium of the coiled cochlear duct. The spiral organ is a continuous strip that lies on the basal side of the duct, basal here signifying in a direction away from the apex of the conical cochlea (Fig. 511). Differentiation begins as an epithelial thickening in the basal turn and advances progressively toward the apex.[20]

The epithelial primordium of the spiral organ soon divides longitudinally into an inner, larger ridge and an outer, smaller ridge (Fig. 512 *A*). The inner cells of the inner ridge become the tall constituents of the *spiral limbus* (*B*); by contrast, the outer part of the ridge undergoes a peculiar autolytic involution until only the thin lining of the *inner spiral sulcus* remains (*C, D*). The outer, smaller ridge is the primordium of the *spiral organ* (of Corti). In it appear the flask-shaped *inner* and *outer hair cells,* while the remaining elements become the various *supporting cells (B–D).* The *spiral tunnel* results from a partial destruction of the supporting cells (*C*). Both ridges are from the beginning covered with the gradually thickening *tectorial membrane.* It is a fibrillar and gelatinous substance secreted by the epithelium.[21] As the spiral sulcus becomes deeper by the cellular dissolution already mentioned, the membrane spans across its trough (*C, D*).

The development of the acoustic nerve and the distribution of its vestibular and cochlear divisions are described on p. 507 and illustrated in Figs. 475, 510. Nerve fibers arborize about the bases of the sensory cells of the cristæ, maculæ and spiral organ. A newborn child hears imperfectly because the external auditory meatus is not entirely free of detritus and the middle ear cavity is filled almost completely with a gelatinous tissue. Following the progressive resorption of this material, normally acute hearing enters in the first weeks after birth.

The mesenchyme surrounding the epithelial labyrinth is differentiated into a fibrous membrane, which lies beneath the epithelium, and into cartilage which envelops the whole labyrinth. At about the tenth week the cartilage immediately bordering the labyrinth begins a secondary reversal of development whereby it returns first to precartilage and then to a syncytial reticulum; the latter becomes the open tissue of the *perilymph spaces* (Fig. 513 *A*).[22] The membranous labyrinth is henceforth suspended in the fluid of the perilymph spaces. The cochlear duct appears triangular in section, for its lateral wall remains attached to the peripheral bony labyrinth while its inner angle is adherent to the bony axis (modiolus) of

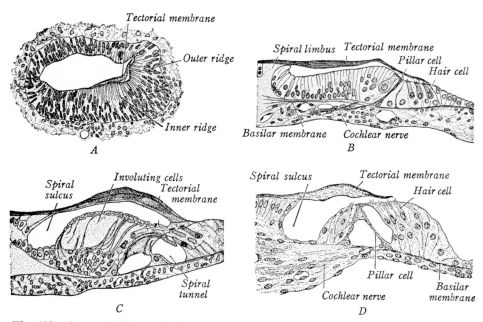

Fig. 512. Stages of differentiation of the human spiral organ, at ten to twenty weeks, in the basal turn of the cochlear duct (after Kolmer and Alexander). × about 150.

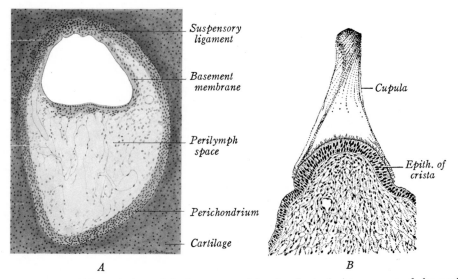

Suspensory
ligament

Basement
membrane

Perilymph
space

Perichondrium

Cartilage

Cupula

Epith. of
crista

A

B

Fig. 513.   Differentiation of the human semicircular duct. *A,* Appearance of the perilymph space, at four months, shown in a transverse section (after Streeter; × 75). *B,* Crista
and its cupula, at five months, sectioned vertically (after Alexander).

the cochlea (Fig. 511). Large perilymph spaces are formed above and
below the cochlear duct. The upper is the *scala vestibuli,* the lower the
scala tympani; both are lined with flattened mesenchymal cells, arranged
like an epithelium. The thin wall separating the cavity of the cochlear duct
from that of the scala vestibuli is the *vestibular membrane* (of Reissner).
Beneath the spiral organ of the cochlear duct, a fibrous *basilar membrane*
is differentiated by the mesenchyme. Its vibrations to sound waves stimulate the receptive hair cells. The *bony labyrinth* is produced in the fifth
fetal month by the replacement of the cartilage capsule by bone. The central
axis of the bony cochlea is exceptional, however, in that it develops
directly from mesenchyme as a membrane bone. At the middle of fetal life
the internal ear has attained its final size.

  The Middle Ear.   Each auditory tube and tympanic cavity represents a
drawn-out first pharyngeal pouch (with which the second perhaps merges, as
well).[23] The entodermal pouches appear in embryos of 3 mm., enlarge
rapidly, flatten dorsoventrally, and are in temporary contact with the ectoderm (Fig. 514 *A*). Toward the close of the second month the proximal
stalk of each pouch undergoes actual constriction to form the more cylindrical *auditory tube.* This canal lengthens and its lumen becomes slit-like
during the fourth month. The blind, outer end of the pouch enlarges into
the *tympanic cavity* (*B*). It is surrounded by loose connective tissue in
which the auditory ossicles develop and for a considerable time lie embedded. In the last fetal months, however, the peculiar spongy tissue that
surrounds the ossicles undergoes degeneration, while the tympanic cavity
expands correspondingly to occupy the new space thus made available; yet
at birth this process is still incomplete. The tympanic epithelium on en-

countering the ossicles wraps itself around them. Even in the adult, the
ossicles, their muscles and the chorda tympani nerve (all of which appear
to have invaded the tympanic cavity) really are outside, since they retain
a covering of mucous epithelium continuous with that lining the cavity.
The pneumatic cells of the mastoid wall result from epithelial invaginations,
which at the close of fetal life begin to invade the simultaneously excavated
temporal bone.

The *auditory ossicles* develop from the condensed mesenchyme of the
first and second branchial arches. When these primordial ossicles are chon-
drifying from single centers, they are still in direct continuation with their
respective cartilaginous arches (Fig. 515). Soon the ear bones lose connection with the rest of the arch, and articulations are developed where the
ossicles touch each other. The *malleus* (hammer) attaches to the ear drum;
the *stapes* (stirrup) fits into the oval window of the perilymph space; the
incus (anvil), intermediate in position, articulates with the other two. Of
these ear bones, the major portions of the malleus and incus are differentiated in serial order from the dorsal end of the first arch (Meckel's car-

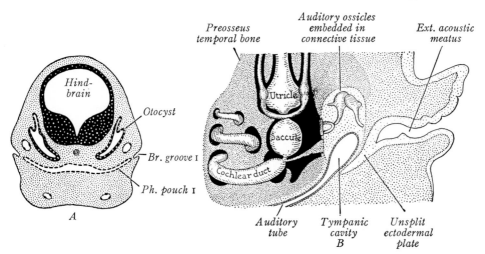

Fig. 514. Progressive association of the primordia of the external, middle and internal
ears, illustrated by partly schematic sections. *A,* At six weeks; *B,* at three months.

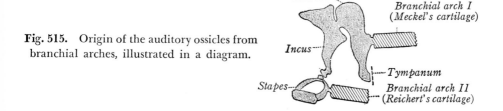

Fig. 515. Origin of the auditory ossicles from
branchial arches, illustrated in a diagram.

tilage).[24] Similarly, the stapes, except its floor plate, is derived from the second branchial arch (Reichert's cartilage) (Fig. 515).[25]

Since the mesenchymal and cartilaginous stages of the stapes are perforated by the stapedial artery, the early shape is that of a ring. This form persists well into the third month when the stapedial artery disappears and the assumption of the final shape is begun. The final size is attained at midfetal life. Substituting for these ossicles in amphibians, reptiles and birds is an unjointed rod, the *columnella;* this rod, however, is equivalent to the stapes alone.

Certain collateral data are correlated with the branchial-arch origin of the auditory ossicles. For instance, the muscle of the malleus, the *tensor tympani,* is derived from the first branchial arch; the *stapedial muscle* of the stapes, from the second arch. These muscles are innervated by the trigeminal and facial nerves, which are respectively the nerves of the first and second arches.

The External Ear.   The external ear is a modification of the first branchial groove, together with additions from the branchial arches bounding that region. In a sense, the *external acoustic meatus* represents the ectodermal groove itself, which for a time is in contact with the entoderm of the first pharyngeal pouch. Later, however, this contact is lost and growth of the head in thickness tends to separate the meatus from the middle ear cavity. Toward the end of the second month the groove deepens centrally to produce a funnel-shaped pit; the whole canal, thus formed, corresponds to the outer portion of the definitive meatus that is surrounded by cartilage. From the bottom of the pit just described, an ectodermal cellular plate grows still deeper until it reaches the wall of the tympanic cavity (Fig. 514 *B*). During the seventh month the plate splits, and the additional cleft acquired in this fashion constitutes the innermost portion of the external meatus. Even at birth a plug of cast-off cells may fill the lumen.

The *tympanic membrane* (ear drum) results from a thinning of the mesodermal tissue in the region where the blind end of the external acoustic meatus is coming to abut against the wall of the tympanic cavity. Hence the permanent membrane is a fibrous sheet covered externally by ectodermal epithelium and internally by entoderm. The area of apposition between these layers does not correspond to any part of the primary tympanic cavity, but is at a region added secondarily through the process of cavity-expansion already described. At birth the ear drum, though of final size, is set so obliquely that it almost lies upon the meatal floor; it erects gradually as the meatus lengthens.

The *auricle* develops around the first branchial groove. Its tissue is furnished both by the first (mandibular) branchial arch and the second (hyoid) arch. During the sixth week six hillocks appear on these parts—three on the caudal border of the first arch and three on the second arch (Fig. 516 *A, B*). For many years it was held that the auricle develops in a rather precise manner from these six elevations and from an auricular fold of the hyoid integument. Later restudies of the problem differ in some details,[26] even to asserting that the entire auricle, except the tragus, is of hyoid origin.[27]

Causal Relations.   The potential ear-field is originally larger than the auditory placode that normally forms within it. The ear-field is very soon polarized in at least two axes, and an otocyst turned upside down will right itself. The occurrence and modeling of an otocyst depend upon the stimulation supplied by the hind-brain, and especially by the mesenchyme of the archenteron roof, during the neurula stage of amphibians. Once its fate is determined, the otocyst can differentiate further when moved to a strange location or placed in culture medium. But the determination for the development of a normal labyrinth occurs gradually and involves several factors brought into play successively. The normal differentiation of the several component parts is also contingent on the presence of an intact medulla oblongata. Hence the otocyst is not self-differentiating and independent in quite the same way, as, for example, is the limb bud (p. 211). Yet the presence of a normal nerve supply is not essential to differentiation, and deprivation of nerve does not cause secondary degeneration. The otocyst, in turn, compels the surrounding mesenchyme to form a cartilaginous capsule in both amphibians and birds. Still further, the element known as the annular tympanic cartilage induces the formation of a tympanic membrane, and without it none forms.

Anomalies.   Agenesis of the inner, middle or outer ear is very rare. Congenital deafness may be the result of imperfect nerve connections, of faulty development of the membranous labyrinth, auditory ossicles or ear drum, and of atresia of the tympanic cavity or external meatus. Deafness from injury to the spiral organ occurs frequently in infants whose mothers experienced German measles at the second month of pregnancy. The auricles may fail to develop appreciably. Ordinarily they are quite variable in form, whereas defective combination of the several primordial parts is responsible for the more serious abnormalities (Fig. 517 *A*). Fetal types of auricle are occasionally seen in adults as the result of inhibited development, but are without further significance. Alleged cases of inherited, pierced ear lobes are really clefts between the incompletely fused tragus and

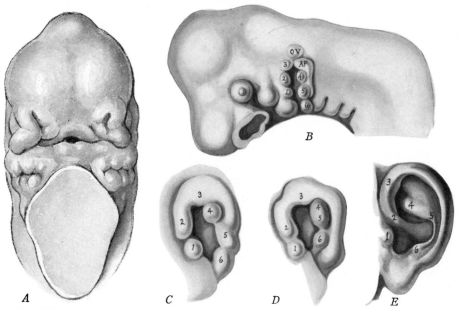

Fig. 516.   Development of the human auricle (partly after His). *A*, Front view of the head, at 12.5 mm. (× 13); below the face (*cf.* Fig. 160 *B*) are the ear hillocks grouped about the first branchial grooves. *B–E*, Side views of the auricle at 11 mm., 13.5 mm., 15 mm. and in adult.

AF, Auricular fold; *OV*, otic vesicle; 1–6, elevations on the mandibular and hyoid arches which respectively become: 1, tragus; 2, 3, helix; 4, 5, anthelix; 6, antitragus.

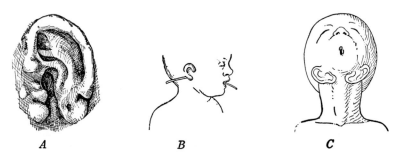

Fig. 517.   Anomalies of the human auricle. *A*, Malformed auricle. *B*, Fistula auris, probed to show its relations. *C*, Synotia, combined with microstomia and agnathia.

antitragus. Pits sometimes occur in regions related to the other primordia, and especially in front of the base of the helix. The whole group is included in the category of fistulæ of the ear. A complete fistula, connecting with the middle ear cavity, is of the greatest rarity (*B*). The almost equally rare condition of *synotia* shows the ears fused, or near the mid-ventral line at the upper part of the neck (*C*); it is associated with agnathia (p. 206), and illustrates the primitive location of the ear primordia before being wedged apart by the growing mandible (*cf.* Fig. 516 *A*).

REFERENCES CITED

1. Jalowy, B. 1939. Z'ts. Anat. u. Entw., *109,* 344–359.
2. Tello, J. F. 1922. Z'ts. Anat. u. Entw., *64,* 348–440.
3. Cuajunco, F. 1927. Carnegie Contr. Embr., *19,* 45–72.
4. Bradley, R. M. & I. B. Stern. 1967. J. Anat., *101,* 743-752.
5. Torrey, T. W. 1941. Proc. Nat. Acad. Sci., *26,* 627–634.
6. Schaeffer, J. P. 1920. The Nose in Man (Blakiston).
7. Torrigiani, C. A. 1914. Arch. Ital. Anat. e. Embr. *12,* 153–253.
8. Smith, D. T. 1920. Johns Hopkins Hosp. Bull., *31,* 239–246.
9. Special-Cirincione, S. 1922. Annali ottal. e. clin. oculist., *50,* 2–49.
10. Van der Stricht, O. 1922. Compt. rend. Soc. biol., *86,* 264–266; 266–269.
11. Parker, G. H. 1908. Am. Nat., *42,* 601–609.
12. Bruesch, S. R. & L. B. Arey. 1942. J. Comp. Neur., *77,* 631–665.
13. Rones, B. 1932. Arch. Ophth., *8,* 568–575.
14. Mann, I. 1931. Trans. Ophth. Soc. Un. King., *51,* 63–85.
15. Duke-Elder, S. 1963. System of Ophthalmology, *3,* Pt. 1 (Kimpton).
16. Fischer, F. 1936. Abhandl. a. d. Augenheilk., *22,* 1–58.
17. Politzer, G. 1952. Z'ts. Anat. u. Entw., *116,* 332–347.
18. Papamiltiades, M. 1947. C. R. Ass'n. Anat., *34,* 418–419.
19. Mann, I. C. 1957. Developmental Anomalies of the Eye (Lippincott).
20. Kolmer, W. 1927. Bd. 3, Teil I in Möllendorff: Handbuch (Springer).
21. Van der Stricht, O. 1918. Carnegie Contr. Embr., *7,* 55–86.
22. Streeter, G. L. 1918. Carnegie Contr. Embr., *7,* 5–54.
23. Frazer, J. E. 1914. J. Anat. & Physiol., *48,* 391–408.
24. Hanson, J. R. & B. J. Anson. 1962. Arch. Otolaryn., *76,* 200–215.
25. Cauldwell, E. W. & B. J. Anson. 1942. Arch. Otolaryn., *36,* 891–925.
26. Streeter, G. L. 1922. Carnegie Contr. Embr., *14,* 111–138.
27. Wood-Jones, F. & I. C. Wen. 1934. J. Anat., *68,* 525–535.

PART III.

A LABORATORY MANUAL OF EMBRYOLOGY[1]

Chapter XXVIII. The Study of Chick Embryos

A. THE UNINCUBATED OVUM AND EMBRYOS OF THE FIRST DAY

Unincubated Egg. The 'yolk' of the hen's egg is a single ovum, enormously expanded with stored food material. When this egg cell is expelled from the ovary at the time of *ovulation* it is enveloped by the *vitelline membrane,* secreted by the cytoplasm of the egg itself (Fig. 518), and by the delicate *zona pellucida* commonly held to be a product of the follicle cells among which the growing egg lay. By the time the liberated ovum passes into the oviduct, the process of *maturation* has progressed to the point where one polar cell is given off. If spermatozoa lie in wait, fertilization ensues; at the same time the second polar cell is extruded, thereby completing maturation. As the egg continues down the oviduct, the viscid *albumen,* papery *shell membrane* and calcareous *shell* are progressively secreted by the epithelial lining of the duct and are added about the yolk as accessory investments (Fig. 518). During this journey, which ends with the laying of the egg, a start has been made toward the formation of a visible embryo. Thus it is that, before external incubation begins, the processes of *cleavage* and entoderm formation are complete; when laid, the embryonic area is represented by the familiar whitish disc to be seen on the surface of the yolk and technically designated the *blastoderm.* The egg is ready to be laid about 24 hours after its discharge from the ovary; at this time the relations of its several components are as indicated in Fig. 518.

[1] More than one-third of the illustrations in this section are copies or adaptations of drawings originally published by Professor C. W. Prentiss in 1915.

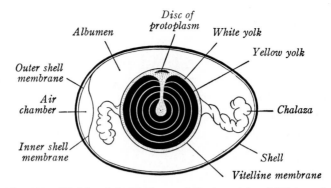

Fig. 518. Highly telolecithal egg of the hen (after Lillie). × 1.

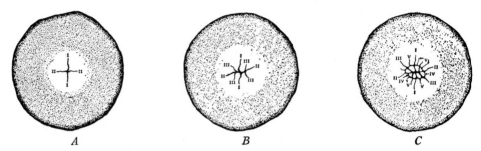

Fig. 519. Cleavage of pigeon's ovum, seen in surface view (Patten, after Blount). × 4. Order of appearance of cleavage furrows on blastoderm is indicated by Roman numerals. *A,* Second cleavage; *B,* third cleavage; *C,* fifth cleavage.

Cleavage and Blastula.   The protoplasmic part of the egg is a tiny disc or cap at its upper pole. Fertilization promptly initiates a series of orderly cell divisions which divide the thin disc into an increasingly large number of cells. This sequence of mitoses comprises the process of *cleavage,* while the component cells are known as *blastomeres* (Fig. 519). The result is a cellular disc, separated from the yolk beneath by a cleft-like space (Fig. 520); the whole makes a highly asymmetrical, hollow sphere which is called a *blastula.*

Stage of Gastrula and Primitive Streak.   Two different processes accomplish gastrulation in birds. The first occurs when an under layer, the *entoderm,* splits away from the blastodermic disc (Figs. 51, 521). In this condition the egg is laid, and without incubation there is no further development. On the commencement of incubation, even though days of dormancy have elapsed since laying, gastrulation continues into its second phase. This consists in the movement of certain cells, destined to become the *mesoderm* and *notochord,* out of the outer layer to a middle-layer position (Figs. 52, 523). The residual outer layer, when these departures are completed, is *ectoderm.*

The crowding toward the midline, as the cells of the future notochord and mesoderm flow and turn beneath, produces an opaque band named the *primitive streak* (Fig. 522 *A*). It is well seen after 18 hours of incubation.

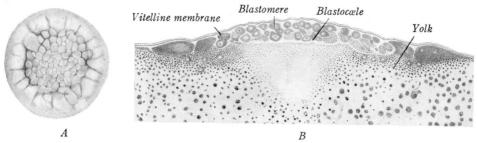

Fig. 520. Early blastula stage in the pigeon (after Blount). *A,* Blastoderm in surface view; *B,* in vertical section.

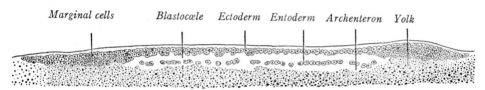

Fig. 521. Entoderm formation in pigeon, shown by sectioned blastoderm. × 50.

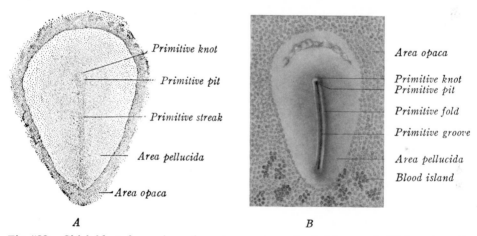

Fig. 522. Chick blastoderms, in surface view, at stage of primitive streak (18 hours). × 17. *A,* Before appearance of primitive groove; *B,* with prominent groove.

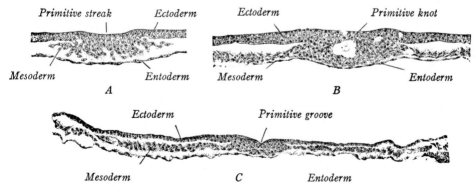

Fig. 523. Transverse sections of chick blastoderms, at stage of primitive streak. × 165. *A,* Through early primitive streak; *B, C,* through later primitive knot and groove.

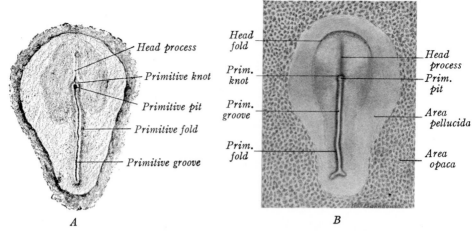

Fig. 524. Chick blastoderm and embryo, in surface view. *A,* Stage of head process (21 hours) (× 16). *B,* Stage of head fold (23 hours) (× 15).

Directly following the earliest appearance of the streak, a *primitive groove* courses lengthwise along its surface (*B*). In the future cephalic direction this gutter ends in the deeper *primitive pit.* At the extreme front end of the streak is a clubbed expansion, known as the *primitive knot* (of Hensen).

Microscopic sections, cut across the primitive streak, show it to be a thickening from which the mesoderm spreads laterad (Fig. 523). The first mesodermal cells are sparse, migratory elements (*A*), but they soon aggregate into distinct plates (*C*) extending both in a lateral and caudal direction. Later the mesoderm invades the region ahead of the streak. At the primitive knot all three germ layers fuse intimately (*B*), but in the caudal half of the streak the entoderm tends to be free (*C*). The primitive groove is the mechanical consequence of this rapid spread of mesoderm which produces a trough through cellular depletion. From the three germ layers, whose origins have been thus briefly reviewed, all the tissues and organs will develop.

Stage of Head Process and Fold. Embryos of about 21 hours' incubation show an axial strand of cells extending forward from the primitive knot (Fig. 524 *A*). This is the so-called *head process;* it is also termed the *notochordal plate* because it becomes the cylindrical notochord, destined to serve as the primitive axis about which the embryo differentiates. The head process results from the turning under of cells, originally located in the outer layer, which pass through the substance of the primitive knot and extend forward in the midline. Their movement is a constituent feature of a second, and later, phase of gastrulation. A longitudinal section shows the relation of head process to primitive knot (Fig. 525); a transverse section demonstrates it as a median, thicker mass, continuous laterally with mesoderm that has grown into this region (Fig. 526). Both sections illustrate the independence of the head process from the ectoderm above, and the temporary fusion which it makes with the entoderm beneath.

After the head process has become prominent, a curved fold begins to show in a position still more cephalad (Fig. 524 *B*). It is the *head fold*, which at first involves ectoderm and entoderm alone (Fig. 525). The further development of this important structure will establish the fore-gut internally and definitely delimit the upper body externally (Fig. 527). The blastoderm as a whole shows a clearer, central region, the *area pellucida*, and a more peripheral *area opaca* (Fig. 524 *B*). The latter is darker because it adheres to the yolk beneath (which it digests).

Stage of Early Neural Groove and Somites.   Even embryos of the previous stage exhibit a broad zone of thickening in the ectoderm overlying and bordering the head process. This region constitutes the *neural plate* (Fig. 526). In an embryo of 23 hours the plate begins to fold lengthwise to form a shallow, gutter-like trough, called the *neural groove* (Fig. 527 *A*). Within the next hour or two this groove becomes flanked by elevated, marginal ridges, the *neural folds* (*B*), which later will unite progressively until the brain and spinal cord are laid down as a continuous tube. The *notochord* is now a definite rod, seen through the transparent ectoderm at the bottom of the neural groove.

The wings of mesoderm, which grew from the sides of the primitive streak, have continued to spread peripherad to the margin of the blastoderm, but have not yet reached the region just in front of the *head fold* (Fig. 527). Alongside the notochord the mesoderm is thick, and in it are appearing pairs of vertical clefts; these separate the mesoderm into successive masses (the first incomplete cranially) which will be seen better in older stages. They are the *somites*, or *mesodermal segments*. The area opaca shows a mottling produced by the consolidation of *blood islands* within its mesoderm.

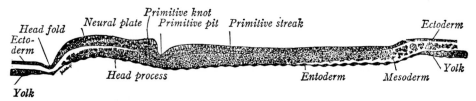

Fig. 525.   Midsagittal section of chick embryo, at stage of the head process and head fold (23 hours). × 100.

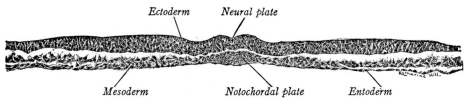

Fig. 526.   Transverse section through head process of 23-hour chick embryo. × 165.

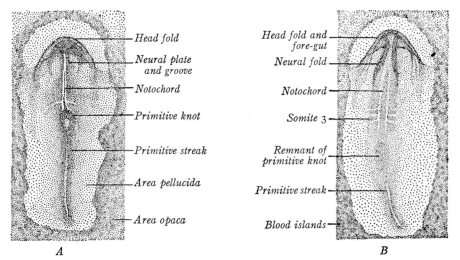

Fig. 527. Chick embryos, in dorsal view, at beginning of somite formation. *A*, Embryo with first intersomitic groove (24 hours) ($\times$ 25). *B*, Embryo of three somites (26 hours) ($\times$ 16).

B. EMBRYO OF FIVE SOMITES (TWENTY-EIGHT HOURS)

By the end of the first day of incubation an embryonic and an extra-embryonic region of the blastoderm are becoming more sharply defined (Fig. 528). The latter is the territory not destined to become a part of the embryo proper. It includes both the *area pellucida,* within which lies the developing embryo, and the more peripheral *area opaca.* In a zone of the opaque area bordering the area pellucida are mottled masses, the *blood islands,* already observed in younger stages but now fusing into a definite network. This mesh is best developed caudally; when complete, it will comprise a distinct subdivision of the area opaca to be called the *area vasculosa.* Mesoderm is still lacking in a clearer region in front of the head; to it the inappropriate name of *proamnion* has been given.

At this period the *head* is growing rapidly. It rises above the blastoderm and projects forward as a somewhat cylindrical part of the embryo which, at its cephalic end, is entirely free (Fig. 528). In accomplishing this result, the shallow head fold of earlier stages appears to have grown caudad and to have liberated the head by undercutting (Fig. 529). A more important factor, however, is a true forward overgrowth on the part of the head itself. Simultaneously with the extension of the head, the entodermal component of the original head fold is elongated as an internal tubular pocket; this is the primitive *fore-gut.* Cranially it is a blind sac; caudally it opens out onto the yolk through an arched aperture which resembles a tunnel-entrance and is termed the *intestinal portal.* In Fig. 528 the lateral limits of the darker fore-gut (labelled 'entoderm') and its relation to the arching intestinal portal are shown plainly. Figure 529 illustrates how the entoderm is reflected into the fore-gut at the level of the portal.

The *neural groove* is both broad and deep (Fig. 528). Midway along its extent the component neural folds have approached and are ready to fuse. Caudally the folds diverge and become increasingly indistinct.

The mesodermal *somites* are clearly defined and block-like. The *notochord* shows through the transparent ectoderm, and the *primitive streak* is shorter, both relatively and actually. Later, when the body form is further indicated by the formation of the tail fold, the primitive streak will disappear. It is a notable fact that the head not only arises soonest but also retains its early advantage over lower levels of the body. The progressive differentiation, leading to the establishment of body form, advances in a caudal direction; it first reaches the end of the trunk at a considerably later period than the stage under consideration (Fig. 558 *C*).

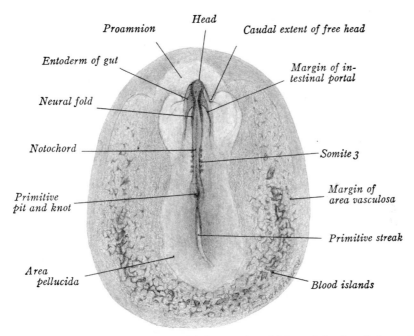

Fig. 528. Chick blastoderm and embryo with five somites (28 hours), in dorsal view. × 14.

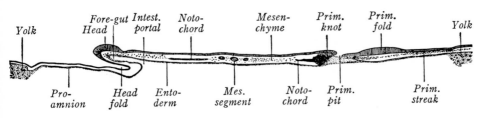

Fig. 529. Longitudinal section of chick embryo with five somites (after Patten). × 25.

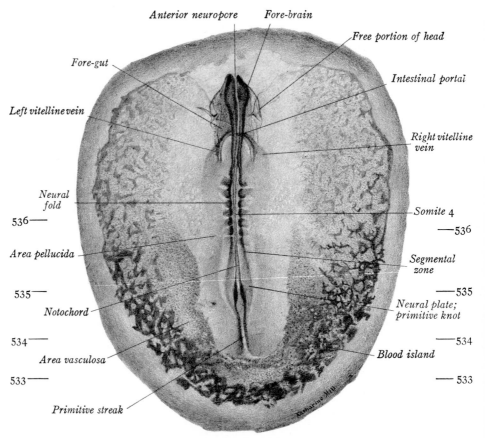

Fig. 530. Chick blastoderm and embryo with seven somites (31 hours), in dorsal view. × 20. Numbered lines indicate levels of sections, Figs. 533–536.

C. EMBRYO OF SEVEN SOMITES (THIRTY-ONE HOURS)

Although a total view of a chick embryo at this stage much resembles the one last described, it does show certain distinct advances (Figs. 530, 531). Nevertheless, the descriptions that follow will apply in all essentials to embryos having between five and ten primitive segments. Among the changes encountered, it is noteworthy that the vascular area of the blastoderm is better organized than previously and extends far cephalad. In front of the head there is a light area, not yet invaded by mesoderm, known by the poorly chosen name *proamnion*. The *primitive streak* is still prominent caudally, but it now measures only about one-fourth the length of the embryo. The *notochord* can be followed cephalad from the primitive knot until it is lost from sight beneath the neural tube.

Neural Tube. The lips of the neural folds have met throughout the cranial two-thirds of the embryo, but have not fused to any extent. The *neural tube,* formed thus by the closing of ectodermal folds, is open at each end; the closure of its cranial opening is characteristically delayed, and this leaves a temporary communication to the outside which has been designated the *anterior neuropore* (Fig. 530). In succeeding stages the more caudal regions of the present neural groove will be rolled progressively together and added to the tube already completed. At the head end the neural tube has begun to expand into the brain; only the *fore-brain* is at all prominent, and from it the *optic vesicles* are bulging laterally (Fig. 531).

Fore-gut. Except for an increase in size, the *fore-gut* is little changed (Fig. 530). Cranially, near its blind end, the floor of the gut is applied to the ectoderm of the under surface of the head; the two comprise the temporary *pharyngeal membrane* (cf. Fig. 545), which later ruptures to make the permanent opening between the mouth and pharynx. The fore-gut opens caudally, through the arched *anterior intestinal portal,* into a space between entoderm and yolk (Fig. 531). It will ultimately specialize into the several divisions of the alimentary canal that extend as far as the middle of the small intestine. The way in which the entoderm is folded up from the blastoderm and carried forward into the head is shown well in Figs. 529, 545.

Mesoderm and Cœlom. The tissue of the middle germ layer assumes two different forms. Throughout most of the head region it makes up a diffuse meshwork of cells that fills in the spaces between the various epithelial layers. This tissue is *mesenchyme* (Fig. 541). In the caudal part of the head and in the remainder of the body, the middle layer at this stage is organizing into more compact *mesoderm*. Nearest the midplane it is already divided by transverse furrows into seven block-like *somites,* four of which belong to the future head (Figs. 530, 532). Caudad, between the somites and the primitive streak, there is the undifferentiated mesoderm of the *segmental zone,* but new pairs of segments will develop progressively in this region. Lateral to each somite is a plate of unsegmented mesoderm, termed the *intermediate cell mass;* it is also called the *nephrotome* because it will

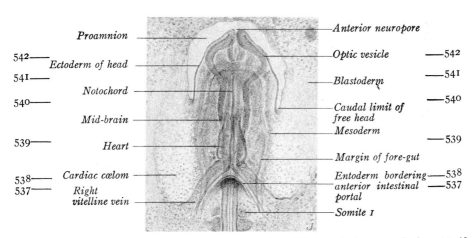

Proamnion

542——
Ectoderm of head
541——
Notochord
540——
Mid-brain
539——
Heart
538—— *Cardiac cœlom*
537—— *Right vitelline vein*

——*Anterior neuropore*
——*Optic vesicle*   ——542
——*Blastoderm*   ——541
——*Caudal limit of free head*   ——540
——*Mesoderm*
——539
——*Margin of fore-gut*
——*Entoderm bordering* ——538
anterior intestinal ——537
portal
——*Somite 1*

Fig. 531. Head of chick embryo with seven somites (31 hours), in ventral view. × 43. Numbered lines indicate levels of sections, Figs. 537–542.

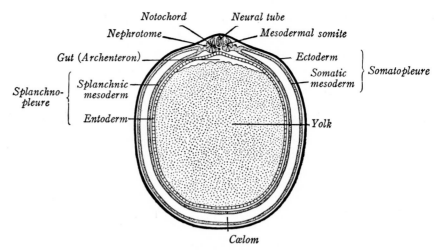

Fig. 532. Diagrammatic transverse section of early vertebrate embryo (Prentiss, after Minot).

play an important rôle in the development of the excretory system (Fig. 532). The nephrotome plate serves as a bridge between the somites and the unsegmented *lateral mesoderm*. When first proliferated, the lateral mesoderm of each side was a solid plate (Fig. 523). However, in stages like the present embryo these have split secondarily into two lamellæ, separated by a space (Figs. 532, 536). The dorsal layer comprises the *somatic mesoderm*, the ventral layer the *splanchnic mesoderm*. It is in the splanchnic layer that the blood vessels are mostly forming. The somatic mesoderm and the ectoderm are closely associated in development, and together are designated the *somatopleure;* it makes up the body wall. Similarly, the splanchnic mesoderm and entoderm are jointly termed the *splanchnopleure;* it is primarily concerned with the development of the gut and its derivatives. Both the somites and the unsegmented mesodermal layers contribute to the loose mesenchymal cells, which play such an important part in development.

The space between the two mesodermal layers first occurs in the form of isolated clefts, but these soon unite on each side into a continuous *body cavity,* or *cœlom.* The originally bilateral cœlomic chambers will later become confluent beneath the gut, thus forming a common cavity (Fig. 532). In the region of the heart the cœlom is already enlarged locally, anticipating its destiny as the pericardial cavity. Other, more caudal portions will subsequently become the pleural cavities of the thorax and the peritoneal cavity of the abdomen.

Heart and Blood Vessels. The *heart* is a straight double tube, lying in the midplane and ventral to the gut (Fig. 531). Traced caudad it is continuous with the converging *vitelline veins,* which enter the body from the area vasculosa by following along the margins of the intestinal portal; the two veins unite as they join the heart. From the cephalic end of the heart are given off short *ventral aortæ.* Dorsal to the gut course paired *dorsal* *aortæ,* just differentiating and not yet linked with the ventral aortæ. The heart is about to begin twitching (at nine somites).

TRANSVERSE SECTIONS

The first embryo to be studied in serial section is most easily understood if the student begins at the caudal end, where differentiation has entered least, and works toward the head. Important facts pertaining to the germ layers, as well as the principles underlying the development of the neural tube, gut, heart and head, are then made simple. The following illustrations and descriptions can be used to interpret sections of chick embryos between the stages of five and ten somites. The level of each section shown can be determined by applying a straight edge across the correspondingly numbered lines on Figs. 530, 531.

Sections through Area Vasculosa (Fig. 533).   The illustrations show, at medium magnification, a sample of the extra-embryonic territory *(area opaca),* peripheral to the area pellucida. In this region the entoderm is associated intimately with the coarsely granular *yolk.* The splanchnic mesoderm contains aggregations of cells known as *blood islands,* many of which are fusing into the network characteristic of the *area vasculosa* (Figs. 530, 533 *A).* The cellular thickenings of the blood islands undergo differentiation into two distinct cell types. Fluid-filled vacuoles first appear within the islands and then expand, so as to set free the innermost cells. These cells soon separate and float about as primitive *blood corpuscles,* while the general process of vacuolation flattens the peripheral cells into the *endothelium* of the developing blood vessel (Fig. 533 *B, C).* The endothelial-lined spaces both coalesce and bud out new vascular sprouts, and in this way the system of extra-embryonic vessels is extended. All blood vessels at first consist of an endothelial layer only.

Section through Primitive Streak (Fig. 534).   The *primitive streak* is a medial thickening of the blastoderm in which the primordial layers of *ectoderm, mesoderm* and *entoderm* all merge. A prominent *primitive groove* indents the streak in its midplane, and this groove is bounded on each side by a *primitive fold.*

Section through Primitive Knot (Fig. 535).   The enlarged cephalic end of the primitive streak is the *primitive knot.* Its common cellular mass separates at higher levels into the three typical germ layers; especially notable is the direct continuity into the notochord. The thickened and grooved *neural plate* of higher levels also extends downward to the

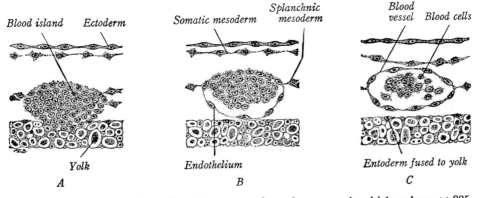

Fig. 533.   Transverse sections through area vasculosa of seven-somite chick embryo. × 225.

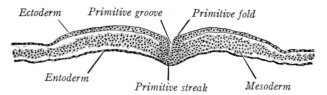

Fig. 534. Transverse section through primitive streak of seven-somite chick embryo. × 90.

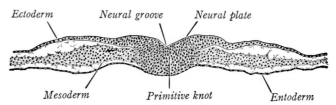

Fig. 535. Transverse section through primitive knot of seven-somite chick embryo. × 90.

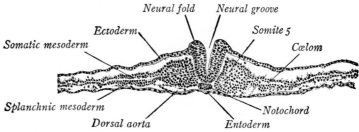

Fig. 536. Transverse section through fifth pair of somites of seven-somite chick embryo. × 90.

region of the knot and even over-rides it. This *neural groove* should not be confused with the smaller and fundamentally different primitive groove of lower levels.

Section through Fifth Somite (Fig. 536). The general level of the primitive segments is characterized by the greater specialization of the mesoderm, the elevation of high *neural folds,* and the presence of a *dorsal aorta* on each side between the mesodermal somites and the entoderm. The neural folds are thick, as is the adjoining ectoderm to a much less degree. The *notochord* is a sharply defined, oval mass of cells which will be observed just below the *neural groove;* it appears in all sections of the series, except those through the tip of the head and the primitive streak. The *somites* are somewhat triangular in outline; each is connected with the *lateral mesoderm* by a short plate, the *nephrotome.* The lateral mesoderm is partially divided by irregular, flattened spaces into two sheets; the dorsal of these is the *somatic layer,* the ventral is the *splanchnic layer.* When the spaces unite to form a definite *cœlom,* or primitive body cavity, the mesodermal lining of the cavity then specializes into a flat epithelium called *mesothelium.*

In the higher somites of the series the differentiation of mesoderm and cœlom is more advanced (*cf.* Fig. 554). Caudal to the seventh somite, in the region of the *segmental zone,* the mesoderm still forms solid plates (*cf.* Fig. 555).

Section Caudal to Intestinal Portal (Fig. 537). The section is characterized: (1) by the meeting of the neural folds preparatory to closing the *neural tube;* (2) by the arching of the entoderm which, a few sections nearer the head end, folds forward into the *fore-gut;* (3) by the presence of *vitelline veins* between the entoderm and folds of the splanchnic mesoderm; (4) by the wide separation of the somatic and splanchnic mesoderm and the consequent increase in the size of the cœlom. In this location the cœlom later surrounds the heart and is converted into the pericardial cavity. The neural tube at this level is transforming into the third brain vesicle, or *hind-brain*. The neural folds have not yet fused, and at their dorsal angles are located the *neural crests,* the forerunners of the spinal ganglia. Somites never develop as far craniad as this region; instead, diffuse masses of mesenchyme occupy comparable positions adjacent to the neural tube. On the left of the section an asterisk marks the point of junction between somatic and splanchnic mesoderm.

Section through Intestinal Portal (Fig. 538). This section passes through a vertical fold of entoderm at the exact point where the latter is reflected into the head as the *fore-gut* (*cf.* Figs. 529, 545). Since the entoderm is here cut on the flat, it appears as a continuous sheet of tissue; it is located between the vitelline veins and closes the fore-gut ventrally. On each side, lateral to the endothelial layer of the *vitelline veins,* the splanchnic mesoderm is thrown into a thick-walled, bulging fold. *Dorsal aortæ,* just beginning to differentiate, are represented inconstantly at these levels (Figs. 537–539).

A few sections craniad, the reflected region of the entodermal layer no longer shows, and the gut is quite separate from the general entoderm; this separation allows first the endothelial heart tubes to meet, and then the flanking folds of splanchnic mesoderm (*cf.* Fig. 539).

Section through Heart (Fig. 539). Passing craniad in the series to a level somewhat above the intestinal portal, one finds that the vitelline veins converge and open into the *heart.* The entoderm of the original head fold can now be identified as the crescentic *pharynx* of the fore-gut; it is separated by the heart, cœlom and splanchnic mesoderm from the entoderm of the general blastoderm. The heart has resulted from the union of two endothelial tubes, continuous with those constituting the vitelline veins in sections already

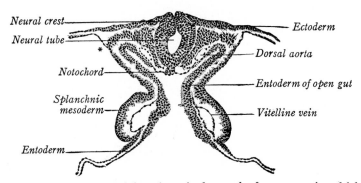

Fig. 537. Transverse section caudal to intestinal portal of seven-somite chick embryo. × 90.

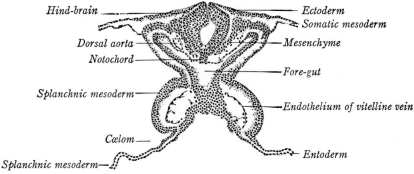

Fig. 538. Transverse section through intestinal portal of seven-somite chick embryo. × 90.

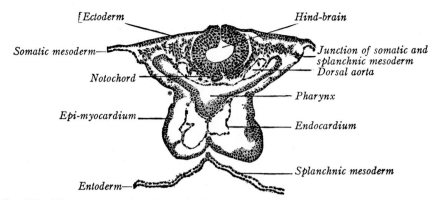

Fig. 539. Transverse section through the heart of seven-somite chick embryo. × 90.

studied. The median walls of these tubes fuse and disappear at a slightly older stage; this merger establishes a single tube, the *endocardium*. Thickened layers of splanchnic meso-derm, which in the preceding section invested each vitelline vein laterally, now form the external wall of the heart; such tissue will give rise to both the *myocardium* and the *epicardium*. In the ventral midplane the layers of cardiac mesoderm of each side have fused and separated from the splanchnic mesoderm of the blastoderm; as a consequence, the two halves of the future pericardial cavity are put in communication. Dorsal to the heart (at asterisk, on left) the paired layers of splanchnic mesoderm approach slightly; this presages the *dorsal mesocardium,* or mesentery of the heart, which is much better developed in slightly older embryos (Fig. 551). Traced still more dorsad, the splanchnic mesoderm extends to a point where the original cœlomic split separated it from the somatic layer; this junction is labeled on the right side of Fig. 539.

Origin of Heart and Embryonic Vessels. From the two sections last described, it is seen that the heart arises as a pair of endothelial tubes which lie in longitudinal folds of the splanchnic mesoderm. These tubes are continuous with paired veins entering from lower levels and paired arteries leaving for higher ones; hence, the vascular system is primi-tively a paired system throughout. Later the endothelial heart-tubes fuse, and the meso-dermal folds are also brought together. The heart then consists of a single endothelial tube within a thick-walled investment of mesoderm. The endothelial cells of the heart often appear to be splitting off from the entoderm (Fig. 538) but this is perhaps a deception, for elsewhere endothelium is mesodermal in origin. Primarily the blood vessels of the body are delicate endothelial channels which originate as clefts in the mesenchyme. Coalescence and budding produce a continuous plexus from which definite vessels are then selected (Fig. 329).

Section through Head Fold (Fig. 540). It will be remembered that an ectodermal *head fold* undercuts the head both from in front and at the sides (Figs. 529, 530). The portion of the embryo cephalad of this fold is necessarily free from the blastoderm. The present section is located just craniad of the heart, at a level into which the central portion of the head fold has not yet extended fully. The inspection of a few sections both in front of and behind this critical region will demonstrate how the embryonic and extra-embryonic terri-tories are related and how they become separate. The cœlom does not extend into the head, which contains much loose mesenchyme. Midway of the blastoderm is a region that still lacks mesoderm; it is the so-called *proamnion.* Ventral to the pharynx occurs the *ventral aorta,* here transitional between a single vessel which is continuous with the heart in a caudal direction and the separate vessels which pass craniad in the opposite direction. Above the pharynx are prominent *dorsal aortæ* and the dilated middle brain vesicle, or *mid-brain.*

Section through Pharyngeal Membrane (Fig. 541). This section shows the head free from the underlying blastoderm (*cf.* Fig. 545). Ectoderm surrounds the head completely. Near the midventral line it is bent dorsad, thickened somewhat and comes in contact with the thick entoderm of the *pharynx*. The area of contact between ectoderm and pharyngeal entoderm constitutes the *pharyngeal membrane*. Later this plate breaks through and establishes continuity between the mouth and pharynx. As in the previous section, the neural tube is closed and entirely separate from the superficial ectoderm. The level is at a narrow neck connecting the more dilated fore- and mid-brain. The *dorsal aortæ* are represented by small vessels just above the lateral wings of the pharynx, but the ventral aortæ have not extended this far forward. The region of the blastoderm directly beneath the head is the still broad *proamnion*. Far laterad may be seen the layers of the mesoderm, as well, and a little of the cœlom.

Section through Fore-brain and Optic Vesicles (Fig. 542). The neural tube is unclosed here, and the section is chiefly made up of a continuous double layer of ectoderm, infolded

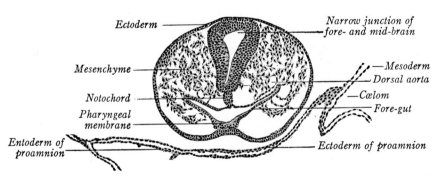

Fig. 540. Transverse section through head fold of seven-somite chick embryo. × 90.

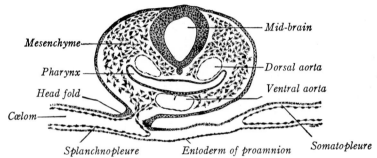

Fig. 541. Transverse section through pharyngeal membrane of seven-somite chick embryo. × 90.

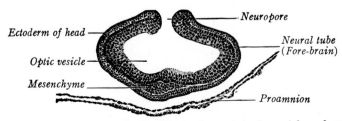

Fig. 542. Transverse section through the fore-brain and optic vesicles of seven-somite chick embryo. × 90.

gastrula-fashion. The opening from the first brain vesicle, or *fore-brain,* to the outside is the temporary *anterior neuropore.* The ectoderm on the surface of the head is continuous at the neuropore with the much thicker wall of the fore-brain. These two ectodermal layers are in contact with each other, except in the midventral region where the mesenchyme is beginning to penetrate and separate them. The paired, lateral expansions of the fore-brain are the *optic vesicles,* which eventually give rise to the retina of the eye.

D. EMBRYO OF SEVENTEEN SOMITES (FORTY-EIGHT HOURS)

The stage selected as a type for illustrating the significant advances since the seven-somite embryo is a chick of about 48 hours' incubation which possesses.17 primitive segments (Fig. 543 *B, C*). Since at this time the somites are developing rapidly, the descriptions that follow will apply satisfactorily to embryos between 40 hours (12 somites) and 49 hours (18 somites). Intermediate conditions between seven and 17 somites are illustrated by the embryo shown in Fig. 543 *A.*

General Features. The long axis of the embryo is still nearly straight, but specimens of full 17 somites should show a flexing of the head ventrad (Fig. 546) and a slight turning of the tip of the head on its left side (Fig. 543 *C*). In these respects the embryo in Fig. 543 *B* is slightly retarded. The area pellucida is dumb-bell shaped and is developing a vascular network. The extra-embryonic vessels of the area opaca are well differentiated and the vascular area ends in a bordering *terminal sinus.* Adjacent to the caudal end of the heart, the vascular networks of the blastoderm converge and become continuous with the stems of the vitelline veins. Connections have been established also between the dorsal aortæ and the vascular area at the level of the lowest somites, but as yet these have not organized as distinct vitelline arteries (Fig. 544). The tubular heart is enlarged and bent to the embryo's right; the head is more prominent than formerly and the three primary vesicles of the brain are easily distinguishable; seen through the brain walls is the notochord which extends in the midplane as far cephalad as the fore-brain; the proamniotic area is reduced to a small region in front of the head; the primitive streak is short and relatively inconspicuous.

Neural Tube and Sensory Primordia. The tardy sealing of the anterior neuropore has occurred and the neural tube is closed, save at its caudal end where the divergent neural folds bound the so-called *rhomboidal sinus* (Fig. 543 *B*). In the head the neural tube has differentiated into three brain vesicles, set off from one another by constrictions. The *fore-brain* (prosencephalon) is characterized by the outgrowing *optic vesicles.* The *mid-brain* (mesencephalon) is a simple dilatation. The elongate *hind-brain* (rhombencephalon) gradually passes into the *spinal cord;* it shows a number of secondary dilatations, the *neuromeres.*

The ectoderm is thickened into a *lens placode* where it overlies the lateral wall of each *optic vesicle* (Fig. 548). The optic vesicle flattens at this point and will soon invaginate to produce the optic cup. Dorsolaterally, in the hind-brain region, the ectoderm is also thickened into *auditory placodes* which are already indented as the *auditory pits* (Fig. 544). Each pit will become an otocyst, or otic vesicle, from which differentiates the sensory epithelium of the internal ear (membranous labyrinth).

Fore-gut. Caudally the entoderm is flat, but nearing the intestinal portal it elevates progressively. In Fig. 544 the greater part of the entoderm is omitted. The broad *fore-gut*, folded inward at the portal, shows indications of three lateral diverticula, the *pharyngeal pouches*, which will be much plainer in the next embryo studied. At its cephalic end the pharynx is closed ventrally by the double-layered *pharyngeal membrane;* the ectodermal depression external to it is the *stomodeum* (Figs. 545, 546).

Heart and Blood Vessels. The heart tube is flexed, yet does not vary in structure throughout its length. Nevertheless, certain portions can be identified as forerunners of later regions (Fig. 544). The caudal end of the

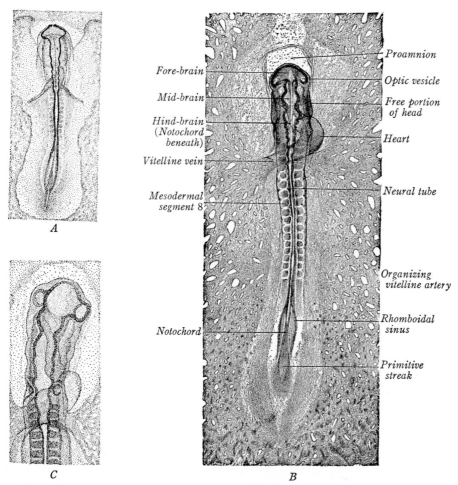

Fig. 543. Chick embryos of 38 to 48 hours, in dorsal view. *A,* At 11 somites (38 hours) (× 13). *B,* At 17 somites (47 hours) (× 20). *C,* At 18 somites, with head slightly rotated and bent (48 hours) (× 25).

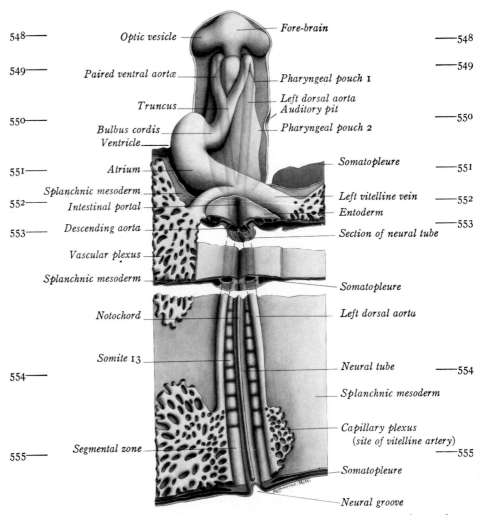

548 —
Optic vesicle —
Fore-brain
— 548

549 —
Paired ventral aortæ —
Pharyngeal pouch 1
— 549

Truncus —
Left dorsal aorta
Auditory pit

550 —
Bulbus cordis —
Ventricle —
Pharyngeal pouch 2
— 550

551 —
Atrium —
Somatopleure
— 551

Splanchnic mesoderm —
Intestinal portal —
Left vitelline vein
Entoderm
— 552

552 —
— 553

Descending aorta —
Section of neural tube

553 —
Vascular plexus —

Splanchnic mesoderm —
Somatopleure

Notochord —
Left dorsal aorta

Somite 13 —
Neural tube
— 554

554 —
Splanchnic mesoderm

Capillary plexus
(site of vitelline artery)
— 555

Segmental zone —
Somatopleure

555 —
Neural groove

Fig. 544. Ventral reconstruction of 17-somite chick embryo. × 38. Ectoderm of ventral surface of head, mesoderm of head and heart regions, and entoderm (except about intestinal portal) have been removed. Numbered lines indicate levels of Figs. 548–555.

tube, where the vitelline and cardinal veins open, is the *sinus venosus.* This dilates into the *atrium,* which bends ventrad and to the embryo's right. The tube then bends dorsad and to the midplane, as the *ventricle,* thereby completing a U-shaped bend. Continuing craniad, the ventricle narrows into the *bulbus* which, in turn, passes over into the *truncus* and *ventral aorta.* The latter vessel lies beneath the pharynx and divides into two trunks. Near the tip of the pharynx these paired ventral aortæ bend dorsad around the sides of the pharynx as the first pair of *aortic arches.* The arches then turn sharply caudad as the paired *dorsal aortæ.* In the region of the intestinal portal they not only lie close together but also have fused for a short distance to form a single vessel, the *descending aorta.* Below this level they separate again, and opposite the lowest somites connect by numerous capillaries with the general vascular network. It is in this connecting region that paired *vitelline arteries* presently will be differentiated. The heart already beats spasmodically and moves blood in the vessels. Blood drains from the vascular area by way of the *vitelline veins* to the heart; here it is pumped around the aortæ and flows through the organizing vitelline arteries back again to the area vasculosa. This circuit constitutes the vitelline circulation; through it the embryo receives nutriment from the yolk for its continued development.

Heretofore the body of the embryo has been without definite veins, but now two pairs of vessels are developing for the purpose of returning blood to the heart (Fig. 546). The *anterior cardinal veins* collect blood from the head region; the *posterior cardinals,* just appearing at this stage, will perform a similar function for the lower body. The two vessels on each side unite into a *common cardinal vein* (duct of Cuvier) which enters the sinus venosus.

Mesoderm and Cœlom. The production of early mesodermal somites and the addition of new ones by a progressive furrowing of the *segmental zone* have been observed in previous stages. The *somites,* thus formed, are block-like with rounded corners when viewed from above; in transverse sections they appear triangular (Fig. 547). In higher vertebrates these primitive segments contain indications of a space that represents a cavity continuous in lower vertebrates with the general cœlom. In the chick this

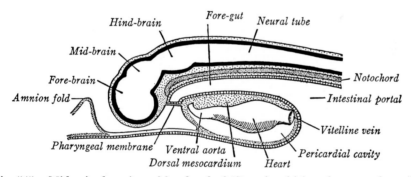

Fig. 545. Midsagittal section of head-end of 17-somite chick embryo. $\times$ about 50.

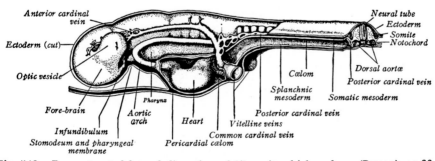

Fig. 546. Reconstructed lateral dissection of 17-somite chick embryo (Patten). $\times$ 33.

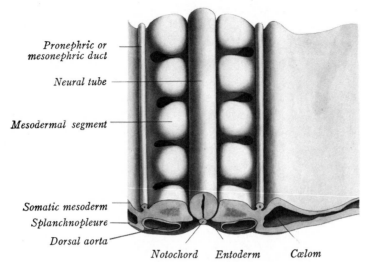

Pronephric or mesonephric duct

Neural tube

Mesodermal segment

Somatic mesoderm
Splanchnopleure
Dorsal aorta

Notochord   *Entoderm*   *Cœlom*

Fig. 547. Reconstruction through lower somites of two-day chick embryo. Ectoderm removed from dorsal surface.

rudiment is a minute, central cleft which is mostly filled with a cellular core; the other cells of the somite form a thick, radially-arranged shell about it (Fig. 554). The ventral wall and a portion of the medial wall of each somite break down into a mass of mesenchyme termed the *sclerotome;* these later surround the notochord and neural tube where they transform into the axial skeleton. The remaining portions of the somite constitute the *dermo-myotome* (Fig. 551). The cells of the dorsomedial wall of this plate, the *myotome,* eventually give rise to the skeletal musculature of the body. The lateral plate is the *dermatome* which contributes to the connective tissue of the integument.

The cellular plate connecting a primitive segment with the lateral mesodermal layers is the *intermediate cell mass,* or *nephrotome* (Fig. 532). In the chick the nephrotomes of the fifth to sixteenth somites give rise to segmental pairs of budlike sprouts which extend dorsad (Fig. 554). These are the *pronephric tubules* of a rudimentary type of kidney. Although functionless as excretory tubules their ends turn caudad and link into a tube, known as the *pronephric duct,* which grows to the cloaca (Fig. 547). More caudal nephrotomes will soon differentiate a temporary functional kidney, the *mesonephros;* its tubules will open into the pronephric duct which thereafter is called the *mesonephric duct.* Later still, the permanent kidney develops partly from the pronephric duct and partly from nephrotome tissue of a lower level. Accordingly, the intermediate cell masses may be regarded as the source of the urogenital glands and ducts—all mesodermal in origin.

In the previous embryo of seven somites the lateral mesoderm was observed to split into two layers, the dorsal somatic and the ventral splanchnic mesoderm. These layers persist as components of the *somatopleure* and *splanchnopleure;* the somatic mesoderm will give rise to the parietal walls of the pericardial, pleural and peritoneal cavities, while the splanchnic layer forms the epi-myocardium, the visceral pleura, and the mesenteries and mesodermal layers of the gut.

The *cœlom* has not progressed much beyond its condition in the previous stage (Fig. 546), although a beginning has been made toward the isolation of a portion of it within the body of the embryo (Fig. 552).

TRANSVERSE SECTIONS

In studying serial sections of an embryo it is not sufficient merely to identify the structures seen. The student should determine also the exact level of each significant section with respect to drawings or models of the total embryo; this has been done along the margins of Fig. 544 for the particular series that follows. It is also important to trace the several organs and parts faithfully from section to section in a series. The novice is then ready to reconstruct mentally the complete picture of a part and to interpret its origin and relations.

The following sections are drawn as if viewed from the cranial surface; hence, the right side of the embryo is at the reader's left. These illustrations and descriptions may be used for guidance in the study of chick embryos between 40 hours (12 somites) and 49 hours (18 somites).

Section through Fore-brain and Optic Vesicles (Fig. 548). The first sections encountered in the series are shavings through the tip of the free head. The brain cavity straightway enlarges, and about midway along the fore-brain the present level is reached. Here the *optic stalks* connect the *optic vesicles* with lateral portions of the *fore-brain*. Dorsally the section passes through the *mid-brain*, due to the somewhat ventral flexion of the head (*cf.* Fig. 545). The *lens placodes* are thickenings of the surface ectoderm over the optic vesicles. Note that there is now a considerable amount of mesenchyme filling in between the ectoderm and the neural tube; the small spaces seen are terminal branches of the

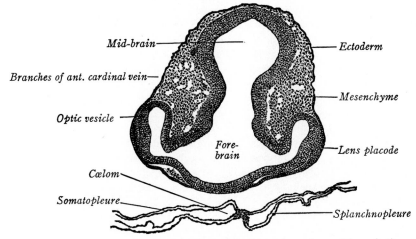

Fig. 548. Transverse section through fore-brain and optic vesicles of 17-somite chick embryo. × 75.

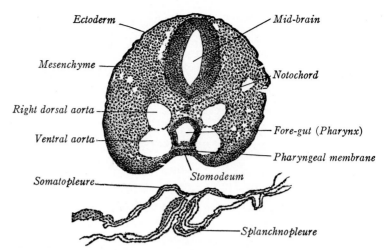

Fig. 549. Transverse section through mid-brain and pharyngeal membrane of 17-somite chick embryo. × 75.

anterior cardinal veins. Layers of extra-embryonic mesoderm extend to the midplane in the underlying blastoderm.

Section through Mid-brain and Pharyngeal Membrane (Fig. 549). At this level the fore-brain has been passed and the *mid-brain* alone is included. In the midventral line the thickened ectoderm bends up into contact with the entoderm of the tubular *pharynx* of the fore-gut. The resulting plate is the *pharyngeal membrane* and the ectodermal pit leading to it is the *stomodeum.* The membrane is transient because at this point the *oral opening* will break through and make the stomodeum continuous with the rest of the mouth cavity (which is entodermal). Lateral to the pharynx two pairs of large vessels are seen. The ventral pair is the *ventral aortæ,* while *dorsal aortæ* make up the dorsal pair. Two sections cephalad in the series, the two sets become continuous around the first *aortic arches.* The caudal end of the *mesencephalon* is the portion of the neural tube showing; its thick walls surround an oval cavity. A large amount of unspecialized mesenchyme is present throughout the section. The structure of the blastoderm is complicated laterally by the presence of collapsed blood vessels in the splanchnopleure.

Section through Hind-brain, Auditory Pits and Heart (Fig. 550). Between the plane of the last section and this one, the head fold ceases to separate the body from the blastoderm. Nevertheless, lateral prolongations of the head fold continue to indent the somatopleure here and for some distance caudad.

The section selected is characterized by: (1) the *auditory placodes,* already deepening into pits which represent the beginnings of the internal ears; (2) the large *hind-brain,* somewhat thin and flattened dorsally; (3) the *neural crests,* destined to give rise to ganglia and other derivatives; (4) the broad *pharynx,* cut through its second pair of pharyngeal pouches, above which on each side lie the *dorsal aortæ;* (5) definite *anterior cardinal veins,* ventrolateral to the brain, which return blood from the head region; (6) the presence of two portions of the *heart,* cut near its cephalic end. Because of its sinuous shape, the heart is sectioned twice. The smaller part is the single *bulbus,* which now replaces the paired ventral aortæ of higher levels. The large *ventricle* lies on the right side of the embryo; a few sections caudad in the series it is continuous with the bulbus (*cf.* Fig. 544). Between the somatic and splanchnic mesoderm is the large potential *pericardial cavity* surrounding the heart.

Section through Atrial End of the Heart (Fig. 551). The section is toward the caudal end of the *pharynx,* but the lower end of the *hind-brain* is still included. The *dorsal aortæ* are separated merely by a thin septum which has ruptured at this level. The *anterior cardinal veins* are cut where they bend ventrad to connect with the common cardinal veins. The outer (epi-myocardial) wall of the *atrium* diverges dorsally, right and left, and continues into the general splanchnic mesoderm of the embryo. Beneath the pharynx these approximated folds constitute the *dorsal mesocardium,* which serves as a transient mesentery to the heart. On the right side of the section there is fusion between the *epi-myocardium* of the heart and the somatic mesoderm; this is separating off an embryonic portion of the cœlom. Mesodermal *somites* were not observed at higher levels, but now they appear alongside the hind-brain. The ventromedial part of the segment is breaking down into the *sclerotome;* the dorsomedial wall represents the *myotome,* while the lateral plate is the *dermatome.*

Section through Intestinal Portal and Venous Stems (Fig. 552). Both heart and brain have been passed but the *spinal cord* now becomes a prominent feature. The dorsal aortæ

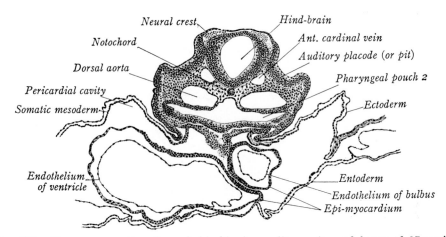

Fig. 550. Transverse section through hind-brain, auditory pits and heart of 17-somite chick embryo. × 75.

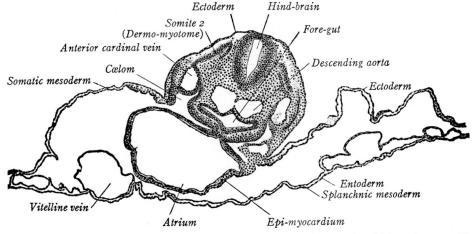

Fig. 551. Transverse section through atrial end of heart of 17-somite chick embryo. × 75.

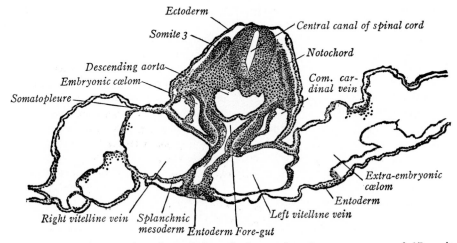

Fig. 552. Transverse section through intestinal portal and venous stems of 17-somite chick embryo. × 90.

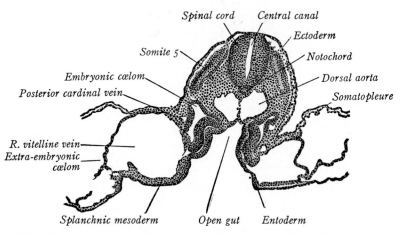

Fig. 553. Transverse section through open gut of 17-somite chick embryo. × 90.

lose their separating median septum and combine into a single vessel, henceforth to be known as the *descending aorta*. The section cuts through the entoderm at the point where it is folded dorsad and craniad into the head as the *fore-gut* (*cf.* Fig. 545). Two sections caudad occurs the opening (*intestinal portal*) where the fore-gut communicates with the progressively flattened, open gut spreading out between the entoderm and the yolk. On each side of the fore-gut is a large *vitelline vein*, sectioned obliquely as it diverges from the heart. The splanchnic mesoderm overlying these veins is pressed by them against the somatic mesoderm, and the cavity of the cœlom is thus interrupted on each side; the portions lying within the embryo's body are the beginnings of an *embryonic cœlom*. The *common cardinal veins* are cut near the level where they join the sinus venosus.

Section through Open Gut (Fig. 553). In general this section resembles the preceding, save that the gut is clearly open and without a ventral wall. Its lining is directly continuous with the splanchnopleure, and in this region one speaks of the *mid-gut*. The *vitelline veins* are still large and may be traced laterad into the vascular plexus of the blastoderm. Lateral to the enclosed cœlom, on each side, are rounded spaces which represent the *pos-*

terior cardinal veins, just differentiating. The *dorsal aortæ* are about to become separate once more.

Section through Fourteenth Pair of Primitive Segments (Fig. 554). The body of the embryo is now flattened on the surface of the yolk, and the section is characterized by its relative simplicity. Here the *dorsal aortæ* are again separate. Other prominent features are the *spinal cord, notochord, somites, nephrotomes,* and layers of *somatic* and *splanchnic mesoderm.* These somites are much less specialized than the older ones at higher levels. Arising from the nephrotomes are sprout-like *pronephric tubules.* The tips of these hollow out and unite to produce the *pronephric duct,* which is the primary excretory duct of the embryo.

Section through Rhomboidal Sinus and Segmental Zone (Fig. 555). The section passes through the *segmental zone,* which is a region of unsegmented mesoderm destined to be cut up into additional somites. Large blood vessels (of the area vasculosa) occur in the splanchnic mesoderm, next to the entoderm. The dorsal aortæ of higher levels lose their identity in this plexus, which precedes the appearance of a definite *vitelline artery.* The lateral mesoderm is separated by narrow, cœlomic clefts. The open neural groove is called the *rhomboidal sinus.* The ectoderm is notable for the columnar form of its cells. At the point where the general ectoderm of each side joins the neural fold there is prospective neural-crest tissue, from which the spinal ganglia will differentiate.

Section through Primitive Knot (Fig. 556). The three germ layers merge at the *primitive knot* (of Hensen) into a common mass of unspecialized tissue. This knob of formative tissue is also to be known as the *end bud,* or *tail bud,* since it gives rise to the lower body. The lateral mesoderm is split into somatic and splanchnic layers; the splanchnic mesoderm contains numerous small blood vessels of the vascular network.

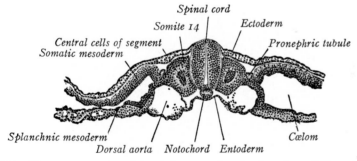

Fig. 554. Transverse section through fourteenth pair of somites of 17-somite chick embryo. × 90.

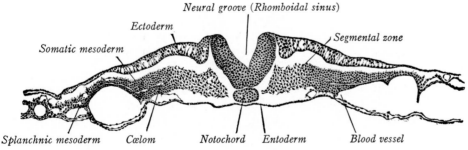

Fig. 555. Transverse section through rhomboidal sinus and segmental zone of 17-somite chick embryo. × 90.

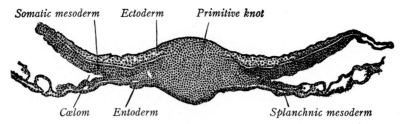

Fig. 556. Transverse section through primitive knot of 17-somite chick embryo. × 90.

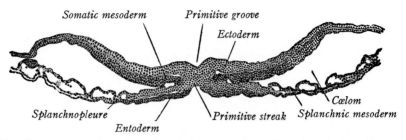

Fig. 557. Transverse section through primitive streak of 17-somite chick embryo. × 90.

Section through Primitive Streak (Fig. 557). In the mid-dorsal line is the *primitive groove.* All of the germ layers can be seen in direct continuity with the undifferentiated tissue of the *primitive streak* which underlies the groove. Laterally blood vessels are present in the splanchnic mesoderm, as in the preceding sections.

E. EMBRYO OF TWENTY-SEVEN SOMITES (TWO DAYS)

Although a chick embryo with 27 somites (53 hours) is chosen as the norm (Fig. 558 *B*), the descriptions that follow are applicable to stages between 51 hours (23 somites) and 58 hours (31 somites). An earlier and a later stage than 50 hours are shown in Fig. 558 *A* and *C*, respectively.

General Features. During the latter half of the second day a remarkable change occurs in the appearance of the embryo and in its position with respect to the blastoderm (Fig. 558 *A, B*). The bending of the head, already begun in the stage last studied, has continued until the fore- and hindbrains are nearly parallel. This marked *cephalic flexure,* which makes the embryo Γ-shaped, occurs at the region of the mid-brain (Fig. 558 *B*). It is manifest that as long as the embryo retained its original prone position with respect to the yolk it would be difficult for the head to bend greatly ventrad. In order to facilitate such flexion, and to allow it to proceed to completion, the upper body has twisted about its long axis until the left side lies flat upon the yolk. In a dorsal view, therefore, one sees the right side of the head but the dorsal side of the lower body. The actual region of torsion, now half way down the trunk, will advance caudad progressively until the whole embryo lies on its left side. Additional curvatures ultimately bend it into the shape of the letter **C**. One of these flexures is already appearing opposite the lower end of the heart, at the junction of head and trunk; for this reason it is named the *cervical flexure* (Fig. 558 *B, C*).

Most of the body is rather sharply delimited from the blastoderm: the head is free; much of the midbody is bounded by deep *lateral folds*. Caudally the *tail bud* indicates the material for the future hind end of the body; it is bordered by a *tail fold*. The further overgrowth of the embryo, beyond the limits of the head-, lateral- and tail folds, will appear to constrict the embryo from extra-embryonic blastoderm.

The head is now covered by folded somatopleure, the so-called head fold of the *amnion;* it envelops the upper half of the body like a veil (Fig. 558 *B*). The primitive *brain vesicles* are plainly seen and the primordia of the *eye* and *ear* are prominent. The *heart* bends in the form of a letter **S**, and distinct *vitelline arteries* and *vitelline veins* connect with a profuse plexus of extra-embryonic vessels. Three ectodermal furrows form *branchial grooves* on the sides of the neck. Additional *somites* extend far down the embryo; they have differentiated from tissue that was in the territory of the *segmental zone* of the previous stage.

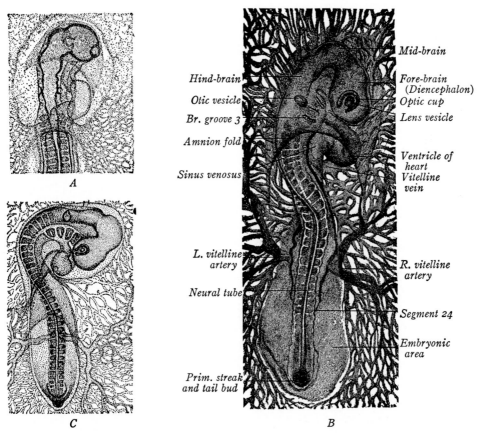

Fig. 558. Chick embryos of 49 to 60 hours. Upper portion is in lateral view, lower in dorsal view. *A*, At 20 somites (49 hours) (× 19). *B*, At 27 somites (53 hours) (× 14). **C**, At 31 somites (60 hours) (× 8).

Central Nervous System and Sense Organs. The brain region of the neural tube is separated by constrictions into five vesicles, but these subdivisions are not so distinct as they will be somewhat later (Fig. 558 *B*). The first subdivision of the primitive fore-brain is the *telencephalon;* the rest constitutes the *diencephalon.* The *mesencephalon* remains undivided, but is bent at its middle by the *cephalic flexure.* The hind-brain shows indistinctly two regions of specialization. A short segment with a thick roof, adjoining the mid-brain, is the *metencephalon;* the thin-roofed remainder is the *myelencephalon.* The *spinal cord* is now closed to its extreme end, and consequently the rhomboidal sinus no longer exists.

Each lens placode has infolded and become a *lens vesicle.* At the same time the outer wall of the optic vesicle also invaginates, thereby making a double-walled structure named the *optic cup* (Fig. 558 *B*). The latter is not a complete cup, for on one side a segment of the wall is missing; this *optic* or *chorioid fissure* gives the cup a horseshoe-shaped outline in surface view (*cf.* Fig. 496 *B*). The auditory placode of earlier stages has become a sac, the *otocyst* or *otic vesicle,* which still remains connected with the body ectoderm.

Digestive System. The entodermal canal shows two regional divisions. Of these, the *fore-gut* is best differentiated; it will be described more fully in the next paragraph. In Fig. 559 most of the entoderm has been removed, so that the open *mid-gut* scarcely shows; it extends from the *anterior intestinal portal* to the tail bud and, lacking a ventral wall, lies directly upon the yolk. At the extreme caudal end of the body a shallow pocket opens ventrally, just caudal to the main mass of the tail bud. This is the first indication of the *allantois,* which is an extra-embryonic sac soon to be used for respiration and the deposit of urinary wastes. The *hind-gut* does not fold off until later in the third day.

The *pharyngeal membrane* lies at the bottom of a deep pit, the *stomodeum,* formed by depressed ectoderm. A median ectodermal sac, just in front of the pharyngeal membrane, is *Rathke's pouch.* It is seen as a Λ-shaped indentation next to the tip of the pharynx in Fig. 559. It extends along the ventral surface of the diencephalon where it will develop into the epithelial portion of the hypophysis. The entodermal *pharynx* bears three pairs of lateral outpocketings known as *pharyngeal pouches* (Fig. 559). They occur opposite the three external *branchial grooves,* and here ectoderm of the groove and entoderm of the pouch are in contact, forming *closing plates* (Fig. 558). At about this age the first pair of plates ruptures, thereby making a free opening, or *branchial cleft,* into the pharynx. These transitory apertures correspond to the gill clefts in lower, aquatic vertebrates. Between the successive pouches lie solid, bar-like portions of the body wall, the *branchial arches;* in animals with aquatic respiration the arches bear gills, and even in higher embryos, like the chick, an artery courses through each. At the level of the second pair of pouches, a broadly open pocket grows away from the median floor of the pharynx; it is the *thyroid gland* (Fig. 566). Beyond the pharynx the fore-gut narrows, but esophagus, stomach and duodenum are not yet clearly distinguishable. Toward the anterior intestinal portal the fore-gut is flattened from side to side, and just before it opens into the mid-gut there is budded off a bilobed diverticulum, the *liver*, which lies between the vitelline veins (Fig. 559).

Cardio-vascular System. The disappearance of the dorsal mesocardium leaves the large, tubular heart attached solely by its two ends. Since the heart tube is growing faster than the surrounding body, it of necessity bends; when viewed from the ventral side it comes to look like the letter **S** (Fig. 559). Four regions can be distinguished: (1) the *sinus venosus,* into which the veins open; (2) a dilated, dorsal chamber, the *atrium;* (3) a tubular, ventral portion, bent in the form of a **U**, of which the left limb is the *ventricle,* the right limb (4) the *bulbus cordis.* From the bulbus is given off the *truncus* and *ventral aortæ.* There are now three pairs of *aortic*

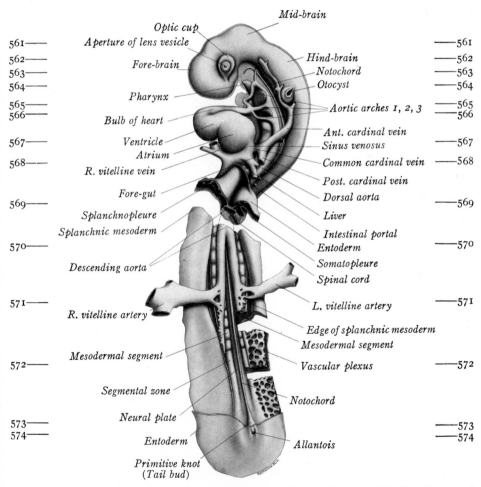

Fig. 559. Ventral reconstruction of 27-somite chick embryo. × 18. Ectoderm and mesoderm of upper body and entoderm of lower body have been mostly removed. Numbered lines indicate levels of Figs. 561–574.

arches, which open into the paired *dorsal aortæ.* The first aortic arch runs axially through the first branchial arch, located just cranial to the first pharyngeal pouch; it is the same vessel that connected ventral and dorsal aortæ in the 48-hour embryo. The second and third aortic arches course in the second and third branchial arches, which stand similarly craniad of the second and third pharyngeal pouches. They are developed by the enlargement of channels in primitive capillary networks between the ventral and dorsal aortæ. At the level of the sinus venosus the paired dorsal trunks fuse to form the single *descending aorta,* which extends as far back as the fifteenth pair of primitive segments. At this point the aortæ again separate; opposite the twenty-second somites each connects with the trunk of a *vitelline artery,* which conveys blood to the vascular area. Caudal to the vitelline arteries the aortæ decrease rapidly in size and soon end.

As in the previous stage, the blood is returned from the vascular area to the heart by the *vitelline veins,* now two large trunks (Fig. 559). In the body of the embryo the *anterior cardinal veins* course ventrolateral to the brain, and already are of large size. The smaller *posterior cardinal veins* are developing caudal to the atrium. They lie in the mesenchyme of the somatopleure, lateral in position. Opposite the sinus venosus the anterior and posterior cardinal veins of each side unite to form the *common cardinal veins* (ducts of Cuvier), which open into the dorsal wall of the sinus venosus. The set of primitive veins is thus paired like the arteries, and like them develops by the enlargement of channels in a network of capillaries.

Differentiation of Mesoderm and Cœlom. The formation of new somites and the progressive differentiation of older ones into sclerotome, myotome and dermatome continue as described for the preceding embryo (p. 570). The nephrotome region shows the beginning of additional features. The *pronephric duct* has continued beyond its original site of formation and extends tailward as a blindly growing cord. A second set of kidney tubules is now starting to differentiate which will ultimately extend between the fourteenth and thirtieth somites. They arise from the intermediate cell masses caudal to the pronephric group. At first taking the shape of vesicles, they later will become *mesonephric tubules* and join the pronephric (hereafter mesonephric) duct. The *mesonephros* constitutes a functional kidney of the embryo, but not the definitive one of the later embryo and adult.

The splanchnopleure of this stage is chiefly involved in gut formation. Below the level of the free head the somatopleure is continuous with the extra-embryonic blastoderm, but it is already being indented deeply by *lateral body folds* whose union will progressively close the ventral body wall (Fig. 569). The establishment of a complete body wall is the chief factor in separating embryonic from extra-embryonic cœlom. Up to the present time this closure has not occurred. The only stretch of embryonic cœlom is due to fusions between somatopleure and splanchnopleure at the caudal level of the heart. Since the lungs will bud out here, these paired cœlomic canals are potentially pleural cavities (Fig. 567).

Amnion and Chorion.   At the end of the second day two extra-embryonic protective membranes have become prominent. They are the *amnion* which will form a membranous, fluid-filled sac about the embryo itself, and the *chorion* which eventually encloses both the embryo and all extra-embryonic structures. Developmentally the two membranes are nothing more than the outer and inner portions of a circular fold, thrown up around the embryo by the extra-embryonic somatopleure. The two membranes arise simultaneously by a single process of folding (Fig. 560). The first indication of them is a fold in front of the embryo, followed later by lateral and caudal ones (*A*). These hood-like, arching folds close in from all sides (*B, E*) until they meet and fuse over the embryo (*C, D, F*). The inner somatopleuric layer is the amnion; the outer somatopleuric layer constitutes the chorion, of little importance to the chick. It should be noted that, while the folding brings the mesodermal components of these membranes facing each other, the two are separated by the extra-embryonic cœlom.

The *head fold* of the amnion had begun in the chick of the previous stage. At the end of the second day it is continuous along a crescentic margin with the *lateral folds* and envelops the upper half of the body (Fig. 558). As yet the *tail fold* of the amnion has scarcely started.

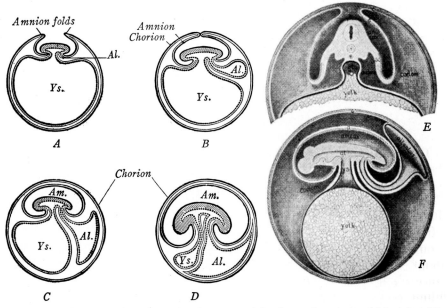

Fig. 560.   Diagrams illustrating development of fetal membranes in chick (after Mc-Murrich and Kingsley). In *A–D* ectoderm, mesoderm and entoderm are represented by heavy, light and dotted lines respectively; in *E, F* ectoderm is hatched, mesoderm gray and entoderm black.

a, Amnion; *al,* allantois; *am,* amniotic cavity; *c,* chorion; *gt,* gut; *so,* somatopleure; *ys,* yolk stalk and sac.

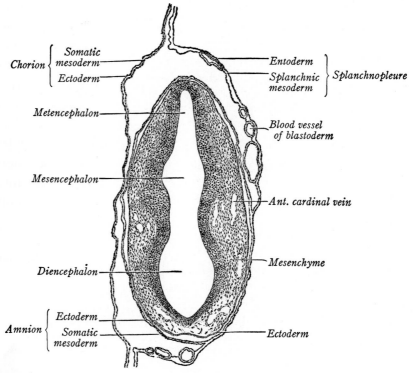

Fig. 561. Transverse section through flexed brain of 27-somite chick embryo. × 50.

TRANSVERSE SECTIONS

The following series of transverse sections from a 27 somite chick embryo shows the fundamentally important structures; the illustrations and descriptions are equally applicable to the study of embryos between 51 hours (23 somites) and 58 hours (31 somites). The sections used are drawn from the cranial surface; hence, the right side of the embryo is at the reader's left. The precise level of each significant section in the student's slides should be determined with respect to Figs. 558 *B* and 559; for the sections about to be described this has been indicated along the margins of Fig. 559. Since the head is bending rapidly during the last hours of the second day, minor variations in the appearance of different series of sections through the head are unavoidable; this, however, is chiefly a question of which particular structures happen to appear together in the fore-brain and hind-brain portions of a section.

Section through Flexed Brain (Fig. 561).   Because of flexion of the head, the first sections encountered pass through the *mesencephalon*. A little farther down the series the *metencephalon* of the hind-brain and the *diencephalon* of the fore-brain are included as well; constrictions mark the boundaries between these divisions, as in Fig. 561. The blood vessels seen in the mesenchyme are branches of the *anterior cardinal veins*. The splanchnopleure is characterized in this and subsequent sections by the presence of blood vessels in its mesodermal layer; these obvious structures make easy the identification of the yolk-side of the blastoderm.

The entire head is enveloped by the *amnion;* by contrast, the *chorion* surrounds both embryo and yolk, and consequently overlies the right side of the head only (*i.e.,* the free side of the head, away from the yolk). In each of these membranes a layer of ectoderm and a separate layer of mesoderm can be identified; the mesodermal components of the amnion and chorion face each other across the extra-embryonic cœlom, but because of collapse in the process of preparation they may be found partly in contact.

Section through Optic Cups and First Aortic Arches (Fig. 562).   Continuing down the series, the mid-brain is passed and the brain then becomes cut twice in each section; the *myelencephalon* is always recognized by its thin roof and its close association with the notochord. Observe that in these sections through the bent head, progress is caudad down the hind-brain half of the section, but rostrad toward the tip of the fore-brain.

Since the section illustrated passes above the level of the optic stalks, the *optic cups* appear unconnected with the fore-brain. The overlying ectoderm has thickened and in-vaginated to form the *lens vesicles.* The thicker wall of the optic cup, next the lens, will give rise to the nervous layer of the retina; the thinner outer wall becomes the pigmented epithelium. Ventrally in the section is the *telencephalon* and *diencephalon.* Dorsally oc-curs the *myelencephalon* of the hind-brain, with its roof a thin *ependymal layer.* In the mesenchyme, between the brain vesicles, are longitudinal sections of the first pair of *aortic arches.* Lateral to the hind-brain are portions of the *anterior cardinal veins,* which convey blood from the head to the heart.

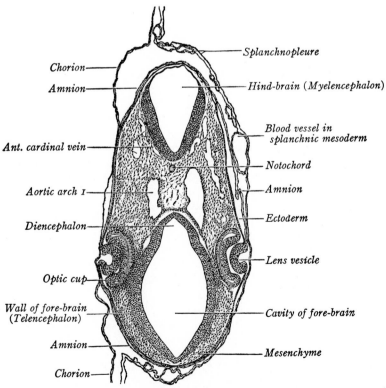

Fig. 562.   Transverse section through optic cups and first aortic arches of 27-somite chick embryo. × 50.

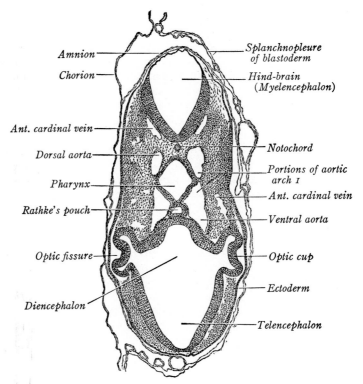

Fig. 563.   Transverse section through Rathke's pouch and optic stalks of 27-somite chick embryo. × 50.

Section through Rathke's Pouch and Optic Stalks (Fig. 563).   The section passes just caudal to the lens, but it includes the caudal margins of the *optic cups*. The shallow concavity on the margin is the *optic fissure*. Each cup is connected with the wall of the fore-brain (specifically the *diencephalon*) by an eccentrically attached *optic stalk;* this stalk will furnish the path through which optic nerve fibers grow from retina to brain. Use Fig. 496 *B* to explain all appearances encountered in sections of the lens, cup and stalk. Both the *ventral* and *dorsal aortæ* are seen. Parts of the first pair of aortic arches, cut along their caudal borders, connect with them. Between the ventral wall of the fore-brain and the *pharynx* is a transverse section through an invagination of the stomodeal ectoderm, *Rathke's pouch;* it will become the epithelial lobe of the hypophysis. The *anterior cardinal veins* have assumed their characteristic position ventrolateral to the mid-brain.

Section through Stomodeum and Pharyngeal Membrane (Fig. 564).   The most important feature of this level is the head, cut in two separate sections. One part includes the *hind-brain* and *pharynx;* the other, the *fore-brain* and end of the bent head. The space between these two parts is the region of the *stomodeum.* Here the *pharyngeal membrane,* composed of fused ectoderm and entoderm, still separates stomodeum from pharynx. Here, also, the mouth of *Rathke's pouch* opens. *Dorsal* and *ventral aortæ* show their characteristic positions with respect to the pharynx. This is about the lowest section to include shavings off the *optic stalks.* Tracing down the series a few sections shows how the fore-brain and tip of the head come to an end.

Section through Otocysts, Bulbus and Second Aortic Arches (Fig. 565).   The bent part of the head of the ſ-shaped embryo has been passed. The *otocysts* are sectioned caudal to their apertures, and so appear as closed sacs alongside the hind-brain. Ventral to the *pharynx,* the *bulbus cordis* is sectioned obliquely. Continuous with the bulbus is the unpaired *truncus,* which gives off the second pair of *aortic arches;* these pass around the sides of the pharynx and connect with the *dorsal aortæ.* Surrounding the bulbus cordis is the *cœlom,* which is not yet enclosed by body wall and for this reason is not yet specifically a pericardial cavity. The *amnion* attaches to the body on each side; on the embryo's right it is folded upon itself. This is because the primitive amniotic folds fuse directly over the original dorsal line, regardless of the turning of the embryo; consequently, on the right there is 'slack.'

Section through Pharyngeal Pouches, Thyroid Gland and Ventricle (Fig. 566). Since the section figured represents a level between the second and third aortic arches, the *dorsal aortæ* and heart are unconnected; nevertheless, the ventral ends of the third pair of *aortic arches* have been grazed and do show. Tangential shavings have also been cut from the caudal walls of the otocysts, just as they are being left behind. Extending laterad from the pharynx is the second pair of *pharyngeal pouches,* which have already come in contact with the ectoderm to form *closing plates;* the complementary, external *branchial grooves* are not well seen in this specimen. A pocket-like depression in the midventral floor of the pharynx indicates the beginning of the *thyroid gland;* later it becomes saccular and loses its connection with the pharyngeal entoderm. The splanchnic mesodermal wall of the heart is destined to give rise to the *epicardium* and *myocardium.* Only the beginning of the *ventricle* appears, but a short distance down the series its large loop is met; the main part of the ventricle is free and no longer suspended by the former dorsal mesocardium.

Section through Atrium, Venous Stems and Pleural Cavities (Fig. 567). Between the previous level and this one, the third aortic arches and much of the heart have been passed. Also the anterior cardinal veins have bent downward to join the posterior cardinal veins

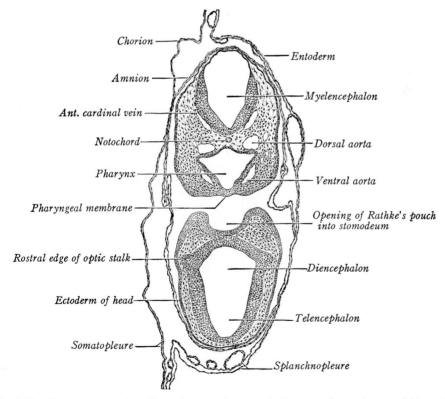

Fig. 564. Transverse section through stomodeum and pharyngeal membrane of 27-somite chick embryo. × 50.

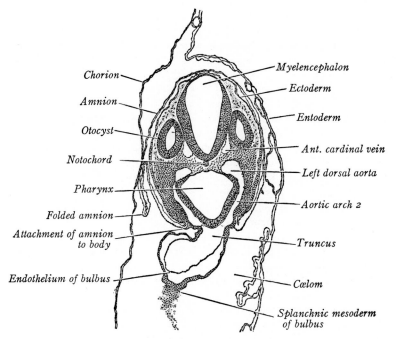

Chorion

Amnion

Otocyst

Notochord

Pharynx

Folded amnion

Attachment of amnion to body

Endothelium of bulbus

Myelencephalon

Ectoderm

Entoderm

Ant. cardinal vein

Left dorsal aorta

Aortic arch 2

Truncus

Cœlom

Splanchnic mesoderm of bulbus

Fig. 565.   Transverse section through otocysts, bulbus and second aortic arches of 27-somite chick embryo. × 50.

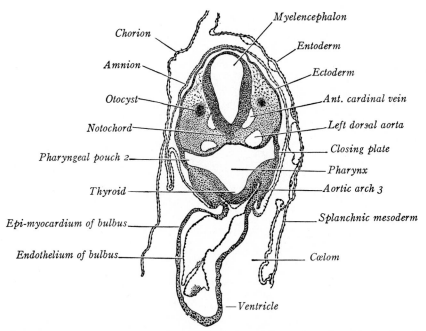

Chorion

Amnion

Otocyst

Notochord

Pharyngeal pouch 2

Thyroid

Epi-myocardium of bulbus

Endothelium of bulbus

Myelencephalon

Entoderm

Ectoderm

Ant. cardinal vein

Left dorsal aorta

Closing plate

Pharynx

Aortic arch 3

Splanchnic mesoderm

Cœlom

Ventricle

Fig. 566.   Transverse section through second pharyngeal pouches, thyroid gland and ventricle of 27-somite chick embryo. × 50.

in a stem known as the common cardinal. The present level shows the *posterior* and *common cardinal veins,* the latter opening into the thin-walled *sinus venosus.* The sinus receives all of the blood being returned to the heart and is separated from the larger *atrium* by a slight constriction only. Passing a few sections lower, the opening of the *vitelline veins* into the sinus may also be demonstrated. The dorsal aortæ have united to form the single *descending aorta.*

On each side of the pharynx is a subdivision of the cœlom that will serve as a *pleural cavity* when the lung buds appear. These canals are partially separated from the pericardial cavity by the *septum transversum* (primitive diaphragm), through which the common cardinal veins cross to the sinus venosus. Between the previous section and this one the myelencephalon has given way to the *spinal cord,* and here the fourth pair of *somites* is seen. These somites have differentiated into a *dermo-myotome* plate and a more diffuse *sclerotome.* At all higher levels the general mesoderm was purely mesenchyme and without visible specialization. The mesodermal components of the two amnion folds come into close approximation but remain separate at this level.

Section through Vitelline Veins and Liver (Fig. 568). The fore-gut is now flattened from side to side and its cavity is narrow; a few sections caudad it bends downward to open through the *anterior intestinal portal* onto the yolk sac. Midventrally there is evaginated from the gut-entoderm a pair of diverticula which constitute the earliest indication of the *liver.* At the side of each bud is a *vitelline vein* (the left, cut as it swings in from the blastoderm); their destination in the sinus venosus has already been traced. The primitive liver bud does not always appear bilobed, as in this specimen. Note the intimate relation between the entodermal epithelium of the liver and the endothelium of the vitelline veins; this is significant since later there will be a mutual intergrowth between the two to give the characteristic relation of hepatic cords and sinuses.

The *septum transversum* is still present at this level; in fact, it was originally produced through the bulging vitelline veins fusing with the somatopleure. Lateral to the fore-gut are small *cœlomic cavities,* and lateral to these, in turn, appear portions of the *posterior cardinal veins.* The *descending aorta* again shows signs of doubleness through the incomplete fusion of its component dorsal aortæ. At this and other levels, *neural crest* cells can be seen.

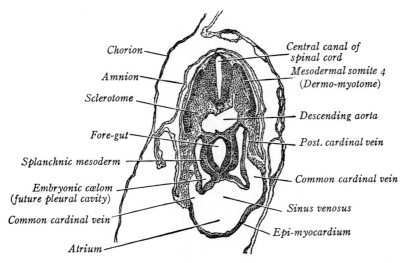

Fig. 567. Transverse section through atrium, venous stems and pleural cavities of 27-somite chick embryo. × 50.

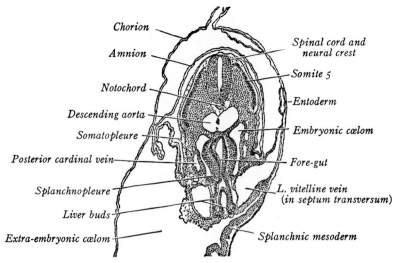

Fig. 568. Transverse section through vitelline veins and liver of 27-somite chick embryo. × 50.

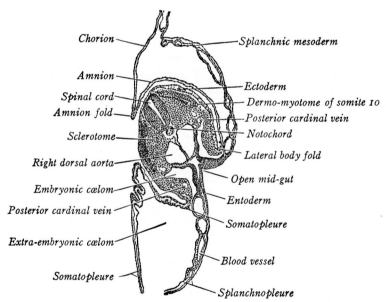

Fig. 569. Transverse section through open gut and amnion folds of 27-somite chick embryo. × 50.

Section through Open Gut and Amnion Folds (Fig. 569). The intestine has opened ventrally as the *mid-gut,* its splanchnopleuric wall passing directly over onto the vascular blastoderm. The descending aorta is again divided by a septum into its primitive components, the right and left *dorsal aortæ.* Lateral to the aortæ and in the somatopleure are the small *posterior cardinal veins.* The embryonic cœlom is in communication with the extra-embryonic cœlom. Deep *lateral body folds* of somatopleure indicate how, by their ventral union (and the similar folding-off of the mid-gut by the splanchnopleure), the body becomes established free from the blastoderm.

The *amnion folds* have not joined at this level, thus leaving the amniotic cavity open; some variation may be found in the exact level where closure is occurring. In such a section as this the somatopleuric components (ectoderm and somatic mesoderm) of the amnion and chorion are easily traced, and a few sections cephalad the manner of union of the two compound folds is demonstrated.

Section through Seventeenth Pair of Somites (Fig. 570). The body of the embryo is no longer rotated. On the left side of the embryo the mesodermal somite is beginning to specialize into a *dermo-myotome* plate and a less compact *sclerotome;* on the right side the section merely grazes the edge of a somite. Lateral to each aorta appears a section of the *pronephric (mesonephric) duct* and a *mesonephric tubule.* The space nearby is the *posterior cardinal vein.* The embryonic somatopleure is grooved by widely separated *lateral body folds.* Still farther laterad are elevations of the somatopleure that represent the caudal extent of the *lateral folds of the amnion (cf.* Fig. 558 B).

Section through Twenty-third Somites and Vitelline Arteries (Fig. 571). In this region the embryo is flatter and simpler in structure, corresponding to the condition at higher levels in younger embryos. Somites, nephrotomes and lateral layers of somatic and splanchnic mesoderm are little changed from their original appearance. On the left side the *vitelline artery* leaves the aorta; on the right side this connection has been passed. The right *posterior cardinal vein* is present just lateral to the *mesonephric tubule* and *duct.* The small clusters of cells dorsolateral to the spinal cord are the *neural crests,* which will differentiate into spinal ganglia.

Section through Segmental Zone (Fig. 572). The somites are replaced by the *segmental zone.* This is a somewhat triangular column of primitive mesoderm, ready to serve as the source from which both somites and nephrotomes will be progressively blocked out. The solid, caudal ends of the free-growing *mesonephric ducts* come to an end just below this level. The *aortæ* are smaller than heretofore, and a short distance caudad they

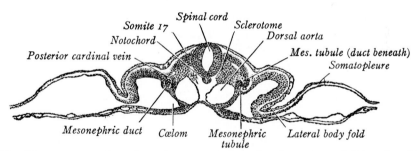

Fig. 570. Transverse section through seventeenth pair of somites of 27-somite chick embryo. × 50.

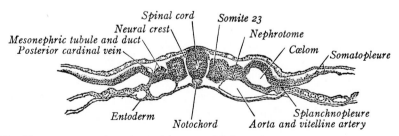

Fig. 571. Transverse section through twenty-third somites and the vitelline arteries of 27-somite chick embryo. × 50.

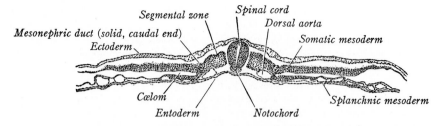

Fig. 572. Transverse section through segmental zone of 27-somite chick embryo. × 50.

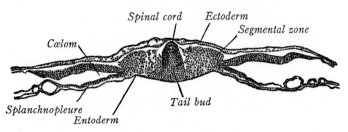

Fig. 573. Transverse section through the tail bud of a 27-somite chick embryo. × 50.

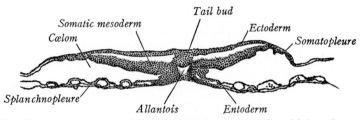

Fig. 574. Transverse section through allantois of 27-somite chick embryo. × 50.

disappear in the plexus of the area vasculosa. Laterally the *somatopleure* and *splanchnopleure* are flat and separated by the slit-like *cœlom*.

Section through Tail Bud (Fig. 573). In embryos of two days the neural groove has rolled into a tube and is cut off from the surface ectoderm throughout its full length. At the present level the caudal tip of the *spinal canal* is seen, and the ventral wall of the neural tube is merged with an undifferentiated mass of dense tissue which is a common meeting ground for the ectoderm, mesoderm and entoderm. This tissue has the essential relations of the primitive knot of earlier stages, and like it is a region of progressive proliferation and differentiation; the entire formative mass has been called the *tail bud,* since the caudal end of the body develops from its substance.

Section through Allantois (Fig. 574). A short pocket, located in the midplane, is cut across at this level; a few sections craniad in the series it opens and becomes continuous with the general entoderm. This diverticulum is the first indication of the sacculating *allantois.* It precedes the establishment of a hind-gut by folding. Also in the midplane may be seen the caudal end of the *tail bud.* It is continuous dorsally with the *ectoderm,* ventrally with the *entoderm* of the allantois and laterally with the *mesoderm.*

F. EMBRYO OF THIRTY-FIVE SOMITES (THREE DAYS)

The descriptions that follow apply specifically to the 72-hour chick, but they can be used satisfactorily in conjunction with the study of stages between 58 hours (31 somites) and 84 hours (38 somites). Examples of these extremes are shown in Figs. 558 *C*, 589 *A*.

Several gross advances that have occurred during the third day attract attention (Fig. 575). A *tail fold* had demarcated a definite, caudal end to the trunk, which now terminates in a conical *end bud* (or *tail bud*). Somites extend down to this terminal bud. The twist that eventually turns the embryo onto its left side has proceeded two-thirds of the way down the body. The *cervical flexure* is now so pronounced that the hind-brain is the most cephalically situated part of the embryo, and the fore-brain is made to point at the heart. As one result of this bending, the *eye* lies much farther caudad than the *ear*. A new feature at the caudal end of the embryo is the *caudal flexure,* and this creates a curve ending in a *tail*. Other new features are the *limb buds,* which make noticeable swellings along the trunk.

Nervous System and Sense Organs.   The *telencephalon* is plainly demarcated from the *diencephalon* and shows prominent swellings, which are the beginnings of the *cerebral hemispheres* (Figs. 575, 576). The diencephalon bears a dorsal evagination which is the *epiphysis*. The *mesencephalon* is sharply set off from the hind-brain, and the latter shows signs of subdivision into a short, thick-roofed *metencephalon* and a long, thin-roofed *myelencephalon*. The caudal end of the brain passes without boundary into the *spinal cord*. The cavities of the neural tube are named as follows:

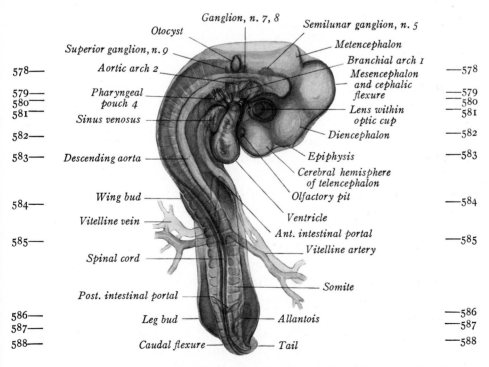

Fig. 575.   Chick embryo with 35 somites (72 hours), in part lateral and part dorsal view. × 14. Numbered lines indicate levels of sections, Figs. 578–588.

lateral ventricles of the cerebral hemispheres; *third ventricle* of the diencephalon; *cerebral aqueduct* of the mesencephalon; *fourth ventricle* of the met- and myelencephalon; and *central canal* of the spinal cord.

Paired masses, occurring along the ventrolateral border of the myelencephalon, are early *cranial ganglia* (Fig. 575). These are: the *semilunar ganglia* of the fifth nerve; the temporarily merged *geniculate* and *acoustic ganglia* of the seventh and eighth nerves; and the *superior* (proximal) and *petrosal* (distal) *ganglia* of the ninth nerves. The ganglia of the tenth nerves are not distinguishable as yet. The pairs of cranial nerves that exist as entities at this period are the third, fifth, seventh and ninth. The *oculomotor* (third) *nerves* are emerging from the under surface of the mesencephalon. The *trigeminal* (fifth) *nerves* subdivide and may be traced into the region of the eyes and toward the subdivided first branchial arches. The *facial* (seventh) *nerves* and *glossopharyngeal* (ninth) *nerves* pass to the second and third branchial arches, respectively. In the region of the spinal cord the *spinal ganglia* are beginning to show as definite, segmental masses.

Among the sense organs a new feature is presented by the shallow *olfactory pits* which lie near the tip of the head (Fig. 575). They come from the invagination of locally thickened ectoderm. The stalked *eye cups* are horseshoe-shaped in surface view, the notch-like defect in the rim representing the temporary *optic fissure*. The *lens* is now a closed-off sac. The *otocysts* are also closed vesicles, but pear-shaped and continued into a short *endolymph duct* at the upper end.

Digestive and Respiratory Systems. The *stomodeum* and *pharynx* are becoming continuous because of a local rupture of the *pharyngeal membrane* (Fig. 576). *Rathke's pouch* is more of a stalked sac than before. A fourth pair of *pharyngeal pouches* has been added to the three previous sets. The first two pairs show partially ruptured *closing plates,* so that there is temporary communication between these pouches and their corresponding *branchial grooves*. The *thyroid* diverticulum is a nearly closed pouch. In line with the thyroid, but at the caudal end of the pharyngeal floor, is a ventral gutter known as the *laryngo-tracheal groove*. At its caudal end are merely the primary bronchi, whereas the smaller air ducts and sacs of the lobed lungs will arise later by further buddings and branchings. Still farther along the fore-gut are a cranial and a caudal *liver bud,* whose branching masses interconnect. The *esophagus* is a short tube, directly caudal to the origin of the lung buds; it continues into the slightly dilated *stomach*. The latter narrows into the short *duodenum,* which gives off the liver primordia and then opens almost immediately into the unclosed mid-gut. On its dorsal wall, at the level of the caudal hepatic diverticulum, is a solid thickening that represents the earliest sign of the dorsal primordium of the *pancreas*.

The *mid-gut* is still a long, unclosed stretch of intestine, overlying the yolk sac and opening into the fore-gut and hind-gut by *intestinal portals* (Fig. 576). The *hind-gut* gives off a ventral diverticulum which is the slightly dilated *allantois*. Just caudal to the allantois is the *cloacal membrane* where ectoderm and entoderm fuse. Still farther caudad the hind-gut terminates in a short *tail-gut*.

Mesoderm and Cœlom. *Somites* extend all the way to the tail bud, and at different levels there are exemplified various stages in the differentiation of the dermatome, myotome and sclerotome (Fig. 575). In the older somites continued growth of the myotome plate has produced a complete layer underneath the dermatome. Sclerotomic masses are somewhat more definite than before.

Of the nephrotomic derivatives, the *mesonephric duct* is prominent. By continued downgrowth it has reached the hind-gut, where just caudal to the allantois, it is about to open into the gut. This caudal region of the primitive gut, now shared by the intestine, allantois, mesonephric ducts (and, soon, by the oviducts), is named the *cloaca*. Representatives of the *mesonephric tubules* occur between the fourteenth and thirteenth somites. Those at higher levels in the series are more advanced in development, but as yet none have completed the differentiation of actual tubules that will presently link up with the duct.

Enlargement of the cœlom toward the midplane has reduced the originally broad mesodermal region above the tubular gut to a thin plate, which is the *dorsal mesentery*. A *ventral mesentery*, brought into being as

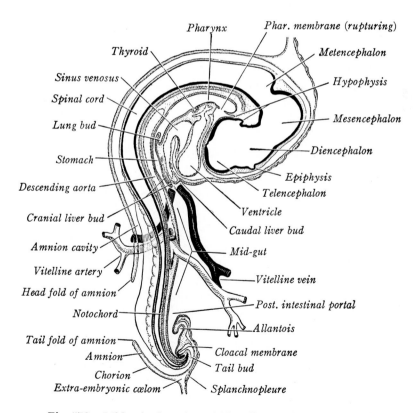

Fig. 576. Midsagittal section of 35-somite chick embryo. × **13.**

the body folds off from the yolk sac, attaches the fore-gut to the midventral body wall; it is so transient as to scarcely have expression in the region of the mid-gut. In the hind-gut region a dorsal and ventral mesentery are created when the cœlom penetrates into this region, after the caudal end of the body has already folded off. At the level of the sinus venosus and liver the ventral mesentery, and the lateral folds in which the common cardinal veins course, together represent the beginning of an important mesodermal mass to be known as the *septum transversum.*

During the third day folding of the lateral body walls has begun to extend the trunk caudad in the region of the heart and thus confine some of the cœlom within the folding-off body. The cœlom then consists of a large *pleuro-pericardial cavity,* containing tiny lungs and the large heart. This region is continuous caudad with the future *peritoneal cavity,* which is partially partitioned in the midplane by the mesentery. Cœlomic extensions are invading the caudal end of the folded trunk, where they lie as blind tubes alongside the hind-gut. Only this region and the more cranial portion of the pericardial space no longer communicate laterally with the extra-embryonic cœlom. The lateral bridges, in which course the common cardinal veins, are important because here will arise two sets of membranous partitions which aid the septum transversum in separating the pleural cavities from the pericardial cavity cranially, and from the peritoneal cavity caudally.

Cardio-vascular System.   The *heart* shows the same divisions as before: *sinus venosus; atrium; ventricle;* and *bulbus* (Figs. 575, 576). But further growth has dropped the *ventricular loop* to a more caudal position than other parts of the heart. The two limbs of this loop represent approximately the future right and left ventricles. The *atrium* now shows a larger left and smaller right expansion. The heart is free except at its attached cranial and caudal ends.

The *truncus arteriosus* lies in the floor of the pharynx and gives off the *aortic arches.* There are now four arches; the first and second are broken, or are about to become interrupted (Fig. 577). The *dorsal aortæ* extend, as *internal carotid arteries,* into the region of the diencephalon. Just caudal to the aortic arches the aortæ unite into a common *descending aorta* which again separates into paired vessels at a mid-gut level; these continue as far as the tail bud. About midway along the mid-gut prominent *vitelline arteries* pass to the yolk sac. Pairs of tiny dorsal branches (*intersegmentals*) are growing off the joined and paired aortæ, and at the level of the wing buds a prominent pair constitutes the primitive *subclavian arteries.*

The *anterior cardinal veins* (internal jugulars) have not advanced greatly except in complexity of branching (Fig. 577). Into their bases may now drain slender *external jugulars* from the floor of the pharynx. The *posterior cardinals* are more prominent than formerly as they overlie the mesonephric ducts, and they show dorsal intersegmental branches. Beneath the mesonephroi are appearing tiny vessels named the *subcardial veins.* The *um-*

bilical veins are primitive vessels coursing in the lateral body wall and emptying into the *common cardinals*. The paired *vitelline veins* are uniting by a slender but important anastomosis close to the anterior intestinal portal. Just cephalad, in the region of the liver, these veins merge bodily before draining into the sinus venosus.

The beating heart sends the blood through two circuits: (1) Oldest is an arc for the yolk sac. It operates by way of the vitelline arteries and veins, which supply and drain the vitelline network and its marginal sinus. (2) Somewhat slower to perfect was the circuit within the body of the embryo itself. Blood is pumped through the aortic arches to the dorsal aortæ, whence it passes to the head by way of the primitive carotid vessels and to the dorsal trunk by way of the serially repeated intersegmental branches. The return is handled by the anterior and posterior cardinal veins, which share a common stem that opens into the sinus venosus.

Extra-embryonic Membranes. Splanchnopleure, spreading over the surface of the yolk, has half-enclosed it as a *yolk sac;* only part of this expanse is vascularized. The *allantois* is a stalked and slightly dilated sac that projects ventrad from the hind-gut and causes a local bulge there (Figs. 575, 576). During the third day the *head fold* and *lateral folds* of the amnion, already present, close in and extend the amnion and chorion far down the embryo. A similar *tail fold* of the amnion appears during the third day, just caudal to the embryo, and extends a short distance cephalad. The *amnion,* however, is still not completely pursed into a closed sac.

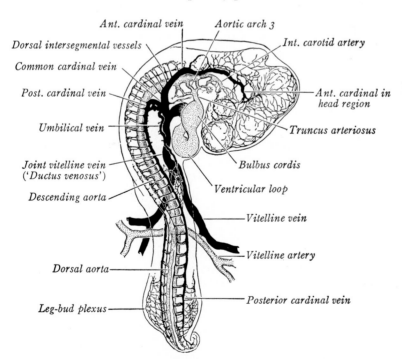

Fig. 577. Principal blood vessels of 35-somite chick embryo, in part lateral and part dorsal view. × 13.

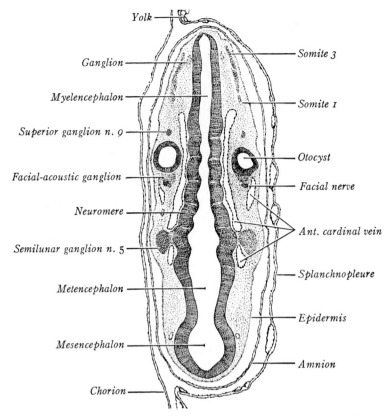

Fig. 578.   Transverse section through otocysts and cranial ganglia of 35-somite chick embryo. × 36.

TRANSVERSE SECTIONS

The following series of sections from a three-day chick can be used successfully to interpret embryos between 58 hours (31 somites) and 84 hours (35 somites). They are drawn from the cephalic surface; hence the right side of the embryo is on the reader's left. The level of each section is indicated along the margins of Fig. 575. Although this series is designated as being cut transversely, actually the highly flexed head makes the plane of section through the fore- and hind-brain more or less frontal.

Sections through the Hind-brain.   The first sections encountered cut the *hind-brain* at a high level through a frontal plane. They contain mostly portions of the thin roof. At lower levels the brain wall thickens and a section of the *mesencephalon* appears, which is at first a separate, globular structure. The head, as most of the body, is enclosed in *amnion.* Beneath the left side of the head is *splanchnopleure,* which is easily distinguishable by the presence in it of blood vessels and, farther laterad, of yolk. Above the head is a part of the *chorion.*

Section through Otocysts and Cranial Ganglia (Fig. 578). It should be noted in Figs. 578–583 that the embryo is lying on its left side, but for convenience of study has been placed in the erect position. Hence the blastoderm is vertical rather than horizontal. The head is bounded by ectoderm which will become *epidermis*. Portions of the three highest *somites* are cut at this level, and medial to the third pair are rudimentary *ganglia*. The hind-brain (*metencephalon* and *myelencephalon*) is sectioned frontally, while the *mesencephalon* is cut more transversely, owing to its flexure. Six *neuromeres* give a beaded appearance to the wall of the hind-brain, but these are only transient features. Conspicuous masses are the *semilunar ganglia,* the right one showing well its root-connection with the brain wall. Traced down the series a few sections each ganglion gives off three nerves, which pass toward the regions of the future eyes and jaws. One or both of the more lateral of the three masses can be traced into continuity with thickenings (*placodes*) of the epidermis.

Also conspicuous are the *otocysts;* followed to a slightly higher level they dwindle into slender *endolymph ducts,* whose tips still attach to the ectoderm. On the rostral side of the otocysts are masses that represent the temporary *facial-acoustic ganglia,* not yet separated into distinct geniculate and acoustic subdivisions. A few sections higher in the series, facial rootlets join the brain wall while at the present level the left *facial nerve* is becoming distinct. Each nerve passes to a second branchial arch and merges with an ectodermal *placode* there. The acoustic portion of the common ganglionic mass is in contact with the

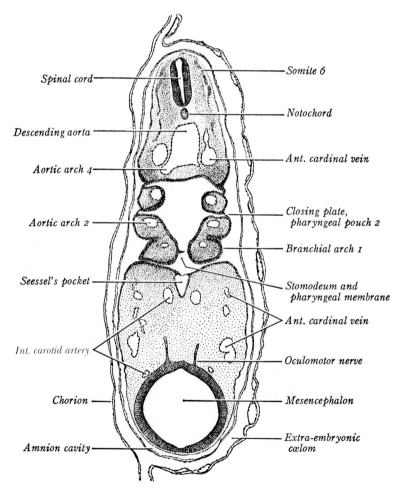

Fig. 579.   Transverse section through pharynx and oculomotor nerves of 35-somite chick embryo. × 36.

otocyst wall, and at one level the two become continuous. Just caudal to the otocysts are small *superior ganglia*. Slightly lower they continue into more cord-like *glossopharyngeal nerves*, while near the third branchial arches they again enlarge into the beginnings of *petrosal ganglia* and fuse with ectodermal placodes. Following alongside the brain are the *anterior cardinal veins*, cut longitudinally.

Section through Pharynx and Oculomotor Nerves (Fig. 579). This section lies at a level where the bent, rostral half of the head appears as a separate section in addition to the main section. The *mesencephalon* is cut transversely, and from its ventral side emerge short *oculomotor nerves*. Most of the hind-brain and cranial portion of the notochord (sectioned lengthwise) have been passed by. At the present level is the beginning of the *spinal cord* which lies between the two highest cervical *somites* (whose dermo-myotome plates and plain) and over the *notochord*—all sectioned somewhat slantingly. About half-way between Fig. 578 and this level the paired *dorsal aortæ* were cut lengthwise as they coursed dorsal to the pharynx. Before reaching the present level the aortæ have: (1) joined into a common *descending aorta;* (2) received four pairs of *aortic arches* which lie within *branchial arches;* and (3) continued as *internal carotid arteries*, into the rostral part of the head. Here branchial arches and aortic arches are cut transversely; the fourth pair of the aortic arches show their union with the aorta. Portions of the internal carotids appear in the rostral half of the head. A more caudal portion of each *anterior cardinal vein* lies lateral to the aorta, while their more rostral extensions occur in the separate section of the head.

The *pharynx* is cut on the flat. Prominent are the second and third pairs of *pharyngeal pouches*, with *closing plates* and adjacent (ectodermal) *branchial grooves*. The first pair of pouches occurs at a slightly higher level than this. The *pharyngeal membrane* is a compound plate separating the entodermal pharynx from the ectodermal *stomodeum*. In the present section it shows beginning perforation. In the section of the rostral head is a cranial tip of the pharynx known as *Seessel's pocket;* it is a part of the pharynx that extends rostrad of the oral opening, but has no further significance (Fig. 590).

Section through Thyroid and Hypophysis (Fig. 580). The *branchial arches* are cut near their bases. The *mandibular processes* of the first pair join in the midplane. Directly across on the more rostral portion of the head are the *maxillary processes*, and the two become confluent at a higher level in the series. Arches 1–4 are plain; arch 5 is only a transitory structure behind the fourth pharyngeal pouch. Three pairs of *aortic arches* are cut just as they leave the *truncus* and enter branchial arches 1–3. The fourth aortic arches were seen in Fig. 579. The second pair of *pharyngeal pouches* show merely as cut-off wings; the third pair has normal relations; the fourth pouches are small and not yet in contact with the ectoderm. The first pouches had their closing plates ruptured locally; in the present section one of the second pouches has opened to the outside (and the other is likewise perforated in adjoining sections); the closing plates of the third pouches are intact, but will rupture presently. Such openings are temporary and normally close during the next 36 hours.

The *thyroid* is a thick-walled sac which is cut almost horizontally; several sections higher in the series, it opens into the pharynx by a narrower mouth. The arched, dorsal portion of the pharynx represents the region where it begins to narrow into the *laryngotracheal groove*. *Spinal ganglia* lie dorsally alongside the *spinal cord*, but they still lack sharply delimited boundaries. The *mesencephalon* is becoming continuous with a shaving off the wall of the *diencephalon*. The rounded, hollow region of this shaving is the *infundibulum*, which will give rise to the neural lobe of the *hypophysis*. Two sections higher in the series *Rathke's pouch* (epithelial lobe of the hypophysis) extended even to the infundibulum, but at the present level it has been largely replaced by brain wall, and only its mouth is seen.

Section through Lung Buds and Optic Cups (Fig. 581). Between the previous level and this one, several notable changes have occurred: (1) The branchial arch-pouch region has been left behind. (2) The broad pharynx gave way to a compressed slit-like tube, known as the *laryngo-tracheal groove*. At this level the groove is broadening into an *esophagus* dorsally. It also is in the process of bifurcating into two so-called *lung buds* (actually primary bronchi); a few sections farther along, these buds are separate from the esophagus and from each other. (3) Esophagus and lung buds are contained in a thick mesentery. This broad attachment will be retained and is later called the *mediastinum.* (4) The truncus has given way to the *bulbus,* which lies at the right, while other portions of the bent heart (*atrium* and *sinus venosus*) have come into view on the left. In the bulbus, as also in the ventricle (Fig. 583), the endocardium seems shrunken and displaced. This is because of the occurrence of an important constituent, the *endocardial jelly,* which does not stain. (5) The *anterior cardinal veins* have dropped to a more ventral position, and at their bases some specimens are receiving *external jugular veins* from the floor of the pharynx. At the present level the left anterior cardinal is being joined by a minor root of the *posterior cardinal.* The combined trunk constitutes the *common cardinal.* (6) From the most cranial position of the truncus down to the present level (and much farther) the heart is enclosed within a completed *body wall;* the cavity thus circumscribed is the *pleuropericardial cœlom.*

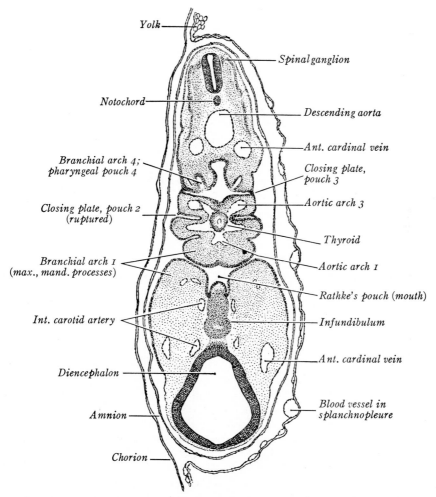

Fig. 580. Transverse section through thyroid and hypophysis of 35-somite chick embryo.
× 36.

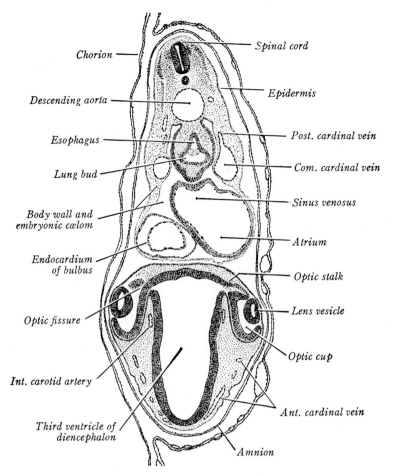

Fig. 581.   Transverse section through lung buds and optic cups of 35-somite chick embryo. × 36.

(7) In reaching this separate section through the rostral head, the mesencephalon has been passed by and a slanting section of the *diencephalon* now appears. Prominent ventrally are the *optic cups* and *lens vesicles*. Each cup is attached to the brain by a short *optic stalk*. The left cup shows well this relation and the continuity of layers. In the double-walled cup the thick lining will become the *nervous layer* of the *retina*, while the thin covering represents the future *pigment layer*. The invagination which creates the cup is continued along the stalk; this also results in a defect or notch at one point in the cup. In the right cup the section passes exactly through this notch (*optic* or *chorioid fissure*), just grazing one edge of its bounding wall. The *lens* is a closed sac, but still is attached to the epidermis in neighboring sections. (8) The *internal carotid arteries*, which were seen in Figs. 579, 580, have passed to positions median to the optic cups; the *anterior cardinal veins* also continue their penetration into the rostral part of the head.

    Section through Sinus Venosus, Olfactory Pits and Epiphysis (Fig. 582).   Changes since Fig. 581 are as follows: (1) Below the level of the lung buds the esophagus-stomach portion of the digestive tube has come to view; although there is no clear difference in appearance between the two, the present level is through the *stomach* and its mesentery (*mesogastrium*). (2) Directly beneath the stomach is a flattened mass that consists of slightly branching liver cords. It is the more cranial of two *hepatic diverticula*. A short distance caudad its connection with the duodenum can be found, appearing essentially like the more caudal evagination seen in Fig. 583. (3) Both *common cardinal veins* bridge across the cœlom and are opening into the *sinus venosus* which is constricting from the more ventral *atrium*. The vessels lateral to the aorta are the *posterior cardinal veins*. Other portions of these vessels (which connect with the common cardinal by several stems) lie in the lateral body wall. (4) The separate section passes through the rostral part of the *diencephalon* which here gives off a dorsal evagination that is the *epiphysis*. The larger part of the brain, however, is the *telencephalon,* with somewhat swollen beginnings of *cerebral hemispheres.* (5) Ventrolaterally the epidermis has thickened into *olfactory placodes.* These now take the form of concave *olfactory pits.*

Section through Hepatic Diverticulum and Ventricle (Fig. 583). Changes since Fig. 582 are as follows: (1) Just below the previous level the body wall separated into *lateral body folds* and the embryonic and extra-embryonic cœlom became continuous. In the present section these folds are far apart and the viscera protrude prominently beyond them. (2) The separate section, now consisting only of shavings from the tip of the head and telencephalon, marks the apical end of this highly flexed portion of the embryo. (3) In the main section the gut (*duodenum*) is bending toward the intestinal portal and is, consequently, cut on the slant. From its lower end there projects a second *liver bud*. Traced craniad, it branches beneath the combined vitelline veins and passes dorsad and to the

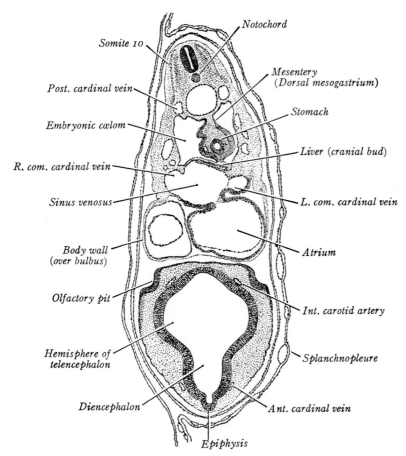

Fig. 582. Transverse section through sinus venosus, olfactory pits and epiphysis of 35-somite chick embryo. × 36.

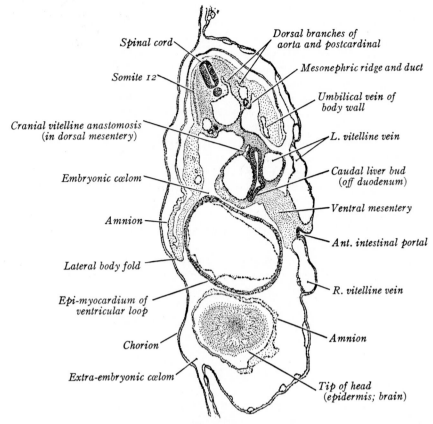

Spinal cord

Dorsal branches of
aorta and postcardinal

Somite 12

Mesonephric ridge and duct

Umbilical vein of
body wall

Cranial vitelline anastomosis
(in dorsal mesentery)

L. vitelline vein

Embryonic cœlom

Caudal liver bud
(off duodenum)

Ventral mesentery

Amnion

Ant. intestinal portal

Lateral body fold

R. vitelline vein

Epi-myocardium of
ventricular loop

Chorion

Amnion

Extra-embryonic cœlom

Tip of head
(epidermis; brain)

Fig. 583. Transverse section through hepatic diverticulum and ventricle of 35-somite chick embryo. × 36.

left of that vessel where it joins branches from the more cranial bud. The relation of these two diverticula to the planes of section will be made clear by consulting Fig. 576. (4) The *dorsal* and *ventral mesentery* have assumed typical relations to the body wall and gut, and both liver buds are burrowing into the ventral mesentery. (5) The extreme cranial end of the *mesonephric ridge* has appeared, with the *mesonephric duct* as its most conspicuous feature. In this region pronephric tubules alone were formed and some traces of them (*nephrostomes*) may still be found.

(6) The bulbus and atrium have been succeeded by a large *ventricular loop,* here cut tangentially. Its thicker myocardial wall is made spongy by blood channels. (7) The *sinus venosus* becomes continuous in a caudal direction with a large channel (*ductus venosus* in part), formed by the fusion of the two vitelline veins. (8) Between the level of the cranial hepatic bud and the present caudal one the gut (*duodenum*) began to bend ventrad on its way to reach its portal. This keeps the *vitelline veins* separate, and they lie right and left of the gut. (9) In this section a communicating *anastomosis* (part of which is seen) joins the two vitelline veins. Farther caudad a second anastomosis is forming close to the intestinal portal. These cross-connections play an important rôle in the development of the portal vein. (10) Prominent vessels in the somatopleure of the body wall are the *umbilical* (or allantoic) *veins,* which drain into the common cardinals near the level of Fig. 582. (11) Features not encountered previously (in the sections drawn) are *dorsal intersegmental vessels,* seen here both as branches of the aorta and of the left postcardinal.

Section through the Wing Buds and Mid-gut (Fig. 584). The remaining sections (Figs. 584–588) pass through the partly turned to unrotated body. They are shown in relation to a horizontal blastoderm. Since the previous level, changes have come in as follows: (1) The body is nearly twice as broad as previously; this is because of the presence of *wing buds* which are produced as local expansions of the somatopleure. (2) The fore-gut has opened at its *intestinal portal,* whose site is indicated by an indentation of the splanchnopleure in Fig. 583. The present section lies some distance below that point, and here the *mid-gut* is an open arch, continuous with the general splanchnopleure. (3) Shortly before it opens, the duodenum bears dorsally a slight thickening, which is the dorsal primordium of the *pancreas.* (4) The *vitelline veins* remain essentially as before, but traced caudad they diverge and become continuous with the area vasculosa of the blastoderm. (5) Fusion of the paired aortæ of lower levels is practically complete. (6) On the left side an important dorsal intersegmental artery extends into the wing bud; this is the future *subclavian artery.*

(7) The differentiating *mesonephroi* produce bulges into the cœlom that are the *mesonephric ridges.* Balls of tissue appear and hollow into vesicles, but the tubules are not completed at this stage and do not yet link up with the prominent *mesonephric duct.* (8) The *posterior cardinal vein* overlies the mesonephric duct and a tiny new vessel, the *subcardinal vein,* is appearing beneath and medial to the duct. (9) Between the mesonephric ridge and the mesentery, the cœlomic epithelium is thickened and bulging; it is the site of the 'germinal epithelium' of the future *genital ridge.* (10) The amnion and chorion are no longer wholly separate; at a region where the amnion folds meet, the ectodermal components are continuous.

Section through Vitelline Arteries (Fig. 585). This section is much like the preceding level but shows a few changes: (1) The wing-buds have been passed. (2) The aorta, which showed the beginning of a septum in the preceding figure, has again become two vessels which have separated slightly. (3) The left dorsal aorta gives off a large ventral vessel; the right aorta did the same a few sections cephalad, but the continuity is no longer seen at this level. These stems are *vitelline arteries;* traced caudad they break up into branches supplying the vascular area of the blastoderm. (4) The *dermo-myotomes* have taken an increasingly slanting position as the body flattens; the myotome plate is not so extensive as at higher levels (where it underlies the entire dermatome); the left *sclerotome* shows well.

Section through Posterior Intestinal Portal and Leg Buds (Fig. 586). New and altered features at this level are the following: (1) The body broadens again because of the presence of *leg buds* as local swellings of the somatopleure. (2) Not far craniad of this

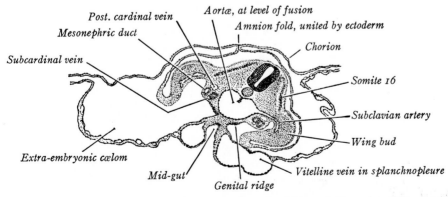

Fig. 584. Transverse section through wing buds and mid-gut of 35-somite chick embryo.
× 36.

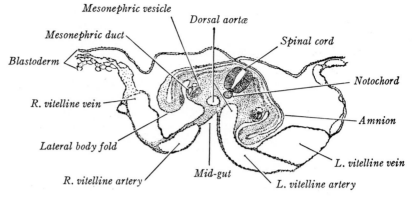

Fig. 585.   Transverse section through vitelline arteries of 35-somite chick embryo ($\times$ 36).

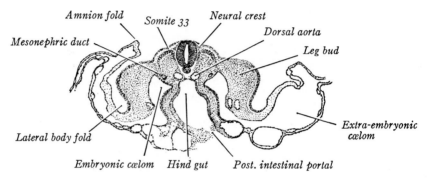

Fig. 586.   Transverse section through posterior intestinal portal and leg buds of 35-somite chick embryo. $\times$ 36.

level the amnion opened and its *amnion folds* are here spread far apart. (3) The open mid-gut is becoming continuous, through a *posterior intestinal portal,* with the more caudal closed-off *hind-gut.* At this level the splanchnopleure is just in the act of completing the fusions that produce this change. The next section in the series shows a broad bridge of entoderm, separating hind-gut from the space beneath the splanchnopleure of the blastoderm. (4) *Somites* are smaller and considerably simpler. (5) *Mesonephric ducts* are still seen but there are no tubule primordia; in fact, somite 30 is the last one that forms mesonephric tubules. (6) Posterior cardinal veins are no longer continuous vessels, but the *aortæ* are well represented. (7) Ganglia of high levels are represented merely by a simple *neural crest.* (8) Caudal to somite 27, the neural tube arises by canalizing a cellular rod laid down by the end bud.

 Section through Hind-gut and Allantois (Fig. 587).   The *lateral body folds* are approaching but still are some distance apart. In spite of this, the hind end of the embryo has become separate from the splanchnopleure of the blastoderm by reason of wedges of mesoderm from the lateral body wall, which here has pushed beneath the gut. In so doing, some of the *cœlom* has been separated off as a blind tunnel which extends a short distance caudad of the present level. The *hind-gut* has produced an expanded diverticulum which protrudes ventrad below the level of the lateral body folds; this is the *allantois.* Even simpler *somites* than before, *mesonephric ducts,* tiny *aortæ* and *leg buds* are still seen. Presently, intersegmental branches off the aortæ will supply the allantois as *allantoic* (or *umbilical*) *arteries.* Veins from the allantois (some of which are seen here) will complete the allantoic circuit by joining with *umbilical veins* already observed at higher levels (Fig. 583).

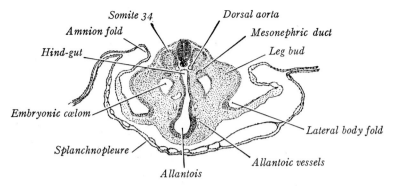

Fig. 587. Transverse section through hind-gut and allantois of 35-somite chick embryo. × 36.

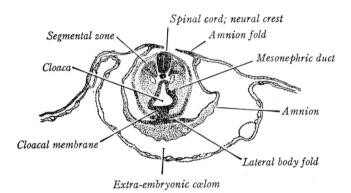

Fig. 588. Transverse section through cloaca and cloacal membrane of 35-somite chick embryo. × 36.

Section through Cloaca and Cloacal Membrane (Fig. 588). The *lateral body folds* have met and are just setting the caudal end of the body free. *Amnion folds* are again about to meet and enclose this caudal end of the embryo. The caudalmost pair of somites at this stage is being blocked out, but they are not cut off on their caudal surfaces; hence this is the region of the *segmental zone*. The left *mesonephric duct* comes to an end in contact with the cloacal wall; the right one has already done so a few sections higher in the series. In a slightly older embryo they actually open into the cloaca. Since this region of the primitive hind-gut also is a common receptacle connecting with the allantois and mesonephric ducts, it is a *cloaca*. Ventrally its thick entoderm is fused with the ectoderm, and this joint plate is the *cloacal membrane*.

Sections through Tail Bud. The gut continues a short distance caudad of the cloacal membrane as the *tail-gut*. The *segmental zone* has the same shape as somites in transverse section, but is not yet broken up into blocks. Slightly farther caudad the *notochord* fuses with the tail-gut and *neural tube;* both of these hollow structures then lose their cavities. A few sections still farther caudad all blend in an undifferentiated mass of formative tissue, which is the *tail bud.*

G. EMBRYO OF FOUR DAYS

During the fourth day of incubation the chick attains a stage of development corresponding to the youngest pig embryos customarily studied. It

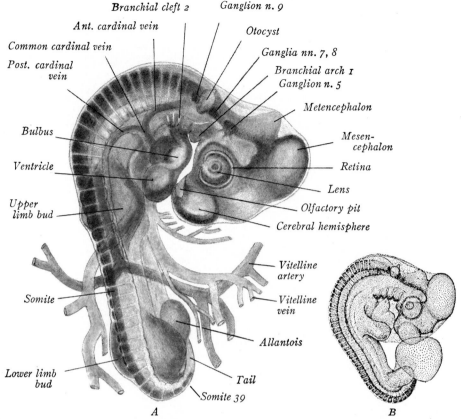

Branchial cleft 2   *Ganglion n. 9*

Ant. cardinal vein

Otocyst

Common cardinal vein

Post. cardinal vein

Ganglia nn. 7, 8

Branchial arch 1

Ganglion n. 5

Metencephalon

Bulbus

Mesen-cephalon

Ventricle

Retina

Lens

Upper limb bud

Olfactory pit

Cerebral hemisphere

Vitelline artery

Somite

Vitelline vein

Allantois

Lower limb bud

Tail

Somite 39

A   *B*

Fig. 589. Chick embryos of fourth day, viewed from right side. *A,* At 85 hours ($\times$ 14); *B,* at 96 hours ($\times$ 6).

will be sufficient, therefore, to describe only such essential features of developmental advance as may prepare for the detailed treatment of pig embryos in the next chapter.

External Form. Continued and completed torsion causes the embryo to lie wholly on its left side (Fig. 589). *Cervical* and *caudal flexures* are increasingly pronounced and a new *dorsal flexure* has appeared; as a result, the embryo becomes so curved into a **C**-shape that its head and tail approach. Continued growth on the part of the embryo as a whole, and under-cutting from all directions by the body folds, have led to an extensive delimitation of embryo from blastoderm. This reduces the region of the open gut greatly and creates a slender *yolk stalk* leading to the *yolk sac.* Caudally the *allantois* is now a prominent bladder-like dilatation, nearby the narrow, tapering tail. Branchial arches are thicker and more conspicuous. The *limb buds* have become elongate, paddle-like appendages. The *eye* is large and the *brain* is swollen regionally.

Nervous System and Sense Organs. The *brain* is featured by prominent *cerebral hemispheres* and a bulging *mesencephalon* (Figs. 589, 590). The primordium of the neural lobe of the *hypophysis* evaginates into contact with Rathke's pouch, which will become the epithelial lobe of that organ. New nerves, arising from the mesencephalon, metencephalon and myelencephalon, respectively, are the *trochlear* (fourth), *abducens* (sixth) and *vagus* (tenth). The fourth and sixth nerves extend to the vicinity of the eyes, while the vagus nerves develop *jugular* and *nodose ganglia* and grow into the fourth branchial arches. *Accessory* (eleventh) and *hypoglossal* (twelfth) nerves are just appearing. The *olfactory pits* are deep; each is bounded by a ∩-shaped elevation, known as a *lateral* and a *medial nasal process.* Except for increased size, the *eye* and *ear* have not changed drastically. *Spinal ganglia* are distinct, and the roots of *spinal nerves* are now beginning to be recognizable. A chain of *autonomic ganglia* lies dorsolateral to the aorta, on each side of the midplane.

Digestive and Respiratory Systems. Rupture and disappearance of the pharyngeal membrane have put the *stomodeum* (mouth) into continuity with the *pharynx,* but the *cloacal membrane* will remain intact until considerably later (Fig. 590). *Pharynx, lung buds, digestive tube* and *cloaca* have not changed greatly except in size and added distinctness. The earlier primordia of the *liver* have consolidated into a prominent organ, and the *gall bladder* and *bile duct* are now plain. Dorsal and ventral *pancreatic primordia* are budding from the intestine and bile duct, respectively.

Urogenital System. All traces of the pronephric tubules have disappeared, but *mesonephric tubules* are still developing and differentiating. The *metanephros,* or permanent kidney, is just making an appearance. Its ureteric bud is growing out of the mesonephric duct, near the cloaca. Its secretory tubules will differentiate from adjacent nephrotome tissue, located opposite somites 31 to 33; this is just caudal to the lowest mesonephric tubules. The mesonephroi, as a whole, bulge in paired longitudinal ridges along each side of the mesentery. In the narrow strip between mesonephros and mesentery, a thickening of the peritoneal epithelium gives the first definite indication of a future *gonad.*

Cardio-vascular System. Externally the *heart* is not greatly changed in appearance (Fig. 589). Internally its ventricle has acquired a thick, spongy wall, and there are beginnings of the atrial and ventricular partitions that eventually produce a four-chambered heart.

Rudimentary, inconstant fifth aortic arches may have formed since the third day, whereas sixth (pulmonary) arches are regularly added to the aortic-arch series. On the other hand, the first two pairs are now broken and disappearing. The dorsal aortæ continue forward from the third aortic arches as *internal carotid arteries* and extend to the eyes and brain. Near the ventral ends of the third arches, *external carotids* are growing toward the jaw region (first branchial arches). The *vitelline veins* have completed a second anastomosis, and thus created a second venous ring, near the liver; parts of both rings are disappearing, leaving a trunk to be known as the *portal vein.* Hepatic sinusoids have connected with the right subcardinal

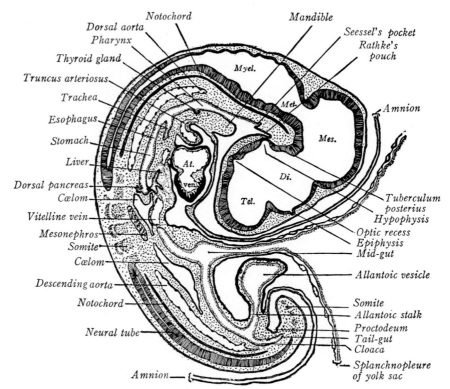

Fig. 590. Midsagittal section of chick embryo of four days, viewed from right side (Patten). (× 14).

vein to establish a venous drainage that will presently become the *inferior vena cava.*

In addition to the vitelline and intra-embryonic vascular circuits, present at three days, a third arc, known as allantoic or umbilical, has been completed. *Allantoic arteries* pass to the allantoic sac and ramify in its wall, while *allantoic veins* return this blood, by way of the lateral body wall and ductus venosus to the heart. These latter vessels are also called *umbilical;* they serve the chick for respiratory purposes but will become still more important as placental vessels in mammals.

Extra-embryonic Membranes. Much of the yolk-mass has become covered by advancing splanchnopleure, which thus creates a vascularized *yolk sac.* It is continuous over a narrow *yolk stalk* with the mid-gut; as the embryo enlarges, this connecting stalk appears relatively narrower (Fig. 590). Through the vitelline vessels the yolk supplies all the food materials for embryonic growth. The *allantois* is a prominent, stalked sac which later occupies the space beneath the shell. Its highly vascularized sac will serve as the principal fetal organ of respiration and excretion. During the fourth day the encircling fold of somatopleure that is responsible for the amnion and chorion has completed its arching growth and, in so doing, has created a closed *amniotic sac.* It is filled with *amniotic fluid,* in which the embryo 'floats,' and thus serves as a protective sac. The *chorion,* formed by the same folding process, but of little functional significance, surrounds the embryo, yolk sac and allantois.

H. EMBRYOS OF SEVEN AND TEN DAYS

Figure 591 illustrates the advances in form acquired up to the middle of the incubation period. By the end of this time the fetus becomes unmis-

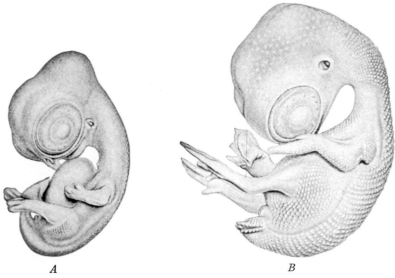

A   *B*

Fig. 591. Later chick embryos, viewed from left side (after Keibel and Abraham). × 3.
A, At seven days; *B,* at nine days.

takably bird-like in its external characteristics, even though the *eye* is disproportionately large. The original cervical flexure has been lost, and a distinct *neck* now separates head from thorax. The first branchial arches remain as the primitive *jaws,* which assume the appearance of a beak, and the first branchial cleft is retained as the *external acoustic meatus,* about which no auricle develops. The other arches and clefts have disappeared. The ventral surface of the body bulges prominently as the viscera become more protuberant. *Feather* primordia appear in definite patterns. The contours of the body, including the head and tail, become recognizably avian, and the fore limbs are wing-like. The amnion narrows to a tube, which contains the allantoic and yolk stalks and attaches to the lower body wall as a sort of umbilical cord.

RECOMMENDED COLLATERAL READING

Hamilton, H. L.   Lillie's Development of the Chick.   Holt.
Patten, B. M.   The Early Embryology of the Chick.   Blakiston.
Romanoff, A. L.   The Avian Embryo.   Macmillan.

Chapter XXIX. The Study of Pig Embryos

The maturing eggs of the sow are expelled from the ovary during the period of heat, following which they become promptly fertilized (*cf.* Fig. 35). Cleavage and the formation of a blastocyst are illustrated in Fig. 42. Gastrulation (the segregation of entoderm and mesoderm) is essentially like the stages shown in Figs. 57, 59, while the organization of a typical blastoderm is illustrated in Fig. 58.

At the completion of germ-layer formation the embryonic disc possesses a typical primitive streak (Fig. 592 *A*). This is quickly followed by the appearance of neural folds (*B*) and then mesodermal somites (*C*, *D*). The

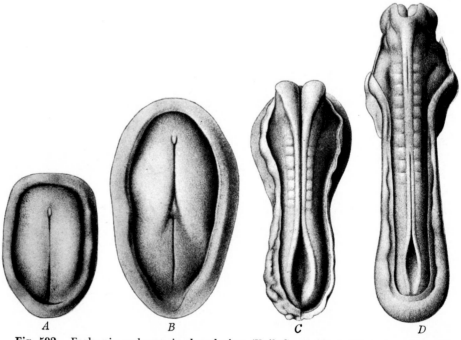

A   B   C   D

Fig. 592. Early pig embryos, in dorsal view (Keibel). × 20. *A*, Blastoderm at twelve days, with primitive streak and knot. *B*, Blastoderm at thirteen days, with primitive streak and neural groove. *C*, Embryo of fourteen days, with seven somites. *D*, Embryo of fifteen days, with 11 somites.

610

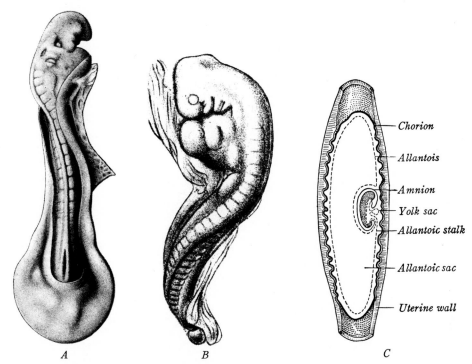

— Chorion

— Allantois

— Amnion

— Yolk sac

— Allantoic stalk

— Allantoic sac

— Uterine wall

A   B   C

Fig. 593. Early pig embryos, in dorsal view (Keibel). *A*, At sixteen days, with seventeen somites ($\times$ 15). *B*, At seventeen days, approximately 4 mm. long, with nearly full number of somites ($\times$ 12). *C*, Hemisection of uterus, showing fetal membranes and placental relations.

fundamental similarity of the stages shown in Figs. 592, 593 *A* to chick embryos of the first two days is apparent. The stages immediately succeeding correspond to those of three-day chick embryos, but are complicated by flexion and spiral twisting (Fig. 593 *B*); this makes sections difficult for the beginner to interpret. However, in embryos about 6 mm. long the twist of the body has disappeared sufficiently so that its structure may be studied to better advantage. At this time the state of development is generally comparable to that of a four-day chick (Fig. 594). Notice the similarity of Figs. 592, 593 to the human stages shown as Figs. 68, 74, 75, 76.

The fetal membranes of the pig stand somewhat intermediate between those of the chick and man. The amnion, chorion and allantois develop very much as in the chick (Fig. 92). The yolk sac for a time grows rapidly, but its functions are soon transferred to the allantois which fuses with the chorion; the two enter into a contact-relation with the lining of the uterus and constitute a placenta, which is the organ of fetal respiration, nutrition and excretion (Fig. 593 *C*). The development and relations of these extraembryonic structures are described on pp. 116–118, 144–147.

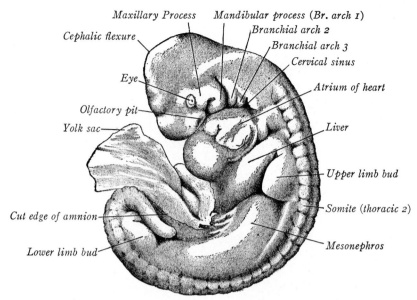

Fig. 594. Pig embryo of 6 mm., with amnion removed, viewed from left side. × 12.

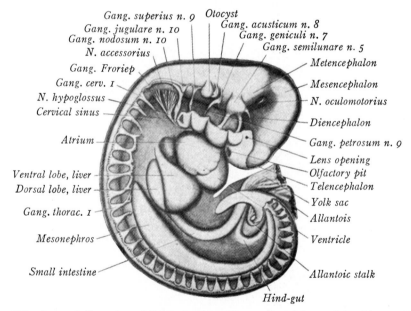

Fig. 595. Lateral dissection of 5.5 mm. pig embryo, viewed from right side. × 12.

A. ANATOMY OF A SIX MM. PIG EMBRYO

The general structure of a 6 mm. pig embryo is illustrated in Figs. 594–597. This should be compared with the chick embryo of four days (Figs. 589, 590) and the 5 and 8 mm. human embryos (Fig. 77). Some familiarity with the gross anatomy of the 6 mm. pig embryo will make the detailed study of the 10 mm. pig embryo, which follows, easier to understand.

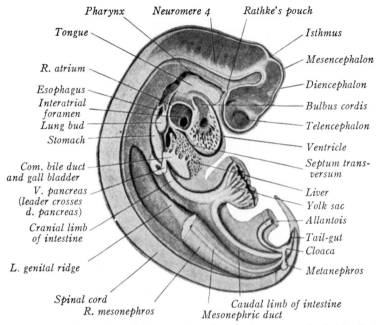

Pharynx Neuromere 4 Rathke's pouch

Tongue

Isthmus

Mesencephalon

R. atrium

Diencephalon

Esophagus

Interatrial foramen

Bulbus cordis

Lung bud

Telencephalon

Stomach

Ventricle

Com. bile duct and gall bladder

Septum transversum

V. pancreas (leader crosses d. pancreas)

Liver

Yolk sac

Allantois

Cranial limb of intestine

Tail-gut

Cloaca

L. genital ridge

Metanephros

Spinal cord

R. mesonephros

Caudal limb of intestine

Mesonephric duct

Fig. 596. Median dissection of 6 mm. pig embryo, after removal of right half. × 12.

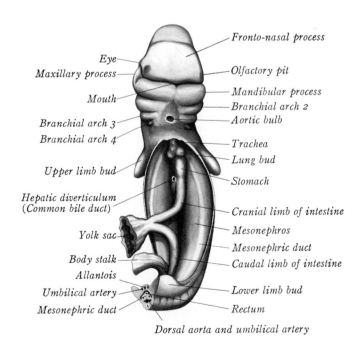

Fronto-nasal process

Eye

Maxillary process

Olfactory pit

Mouth

Mandibular process

Branchial arch 2

Branchial arch 3

Aortic bulb

Branchial arch 4

Trachea

Lung bud

Upper limb bud

Stomach

Hepatic diverticulum (Common bile duct)

Cranial limb of intestine

Mesonephros

Yolk sac

Mesonephric duct

Body stalk

Caudal limb of intestine

Allantois

Umbilical artery

Lower limb bud

Mesonephric duct

Rectum

Dorsal aorta and umbilical artery

Fig. 597. Ventral dissection of 6 mm. pig embryo. × 11. Head is bent dorsad.

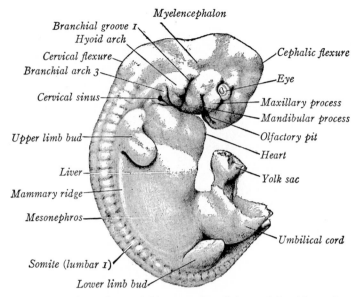

Fig. 598. Pig embryo of 10 mm., viewed from right side. × 7.

B. ANATOMY OF A TEN MM. PIG EMBRYO

This is the most instructive single stage of later development. Nearly all the important organs are represented, and yet the embryo is not so complex as to confuse unduly a beginner. Embryos between 8 and 14 mm. long may be used satisfactorily in conjunction with the descriptions that follow. Human embryos of 8 mm. and 12 mm. are shown in Figs. 77 *B* and 78, respectively. At this period a human embryo is slightly further advanced than a pig embryo of equal size, but at corresponding stages of development they are fundamentally alike.

External Form.   The *head,* which is relatively large on account of the dominance of the brain, makes a right-angled bend at the *cephalic flexure* (Fig. 598). On the under surface of the head are the *olfactory pits,* now drawn into elongate grooves and bounded by *lateral* and *median nasal processes.* The lens of the eye is prominent as it lies beneath the ectoderm, surrounded by the optic cup. At the sides of the head are four *branchial arches,* separated by three *branchial grooves.* The first branchial arch of each side forks ventrally into two parts. The smaller *maxillary processes* show signs of fusing with the median nasal processes to form the upper jaw, while the larger *mandibular processes* have united already into the lower jaw (*cf.* Fig. 604 *A*). Next caudad is the prominent second, or hyoid arch. Small tubercles, which will combine into the *auricle* of the external ear, bound the first branchial groove; the groove itself will become the *external acoustic meatus.* The third branchial arch is still visible in the future neck region, but the fourth arch has sunk into the *cervical sinus;* both disappear at a slightly later stage.

At the *cervical flexure* the head is bent at right angles to the body, thus bringing the ventral surface of the head close to the trunk (Fig. 598); it is probably owing to this flexure that the third and fourth branchial arches buckle inward to give rise to the *cervical sinus* (Fig. 604 *A*). Along its dorsal surface the trunk curves convexly, but this feature is not so prominent as at 6 mm. The reduction in trunk curvature results from the increased size of the heart, liver and mesonephroi. These organs are plainly indicated through the translucent body wall, while the position of the septum trans-versum may be noted between the heart and the liver (*cf.* Fig. 599). The *limb buds* are growing rapidly; the upper limbs are at a level between heart and liver. The *umbilical cord* is relatively large; it attaches at the lower end of the trunk. Dorsally the *somites* occur in serial order; toward the tail they become progressively smaller. Paralleling them and extending in a curve between the bases of the limb buds is the *mammary ridge;* on this thickened band of ectoderm will differentiate the mammary glands. The *tail* is long and tapering. Between its base and the umbilical cord is the *genital tubercle* (Fig. 599).

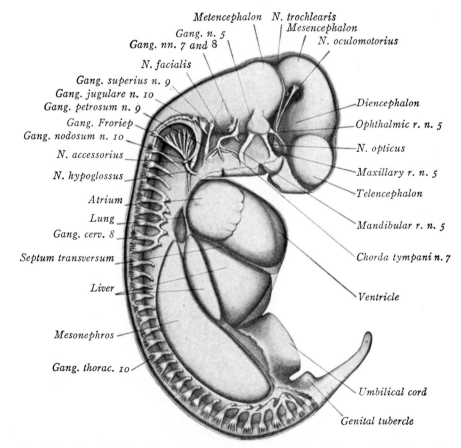

Fig. 599. Lateral dissection of 10 mm. pig embryo, viewed from right side. × 11.

Nervous System. BRAIN AND SPINAL CORD. Five distinct regions of the brain can be distinguished (Figs. 599, 601): (1) The *telencephalon* exhibits its rounded lateral outgrowths, the *cerebral hemispheres.* Their cavities, the *lateral ventricles,* communicate by interventricular foramina with the third ventricle. (2) The *diencephalon* shows a laterally flattened cavity, the *third ventricle.* From the ventrolateral side of the diencephalon pass off the optic stalks, while an evagination of the midventral wall (the infundibulum) will produce the neural lobe of the *hypophysis.* (3) The *mesencephalon* never subdivides, and its cavity becomes the *cerebral aqueduct* leading caudad into the fourth ventricle. (4) The *metencephalon* is separated from the mesencephalon by a constriction, the *isthmus.* Dorsolaterally it becomes the *cerebellum,* ventrally the *pons.* (5) The elongate *myelencephalon* is roofed over by a thin and non-nervous ependymal layer. Its ventrolateral wall is thickened and still gives internal indication of the *neuromeres.* The cavity of the metencephalon and myelencephalon is the *fourth ventricle.*

The *spinal cord* begins without specific demarcation and extends into the tapering tail. Just beneath the hind-brain and spinal cord lies the *notochord.*

CRANIAL NERVES. Of the twelve pairs of cranial nerves, all but the olfactory and abducent are represented in Fig. 599, where they occur in the order listed: (1) The *olfactory nerve* is not grossly demonstrable at this stage. (2) Fibers of the *optic nerve* are growing brainward within the optic stalk, cut through in this illustration. (3) The *oculomotor nerve,* a motor nerve to four of the eye muscles, takes origin from the ventrolateral wall of the mesencephalon and passes downward between the two parts of the bent brain. (4) The *trochlear nerve,* motor and destined for the superior oblique muscle of the eye, really arises from the ventral wall of the mesencephalon but emerges dorsally at the isthmus. The next eight pairs of nerves pass off from the rhombencephalon; four of these are rostral to the otocyst in the metencephalon and four lie caudally in the myelencephalon. (5) The *trigeminal nerve* is conspicuous because of its large *semilunar ganglion* and three branches (ophthalmic, maxillary and mandibular rami) which carry motor impulses to the jaw muscles and bring sensory impulses from the head. (6) The *abducent nerve* originates from the ventral brain wall and passes to the eye, where it will innervate the external rectus muscle. (7) The *facial nerve* is mixed, sensory and motor; it bears the *geniculate ganglion* and divides into chorda tympani, facial and superficial petrosal rami in the order named; most of the nerve has to do with the motor innervation of the face, whereas the sensory supply goes to the tongue. (8) The *acoustic nerve* arises just rostral to the otocyst; it bears the *acoustic ganglion* which will send sensory fibers to the internal ear. The postotic nerves are displayed in greater detail in Fig. 600. (9) Caudal to the otocyst is the *glossopharyngeal nerve,* showing a proximal *superior* and a more distal *petrosal ganglion;* its sensory and motor fibers innervate both tongue and pharynx. (10) The *vagus nerve* is mixed in function and has both a *jugular* and a *nodose ganglion;* its fibers innervate chiefly the viscera. (11) The *accessory nerve* has motor fibers which take origin both from the lateral wall of the myelencephalon and from the spinal cord as far caudad as the sixth cervical ganglion; an internal branch accompanies the vagus, while the external branch is distributed to the sterno-mastoid and trapezius muscles. (12) The *hypoglossal nerve* arises by five or six rootlets from the ventral wall of the myelencephalon; it is purely motor and supplies the muscles of the tongue.

It should be noted that the fifth, seventh, ninth and tenth cranial nerves pass into the four branchial arches in the order named. This primitive relation, better seen in Fig. 595, is maintained in the adult when the nerves innervate the derivatives of these arches.

A nodular chain of ganglion cells extends caudad from the jugular ganglion of the vagus (Fig. 600). These have usually been interpreted as *accessory vagus ganglia.* They may, however, be continuous with *Froriep's ganglion* which sends sensory fibers to the n. hypoglossus. In pig embryos of 15 mm. this chain is frequently divided into four or five ganglionic masses, of which occasionally two or three (including Froriep's ganglion) contribute fibers to the root fascicles of the hypoglossal nerve (Fig. 600).

SPINAL NERVES. Each nerve has a single *spinal ganglion,* from which the sensory *dorsal root* fibers are developed (Fig. 599, 619). The motor fibers take origin from the ventral cells of the neural tube; they form the *ventral roots* which join the dorsal roots in the common *nerve trunk.* In the region of the upper and lower limb buds the spinal nerves unite and give rise respectively to the *brachial* and *lumbo-sacral plexuses.*

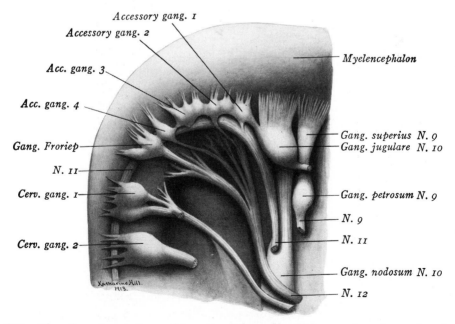

Fig. 600. Dissection of postotic cranial nerves and ganglia of 15 mm. pig embryo, viewed from right side. × 25.

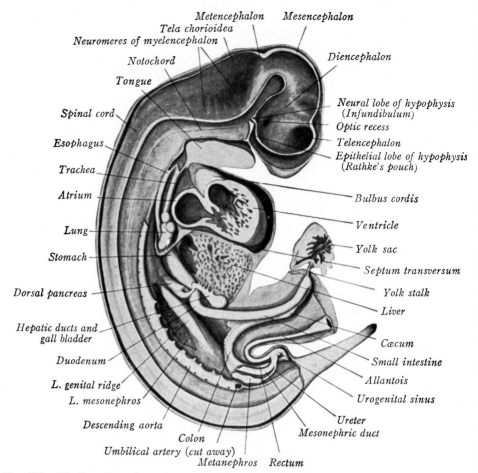

Fig. 601.   Median dissection of 10 mm. pig embryo, after removal of right half. × 10.

AUTONOMIC SYSTEM.   Paired masses of ganglion cells, located dorsolateral to the descending aorta, are the forerunners of this division of the nervous system. These segmental masses are *autonomic ganglia* and, as ganglionated chains, become the *sympathetic trunks.* Fibers pass from the spinal cord to the ganglia and these fibers comprise a branch of the spinal nerve known as its *communicating ramus* (r. communicans).

Sense Organs.   The *olfactory pits* are deep fossæ, flanked by the median and lateral nasal processes (Fig. 598). The stalked *optic cups* are prominent and the *lens vesicles* have detached from the ectoderm. Each *otocyst* is a compressed oval vesicle with a tubular *endolymph duct* growing from its medial side.

Digestive and Respiratory Systems.   MOUTH AND PHARYNX.   The pharyngeal membrane disappeared at a considerably earlier stage and the *stomodeum,* or ectodermal mouth, then becomes continuous with the pharynx. From the dorsal wall of the stomodeum, *Rathke's pouch* (epithelial hypophysis) extends as a long, stalked sac which forks at its end near the primordium of the neural lobe (Fig. 601). The floor of the mouth and pharynx is occupied by the *tongue* and *epiglottis* (Fig. 602). From the mandibular arches arise paired *lateral swellings* that become the body of the tongue. Lying between these thickenings is the transient *tuberculum impar.* The thyroglossal duct, which formerly opened just caudal to the tuberculum impar, is already obliterated; the *thyroid gland* itself, composed of branching epithelial cords, is now located in the midplane between the second and third branchial arches (Fig. 603). A median ridge, named the *copula,* unites the second arches and represents the primitive root of the tongue; it connects the tuberculum impar with the epiglottis which develops from the bases of the third and fourth branchial arches (Fig. 602). On each side of the slit-like *glottis* is an *arytenoid fold* of the larynx.

The pharynx is flattened dorsoventrally; it is broad at the oral end. Opposite the third branchial arch the pharynx bends sharply in conformity with the cervical flexure. The paired *pharyngeal pouches* are large, and each bears a dorsal and a ventral wing (Fig. 603); these relations are plainer in Fig. 195 *B.* The first pouch on each side persists as the *auditory tube* and *tympanic cavity;* the 'closing plate' between it and the first branchial groove forms the *tympanic membrane,* while the ectodermal groove becomes the *external acoustic meatus.* The second pouches are destined largely to disappear; about each develops a *palatine tonsil.* The dorsal wing of each tubular third pouch forms a *parathyroid gland;* the ventral wings differentiate into the *thymus.* The fourth pouches are smaller; their dorsal wings give rise to another pair of parathyroids, while the ventral wings are rudimentary. A tubular outgrowth, just caudal to each fourth pouch, is sometimes regarded as a fifth pharyngeal pouch; it forms an *ultimobranchial body.*

LARYNX, TRACHEA AND LUNGS.   The *larynx* and *epiglottis* are appearing (Fig. 602), and the *trachea* is a definite tube (Figs. 601, 603). Terminally the trachea bifurcates into *primary bronchi;* each of these has already divided again into secondary bronchial buds which indicate the two lobes of the *left lung* and the middle and lower lobes of the *right lung* (Fig. 604 *A*). From the right side of the trachea itself appears another bud which, in the pig, represents the upper lobe of the right lung.

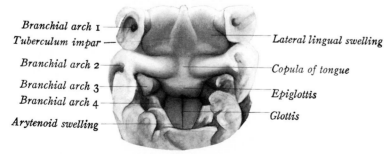

Fig. 602.   Floor of mouth and pharynx of 10 mm. pig embryo, after removing upper half of head. × 12.

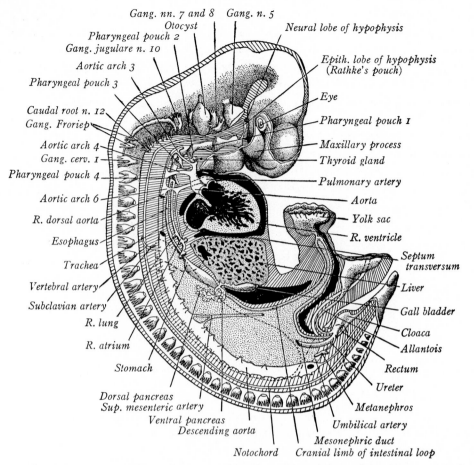

Fig. 603. Reconstruction of 10 mm. pig embryo, viewed from right side. × 10. Veins are not included; broken lines indicate outline of left mesonephros and positions of limb buds.

ESOPHAGUS AND STOMACH. The *esophagus* extends as a narrow tube past the lungs, whereupon it dilates into the laterally flattened *stomach* (Figs. 601, 603). The entire stomach has rotated about its axis so that the original dorsal border, now the convex greater curvature, lies to the left and the primitive ventral border (lesser curvature) to the right (Fig. 604 *A*). At this stage the rotation is incomplete.

INTESTINE. The pyloric end of the stomach opens into the *duodenum* which also shows the effect of stomach rotation; the stem of the *hepatic diverticulum,* originating from it ventrally, now lies to the right (Fig. 604 *A*). The diverticulum itself has differentiated into various things. From its tip has come the four-lobed *liver,* filling in the space between the heart, stomach and duodenum (Fig. 601). One of the several ducts now connecting the liver with the parent diverticulum will persist as the *hepatic duct.* The main stem of the diverticulum is the *common bile duct,* while a side sacculation is the cystic duct and gall bladder (Figs. 603, 604 *B*). The *ventral pancreas* springs from the common bile duct near its point of origin. It is directed dorsad and caudad, to the right of the duodenum. The *dorsal pancreas* arises a little more caudally (in man, cranially) from the dorsal wall of the duodenum; its larger, lobulated body grows dorsad and craniad (Figs. 603, 604 *B*). The two glands will interlock into a single organ; in the pig it is the duct of the dorsal pancreas that persists as the functional duct.

Beyond the duodenum the intestine is thrown into a loop, which extends well into the umbilical cord and connects with the *yolk stalk* there (Figs. 601, 603). Owing to rotation in the entire loop, the cranial limb of the intestine lies to the right, the caudal limb to the left. The small intestine (jejunum and ileum) extends as far as a slight enlargement on the caudal limb of the loop (Fig. 601). This is the *cæcum* which marks the beginning of the *large intestine* (colon and rectum). The cloaca is now subdividing into the *rectum* and *urogenital sinus*.

Cœlom and Mesenteries. The cœlom is a continuous, communicating system which includes the single *pericardial* and *peritoneal cavities*, still connected by paired *pleural canals*. Between the heart and liver is a prominent partition, the *septum transversum;* the liver is broadly fused to this

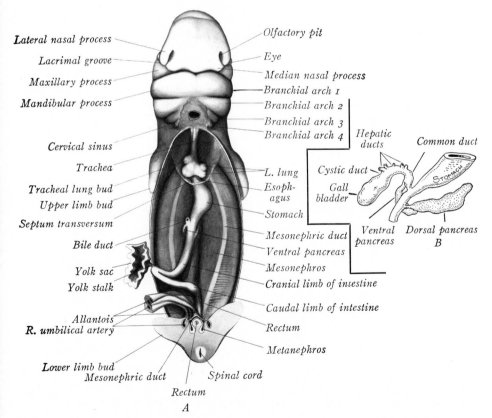

Fig 604. Ventral dissections of 10 mm. pig embryo. *A,* General view, with head bent dorsad ($\times$ 9). *B,* Detail of duodenal region ($\times$ 20).

septum which will comprise much of the diaphragm (Fig. 601). The double sheet of splanchnic mesoderm that serves as the primitive dorsal mesentery receives special names at different levels. Where it suspends the stomach it is known as the *mesogastrium,* or *greater omentum.* Then in order come the *mesoduodenum, mesentery* proper of the small intestine, *mesocolon* and *mesorectum.* The mesentery proper and some of the mesocolon follow the intestinal loop out into the umbilical cord (Fig. 603). The ventral mesentery is limited in extent. It persists as the *lesser omentum* between stomach and liver, encloses the liver, and continues as the *falciform ligament* between the liver and ventral body wall. A saccular recess, between the caval mesentery and liver ón the right and the stomach and its mesenteries on the left, is the *vestibule* and *omental bursa.* It opens through a narrowed *epiploic foramen* (of Winslow) (*cf.* Fig. 238 *A*).

Urogenital System.   The *mesonephroi* are large and complex in the pig (Figs. 599, 604 *A*). Along the middle of their ventromedial surfaces *genital ridges* have become prominent (Fig. 601). In a ventral dissection the course of the *mesonephric ducts* can be traced along the ventral margins of the mesonephroi and into the *urogenital sinus* (Fig. 604 *A*). The allantois is a conspicuous, stalked sac which communicates with the ventral part of the urogenital sinus (Fig. 601).

The *metanephroi,* or permanent kidneys, lie far caudad between the roots of the umbilical arteries (Figs. 601, 603). At the present stage each consists of a tubular epithelial portion, surrounded by a mass of condensed mesenchyme. The epithelial tube has budded off the mesonephric duct, near its ending; proximally there is a slender duct, the *ureter,* while a distal dilatation is the *renal pelvis.* From the pelvis grow out later the calyces and collecting tubules of the kidney. Encasing the pelvic primordium is a layer of condensed mesenchyme, derived from the lower nephrotomes and destined to differentiate into the secretory tubules, or *nephrons.*

Heart.   This organ lies within the pericardial cavity. Its general form and relations are illustrated in Figs. 599, 603. There are two *atria* and two thicker-walled *ventricles.* In addition, a small chamber, the *sinus venosus,* receives all the blood returned to the heart and directs it into the right atrium while the *bulbus cordis* still serves as a common arterial outlet (Fig. 605 *B*).

From the two hemisections shown in Fig. 605 the internal structure of the heart can be understood. The entrance from the sinus venosus into the right atrium is a nearly sagittal slit, guarded by right and left *valves of the sinus venosus* (*B*). Dorsally the two valves join and continue a short distance as the temporary *septum spurium* (*A*). Somewhat later, the sinus largely loses its identity by merging with the right atrium, although its middle part does persist as the *coronary sinus.* The dorsal wall of the left atrium is receiving a single *pulmonary vein* (not shown; *cf.* Fig. 346 *A*). The two *atria* are incompletely partitioned by the *septum primum* which contains an opening, the *interatrial foramen* (Fig. 605 *A*). On the right side of this partition a second sickle-like fold, the *septum secundum,* is forming. It also becomes an incomplete septum which bears an opening, known as the *foramen ovale.* After birth these two septa, together with the left valve of the sinus venosus, will fuse to complete the final atrial septum.

In a slightly younger embryo the atria and ventricles communicated through a common canal, bounded by two thickenings named *endocardial cushions.* At the present stage the two cushions have joined midway, have received the septum primum, and now subdivide this passage into two *atrio-ventricular canals* (Fig. 605 *A*). About the right canal the endocardium is already undermined, and in the process of forming the *tricuspid valve* (*B*); similarly, on the left is the developing *bicuspid valve.* The two *ventricles* are separated by a *ventricular septum;* it is complete except for the *inter-ventricular foramen* which connects the left ventricle with the bulbar part of the right (*A, B*). The *bulbus cordis* separates distally into *ascending aorta* and *pulmonary artery,* but proximally it is still undivided (*B*). The ventricular walls are thick and spongy, forming a meshwork of muscular trabeculæ, separated by sinusoids. Until later, when coronary vessels are developed, the heart receives all its nourishment from the blood circulating in the sinusoids.

Vascular System. ARTERIES. The aortic-arch system is still represented, although somewhat modified and in process of transforming into its permanent derivatives (Fig. 606). The first two pairs of arches have disappeared. The *ascending aorta* continues into the third and fourth pairs of arches. The third pair and the extensions of the dorsal aortæ into the head constitute continuous channels, to be known as the *internal carotid arteries.* Near their bases arise the *external carotid arteries,* which extend into the region of the lower jaw. The fourth aortic arch is largest, and on the left side it will form the permanent *arch of the aorta.* The sixth (fifth?) aortic arches connect with

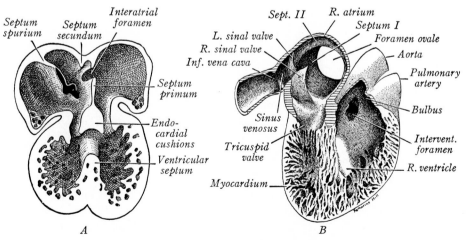

Fig. 605. Dissections of heart of pig embryos. × 20. *A,* At 10 mm., with ventral wall removed (after Patten). *B,* At 12 mm., with right wall removed.

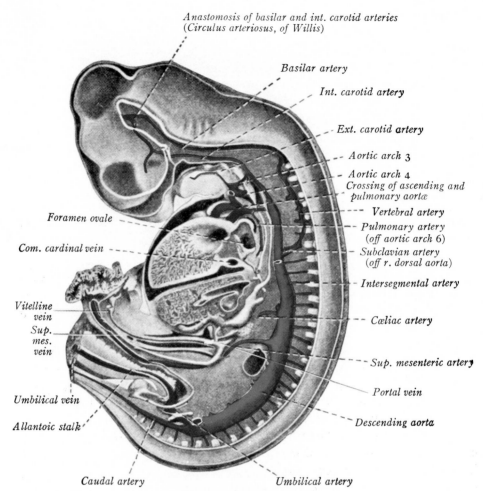

Anastomosis of basilar and int. carotid arteries
(Circulus arteriosus, of Willis)

Basilar artery

Int. carotid artery

Ext. carotid artery

Aortic arch 3

Aortic arch 4
Crossing of ascending and
pulmonary aortæ

Vertebral artery

Pulmonary artery
(off aortic arch 6)

Subclavian artery
(off r. dorsal aorta)

Intersegmental artery

Cœliac artery

Sup. mesenteric artery

Portal vein

Descending aorta

Foramen ovale

Com. cardinal vein

Vitelline vein

Sup. mes. vein

Umbilical vein

Allantoic stalk

Caudal artery

Umbilical artery

Fig. 606.   Reconstruction of midplane and right arteries of 12 mm. pig embryo (after
Patten, from Lewis). × 9.

the pulmonary trunk, and from them small *pulmonary arteries* pass to the
lungs; the left arch continues until birth as the *ductus arteriosus.* (Some
number the pulmonary arches as fifth, because of an inconstant pair.)

The paired *dorsal aortæ* unite opposite the eighth somites and continue
caudad as the median *descending aorta* (Fig. 606). The aorta shows dorsal,
lateral and ventral branches. The dorsal branches pass upward between the
somites, and accordingly can be called *intersegmental arteries.* From the
sixth cervical pair, which is located just where the dorsal aortæ combine, the
subclavian arteries pass off to the upper limb buds, and *vertebral arteries*
run cephalad into the head. The latter vessels are formed by longitudinal
anastomoses between the first six pairs on each side, after which the stems of
the first five atrophy. Under the brain the vertebrals are continuous with
the unpaired *basilar artery;* the latter connects with the internal carotids
beneath the diencephalon. Lateral branches of the descending aorta supply

the mesonephroi and genital ridges. Ventral branches form the *cœliac artery* to the stomach region, the *superior mesenteric artery* (primitive vitelline) to the small intestine and the *inferior mesenteric artery* to the large intestine. The *umbilical arteries* (to the allantois and placenta) belong in this ventral series, but they now arise laterally from secondary trunks which persist as the *common iliacs*. Beyond this point the aorta narrows into the diminutive *caudal artery* extending into the tail.

VEINS. Three sets of plexuses, which are the forerunners of the *dural sinuses,* occur alongside the brain. They drain into the *anterior cardinal veins,* now becoming the *internal jugular veins* (Fig. 607). After receiving the newer *external jugular veins* from the mandibular region, the anterior cardinals open into the *common cardinal veins* (ducts of Cuvier). The latter vessels also receive the *subclavian veins* from the upper limb-buds and the *posterior cardinals* from the lower body. They empty into the sinus venosus. The *posterior cardinal veins* are the oldest veins caudal to the level of the heart. They course dorsal to the mesonephroi and drain the mesonephric sinusoids (Fig. 607). However, the posterior cardinal veins are already beginning to decline, and midway along their lengths an interruption occurs; for

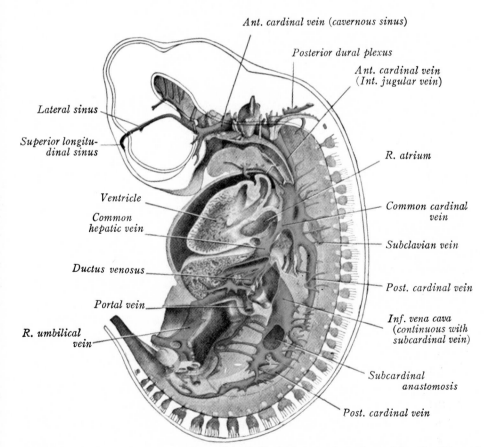

Fig. 607. Reconstruction of veins in right half of 12 mm. pig embryo (after Lewis). × 9.

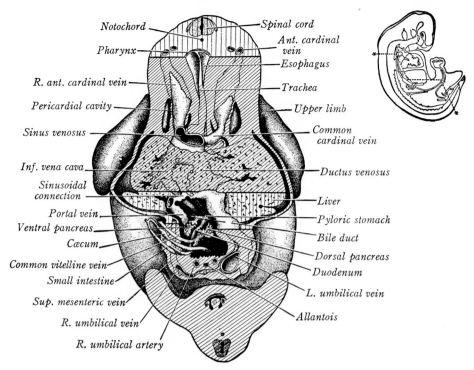

Notochord

Pharynx

R. ant. cardinal vein

Pericardial cavity

Sinus venosus

Inf. vena cava

Sinusoidal connection

Portal vein

Ventral pancreas

Cæcum

Common vitelline vein

Small intestine

Sup. mesenteric vein

R. umbilical vein

R. umbilical artery

Spinal cord

Ant. cardinal vein

Esophagus

Trachea

Upper limb

Common cardinal vein

Ductus venosus

Liver

Pyloric stomach

Bile duct

Dorsal pancreas

Duodenum

L. umbilical vein

Allantois

Fig. 608. Ventral reconstruction of 10 mm. pig embryo, especially to show umbilical and vitelline veins. × 15. In the small orientation figure (*cf.* Fig. 603) various planes are indicated by broken lines—*- - - - - -*.

this reason only the cranial halves communicate with the common cardinal stems.

Considerable diversion of blood from the posterior cardinal veins has been brought about by the development of *subcardinal veins* along the ventromedial surfaces of the mesonephroi. These vessels arose as longitudinal chanels in a mesonephric plexus that was originally tributary to the posterior cardinal veins. Connections between the post- and subcardinal systems of each side still exist, while the two subcardinals also communicate by a prominent anastomosis across the midplane of the body. Their drainage is now shifting into the just organizing inferior vena cava. A fairly prominent *ventral vein* of the mesonephros follows the ventral border of this organ, but it soon disappears.

The *inferior vena cava* is becoming established at this stage. It has a compound origin, some of which is now identifiable. In the mesonephric region the larger right subcardinal is an important component (Fig. 607). More cephalad a vein has developed in a specialized portion (caval mesentery) of the mesogastrium. This vessel connects the subcardinal with the hepatic (vitelline) sinusoids. The blood-flow through the sinusoids is already consolidating into a definite channel, and this is the hepatic part of the inferior vena cava (Fig. 608). It empties into the common hepatic vein (primitive right vitelline), which constitutes the stem of the vena cava.

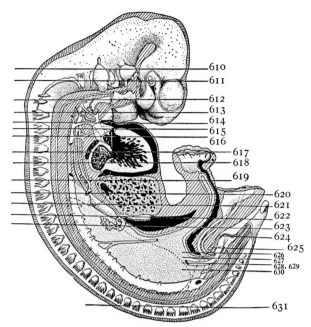

Fig. 609. Reconstruction of 10 mm. pig embryo (*cf.* Fig. 603). × 8. Numbered lines indicate levels of transverse sections shown as Figs. 610–631.

The *umbilical veins* follow the allantoic stalk back from the placenta. In the umbilical cord they have merged into a common vessel, but they separate again on entering the embryo where they course in the ventrolateral body wall of each side to the level of the liver (Fig. 607). The left umbilical is the larger of the two, and it alone persists in older fetuses. Cephalad of the liver the original stems that connected with the sinus venosus have disappeared, and umbilical blood is now routed through the liver in sinusoidal channels. Most important of these is an enlarged fetal passage, connecting the left umbilical with the inferior vena cava; it is the *ductus venosus* (Fig. 608).

Distally the two *vitelline veins* are fused. Passing inward from the regressive yolk sac, they course cephalad of the intestinal loop (Figs. 606, 608). In the pancreas-region the left vein receives the *superior mesenteric vein* which is a new vessel arising in the mesentery of the intestinal loop. Above this junction a cross anastomosis and a continuation of the right vein make a new channel which is known as the *portal vein*. It gives off branches to the hepatic sinusoids, which have arisen much earlier from a breaking up of the vitelline veins in this region, and connects with the left umbilical vein within the liver. Beyond the sinusoids the vitelline vessels are retained as *hepatic veins* and the stem of the *inferior vena cava*.

TRANSVERSE SECTIONS OF A TEN MM. PIG EMBRYO

The more important levels, as indicated by guide lines on Fig. 609, are illustrated and described. These are useful for the identification of organs,

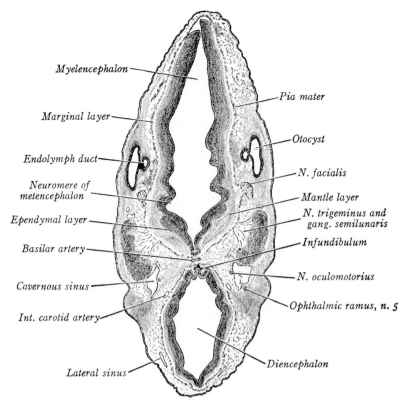

Myelencephalon

Marginal layer

Endolymph duct

Neuromere of
metencephalon

Ependymal layer

Basilar artery

Cavernous sinus

Int. carotid artery

Lateral sinus

Pia mater

Otocyst

N. facialis

Mantle layer

N. trigeminus and
gang. semilunaris

Infundibulum

N. oculomotorius

Ophthalmic ramus, n. 5

Diencephalon

Fig. 610.   Transverse section through semilunar ganglia and otocysts of 10 mm. pig
embryo. × 23.

but the student must interpret his sections with reference to the dissections
and reconstructions, and especially Fig. 603. All sections are drawn from the
cranial surface; accordingly, the right side of the embryo is at the reader's
left.

   Sections through Cephalic Flexure.   Since the head is flexed, the sections first encountered pass through the *mesencephalon* and *metencephalon*. At a slightly lower level
the metencephalon also becomes continuous with the thin-roofed *myelencephalon,* but
presently the mid-brain gives way to the *diencephalon* as the brain becomes cut twice.
Several important structures should be identified in the mesenchyme between these two
portions of the brain. In the midplane, but nearer the metencephalon, is the single
basilar artery; ventrolateral to the diencephalon are the paired *internal carotids* (*cf.* Fig.
610). These three vessels unite at the location of the future *arterial circle* (of Willis).
About halfway between the midplane and the lateral wall appear branches of the *anterior
cardinal veins* and the *oculomotor* and *trochlear nerves.* Of the two nerves, the trochlear
is smaller and slightly more lateral in position; in some series at this stage it extends only
a slight distance. The origin and relations of these nerves show plainly in Fig. 599.
   Section through Infundibulum, Semilunar Ganglia and Otocysts (Fig. 610).   The
brain is sectioned twice. At the bottom of the section is the *diencephalon,* cut transversely;
its cavity is the *third ventricle.* Midventrally the diencephalon gives off the *infundibulum*
which furnishes the neural lobe of the *hypophysis.* The *metencephalon* and *myelencephalon* are sectioned frontally; there is no clear demarcation between the two. Their walls

bear the prominent scalloping of the *neuromeres,* while the common cavity is the *fourth ventricle.* The wall of the entire neural tube now shows a differentiation into three layers: (1) an inner *ependymal layer,* densely cellular, next the central canal; (2) a middle *mantle layer,* of nerve cells and fibers; and (3) an outer *marginal layer,* chiefly fibrous. A thin, vascular layer surrounds the brain wall as the primitive *pia mater.*

The interval between the two portions of the brain contains several structures, sectioned transversely. Next the metencephalon is the unpaired *basilar artery;* ventrolateral to the diencephalon are the paired *internal carotid arteries.* Near the latter are *oculomotor nerves.* In this embryo the trochlear nerves had not grown down to this level. On the left side is a part of the ophthalmic branch of the *trigeminal nerve.* Tributaries of the *anterior cardinal veins* occur, the largest rostral to the semilunar ganglia. This is a portion of the *cavernous sinus,* while the stem of the *middle dural plexus* is just caudal to the semilunar ganglion; alongside the diencephalon the *lateral sinus* is cut.

Near the beginning of the hind-brain are the large *semilunar ganglia;* from their medial sides nerve fibers of the *trigeminal nerves* join the brain wall. This ganglion, situated at the pontine flexure of the metencephalon, constitutes one of the most important landmarks of the embryonic head. Slightly caudad lie the *facial nerves;* the left is cut as it leaves the brain wall. Midway along each side of the hind-brain will be seen the apex of an *otocyst,* and medial to it, the *endolymph duct;* on the left side the latter is cut near its origin from the otocyst.

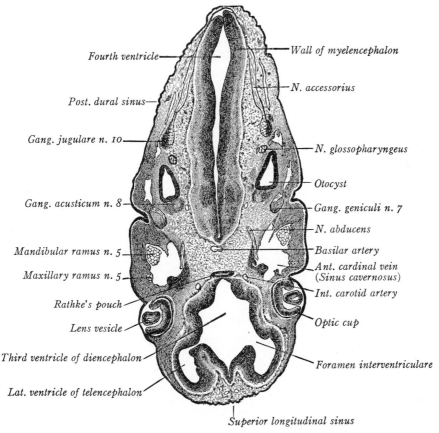

Fig. 611. Transverse section through cerebral hemispheres and eyes of 10 mm. pig embryo. × 23.

Section through Cerebral Hemispheres and Eyes (Fig. 611). This level shows some important new features. The *diencephalon* is now continuous with the *telencephalon*. The latter consists of a medial region which has evaginated paired *cerebral hemispheres;* their cavities, the *lateral ventricles,* connect through the *interventricular foramina* with the *third ventricle* of the diencephalon. Close to the ventral wall of the diencephalon is a section of the epithelial lobe of the hypophysis (*Rathke's pouch*), near which are the *internal carotid arteries.* Lateral to the diencephalon are the *optic cups,* sectioned just caudal to their stalks. The double wall of the optic cup comprises the *retina;* the thin outer layer is the pigmented epithelium, the inner and thicker coat is the neural layer. The *lens* is now a closed vesicle, distinct from the overlying *corneal ectoderm.*

The irregular vascular spaces are tributaries of the *anterior cardinal veins.* The largest space is the *cavernous sinus,* in the vicinity of the fifth nerve. The upper half of the section contains portions of the *posterior dural plexus,* while the two small vessels between the cerebral hemispheres at the bottom of the section represent the *superior longitudinal sinus.*

By working above and below the present level all the cranial nerves and ganglia, as well as the central connections and peripheral courses of these nerves, will be observed. In Fig. 611 transverse sections of the maxillary and mandibular branches of the *trigeminal nerve* are seen, while the *abducent nerve* is sectioned longitudinally as it passes from the under surface of the hind-brain toward the eyes. On the rostral side of each otocyst occur the *geniculate ganglion* of the *facial nerve* and the *acoustic ganglion* of the *acoustic nerve.* The *otocyst* is a sharply defined, epithelial sac that lies at the junction of the metencephalon and myelencephalon and makes a convenient landmark in identifying ganglia and nerves. Caudal to the otocyst, the *glossopharyngeal nerve* and the *jugular ganglion* of the *vagus nerve* are cut transversely, while the trunk of the *accessory nerve* is sectioned lengthwise as it curves forward from the level of the spinal cord.

Section through Mouth, Tongue and First and Second Pouches (Fig. 612). The tip of the head, with parts of the *telencephalon* and *olfactory pits,* is now separate from the rest of the section. Since the level last described, Rathke's pouch has opened into the ectodermal *stomodeum* between the jaws; the present section is at the actual *oral opening,* bounded by the *maxillary* and *mandibular processes* of the *first branchial arches.* At a slightly lower level, the mandibular processes merge in *lateral lingual swellings* which will become the body of the *tongue.* With the disappearance of the pharyngeal membrane, the stomodeum and entodermal mouth cavity have become continuous. The *pharynx* shows ventral portions of the *first* and *second pharyngeal pouches,* destined to be utilized as auditory tubes and tonsillar fossæ, respectively. Opposite the first pouch, externally, is the *first branchial groove,* or future external acoustic meatus. A shaving has been cut from the *tuberculum impar* of the *tongue* as it rises above the floor of the pharynx; the upper (caudal) end traces into connection with the second arches which unite as the *copula* (Fig. 602). The *facial nerves* of the *second branchial arches* are cut across. A little craniad of the present level the *trigeminal nerves* have ended in the maxillary and mandibular processes of the first arches.

The *myelencephalon* is sectioned close to its continuation into the spinal cord; *Froriep's ganglion* and some of the *accessory nerve* are included. Between the myelencephalon and the pharynx are seen on each side the several rootlets of the *hypoglossal nerve,* fibers of the *vagus* and *accessory nerves,* and the *petrosal ganglion* of the *glossopharyngeal nerve* which stands in relation to the third arch. Medial to the petrosal ganglia are the *dorsal aortæ,* and lateral to the vagi are the *anterior cardinal veins.* In the midplane is a bit of the *notochord,* cut lengthwise. The *basilar artery* still lies beneath the myelencephalon, but a short distance caudad it is replaced by the paired *vertebral arteries.*

Section through Third Pouches and Olfactory Pits (Fig. 613). The tip of the head is now small and includes on either side an open *olfactory pit,* lined with thickened ectodermal epithelium. Each pit is bordered by a *lateral* and a *median nasal process,* which assist in the formation of the nose and upper jaw. In preceding sections the epithelium within each pit fuses with that of the mouth; this *oro-nasal membrane* ruptures later and produces a *primitive choana.* The first three pairs of *branchial arches* show; the first is fused as the *mandible,* and the third is slightly sunken in the *cervical sinus.* The dorsal wings of the *third pharyngeal pouches* extend toward the ectoderm of the *third branchial grooves;* attached to these wings are prominent *parathyroid* primordia. The ventral wings are large, epithelial sacs that can be followed into succeeding sections; one is shown here as a separate drawing. They give rise to the *thymus.* The floor of the pharynx is sectioned through the *epiglottis.*

Ventral to the pharynx are portions of the *third aortic arches* (internal carotids) and the solid cords of the *thyroid gland.* (*External carotid arteries* arise at a slightly higher level, close to the ventral origins of the third aortic arches; they can be traced for a variable distance into the substance of the lower jaw.) Beneath the thyroid and in the mandible are portions of the *external jugular veins.* Dorsally the section passes through the *spinal cord* and the first pair of *cervical ganglia.* Between the spinal cord and pharynx, laterally, are the *anterior cardinal veins,* the *hypoglossal nerves* and the *nodose ganglia* of the *vagi.* Lateral to each ganglion is the external branch of the *accessory nerve* and medial to the ganglia are the small *dorsal aortæ.* The *notochord* is conspicuous in the midplane.

Section through Glottis and Fourth Aortic Arches (Fig. 614). At this level the first three branchial arches have been passed and the cephalic border of the heart is coming

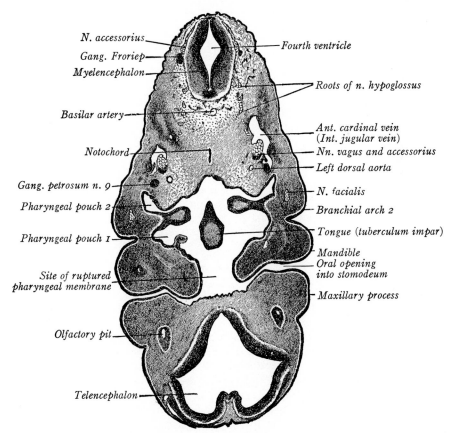

Fig. 612. Transverse section through mouth, tongue and first and second pharyngeal pouches of 10 mm. pig embryo. × 23.

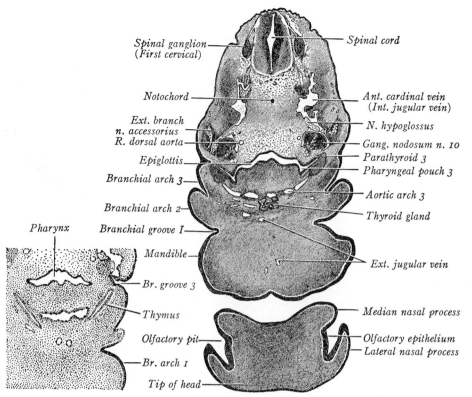

Fig. 613. Transverse sections through third pharyngeal pouches, thyroid gland and olfactory pits of 10 mm. pig embryo. × 23. Section at left is slightly caudad.

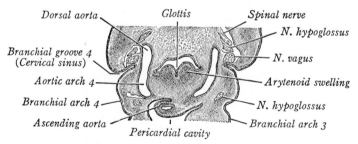

Fig. 614. Transverse section through glottis and fourth aortic arches of 10 mm. pig embryo. × 23.

into view. The illustration includes only the region of the *fourth branchial arches,* and the fourth pair of *aortic arches* which course through them. The left aortic arch will become the permanent *arch of the aorta;* the right serves as the stem of the right *subclavian artery.* The *ascending aorta* connects with these aortic arches a little craniad in the series. The *fourth branchial groove* now begins at the *cervical sinus* which is seen here. This duct extends for some distance before coming in contact with the fourth pharyngeal pouch of Fig. 615 (*cf.* Fig. 157). The section cuts across the *pharynx,* the *glottis* (entrance to the larynx), and its bordering *arytenoid swellings.* In addition to a spinal nerve, sections of the *vagus* and *hypoglossal nerves* are encountered.

Section through Fourth Pouches and Larynx (Fig. 615).   The head has been passed and the section is now dominated by the *heart,* lying within its *pericardial cavity.* The tips of the *atria* are sectioned as they project at the sides of the *bulbus cordis.* The bulbus is dividing into the aortic stem and the pulmonary trunk. The small section of the *ascending aorta* traces cephalad into the third (Fig. 613) and fourth aortic arches (Fig. 614), and caudad into the ventricle (Fig. 616). The *pulmonary trunk* is cut twice; its distal portion can be followed caudad into connection with the sixth aortic arches (Fig. 616).

The crescentic *pharynx* is continued laterad as the small *fourth pharyngeal pouches.* Each gives origin to a dorsal wing *(parathyroid),* encountered a few sections cephalad in the series and shown here as a separate drawing. At the level of the main section the pharynx is also continuous with a saccular *ultimobranchial body* (pouch 5?). From the midventral wall of the pharynx arises the solid epithelial plate of the *larynx.* A section of the *vagus nerve* is located between the dorsal aorta and the *anterior cardinal vein* of each side. Ventral to the anterior cardinals (soon to be called the *internal jugular veins)* are small *external jugular veins.* The left *dorsal aorta* is larger than the right in anticipation of its conversion into the permanent descending aorta of this level.

Section through Pulmonary Arches and Bulbus Cordis (Fig. 616).   The heart is little changed from the last level. However, the *ascending aorta* communicates with the *bulbus,* while the latter shows on each side the thickened internal *ridges* that will progressively meet and continue the separation of the aortic and pulmonary trunks. The *sixth aortic arches* connect the dorsal aortæ with the main *pulmonary trunk,* whose origin was traced in the preceding section. On the left side of the embryo the arch is complete; it represents the *ductus arteriosus,* which remains patent until birth. From these pulmonary arches small *pulmonary arteries* may be traced caudad in the series toward the lungs. The *esophagus* is now separate from the *trachea;* both are cut through their extreme cephalic ends. Ventrolateral to the spinal cord are somewhat diffuse *myotomes,* while *sclerotome*

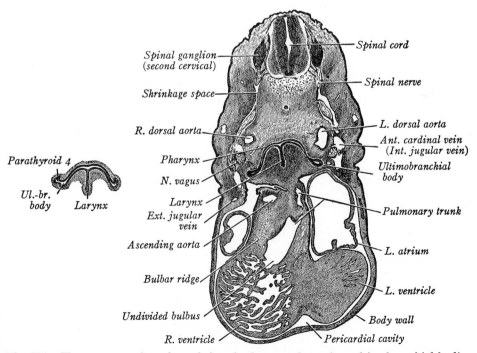

Fig. 615.   Transverse sections through fourth pharyngeal pouches, ultimobranchial bodies and larynx of 10 mm. pig embryo. × 23. Section at left is slightly craniad.

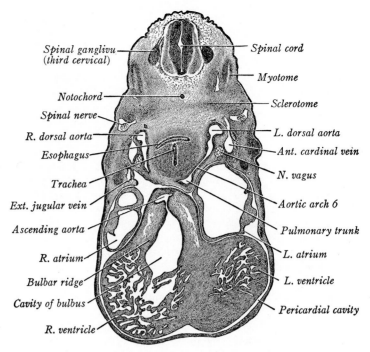

Spinal ganglivu
(third cervical)

Notochord

Spinal nerve

R. dorsal aorta

Esophagus

Trachea

Ext. jugular vein

Ascending aorta

R. atrium

Bulbar ridge

Cavity of bulbus

R. ventricle

Spinal cord

Myotome

Sclerotome

L. dorsal aorta

Ant. cardinal vein

N. vagus

Aortic arch 6

Pulmonary trunk

L. atrium

L. ventricle

Pericardial cavity

Fig. 616.   Transverse section through pulmonary arches and bulbus cordis of 10 mm. pig embryo. × 23.

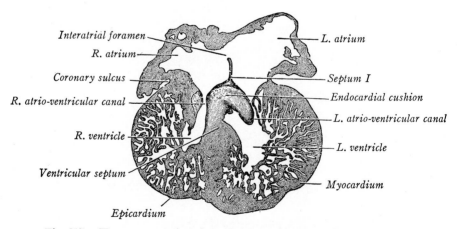

Interatrial foramen

R. atrium

Coronary sulcus

R. atrio-ventricular canal

R. ventricle

Ventricular septum

Epicardium

L. atrium

Septum I

Endocardial cushion

L. atrio-ventricular canal

L. ventricle

Myocardium

Fig. 617.   Transverse section through heart of 10 mm. pig embryo. × 23.

masses surround the notochord. The *vagus nerves* are prominent. At about this level the *external jugular veins* join the *anterior cardinals*.

   Section through Heart and Interatrial Foramen (Fig. 617).   Only the heart is figured, taken from a total section much like Fig. 618. All four chambers are shown. The *atria* are partially separated by the *septum primum,* which is incomplete because of the *interatrial foramen;* this foramen will remain open until birth. A septum secundum is not seen satisfactorily in embryos of this size. Each atrium communicates with the ventricle of the same side through an *atrio-ventricular canal.* Between these openings is the fused portion of the *endocardial cushions.* At an earlier stage these cushions were double, but they have fused midway and thus divide the originally single canal into two; they will also help in the formation of the *bicuspid* and *tricuspid valves*. The atria are marked off externally from the ventricles by the *coronary sulcus*. Between the two ventricles is the *ventricular septum;* a little higher in the series of sections the thin region (above leader) gives way to a temporary *interventricular foramen* (*cf.* Fig. 605 *B*). The ventricular walls are thick and spongy, forming a network of muscular *trabeculæ* surrounded by labyrinthine blood spaces. This muscular layer constitutes the *myocardium*. It is lined by an endothelial layer, the *endocardium,* while the entire heart is surrounded by a layer of mesothelium, the *epicardium,* or visceral pericardium. The latter sac is continuous with the parietal pericardium, which lines the body wall in the region of the heart.

Section through Common Cardinals and Sinus Venosus (Fig. 618). The section is also marked by the large *heart* and the bases of the upper *limb buds*. Dorsal to the atria are the *common cardinal veins*. The right vein empties into the *sinus venosus;* the left crosses the midplane and connects with the sinus at a lower level. Just above the plane of this section the right common cardinal has received the right *subclavian vein* from the limb bud; the left subclavian is still separate. The sinus venosus drains into the right atrium through a slit-like opening in the dorsal and caudal atrial wall. The opening is guarded by the right and left *valves of the sinus venosus,* both of which project into the atrium. The *septum primum* appears like a complete membrane at this level, which is a little caudal to the interatrial foramen and the atrioventricular canals. The septum joins the fused *endocardial cushions,* as does the *ventricular septum* from below.

The *esophagus* and *trachea* are tubular. Around the epithelium of both are condensations of mesenchyme, from which their fibrous and muscular layers are to be differentiated; laterally in this mass lie the *vagus nerves,* unlabeled in the illustration. Ventral to the trachea are the *pulmonary arteries*. The left *dorsal aorta* is large and is here continuing

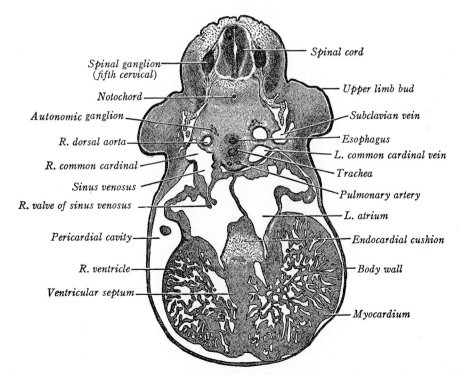

Fig. 618. Transverse section through common cardinals and sinus venosus of 10 mm. pig embryo. × 23.

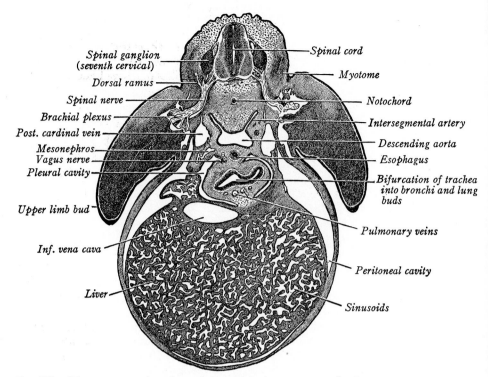

Fig. 619.   Transverse section through brachial plexus and tracheal bifurcation of 10 mm. pig embryo. × 23.

the arch of the aorta caudad; the right dorsal aorta at this level will serve as the stem of the right subclavian artery. Dorsolateral to the aortæ are *autonomic ganglia.* The condensation of mesenchyme about the notochord foreshadows a future *vertebra.*

Section through Brachial Plexus and Tracheal Bifurcation (Fig. 619).   Other distinctive features are the presence of upper *limb buds* and the *liver.* The seventh pair of cervical *spinal nerves* is cut lengthwise in diagrammatic fashion. The spindle-shaped ganglion is associated with the *dorsal root* whose fibers grow out from its cells; the fibers of the *ventral root* arise from the middle, cellular layer of the cord and join the dorsal root in the common *nerve trunk.* On the right side, a short *dorsal ramus* supplies the dorsal muscle mass close by the ganglion. The much larger *ventral ramus* unites with similar rami of several other nerves and forms the *brachial plexus,* a part of which is seen. Adjacent to the esophagus are the cut *vagus nerves,* which follow this tube laterally.

The *descending aorta* shows its manner of origin from paired vessels. From the sixth pair of cervical *intersegmental arteries,* which arise dorsally from it, the *subclavian arteries* are given off two sections caudad in the series. Traced cephalad these seventh intersegmentals become continuous with the *vertebral arteries.* The latter lie medial to the stem portions of the spinal nerves. In some embryos they are imperfectly developed at this stage. Lateral to the aorta are the *posterior cardinal veins,* easily traced to the common cardinals of the previous figure.

Beneath the *esophagus* the trachea branches into short *primary bronchi.* These continue laterad into buds that represent the upper left and middle right lobes of the *lungs.* The upper right bud (unpaired) comes off the trachea a little cephalad in the series (*cf.* Fig. 604); the paired lower lobes are more caudad (Fig. 620). Crescentic *pleural cavities* bound the bulging pulmonary tissue laterally. On the embryo's left this cavity is separated from the peritoneal cavity by a *pleuro-peritoneal membrane,* and here the parietal and visceral pleura assume typical relations. The broad medial mass of mesenchyme containing the esophagus and lungs is the *mediastinum.* Ventral to the bifurcating trachea are sections of the *pulmonary veins,* which can be traced into the left atrium at a slightly higher level. The liver, with its close network of trabeculæ and sinusoids, is large and nearly fills the *peritoneal cavity.* A little craniad in the series its attachment to the septum transversum is seen better. In the present section the *inferior vena cava* is leaving the liver and entering the septum. This portion of the vena cava is, in reality, the *common hepatic vein* (stem of the primitive right vitelline).

Section through **Lungs, Upper Limbs and Autonomic Ganglia** (Fig. 620).   The *limb buds* are ectodermal sacs stuffed with dense, undifferentiated mesenchyme. Ectodermal thickenings, at the tips, exert control over their progressive organization. Flanking the now circular *descending aorta* are the cranial ends of the *mesonephroi.* Above the aorta is a pair of *autonomic ganglia;* the *communicating rami* connecting the spinal nerve trunks do not show, but can be seen in neighboring sections, like Figs. 622, 623. The esophagus is just beginning to dilate into the *stomach,* and at this level the *vestibule* of the omental bursa appears as a crescentic slit to the right and below it. The mediastinum of higher levels has changed into a typical mesentery of the lower esophagus (*meso-esophagus*) and stomach (*mesogastrium,* or *omentum*). The *lungs* are sectioned through their paired lower lobes. Both *pleural cavities* still communicate freely with the peritoneal cavity. In the right dorsal lobe of the liver is located more of the intrahepatic portion of the *inferior vena cava;* this particular segment is organizing from enlarged hepatic sinusoids. Near the midplane is the large *ductus venosus,* which traces into union with the vena cava a short distance cephalad. The *posterior cardinal veins* are coming into intimate relation with the *mesonephroi;* these temporary kidneys are cut near their cranial ends.

Section through **Stomach and Liver** (Fig. 621).   It is the *stomach* and lobate *liver* that feature this level. The stomach has rotated partially, so that its original dorsal margin is now dropping to a position on the embryo's left, whereas the primitive ventral margin is rising correspondingly on the right. These margins are to be the *greater* and *lesser curvatures,* respectively. The stomach is attached dorsally by the *greater omentum;* ventrally

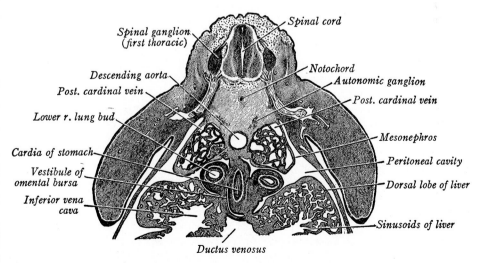

Fig. 620.   Dorsal half of transverse section through lungs, upper limbs and autonomic ganglia of 10 mm. pig embryo. × 23.

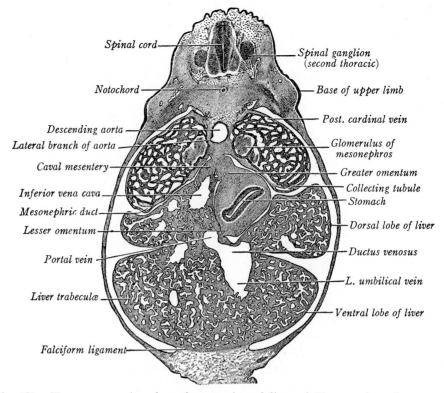

Spinal cord

Notochord

Descending aorta

Lateral branch of aorta

Caval mesentery

Inferior vena cava

Mesonephric duct

Lesser omentum

Portal vein

Liver trabeculæ

Falciform ligament

Spinal ganglion (second thoracic)

Base of upper limb

Post. cardinal vein

Glomerulus of mesonephros

Greater omentum

Collecting tubule

Stomach

Dorsal lobe of liver

Ductus venosus

L. umbilical vein

Ventral lobe of liver

Fig. 621. Transverse section through stomach and liver of 10 mm. pig embryo. × 23.

the *lesser omentum* passes to the liver. This ventral mesentery splits into halves and is continued as a peritoneal reflection around the liver; the component layers then come together again as the *falciform ligament,* attaching the midventral border of the liver to the body wall. Both the body wall and the abdominal viscera are thus seen to be surfaced with a continuous sheet of mesothelium underlaid by mesenchyme; this serous investment is the *peritoneum,* in which a parietal and a visceral division are recognized.

The liver shows paired dorsal and ventral lobes. The right dorsal lobe is fused dorsally to the greater omentum. This connection forms the *caval mesentery,* in which courses the *inferior vena cava.* Between the attachments of the stomach and liver, and to the right of the stomach, is the *vestibule* of the omental bursa. Midventrally in the liver is the *ductus venosus,* sectioned just at the point where it receives the intrahepatic continuation of the left *umbilical vein* and a branch from the *portal vein.* The liver tissue is a complicated network of trabeculæ and sinusoids; the component *liver cords* are composed of hepatic cells, surrounded by the endothelium of the sinusoids; red blood cells are developing here at this stage.

The *mesonephroi* are becoming prominent organs. Along their ventral margins course the *mesonephric ducts;* each shows a connection laterally with the terminal segment of a *collecting tubule.*

Section through Pyloric Stomach and Gall Bladder (Fig. 622). The section grazes the third thoracic *ganglia,* passes slantingly through the pyloric end of the *stomach,* then cuts the *common bile duct,* and finally passes lengthwise through the *gall bladder* superficially embedded in the substance of the *liver.* Within the distance of a few sections it is easy to demonstrate the continuity of these component parts of the digestive system. The *greater*

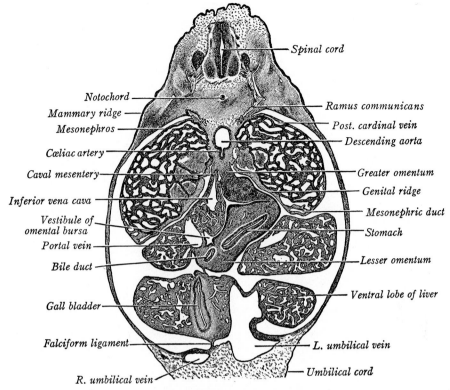

Notochord
Mammary ridge
Mesonephros
Cœliac artery
Caval mesentery
Inferior vena cava
Vestibule of omental bursa
Portal vein
Bile duct
Gall bladder
Falciform ligament
R. umbilical vein

Spinal cord
Ramus communicans
Post. cardinal vein
Descending aorta
Greater omentum
Genital ridge
Mesonephric duct
Stomach
Lesser omentum
Ventral lobe of liver
L. umbilical vein
Umbilical cord

Fig. 622. Transverse section through pyloric stomach and gall bladder of 10 mm. pig embryo. × 23.

omentum of the stomach is larger and more folded than in the previous illustration, and the *omental bursa* (lying horizontally over the stomach) is correspondingly expansive. It opens into a vertical space, which is the *vestibule;* traced caudad a short distance the region of the vestibule, labeled in Fig. 622, opens into the peritoneal cavity through the *epiploic foramen* (of Winslow). Beneath the stomach is a caudal portion of the *lesser omentum.* The blood supply of the stomach-pancreas region comes from the *cœliac artery,* which is seen emerging as a ventral branch of the aorta.

The liver has nearly been left behind, and its dorsal and ventral sets of lobes are now separate. Associated with the liver are several veins. Dorsally is the *inferior vena cava,* about to leave the liver within a lip-like fold, the *caval mesentery.* Also in the right upper lobe is the *portal vein.* Ventrally, where the umbilical cord joins the body wall, are the *umbilical veins;* the larger left vessel is entering the left ventral lobe of the liver on its way to the ductus venosus. On each dorsolateral surface of the trunk is a thickened ectodermal ridge, poorly shown in Fig. 622, which represents the *mammary ridge.* At the same horizontal level a *ramus communicans* passes from the left nerve trunk to an *autonomic ganglion.* The bottom of the figure includes a little of the insertion of the *umbilical cord* on the abdominal wall.

Section through Pancreas, Intestinal Loop and Tail (Fig. 623). The bulging *mesonephroi* are conspicuous. *Mesonephric corpuscles,* with vascular *glomeruli* indenting them, are medial in position. The glomeruli receive *mesonephric arteries,* arising as lateral branches from the aorta; one of these is shown. The *mesonephric tubules* are contorted and variously sectioned; they are lined with a cuboidal epithelium and empty into the *mesonephric duct* coursing along the ventral margin of the gland. A reconstruction of

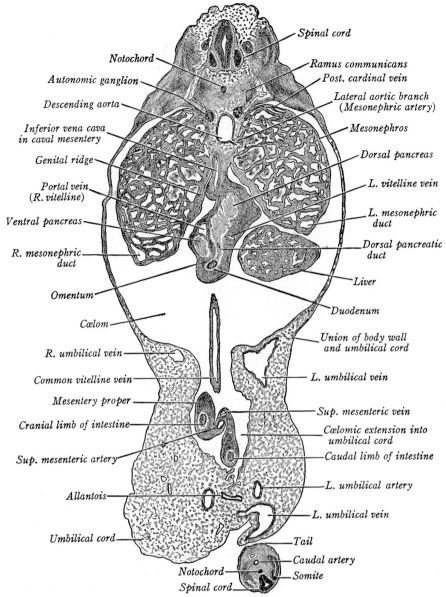

Fig. 623. Transverse section through pancreas, intestinal loop and tail of 10 mm. pig embryo. × 23.

a complete tubule is shown in Fig. 267 C. On the medial surface of each mesonephros the epithelium is thickened; from these *genital ridges* the sex glands will differentiate. The *posterior cardinal veins* lie on the dorsal surfaces of the mesonephroi, and transient *ventral veins* of the mesonephroi course along the ventral borders. The *inferior vena cava* is a vertical slit in the caval mesentery. Traced caudad a short distance it joins the right sub-cardinal vein, which continues this important venous channel down the trunk. The *sub-cardinal veins* are prominent vessels, seen well at a slightly lower level on the medial surfaces of the *mesonephroi*. There they interconnect by a large anastomosis (*cf.* Fig. 607).

The *duodenum* lies within its dorsal mesentery (*mesoduodenum*). The duct of the lobulated *dorsal pancreas* is shown arising directly from the duodenal wall. More to the right is a section of the *ventral pancreas* which traces craniad to its origin from the stem of the common bile duct. On each side of the dorsal pancreas are portions of the *vitelline veins,* the right at this level being a part of the *portal vein.* A few sections caudad this vessel can be followed across a transverse anastomosis to the left vitelline vein where the *superior mesenteric vein* from the mesentery attaches. Beyond the duodenum the vitelline veins have fused into a common vessel, cut lengthwise in the present section; it can be followed to the yolk sac, where the right and left components again separate (*cf.* Fig. 606).

The ventral body wall is continuous with the *umbilical cord.* This contains an extension of the embryonic *cœlom,* and a portion of the *intestinal loop* within its *mesentery.* Between the two intestinal limbs are the *superior mesenteric artery* and *vein.* The former is a ventral branch of the aorta; the latter joins the portal vein. The *allantois* is flanked by *umbilical arteries,* while *umbilical veins* are cut in the cord and again as they enter the body wall. The tip of the recurved *tail* shows as a separate section; its structure is simple.

Section through Cloacal Membrane (Fig. 624).   To maintain the proper relations with sections already studied, this and succeeding sections through the curved, caudal region are shown dorsal side down. The caudal end of the embryo is small. Its laggard differentiation, in comparison to higher levels, is reflected in the less specialized *spinal cord* and *somites.* The slender *tail-gut* is cut across. Between the *notochord* and tail-gut is the continuation of the aorta, known here as the *caudal artery.* On each side of the latter lies the termination of a *posterior cardinal vein.* The ventral half of the section is featured by an epithelial plate that represents the solid *cloacal membrane.* The plane of section is such that the fusion of ectoderm with the entoderm of the plate is shown only at one end.

Section through Subdividing Cloaca (Fig. 625).   Tracing a short distance down the series from the previous level, the tail-gut joins the *cloaca* and the latter gains a cavity. Still farther, at the present level, the cloaca is separating into a dorsal *rectum* and a ventral *urogenital sinus.* (It should be understood that the recurvation of the tail-end of the embryo makes caudal progress in the sectioned series actually craniad on this part of the embryo; *cf.* Fig. 609.)

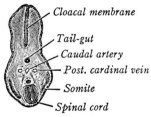

Fig. 624. Transverse section through cloacal membrane of 10 mm. pig embryo. × 23.

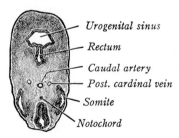

Fig. 625. Transverse section through subdividing cloaca of 10 mm. pig embryo. × 23.

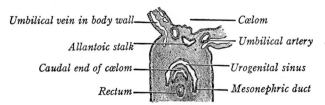

Fig. 626. Transverse section through allantois, urogenital sinus and rectum of 10 mm. pig embryo. × 23.

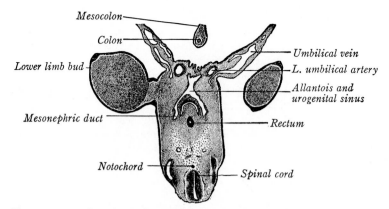

Fig. 627. Transverse section through stems of mesonephric ducts and allantois of 10 mm. pig embryo. × 23.

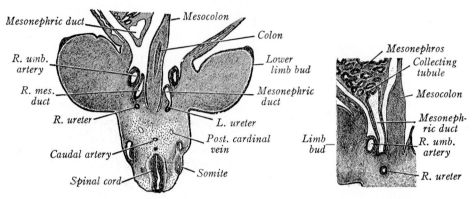

Fig. 628. Transverse section through lower limb buds and ureteric stems of 10 mm. pig embryo. × 23.

Fig. 629. Transverse section through right mesonephric duct and ureter of 10 mm. pig embryo. × 23.

Section through Allantois, Urogenital Sinus and Rectum (Fig. 626). This illustration (and the four that follow) include only the caudal, recurved part of the embryo (*cf.* Fig. 609). It is but a few sections craniad of Fig. 627 and resembles it closely. In the ventral body wall is seen the *allantoic stalk,* accompanied by the *umbilical arteries;* slightly craniad in the series they all enter the umbilical cord. More dorsad are the crescentic *urogenital sinus* and the *rectum,* now separate. The caudalmost portion of the *cœlom* tapers to an end between the two.

Section through Stems of Mesonephric Ducts and Allantois (Fig. 627). The *colon* is contained within a portion of the mesentery that is specifically named the *mesocolon.* In the body wall are tributaries of the *umbilical veins,* and medial to them are the *umbilical arteries.* The *allantoic stalk* is sectioned as it opens into the *urogenital sinus.* Dorsal to the sinus is a section of *rectum,* separated from the sinus by a crescentic prolongation of the *cœlom.* The rectum has no mesentery. The horns of the urogenital sinus receive the *mesonephric ducts.*

Section through Lower Limb Buds and Ureteric Stems (Fig. 628). The section cuts through the middle of both lower *limb buds.* Like the upper set, already studied, they consist of undifferentiated mesenchyme contained within an epithelial sac whose apex shows marked thickening. Medial to the limb buds are the *umbilical arteries,* which in turn lie lateral to the *mesonephric ducts.* The left mesonephric duct is cut at just the proper plane to show the *ureter,* or duct of the metanephros, being given off dorsally. The right *ureter* is sectioned transversely and appears as a separate tube. The *colon* is cut lengthwise; tracing craniad in the series, it becomes continuous with the colon and rectum, as seen in Fig. 627.

Section through Mesonephric Ducts and Ureters (Fig. 629). Continuing down the series the ureters can be traced for some distance. The present illustration is a small part of a section close to Fig. 628. In it the right *ureter* is seen. Also the right *mesonephric duct* is cut lengthwise (frontally) as it leaves the mesonephros on its way to connect with the urogenital sinus.

Section through Metanephroi (Fig. 630). The ureters are found to terminate in the *metanephroi,* figured here. Each of these kidney primordia consists of two parts. Internally there is a dilated expansion of the ureter that represents the *renal pelvis;* from it, first the *calyces* and then the system of *collecting tubules* will bud and grow. The periphery of the double primordium is a mass of condensed mesenchyme, derived from nearby nephro-

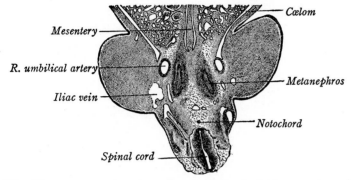

Fig. 630. Transverse section through metanephroi of 10 mm. pig embryo. × 23.

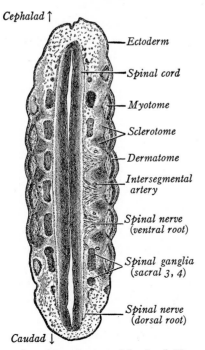

Fig. 631. Transverse section through curved back of 10 mm. pig embryo. × 23.

tomes; this tissue will differentiate into the *secretory tubules* of the kidney. *Iliac veins,* branches of the posterior cardinals, are passing into the limb buds.

Section through Curved Back (Fig. 631). Owing to the lumbar curvature, this section is actually frontal. The *spinal cord* is cut lengthwise. It is flanked by *spinal ganglia,* except midway on the left side where a slightly deeper plane includes several *spinal nerves.* The ganglia include the third lumbar to the fourth sacral. The somites show spindle-shaped *myotomes* and more laterally located so-called *dermatomes.* The medial side of each somite is a *sclerotome;* this shows subdivision into a caudal denser and a cranial less dense half. Recombination of the dense half of one somite with the sparser half of the somite next caudad will produce a definitive *vertebra* (*cf.* Fig. 365). *Intersegmental arteries* appear between some of the somites on the left side.

C. ANATOMY OF AN EIGHTEEN MM. PIG EMBRYO

Most of the important organs are laid down in 10 mm. embryos. Older stages are chiefly instructive, therefore, to demonstrate the growth and differentiation of parts already present, rather than the introduction of new ones. Dissections show perfectly the form and relations of organs, their relative rates of growth and changes of position. Since the illustrations indicate better than descriptions the several structures and their states of development, only certain selected features will be mentioned.

External Form. The neck and back are much straighter than before, but the ventral body is still highly convex. The head is relatively larger, the umbilical cord smaller. The sense organs are prominent, and the face, with snout and jaws, is plain. The branchial grooves and cervical sinus have disappeared from the neck. The limbs show indications of proximal and distal divisions, and the hand and foot are paddle-like. Several mammary gland primordia occur along the mammary ridges, now located more ventrally. The genital tubercle has become a distinct phallus.

Lateral Dissection (Fig. 632). The cerebral hemispheres are larger and the cerebellum is appearing. Beneath the cerebellum is the prominent pontine flexure of the brain, pointing ventrad. Nerves and ganglia show clearly; the brachial and lumbo-sacral plexuses, opposite the limbs, are noteworthy. The liver and lungs are relatively larger and more plainly lobed than before; the heart and mesonephroi are smaller.

Midsagittal Dissection (Fig. 633). The corpus striatum has developed in the floor of the cerebral hemisphere, a chorioid plexus invades the fourth ventricle, and the neural (posterior) lobe of the hypophysis is growing into association with the detached Rathke's pouch. Sclerotomic primordia of vertebræ condense about the notochord. The viscera show only quantitative changes from the 10 mm. stage, but the urogenital sinus and rectum are now separate, as are the aorta and pulmonary artery. The intestinal loop has rotated until the cranial and caudal limbs lie right and left, respectively. The cæcum is conspicuous and a urinary bladder has developed between the allantois and urogenital sinus.

Ventral Dissections. The lungs, septum transversum, stomach, intestine

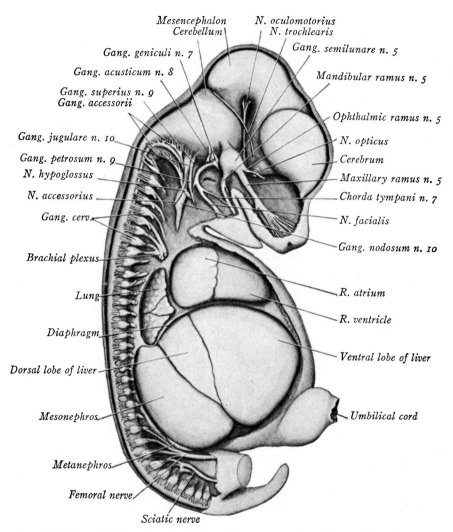

Mesencephalon
Cerebellum
N. oculomotorius
N. trochlearis
Gang. geniculi n. 7
Gang. semilunare n. 5
Gang. acusticum n. 8
Mandibular ramus n. 5
Gang. superius n. 9
Gang. accessorii
Ophthalmic ramus n. 5
Gang. jugulare n. 10
N. opticus
Gang. petrosum n. 9
Cerebrum
N. hypoglossus
Maxillary ramus n. 5
N. accessorius
Chorda tympani n. 7
Gang. cerv.
N. facialis
Brachial plexus
Gang. nodosum n. 10
Lung
R. atrium
Diaphragm
R. ventricle
Dorsal lobe of liver
Ventral lobe of liver
Mesonephros
Metanephros
Umbilical cord
Femoral nerve
Sciatic nerve

Fig. 632. Lateral dissection of 18 mm. pig embryo, viewed from right side. × 8.

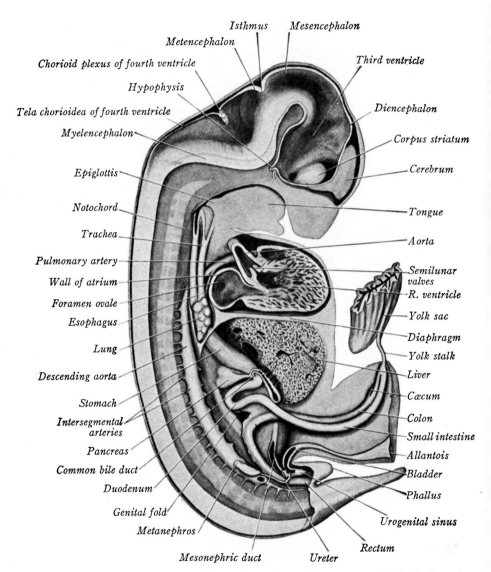

Fig. 633.   Median dissection of 18 mm. pig embryo, after removal of right half. × 8.

and mesonephroi are the chief organs seen in the 15 mm. embryo shown as Fig. 634. Of special interest are the beginnings of the Müllerian ducts.

Figure 635 is a dissection of a slightly larger embryo (18 mm.). From it the stomach and small intestine have been removed to display the genital glands. These organs have advanced rapidly since the 10 mm. stage. They are now definitely established and localized as recognizable gonads. Each Müllerian duct opens cranially by a funnel-shaped ostium; the duct proper is growing caudad as a blind tube that parallels the course of the mesonephroic duct.

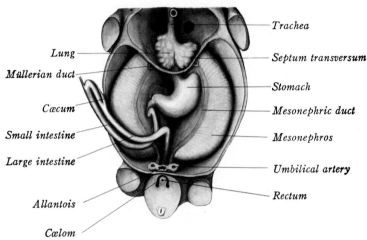

Fig. 634. Ventral dissection of 15 mm. pig embryo. × 6. Heart and liver have been removed and lungs are viewed through transparent pericardium.

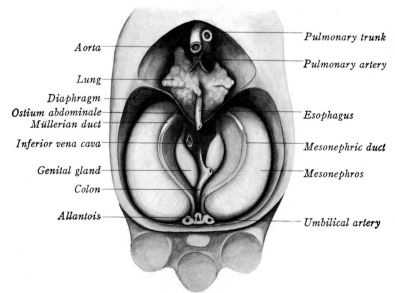

Fig. 635. Ventral dissection of 18 mm. pig embryo. × 7. Heart, liver, stomach and small intestine have been removed.

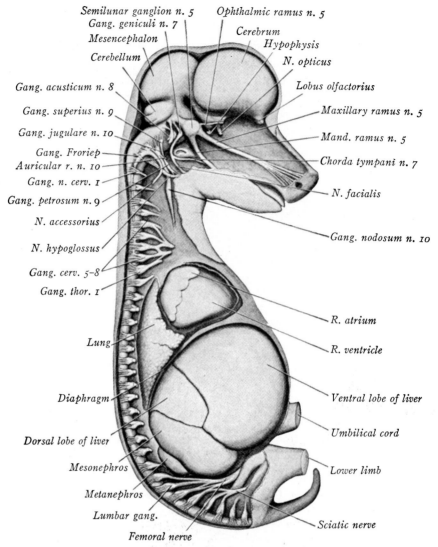

Fig. 636. Lateral dissection of 35 mm. pig embryo, viewed from right side. × 4.

D. ANATOMY OF A THIRTY-FIVE MM. PIG EMBRYO

External Form. The embryo is straighter, slenderer and its ventral surface less protuberant. The head, with its prominent snout, is shaping like that of a lower mammal, and the neck becomes distinct. Digits have appeared on the elongate extremities. The umbilical cord and tail are losing rapidly in relative size.

Lateral Dissection (Fig. 636). The spinal cord and brain are relatively smaller, but the latter is becoming highly specialized and folded. The cerebral hemispheres are large, and olfactory lobes extend forward from the rhinencephalon. The body of the embryo elongates faster than the spinal cord, so that the spinal nerves, at first directed at right angles, course ob-

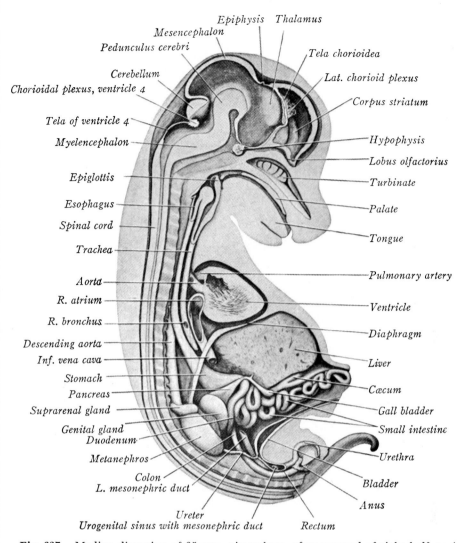

Fig. 637. Median dissection of 35 mm. pig embryo, after removal of right half. × 4.

liquely in the lumbo-sacral region. Note especially how the viscera have 'receded' caudad (*cf.* Figs. 595, 599), and how the liver dominates the abdomen as the mesonephros loses prominence. The kidney is exceptional in that it shifts cephalad.

Midsagittal Section (Fig. 637).   New features of the brain are the olfactory lobes, the chorioid plexus of the third and lateral ventricles, the thalami, the epiphysis and the consolidated hypophysis. The primitive mouth cavity is now divided by the palatine folds into the upper nasal passages and lower oral cavity. Of the viscera, the distinct genital and suprarenal glands and the enlarged metanephros command attention, as does the coiling of the intestine. The ureters have acquired separate openings at the base of the bladder, and the urethra extends to the tip of the phallus.

Ventral Dissection (Fig. 638).   The chief new features are the markedly

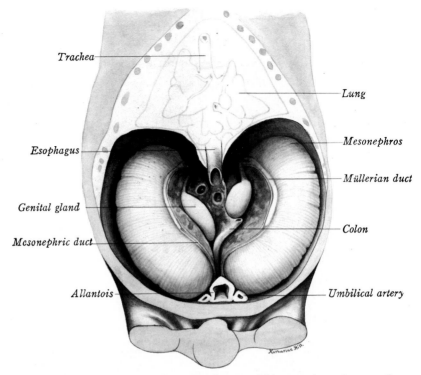

Fig. 638.   Ventral dissection of abdomen of 24 mm. pig embryo. × 6.

lobate lungs and the longer Müllerian ducts with expanded upper ends. The mesonephros is nearing its maximum absolute size.

RECOMMENDED COLLATERAL READING

Patten, B. M.   The Embryology of the Pig.   Blakiston.

INDEX

Epithelium(a) (*Continued*)
 lens, 535
 of palatine tonsils, 238
 olfactory, 526
Epitrichium, 441
Eponychium, 443
Epoöphoron, 325
 duct of, 325
Equational meiosis, 36
Erosion, endometrial, 125–126
Eruption of teeth, 222–223
Erythroblast, 345
Erythrocytes, 346. See also *Blood cells.*
Esophagus, 246
 anomalies, 247
 chick, 592
 glands, deep, 247
  superficial, 247
 pig, 620, 633, 635, 636
Estradiol, 158
Estriol, 161
Estrogen, 158–159
 chorionic, progesterone and, 161–162
 effect on uterine growth, 327
Estrogenic hormone, 157
Estrone, 159, 161
Estrous cycle, 150–152
 comparison with menstrual cycle, 154–155
Estrus, 48, 150
Ethmoid bone, 415–416
Ethmoidal cells, 416, 528
Ethmoidal labyrinths, 416
Ethmo-turbinals, 527
Eustachian tube, 237
Eustachian valve, 384
Evocator, 177, 178
Excitation, nervous, 182
Excretion, placental, 139
Excretory duct, 296, 297
Exoccipitals, 415
Experimental embryology. See *Embryology, experimental.*
Explant, 168
Exstrophy of bladder, 313
Extensor muscles, 434
External acoustic meatus, 238, 548
External body form, 199–212
External ear, 207, 548
External factors in experimental embryology, 167
External genitalia. See *Genitalia, external.*
External jugular veins, chick, 594, 599
Extra-embryonic cœlom, 87, 94, 114
Extra-embryonic ectoderm, 90
Extra-embryonic membranes, chick, 595, 608–609
Extra-embryonic mesoderm, 81, 94
Eye(s), 98, 206, 207, 529–541
 accessory tunics, 529
 albinism, 541
 anomalies, 540–541
 associated organs, 539–540
 chick, 591, 606, 609
 cilia, 539
 pig, 630

Eyelid(s), 206, 539
 cleft, 541

Face, 98, 204–207
 anomalies, 206–207
Facial-acoustic ganglion, 507
 chick, 597
Facial cleft, oblique, 206
Facial components, fates, 207
Facial nerve, 510–511, 512
 chick, 592, 597
 pig, 616, 629, 630
Factors, in experimental embryology, 167
Falciform ligament, 283
 pig, 622, 638
Fallot, tetrology of, 391
Fascia(æ), 397, 437
Fasciculus cuneatus, 471
Fasciculus gracilis, 471
Fasciculi proprii, 471
Feather primordia, chick, 609
Female pronucleus, 35, 57
Femoral artery, 359
Femoral vein, 369
Femur, 422
Fertilization, 48, 55–59, 62, 169
 age, 105
 events of, 55–58
 human, 58
 results, 58
Fetal and embryonic stages, 85–106
Fetal circulation, 369
 birth changes and, 392–394
Fetal heart, 380
Fetal membranes. See *Membranes, fetal.*
Fetal placenta, 132–135
Fetus, 98
 amorphous, 197
 anomalies, 101–102
 period of, 100–102
  in prenatal development, 85
 third through tenth month, 100–102
Fetus papyraceus, 102
Fiber(s)
 afferent, 500, 501
 collagenous, 397, 441
 dentinal, 221
 efferent, 500, 501
 elastic, 397, 441
 functional classification, 501
 lens, 535
 motor, 501
 muscle, 426
 myelinated, 463
 nerve, 500
  olfactory, 526
 of Müller, 534
 preganglionic, 516
 postganglionic, 516
 Purkinje, 427
 root, dorsal, 502
  ventral, 501
 sensory, 501
 unmyelinated, 463